PROBLEM SOLVERS®

Algebra &
Trigonometry

Staff of Research & Education Association

With Contributions By
Jerry Shipman, Ph.D.
Chairperson and Professor of Mathematics
Alabama A & M University
Normal, Alabama

Research & Education Association
Visit our website at
www.rea.com

Research & Education Association
61 Ethel Road West
Piscataway, New Jersey 08854
E-mail: info@rea.com

THE ALGEBRA & TRIGONOMETRY PROBLEM SOLVER®

Printed in the United States of America

Library of Congress Control Number 2006934768

International Standard Book Number 0-87891-508-7

Let REA's Problem Solvers®
work for you

REA's *Problem Solvers* are for anyone—from student to seasoned professional—who seeks a thorough, practical resource. Each *Problem Solver* offers hundreds of problems and clear step-by-step solutions not found in any other publication.

Perfect for self-paced study or teacher-directed instruction, from elementary to advanced academic levels, the *Problem Solvers* can guide you toward mastery of your subject.

Whether you are preparing for a test, are seeking to solve a specific problem, or simply need an authoritative resource that will pick up where your textbook left off, the *Problem Solvers* are your best and most trustworthy solution.

Since the problems and solutions in each *Problem Solver* increase in complexity and subject depth, these references are found on the shelves of anyone who requires help, wants to raise the academic bar, needs to verify findings, or seeks a challenge.

For many, *Problem Solvers* are homework helpers. For others, they're great research partners. What will *Problem Solvers* do for you?

- Save countless hours of frustration groping for answers
- Provide a broad range of material
- Offer problems in order of capability and difficulty
- Simplify otherwise complex concepts
- Allow for quick lookup of problem types in the index
- Be a valuable reference for as long as you are learning

Each *Problem Solver* book was created to be a reference for life, spanning a subject's entire breadth with solutions that will be invaluable as you climb the ladder of success in your education or career.

—Staff of Research & Education Association

How to Use This Book

The genius of the *Problem Solvers* lies in their simplicity. The problems and solutions are presented in a straightforward manner, the organization of the book is presented so that the subject coverage will easily line up with your coursework, and the writing is clear and instructive.

Each chapter opens with an explanation of principles, problem-solving techniques, and strategies to help you master entire groups of problems for each topic.

The chapters also present progressively more difficult material. Starting with the fundamentals, a chapter builds toward more advanced problems and solutions—just the way one learns a subject in the classroom. The range of problems takes into account critical nuances to help you master the material in a systematic fashion.

Inside, you will find varied methods of presenting problems, as well as different solution methods, all of which take you through a solution in a step-by-step, point-by-point manner.

There are no shortcuts in *Problem Solvers*. You are given no-nonsense information that you can trust and grow with, presented in its simplest form for easy reading and quick comprehension.

As you can see on the facing page, the key features of this book are:

- Clearly labeled chapters
- Solutions presented in a way that will equip you to distinguish key problem types and solve them on your own more efficiently
- Problems numbered and conveniently indexed by the problem number, not by page number

Get smarter....Let *Problem Solvers* go to your head!

Anatomy of a Problem Solver®

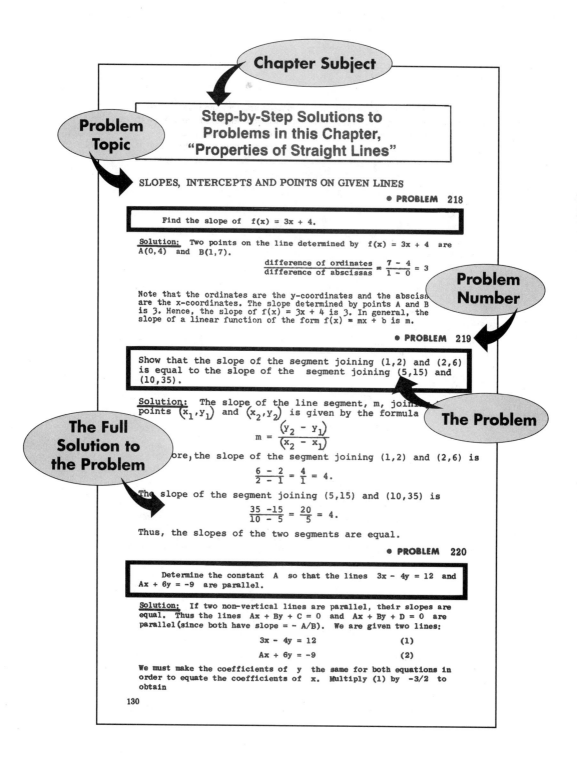

Chapter Subject

Problem Topic

Step-by-Step Solutions to Problems in this Chapter, "Properties of Straight Lines"

SLOPES, INTERCEPTS AND POINTS ON GIVEN LINES

● **PROBLEM** 218

Find the slope of $f(x) = 3x + 4$.

Solution: Two points on the line determined by $f(x) = 3x + 4$ are A(0,4) and B(1,7).

$$\frac{\text{difference of ordinates}}{\text{difference of abscissas}} = \frac{7 - 4}{1 - 0} = 3$$

Note that the ordinates are the y-coordinates and the abscis are the x-coordinates. The slope determined by points A and B is 3. Hence, the slope of $f(x) = 3x + 4$ is 3. In general, the slope of a linear function of the form $f(x) = mx + b$ is m.

Problem Number

● **PROBLEM** 219

Show that the slope of the segment joining (1,2) and (2,6) is equal to the slope of the segment joining (5,15) and (10,35).

Solution: The slope of the line segment, m, join points (x_1, y_1) and (x_2, y_2) is given by the formula

The Problem

$$m = \frac{(y_2 - y_1)}{(x_2 - x_1)}$$

ore, the slope of the segment joining (1,2) and (2,6) is

$$\frac{6 - 2}{2 - 1} = \frac{4}{1} = 4.$$

The Full Solution to the Problem

The slope of the segment joining (5,15) and (10,35) is

$$\frac{35 - 15}{10 - 5} = \frac{20}{5} = 4.$$

Thus, the slopes of the two segments are equal.

● **PROBLEM** 220

Determine the constant A so that the lines $3x - 4y = 12$ and $Ax + 6y = -9$ are parallel.

Solution: If two non-vertical lines are parallel, their slopes are equal. Thus the lines $Ax + By + C = 0$ and $Ax + By + D = 0$ are parallel (since both have slope = - A/B). We are given two lines:

$$3x - 4y = 12 \qquad (1)$$

$$Ax + 6y = -9 \qquad (2)$$

We must make the coefficients of y the same for both equations in order to equate the coefficients of x. Multiply (1) by -3/2 to obtain

CONTENTS

x

CHAPTER 1

FUNDAMENTAL ALGEBRAIC LAWS AND OPERATIONS

> **Basic Attacks and Strategies for Solving Problems in this Chapter. See pages 1 to 8 for step-by-step solutions to problems.**

There are fundamental algebraic properties that describe the way arithmetic and algebraic operations involving addition and multiplication of numbers and variable expressions are handled. These properties for the addition operation include the

commutative property of addition

$$[a + b = b + a],$$

associative property of addition

$$[a + (b + c) = (a + b) + c],$$

inverse property of addition

$$[a + (- a) = 0],$$

and identity property of addition

$$[a + 0 = a].$$

For the multiplication operation, the properties are the

commutative property of multiplication

$$[ab = ba],$$

associative property of multiplication

$$[a(bc) = (ab)c],$$

multiplicative inverse property

$$[a(1/a) = (1/a)a = 1, a \neq 0],$$

and the multiplicative identity property

$[a \cdot 1 = 1 \cdot a = a]$.

One general law that governs the operation of multiplication over addition is the distributive property

$[a(b + c) = ab + ac]$.

In determining the sum of two real numbers observe whether they have the same or different signs. If the signs of the numbers are the same, add the absolute values of the numbers and attach the sign of the addends. If the signs are unlike, then find the difference between the absolute values of the numbers and attach the sign of the number with the greatest absolute value. If subtraction is involved, the first step is to rewrite the subtraction process by using addition of the opposite, that is, $x - y = x + (-y)$. Then, proceed as indicated with addition.

With respect to multiplication and division of two real numbers, the pattern is simply that the product or quotient is positive if both numbers involved have like signs. On the other hand, the product or quotient is negative if the sign of the numbers involved are unlike.

When evaluating a mathematical expression, it is important to perform the operations in an order that begins with the expression in the innermost parentheses, brackets, or braces first and working outward. In the process of working outward, a hierarchical order must be observed. First, simplify all numbers or expressions with exponents, working from left to right if more than one of these expressions is present. Second, do all multiplications and divisions from left to right; and then perform all additions and subtractions from left to right. For instance, to simplify the expression given by

$$3[-4(2 + 3)^2 + 2(2 - 4)^3] + 10,$$

first perform the exponential operations of the expressions in the innermost symbols of grouping, from left to right. Thus,

$$3[-4(2 + 3)^2 + 2(2 - 4)^3] + 10 = 3[-4(5)^2 + 2(-2)^3] + 10$$
$$= 3[-4(25) + 2(-8)] + 10.$$

Next, simplify within the brackets, from left to right, the multiplication operations to obtain

$$= 3[-100 + (-16)] + 10.$$

Next, the distributive property can be used to multiply 3 by the terms in the brackets to obtain

$$= -300 + (-48) + 10.$$

Finally, add the above terms, from left to right, to get the results as follows:

$$= -348 + 10 = -338$$

Step-by-Step Solutions to Problems in this Chapter, "Fundamental Algebraic Laws and Operations"

● PROBLEM 1

Find the sum 8 + (- 3).

Solution: The sum of 8 + (- 3) can be obtained by using facts from arithmetic and the associative law:

$$8 + (- 3) = (5 + 3) + (- 3)$$

Use the associative law of addition (a + b) + c = a + (b + c):

$$= 5 + [3 + (- 3)]$$

Using the additive inverse property, a + (- a) = 0:

$$= 5 + 0$$

Using the additive identity property, a + 0 = a

$$= 5.$$

● PROBLEM 2

Show that (- 2) + (- 3) = - 5.

Solution: This small problem illustrates some of the basic ideas involved in mathematical proof. We know that (- 2) + (- 3) is an integer because the integers are closed under addition. To show that this integer is - 5, we ask ourselves what property is characteristic of - 5.

$$5 + (- 5) = 0,$$

by the additive inverse property, a + (- a) = 0. Moreover, - 5 is the only number which when added to 5 gives 0; for if (5 + b) = 0, by the additive identity, a = a + 0, - 5 = - 5 + 0 = - 5 + (5 + b), by our hypothesis, 5 + b = 0

$= (-5 + 5) + b$	by associative law of addition $a + (b+c) = (a+b)+c$.
$= 0 + b$	by additive inverse property, $a + (-a) = 0$
$= b + 0$	by commutative law of addition, $a + b = b + a$
$= b$	by additive identity, $a + 0 = a$

Thus, $- 5 = b$, proving that $- 5$ is the only number which when added to 5 gives 0.

We therefore see that $(- 2) + (- 3) = - 5$ if and only if $5 + [(- 2) + (- 3)] = 0$. We show below that this sum is zero.

$$5 + [(- 2) + (- 3)] = [3 + 2] + [(- 2) + (- 3)]$$

$$= 3 + \{2 + [(- 2) + (- 3)]\},$$

by associative law of addition, $(a + b) + c = a + (b + c)$:
$$= 3 + [0 + (- 3)] ,$$

by additive inverse property, $a + (- a) = 0$, $= 3 + [(- 3) + 0]$, by commutative law of addition, $a + b = b + a$;

$$= 3 + (- 3),$$

by additive identity property, $a + 0 = a$; $= 0$, by additive inverse property, $a + (- a) = 0$.

Thus we have shown (a) $5 + (- 5) = 0$

(b) $(- 5)$ is the only number which when added to 5 equals 0.

(c) $5 + (- 5) = 0 = 5 + [(- 2) + (- 3)]$ and therefore

$$(- 5) = (- 2) + (- 3),$$

completing our proof.

● **PROBLEM 3**

Find the quotient q and remainder r upon dividing 575 by 21.

<u>Solution:</u> If we divide 575 by 21, we obtain

$$
\begin{array}{r}
27 \\
21{\overline{\smash{\big)}\,575}} \\
\underline{42} \\
155 \\
\underline{147} \\
8
\end{array}
$$

The quotient is 27 and the remainder is 8.

Check: To check the quotient and the remainder obtained, multiply the quotient by the divisor and then add the remainder to this product. The sum should be equal to the dividend.

$$(27)(21) + 8 = 567 + 8 = 575.$$

Hence, the sum is equal to the dividend, 575. Therefore, the quotient and the remainder obtained are correct.

● **PROBLEM 4**

Evaluate $2 - \{5 + (2 - 3) + [2 - (3 - 4)]\}$

<u>Solution:</u> When working with a group of nested parentheses, we evaluate the innermost parenthesis first.

Thus, $2 - \{5 + (2 - 3) + [2 - (3 - 4)]\}$

$$= 2 - \{5 + (2 - 3) + [2 - (-1)]\}$$

$$= 2 - \{5 + (-1) + [2 + 1]\}$$

$$= 2 - \{5 + (-1) + 3\}$$

$$= 2 - \{4 + 3\}$$

$$= 2 - 7$$

$$= -5.$$

● **PROBLEM** 5

Simplify $4[-2(3 + 9) \div 3] + 5$.

<u>Solution:</u>   To simplify means to find the simplest expression. We perform the operations within the innermost grouping symbols first. That is $3 + 9 = 12$.

Thus, $4[-2(3 + 9) \div 3] + 5 = 4[-2(12) \div 3] + 5$

Next we simplify within the brackets:

$$= 4[-24 \div 3] + 5$$

$$= 4 \cdot (-8) + 5$$

We now perform the multiplication, since multiplication is done before addition:

$$= -32 + 5$$

$$= -27$$

Hence, $4[-2(3 + 9) \div 3] + 5 = -27$.

● **PROBLEM** 6

Is the set of all natural numbers from 1 to 10 a closed system under addition?

<u>Solution:</u> For $\{1,2,3,\ldots 10\}$ to be closed with respect to addition, the sum of any two numbers in this set must also be a member of this set. The set of all natural numbers from 1 to 10, inclusive, is therefore not a closed system under addition for it would not be correct to say that given any two numbers in the set there is a number in the set called their sum. For instance, 4 and 7 are in the set but their sum, 11, is not.

● **PROBLEM** 7

Simplify the following expressions, removing the parentheses.

1) $a + (b - c)$

2) $ax - (by - c)$

3) $2 - (-x + y)$.

3

Solution: 1) Place a factor of 1 between the + (plus) sign and the term in the parenthesis. This procedure does not change the value of the entire expression. Hence, $a + (b-c) = a + 1(b-c)$

$$= a + 1(b) + 1(-c) \text{ distributing}$$
$$= a + b - c$$

2) Again, place a factor of 1 between the - (minus) sign and the term in the parenthesis. Again, this procedure does not change the value of the entire expression. Hence,

$$ax - (by - c) = ax - 1(by - c)$$
$$= ax - 1(by) - 1(-c) \text{ distributing}$$
$$= ax - by + c$$

3) Again, place a factor of 1 between the - (minus) sign and the term in parenthesis. Again, this procedure does not change the value of the entire expression. Hence,

$$2 - (-x + y) = 2 - 1(-x + y)$$
$$= 2 - 1(-x) - 1(y) \text{ distributing}$$
$$= 2 + x - y$$

● **PROBLEM 8**

Evaluate $3s - [5t + (2s - 5)]$

Solution: We always evaluate the expression within the innermost parentheses first, when working with a group of nested parentheses. Thus,

$$3s - [5t + (2s - 5)] = 3s - [5t + 2s - 5]$$
$$= 3s - 5t - 2s + 5$$
$$= 3s - 2s - 5t + 5$$
$$= s - 5t + 5.$$

● **PROBLEM 9**

Simplify
$$8x^2 - \left[7x - \left(x^2 - x + 5y\right)\right] + \left(2x - 3y\right).$$

Solution: Where a succession of arithmetic operations is involved, appropriate grouping symbols indicate clearly how these algebraic operations are to be performed; that is, we perform them before other operations. In this problem we also have grouping symbols within grouping symbols. Therefore, we perform the operation in the innermost parentheses first. Hence, multiply the terms in the parentheses by minus one in order to remove the parentheses. Furthermore, they can be removed from the last two terms.

$$8x^2 - \left[7x - \left(x^2 - x + 5y\right)\right] + \left(2x - 3y\right) = 8x^2 - \left[7x - x^2 + x - 5y\right]$$
$$+ 2x - 3y$$

Remove the brackets by multiplying the terms inside by minus one. Then,

$$8x^2 - \left[7x - \left(x^2 - x + 5y\right)\right] + \left(2x - 3y\right) = 8x^2 - 7x + x^2 - x$$
$$+ 5y + 2x - 3y$$

Now group like terms. Then perform the indicated operations from left to right. Thus, we obtain:

$$8x^2 - \left[7x - \left(x^2 - x + 5y\right)\right] + \left(2x - 3y\right) = 8x^2 + x^2 - 7x - x + 2x$$
$$+ 5y - 3y$$

$$= 9x^2 - 6x + 2y.$$

In this example, we have found the algebraic sum of these three quantities: $8x^2$ and $-\left[7x - \left(x^2 - x + 5y\right)\right]$ and $\left(2x - 3y\right)$.

● **PROBLEM** 10

Simplify: $3a - 2\{3a - 2[1 - 4(a - 1)] + 5\}$.

Solution: When working with several sets of brackets and, or parentheses, we work from the inside out. That is, we use the law of distribution throughout the expression, starting from the innermost parentheses, and working our way out. Hence in this case we have: $2\{3a - 2[1 - 4(a - 1)] + 5\}$ and we note that $(a - 1)$ is the innermost parenthesis, so our first step is to distribute the (-4). Thus, we obtain:

$$3a - 2[3a - 2(1 - 4a + 4) + 5].$$

We now find that $(1 - 4a + 4)$ is in our innermost parentheses. Combining terms we obtain:

$$(1 - 4a + 4) = (5 - 4a) ;$$

hence, $3a - 2[3a - 2(1 - 4a + 4) + 5] = 3a - 2[3a - 2(5 - 4a) + 5]$ and since $(5 - 4a)$ is in the innermost parentheses we distribute the (-2), obtaining:

$$3a - 2(3a - 10 + 8a + 5).$$

We are now left with the terms in our last set of parentheses, $(3a - 10 + 8a + 5)$. Combining like terms we obtain:

$$(3a - 10 + 8a + 5) = (11a - 5)$$

hence, $3a - 2(3a - 10 + 8a + 5) = 3a - 2(11a - 5)$.

Distributing the (-2), $= 3a - 22a + 10$

combining terms, $= -19a + 10$.

Hence $3a - 2\{3a - 2[1 - 4(a - 1)] + 5\} = -19a + 10$.

● **PROBLEM** 11

(a) Add, $3a + 5a$

(b) Factor, $5ac + 2bc$.

Solution: (a) To add $3a + 5a$, factor out the common factor a. Then,

$$3a + 5a = (3 + 5)a = 8a.$$

5

(b) To factor $5ac + 2bc$, factor out the common factor c. Then,

$$5ac + 2bc = (5a + 2b)c.$$

● **PROBLEM 12**

Express each of the following as a single term.

(a) $3x^2 + 2x^2 - 4x^2$ (b) $5axy^2 - 7axy^2 - 3xy^2$

Solution: (a)   Factor x^2 in the expression.

$$3x^2 + 2x^2 - 4x^2 = (3 + 2 - 4)x^2 = 1x^2 = x^2.$$

(b)   Factor xy^2 in the expression and then factor a.

$$5axy^2 - 7axy^2 - 3xy^2 = (5a - 7a - 3)xy^2$$
$$= [(5-7)a - 3]xy^2$$
$$= (-2a - 3)xy^2.$$

● **PROBLEM 13**

Simplify $x = a + 2[b - (c - a + 3b)]$.

Solution: When working with several groupings, we perform the operations in the innermost parenthesis first, and work outward. Thus, we first subtract $(c - a + 3b)$ from b:

$$x = a + 2[b - (c - a + 3b)] = a + 2(b - c + a - 3b)$$

Combining terms,

$$= a + 2(-c + a - 2b)$$

distributing the 2,

$$= a - 2c + 2a - 4b$$

combining terms,

$$= 3a - 2c - 4b$$

To check that $a + 2[b - (c - a + 3b)]$ is equivalent to $3a - 2c - 4b$, replace a,b, and c by any values. Letting $a = 1$, $b = 2$, $c = 3$, the original form $a + 2[b - (c - a + 3b)] = 1 + 2[2 - (3 - 1 + 3 \cdot 2)]$

$$= 1 + 2[2 - (3 - 1 + 6)]$$
$$= 1 + 2[2 - 8]$$
$$= 1 + 2(-6)$$
$$= 1 + (-12)$$
$$= -11$$

The final form, $3a - 2c - 4b = 3(1) - 2(3) - 4(2) = 3 - 6 - 8$

$$= -11$$

Thus, both forms yield the same result.

● **PROBLEM 14**

Use the field properties to derive the equation $x = 5$ from the equation $5x - 3 = 2(x + 6)$.

Solution: $5x - 3 = 2(x + 6)$ Given

6

$$5x - 3 = 2x + 12 \qquad \text{distributive property of}$$
$$\text{multiplication over addition}$$
$$(5x - 3) + (-2x) = 2x + 12 + (-2x) \qquad \text{Additive Property } (-2x)$$
$$3x - 3 = 12 \qquad \text{Simplifying}$$
$$(3x - 3) + 3 = 12 + 3 \qquad \text{additive property } (+3)$$
$$3x = 15 \qquad \text{Simplifying}$$
$$\tfrac{1}{3} \cdot (3x) = \tfrac{1}{3} \cdot 15 \qquad \text{Multiplicative Property} (\tfrac{1}{3})$$
$$x = 5 \qquad \text{Simplifying}$$

We could also derive $5x - 3 = 2(x + 6)$ from $x = 5$ by reversing the steps in the solution. Let us see if 5 will make the equation $5x - 3 = 2(x + 6)$ true.

$$5(5) - 3 \overset{?}{=} 2(5 + 6)$$

$$22 = 22$$

Two equations are equivalent if and only if they have the same solution set. Since $5x - 3 = 2(x + 6)$ and $x = 5$ have the same solution set, $\{5\}$, the two equations are equivalent.

● **PROBLEM** 15

Approximate:

$$A = \frac{\pi \times \sqrt{2} \times 2.17}{(6.83)^2 + (1.07)^2}$$

Solution: We use the following approximate values:

$$\pi = 3.1416 \cong 3$$

$$\sqrt{2} = 1.414 \cong 1.5$$

$$2.17 \cong 2$$

$$(6.83)^2 \cong 7^2 = 49 \cong 50$$

$$(1.07)^2 \cong 1^2 = 1$$

Then,

$$A \cong \frac{3 \times 1.5 \times 2}{50 + 1} = \frac{9}{51} \cong \frac{10}{50} = .2$$

● **PROBLEM** 16

Evaluate $p = \dfrac{(a - b)(ab + c)}{(cb - 2a)}$

when $a = +2$, $b = -\tfrac{1}{2}$, and $c = -3$.

Solution: Inserting the given values of a, b, and c

$$p = \frac{[+ 2 - (-\tfrac{1}{2})][(+ 2)(-\tfrac{1}{2}) + (- 3)]}{[(- 3)(-\tfrac{1}{2}) - 2(+ 2)]}$$

$$= \frac{[+ 2 + \tfrac{1}{2}][- 1 - 3]}{[+ 1\tfrac{1}{2} - 4]}$$

7

$$= \frac{(2\frac{1}{2})(-4)}{-(2\frac{1}{2})}$$

The $2\frac{1}{2}$ in the numerator cancels the $2\frac{1}{2}$ in the denominator.

$$p = \frac{-4}{-1}$$

Multiplying numerator and denominator by -1

$$p = \frac{+4}{+1}$$

$$p = +4.$$

CHAPTER 2

LEAST COMMON MULTIPLE/ GREATEST COMMON DIVISOR

Basic Attacks and Strategies for Solving Problems in this Chapter. See pages 9 to 11 for step-by-step solutions to problems.

The *Least Common Multiple* (LCM) of a set of two or more integers is the smallest integer that can be found such that it is divisible by each integer from the given set. To determine the Least Common Multiple of two or more integers, the first step is to factor each of the numbers into their prime factors. Then, select among all the factors a list of those which are unique prime factors of the numbers. Next, determine the exponent to be used for each selected factor by finding the highest number of times the factor appeared in either of the factorizations. The product of the unique prime factors with the appropriate highest exponents is the LCM. For example, the LCM of 12 and 21 is 84. This is determined as follows: prime factors of 12 are

$2 \times 2 \times 3$

and for 21 are

$3 \times 7.$

The unique prime factors are 2, 3, and 7, respectively, of which the highest number of times 2 appears is twice, 3 appears once, and 7 appears once. So, the product or LCM is:

$2^2 \times 3 \times 7 = 4 \times 21 = 84.$

The LCM can be very useful in determining the least common denominator in the addition and/or subtraction of two or more fractions.

The *Greatest Common Divisor* (GCD) of two or more integers is the greatest integer that will divide into all the integers. When finding the GCD, the first step is to factor each number involved into their prime factors. The second step is to select among each of the factorizations the unique common factors. Now determine the exponent to be used for each selected factor by finding the smallest number of times the factor appears. The final step is to find the product of the selected factors with appropriate exponents which is the GCD. For instance, the GCD of 24 and 60 is 12. This solution is found by first factoring 24 and 60 into prime factors as follows:

$$24 = 2 \times 2 \times 2 \times 3 \quad \text{and} \quad 60 = 2 \times 2 \times 3 \times 5.$$

The unique common factors among the two factorizations are 2 and 3 only. The smallest number of times 2 appears among the factorizations is twice and for 3 is once. Thus, the final product is composed of:

$$2^2 \times 3^1 = 4 \times 3 = 12$$

which is the GCD of 24 and 60.

The GCD can be very useful in reducing a fraction to the lowest terms.

Step-by-Step Solutions to Problems in this Chapter, "Least Common Multiple/ Greatest Common Divisor"

● **PROBLEM** 17

Find the least common multiple (lcm) of 15 and 18.

Solution: Some of the integers divisible by 15 are

$$15,30,45,60,75,90,105,\ldots$$

Some of the integers divisible by 18 are

$$18,36,54,72,90,108,\ldots$$

The smallest positive integer divisible by both 15 and 18 is 90. Thus,

$$\mathrm{lcm}\{15,18\} = 90$$

Another method for finding lcm{15,18} is the following:
    Factor 15 and 18 into their prime factors.

$$15 = 3 \cdot 5$$

$$18 = 2 \cdot 3 \cdot 3$$

Now, take the different factors of the two numbers and multiply them together. The exponent to be used for each factor is the highest number of times that the factor appears in either number (15 or 18). The product obtained will be the lcm{15,18}. Hence:

$$\mathrm{lcm}\{15,18\} = 2^1 \cdot 3^2 \cdot 5^1 = 2(9)(5) = 90.$$

● **PROBLEM** 18

Find the Least Common Multiple, LCM, of 26, 39, and 66.

Solution: Write each number as the product of primes:

$$26 = 2(13), \quad 39 = 3(13), \quad 66 = 2(3)(11)$$

The LCM is obtained by using the greatest power of each prime, only once, to form a product. Thus,

$$\mathrm{LCM} = 2(3)(11)(13)$$

$$= 858.$$

● **PROBLEM** 19

Find the Least Common Multiple, LCM, of 12, 18, 21, 25 and 35.

Solution: We want to express each number as a product of prime factors:

9

$$12 = 2^2(3), \quad 18 = 2(3^2), \quad 21 = 3(7),$$

$$25 = 5^2, \quad 35 = 5(7)$$

Find the LCM by retaining the highest power of each distinct factor and multiplying them together, making sure to use each factor only once regardless of the number of times it appears. Thus,

$$\text{LCM} = (2^2)(3^2)(5^2)(7) = (4)(9)(25)(7)$$

$$= 6300$$

● PROBLEM 20

Find the greatest common divisor $\{15,28\}$.

Solution: If 15 and 28 are factored completely into their respective prime factors, $15 = 3 \cdot 5$ and $28 = 2 \cdot 2 \cdot 7$

Since 1 divides every integer, and since 15 and 28 possess no common prime factors, it follows that

$$\gcd\{15,28\} = 1.$$

If the gcd of two integers is 1, then the two integers are said to be relatively prime. Since $\gcd\{15,28\} = 1$, the integers 15 and 28 are relatively prime.

● PROBLEM 21

Find the greatest common divisor and the least common multiple of 16 and 12.

Solution: Factor the two given numbers into their prime factors.

$$12 = 2 \cdot 2 \cdot 3$$

$$16 = 2 \cdot 2 \cdot 2 \cdot 2$$

The greatest common divisor of 16 and 12, or $\gcd\{12,16\}$, is the largest number which divides both 16 and 12, (largest common factor).

$$\frac{12}{16} = \frac{2 \cdot 2 \cdot 3}{2 \cdot 2 \cdot 2 \cdot 2} = \frac{(4)(3)}{(4)(2)(2)}$$

Hence, $\gcd\{12,16\} = 4$.

The least common multiple of 16 and 12, or $\text{lcm}\{12,16\}$, is obtained in the following way. Take the different factors of the two numbers and multiply them together. The exponent to be used for each factor is the highest number of times that the factor appears in either number (12 or 16). The product obtained will be the $\text{lcm}\{12,16\}$. Hence:

$$\text{lcm}\{16,12\} = 2^4 \cdot 3^1 = (16)(3) = 48 \ .$$

● PROBLEM 22

Find the greatest common divisor of 24 and 40. Also, find the least common multiple of 24 and 40.

Solution: To find the greatest common divisor of 24 and 40, or $\gcd\{24,40\}$, we write down the set of all positive integers which divide both 24 and 40. Thus we obtain the two sets

$$\{1,2,3,4,6,8,12,24\} \quad \text{for} \quad 24$$
$$\{1,2,4,5,8,10,20,40\} \quad \text{for} \quad 40$$

Those integers dividing both 24 and 40 are in the intersection of these two sets. Thus,

$$\{1,2,3,4,6,8,12,24\} \cap \{1,2,4,5,8,10,20,40\} = \{1,2,4,8\}$$

The largest element in this last set is 8. Thus,

$$8 = \gcd\{24,40\} \ .$$

Another method for finding the $\gcd\{24,40\}$ is called the factoring technique. Factor the two given numbers into their prime factors.

$$24 = 2 \cdot 2 \cdot 2 \cdot 3$$
$$40 = 2 \cdot 2 \cdot 2 \cdot 5$$

The greatest common divisor of any two numbers is the largest number which divides both of those numbers. Therefore,

$$\frac{24}{40} = \frac{2 \cdot 2 \cdot 2 \cdot 3}{2 \cdot 2 \cdot 2 \cdot 5} = \frac{(8)(3)}{(8)(5)} \ . \quad \text{Hence,}$$

$$\gcd\{24,40\} = 8 \ .$$

The following technique is the definition for finding the least common multiple of 24 and 40, or lcm $\{24,40\}$. To find the lcm$\{24,40\}$, we write down the set of all positive integer multiples of both 24 and 40. Then we obtain

$$\{24,48,72,96,120,144,168,192,216,240,264,\dots\} \quad \text{for} \quad 24$$

$$\{40,80,120,160,200,240,280,\dots\} \qquad\qquad \text{for} \quad 40$$

The integers which are multiples of both 24 and 40; that is, common multiples of 24 and 40, are in the intersection of these two sets. This is the set $\{120,240,\dots\}$. The smallest element of this set is 120. Hence, lcm$\{24,40\}$ = 120. Another method for finding the lcm$\{24,40\}$ is called the factoring technique. Factor the two given numbers into their prime factors.

$$24 = 2 \cdot 2 \cdot 2 \cdot 3$$
$$40 = 2 \cdot 2 \cdot 2 \cdot 5$$

Now, take the different factors of the two numbers and multiply them together. The exponent to be used for each factor is the highest number of times that the factor appears in either number (24 or 40). The produce obtained will be the lcm$\{24,40\}$. Hence:

$$\text{lcm}\{24,40\} = 2^3 \cdot 3^1 \cdot 5^1 = (8)(3)(5) = (24)(5) = 120.$$

CHAPTER 3

SETS AND SUBSETS

> **Basic Attacks and Strategies for Solving Problems in this Chapter. See pages 12 to 16 for step-by-step solutions to problems.**

There are two basic operations used to combine sets — *union* and *intersection*. Other operations involving sets include the *complement* and *cartesian* product.

When finding the union of two or more sets, the first step is to examine the sets and determine all common elements that belong to the individual sets and all the unique elements (not common among the sets). Then, the union of the sets, using the roster or list method, is the set of all the common elements written with no repetitions together with all of the unique elements. For example, the union of set $A = \{3, 4, 5\}$ and set $B = \{4, 5, 6\}$ is given by the following set:

$A \cup B = \{3, 4, 5, 6\}$.

To find the intersection of two or more sets, simply identify among the sets all the elements which are common. Then, the intersection, using the roster method, is the set of only the common elements among the sets written with no repetitions. For example, the intersection of sets A and B above is given by:

$A \cap B = \{4, 5\}$.

In addition to the roster or list method of representing the union and intersection of sets, set builder notation and Venn diagrams are used.

The procedure for finding the complement of a set is to first determine a universal set U and a set A whose elements are a part of U. Then, all of the elements of U which do not belong to A form the complement set of A, given by A'.

The Cartesian product of two sets, say A and B and denoted by $A \times B$, is a set formed by all possible ordered pairs where the first component comes from set A and the second component from set B. For example, if set $A = \{1, 2\}$ and set $B = \{1, 4\}$, then $A \times B$ is given as follows:

$A \times B = \{(1, 1), (1, 4), (2, 1), (2, 4)\}$.

A subset of a set A is a set in which each of its members belong to the original set. In addition, the empty set is a subset of any set. For example, the set $\{x\}$ and the empty set are two of the subsets of set $A = \{x, y\}$. To find all the possible subsets N of any set, one can use the formula

$$N = 2^n,$$

where n is the number of elements in the original set. The actual listing of subsets is found by forming a set of each possible pairing of the elements in the original set plus the empty set. For example, the set

$$A = \{x, y\}$$

has 4 subsets since

$$N = 2^2 = 4.$$

The subsets are $\{x\}$, $\{y\}$, $\{x, y\}$, and the empty set.

Step-by-Step Solutions to Problems in this Chapter, "Sets and Subsets"

● **PROBLEM** 23

If A = { 1, 2, 3, 4, 5} and B = { 2, 3, 4, 5, 6}, find A ∪ B.

Solution: The symbol ∪ is used to denote the union of sets. Thus A ∪ B (which is read "the union of A and B") is the set of all elements that are in either A or B or both. In this problem, if,

then   A = { 1, 2, 3, 4, 5} and B = { 2, 3, 4, 5, 6},

A ∪ B = { 1, 2, 3, 4, 5, 6}.

● **PROBLEM** 24

If A = { 1, 2, 3, 4, 5) and B = { 2, 3, 4, 5, 6}, find A ∩ B.

Solution: The intersection of two sets A and B is the set of all elements that belong to both A and B; that is, all elements common to A and B. In this problem, if

A = { 1, 2, 3, 4, 5} and B = { 2, 3, 4, 5, 6},

then
A ∩ B = { 2, 3, 4, 5}.

● **PROBLEM** 25

If A = {2,3,5,7} and B = {1,-2,3,4,-5,√6}, find
(a) A ∪ B and (b) A ∩ B.

Solution: (a) A ∪ B is the set of all elements in A or in B or in both A and B, with no element included twice in the union set.

A ∪ B = {1,2,-2,3,4,5,-5,√6,7}

(b) A ∩ B is the set of all elements in both A and B.

A ∩ B = {3}

Sometimes two sets have no elements in common. Let S = {3,4,7} and T = {2,-4,6}. What is the intersection of S and T? In this case S ∩ T has no elements. Hence S ∩ T = ∅, the empty set. In that case, the sets are said to be disjoint.

The set of all elements entering a discussion is called the universal set, U, When the universal set is not given, we assume it to be the set of real numbers. The set of all elements in the universal set that are not elements of A is called the complement of A, written Ā .

Show that $(A \cap B) \cap C = A \cap (B \cap C)$.

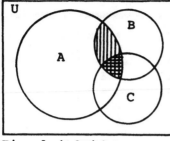
Fig. 1 $(A \cap B) \cap C$

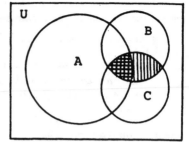
Fig. 2 $A \cap (B \cap C)$

Solution: In Figure 1 the vertically shaded area represents $A \cap B$, and the horizontally shaded area represents the points common to the set $(A \cap B)$ and the set C, that is $(A \cap B) \cap C$. Similarly, in Figure 2, the vertically shaded area represents $B \cap C$, and the horizontally shaded area represents the points common to the set $(B \cap C)$ and the set A, that is $A \cap (B \cap C)$. Since the two horizontally shaded areas in the two figures are the same,

$$(A \cap B) \cap C = A \cap (B \cap C).$$

If $U = \{1, 2, 3, 4, 5\}$ and $A = \{2, 4\}$, find A'.

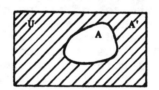

Solution: The complement of a set A in U is the set of all elements of U that do not belong to A. The symbol A' (or, sometimes, $\bar{A}$, $\sim A$, or $\tilde{A}$) denotes the complement of A in U. The figure gives a representation of A', the complement of A in U. In this problem, since,

$$U = \{1, 2, 3, 4, 5\}$$

and

$$A = \{2, 4\},$$

$$A' = \{1, 3, 5\}.$$

$U = \{1,2,3,4,5,6,7,8,9,10\}$, $P = \{2,4,6,8,10\}$, $Q = \{1,2,3,4,5\}$.
Find (a) $\bar{P}$ and (b) $\bar{Q}$.

Solution: $\bar{P}$ and $\bar{Q}$ are the complements of P and Q respectively.

That is, $\bar{P}$ is the set of all elements in the universal set, U, that are not elements of P, and $\bar{Q}$ is the set of elements in U that are not in Q. Therefore,

(a) $\bar{P} = \{1,3,5,7,9\}$; (b) $\bar{Q} = \{6,7,8,9,10\}$

● **PROBLEM** 29

If U = the set of whole numbers and E = the set of even whole numbers: find $\bar{\bar{E}}$.

<u>Solution:</u> $\bar{E}$ is called the complement of E. $\bar{E}$ is the set of all elements in the universal set, U, that are not elements of E. Therefore,

$$\bar{\bar{E}} = \{1,3,5,\ldots\},$$

the set of odd whole numbers.

● **PROBLEM** 30

Show that the complement of the complement of a set is the set itself.

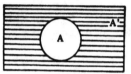

<u>Solution:</u> The complement of set A is given by A'. Therefore, the complement of the complement of a set is given by (A')'. This set, (A')', must be shown to be the set A; that is, that (A')' = A. In the figure the complement of the set A, or A', is the set of all points not in set A; that is, all points in the rectangle that are not in the circle. This is the shaded area in the figure. Therefore, this shaded area is A'. The complement of this set, or (A')', is the set of all points of the rectangle that are not in the shaded area; that is, all points in the circle, which is the set A. Therefore, the set (A')' is the same as set A; that is,

$$(A')' = A.$$

● **PROBLEM** 31

If A = {1, 2, 3} and B = {5, 6}, find A x B and B x A.

<u>Solution:</u> The Cartesian product of two sets A and B, denoted by A x B, is the set of all ordered pairs (x, y) such that x ϵ A and y ϵ B. In this problem, if A = {1, 2, 3} and B = {5, 6}, then the Cartesian product A x B is:

A x B = { (1, 5), (1, 6), (2,5), (2, 6), (3, 5),

(3, 6)}.

The Cartesian product B x A is the set of all ordered pairs (x, y) such that x ϵ B and y ϵ A. Hence, the Cartesian product B x A is:

14

B x A = { (5, 1), (5, 2), (5, 3), (6, 1), (6, 2),

(6, 3) }.

Let M = {1, 2} and N = {p, q}. Find (a) M × N, (b) N × M, and (c) M × M.

<u>Solution:</u> (a) M×N is the set of all ordered pairs in which the first component is a member of M and the second component is a member of N. Thus,

M × N = {(1,p), (1, q), (2, p), (2, q)}.

Note that the number of elements in M is 2,

the number of elements in N is 2,

and the number of elements in M × N = 2 × 2 = 4.

(b) N ×M is the set of all ordered pairs in which the first component is a member of N and the second component is a member of M. Thus,

N × M = {(p, 1), (q, 1), (p, 2), (q, 2)}.

Once again note that the number of elements in N × M is 2 × 2 = 4.

(c) M × M is the set of all ordered pairs in which both components are members of M. Thus,

M × M = {(1, 1),(1, 2), (2, 1), 2, 2)}.

Here too, the number of elements in M × M is 2 × 2 = 4.

List all the subsets of C = {1,2}.

<u>Solution:</u> {1}, {2}, {1,2}, $\emptyset$, where $\emptyset$ is the empty set. Each set listed in the solution contains at least one element of the set C. The set {2,1} is identical to {1,2} and therefore is not listed. $\emptyset$ is included in the solution because $\emptyset$ is a subset of every set.

Find four proper subsets of P = {n: n ε I, -5 < n ≤ 5}.

<u>Solution:</u> P = {-4, -3, -2, -1, 0, 1, 2, 3, 4, 5}. All . these elements are integers that are either less than or equal to 5 or greater than -5. A set A is a proper subset of P if every element of A is an element of B and in addition there is an element of B which is not in A.

(a) B = {-4, -2, 0, 2, 4} is a subset because each element of B is an integer greater than -5 but less than or equal to 5. B is a proper subset because 3 is an element of P but not an element of B. We can write $3 \in P$ but $3 \notin B$.

 (b) C = {3} is a subset of P, since $3 \in P$. However, $5 \in P$ but $5 \notin C$. Hence, $C \subset P$.

 (c) D = {-4, -3, -2, -1, 1, 2, 3, 4, 5} is a proper subset of P, since each element of D is an element of P, but $0 \in P$ and $0 \notin D$.

 (d) $\phi \subset P$, since ϕ has no elements. Note that ϕ is the empty set. ϕ is a proper subset of every set except itself.

CHAPTER 4

ABSOLUTE VALUES

Basic Attacks and Strategies for Solving Problems in this Chapter. See pages 17 to 18 for step-by-step solutions to problems.

When solving an equation containing one or more absolute values, the fundamental procedure is the use of the definition of absolute value. The definition states that:

(1) The absolute value of a positive number or positive expression is the positive of the number or expression itself,

(2) The absolute value of a negative number or negative expression is the opposite of the negative number or expression, and

(3) The absolute value of zero is zero.

If only one absolute value is in the given equation, then the application of the definition yields two equations. The next step is to solve each of these equations for the variable and check the results. For instance, to solve the equation

$$|x + a| = c,$$

where a and c are constants, we can write, according to the definition of absolute value, the following two equations:

$$x + a = c \quad \text{and} \quad x + a = -c.$$

Then, to get the solution of the original equation, solve each of the equations and check the results.

Suppose the equation contains two absolute values, such as

$$|ax + b| = |cx + d| + e,$$

where a, b, c, and d are constants and a and c are not zero. Then, there are four possibilities for equations:

$$ax + b = cx + d + e, \qquad (1)$$

$$-(ax + b) = cx + d + e, \qquad (2)$$

$$ax + b = -(cx + d) + e, \qquad (3)$$

$$-(ax + b) = -(cx + d) + e \qquad (4)$$

Solve the appropriate equations and check the results in order to get the solution of the original equation.

Step-by-Step Solutions to Problems in this Chapter, "Absolute Values"

● **PROBLEM** 35

Solve for x when $|x - 7| = 3$.

Solution: This equation, according to the definition of absolute value, expresses the conditions that x - 7 must be 3 or -3, since in either case the absolute value is 3. If x - 7 = 3, we have x = 10; and if x - 7 = -3, we have x = 4. We see that there are two values of x which solve the equation.

● **PROBLEM** 36

Solve for x when $|3x + 2| = 5$.

Solution: First we write expressions which replace the absolute symbols in forms of equations that can be manipulated algebraically. Thus this equation will be satisfied if either

$$3x + 2 = 5 \quad or \quad 3x + 2 = -5.$$

Considering each equation separately, we find

$$x = 1 \quad and \quad x = -\frac{7}{3}.$$

Accordingly, the given equation has two solutions.

● **PROBLEM** 37

Solve for x when $|5x + 4| = -3$.

Solution: In examining the given equation, it is seen that the absolute value of a number is set equal to a negative value. By definition of an absolute number, however, the number cannot be negative. Therefore the given equation has no solution.

● **PROBLEM** 38

Solve for x when $|5 - 3x| = -2$.

Solution: This problem has no solution, since the absolute value can never be negative and we need not proceed further.

● **PROBLEM** 39

Solve for x in $|2x - 6| = |4 - 5x|$.

Solution: There are four possibilities here. 2x - 6 and 4 - 5x

17

can be either positive or negative. Therefore,

$$2x - 6 = 4 - 5x \qquad (1)$$

$$-(2x - 6) = 4 - 5x \qquad (2)$$

and

$$2x - 6 = -(4 - 5x) \qquad (3)$$

$$-(2x - 6) = -(4 - 5x) \qquad (4)$$

Equations (2) and (3) result in the same solution, as do
equations (1) and (4). Therefore it is necessary to solve only
for equations (1) and (2). This gives:

$$x = \frac{10}{7} \, , \, -\frac{2}{3} \, .$$

● **PROBLEM** 40

Solve for x when $|2x - 1| = |4x + 3|$.

Solution: Replacing the absolute sysmbols with equations that
can be handled algebraically according to the conditions implied
by the given equation, we have:

$$2x - 1 = 4x + 3 \quad \text{or} \quad 2x - 1 = -(4x + 3) \, .$$

Solving the first equation, we have x =-2; solving the second,
we obtain $x = -\frac{1}{3}$, thus giving us two solutions to the original
equation. (We could also write: $-(2x - 1) = -(4x + 3)$, but this
is equivalent to the first of the equations above.)

18

CHAPTER 5

OPERATIONS WITH FRACTIONS

> **Basic Attacks and Strategies for Solving Problems in this Chapter. See pages 19 to 34 for step-by-step solutions to problems.**

When simplifying a complex algebraic fraction, the usual procedure is to first combine or eliminate the sum and difference of fractions in the numerator and denominator. This can be achieved by multiplying both numerator and denominator by the least common multiple of the denominators of all fractions involved and then simplify by reducing the results to the lowest terms. This is the premier procedure for simplifying, especially complex algebraic fractions. For example, simplify the following expression by first noting that the LCD is xy, which is multiplied by both the numerator and denominator in the rational expression. Thus,

$$\frac{\dfrac{2}{y}+2}{\dfrac{1}{x}-3} = \frac{xy\left(\dfrac{2}{y}\right)+xy(2)}{xy\left(\dfrac{1}{x}\right)-xy(3)} = \frac{2x+2xy}{y-3xy}.$$

Another procedure for simplifying a complex algebraic fraction is to first find the LCD, then add and/or subtract the fractions in the numerator. Repeat the same procedure for the fractions in the denominator. Then, multiply the resulting fraction in the numerator by the reciprocal of the resulting fraction in the denominator and simplify by reducing the results to the lowest terms. In the example below notice that the expressions in the numerator have an LCD of y and those in the denominator have an LCD of x. Thus,

$$\frac{\dfrac{5}{y}+2}{3-\dfrac{2}{x}} = \frac{\dfrac{5}{y}+2\left(\dfrac{y}{y}\right)}{3\left(\dfrac{x}{x}\right)-\dfrac{2}{x}} = \frac{\dfrac{(5+2y)}{y}}{\dfrac{(3x-2)}{x}}$$

Now multiply the result in the numerator by the reciprocal of the result in the denominator and reduce to the lowest terms, if necessary, in order to get the final result as follows:

$$= \frac{5+2y}{y} \cdot \frac{x}{3x-2} = \frac{x(5+2y)}{y(3x-2)}.$$

When multiplying two or more algebraic fractions, the procedure is simply to first factor completely the expressions in the numerator and denominator and cancel all possible common factors. The second step is to multiply the numerators together and then the denominators together. Finally, simplify by reducing to the lowest terms. For example, to multiply the following algebraic fractions easily we first factor, cancel, and then reduce to the lowest terms:

$$\frac{3x-6}{5x-20} \cdot \frac{10x-40}{27x-54} = \frac{3(x-2)}{5(x-4)} \cdot \frac{10(x-4)}{27(x-2)} = \frac{3(10)}{5(27)} = \frac{2}{9}$$

Division of algebraic fractions involves multiplying fractions, except take the reciprocal of the fraction designated as the divisor before multiplying. The result after multiplying is simplified by reducing to the lowest terms.

Step-by-Step Solutions to Problems in this Chapter, "Operations with Fractions"

● PROBLEM 41

Simplify $\dfrac{\frac{1}{2} + \frac{1}{3}}{\frac{1}{6}}$.

__Solution:__ $\dfrac{\frac{1}{2} + \frac{1}{3}}{\frac{1}{6}}$ means $\left(\dfrac{1}{2} + \dfrac{1}{3}\right) \div \dfrac{1}{6}$

Since division by a fraction is equivalent to multiplication by its reciprocal:

$$\left(\frac{1}{2} + \frac{1}{3}\right) \div \frac{1}{6} = \left(\frac{1}{2} + \frac{1}{3}\right) \times 6$$

By the distributive law:

$$\left(\frac{1}{2} + \frac{1}{3}\right) \times 6 = \left(\frac{1}{2}\right)(6) + \left(\frac{1}{3}\right)(6)$$

$$= \frac{6}{2} + \frac{6}{3}$$

$$= 3 + 2$$

$$= 5$$

● PROBLEM 42

Simplify the following expression: $1 - \dfrac{1}{2 - \frac{1}{3}}$.

__Solution:__ In order to combine the denominator, $2 - \frac{1}{3}$, we must convert 2 into thirds. $2 = 2 \cdot 1 = 2 \cdot \dfrac{3}{3} = \dfrac{6}{3}$. Thus

$$1 - \frac{1}{2 - \frac{1}{3}} = 1 - \frac{1}{\frac{6}{3} - \frac{1}{3}} = 1 - \frac{1}{\frac{5}{3}}$$

Since division by a fraction is equivalent to multiplication by that fraction's reciprocal

$$1 - \frac{1}{\frac{5}{3}} = 1 - (1)\left(\frac{3}{5}\right) = 1 - \frac{3}{5} = \frac{5}{5} - \frac{3}{5} = \frac{2}{5}$$

Therefore, $1 - \dfrac{1}{2 - \frac{1}{3}} = \dfrac{2}{5}$.

● PROBLEM 43

Simplify the complex fraction $\dfrac{\frac{1}{2} + \frac{1}{3}}{\frac{1}{4} + \frac{1}{5}}$.

<u>Solution:</u> First simplify the expressions in the numerator and denominator by adding the fractions together according to the rule

$$\frac{a}{b} + \frac{c}{d} = \frac{ad + bc}{bd}$$

We now obtain $\dfrac{1}{2} + \dfrac{1}{3} = \dfrac{3 + 2}{6} = \dfrac{5}{6}$

$$\frac{1}{4} + \frac{1}{5} = \frac{4 + 5}{20} = \frac{9}{20}$$

therefore,

$$\frac{\dfrac{1}{2} + \dfrac{1}{3}}{\dfrac{1}{4} + \dfrac{1}{5}} = \frac{\dfrac{5}{6}}{\dfrac{9}{20}}$$

To divide this complex fraction invert the fraction in the denominator and multiply the resulting fraction by the fraction in the numerator:

$$\frac{\dfrac{1}{2} + \dfrac{1}{3}}{\dfrac{1}{4} + \dfrac{1}{5}} = \frac{5}{6} \times \frac{20}{9} = \frac{100}{54} = \frac{50}{27}$$

● **PROBLEM 44**

Simplify $\dfrac{\dfrac{2}{3} + \dfrac{1}{2}}{\dfrac{3}{4} - \dfrac{1}{3}}$.

<u>Solution:</u> A first method is to just add the terms in the numerator and denominator. Since 6 is the least common denominator of the numerator, $\left(\dfrac{2}{3} + \dfrac{1}{2}\right)$, we convert $\dfrac{2}{3}$ and $\dfrac{1}{2}$ into sixths:

$$\frac{2}{3} = \frac{2}{3} \cdot 1 = \frac{2}{3} \cdot \frac{2}{2} = \frac{4}{6} \text{ and } \frac{1}{2} = \frac{1}{2} \cdot 1 = \frac{1}{2} \cdot \frac{3}{3} = \frac{3}{6}$$

Therefore $\dfrac{2}{3} + \dfrac{1}{2} = \dfrac{4}{6} + \dfrac{3}{6} = \dfrac{7}{6}$

Since 12 is the least common denominator of the denominator, $\left(\dfrac{3}{4} - \dfrac{1}{3}\right)$, we convert $\dfrac{3}{4}$ and $\dfrac{1}{3}$ into twelfths:

$$\frac{3}{4} = \frac{3}{4} \cdot 1 = \frac{3}{4} \cdot \frac{3}{3} = \frac{9}{12} \text{ and } \frac{1}{3} = \frac{1}{3} \cdot 1 = \frac{1}{3} \cdot \frac{4}{4} = \frac{4}{12}$$

Therefore $\dfrac{3}{4} - \dfrac{1}{3} = \dfrac{9}{12} - \dfrac{4}{12} = \dfrac{5}{12}$

Thus, $\dfrac{\dfrac{2}{3} + \dfrac{1}{2}}{\dfrac{3}{4} - \dfrac{1}{3}} = \dfrac{\dfrac{7}{6}}{\dfrac{5}{12}}$

Division by a fraction is equivalent to multiplication by the reciprocal hence $\dfrac{\dfrac{7}{6}}{\dfrac{5}{12}} = \dfrac{7}{6} \cdot \dfrac{12}{5}$

Cancelling 6 from the numerator and denominator:

$$= \dfrac{7}{1} \cdot \dfrac{2}{5} = \dfrac{14}{5}$$

A second method is to multiply both numerator and denominator by the least common denominator of the entire fraction. Since we have already seen that L.C.D. of the numerator is 6 and the L.C.D. of the denominator is 12, and 12 is divisible by 6, we use 12 as the L.C.D. of the entire fraction. Thus

$$\dfrac{\dfrac{2}{3} + \dfrac{1}{2}}{\dfrac{3}{4} - \dfrac{1}{3}} = \dfrac{12\left(\dfrac{2}{3} + \dfrac{1}{2}\right)}{12\left(\dfrac{3}{4} - \dfrac{1}{3}\right)}$$

Distribute:

$$= \dfrac{12\left(\dfrac{2}{3}\right) + 12\left(\dfrac{1}{2}\right)}{12\left(\dfrac{3}{4}\right) - 12\left(\dfrac{1}{3}\right)}$$

$$= \dfrac{4 \cdot 2 + 6}{3 \cdot 3 - 4} = \dfrac{8 + 6}{9 - 4} = \dfrac{14}{5}$$

● **PROBLEM** 45

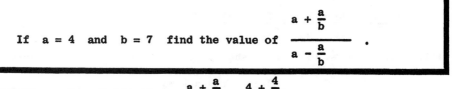

If $a = 4$ and $b = 7$ find the value of $\dfrac{a + \dfrac{a}{b}}{a - \dfrac{a}{b}}$.

<u>Solution:</u> By substitution, $\dfrac{a + \dfrac{a}{b}}{a - \dfrac{a}{b}} = \dfrac{4 + \dfrac{4}{7}}{4 - \dfrac{4}{7}}$.

In order to combine the terms we convert 4 into sevenths:

$$4 = 4 \cdot 1 = 4 \cdot \dfrac{7}{7} = \dfrac{28}{7} .$$

Thus, we have:

$$\dfrac{\dfrac{28}{7} + \dfrac{4}{7}}{\dfrac{28}{7} - \dfrac{4}{7}} = \dfrac{\dfrac{32}{7}}{\dfrac{24}{7}} .$$

Dividing by $\frac{24}{7}$ is equivalent to multiplying by $\frac{7}{24}$. Therefore,

$$\frac{4 + \frac{4}{7}}{4 - \frac{4}{7}} = \frac{32}{7} \cdot \frac{7}{24}$$

Now, the 7 in the numerator cancels with the 7 in the denominator.

Thus, we obtain: $\frac{32}{24}$, and dividing numerator and denominator by 8,

we obtain: $\frac{4}{3}$.

Therefore $\dfrac{a + \frac{a}{b}}{a - \frac{a}{b}} = \frac{4}{3}$ when $a = 4$ and $b = 7$.

● **PROBLEM** 46

Perform the following division: $1 \bigg/ \dfrac{x + y}{x^2}$.

Solution: Division by a fraction is equivalent to multiplication by that fraction's reciprocal. Hence,

$$\frac{1}{\frac{x + y}{x^2}} = 1 \cdot \frac{x^2}{x + y} = \frac{x^2}{x + y}.$$

● **PROBLEM** 47

Perform the indicated operation:

$$\frac{3a - 9b}{x - 5} \cdot \frac{xy - 5y}{ax - 3bx}$$

Solution: According to our definition of multiplication, we need only to write the product of the numerators over the product of the denominators. The only remaining step is that of reducing the fraction to lowest terms by factoring the numerator and denominator of the answer and simplifying the result.

$$\frac{3a - 9b}{x - 5} \cdot \frac{xy - 5y}{ax - 3bx} = \frac{(3a - 9b)(xy - 5y)}{(x - 5)(ax - 3bx)}$$

Factor out 3 from (3a - 9b) and y from (xy - 5y). Also, factor out x from ax - 3bx.

$$= \frac{3(a - 3b)y(x - 5)}{(x - 5)x(a - 3b)}$$

Group the same terms in numerator and the denominator.

$$= \frac{3y}{x} \cdot \frac{a - 3b}{a - 3b} \cdot \frac{x - 5}{x - 5}$$

Cancel like terms.

$$= \frac{3y}{x} \cdot 1 \cdot 1$$

$$= \frac{3y}{x}$$

This procedure could have been abbreviated in the following manner:

$$\frac{3a - 9b}{x - 5} \cdot \frac{xy - 5y}{ax - 3bx} = \frac{3\cancel{(a - 3b)}}{\cancel{x - 5}} \cdot \frac{y\cancel{(x - 5)}}{x\cancel{(a - 3b)}} = \frac{3y}{x}$$

● **PROBLEM   48**

Simplify $\dfrac{4x^3 + 6x^2}{2x}$

__Solution:__ Since the denominator of a fraction cannot equal zero, $2x \neq 0$ or, dividing by 2 we obtain the restriction $x \neq 0$.

Now we proceed to simplify the given expression. First we note that $2x$ may be factored out of the numerator; thus

$$\frac{4x^3 + 6x^2}{2x} = \frac{2x(2x^2 + 3x)}{2x}$$

$$= \frac{2x}{2x} \cdot (2x^2 + 3x)$$

$$= 2x^2 + 3x.$$

Thus, $\dfrac{4x^3 + 6x^2}{2x} = 2x^2 + 3x$, and $x \neq 0$.

● **PROBLEM   49**

Simplify $\dfrac{1 + \dfrac{1}{x}}{1 - \dfrac{1}{x}}$.

__Solution:__ In order to eliminate the fractions in the numerator and denominator we multiply numerator and denominator by x.

$$\frac{1 + \frac{1}{x}}{1 - \frac{1}{x}} = \frac{x\left(1 + \frac{1}{x}\right)}{x\left(1 - \frac{1}{x}\right)} = \frac{x + \frac{x}{x}}{x - \frac{x}{x}} = \frac{x + 1}{x - 1} .$$

● **PROBLEM   50**

Combine into a single fraction in lowest terms.

(a) $\dfrac{6(a + 1)}{a + 8} - \dfrac{3(a - 4)}{a + 8} - \dfrac{2(a + 5)}{a + 8}$

(b) $\dfrac{7x - 3y + 6}{x + y} - \dfrac{2(x - 4y + 3)}{x + y}$

(c) $\dfrac{5x + 2}{x - 6} - \dfrac{3(x + 4)}{x - 6} - \dfrac{x - 7}{x - 6}$

__Solution:__ Noting $\dfrac{a}{x} + \dfrac{b}{x} + \dfrac{c}{x} = \dfrac{a+b+c}{x}$ (where a,b,c are any real numbers and x any non-zero real number), we proceed to evaluate these expressions:

(a) $\dfrac{6(a + 1)}{a + 8} - \dfrac{3(a - 4)}{a + 8} - \dfrac{2(a + 5)}{a + 8} = \dfrac{6(a+1) - 3(a-4) - 2(a+5)}{a+8}$

__Distributing,__ $= \dfrac{6a + 6 - 3a + 12 - 2a - 10}{a+8} = \dfrac{6a-3a-2a+6+12-10}{a+8}$

$$= \frac{a + 8}{a + 8} = 1.$$

(b) $\frac{7x - 3y + 6}{x + y} - \frac{2(x - 4y + 3)}{x + y} = \frac{7x - 3y + 6 - 2(x-4y+3)}{x + y}$

Distributing, $= \frac{7x - 3y + 6 - 2x + 8y - 6}{x + y} = \frac{7x-2x-3y+8y+6-6}{x + y}$

$$= \frac{5x + 5y}{x + y} = \frac{5(x + y)}{x + y} = 5.$$

(c) $\frac{5x + 2}{x - 6} - \frac{3(x + 4)}{x - 6} - \frac{x - 7}{x - 6} = \frac{5x + 2 - 3(x+4) - (x-7)}{x-6}$

Distributing, $= \frac{5x+2-3x-12-x+7}{x-6} = \frac{5x-3x-x+2-12+7}{x-6} = \frac{x - 3}{x - 6}.$

● **PROBLEM** 51

Combine and simplify $1 + \dfrac{1}{1 + \dfrac{1}{1 - x}}$.

Solution: First combine the terms in the denominator. Recall $1 = (1-x)/(1-x)$. Thus,

$$1 + \frac{1}{1 + \frac{1}{1 - x}} = 1 + \frac{1}{\frac{1 - x}{1 - x} + \frac{1}{1 - x}}$$

$$= 1 + \frac{1}{\frac{1 - x + 1}{1 - x}}$$

$$= 1 + \frac{1}{\frac{2 - x}{1 - x}}$$

Division by a fraction is equivalent to multiplication by its reciprocal, thus

$$= 1 + 1 \cdot \frac{(1 - x)}{(2 - x)}$$

$$= 1 + \frac{1 - x}{2 - x}$$

Recall $1 = \dfrac{2 - x}{2 - x}$, therefore,

$$1 + \frac{1}{1 + \frac{1}{1 - x}} = \frac{2 - x}{2 - x} + \frac{1 - x}{2 - x}$$

$$= \frac{2 - x + 1 - x}{2 - x}$$

$$= \frac{3 - 2x}{2 - x} .$$

Simplify the complex fraction $\dfrac{\dfrac{1}{x} - \dfrac{1}{y}}{\dfrac{1}{x^2} - \dfrac{1}{y^2}}$.

Solution: Add both fractions of the numerator together using the rule: $\dfrac{a}{b} + \dfrac{c}{d} = \dfrac{ad + bc}{bd}$; and obtain

$\dfrac{y - x}{xy}$. Similarly for the denominator, obtain:

$$\dfrac{y^2 - x^2}{x^2y^2}$$

Now invert the fraction in the denominator and multiply by the numerator:

$$\dfrac{y - x}{xy} \cdot \dfrac{x^2y^2}{y^2 - x^2} = \dfrac{(y - x)}{xy} \cdot \dfrac{(xy)(xy)}{(y - x)(y + x)}$$

$$= \dfrac{(y - x)(xy)(xy)}{xy(y - x)(y + x)}$$

$$= \dfrac{xy}{y + x} \quad .$$

Simplify $\qquad \dfrac{x + \dfrac{1}{y}}{x - \dfrac{1}{y}}$.

Solution: Obtain the least common denominator, ℓ.c.d., of the two terms in the numerator and of those that appear in the denominator. This is done by writing down the different factors that appear in the denominators of the terms. The exponent to be used for each factor is the smallest number of times that the factor appears in either of the denominators of the terms. Hence, the ℓ.c.d. of the two terms in the denominator $= (1)^1(y)^1 = 1y = y$. Also, the ℓ.c.d. of the two terms in the numerator is obtained in the same way. Therefore, the ℓ.c.d. of the two terms in the numerator $= (1)^1(y)^1 = 1y = y$. Therefore:

$$\dfrac{x + \dfrac{1}{y}}{x - \dfrac{1}{y}} = \dfrac{\dfrac{y(x)}{y} + \dfrac{1}{y}}{\dfrac{y(x)}{y} - \dfrac{1}{y}} = \dfrac{\dfrac{yx + 1}{y}}{\dfrac{yx - 1}{y}}$$

Division is the same as multiplying the numerator by the multiplicative inverse of the denominator. (The multiplicative inverse of a number a is the number n, such that $a \cdot n = 1$. This number n is $1/a$. Hence, $a \cdot 1/a = 1$.) Therefore, the multiplicative inverse of

$$\dfrac{yx - 1}{y} \quad \text{is} \quad \dfrac{y}{yx - 1}.$$

Hence,

$$\dfrac{x + \dfrac{1}{y}}{x - \dfrac{1}{y}} = \dfrac{\dfrac{yx + 1}{y}}{\dfrac{yx - 1}{y}} = \left(\dfrac{yx + 1}{\cancel{y}}\right)\left(\dfrac{\cancel{y}}{yx - 1}\right) = \dfrac{yx + 1}{yx - 1} \ .$$

● **PROBLEM** 54

Simplify the following expressions:

(a) $\quad \dfrac{\dfrac{a}{b} + \dfrac{a}{c}}{ab + ac}$ $\qquad$ (b) $\quad \dfrac{2 - \dfrac{1}{4}}{\dfrac{3}{5} + 1}$

Solution:

(a) In order to combine the fractions $\dfrac{a}{b}$ and $\dfrac{a}{c}$ in the numerator, we convert them into fractions with the same denominator by multiplying a/b by c/c (which is equal to 1) and a/c by b/b (also equal to 1). Multiplication by a fraction equal to 1 does not change the value of the original fractions. Thus,

$$\dfrac{\left(\dfrac{a}{b}\right)\left(\dfrac{c}{c}\right) + \left(\dfrac{a}{c}\right)\left(\dfrac{b}{b}\right)}{ab + ac} = \dfrac{\dfrac{ac}{bc} + \dfrac{ab}{bc}}{ab + ac} = \dfrac{\dfrac{ac + ab}{bc}}{ab + ac}$$

Multiplying numerator and denominator by bc,

$$= \dfrac{\dfrac{ac + ab}{bc}(bc)}{(ab + ac)(bc)} = \dfrac{ac + ab}{(ac + ab)(bc)} = \dfrac{1}{bc} \ .$$

(b) $\dfrac{2 - \dfrac{1}{4}}{\dfrac{3}{5} + 1} = \dfrac{(2)\left(\dfrac{4}{4}\right) - \dfrac{1}{4}}{\dfrac{3}{5} + (1)\left(\dfrac{5}{5}\right)} = \dfrac{\dfrac{8}{4} - \dfrac{1}{4}}{\dfrac{3}{5} + \dfrac{5}{5}} = \dfrac{\dfrac{7}{4}}{\dfrac{8}{5}}$

Since division by a fraction is equivalent to multiplication by that fraction's reciprocal

$$= \dfrac{7}{4} \cdot \dfrac{5}{8}$$

$$= \dfrac{35}{32} \ .$$

● **PROBLEM** 55

Simplify: $\quad \dfrac{x - \dfrac{2}{y}}{x + \dfrac{3}{y}} \ .$

Solution: The Lowest Common Multiple of the denominators is y .

Since $\dfrac{y}{y} = 1$, multiply numerator and denominator by y . Thus,

we obtain:

$$\dfrac{y\left(x - \dfrac{2}{y}\right)}{y\left(x + \dfrac{3}{y}\right)}$$

Distribute:

$$\frac{yx - y\left(\frac{2}{y}\right)}{yx + y\left(\frac{3}{y}\right)}$$

$$= \frac{yx - 2}{yx + 3}$$

therefore,

$$\frac{x - \frac{2}{y}}{x + \frac{3}{y}} = \frac{yx - 2}{yx + 3} \quad .$$

Simplify this expression:

$$\frac{\frac{3}{x} - \frac{2}{y}}{\frac{5}{x} + \frac{6}{y}}$$

Solution: There are two ways to approach this problem. One is to consider it as a division problem:

$$\left(\frac{3}{x} - \frac{2}{y}\right) \div \left(\frac{5}{x} + \frac{6}{y}\right) \quad .$$

Use the least common denominator xy :

$$= \left(\frac{3y}{xy} - \frac{2x}{xy}\right) \div \left(\frac{5y}{xy} + \frac{6x}{xy}\right) \quad .$$

Combine fractions:

$$= \frac{3y-2x}{xy} \div \frac{5y+6x}{xy} .$$

Dividing by a fraction is equivalent to multiplying by its reciprocal:

$$= \left(\frac{3y-2x}{xy}\right) \cdot \left(\frac{xy}{5y+6x}\right) \quad .$$

Cancelling out xy : $= \dfrac{3y-2x}{5y+6x} \quad .$

The second approach is to multiply both numerator and denominator by xy; this is equivalent to multiplying the fraction by 1:

$$\frac{\frac{3}{x} - \frac{2}{y}}{\frac{5}{x} + \frac{6}{y}} \cdot \frac{xy}{xy} = \frac{3y - 2x}{5y + 6x} \quad .$$

Simplify $\dfrac{\frac{2}{x} + \frac{3}{y}}{1 - \frac{1}{x}} \quad .$

Solution: A first method is to just add the terms in the numerator and denominator, obtaining

$$\frac{\frac{2}{x} + \frac{3}{y}}{1 - \frac{1}{x}} = \frac{\frac{2y}{xy} + \frac{3x}{xy}}{\frac{x}{x} - \frac{1}{x}} = \frac{\frac{2y + 3x}{xy}}{\frac{x - 1}{x}}$$

Since dividing by fraction is equivalent to multiplying by its reciprocal,

$$= \frac{2y + 3x}{xy} \cdot \frac{x}{x - 1} = \frac{2y + 3x}{y(x - 1)}$$

A second method is to multiply both numerator and denominator by the least common denominator of the entire fraction, in this case xy:

$$\frac{\frac{2}{x} + \frac{3}{y}}{1 - \frac{1}{x}} = \frac{xy\left(\frac{2}{x} + \frac{3}{y}\right)}{xy\left(1 - \frac{1}{x}\right)}$$

Distributing,
$$= \frac{xy\left(\frac{2}{x}\right) + xy\left(\frac{3}{y}\right)}{xy(1) - xy\left(\frac{1}{x}\right)}$$

Cancelling like terms,

$$= \frac{2y + 3x}{xy - y}$$

Using distributive law,

$$= \frac{2y + 3x}{y(x - 1)} .$$

● **PROBLEM** 58

Combine $\quad a + b - \dfrac{2ab}{a + b}$.

Solution: In order to combine fractions we must transform them into equivalent fractions with a common denominator. In our case we will use a + b as our least common denominator (LCD). Thus

$$a + b - \frac{2ab}{a+b} = \frac{a+b}{a+b}\left(\frac{a+b}{1}\right) - \frac{2ab}{a+b}$$

$$= \frac{(a+b)(a+b)}{a + b} - \frac{2ab}{a+b}$$

$$= \frac{a^2 + 2ab + b^2}{a + b} - \frac{2ab}{a+b}$$

$$= \frac{a^2 + 2ab + b^2 - 2ab}{a + b}$$

$$= \frac{a^2 + b^2}{a + b}$$

● **PROBLEM** 59

Add $\dfrac{2}{x - 3}$ and $\dfrac{5}{x + 2}$.

Solution: Change these fractions to fractions with a common denominator, then add fractions by adding numerators and placing over the common denominator. Neither of the denominators is factorable; therefore, the LCD is the product of the denominators. LCD = $(x - 3)(x + 2)$. To change $\frac{2}{x - 3}$ to an equivalent fraction with $(x - 3)(x + 2)$ as its denominator, multiply by the unit fraction $\frac{(x + 2)}{(x + 2)}$. To change $\frac{5}{x + 2}$ to an equivalent fraction with the LCD as its denominator, multiply by $\frac{x - 3}{x - 3}$. Then add the resulting fractions as follows:

$$\frac{2}{x - 3} + \frac{5}{x + 2} = \frac{2(x + 2)}{(x - 3)(x + 2)} + \frac{5(x - 3)}{(x - 3)(x + 2)}$$

$$= \frac{2(x + 2) + 5(x - 3)}{(x - 3)(x + 2)}$$

$$= \frac{2x + 4 + 5x - 15}{(x - 3)(x + 2)}$$

$$= \frac{7x - 11}{(x - 3)(x + 2)} \qquad \text{Combining Terms}$$

The numerator is not factorable, so the result can not be reduced.

● **PROBLEM** 60

Combine $\frac{1}{6x} + \frac{1}{3y} - \frac{3x + 2y}{12xy}$ into a single fraction.

Solution: Since both 6x and 3y are factors of 12xy, the least common denominator (the LCD) of the given fractions is 12xy. Thus, we wish to convert the given fractions to equal fractions having 12xy as a denominator. We can accomplish this by multiplying each member of the first fraction by 2y and each member of the second by 4x. We thereby obtain

$$\frac{1}{6x} + \frac{1}{3y} - \frac{3x + 2y}{12xy} = \frac{2y \cdot 1}{2y(6x)} + \frac{4x \cdot 1}{4x(3y)} - \frac{3x + 2y}{12xy}$$

$$= \frac{2y}{12xy} + \frac{4x}{12xy} - \frac{3x + 2y}{12xy}$$

$$= \frac{2y + 4x - (3x + 2y)}{12xy}$$

$$= \frac{2y + 4x - 3x - 2y}{12xy}$$

$$= \frac{2y + x - 2y}{12xy}$$

$$= \frac{2y - 2y + x}{12xy}$$

$$= \frac{x}{12xy} .$$

Cancelling x from numerator and denominator,

$$= \frac{1}{12y}$$

Thus, $\frac{1}{6x} + \frac{1}{3y} - \frac{3x + 2y}{12xy} = \frac{1}{12y}$.

● PROBLEM 61

Perform the indicated operations:

$$\frac{3a}{2xy} - \frac{2 - 5x}{y^3} + 6$$

Solution: The first step in adding or subtracting fractions is to convert them into equivalent fractions having like denominators. The
simplest method is to find the least common denominator (LCD). The
LCD is the product of the unique prime factors of all the original
denominators, each factor having as its exponent the positive integer
representing the largest number of times the factor appeared in an
original denominator. In this example the first denominator is
$2 \cdot x \cdot y$, and the second is y^3. Since the factor y appears to the
third power in the second denominator, the LCD is $2xy^3$. Hence, we
multiply the numerator and denominator of each term by the factor
necessary to make the denominator equal to $2xy^3$.

$$\frac{3a}{2xy} - \frac{2 - 5x}{y^3} + \frac{6}{1} = \frac{3a\left(y^2\right)}{2xy\left(y^2\right)} - \frac{(2 - 5x)(2x)}{y^3 \quad (2x)} + \frac{6\left(2xy^3\right)}{1\left(2xy^3\right)}$$

$$= \frac{3ay^2}{2xy^3} - \frac{\left(4x - 10x^2\right)}{2xy^3} + \frac{12xy^3}{2xy^3}$$

$$= \frac{3ay^2 - 4x + 10x^2 + 12xy^3}{2xy^3}$$

● PROBLEM 62

Simplify $\dfrac{\dfrac{1}{x - 1} - \dfrac{1}{x - 2}}{\dfrac{1}{x - 2} - \dfrac{1}{x - 3}}$.

Solution: Simplify the expression in the numerator by
using the addition rule:

$$\frac{a}{b} + \frac{c}{d} = \frac{ad + bc}{bd}$$

Notice bd is the Least Common Denominator, LCD.

We obtain $\dfrac{x - 2 - (x - 1)}{(x - 1)(x - 2)} = \dfrac{-1}{(x - 1)(x - 2)}$ in the

numerator.

Repeat this procedure for the expression in the
denominator:

$$\frac{x - 3 - (x - 2)}{(x - 2)(x - 3)} = \frac{-1}{(x - 2)(x - 3)}$$

We now have

$$\frac{\dfrac{-1}{(x - 1)(x - 2)}}{\dfrac{-1}{(x - 2)(x - 3)}} ,$$

which is simplified by inverting the fraction in the denominator and multiplying it by the numerator and cancelling like terms

$$\frac{-1}{(x - 1)(x - 2)} \cdot \frac{(x - 2)(x - 3)}{-1} = \frac{x - 3}{x - 1} .$$

● **PROBLEM** 63

Simplify:
$$\frac{\dfrac{1}{a} - \dfrac{1}{a - b}}{\dfrac{1}{a} + \dfrac{1}{a - b}}$$

Solution: The Lowest Common Multiple (L.C.M) of the denominators is

$a(a - b)$. Since $\dfrac{a(a-b)}{a(a-b)} = 1$, multiply numerator and denominator by

$a(a - b)$. Thus, we have:

$$\frac{a(a - b)\left(\dfrac{1}{a} - \dfrac{1}{a - b}\right)}{a(a - b)\left(\dfrac{1}{a} + \dfrac{1}{a - b}\right)} .$$

Distribute $a(a - b)$ in numerator and denominator:

$$\frac{a(a - b)\left(\dfrac{1}{a}\right) - a(a - b)\left(\dfrac{1}{a - b}\right)}{a(a - b)\left(\dfrac{1}{a}\right) + a(a - b)\left(\dfrac{1}{a - b}\right)} .$$

Perform the multiplication:

$$\frac{\dfrac{a}{a}(a - b) - a\left(\dfrac{a - b}{a - b}\right)}{\dfrac{a}{a}(a - b) + a\left(\dfrac{a - b}{a - b}\right)}$$

Since $\dfrac{a}{a} = 1$ and $\dfrac{a - b}{a - b} = 1$, we have: $\dfrac{(a - b) - a}{(a - b) + a}$.

Using Associative and Commutative laws of addition, we obtain:

$$\frac{\dfrac{1}{a} - \dfrac{1}{a - b}}{\dfrac{1}{a} + \dfrac{1}{a - b}} = \frac{(a - a) - b}{(a + a) - b} = \frac{-b}{2a - b} .$$

31

Simplify $\dfrac{\dfrac{x}{x+y} + \dfrac{y}{x-y}}{\dfrac{y}{x+y} - \dfrac{x}{x-y}}$.

Solution: To eliminate the fractions in this expression, we multiply numerator and denominator by the least common multiple (L.C.M.), the expression of lowest degree into which each of the original expressions can be divided without a remainder. The L.C.M. is the product obtained by taking each factor to the highest degree. In our case the L.C.M. is $(x + y)(x - y)$. Thus, multiplying numerator and denominator by $(x + y)(x - y)$,

$$\frac{\dfrac{x}{x+y} + \dfrac{y}{x-y}}{\dfrac{y}{x+y} - \dfrac{x}{x-y}} = \frac{\dfrac{x}{x+y} + \dfrac{y}{x-y}}{\dfrac{y}{x+y} - \dfrac{x}{x-y}} \cdot \frac{(x+y)(x-y)}{(x+y)(x-y)} .$$

Distributing, $= \dfrac{\left(\dfrac{x}{x+y}\right)(x+y)(x-y) + \left(\dfrac{y}{x-y}\right)(x+y)(x-y)}{\left(\dfrac{y}{x+y}\right)(x+y)(x-y) - \left(\dfrac{x}{x-y}\right)(x+y)(x-y)}$

Cancelling like terms, $= \dfrac{x(x-y) + y(x+y)}{y(x-y) - x(x+y)}$

Distributing, $= \dfrac{x^2 - xy + yx + y^2}{yx - y^2 - x^2 - xy}$

Using the commutative law, $= \dfrac{x^2 - xy + xy + y^2}{xy - y^2 - x^2 - xy}$

Combining terms, $= \dfrac{x^2 + y^2}{-x^2 - y^2}$

Factoring (-1) from the denominator $= \dfrac{x^2 + y^2}{(-1)(x^2 + y^2)}$

Cancelling $x^2 + y^2$, $= \dfrac{1}{-1}$

$= -1.$

Thus, $\dfrac{\dfrac{x}{x+y} + \dfrac{y}{x-y}}{\dfrac{y}{x+y} - \dfrac{x}{x-y}} = -1.$

● **PROBLEM** 65

A) If $x = \dfrac{c - ab}{a - b}$, find the value of the expression $a(x + b)$.

B) Also, if $x = \dfrac{c - ab}{a - b}$, find the value of the expression $bx + c$.

<u>Solution:</u> A) Substituting $x = \dfrac{c - ab}{a - b}$ for x in the expression $a(x + b)$,

$$a(x + b) = a\left(\frac{c - ab}{a - b} + b\right) \qquad (1)$$

Obtaining a common denominator of a - b for the two terms in parenthesis; equation (1) becomes:

$$a(x + b) = a\left[\frac{c - ab}{a - b} + \frac{(a-b)b}{a - b}\right]$$

Distributing the numerator of the second term in brackets:

$$a(x + b) = a\left[\frac{c-ab}{a-b} + \frac{ab-b^2}{a-b}\right] = a\left[\frac{c-ab+ab-b^2}{a-b}\right]$$

$$= a\left[\frac{c-b^2}{a-b}\right]$$

$$a(x + b) = \frac{a\left(c-b^2\right)}{a-b}$$

B) Substituting $x = \dfrac{c-ab}{a-b}$ for x in the expression bx + c,

$$bx + c = b\left(\frac{c-ab}{a-b}\right) + c$$

$$= \frac{b(c-ab)}{a-b} + c \qquad (2)$$

Obtaining a common denominator of a - b for the two terms on the right side of equation (2):

$$bx + c = \frac{b(c-ab)}{a-b} + \frac{(a-b)c}{a-b}$$

Distributing the numerator of each term on the right side:

$$bx + c = \frac{bc-ab^2}{a-b} + \frac{ac-bc}{a-b}$$

$$= \frac{bc-ab^2+ac-bc}{a-b} = \frac{-ab^2+ac}{a-b}$$

$$= \frac{ac-ab^2}{a-b}$$

Factoring out the common factor of a from the numerator of the right side:

$$bx + c = \frac{a\left(c-b^2\right)}{a-b}$$

● **PROBLEM** 66

When two resistances are installed in an electric circuit in parallel, the reciprocal of the resistance of the system is equal to the sum of the reciprocals of the parallel resistances. If r_1 and r_2 represent the resistances installed and R the resistance of the system, then

$$\frac{1}{R} = \frac{1}{r_1} + \frac{1}{r_2}$$

What single resistance is the equivalent of resistances of 10 ohms and 25 ohms wired in parallel?

<u>Solution:</u> Let r_1 = 10 ohms and r_2 = 25 ohms. We are looking for the single resistance R, which is equivalent to r_1 and r_2.

Here the reciprocal of R $= \dfrac{1}{R}$

Here the reciprocal of $r_1 = \dfrac{1}{r_1}$

and the reciprocal of $r_2 = \dfrac{1}{r_2}$

Now substitute the values for r_1 and r_2 respectively into the equation. Thus,

$$\frac{1}{R} = \frac{1}{10} + \frac{1}{25}$$

Add the fractions according to the rule

$$\frac{a}{b} + \frac{c}{d} = \frac{ad + bc}{bd}$$

$$\frac{1}{R} = \frac{25 + 10}{250} = \frac{35}{250}$$

$$R = \frac{250}{35} = \frac{50}{7} = 7.14 \text{ ohms.}$$

CHAPTER 6

BASE, EXPONENT, POWER

Basic Attacks and Strategies for Solving Problems in this Chapter. See pages 35 to 52 for step-by-step solutions to problems.

For any real number X and any positive integer n, the nth power of X, denoted by X^n, is defined to be the product of n factors of X. In X^n, X is referred to as the base and n is referred to as the exponent. Rules for performing certain operations with powers of any real numbers, say X and Y, and any positive integers, n and m, are as follows:

(1) To multiply two powers of the same base, simply add their exponents and use this sum as the exponent of the common base

$$(\text{e.g.}, X^n * X^m = X^{n+m});$$

(2) To divide two powers with the same base, simply subtract the exponent of the divisor from the exponent of the dividend and use this difference as the exponent on the common base

$$(\text{e.g.}, X^n / X^m = X^{n-m}, \quad n > m > 0);$$

(3) To determine the power of a power, simply multiply the exponents and use this product as the exponent of the base

$$(\text{e.g.}, (X^n)^m = X^{nm});$$

(4) The power of a product is the product of the separate powers

$$(\text{e.g.}, (XY)^n = X^n Y^n); \text{ and}$$

(5) The power of a quotient is the quotient of the separate powers

$$(\text{e.g.}, (X/Y)^n = X^n / Y^n).$$

Note that powers of sums and differences cannot be taken term by term, rather binomial expansion must be done.

If the power of a nonzero real number X is negative, then the first step before applying the above rules is to change all negative exponents to positive exponents by using the definition

$$X^{-n} = 1/X^n.$$

For any real number $X > 0$ and any positive rational number, p/q exponent, the aforementioned rules for performing operations with powers apply. On the other hand, if the exponent is a negative rational number, then first use the definition of negative exponents $X^{-p/q}$ and then simplify the resulting fraction.

For any nonzero real number X, $X^0 = 1$ but 0^0 is undefined.

Step-by-Step Solutions to Problems in this Chapter, "Base, Exponent, Power"

● PROBLEM 67

Simplify: (a) 3^{-2} (b) $\dfrac{1}{5^{-2}}$

Solution: (a) Since $x^{-a} = \dfrac{1}{x^a}$, $3^{-2} = \dfrac{1}{3^2} = \dfrac{1}{3 \cdot 3} = \dfrac{1}{9}$.

(b) Again, recall $\dfrac{1}{x^a} = x^{-a}$; hence,

$$\dfrac{1}{5^{-2}} = 5^{-(-2)} = 5^2 = 5 \cdot 5 = 25.$$

● PROBLEM 68

Simplify the following expressions:

(a) -3^{-2} (b) $(-3)^{-2}$ (c) $\dfrac{-3}{4^{-1}}$

Solution:

(a) Here the exponent applies only to 3.
Since $x^{-y} = \dfrac{1}{x^y}$, $-3^{-2} = -(3^{-2}) = -\dfrac{1}{3^2} = -\dfrac{1}{9}$.

(b) In this case the exponent applies to the negative base.
Thus, $(-3)^{-2} = \dfrac{1}{(-3)^2} = \dfrac{1}{(-3)(-3)} = \dfrac{1}{9}$.

(c) $\dfrac{-3}{4^{-1}} = \dfrac{-3}{(\frac{1}{4})^1} = \dfrac{-3}{\frac{1}{4^1}} = \dfrac{-3}{\frac{1}{4}}$.

Division by a fraction is equivalent to multiplication by that fraction's reciprocal, thus

$$\dfrac{-3}{\frac{1}{4}} = -3 \cdot \dfrac{4}{1} = -12,$$

and $\dfrac{-3}{4^{-1}} = -12.$

Evaluate:

(a) $8\left(-\frac{1}{4}\right)^0$ (b) $6^0 + (-6)^0$ (c) $-7(-3)^0$ (d) 9^{-1} (e) 7^{-2}.

Solution: Note $x^0 = 1$ and $x^{-a} = \dfrac{1}{x^a}$ for all non-zero real numbers x,

(a) $8\left(-\frac{1}{4}\right)^0 = 8(1) = 8$

(b) $6^0 + (-6)^0 = 1 + 1 = 2$

(c) $-7(-3)^0 = -7(1) = -7$

(d) $9^{-1} = \dfrac{1}{9^1} = \dfrac{1}{9}$

(e) $7^{-2} = \dfrac{1}{7^2} = \dfrac{1}{49}$

Simplify the expression $\left(3^{-1} + 2^{-1}\right)^{-2}$.

Solution: Since $x^{-y} = \dfrac{1}{x^y}$, $3^{-1} = \dfrac{1}{3^1} = \dfrac{1}{3}$ and $2^{-1} = \dfrac{1}{2^1} = \dfrac{1}{2}$. Thus,

$$\left(3^{-1} + 2^{-1}\right)^{-2} = \left(\tfrac{1}{3} + \tfrac{1}{2}\right)^{-2}.$$

Now, we combine fractions, using 6 as our least common denominator:

$$= \left[\tfrac{2}{2}(\tfrac{1}{3}) + \tfrac{3}{3}(\tfrac{1}{2})\right]^{-2}$$

$$= \left(\tfrac{2}{6} + \tfrac{3}{6}\right)^{-2}$$

$$= \left(\tfrac{5}{6}\right)^{-2}$$

$$= \dfrac{1}{\left(\tfrac{5}{6}\right)^2}$$

$$= \dfrac{1}{\dfrac{25}{36}}$$

and since division by a fraction is equivalent to multiplying the numerator by the reciprocal of the denominator, we have:

$$= 1 \times \dfrac{36}{25}$$

$$= \dfrac{36}{25}.$$

Perform the indicated operations:

$(7 \cdot 10^5)^3 \cdot (3 \cdot 10^{-3})^4.$

<u>Solution:</u> Since $(ab)^x = a^x b^x$,

$$(7 \cdot 10^5)^3 \cdot (3 \cdot 10^{-3})^4 = (7)^3 (10^5)^3 \cdot (3)^4 (10^{-3})^4.$$

Recall that $\left(a^x\right)^y = a^{xy}$. Thus,

$$= (7^3)(10^{5 \cdot 3}) \cdot (3^4)(10^{-3 \cdot 4})$$

$$= (7^3)(10^{15}) \cdot (3^4)(10^{-12})$$

$$= (7^3)(3^4)(10^{15})(10^{-12}).$$

Since $a^x \cdot a^y = a^{x+y}$,
$$= (7^3)(3^4)\left[10^{15+(-12)}\right]$$

$$= 7^3 3^4 10^3.$$

● **PROBLEM 72**

Simplify:

(a) $2^3 \cdot 2^2$ (b) $a^3 \cdot a^5$ (c) $x^6 \cdot x^4$

<u>Solution:</u> If a is any number and n is any positive integer, the product of the n factors a·a·a ... a is denoted by a^n. a is called the base and n is called the exponent. Also, for base a and exponents m and n, m and n being positive integers, we have the law:
$$a^m \cdot a^n = a^{m+n}.$$

Therefore,

(a) $2^3 \cdot 2^2 = (2 \cdot 2 \cdot 2)(2 \cdot 2) = 8 \cdot 4 = 32$

or $2^3 \cdot 2^2 = 2^{3+2} = 2^5 = 32$

(b) $a^3 \cdot a^5 = (a \cdot a \cdot a)(a \cdot a \cdot a \cdot a \cdot a)$

$$= (a \cdot a \cdot a \cdot a \cdot a \cdot a \cdot a \cdot a) = a^8$$

or $a^3 \cdot a^5 = a^{3+5} = a^8$

(c) $x^6 \cdot x^4 = x^{6+4} = x^{10}.$

● **PROBLEM 73**

Use the laws of exponents to perform the indicated operations:

(a) $5x^5 \cdot 2x^2$ (b) $\left(x^4\right)^6$ (c) $\dfrac{8y^8}{2y^2}$ (d) $\dfrac{x^3}{x^6}\left(\dfrac{7}{x}\right)^2.$

<u>Solution:</u> Noting the following properties of exponents:

(1) $a^b \cdot a^c = a^{b+c}$ (2) $\left(a^b\right)^c = a^{b \cdot c}$ (3) $\dfrac{a^b}{a^c} = a^{b-c}$ (4) $\left(\dfrac{a}{b}\right)^c = \dfrac{a^c}{b^c}$

we proceed to evaluate these expressions.

(a) $5x^5 \cdot 2x^2 = 5 \cdot 2 \cdot x^5 x^2 = 10 \cdot x^5 \cdot x^2 = 10x^{5+2} = 10x^7$

(b) $\left(x^4\right)^6 \quad = x^{4 \cdot 6} \quad = x^{24}$

(c) $\dfrac{8y^8}{2y^2} \quad = \dfrac{8}{2} \cdot \dfrac{y^8}{y^2} \quad = 4 \cdot y^{8-2} \quad = 4y^6$

(d) $\left(\dfrac{x^3}{x^6}\right)\left(\dfrac{7}{x}\right)^2 = \left(\dfrac{x^3}{x^6}\right)\left(\dfrac{7^2}{x^2}\right) = \dfrac{x^3 \cdot 49}{x^6 \cdot x^2} \quad = \dfrac{49x^3}{x^{6+2}} = \dfrac{49x^3}{x^8} = \dfrac{49x^3}{x^{5+3}}$

$$= \frac{49x^3}{x^5 \cdot x^3} \qquad = \frac{49}{x^5}$$

Write $5x^{-3}y^0$ without zero or negative exponents.

Solution: Since $a^{-b} = \frac{1}{a^b}$ by definition,

$$x^{-3} = \frac{1}{x^3}$$

and since $x^0 = 1$ by definition (any real non-zero base raised to an exponent of zero equals one),

$$y^0 = 1 .$$

Substituting these values for x^{-3} and y^0 we obtain

$$5x^{-3}y^0 = 5 \cdot \frac{1}{x^3} \cdot 1 = \frac{5}{x^3} .$$

Simplify the quotient $\frac{2x^0}{(2x)^0}$.

Solution: The following two laws of exponents can be used to simplify the given quotient:

1) $a^0 = 1$ where a is any non-zero real number, and

2) $(ab)^n = a^n b^n$ where a and b are any two numbers.

In the given quotient, notice that the exponent in the numerator applies only to the letter x. However, the exponent in the denominator applies to both the number 2 and the letter x; that is, the exponent in the denominator applies to the entire term (2x). Using the first law, the numerator can be rewritten as:

$$2x^0 = 2(1) = 2$$

Using the second law with n = 0, the denominator can be rewritten as:

$$(2x)^0 = 2^0 x^0$$

Using the first law again to further simplify the denominator:

$$(2x)^0 = 2^0 x^0$$
$$= (1)(1)$$
$$= 1$$

Therefore,

$$\frac{2x^0}{(2x)^0} = \frac{2}{1} = 2$$

Write the expression $\left(x + y^{-1}\right)^{-1}$ without using negative exponents.

Solution: Since $x^{-a} = \dfrac{1}{x^a}$, $y^{-1} = \dfrac{1}{y^1} = \dfrac{1}{y}$,

$$\left(x + y^{-1}\right)^{-1} = \left(x + \frac{1}{y}\right)^{-1}$$

$$= \frac{1}{x + \dfrac{1}{y}}$$

Multiply numerator and denominator by y in order to eliminate the fraction in the denominator,

$$\frac{y(1)}{y\left(x + \dfrac{1}{y}\right)} = \frac{y}{yx + \dfrac{y}{y}} = \frac{y}{yx + 1}$$

Thus

$$\left(x + y^{-1}\right)^{-1} = \frac{y}{yx + 1}$$

Simplify the expression $xy\left(x^{-1} + y^{-1}\right)$.

Solution: The following two laws of exponents can be used to simplify the given expression:

1) $a^{-n} = \dfrac{1}{a^n}$ and

2) $a^m \cdot a^n = a^{m+n}$.

Using the first law,

$$xy\left(x^{-1} + y^{-1}\right) = xy\left(\frac{1}{x^1} + \frac{1}{y^1}\right)$$

$$= xy\left(\frac{1}{x} + \frac{1}{y}\right)$$

Using the distributive property, this last equation becomes:

$$= xy\left(\frac{1}{x}\right) + xy\left(\frac{1}{y}\right)$$

$$= x\left(\frac{1}{x}\right)y + xy\left(\frac{1}{y}\right)$$

$$= y + x$$

Using the second law, we can solve this problem in another way.

$$xy\left(x^{-1} + y^{-1}\right) = xyx^{-1} + xyy^{-1}$$

$$= x^{1+(-1)}y + xy^{1+(-1)}$$

$$= x^0y + xy^0$$
$$= (1 \cdot y) + (x \cdot 1)$$
$$= y + x$$

● **PROBLEM** 78

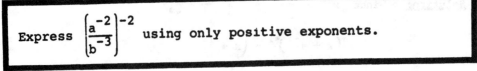

Express $\left(\dfrac{a^{-2}}{b^{-3}}\right)^{-2}$ using only positive exponents.

<u>Solution A:</u> By the law of exponents which states that $(x)^{-n} = \dfrac{1}{x^n}$ where n is a positive integer,

$$\left(\frac{a^{-2}}{b^{-3}}\right)^{-2} = \frac{1}{\left(\dfrac{a^{-2}}{b^{-3}}\right)^{2}}.$$

Since $\left(\dfrac{x}{y}\right)^n = \dfrac{x^n}{y^n}$, $\left(\dfrac{a^{-2}}{b^{-3}}\right)^2 = \dfrac{(a^{-2})^2}{(b^{-3})^2}$. Also, since $(x^m)^n = x^{mn}$,

$(a^{-2})^2 = a^{(-2)(2)} = a^{-4}$, $(b^{-3})^2 = b^{(-3)(2)} = b^{-6}$. Hence,

$$\left(\frac{a^{-2}}{b^{-3}}\right)^{-2} = \frac{1}{\left(\dfrac{a^{-2}}{b^{-3}}\right)^{2}}$$

$$= \frac{1}{\dfrac{a^{-4}}{b^{-6}}}$$

$$= \frac{1}{\dfrac{(a^4)^{-1}}{(b^6)^{-1}}}$$

$$= \frac{1}{\left(\dfrac{a^4}{b^6}\right)^{-1}}$$

$$= \frac{1}{\left[\dfrac{1}{\dfrac{a^4}{b^6}}\right]}$$

Note that division is the same as multiplying the numerator by the reciprocal of the denominator. This principle is applied to the term in brackets.

$$\left(\frac{a^{-2}}{b^{-3}}\right)^{-2} = \frac{1}{(1)\left[\dfrac{b^6}{a^4}\right]} = \left(\dfrac{1}{\dfrac{b^6}{a^4}}\right).$$

Applying the same principle to the term in parenthesis on the right side of the equation:

$$\left(\frac{a^{-2}}{b^{-3}}\right)^{-2} = \left(\frac{1}{\frac{b^6}{a^4}}\right) = (1)\left(\frac{a^4}{b^6}\right) = \frac{a^4}{b^6}.$$

__Solution B:__ Since $\left(\frac{x}{y}\right)^n = \frac{x^n}{y^n}$, $\left(\frac{a^{-2}}{b^{-3}}\right)^{-2} = \frac{(a^{-2})^{-2}}{(b^{-3})^{-2}}$. Also, since $(x^m)^n = x^{mn}$, $(a^{-2})^{-2} = a^{(-2)(-2)} = a^4$, and $(b^{-3})^{-2} = b^{(-3)(-2)} = b^6$. Hence,

$$\left(\frac{a^{-2}}{b^{-3}}\right)^{-2} = \frac{a^4}{b^6}.$$

● **PROBLEM**   79

Express $2c^{-2}d^{-1}/3x^{-1}y^3$ as an equal fraction involving only positive exponents.

__Solution:__ Since $a^{-b} = \frac{1}{a^b}$ for all real b,

$$c^{-2} = \frac{1}{c^2}, \ d^{-1} = \frac{1}{d}, \ x^{-1} = \frac{1}{x}.$$

Hence $\dfrac{2c^{-2}d^{-1}}{3x^{-1}y^3} = \dfrac{2\left(\frac{1}{c^2}\right)\left(\frac{1}{d}\right)}{3\left(\frac{1}{x}\right)(y^3)} = \dfrac{\frac{2}{c^2 d}}{\frac{3y^3}{x}}$. Division by a fraction is

equivalent to multiplication by its reciprocal, thus

$$\frac{\frac{2}{c^2 d}}{\frac{3y^3}{x}} = \left(\frac{2}{c^2 d}\right) \times \left(\frac{x}{3y^3}\right) = \frac{2x}{c^2 d 3 y^3} = \frac{2x}{3y^3 c^2 d} \ .$$

● **PROBLEM**   80

Convert $a^2 b^{-3} c^{-2}/x^{-1}y^3 z^{-3}$ into an equal fraction in which all exponents are positive.

__Solution:__ Since $m^{-n} = \frac{1}{m^n}$ for all real n, $b^{-3} = \frac{1}{b^3}$, $c^{-2} = \frac{1}{c^2}$, $x^{-1} = \frac{1}{x}$, and $z^{-3} = \frac{1}{z^3}$. Hence,

$$\frac{a^2 b^{-3} c^{-2}}{x^{-1}y^3 z^{-3}} = \frac{(a^2)\left(\frac{1}{b^3}\right)\left(\frac{1}{c^2}\right)}{\left(\frac{1}{x}\right)(y^3)\left(\frac{1}{z^3}\right)} = \frac{\frac{a^2}{b^3 c^2}}{\frac{y^3}{xz^3}} \ .$$

Division by a fraction is equivalent to multiplication by its reciprocal, thus

$$\frac{\dfrac{a^2}{b^3c^2}}{\dfrac{y^3}{xz^3}} = \left(\frac{a^2}{b^3c^2}\right) \times \left(\frac{xz^3}{y^3}\right) = \frac{a^2xz^3}{b^3c^2y^3}$$

Simplify the quotient $\dfrac{(x^{-2}y^4)^3}{(xy)^{-3}}$.

<u>Solution</u>: The following six laws of exponents will be used to simplify the given quotient:

(1) $x^{-n} = \dfrac{1}{x^n}$, where n is any positive integer

(2) $\left(\dfrac{x}{y}\right)^m = \dfrac{x^m}{y^m}$,

(3) $(x^m)^n = x^{m \cdot n}$,

(4) $(xy)^m = x^m y^m$,

(5) $x^m \cdot x^n = x^{m+n}$,

(6) $\dfrac{x^m}{x^n} = x^{m-n}$

Using the first law to simplify the quotient:

$$\frac{\left(x^{-2}y^4\right)^3}{(xy)^{-3}} = \frac{\left[\left(\dfrac{1}{x^2}\right)y^4\right]^3}{\dfrac{1}{(xy)^3}}$$

$$= \frac{\left(\dfrac{y^4}{x^2}\right)^3}{\dfrac{1}{(xy)^3}}$$

Using the second law to simplify the numerator,

$$\frac{\left(x^{-2}y^4\right)^3}{(x-y)^{-3}} = \frac{\left(\dfrac{y^4}{x^2}\right)^3}{\dfrac{1}{(xy)^3}}$$

$$= \frac{\dfrac{(y^4)^3}{(x^2)^3}}{\dfrac{1}{(xy)^3}}$$

Using the third and fourth laws to simplify both the numerator and the denominator,

$$\frac{\left(x^{-2}y^4\right)^3}{(xy)^{-3}} = \frac{\dfrac{\left(y^4\right)^3}{\left(x^2\right)^3}}{\dfrac{1}{(xy)^3}}$$

$$= \frac{\dfrac{y^{4\cdot3}}{x^{2\cdot3}}}{\dfrac{1}{x^3y^3}}$$

$$= \frac{\dfrac{y^{12}}{x^6}}{\dfrac{1}{x^3y^3}}$$

Since multiplying the numerator by the reciprocal of the denominator is equivalent to division, the equation becomes:

$$\frac{\left(x^{-2}y^4\right)^3}{(xy)^{-3}} = \frac{y^{12}}{x^6} \cdot \frac{x^3y^3}{1}$$

$$= \frac{y^{12}x^3y^3}{x^6}$$

Using the fifth law to simplify the numerator:

$$\frac{\left(x^{-2}y^4\right)^3}{(xy)^{-3}} = \frac{y^{12+3}x^3}{x^6}$$

$$= \frac{y^{15}x^3}{x^6}$$

Using the sixth law to make the last simplification:

$$\frac{\left(x^{-2}y^4\right)^3}{(xy)^{-3}} = \frac{y^{15}x^3}{x^6} = y^{15}\left(\frac{x^3}{x^6}\right) = y^{15}x^{3-6} = y^{15}x^{-3}$$

Hence, $\dfrac{\left(x^{-2}y^4\right)^3}{(xy)^{-3}} = y^{15}x^{-3}$ or $\dfrac{y^{15}}{x^3}$.

● **PROBLEM 82**

Evaluate the following expression: $\dfrac{12x^7y}{3x^2y^3}$

Solution: Noting (1) $\dfrac{abc}{def} = \dfrac{a}{d}\cdot\dfrac{b}{e}\cdot\dfrac{c}{f}$, (2) $a^{-b} = \dfrac{1}{a^b}$ and
(3) $\dfrac{a^b}{a^c} = a^{b-c}$ for all non-zero real values of a,d,e,f, we

43

proceed to evaluate the expression:

$$\frac{12x^7y}{3x^2y^3} = \frac{12}{3} \cdot \frac{x^7}{x^2} \cdot \frac{y}{y^3} = 4 \cdot x^{7-2} \cdot y^{1-3} = 4x^5y^{-2} = \frac{4x^5}{y^2}.$$

● **PROBLEM 83**

Simplify the quotient $\dfrac{\left(x^{-2}y^4\right)^3}{(xy)^{-3}}$.

<u>Solution:</u> Since $(ab)^x = a^x b^x$, and $\left(a^x\right)^y = a^{xy}$:

$$\frac{\left(x^{-2}y^4\right)^3}{(xy)^{-3}} = \frac{\left(x^{-2}\right)^3\left(y^4\right)^3}{x^{-3}y^{-3}} = \frac{x^{(-2)(3)}y^{4 \cdot 3}}{x^{-3}y^{-3}} = \frac{x^{-6}y^{12}}{x^{-3}y^{-3}} .$$

When dividing common bases with different exponents we subtract the exponent of the divisor from the exponent of the dividend $\left(\dfrac{a^x}{a^y} = a^{x-y}\right)$;

thus:

$$= x^{-6-(-3)}y^{12-(-3)}$$

$$= x^{-6+3}y^{12+3}$$

$$= x^{-3}y^{15} .$$

● **PROBLEM 84**

Simplify (rewrite without negative exponents, and reduce to a fraction in lowest terms) $\dfrac{7x^{-1}}{x^{-3}+y^{-4}}$.

<u>Solution 1:</u> We see that all negative exponents can be eliminated by multiplying numerator and denominator by x^3y^4. Hence,

$$\frac{7x^{-1}}{x^{-3}+y^{-4}} = \frac{7x^{-1}}{x^{-3}+y^{-4}} \cdot \frac{x^3y^4}{x^3y^4} = \frac{7x^{-1}x^3y^4}{x^{-3}x^3y^4+y^{-4}x^3y^4} = \frac{7x^{-1+3}y^4}{x^{-3+3}y^4+x^3y^{-4+4}}$$

$$= \frac{7x^2y^4}{y^4+x^3} .$$

<u>Solution 2:</u> Another way to solve this problem is to apply the definition $a^{-n} = 1/a^n$ where $a \neq 0$.

$$\frac{7x^{-1}}{x^{-3}+y^{-4}} = \frac{7\left(\frac{1}{x}\right)}{\frac{1}{x^3}+\frac{1}{y^4}} . \text{ Combine } \frac{1}{x^3} + \frac{1}{y^4} \text{ into one term. The least}$$

common denominator is x^3y^4.

$$\frac{1}{x^3} + \frac{1}{y^4} = \frac{1}{x^3}\left(\frac{y^4}{y^4}\right) + \frac{1}{y^4}\left(\frac{x^3}{x^3}\right) = \frac{y^4 + x^3}{y^4x^3}$$

Therefore,

$$\frac{7\left(\frac{1}{x}\right)}{\frac{1}{x^3} + \frac{1}{y^4}} = \frac{\frac{7}{x}}{\frac{y^4 + x^3}{x^3 y^4}} .$$

Division by a fraction is equivalent to multiplication by that fraction's reciprocal. Thus

$$\frac{\frac{7}{x}}{\frac{y^4 + x^3}{x^3 y^4}} = \frac{7}{x} \cdot \frac{x^3 y^4}{y^4 + x^3} = \frac{7x^2 y^4}{y^4 + x^3}$$

• **PROBLEM** 85

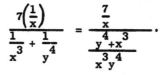

Express $\dfrac{3x^{-1} - y^{-2}}{x^{-2} + 2y^{-1}}$ without negative exponents.

Solution: Since $x^{-a} = \dfrac{1}{x^a}$ for all real $x \neq 0$,

$$x^{-1} = \frac{1}{x}, \qquad y^{-2} = \frac{1}{y^2}$$

$$x^{-2} = \frac{1}{x^2} , \text{ and } y^{-1} = \frac{1}{y}; \text{ thus,}$$

$$\frac{3x^{-1} - y^{-2}}{x^{-2} + 2y^{-1}} = \frac{3\left(\frac{1}{x}\right) - \frac{1}{y^2}}{\frac{1}{x^2} + 2\left(\frac{1}{y}\right)}$$

$$= \frac{\frac{3}{x} - \frac{1}{y^2}}{\frac{1}{x^2} + \frac{2}{y}}$$

Multiplying numerator and denominator by the least common multiple, $x^2 y^2$,

$$= \frac{x^2 y^2 \left(\frac{3}{x} - \frac{1}{y^2}\right)}{x^2 y^2 \left(\frac{1}{x^2} + \frac{2}{y}\right)}$$

Distributing, $= \dfrac{x^2 y^2 \left(\frac{3}{x}\right) - x^2 y^2 \left(\frac{1}{y^2}\right)}{x^2 y^2 \left(\frac{1}{x^2}\right) + x^2 y^2 \left(\frac{2}{y}\right)}$

Cancelling like terms, $= \dfrac{3xy^2 - x^2}{y^2 + 2x^2 y}$

Factoring x from numerator and y from denominator,

$$= \frac{x(3y^2 - x)}{y(y + 2x^2)}$$

Thus, $\dfrac{3x^{-1} - y^{-2}}{x^{-2} + 2y^{-1}} = \dfrac{x(3y^2 - x)}{y(y + 2x^2)}$

● **PROBLEM** 86

Simplify: (a) $\dfrac{a^{-3}b^2}{a^{-2}b^4}$ (b) $\dfrac{3x^{-4}}{y^2} \cdot \dfrac{4x}{9x^2y^{-1}}$

Solution: (a) Since $a^{-n} = \dfrac{1}{a^n}$,

$$\frac{a^{-3}b^2}{a^{-2}b^4} = \frac{\dfrac{b^2}{a^3}}{\dfrac{b^4}{a^2}} = \frac{b^2}{a^3} \cdot \frac{a^2}{b^4}$$

Dividing by a fraction is equivalent to multiplying by the reciprocal of the fraction.

$$\frac{b^2}{a^3} \cdot \frac{a^2}{b^4} = \frac{b^2 a^2}{a^3 b^4} = \frac{a^2}{a^3} \cdot \frac{b^2}{b^4} = \frac{1}{ab^2}$$

(b) $\dfrac{3x^{-4}}{y^2} \cdot \dfrac{4x}{9x^2y^{-1}} = \dfrac{3}{y^2x^4} \cdot \dfrac{4x}{\dfrac{9x^2}{y}} = \dfrac{3}{y^2x^4} \cdot \dfrac{4xy}{9x^2} = \dfrac{3 \cdot 4}{9yx^5} = \dfrac{4}{3yx^5}$

● **PROBLEM** 87

Find the following products.

   a. $(3x^2y)(2xy^2)$

   b. $(-xy^3)(4xyz)(2yz)$

Solution: Use the following law of exponents to find the indicated products:

$$a^m \cdot a^n \cdot a^x \cdot \ldots = a^{m+n+x+\ldots}$$

a) $(3x^2y)(2xy^2) = 6(x^2 \cdot x)(y \cdot y^2)$

$$= 6(x^{2+1})(y^{1+2})$$

$$= 6x^3y^3$$

b) $(-xy^3)(4xyz)(2yz) = -8(x \cdot x)(y^3 \cdot y \cdot y)(z \cdot z)$

$$= -8\left(x^{1+1}\right)\left(y^{3+1+1}\right)\left(z^{1+1}\right)$$

$$= -8(x^2)(y^5)(z^2)$$

$$= -8x^2y^5z^2$$

● **PROBLEM** 88

Use the properties of exponents, to perform the indicated operations in

$$(2^3x^45^2y^7)^5.$$

Solution: Since the product of several numbers raised to the same exponent equals the product of each number raised to that exponent (i.e., $(abcd)^x = a^xb^xc^xd^x$) we obtain,

$$(2^3x^45^2y^7)^5 = (2^3)^5(x^4)^5(5^2)^5(y^7)^5.$$

Recall that $(x^a)^b = x^{a\cdot b}$; thus

$$(2^3x^45^2y^7)^5 = (2^3)^5(x^4)^5(5^2)^5(y^7)^5$$

$$= (2^{3\cdot5})(x^{4\cdot5})(5^{2\cdot5})(y^{7\cdot5})$$

$$= 2^{15}x^{20}5^{10}y^{35}.$$

● **PROBLEM** 89

Perform the indicated operations, and simplify. (Write without negative or zero exponents.) Each letter represents a positive real number.

(a) $\left(7x^{-3}y^5\right)^{-2}$ (b) $\left(5x^7y^{-8}\right)^{-3}.$

Solution: Note that: (1) $(abc)^x = a^xb^xc^x$ (for all real a,b,c), (2) $a^{-x} = \dfrac{1}{a^x}$ (for all non-zero real a) and (3) $\left(a^b\right)^c = a^{bc}$ (for all real a,b,c). These will be useful in evaluating the given expressions.

(a) $\left(7x^{-3}y^5\right)^{-2} = 7^{-2}\left(x^{-3}\right)^{-2}\left(y^5\right)^{-2} = 7^{-2}(x)^{(-3)(-2)}(y)^{(5)(-2)}$

$$= 7^{-2}(x)^6(y)^{-10} = \frac{x^6}{(7^2)(y^{10})} = \frac{x^6}{49y^{10}}.$$

(b) $\left(5x^7y^{-8}\right)^{-3} = 5^{-3}\left(x^7\right)^{-3}\left(y^{-8}\right)^{-3} = 5^{-3}(x)^{(7)(-3)}(y)^{(-8)(-3)}$

$$= 5^{-3}(x)^{-21}(y)^{24} = \frac{y^{24}}{5^3x^{21}} = \frac{y^{24}}{125x^{21}}.$$

● **PROBLEM** 90

Evaluate the following expressions:

(a) $\dfrac{-12x^{10}y^9z^5}{3x^2y^3z^6}$ (b) $\dfrac{-16x^{16}y^6z^4}{-4x^4y^2z^7}$.

Solution: Noting (a) $\dfrac{abcd}{efgh} = \dfrac{a}{e}\cdot\dfrac{b}{f}\cdot\dfrac{c}{g}\cdot\dfrac{d}{h}$, (2) $a^{-b} = \dfrac{1}{a^b}$

and (3) $\dfrac{a^b}{a^c} = a^{b-c}$ for all non-zero real values of a,e,f,g,h,

we proceed to evaluate these expressions:

(a) $\dfrac{-12x^{10}y^9z^5}{3x^2y^3z^6} = \dfrac{-12}{3}\cdot\dfrac{x^{10}}{x^2}\cdot\dfrac{y^9}{y^3}\cdot\dfrac{z^5}{z^6} = -4\cdot x^{10-2}\cdot y^{9-3}\cdot z^{5-6}$

$= -4x^8y^6z^{-1} = \dfrac{-4x^8y^6}{z^1}$.

Thus $\dfrac{-12x^{10}y^9z^5}{3x^2y^3z^6} = \dfrac{-4x^8y^6}{z}$.

(b) $\dfrac{-16x^{16}y^6z^4}{-4x^4y^2z^7} = \dfrac{-16}{-4}\cdot\dfrac{x^{16}}{x^4}\cdot\dfrac{y^6}{y^2}\cdot\dfrac{z^4}{z^7}$

$= 4x^{16-4}\cdot y^{6-2}\cdot z^{4-7} = 4x^{12}y^4z^{-3} = \dfrac{4x^{12}y^4}{z^3}$.

● **PROBLEM 91**

Perform the indicated operations and simplify:

$$\left(\dfrac{-5b^y}{3^2x^5}\right)^3 \left(\dfrac{3x^7}{5b^y}\right)^2 \quad .$$

Solution: $\left(\dfrac{-5b^y}{3^2x^5}\right)^3 \left(\dfrac{3x^7}{5b^y}\right) = \dfrac{(-5b^y)^3}{(3^2x^5)^3} \cdot \dfrac{(3x^7)^2}{(5b^y)^2}$ since $\left(\dfrac{a}{b}\right)^x = \dfrac{a^x}{b^x}$

$= \dfrac{(-5)^3(b^y)^3}{(3^2)^3(x^5)^3} \cdot \dfrac{3^2(x^7)^2}{5^2(b^y)^2}$ since $(ab)^x = a^xb^x$

$= \dfrac{-5^3b^{3y}}{3^6x^{15}} \cdot \dfrac{3^2x^{14}}{5^2b^{2y}}$ since $(a^x)^y = a^{x\cdot y}$

$= \dfrac{\left(-5^3b^{3y}\right)\left(3^2x^{14}\right)}{\left(3^6x^{15}\right)\left(5^2b^{2y}\right)}$

$= \dfrac{\left(3^2x^{14}\right)\left(-5^3b^{3y}\right)}{\left(3^6x^{15}\right)\left(5^2b^{2y}\right)}$ using the commutative law of multiplication

$= \left(3^{2-6}\right)\left(x^{14-15}\right)\left[-\left(5^{3-2}\right)\left(6^{3y-2y}\right)\right]$ because

$\dfrac{x^a}{x^b} = x^{a-b}$,

$$= (3^{-4})(x^{-1})(-5^{1})(b^{y})$$

$$= \frac{-5b^{y}}{3^{4}x} \quad \text{because} \quad x^{-a} = \frac{1}{x^{a}}$$

$$= \frac{-5b^{y}}{3 \cdot 3 \cdot 3 \cdot 3x}$$

$$= \frac{-5b^{y}}{81x}$$

● **PROBLEM 92**

Determine the value of $(0.0081)^{-3/4}$.

Solution: $(0.0081) = .3 \times .3 \times .3 \times .3 = (.3)^{4}$,

therefore, $(0.0081)^{-3/4} = (.3^{4})^{-3/4}$

Recalling the property of exponents,

$$(a^{x})^{y} = a^{x \cdot y}$$

we have,

$$(.3^{4})^{-3/4} = .3^{(4)(-3/4)} = .3^{-3}.$$

Since $\quad a^{-x} = \frac{1}{a^{x}}, \quad .3^{-3} = \frac{1}{.3^{3}} = \frac{1}{0.027} = \frac{1}{\frac{27}{1000}}$

Division by a fraction is equivalent to multiplication by its reciprocal, thus,

$$\frac{1}{\frac{27}{1000}} = \frac{1000}{27}.$$

Hence,

$$(0.0081)^{-3/4} = \frac{1000}{27}.$$

● **PROBLEM 93**

Simplify $\left[\dfrac{1600 \times 10,000}{2000}\right]^{1/3}$.

Solution: Observe $1600 = 16 \times 100 = 16 \times 10^{2}$

$$10,000 = 10^{4}$$

$$2,000 = 2 \times 10^{3}.$$

Thus, $\left[\dfrac{1600 \times 10,000}{2000}\right]^{1/3} = \left[\dfrac{(16 \times 10^{2})(10^{4})}{2 \times 10^{3}}\right]^{1/3}$.

Using the associative property,

$$= \left[\frac{16 \times (10^{2} \times 10^{4})}{2 \times 10^{3}}\right]^{1/3}$$

Recall: $a^{x} \cdot a^{y} = a^{x+y}$,

$$= \left[\frac{16 \times 10^{6}}{2 \times 10^{3}}\right]^{1/3}$$

49

Since $\dfrac{ab}{cd} = \dfrac{a}{c} \cdot \dfrac{b}{d}$,

$$= \left[\dfrac{16}{2} \times \dfrac{10^6}{10^3}\right]^{1/3}$$

Recall $\dfrac{a^x}{a^y} = a^{x-y}$,

$$= \left(8 \times 10^3\right)^{1/3}$$

Since $(ab)^x = a^x b^x$,

$$= 8^{1/3} \times 10^{3(1/3)}$$
$$= 2 \times 10^1$$
$$= 20$$

● PROBLEM 94

Determine the value of

$$\dfrac{5^{3/4}\, 5^{2/3}\, 5^{-5/2}\, 5^{5/3}}{5^{1/3}\, 5^{-5/2}\, 5^{7/4}} .$$

Solution: Since $a^x \cdot a^y \cdot a^z = a^{x+y+z}$ for any real base, then,

$$\dfrac{5^{\frac{3}{4}} \cdot 5^{\frac{2}{3}} \cdot 5^{-\frac{5}{2}} \cdot 5^{\frac{5}{3}}}{5^{\frac{1}{3}} \cdot 5^{-\frac{5}{2}} \cdot 5^{\frac{7}{4}}} = \dfrac{5^{\frac{3}{4} + \frac{2}{3} - \frac{5}{2} + \frac{5}{3}}}{5^{\frac{1}{3} - \frac{5}{2} + \frac{7}{4}}}$$

The fractional exponents have denominators 4, 3, and 2. Their least common denominator (L.C.D.) is the least common multiple of the denominators, 12. Converting the fractional exponents to twelfths we obtain

$$\dfrac{5^{\frac{3}{4} + \frac{2}{3} - \frac{5}{2} + \frac{5}{3}}}{5^{\frac{1}{3} - \frac{5}{2} + \frac{7}{4}}} = \dfrac{5^{\frac{9}{12} + \frac{8}{12} - \frac{30}{12} + \frac{20}{12}}}{5^{\frac{4}{12} - \frac{30}{12} + \frac{21}{12}}}$$

$$= \dfrac{5^{\frac{7}{12}}}{5^{-\frac{5}{12}}}$$

Since $\dfrac{a^x}{a^y} = a^{x-y}$ for any real base $(a \neq 0)$

$$\dfrac{5^{\frac{7}{12}}}{5^{-\frac{5}{12}}} = 5^{\frac{7}{12} - \left(-\frac{5}{12}\right)}$$

$$= 5^{\frac{7}{12} + \frac{5}{12}}$$

$$= 5^{\frac{12}{12}}$$

$$= 5^1$$

$$= 5.$$

Therefore, $\dfrac{5^{\frac{3}{4}}\, 5^{\frac{2}{3}}\, 5^{-\frac{5}{2}}\, 5^{\frac{5}{3}}}{5^{\frac{1}{3}}\, 5^{-\frac{5}{2}}\, 5^{\frac{7}{4}}} = 5.$

Express $\left(5^{\frac{1}{2}} + 9^{\frac{1}{8}}\right) \div \left(5^{\frac{1}{2}} - 9^{\frac{1}{8}}\right)$

as an equivalent fraction with a rational denominator.

Solution: The given expression can be rewritten as,

$$\frac{5^{\frac{1}{2}} + 9^{\frac{1}{8}}}{5^{\frac{1}{2}} - 9^{\frac{1}{8}}} \quad , \qquad \text{and since}$$

$$9^{\frac{1}{8}} = \left(3^2\right)^{\frac{1}{8}} = 3^{\frac{1}{4}} \qquad \text{we write:}$$

$$\frac{5^{\frac{1}{2}} + 3^{\frac{1}{4}}}{5^{\frac{1}{2}} - 3^{\frac{1}{4}}} \quad .$$

To rationalize the denominator, put

$5^{\frac{1}{2}} = x, \quad 3^{\frac{1}{4}} = y$; then since

$$x^4 - y^4 = \left(5^{\frac{1}{2}}\right)^4 - \left(3^{\frac{1}{4}}\right)^4 = 5^2 - 3 = 25 - 3 = 22,$$

which is rational, we can write

$$x^4 - y^4 = (x - y)\left(x^3 + x^2y + xy^2 + y^3\right),$$

and the factor which rationalizes x - y, or
$5^{\frac{1}{2}} - 3^{\frac{1}{4}}$ is $x^3 + x^2y + xy^2 + y^3$, and substituting for x and y:

$$\left(5^{\frac{1}{2}}\right)^3 + \left(5^{\frac{1}{2}}\right)^2 \cdot 3^{\frac{1}{4}} + 5^{\frac{1}{2}} \cdot \left(3^{\frac{1}{4}}\right)^2 + \left(3^{\frac{1}{4}}\right)^3$$

$$= \quad 5^{\frac{3}{2}} + 5^{\frac{2}{2}} \cdot 3^{\frac{1}{4}} + 5^{\frac{1}{2}} \cdot 3^{\frac{2}{4}} + 3^{\frac{3}{4}} \ ;$$

and the rational denominator is

$$x^4 - y^4 = 5^{\frac{4}{2}} - 3^{\frac{4}{4}} = 5^2 - 3 = 22.$$

Now, since

$$x^4 - y^4 = (x - y)\left(x^3 + x^2y + xy^2 + y^3\right), \qquad \text{then}$$

$$(x - y) = \frac{x^4 - y^4}{x^3 + x^2y + xy^2 + y^3}, \qquad \text{and substituting:}$$

$$5^{\frac{1}{2}} - 3^{\frac{1}{4}} = \frac{22}{5^{\frac{3}{2}} + 5^{\frac{2}{2}} \cdot 3^{\frac{1}{4}} + 5^{\frac{1}{2}} \cdot 3^{\frac{2}{4}} + 3^{\frac{3}{4}}}$$

Therefore, the given expression

$$= \frac{\dfrac{5^{\frac{1}{2}} + 3^{\frac{1}{4}}}{22}}{5^{\frac{3}{2}} + 5^{\frac{2}{2}} \cdot 3^{\frac{1}{4}} + 5^{\frac{1}{2}} \cdot 3^{\frac{2}{4}} + 3^{\frac{3}{4}}}$$

$$= \frac{\left(5^{\frac{1}{2}} + 3^{\frac{1}{4}}\right)\left[5^{\frac{3}{2}} + \left(5^{\frac{2}{2}} \cdot 3^{\frac{1}{4}}\right) + \left(5^{\frac{1}{2}} \cdot 3^{\frac{2}{4}}\right) + 3^{\frac{3}{4}}\right]}{22}$$

$$= 5^{\frac{4}{2}} + 5^{\frac{3}{2}} 3^{\frac{1}{4}} + 5^{\frac{3}{2}} 3^{\frac{1}{4}} + 5^{\frac{2}{2}} 3^{\frac{2}{4}} + 5^{\frac{2}{2}} 3^{\frac{2}{4}} +$$

$$+ 5^{\frac{1}{2}} 3^{\frac{3}{4}} + 5^{\frac{1}{2}} 3^{\frac{3}{4}} + 3^{\frac{4}{4}}$$

$$= \frac{5^{\frac{4}{2}} + \left(2 \cdot 5^{\frac{3}{2}} \cdot 3^{\frac{1}{4}}\right) + \left(2 \cdot 5^{\frac{2}{2}} \cdot 3^{\frac{2}{4}}\right) + \left(2 \cdot 5^{\frac{1}{2}} \cdot 3^{\frac{3}{4}}\right) + 3^{\frac{4}{4}}}{22}$$

$$= \frac{5^2 + 2\left[5^{\frac{3}{2}} \cdot 3^{\frac{1}{4}} + 5^{\frac{2}{2}} \cdot 3^{\frac{2}{4}} + 5^{\frac{1}{2}} \cdot 3^{\frac{3}{4}}\right] + 3}{22}$$

$$= \frac{28 + 2\left[5^{\frac{3}{2}} \cdot 3^{\frac{1}{4}} + 5^{\frac{2}{2}} \cdot 3^{\frac{2}{4}} + 5^{\frac{1}{2}} \cdot 3^{\frac{3}{4}}\right]}{22}$$

$$= \frac{14 + 5^{\frac{3}{2}} \cdot 3^{\frac{1}{4}} + 5 \cdot 3^{\frac{1}{2}} + 5^{\frac{1}{2}} \cdot 3^{\frac{3}{4}}}{11}$$

CHAPTER 7

ROOTS AND RADICALS

> ## Basic Attacks and Strategies for Solving Problems in this Chapter. See pages 53 to 78 for step-by-step solutions to problems.

When simplifying and evaluating roots and radicals, it is essential to first observe the fundamental relation between roots, fractional exponents, and radicals. In particular, for any positive real number x and positive rational number p/q (q is not zero),

$$x^{p/q} = \sqrt[q]{x^p} = \left(\sqrt[q]{x}\right)^p = \left(x^{1/q}\right)^p$$

is the fundamental relation on which all simplifications and evaluations of roots and radicals rest. Also, all the properties involving multiplication and division involving exponents are essential from Chapter 6.

When dealing with negative exponents, it is usually helpful to write the expression as a reciprocal in order to get a positive exponent before further simplification and evaluation of roots. For example, to simplify the expression below we first change the negative exponent to a positive exponent and then apply the aforementioned relationship between the fractional exponents and radicals as follows:

$$8^{-2/3} = 1/8^{2/3} = 1/\left(\sqrt[3]{8}\right)^2 = 1/(2)^2 = 1/4.$$

When dealing with negative values of x, it is important to recognize that this leads to complex numbers. In particular, it is important to observe that $\sqrt{-1} = i$, the imaginary part of a complex number. Also, if the index of the radical is an even integer and x is negative, then the results will always include $\sqrt{-1}$. For instance,

$$\sqrt{-4} = \sqrt{4} * \sqrt{-1} = 2i.$$

In the simplification of radicals, it is usually desirable to arrange radicals in a simplest form. A radical with index n is in simplest form if the following conditions hold:

(1)   The nth roots of any n^{th} powers in the radicand have been taken. When this is done every exponent in the radicand will be less than n.

(2) All denominators have been rationalized; that is, all denominators are free of radicals. (See procedure below.)

(3) The index of the radical is reduced as much as possible.

If a fraction contains one or more radicals in the denominator, then the denominator must be rationalized in order to put it in simplest form. To rationalize a denominator which contains a monomial radical, multiply both the numerator and denominator of the given fraction by an appropriate monomial radical which will enable the results in the denominator to be the nth root of a nth power, which is a result that is free of the radical in the denominator. To rationalize a binominal denominator in which one or both terms is a square root, multiply the numerator and denominator of the fraction by the conjugate of the denominator, i.e., by the same two terms but with the opposite sign between them. For example, to simplify the radical expression below, we need to rationalize the denominator as follows:

$$\frac{\sqrt{2}}{\sqrt{2}-\sqrt{x}} = \frac{\sqrt{2}}{\sqrt{2}-\sqrt{x}} * \frac{\sqrt{2}+\sqrt{x}}{\sqrt{2}+\sqrt{x}} = \frac{2+\sqrt{2x}}{2-x}$$

To simplify a radical in which the index is a multiple of the exponents in the radicand, rewrite the expression by simply dividing the index and all of the exponents under the radical by the GCD. Then, simplify the result.

Step-by-Step Solutions to Problems in this Chapter, "Roots and Radicals"

SIMPLIFICATION AND EVALUATION OF ROOTS

● **PROBLEM** 96

Evaluate $\sqrt{400}$.

Solution: $\quad 400 = 4 \times 100$

Thus, $\quad\quad \sqrt{400} = \sqrt{4 \times 100}$

Since $\quad\quad \sqrt{ab} = \sqrt{a}\ \sqrt{b},$

$\quad\quad\quad\quad \sqrt{400} = \sqrt{4}\ \sqrt{100}$

$\quad\quad\quad\quad\quad\quad = 2 \cdot 10$

$\quad\quad\quad\quad\quad\quad = 20$

Check: If $\sqrt{400}$ is 20, then 20^2 must equal 400, which is true. Hence, 20 is the solution.

● **PROBLEM** 97

Find the value of $\sqrt[4]{-64a^4}$.

Solution: We can rewrite $\sqrt[4]{-64^4}$ as,

$$\left[64a^4 \cdot (-1)\right]^{\frac{1}{4}} = \left[\left(\underline{+}\ 8a^2\right)^2 \cdot (-1)\right]^{\frac{1}{4}},$$

by first factoring -1 from the expression under the radical, and then using the fact that $64a^4 = \left(\underline{+}\ 8a^2\right)^2$. Also recall that $\sqrt[4]{x} = x^{\frac{1}{4}}$. Now, since $(ab)^x = a^x \cdot b^x$, and $\left(a^x\right)^y = a^{xy}$, we write:

$$\left[\left(8a^2\right)^2 \cdot (-1)\right]^{\frac{1}{4}} = \left[\left(8a^2\right)^2\right]^{\frac{1}{4}} \cdot (-1)^{\frac{1}{4}}$$

$$= \left[8^2 \cdot \left(a^2\right)^2\right]^{\frac{1}{4}} \cdot (-1)^{\frac{1}{4}}$$

$$= \left(8^2\right)^{\frac{1}{4}} \cdot \left[\left(a^2\right)^2\right]^{\frac{1}{4}} \cdot (-1)^{\frac{1}{4}}$$

$$= 8^{\frac{1}{2}} \cdot \left(a^2\right)^{\frac{1}{2}} \cdot (-1)^{\frac{1}{2} \cdot \frac{1}{2}}$$

$$= 8^{\frac{1}{2}} \cdot \left(a^2\right)^{\frac{1}{2}} \cdot \left[(-1)^{\frac{1}{2}}\right]^{\frac{1}{2}}$$

Since $x^{\frac{1}{2}} = \sqrt{x}$, and $\left(x^{\frac{1}{2}}\right)^{\frac{1}{2}} = \sqrt{\sqrt{x}}$, we write:

$$\sqrt[4]{-64a^4} = \sqrt{\underline{+}\ 8a^2\ \sqrt{-1}} = \sqrt{4a^2 \cdot 2 \cdot \underline{+}\ \sqrt{-1}},$$

and since

53

$$\sqrt{a \cdot b \cdot c} = \sqrt{a}\sqrt{b}\sqrt{c} \, , \quad \sqrt[4]{-64a^4} = \sqrt{4a^2} \cdot \sqrt{2} \cdot \sqrt{\pm\sqrt{-1}}$$
$$= 2a\sqrt{2}\sqrt{\pm\sqrt{-1}} \, .$$

It remains to find the value of $\sqrt{\pm\sqrt{-1}}$. $\sqrt{+\sqrt{-1}}$ is a complex number and can be written as the sum of real and imaginery parts eg. $x+y\sqrt{-1}$ where x is the real part and y is the imaginary part.

Squaring both sides of the equation we obtain:
$$+\sqrt{-1} = (x + y\sqrt{-1})(x + y\sqrt{-1})$$

or, by performing the multiplication,
$$+\sqrt{-1} = x^2 + xy\sqrt{-1} + xy\sqrt{-1} + y\sqrt{-1} \cdot y\sqrt{-1} \, .$$

Combining like terms, and recalling that $\sqrt{-1} \cdot \sqrt{-1} = -1$, we have:
$$+\sqrt{-1} = x^2 - y^2 + 2xy\sqrt{-1} \, .$$

Let us examine the equation,
$$+\sqrt{-1} = x^2 - y^2 + 2xy\sqrt{-1} \, .$$

This equation is only true if $x^2 - y^2 = 0$, and $2xy = 1$, because then the equation becomes:
$$+\sqrt{-1} = 0 + {}^-1\sqrt{-1}$$
$$+\sqrt{-1} = +\sqrt{-1}$$

Therefore, we have the following system of equations:
$$x^2 - y^2 = 0 \quad \text{and} \quad 2xy = 1 \, .$$

To solve for x and y we use the method of substitution. Solving for y in the second equation, $2xy = 1$, we have:
$$y = \frac{1}{2x} \, ;$$

and substituting this value into equation one, $x^2 - y^2 = 0$, we obtain:
$$x^2 - \left(\frac{1}{2x}\right)^2 = 0$$
$$x^2 - \frac{1}{4x^2} = 0$$

Multiply both sides by $4x^2$: $\quad 4x^2\left(x^2 - \frac{1}{4x^2}\right) = 4x^2(0)$

Distribute: $\qquad\qquad\qquad\qquad 4x^4 - 1 = 0$

Add 1 to both sides: $\qquad\qquad\quad 4x^4 = 1$

Divide both sides by 4: $\qquad\qquad x^4 = \tfrac{1}{4}$

Now, taking the fourth root of both sides we obtain:
$$x = \sqrt[4]{\tfrac{1}{4}}$$
$$= \left(\tfrac{1}{4}\right)^{\tfrac{1}{4}} \, , \text{ since } \sqrt[4]{x} = x^{\tfrac{1}{4}}$$
$$= \frac{1^{\tfrac{1}{4}}}{4^{\tfrac{1}{4}}} \, , \text{ since } \left(\frac{a}{b}\right)^x = \frac{a^x}{b^x}$$
$$= \frac{1}{(2^2)^{\tfrac{1}{4}}}$$
$$= \frac{1}{2^{\tfrac{1}{2}}} \, , \text{ since } \left(a^b\right)^x = a^{bx}$$
$$= \frac{1}{\sqrt{2}} \, , \text{ since } x^{\tfrac{1}{2}} = \sqrt{x}$$

Now, since $y = \frac{1}{2x}$, by substitution:

$$y = \frac{1}{2\left(\frac{1}{\sqrt{2}}\right)} = \frac{1}{\frac{2}{\sqrt{2}}} = \frac{\sqrt{2}}{2}$$

Observing the y-value, $\frac{\sqrt{2}}{2}$, closely we see that it is equivalent to the x-value, $\frac{1}{\sqrt{2}}$. This can be seen by multiplying $\frac{\sqrt{2}}{2}$ by $\frac{\sqrt{2}}{\sqrt{2}}$ (which equals 1 and therefore does not alter the value of the fraction). Thus,

$$\frac{\sqrt{2}}{2} \cdot \frac{\sqrt{2}}{\sqrt{2}} = \frac{2}{2\sqrt{2}} = \frac{1}{\sqrt{2}}$$

Therefore,

$$x = \frac{1}{\sqrt{2}} , \; y = \frac{1}{\sqrt{2}} ; \text{ or } x = -\frac{1}{\sqrt{2}} , \; y = -\frac{1}{\sqrt{2}} ;$$

We must include the negative values for x and y as solutions to the equations, because these give us the same results in the equations $x^2 - y^2 = 0$ and $2xy = 1$ as do the positive values for x and y, since a negative value squared is positive (equation 1), and a negative multiplied by a negative is positive (equation 2). Therefore, substituting into the equation, $\sqrt{+\sqrt{-1}} = x + y\sqrt{-1}$, we have:

$$\sqrt{+ \sqrt{-1}} = \pm \frac{1}{\sqrt{2}} + \left(\pm \frac{1}{\sqrt{2}} \right)\sqrt{-1} ,$$

and factoring $\pm \frac{1}{\sqrt{2}}$ from both terms on the right side:

$$\sqrt{+ \sqrt{-1}} = \pm \frac{1}{\sqrt{2}}(1 + \sqrt{-1}).$$

Similarly, we assume $\sqrt{-\sqrt{-1}} = x - y\sqrt{-1}$, and proceeding as in the case when $\sqrt{+\sqrt{-1}} = x + y\sqrt{-1}$, we find that the x and y values are again $\pm \frac{1}{\sqrt{2}}$, thus:

$$\sqrt{- \sqrt{-1}} = \pm \frac{1}{\sqrt{2}}(1 - \sqrt{-1})$$

Therefore,

$$\sqrt{\pm \sqrt{-1}} = \pm \frac{1}{\sqrt{2}}(1 \pm \sqrt{-1}) ;$$

and finally, from the fact that,

$$\sqrt[4]{-64a^4} = 2a \sqrt{2} \sqrt{\pm \sqrt{-1}} , \text{ and}$$

$$\sqrt{\pm \sqrt{-1}} = \pm \frac{1}{\sqrt{2}}(1 \pm \sqrt{-1}) , \text{ we have:}$$

$$\sqrt[4]{-64a^4} = 2a\sqrt{2} \cdot \left[\pm \frac{1}{\sqrt{2}} \left(1 \pm \sqrt{-1} \right) \right]$$

$$= 2a\left[\pm (1 \pm \sqrt{-1}) \right] , \text{ by cancelling } \frac{\sqrt{2}}{\sqrt{2}} .$$

Therefore,

$$\sqrt[4]{-64a^4} = \pm 2a(1 \pm \sqrt{-1}) .$$

Evaluate $16^{-\frac{3}{4}}$.

Solution: $$16^{-\frac{3}{4}} = \frac{1}{16^{\frac{3}{4}}}$$

$$= \frac{1}{\left(\sqrt[4]{16}\right)^3} \quad .$$

Note that $2^4 = 2 \cdot 2 \cdot 2 \cdot 2 = 16$, hence $\sqrt[4]{16} = 2$. Thus, $16^{-\frac{3}{4}}$

$$= \frac{1}{2^3} = \frac{1}{2 \cdot 2 \cdot 2} = \frac{1}{8} \quad .$$

Show that $\sqrt[3]{(-8)^3} = \left(\sqrt[3]{-8}\right)^3$.

Solution: $\sqrt[3]{(-8)^3} = \sqrt[3]{-512} = -8$. Since $(-8)^3 = -512$, $\sqrt[3]{-512} = -8$. $\sqrt[3]{-8} = -2$ since $(-2)^3 = -8$. Therefore $\left(\sqrt[3]{-8}\right)^3 = (-2)^3 = -8$. $\sqrt[3]{(-8)^3} = -8 = \left(\sqrt[3]{-8}\right)^3$, hence, $\sqrt[3]{(-8)^3} = \left(\sqrt[3]{-8}\right)^3$.

Find the indicated roots.

(a) $\sqrt[5]{32}$ (b) $\pm \sqrt[4]{625}$ (c) $\sqrt[3]{-125}$ (d) $\sqrt[4]{-16}$.

Solution: The following two laws of exponents can be used to solve these problems: 1) $\left(\sqrt[n]{a}\right)^n = \left(a^{1/n}\right)^n = a^1 = a$, and 2) $\left(\sqrt[n]{a}\right)^n = \sqrt[n]{a^n}$.

(a) $\sqrt[5]{32} = \sqrt[5]{2^5} = \left(\sqrt[5]{2}\right)^5 = 2$. This result is true because $(2)^5 = 32$, that is, $2 \cdot 2 \cdot 2 \cdot 2 \cdot 2 = 32$.

(b) $\sqrt[4]{625} = \sqrt[4]{5^4} = \left(\sqrt[4]{5}\right)^4 = 5$. This result is true because $\left(5^4\right) = 625$, that is, $5 \cdot 5 \cdot 5 \cdot 5 = 625$.

$-\sqrt[4]{625} = -\left(\sqrt[4]{5^4}\right) = -\left[\left(\sqrt[4]{5}\right)^4\right] = -\left[5\right] = -5$. This result is true because $(-5)^4 = 625$, that is, $(-5) \cdot (-5) \cdot (-5) \cdot (-5) = 625$.

(c) $\sqrt[3]{-125} = \sqrt[3]{(-5)^3} = \left(\sqrt[3]{-5}\right)^3 = -5$. This result is true because $(-5)^3 = -125$, that is, $(-5) \cdot (-5) \cdot (-5) = -125$.

(d) There is no solution to $\sqrt[4]{-16}$ because any number raised to the fourth power is a positive number, that is, $N^4 = (N) \cdot (N) \cdot (N) \cdot (N) = $ a positive number $\neq$ a negative number, -16.

Simplify: (a) $\sqrt[3]{-512}$ (b) $\sqrt[4]{\dfrac{81}{16}}$ (c) $\sqrt[3]{-16} \div \sqrt[3]{-2}$.

Solution: (a) By the law of radicals which states that $\sqrt[n]{ab} = \sqrt[n]{a}\,\sqrt[n]{b}$ where a and b are any two numbers, $\sqrt[3]{-512} = \sqrt[3]{8(-64)} = \sqrt[3]{8}\sqrt[3]{-64}$. Therefore, $\sqrt[3]{-512} = \sqrt[3]{8}\sqrt[3]{-64} = (2)(-4) = -8$. The last result is true because $(2)^3 = 8$ and $(-4)^3 = -64$.

(b) By another law of radicals which states that $\sqrt[n]{\dfrac{a}{b}} = \dfrac{\sqrt[n]{a}}{\sqrt[n]{b}}$ where a and b are any two numbers, $\sqrt[4]{\dfrac{81}{16}} = \dfrac{\sqrt[4]{81}}{\sqrt[4]{16}}$.

Therefore, $\sqrt[4]{\dfrac{81}{16}} = \dfrac{\sqrt[4]{81}}{\sqrt[4]{16}} = \dfrac{3}{2}$. The last result is true because $(3)^4 = 81$ and $(2)^4 = 16$.

(c) By the law of radicals stated in example (b), $\sqrt[3]{-16} \div \sqrt[3]{-2} = \dfrac{\sqrt[3]{-16}}{\sqrt[3]{-2}} = \sqrt[3]{\dfrac{-16}{-2}} = \sqrt[3]{8} = 2$. The last result is true because $(2)^3 = 8$.

Show that: (a) $(-8)^{2/3} = (-8^{1/3})^2$

(b) $\left(\dfrac{1}{64}\right)^{4/3} = \left[\left(\dfrac{1}{64}\right)^{1/3}\right]^4$

Solution: (a) By the law of exponents which states that $(N)^{a/b} = \sqrt[b]{N^a}$ where N is any number, $(-8)^{2/3} = \sqrt[3]{(-8)^2}$. Therefore, $(-8)^{2/3} = \sqrt[3]{(-8)^2} = \sqrt[3]{64} = 4$. The last result is true because $(4)^3 = 64$. Also, by the law of exponents which states that $\left[N_1^{a_1} N_2^{a_2}\right]^b = N_1^{a_1 b} N_2^{a_2 b}$ where N_1 and N_2 are any numbers. $\left(-8^{1/3}\right)^2 = \left(-1 \cdot 8^{1/3}\right)^2 = (-1)^2\left(8^{1/3}\right)^2 = 1\left(8^{2/3}\right) = \sqrt[3]{8^2} = \sqrt[3]{64} = 4$. Again, the last result is true because $(4)^3 = 64$. Hence, $(-8)^{2/3} = 4 = \left(-8^{1/3}\right)^2$.

(b) By the first law of exponents stated above, $\left(\dfrac{1}{64}\right)^{4/3} = \sqrt[3]{\left(\dfrac{1}{64}\right)^4}$. By another law of exponents which states that $\left(\dfrac{a}{b}\right)^n = \dfrac{a^n}{b^n}$ where a and b are any two numbers, $\left(\dfrac{1}{64}\right)^4 = \dfrac{(1)^4}{(64)^4} = \dfrac{1}{(4^3)^4}$. (Note the last result is true since

$4^3 = 64$.) Hence, $\left(\frac{1}{64}\right)^4 = \frac{1}{(4^3)^4} = \frac{1}{4^{12}}$. Therefore,

$\left(\frac{1}{64}\right)^{4/3} = \sqrt[3]{\left(\frac{1}{64}\right)^4} = \sqrt[3]{\frac{1}{4^{12}}}$. By a law of radicals which states

that $\sqrt[n]{\frac{a}{b}} = \frac{\sqrt[n]{a}}{\sqrt[n]{b}}$ where a and b are any two numbers, $\sqrt[3]{\frac{1}{4^{12}}} =$

$\frac{\sqrt[3]{1}}{\sqrt[3]{4^{12}}} = \frac{\sqrt[3]{1}}{\sqrt[3]{(4^4)^3}} = \frac{1}{4^4} = \frac{1}{256}$. The expression $\left[\left(\frac{1}{64}\right)^{1/3}\right]^4 =$

$= \left[\sqrt[3]{\frac{1}{64}}\right]^4 = \left[\frac{\sqrt[3]{1}}{\sqrt[3]{64}}\right]^4 = \left[\frac{\sqrt[3]{1}}{\sqrt[3]{(4)^3}}\right]^4 = \left[\frac{1}{4}\right]^4 = \frac{(1)^4}{(4)^4} = \frac{1}{256}$. Hence,

$\left(\frac{1}{64}\right)^{4/3} = \sqrt[3]{\frac{1}{4^{12}}} = \frac{1}{256} = \left[\left(\frac{1}{64}\right)^{1/3}\right]^4$. Therefore, $\left(\frac{1}{64}\right)^{4/3} =$

$\left[\left(\frac{1}{64}\right)^{1/3}\right]^4$.

● **PROBLEM 103**

Find the numerical value of each of the following.
(a) $8^{2/3}$ (b) $25^{3/2}$

Solution:

(a) Since $x^{a/b} = \left(x^{1/b}\right)^a$, $8^{2/3} = \left(8^{1/3}\right)^2 = \left(\sqrt[3]{8}\right)^2 = (2)^2 = 4$

(b) Similarly, $25^{3/2} = \left(25^{1/2}\right)^3 = 5^3 = 125$.

● **PROBLEM 104**

Simplify $\sqrt{12} - \sqrt{27}$.

Solution: Here we have two different radicals, yet
when each is simplified, the distributive law gives a
simpler form for the expression. Note that 12 and 27 both
have a factor 3, hence

$\sqrt{12} - \sqrt{27} = \sqrt{4 \cdot 3} - \sqrt{9 \cdot 3}$

Because $\sqrt{ab} = \sqrt{a} \cdot \sqrt{b}$, $= \sqrt{4} \cdot \sqrt{3} - \sqrt{9} \cdot \sqrt{3}$

$= 2\sqrt{3} - 3\sqrt{3}$

Now, we use the distributive law, $= (2 - 3)\sqrt{3}$

$= (-1)\sqrt{3}$

$= -\sqrt{3}$

● **PROBLEM 105**

Simplify $5\sqrt{12} + 3\sqrt{75}$.

Solution: Express 12 and 75 as the product of perfect squares if possible. Thus, $12 = 4 \cdot 3$ and $75 = 25 \cdot 3$; and $5\sqrt{12} + 3\sqrt{75} = 5\sqrt{4 \cdot 3} + 3\sqrt{25 \cdot 3}$.

Since $\sqrt{a \cdot b} = \sqrt{a} \cdot \sqrt{b}$:
$$= [5 \cdot \sqrt{4} \cdot \sqrt{3}] + [3\sqrt{25} \cdot \sqrt{3}]$$
$$= [(5 \cdot 2)\sqrt{3}] + [(3 \cdot 5)\sqrt{3}]$$
$$= 10\sqrt{3} + 15\sqrt{3}.$$

Using the distributive law:
$$= (10 + 15)\sqrt{3}$$
$$= 25\sqrt{3}.$$

● **PROBLEM** 106

Approximate $\sqrt{23} \times \sqrt{40}$ and $\sqrt{23} \div \sqrt{40}$.

Solution: A four-place table of square roots gives

$$\sqrt{23} = 4.796 \quad \text{and} \quad \sqrt{40} = 6.325.$$

The product $\sqrt{23} \times \sqrt{40} = (4.796)(6.325)$

$$= 30.334700$$

Thus, rounding off to the nearest one hundredth, we obtain

$\sqrt{23} \sqrt{40} = 30.33$, approximately.

For the division: $\sqrt{23} \div \sqrt{40} = 4.796 \div 6.325$

$$= 6.325\overline{)4.796}$$

Thus,

```
            .7582
  6.325)4.7960000
        4 4275
          36850
          31625
          52250
          50600
          16500
          12650
           3850
```

Hence, $\sqrt{23} \div \sqrt{40}$ is approximately 0.7582.

● **PROBLEM** 107

If $a = 3$ and $b = 2$, find $(6a - b)^{-5/4}$.

Solution: Substitute $a = 3$ and $b = 2$: $(6 \cdot 3 - 2)^{-5/4}$

59

Perform the indicated multiplication: $(18-2)^{-5/4}$

$$= 16^{-5/4}$$

Since $x^{-y} = \dfrac{1}{x^y}$: $= \dfrac{1}{16^{5/4}}$

$$= \dfrac{1}{\left(\sqrt[4]{16}\right)^5}$$

Since $2^4 = 2\cdot2\cdot2\cdot2 = 16$,

$\sqrt[4]{16} = 2$. Hence: $= \dfrac{1}{2^5}$

$$= \dfrac{1}{32}$$

● **PROBLEM** 108

Simplify the quotient $\sqrt{x}/\sqrt[4]{x}$. Write the result in exponential notation.

<u>Solution:</u> Since $\sqrt[b]{n^a} = n^{a/b}$, the numerator and the denominator can be rewritten as:

$$\sqrt{x} = x^{1/2} \quad \text{and}$$
$$\sqrt[4]{x} = x^{1/4}$$

Therefore,

$$\frac{\sqrt{x}}{\sqrt[4]{x}} = \frac{x^{1/2}}{x^{1/4}} \qquad\qquad (1)$$

According to the law of exponents which states that $\dfrac{n^a}{n^b} = n^{a-b}$, equation (1) becomes:

$$\frac{\sqrt{x}}{\sqrt[4]{x}} = \frac{x^{1/2}}{x^{1/4}}$$

$$= x^{\frac{1}{2}-\frac{1}{4}}$$

$$= x^{\frac{2}{4}-\frac{1}{4}}$$

$$= x^{\frac{1}{4}}$$

● **PROBLEM** 109

Simplify: (a) $\sqrt{8x^3y}$ (b) $\sqrt{\dfrac{2a}{4b^2}}$ (c) $\sqrt[4]{25x^2}$.

<u>Solution:</u>(a) $\sqrt{8x^3y}$ contains the perfect square $4x^2$. Factoring

out $4x^2$ we obtain,

$$\sqrt{8x^3y} = \sqrt{4x^2 \cdot 2xy} \quad .$$

Recall that $\sqrt{ab} = \sqrt{a} \cdot \sqrt{b}$. Thus,

$$= \sqrt{4x^2} \cdot \sqrt{2xy}$$

$$= \sqrt{4}\sqrt{x^2}\sqrt{2xy} \quad .$$

Since $\sqrt{x^2} = |x|$,

$$\sqrt{8x^3y} = 2|x|\sqrt{2xy} \quad .$$

  (b) $\sqrt{\dfrac{2a}{4b^2}}$ has a fraction for the radicand, but the denominator

is a perfect square.

$$\sqrt{\frac{2a}{4b^2}} = \frac{\sqrt{2a}}{\sqrt{4b^2}} \quad , \text{ since } \sqrt{\frac{a}{b}} = \frac{\sqrt{a}}{\sqrt{b}} \; ; \; \frac{\sqrt{2a}}{\sqrt{4b^2}} = \frac{\sqrt{2a}}{2|b|} \quad .$$

  (c) $\sqrt[4]{25x^2}$ has a perfect square for the radicand.

$$\sqrt[4]{25x^2} = \sqrt[4]{(5x)^2} \quad .$$

Recall that $\sqrt[4]{x} = \sqrt[2]{\sqrt[2]{x}}$; hence $\sqrt[4]{(5x)^2} = \sqrt[2]{\sqrt[2]{(5x)^2}}$. Now, since

$\sqrt[2]{(5x)^2} = |5x|$, $\qquad = \sqrt[2]{|5x|}$. Since

$$|ab| = |a||b| \; , \; = \sqrt[2]{|5||x|} = \sqrt[2]{5|x|} = \sqrt{5|x|} \quad .$$

  Radicals with the same index can be multiplied by finding the product of the radicands, the index of the product being the same as the factors.

● **PROBLEM** 110

Find the product $\sqrt[4]{x^3y} \cdot \sqrt[4]{xy^2}$ and simplify.

<u>Solution:</u> Note that $\sqrt[x]{a} \cdot \sqrt[x]{b} = \sqrt[x]{ab}$; thus,

$$\sqrt[4]{x^3y} \cdot \sqrt[4]{xy^2} = \sqrt[4]{(x^3y)(xy^2)} \quad .$$

Recall that when multiplying, we add exponents; hence

$$\left(x^3y^1\right)\left(x^1y^2\right) = \left(x^{3+1} \, y^{1+2}\right), \quad \text{and}$$

we obtain,

$$= \sqrt[4]{x^4y^3}$$

$$= \sqrt[4]{x^4}\left(\sqrt[4]{y^3}\right)$$

Now, since $\sqrt[4]{x^4} = \left(x^{\frac{1}{4}}\right)^4 = x^1 = x$, $\sqrt[4]{x^3y} \cdot \sqrt[4]{xy^2} = x\sqrt[4]{y^3}$.

● **PROBLEM** 111

  Perform the indicated operations in the following expression and write the final result without negative or zero exponents:

$$\left(\frac{64a^{-3}b^{4/3}}{27a^{-9}b^{-14/3}}\right)^{-2/3}.$$

Solution: Since $\left(\frac{a}{b}\right)^n = \frac{a^n}{b^n}$

$$\left(\frac{64a^{-3}b^{4/3}}{27z^{-9}b^{-14/3}}\right)^{-2/3} = \frac{\left(64a^{-3}b^{4/3}\right)^{-2/3}}{\left(27a^{-9}b^{-14/3}\right)^{-2/3}}$$

and $(abc)^n = a^n b^n c^n$. Thus

$$\frac{(64a^{-3}b^{4/3})^{-2/3}}{\left(27a^{-9}b^{-14/3}\right)^{-2/3}} = \frac{\left(64\right)^{-2/3}\left(a^{-3}\right)^{-2/3}\left(b^{4/3}\right)^{-2/3}}{\left(27\right)^{-2/3}\left(a^{-9}\right)^{-2/3}\left(b^{-14/3}\right)^{-2/3}}$$

Recall $(x^y)^z = x^{yz}$ thus

$$\frac{(64)^{-2/3}(a^{-3})^{-2/3}(b^{4/3})^{-2/3}}{(27)^{-2/3}(a^{-9})^{-2/3}(b^{-14/3})^{-2/3}} = \frac{(64)^{-2/3}(a^2)(b^{-8/9})}{(27)^{-2/3}(a^6)(b^{28/9})}$$

Since $a^6 = a^4 \cdot a^2$, cancel a^2 from numerator and denominator,

$$= \frac{(64)^{-2/3}(b^{-8/9})}{(27)^{-2/3}(a^4)(b^{28/9})}$$

and since $\frac{a^x}{a^y} = a^{x-y}$, $\frac{b^{-8/9}}{b^{28/9}} = b^{-8/9 - 28/9} = b^{-36/9} = \frac{1}{b^{36/9}} = \frac{1}{b^4}$

thus

$$\frac{(64)^{-2/3}(b^{-8/9})}{(27)^{-2/3}(a^4)(b^{28/9})} = \frac{(64)^{-2/3}}{(27)^{-2/3}(a^4)(b^4)}$$

Since

$$x^{a/b} = \left(b\sqrt{x}\right)^a$$

$$(64)^{-2/3} = \left(3\sqrt{64}\right)^{-2} = (4)^{-2}$$

$$(27)^{-2/3} = \left(3\sqrt{27}\right)^{-2} = (3)^{-2}$$

Thus

$$\frac{(64)^{-2/3}}{(27)^{-2/3}a^4 b^4} = \frac{(4)^{-2}}{(3)^{-2}a^4 b^4}$$

Recall $x^{-a} = \frac{1}{x^a}$

therefore $(4)^{-2} = \frac{1}{4^2} = \frac{1}{16}$

and $(3)^{-2} = \frac{1}{3^2} = \frac{1}{9}$

hence $\frac{(4)^{-2}}{(3)^{-2}a^4 b^4} = \frac{1/16}{\frac{1}{9}a^4 b^4}$

Multiply numerator and denominator by $16 \cdot 9$,

$$= \frac{16 \cdot 9(1/16)}{16 \cdot 9\left(\frac{1}{9}\right)a^4 b^4}$$

$$= \frac{9}{16 \, a^4 b^4}$$

thus $\left(\frac{64a^{-3} \, b^{4/3}}{27a^{-9} \, b^{-14/3}} \right)^{-2/3} = \frac{9}{16 \, a^4 b^4}$

● **PROBLEM** 112

Simplify $\sqrt[3]{-81x^3} - 2x\sqrt[3]{3} + 5x\sqrt[3]{24}$.

<u>Solution:</u> Rewrite the expression so it contains similar radicals.

$\sqrt[3]{-81x^3} = \sqrt[3]{(-3)^3 x^3 \cdot 3} = \sqrt[3]{(-3x)^3 \cdot 3}$ by the law $(ab)^n = a^n b^n$. Also, since $\sqrt[n]{ab} = \sqrt[n]{a} \sqrt[n]{b}$, $\sqrt[3]{(-3x)^3 \cdot 3} = \sqrt[3]{(-3x)^3} \sqrt[3]{3}$. Hence, $\sqrt[3]{-81x^3} = \sqrt[3]{(-3x)^3} \sqrt[3]{3}$. Since $(\sqrt[n]{a})^n = (a^{1/n})^n = a^{\frac{1}{n} \cdot n} = a^1 = a$, $\sqrt[3]{(-3x)^3} = -3x$. Therefore, $\sqrt[3]{-81x^3} = -3x\sqrt[3]{3}$. By the same laws, $5x\sqrt[3]{24} = 5x\sqrt[3]{2^3 \cdot 3} = 5x\sqrt[3]{(2)^3}\sqrt[3]{3} = 5x(2)\sqrt[3]{3} = 10x\sqrt[3]{3}$. Therefore, $\sqrt[3]{-81x^3} - 2x\sqrt[3]{3} + 5x\sqrt[3]{24} = -3x\sqrt[3]{3} - 2x\sqrt[3]{3} + 10x\sqrt[3]{3}$

$$= -5x\sqrt[3]{3} + 10x\sqrt[3]{3} = 5x\sqrt[3]{3}.$$

Hence, $\sqrt[3]{-81x^3} - 2x\sqrt[3]{3} + 5x\sqrt[3]{24} = 5x\sqrt[3]{3}$. Note that the radical used to simplify the given expression was $\sqrt[3]{3}$.

● **PROBLEM** 113

Find the square root of

$$\frac{3}{2} (x - 1) + \sqrt{2x^2 - 7x - 4}.$$

<u>Solution:</u> The given expression can be rewritten as

$$\frac{3}{2} x - \frac{3}{2} + \sqrt{2x^2 - 7x - 4}.$$

We can eliminate the fractions by factoring $\frac{1}{2}$ from all the terms. To do this we must first multiply the third term by 2 so as not to change the value of this term. Thus, we obtain:

$$\frac{1}{2} \left(3x - 3 + 2\sqrt{2x^2 - 7x - 4} \right)$$

Let us examine the expression under the radical. Notice that this can be rewritten in factored form since

$$2x^2 - 7x - 4 = (2x + 1)(x - 4).$$

Substituting this in the given expression we have:

$$\frac{1}{2} \left(3x - 3 + 2\sqrt{(2x + 1)(x - 4)} \right).$$

Our aim now is to transform the expression into one which is a perfect square. This can be accomplished as follows: Rewrite the first two terms, 3x − 3 as: (2x + 1) + (x − 4), and substitute this into the expression. Thus, we obtain,

$$\frac{1}{2}\left[(2x + 1) + (x - 4) + 2\sqrt{(2x + 1)(x - 4)}\right],$$

and we are looking for:

$$\sqrt{\frac{1}{2}\left[(2x + 1)+(x - 4) + 2\sqrt{(2x + 1)(x - 4)}\right]}$$

But,

$$(2x + 1) + (x - 4) + 2\sqrt{(2x + 1)(x - 4)}$$

$$= (2x + 1) + (x - 4) + 2\sqrt{2x + 1}\sqrt{x - 4}$$

$$= (\sqrt{2x + 1} + \sqrt{x - 4})(\sqrt{2x + 1} + \sqrt{x - 4})$$

$$= (\sqrt{2x + 1} + \sqrt{x - 4})^2;$$

Therefore, our expression becomes:

$$\sqrt{\frac{1}{2}(\sqrt{2x + 1} + \sqrt{x - 4})^2}$$

$$= \sqrt{\frac{1}{2}(\sqrt{2x + 1} + \sqrt{x - 4})}$$

$$= \frac{1}{\sqrt{2}}(\sqrt{2x + 1} + \sqrt{x - 4}).$$

● **PROBLEM** 114

Find the cube root of $9\sqrt{3} + 11\sqrt{2}$.

Solution: In this problem we wish to find:

$$\sqrt[3]{9\sqrt{3} + 11\sqrt{2}}$$

Factoring $3\sqrt{3}$ from both terms under the radical, $9\sqrt{3}$ and $11\sqrt{2}$, we obtain:

$$\sqrt[3]{3\sqrt{3}\left[3 + \frac{11}{3}\frac{\sqrt{2}}{\sqrt{3}}\right]}, \qquad \text{or}$$

$$\sqrt[3]{9\sqrt{3} + 11\sqrt{2}}$$

$$= \sqrt[3]{3\sqrt{3}\left[3 + \frac{11}{3}\sqrt{\frac{2}{3}}\right]}.$$

Now, since $\sqrt[3]{3\sqrt{3}} = \sqrt{3}$ (that is, $\sqrt[3]{3\sqrt{3}} = \sqrt[3]{3(3^{\frac{1}{2}})} = \sqrt[3]{3^{(1+\frac{1}{2})}} =$

$\sqrt[3]{3^{\frac{3}{2}}} = (3^{\frac{3}{2}})^{\frac{1}{3}} = 3^{\frac{1}{2}} = \sqrt{3}$) we can write the expression as:

$$= \sqrt{3}\left(\sqrt[3]{3 + \frac{11}{3}\sqrt{\frac{2}{3}}}\right)$$

Observe that $3 + \frac{11}{3}\sqrt{\frac{2}{3}}$ is a perfect cube, since:

$$\left(1 + \sqrt{\frac{2}{3}}\right)\left(1 + \sqrt{\frac{2}{3}}\right)\left(1 + \sqrt{\frac{2}{3}}\right)$$

$$= \left(1 + 2\sqrt{\frac{2}{3}} + \frac{2}{3}\right)\left(1 + \sqrt{\frac{2}{3}}\right)$$

$$= 1 + 2\sqrt{\frac{2}{3}} + \frac{2}{3} + \sqrt{\frac{2}{3}} + 2 \cdot \frac{2}{3} + \frac{2}{3}\sqrt{\frac{2}{3}}$$

$$= \left(1 + \frac{4}{3} + \frac{2}{3}\right) + \left(2\sqrt{\frac{2}{3}} + \sqrt{\frac{2}{3}} + \frac{2}{3}\sqrt{\frac{2}{3}}\right)$$

$$= 3 + \frac{11}{3}\sqrt{\frac{2}{3}} .$$

Thus,

$$\sqrt[3]{3 + \frac{11}{3}\sqrt{\frac{2}{3}}} = 1 + \sqrt{\frac{2}{3}} , \text{ and}$$

the required cube root $= \sqrt{3}\left(1 + \sqrt{\frac{2}{3}}\right)$

$$= \sqrt{3} + \sqrt{2}.$$

RATIONALIZING THE DENOMINATOR

● **PROBLEM** 115

Write in fractional exponent form with no denominators.

(a) $\sqrt[3]{\frac{x}{y}}$ (b) $\sqrt[3]{3}$ (c) $\sqrt[4]{x} \ \sqrt[3]{xy^{-1}}$

Solution: Noting that $\sqrt[b]{a} = a^{1/b}$, $\left(\frac{a}{b}\right)^c = \frac{a^c}{b^c}$, and $a^{-b} = \dfrac{1}{a^b}$, we proceed to evaluate these expressions.

(a) $\sqrt[3]{\dfrac{x}{y}} = \left(\dfrac{x}{y}\right)^{1/3} = \dfrac{x^{1/3}}{y^{1/3}} = x^{\frac{1}{3}} y^{-\frac{1}{3}}$

(b) $\sqrt[3]{3} = 3^{\frac{1}{3}}$

(c) $\sqrt[4]{x^2}\ \sqrt[3]{xy^{-1}} = (x^2)^{\frac{1}{4}} \left(xy^{-1}\right)^{\frac{1}{3}}$

$\qquad = (x^{\frac{2}{4}})(x^{\frac{1}{3}})(y^{-\frac{1}{3}})$, since $(a^b)^c = a^{bc}$

$\qquad\qquad\qquad\qquad\qquad$ and $(ab^c)^d = a^d b^{cd}$

$\qquad = (x^{\frac{1}{2}})(x^{\frac{1}{3}})(y^{-\frac{1}{3}})$

$\qquad = x^{\frac{1}{2}+\frac{1}{3}}\ y^{-\frac{1}{3}}$, since $(x^a)(x^b) = x^{a+b}$

$\qquad = x^{\frac{3}{6}+\frac{2}{6}}\ y^{-\frac{1}{3}}$

$\qquad = x^{\frac{5}{6}}\ y^{-\frac{1}{3}}$

● **PROBLEM** 116

Rationalize the denominator in the quotient $1\Big/ \sqrt[5]{x^2}$.

Solution: $\sqrt[y]{a^x} = a^{\frac{x}{y}}$, thus $\dfrac{1}{\sqrt[5]{x^2}} = \dfrac{1}{x^{\frac{2}{5}}}$. Rationalizing a

denominator means eliminating the radical from the denominator, thus we want to eliminate the fractional exponent.
When multiplying numbers with the same base, exponents are
added, hence multiplying numerator and denominator by

$x^{\frac{3}{5}}$ will eliminate the fractional exponent in the denominator:

$$\dfrac{1}{\sqrt{x^2}} = \dfrac{1}{x^{\frac{2}{5}}} = \dfrac{1}{x^{\frac{2}{5}}}\ \dfrac{x^{\frac{3}{5}}}{x^{\frac{3}{5}}} = \dfrac{x^{\frac{3}{5}}}{x^{\frac{2}{5}+\frac{3}{5}}} = \dfrac{x^{\frac{3}{5}}}{x^1} = \dfrac{\sqrt[5]{x^3}}{x}.$$

● **PROBLEM** 117

Write in radical form without negative exponents, rationaliz-
ing denominators.　(a) $\left(x^{1/3}\right)^{-3/4}$　(b) $x^{1/6}\Big/x^{-2/3}$.

Solution: Noting $\left(a^b\right)^c = a^{bc}$, $\dfrac{a^b}{a^c} = a^{b-c}$, $a^{-b} = \dfrac{1}{a^b}$,
$a^{1/b} = \sqrt[b]{a}$, and $\sqrt[b]{a}\ \sqrt[c]{c} = \sqrt[bc]{ac}$, we proceed to evaluate these
expressions.

66

(a) $\left(x^{1/3}\right)^{-3/4} = x^{\left(\frac{1}{3}\right)\left(\frac{-3}{4}\right)} = x^{\frac{-3}{12}} = x^{-\frac{1}{4}} = \frac{1}{x^{1/4}} = \frac{1}{\sqrt[4]{x}}$

$$= \frac{1}{\sqrt[4]{x}} \cdot \frac{\sqrt[4]{x^3}}{\sqrt[4]{x^3}} \qquad \left(\text{since } \frac{\sqrt[4]{x^3}}{\sqrt[4]{x^3}} = 1\right)$$

$$= \frac{\sqrt[4]{x^3}}{\sqrt[4]{x}\,\sqrt[4]{x^3}} = \frac{\sqrt[4]{x^3}}{\sqrt[4]{x \cdot x^3}} = \frac{\sqrt[4]{x^3}}{\sqrt[4]{x^4}}.$$

Since $\sqrt[n]{a^n} = (a^n)^{1/n} = a^{n \cdot \frac{1}{n}} = a^{n/n} = a^1 = a$, $\sqrt[4]{x^4} = x$. Thus,

$$\left(x^{\frac{1}{3}}\right)^{-\frac{3}{4}} = \frac{\sqrt[4]{x^3}}{x}$$

(b) $\dfrac{x^{1/6}}{x^{-2/3}} = x^{\frac{1}{6}-\left(-\frac{2}{3}\right)} = x^{\frac{1}{6}+\frac{2}{3}} = x^{\frac{1}{6}+\frac{4}{6}} = x^{\frac{5}{6}} = \sqrt[6]{x^5}$

● **PROBLEM** 118

Rationalize $\dfrac{\sqrt[3]{3ax}}{\sqrt[3]{4a^2}}$.

Solution: Multiply the numerator and the denominator by the radical $\sqrt[3]{(4a^2)^2}$ to eliminate the radical in the denominator:

$$\frac{\sqrt[3]{3ax}}{\sqrt[3]{4a^2}} = \frac{\left[\sqrt[3]{(4a^2)^2}\right]\sqrt[3]{3ax}}{\left(\sqrt[3]{(4a^2)^2}\right)\sqrt[3]{(4a^2)}} = \frac{\sqrt[3]{(4a^2)^2}\,\sqrt[3]{3ax}}{\sqrt[3]{(4a^2)^3}}$$

Note the last result is true because of the law involving radicals which states that $\sqrt[3]{a} \cdot \sqrt[3]{b} = \sqrt[3]{ab}$. Also, since $\sqrt[3]{a^3} = (\sqrt[3]{a})^3 = (a^{1/3})^3 = a^1 = a$, $\sqrt[3]{(4a^2)^3} = \left(\sqrt[3]{4a^2}\right)^3 = 4a^2$. Hence,

$$\frac{\sqrt[3]{3ax}}{\sqrt[3]{4a^2}} = \frac{\sqrt[3]{(4a^2)^2}\,\sqrt[3]{3ax}}{4a^2} = \frac{\sqrt[3]{16a^4}\,\sqrt[3]{3ax}}{4a^2} .$$

Since $\sqrt[3]{ab} = \sqrt[3]{a}\sqrt[3]{b}$, $\sqrt[3]{16a^4} = \sqrt[3]{(8a^3)(2a)} = \sqrt[3]{8a^3}\sqrt[3]{2a}$

$$= \sqrt[3]{(2a)^3}\sqrt[3]{2a}.$$

Note that the last result is true because $(ab)^x = a^x b^x$; that is, $8a^3 = 2^3 a^3 = (2a)^3$. Hence:

$$\frac{\sqrt[3]{3ax}}{\sqrt[3]{4a^2}} = \frac{\sqrt[3]{(2a)^3}\sqrt[3]{2a}\sqrt[3]{3ax}}{4a^2}$$

$$= \frac{2a\sqrt[3]{2a}\sqrt[3]{3ax}}{4a^2}$$

$$= \frac{2a\sqrt[3]{(2a)(3ax)}}{4a^2}$$

$$= \frac{2a\sqrt[3]{6a^2x}}{4a^2} \ .$$

Therefore, $\dfrac{\sqrt[3]{3ax}}{\sqrt[3]{4a^2}} = \dfrac{\sqrt[3]{6a^2x}}{2a}.$

● **PROBLEM** 119

Express the product $1/\sqrt[3]{x^2} \cdot 1/\sqrt[4]{x}$ in simplest radical form, rationalizing the denominator.

<u>Solution:</u> Since $\dfrac{1}{b\sqrt{x^a}} = \dfrac{1}{x^{a/b}}$

$$\frac{1}{\sqrt[3]{x^2}} = \frac{1}{x^{2/3}} \quad \text{and} \quad \frac{1}{\sqrt[4]{x}} = \frac{1}{x^{1/4}}$$

thus,

$$\frac{1}{\sqrt[3]{x^2}} \cdot \frac{1}{\sqrt[4]{x}} = \frac{1}{x^{2/3}} \cdot \frac{1}{x^{1/4}} \ .$$

Since $a^x \cdot a^y = a^{x+y}$

$$\frac{1}{x^{2/3}} \cdot \frac{1}{x^{1/4}} = \frac{1}{x^{2/3 + 1/4}} = \frac{1}{x^{8/12 + 3/12}} = \frac{1}{x^{11/12}}$$

To rationalize the denominator we wish to obtain an integral exponent of x in the denominator, thus we multiply numerator and denominator by $x^{1/12}$,

$$\frac{1}{x^{11/12}} \cdot \frac{x^{1/12}}{x^{1/12}} = \frac{x^{1/12}}{x^{11/12 + 1/12}} = \frac{x^{1/12}}{x^{12/12}}$$

$$= \frac{\sqrt[12]{x}}{x^1} = \frac{\sqrt[12]{x}}{x}$$

Thus

$$\frac{1}{\sqrt[3]{x^2}} \cdot \frac{1}{\sqrt[4]{x}} = \frac{\sqrt[12]{x}}{x} \ .$$

● **PROBLEM** 120

Find the factor which will rationalize $\sqrt{3} + \sqrt[3]{5}$.

<u>Solution:</u> We can rewrite $\sqrt{3} + \sqrt[3]{5}$ as,

$$3^{\frac{1}{2}} + 5^{\frac{1}{3}}$$

Observe that both of the above irrational numbers, when raised to the sixth power, become rational

$$\left[\left(3^{\frac{1}{2}}\right)^6 = 3^3 = 27, \ \left(5^{\frac{1}{3}}\right)^6 = 5^2 = 25 \right].$$

Let $x = 3^{\frac{1}{2}}$, $y = 5^{\frac{1}{3}}$; then x^6 and y^6 are both rational.

Since x^6 and y^6 are rational, so is $x^6 - y^6$ (and in fact, is equal to $27 - 25 = 2$). To find the factor which rationalizes $x + y = (\sqrt{3} + \sqrt[3]{5})$, we divide $x^6 - y^6$ by $x + y$, and find the quotient to be,

$$x^5 - x^4y + x^3y^2 - x^2y^3 + xy^4 - y^5.$$

Thus

$$x^6 - y^6 = (x + y)(x^5 - x^4y + x^3y^2 - x^2y^3 + xy^4 - y^5);$$

and substituting for x and y, the required factor is

$$x^5 - x^4y + x^3y^2 - x^2y^3 + xy^4 - y^5 =$$

$$\left(3^{\frac{1}{2}}\right)^5 - \left(3^{\frac{1}{2}}\right)^4 5^{\frac{1}{3}} + \left(3^{\frac{1}{2}}\right)^3 \left(5^{\frac{1}{3}}\right)^2 - \left(3^{\frac{1}{2}}\right)^2 \left(5^{\frac{1}{3}}\right)^3$$

$$+ 3^{\frac{1}{2}} \left(5^{\frac{1}{3}}\right)^4 - \left(5^{\frac{1}{3}}\right)^5 =$$

$$3^{\frac{5}{2}} - 3^{\frac{4}{2}} \cdot 5^{\frac{1}{3}} + 3^{\frac{3}{2}} \cdot 5^{\frac{2}{3}} - 3^{\frac{2}{2}} \cdot 5^{\frac{3}{3}}$$

$$+ 3^{\frac{1}{2}} \cdot 5^{\frac{4}{3}} - 5^{\frac{5}{3}} , \qquad \text{or}$$

$$3^{\frac{5}{2}} - 9 \cdot 5^{\frac{1}{3}} + 3^{\frac{3}{2}} \cdot 5^{\frac{2}{3}} - 15 + 3^{\frac{1}{2}} \cdot 5^{\frac{4}{3}} - 5^{\frac{5}{3}} ;$$

and the rational product, $\left(3^{\frac{1}{2}}\right)^6 - \left(5^{\frac{1}{3}}\right)^6$, of the above factor and $3^{\frac{1}{2}} + 5^{\frac{1}{3}}$ is

$$3^{\frac{6}{2}} - 5^{\frac{6}{3}} = 3^3 - 5^2 = 2.$$

● **PROBLEM** 121

Rationalize the denominator in the quotient:

$$\frac{1}{\sqrt{x} - \sqrt{y}} .$$

<u>Solution</u>: To rationalize a denominator we multiply numerator and denominator by the conjugate of the denominator (recall $a + bi$ and $a - bi$ are complex conjugates). In our example the conjugate of $\sqrt{x} - \sqrt{y}$ is $\sqrt{x} + \sqrt{y}$. Hence, multiplying numerator and denominator by $\sqrt{x} + \sqrt{y}$:

$$\frac{1}{\sqrt{x} - \sqrt{y}} \cdot \frac{\sqrt{x} + \sqrt{y}}{\sqrt{x} + \sqrt{y}} = \frac{\sqrt{x} + \sqrt{y}}{\sqrt{x} \cdot \sqrt{x} + \sqrt{x}\sqrt{y} - \sqrt{x}\sqrt{y} - \sqrt{y}\sqrt{y}} .$$

Recall that $\sqrt{a} \cdot \sqrt{a} = \sqrt{a^2} = a$, thus $\sqrt{x}\sqrt{x} = x$ and $\sqrt{y}\sqrt{y} = y$; and we obtain:

$$= \frac{\sqrt{x} + \sqrt{y}}{x - y} .$$

Rationalize $\dfrac{\sqrt{3xy}}{\sqrt{2x} - \sqrt{3y}}$.

Solution: To rationalize a fraction, we multiply numerator and denominator by the conjugate of the denominator (where the conjugate of a + b is a − b). In our example, we multiply numerator and denominator by the conjugate of $\sqrt{2x} - \sqrt{3y}$, which is $\sqrt{2x} + \sqrt{3y}$. Thus,

$$\frac{\sqrt{3xy}}{\sqrt{2x} - \sqrt{3y}} = \frac{\sqrt{3xy}}{\sqrt{2x} - \sqrt{3y}} \cdot \frac{\sqrt{2x} + \sqrt{3y}}{\sqrt{2x} + \sqrt{3y}}$$

$$= \frac{\sqrt{3xy} \, (\sqrt{2x} + \sqrt{3y})}{(\sqrt{2x})^2 - (\sqrt{3y})^2} .$$

Since $(\sqrt{a})^2 = \sqrt{a} \cdot \sqrt{a} = \sqrt{a^2} = a$,

$$(\sqrt{2x})^2 = 2x$$

$$(\sqrt{3y})^2 = 3y.$$

Making these substitutions,

$$\frac{\sqrt{3xy} \, (\sqrt{2x} + \sqrt{3y})}{(\sqrt{2x})^2 - (\sqrt{3y})^2} = \frac{\sqrt{3xy} \, (\sqrt{2x} + \sqrt{3y})}{2x - 3y}$$

$$= \frac{\sqrt{3xy} \cdot \sqrt{2x} + \sqrt{3xy} \cdot \sqrt{3y}}{2x - 3y}$$

Since $\sqrt{a} \sqrt{b} = \sqrt{ab}$ and $\sqrt{a} \sqrt{b} \sqrt{c} = \sqrt{abc}$,

$$\frac{\sqrt{3xy} \cdot \sqrt{2x} + \sqrt{3xy} \cdot \sqrt{3y}}{2x - 3y} = \frac{\sqrt{6x^2 y} + \sqrt{9xy^2}}{2x - 3y}$$

$$= \frac{\sqrt{x^2} \sqrt{6y} + \sqrt{9} \sqrt{y^2} \sqrt{x}}{2x - 3y}$$

$$= \frac{x \sqrt{6y} + 3y \sqrt{x}}{2x - 3y} .$$

Reduce $\dfrac{\left(2 + 3\sqrt{-1}\right)^2}{2 + \sqrt{-1}}$ to the form $A + B\sqrt{-1}$.

Solution: Expanding the numerator, we can rewrite the fraction as,

$$\frac{(2 + 3\sqrt{-1})(2 + 3\sqrt{-1})}{2 + \sqrt{-1}} = \frac{4 + 6\sqrt{-1} + 6\sqrt{-1} + 3\sqrt{-1} \cdot 3\sqrt{-1}}{2 + \sqrt{-1}} =$$

$$\frac{4 + 12\sqrt{-1} + 9(-1)}{2 + \sqrt{-1}} = \frac{-5 + 12\sqrt{-1}}{2 + \sqrt{-1}} .$$

We now rationalize the denominator by multiplying both numerator and denominator by $2 - \sqrt{-1}$. This does not change the value of the fraction, since

$$\frac{2 - \sqrt{-1}}{2 - \sqrt{-1}} = 1 \; ,$$

and multiplication by 1 does not change the value of any expression. Thus, we have:

$$\frac{-5 + 12\sqrt{-1}}{2 + \sqrt{-1}}\left(\frac{2 - \sqrt{-1}}{2 - \sqrt{-1}}\right) = \frac{-10 + 24\sqrt{-1} + 5\sqrt{-1} - 12\sqrt{-1} \cdot \sqrt{-1}}{4 + 2\sqrt{-1} - 2\sqrt{-1} - \sqrt{-1} \cdot \sqrt{-1}}$$

$$= \frac{-10 + 29\sqrt{-1} - 12(-1)}{4 - (-1)} = \frac{2 + 29\sqrt{-1}}{5}$$

$$= \frac{2}{5} + \frac{29}{5}\sqrt{-1} \; ;$$

which is of the required form, $A + B\sqrt{-1}$.

● **PROBLEM** 124

Express with rational denominator $\dfrac{4}{\sqrt[3]{9} - \sqrt[3]{3} + 1}$.

<u>Solution:</u> The given expression can be written as,

$$\frac{4}{9^{\frac{1}{3}} - 3^{\frac{1}{3}} + 1} = \frac{4}{\left(3^2\right)^{\frac{1}{3}} - 3^{\frac{1}{3}} + 1} = \frac{4}{3^{\frac{2}{3}} - 3^{\frac{1}{3}} + 1} \; .$$

To ratonalize the denominator, multiply both numerator and denominator by $\left(3^{\frac{1}{3}} + 1\right)$. This will not change the value of the fraction because

$$\frac{3^{\frac{1}{3}} + 1}{3^{\frac{1}{3}} + 1} = 1,$$

and multiplication by 1 does not change the value of any given expression.

Thus, multiplying, we obtain:

$$\frac{4\left(3^{\frac{1}{3}} + 1\right)}{\left(3^{\frac{1}{3}} + 1\right)\left(3^{\frac{2}{3}} - 3^{\frac{1}{3}} + 1\right)}$$

$$= \frac{4\left(3^{\frac{1}{3}} + 1\right)}{\left(3^{\frac{1}{3}}\right)\left(3^{\frac{2}{3}}\right) + 3^{\frac{2}{3}} - \left(3^{\frac{1}{3}}\right)\left(3^{\frac{1}{3}}\right) - 3^{\frac{1}{3}} + 3^{\frac{1}{3}} + 1}$$

$$= \frac{4\left(3^{\frac{1}{3}} + 1\right)}{3^{\frac{3}{3}} + 3^{\frac{2}{3}} - 3^{\frac{2}{3}} - 3^{\frac{1}{3}} + 3^{\frac{1}{3}} + 1}$$

$$\frac{4\left(3^{\frac{1}{3}} + 1\right)}{3 + 1} = \frac{4\left(3^{\frac{1}{3}} + 1\right)}{4}$$

$$= 3^{\frac{1}{3}} + 1.$$

OPERATIONS WITH RADICALS

● **PROBLEM** 125

Find the product $\sqrt[3]{2a^2x} \cdot \sqrt{2x}$

Solution: Since $\sqrt[n]{N} = N^{1/n}$, $\sqrt[3]{2a^2x} = \left(2a^2x\right)^{1/3}$ and $\sqrt{2x} = (2x)^{1/2}$. Therefore, $\sqrt[3]{2a^2x} \cdot \sqrt{2x} = \left(2a^2x\right)^{1/3}(2x)^{1/2}$. Also, since $\left(N_1 N_2\right)^{1/n} = \left(N_1\right)^{1/n}\left(N_2\right)^{1/n}$ where N_1 and N_2 are any two numbers and n is any positive integer, $\left(2a^2x\right)^{1/3} = \left[\left(2a^2\right)(x)\right]^{1/3} = \left(2a^2\right)^{1/3}(x)^{1/3}$ and $(2x)^{1/2} = 2^{1/2}x^{1/2}$. Therefore:

$$\sqrt[3]{2a^2x} \cdot \sqrt{2x} = \left(2a^2\right)^{1/3}(x)^{1/3}(2)^{1/2}(x)^{1/2}$$

$$= (2)^{1/3}\left(a^2\right)^{1/3}(x)^{1/3}(2)^{1/2}(x)^{1/2}.$$

By the law of exponents which states that $a^m \cdot a^n = a^{m+n}$, $(2)^{1/3}(2)^{1/2} = 2^{1/3+1/2}$ and $(x)^{1/3}(x)^{1/2} = x^{1/3+1/2}$. Obtaining a least common denominator of 6 for the two fractions in the exponents:

$\frac{1}{3} + \frac{1}{2} = \frac{2(1)}{2(3)} + \frac{3(1)}{3(2)} = \frac{2}{6} + \frac{3}{6} = \frac{5}{6}$. Hence, $(2)^{1/3}(2)^{1/2} = 2^{1/3 + 1/2} = 2^{5/6}$ and $(x)^{1/3}(x)^{1/2} = x^{1/3 + 1/2} = x^{5/6}$.

Therefore, $\sqrt[3]{2a^2x} \cdot \sqrt{2x} = (2)^{1/3}\left(a^2\right)^{1/3}(x)^{1/3}(2)^{1/2}(x)^{1/2}$

$$= (2)^{1/3}(2)^{1/2}(x)^{1/3}(x)^{1/2}\left(a^2\right)^{1/3}$$

$$= \left(2^{5/6}\right)\left(x^{5/6}\right)\left(a^2\right)^{1/3}$$

$$= (2x)^{5/6}\left(a^2\right)^{1/3}.$$

Since $N^{m/n} = \sqrt[n]{N^m}$, $(2x)^{5/6} = \sqrt[6]{(2x)^5}$. Also, since $\left(N^m\right)^n = N^{mn}$, $\left(a^2\right)^{1/3} = a^{2/3}$. Therefore,

$$\sqrt[6]{2a^2x} \cdot \sqrt{2x} = \sqrt[6]{(2x)^5}\, a^{2/3} = \sqrt[6]{(2x)^5}\, a^{4/6} = \sqrt[6]{(2x)^5}\, \sqrt[6]{a^4}$$

Since $(xy)^a = x^a y^a$,
$$= \sqrt[6]{(2)^5 x^5}\, \sqrt[6]{a^4}$$
$$= \sqrt[6]{32x^5}\, \sqrt[6]{a^4}$$
$$= \sqrt[6]{32a^4 x^5} \ .$$

● **PROBLEM** 126

Simplify $(\sqrt{3} + \sqrt{2}) \cdot (\sqrt{2} - \sqrt{6})$.

<u>Solution:</u> Using the distributive property the expression on the right can be multiplied by each term in the expression on the left or vice versa. Hence,

$$(\sqrt{3} + \sqrt{2})(\sqrt{2} - \sqrt{6}) = \sqrt{3}(\sqrt{2} - \sqrt{6}) + \sqrt{2}(\sqrt{2} - \sqrt{6})$$

The distributive property again enables us to multiply the terms on the left by their respective right hand members;

$$= (\sqrt{3} \cdot \sqrt{2}) - (\sqrt{3} \cdot \sqrt{6}) + (\sqrt{2} \cdot \sqrt{2})$$
$$- (\sqrt{2} \cdot \sqrt{6})$$

Then since,
$$\sqrt{a} \cdot \sqrt{6} = \sqrt{ab},$$
$$(\sqrt{3} + 2)(\sqrt{2} - \sqrt{6}) = (\sqrt{6}) - (\sqrt{18}) + (2) - (\sqrt{12})$$
$$= \sqrt{6} - 3\sqrt{2} + 2 - 2\sqrt{3}.$$

● **PROBLEM** 127

Multiply $(2\sqrt{3} + 3\sqrt{2})$ by $(3\sqrt{3} - 2\sqrt{2})$.

<u>Solution:</u> Using the following method (foil method) of polynomial multiplication:
$$(x + y)(a + b) = xa + xb + ya + yb,$$
we obtain
$$(2\sqrt{3} + 3\sqrt{2})(3\sqrt{3} - 2\sqrt{2}) = (2\sqrt{3})(3\sqrt{3}) + (2\sqrt{3})(-2\sqrt{2}) + (3\sqrt{2})(3\sqrt{3})$$
$$+ (3\sqrt{2})(-2\sqrt{2})$$
$$= (6)(\sqrt{3})^2 - 4(\sqrt{3}\sqrt{2}) + (9)(\sqrt{3}\sqrt{2}) - 6(\sqrt{2})^2 \ .$$

Since
$$(\sqrt{a})^2 = \left(a^{\frac{1}{2}}\right)^2 = a^{2/2} = a^1 = a$$
$$(\sqrt{3})^2 = 3 \quad \text{and} \quad (\sqrt{2})^2 = 2$$

Making these substitutions we obtain,

$$(2\sqrt{3} + 3\sqrt{2})(3\sqrt{3} - 2\sqrt{2}) = (6)(3) + 5(\sqrt{3})(\sqrt{2}) - 6(2)$$
$$= 18 + 5(\sqrt{3})(\sqrt{2}) - 12$$
$$= 6 + 5\sqrt{3}\sqrt{2}$$

Since $\sqrt{a} \cdot \sqrt{b} = \sqrt{ab}$

$$\sqrt{3} \cdot \sqrt{2} = \sqrt{3 \cdot 2} = \sqrt{6}$$

Therefore $(2\sqrt{3} + 3\sqrt{2})(3\sqrt{3} - 2\sqrt{2}) = 6 + 5\sqrt{6}$

● **PROBLEM** 128

Find the product by inspection: $\sqrt{3}(x - \sqrt{5})(x + \sqrt{5})$.

<u>Solution:</u> The formula for the difference of two squares can be used to obtain the product. This formula is: $(x^2 - y^2) = (x + y)(x - y)$. The product $(x - \sqrt{5})(x + \sqrt{5})$ corresponds to the right side of the formula where x is replaced by x and y is replaced by $\sqrt{5}$. Therefore,

$$3(x - \sqrt{5})(x + \sqrt{5}) = 3\left(x^2 - (\sqrt{5})^2\right)$$

Since $(\sqrt{a})^2 = \sqrt{a} \ \sqrt{a} = \sqrt{a^2} = a$, $(\sqrt{5})^2 = 5$. Thus

$$\sqrt{3}(x - \sqrt{5})(x + \sqrt{5}) = \sqrt{3}\left(x^2 - 5\right).$$

Distributing $\sqrt{3}$,

$$= \sqrt{3} \ x^2 - \sqrt{3}(5)$$

$$= \sqrt{3} \ x^2 - 5\sqrt{3}.$$

Thus $\sqrt{3}(x - \sqrt{5})(x + \sqrt{5}) = \sqrt{3} \ x^2 - 5\sqrt{3}$.

● **PROBLEM** 129

Find the following products:

    a. $\sqrt{x} \ (\sqrt{2x} - \sqrt{x})$

    b. $(\sqrt{x} - 2 \sqrt{y})(2\sqrt{x} + \sqrt{y})$

<u>Solution:</u> The following two laws concerning radicals can be used to find the indicated products:

    1) $\sqrt{a} \ \sqrt{b} = \sqrt{ab}$

where a and b are any two numbers

    2) $\sqrt{a^2} = (\sqrt{a})^2 \ \left(a^{1/2}\right)^2 = a^{(1/2) \ 2} = a^1 = a$

    <u>or</u> $\sqrt{a^2} = a$

a) Using the distributive property,

$$\sqrt{x} \ (\sqrt{2x} - \sqrt{x}) = \sqrt{x} \ (\sqrt{2x}) - \sqrt{x} \ (\sqrt{x})$$

Using the first law concerning radicals to further simplify this equation,

$$\sqrt{x}\ (\sqrt{2x} - \sqrt{x}) = \sqrt{x \cdot 2x} - \sqrt{x \cdot x}$$

$$= \sqrt{2x^2} - \sqrt{x^2}$$

$$= \sqrt{2}\sqrt{x^2} - \sqrt{x^2}$$

Using, the second law to simplify this equation,

$$\sqrt{x}\ \ (\sqrt{2x} - \sqrt{x}) = \sqrt{2}\ (x) - x$$

$$= \sqrt{2}x - x$$

b) $(\sqrt{x} - 2\sqrt{y})(2\sqrt{x} + \sqrt{y}) = \sqrt{x}\ (2\sqrt{x}) - 2\sqrt{y}(2\sqrt{x}) + \sqrt{x}(\sqrt{y})$

$- 2\sqrt{y}\ (\sqrt{y})$

Using the first law concerning radicals to simplify this equation,

$$(\sqrt{x} - 2\sqrt{y})(2\sqrt{x} + \sqrt{y}) = 2\sqrt{x \cdot x} - 4\sqrt{x \cdot y} + \sqrt{x \cdot y} - 2\sqrt{y \cdot y}$$

$$= 2\sqrt{x^2} - 4\sqrt{xy} + \sqrt{xy} - 2\sqrt{y^2}$$

$$= 2\sqrt{x^2} - 3\sqrt{xy} - 2\sqrt{y^2}$$

Using the second law to simplify this equation,

$$(\sqrt{x} - 2\sqrt{y})(2\sqrt{x} + \sqrt{y}) = 2(x) - 3\sqrt{xy} - 2(y)$$

$$= 2x - 3\sqrt{xy} - 2y$$

● **PROBLEM** 130

When $x = \dfrac{3 + 5\sqrt{-1}}{2}$, find the value of $2x^3 + 2x^2 - 7x + 72$; and show that it will be unaltered if $\dfrac{3 - 5\sqrt{-1}}{2}$ be substituted for x.

Solution: When $x = \dfrac{3 + 5\sqrt{-1}}{2}$,

(eq. 1) $2x^3 + 2x^2 - 7x + 72 = 2\left(\dfrac{3+5\sqrt{-1}}{2}\right)^3 + 2\left(\dfrac{3+5\sqrt{-1}}{2}\right)^2 - 7\left(\dfrac{3+5\sqrt{-1}}{2}\right) + 72.$

Simplifying the right side of this equation:

$\left(\dfrac{3+5\sqrt{-1}}{2}\right)^2 = \left(\dfrac{3+5\sqrt{-1}}{2}\right)\left(\dfrac{3+5\sqrt{-1}}{2}\right) = \dfrac{9 + 30\sqrt{-1} + 25(-1)}{4}$

$= \dfrac{30\sqrt{-1} - 16}{4}$

$\left(\dfrac{3+5\sqrt{-1}}{2}\right)^3 = \left(\dfrac{3+5\sqrt{-1}}{2}\right)\left(\dfrac{3+5\sqrt{-1}}{2}\right)^2 = \left(\dfrac{3+5\sqrt{-1}}{2}\right)\left(\dfrac{30\sqrt{-1} - 16}{4}\right)$

75

$$= \frac{90\sqrt{-1} + 150(-1) - 48 - 80\sqrt{-1}}{8}$$

$$= \frac{10\sqrt{-1} - 198}{8}$$

Therefore, equation (1) becomes:

$$2x^3 + 2x^2 - 7x + 72 = 2\left(\frac{10\sqrt{-1} - 198}{8}\right) + 2\left(\frac{30\sqrt{-1} - 16}{4}\right) - 7\left(\frac{3+5\sqrt{-1}}{2}\right) + 72$$

$$= \frac{10\sqrt{-1} - 198}{4} + \frac{30\sqrt{-1} - 16}{2} - \frac{21 + 35\sqrt{-1}}{2} + 72$$

$$= \frac{10\sqrt{-1}}{4} - \frac{198}{4} + 15\sqrt{-1} - 8 - \frac{21}{2} - \frac{35}{2}\sqrt{-1} + 72$$

$$= 2\tfrac{1}{2}\sqrt{-1} - 49\tfrac{1}{2} + 15\sqrt{-1} - 8 - 10\tfrac{1}{2} - 17\tfrac{1}{2}\sqrt{-1} + 72$$

Associating,
$$= (2\tfrac{1}{2}\sqrt{-1} + 15\sqrt{-1} - 17\tfrac{1}{2}\sqrt{-1}) - 49\tfrac{1}{2} - 8 - 10\tfrac{1}{2} + 72$$

$$= (17\tfrac{1}{2}\sqrt{-1} - 17\tfrac{1}{2}\sqrt{-1}) - 68 + 72$$

$$= 0 - 68 + 72$$

$$= 4$$

When $x = \dfrac{3 - 5\sqrt{-1}}{2}$,

(eq. 2) $2x^3 + 2x^2 - 7x + 72 = 2\left(\dfrac{3-5\sqrt{-1}}{2}\right)^3 + 2\left(\dfrac{3-5\sqrt{-1}}{2}\right)^2 - 7\left(\dfrac{3-5\sqrt{-1}}{2}\right) + 72.$

Simplifying the right side of this equation:

$$\left(\frac{3-5\sqrt{-1}}{2}\right)^2 = \left(\frac{3-5\sqrt{-1}}{2}\right)\left(\frac{3-5\sqrt{-1}}{2}\right) = \frac{9 - 30\sqrt{-1} + 25(-1)}{4}$$

$$= \frac{-30\sqrt{-1} - 16}{4}$$

$$\left(\frac{3-5\sqrt{-1}}{2}\right)^3 = \left(\frac{3-5\sqrt{-1}}{2}\right)\left(\frac{3-5\sqrt{-1}}{2}\right)^2 = \left(\frac{3-5\sqrt{-1}}{2}\right)\left(\frac{-30\sqrt{-1} - 16}{4}\right)$$

$$= \frac{-90\sqrt{-1} - 48 + 150(-1) + 80\sqrt{-1}}{8}$$

$$= \frac{-10\sqrt{-1} - 198}{8}$$

Therefore, equation (2) becomes:

$$2x^3 + 2x^2 - 7x + 72 = 2\left(\frac{-10\sqrt{-1} - 198}{8}\right) + 2\left(\frac{-30\sqrt{-1} - 16}{4}\right) - 7\left(\frac{3-5\sqrt{-1}}{2}\right) + 72$$

$$= \frac{-10\sqrt{-1}}{4} - \frac{198}{4} - 15\sqrt{-1} - 8 - \frac{21}{2} + \frac{35}{2}\sqrt{-1} + 72$$

$$= -2\tfrac{1}{2}\sqrt{-1} - 49\tfrac{1}{2} - 15\sqrt{-1} - 8 - 10\tfrac{1}{2} + 17\tfrac{1}{2}\sqrt{-1} + 72$$

Associating, $= \left(-2\tfrac{1}{2}\sqrt{-1} - 15\sqrt{-1} + 17\tfrac{1}{2}\sqrt{-1}\right) - 49\tfrac{1}{2} - 8 - 10\tfrac{1}{2} + 72$

$$= \left(-17\tfrac{1}{2}\sqrt{-1} + 17\tfrac{1}{2}\sqrt{-1}\right) - 68 + 72$$

$$= 0 - 68 + 72$$

$$= 4$$

Therefore, when $x = \dfrac{3+5\sqrt{-1}}{2}$, the value of the equation is 4. The equation is unaltered when $3-5\sqrt{-1}/2$ is substituted for x, since the value of the equation is also 4 for this substitution.

Find the value of $y^2 - 3y + 1$, for $y = 3 - \sqrt{2}$.

Solution: Substituting $3 - \sqrt{2}$ for y in the expression

$$y^2 - 3y + 1,$$

$$y^2 - 3y + 1 = (3 - \sqrt{2})^2 - 3(3 - \sqrt{2}) + 1$$

$$= (3 - \sqrt{2})(3 - \sqrt{2}) - 3(3 - \sqrt{2}) + 1$$

$$= (9 - 6\sqrt{2} + 2) - 3(3 - \sqrt{2}) + 1$$

$$= (11 - 6\sqrt{2}) - 9 + 3\sqrt{2} + 1$$

$$= 11 - 6\sqrt{2} - 9 + 3\sqrt{2} + 1$$

$$= 11 - 9 + 1 - 6\sqrt{2} + 3\sqrt{2}$$

$$= 3 - 3\sqrt{2}$$

Find the product by inspection:
$$\left(\sqrt[3]{2}a + \sqrt[3]{4}b\right)\left(\sqrt[3]{4}a^2 - 2ab + 2\sqrt[3]{2}b^2\right)$$

Solution: The formula for the sum of two cubes can be used to find the product. This formula is:

$$x^3 + y^3 = \left(x + y\right)\left(x^2 - xy + y^2\right).$$

The product $\left(\sqrt[3]{2}a + \sqrt[3]{4}b\right)\left(\sqrt[3]{4}a^2 - 2ab + 2\sqrt[3]{2}b^2\right)$ corresponds to the right side of the formula for the sum of two cubes where x is replaced by $\sqrt[3]{2}a$ and y is replaced by $\sqrt[3]{4}b$. Hence,

$$\left(\sqrt[3]{2}a + \sqrt[3]{4}b\right)\left(\sqrt[3]{4}a^2 - 2ab + 2\sqrt[3]{2}b^2\right) = \left(\sqrt[3]{2}a\right)^3 + \left(\sqrt[3]{4}b\right)^3$$

$$= \left(\sqrt[3]{2}\right)^3 a^3 + \left(\sqrt[3]{4}\right)^3 b^3$$

since $(ab)^x = a^x b^x$. Also,

$$\left(\sqrt[n]{x}\right)^n = \left[x^{\frac{1}{n}}\right]^n = x^{\frac{n}{n}} = x^1 = x, \text{ hence}$$

$$\left(\sqrt[3]{2}\right)^3 = 2 \text{ and } \left(\sqrt[3]{4}\right)^3 = 4.$$

Therefore $\left(\sqrt[3]{2}a + \sqrt[3]{4}b\right)\left(\sqrt[3]{4}a^2 - 2ab + 2\sqrt[3]{2}b^2\right) = 2a^3 + 4b^3.$

Find the value of $3x^2 - 4x - 2$, for

$$x = \frac{2 - \sqrt{10}}{3}$$

Solution: Substituting $\frac{2 - \sqrt{10}}{3}$ for x in the expression

$3x^2 - 4x - 2$

$$3x^2 - 4x - 2 = 3\left(\frac{2 - \sqrt{10}}{3}\right)^2 - 4\left(\frac{2 - \sqrt{10}}{3}\right) - 2$$

$$= \frac{3(2 - \sqrt{10})^2}{(3)^2} - 4\left(\frac{2 - \sqrt{10}}{3}\right) - 2$$

$$= \frac{3(2 - \sqrt{10})(2 - \sqrt{10})}{9} - 4\left(\frac{2 - \sqrt{10}}{3}\right) - 2$$

$$= \frac{3(4 - 4\sqrt{10} + 10)}{9} - 4\left(\frac{2 - \sqrt{10}}{3}\right) - 2$$

$$= \frac{\overset{1}{\cancel{3}}\, 14 - 4\sqrt{10}}{\underset{3}{\cancel{9}}} - 4\left(\frac{2 - \sqrt{10}}{3}\right) - 2$$

$$= \frac{14 - 4\sqrt{10}}{3} - \frac{(8 - 4\sqrt{10})}{3} - 2$$

$$= \frac{14 - 4\sqrt{10}}{3} - \frac{(8 - 4\sqrt{10})}{3} - \frac{6}{3}$$

$$= \frac{14 - 4\sqrt{10} - 8 + 4\sqrt{10} - 6}{3}$$

$$= \frac{14 - 8 - 6 - 4\sqrt{10} + 4\sqrt{10}}{3}$$

$$= \frac{6 - 6}{3}$$

$$= \frac{0}{3}$$

$$= 0$$

CHAPTER 8

ALGEBRAIC ADDITION, SUBTRACTION, MULTIPLICATION, AND DIVISION

> **Basic Attacks and Strategies for Solving Problems in this Chapter. See pages 79 to 89 for step-by-step solutions to problems.**

An algebraic expression is one that contains one or more variables. To evaluate an algebraic expression means to replace the variable or variables in the expression with given numbered values and then simplify the resulting numerical expression.

One type of algebraic expression, in which the terms are all monomials, is called a polynomial. The four fundamental operations can be performed using polynomials. The addition and subtraction of polynomial expressions rely on essentially the same ideas. The commutative, associative, and distributive properties provide the basis for rearranging, regrouping, and combining similar terms in the addition and subtraction of polynomial expressions. To add polynomials, simply combine like terms. To subtract polynomials, simply add the opposite of the second polynomial to the first and combine like terms. The final resulting polynomial in each case should be written in decreasing order according to the power of the variable. Either a vertical or horizontal format can be used for either operation. For example, we can combine each of the following expressions using the indicated operations in vertical or horizontal format.

Simplify

$$(3x^2 - 7x + 1) + (7x^2 - 5) \text{ and} \tag{1}$$

$$(7x^2 - 5) - (3x^2 - 7x + 1). \tag{2}$$

Vertical Format (Addition)

$$
\begin{array}{r}
3x^2 - 7x + 1 \\
7x^2 \qquad -5 \\
\hline
10x^2 - 7x - 4
\end{array}
$$

Horizontal Format (Subtraction)

$$(7x^2 - 5) - (3x^2 - 7x + 1) = 7x^2 - 5 - 3x^2 + 7x - 1$$

$$= 4x^2 + 7x - 6$$

Multiplying polynomials involves the use of the properties of exponents and the distributive property. In general, to find the product of two polynomials, multiply each term of the first polynomial by each term of the second polynomial, combine similar terms, and write the resulting polynomial in decreasing order. If there are more than two polynomials to be multiplied, multiply the product of the first two by the next polynomial, combine like terms, and write the final product in decreasing order.

To divide a polynomial by a monomial, divide each term in the numerator by the denominator and write the sum of the quotients. To divide a polynomial by a polynomial (degree 2 or more), use a method similar to that used for division of whole numbers. In particular, first arrange the terms in the divisor and dividend in descending powers of the variable, filling in any missing power of the variables in the dividend with +0 times the missing power of the variable. Then, the next step is to divide the first term of the divisor into the first term of the dividend. Multiply the quotient from the division by each term in the divisor and subtract the products of each term from the dividend. This result (difference) is a new dividend. Repeat the last step using the divisor and the new dividend again until we obtain a remainder which is of degree less than that of the divisor or zero. The final quotient is the sum of the quotients obtained from each step plus the remainder expressed as a fraction.

● **PROBLEM** 134

Evaluate the expression $3x^2y - 2xy^2z + xyz$ when $x = 2$, $y = -1$, and $z = 3$.

<u>Solution:</u> There are two ways to find the value.

(a) We can simplify $3x^2y - 2xy^2z + xyz$ and substitute. Simplifying can be done by applying the distributive property since each term of the expression has a common factor of xy. Thus, we have

$$(3x - 2yz + z)xy$$

We now substitute in the expression:

$$[3(2) - 2(-1)(3) + 3](2)(-1) = (6 + 6 + 3)(-2) = -30.$$

(b) We can substitute in the original expression:

$$3(2)^2(-1) - 2(2)(-1)^2(3) + (2)(-1)(3) = -12 - 12 - 6 = -30.$$

In the example above we found that both $(3x - 2yz + z)xy$ and $3x^2y - 2xy^2z + xyz$ have the same value when $x = 2$, $y = -1$, and $z = 3$. It is apparent that both expressions will have equal values for any set of replacements of the variable; they are called equivalent expressions.

● **PROBLEM** 135

Find the value of the polynomial $3x^2y - 2xy^2 + 5xy$ when $x = 1$ and $y = -2$.

<u>Solution:</u> Replace x by 1 and y by -2 in the given polynomial to obtain,

$$3x^2y - 2xy^2 + 5xy = \left[3(1)^2 - (-2)\right] - \left[2(1)(-2)^2\right]$$
$$+ [5(1)(-2)]$$
$$= [(3)(-2)] - [(2)(4)]$$
$$+ [(5)(-2)]$$
$$= -6 - 8 - 10$$
$$= -24.$$

Thus, when $x = 1$ and $y = -2$, the polynomial

$$3x^2y - 2xy^2 + 5xy = -24.$$

● **PROBLEM** 136

Combine the expressions $2a - 5b - c$ and $8a + 4b - 3c$.

Solution: Whenever two expressions are combined, those two expressions should be added. Therefore, the two expressions 2a-5b-c and 8a+4b-3c to be combined will be added. (2a-5b-c) + (8a+4b-3c) is the problem. Place all similiar terms together; that is, place the a terms together, the b terms together and the c terms together. Therefore:

$$(2a - 5b - c) + (8a + 4b - 3c) = (2a + 8a) + (-5b + 4b)$$
$$+ (-c - 3c)$$
$$= 10a - b - 4c.$$

This solution can be written directly, as

$$(2a - 5b - c) + (8a + 4b - 3c) = 10a - b - 4c,$$

or it can be arranged so that like terms appear in columns, as

$$2a - 5b - c$$
$$\underline{8a + 4b - 3c}$$
$$10a - b - 4c$$

in which the columns are added.

● **PROBLEM** 137

Add $\left(3xy^2 + 2xy + 5x^2y\right) + \left(2xy^2 - 4xy + 2x^2y\right)$.

Solution: Use the vertical form, align all like terms, and apply the distributive property.

$$3xy^2 + 2xy + 5x^2y$$
$$\underline{2xy^2 - 4xy + 2x^2y}$$
$$(3 + 2)xy^2 + (2 - 4)xy + (5 + 2)x^2y$$

Thus, the sum is $5xy^2 - 2xy + 7x^2y$.

● **PROBLEM** 138

Add the expression $4a^2 - 3 + 5a$, $6a - 2a^2 + 2$, and $2a^2 - 3a + 8$.

Solution: Arrange each polynomial in decending order of exponents of a; by commutation, $4a^2 - 3 + 5a = 4a^2 + 5a - 3$

similarly, $6a - 2a^2 + 2 = -2a^2 + 6a + 2$

and, $2a^2 - 3a + 8 = 2a^2 - 3a + 8$.

Since in each column we have the same power of a we may simply add columns to obtain:

$$4a^2 + 5a - 3$$
$$-2a^2 + 6a + 2$$
$$\underline{2a^2 - 3a + 8}$$
$$4a^2 + 8a + 7$$

Subtract $4y^2 - 5y + 2$ from $7y^2 - 6$.

<u>Solution:</u> The problem is the following:

$$\left(7y^2 - 6\right) - \left(4y^2 - 5y + 2\right).$$

Whenever a minus sign (-) appears before an expression in parentheses, change the sign of every term in the paren-theses.

In this problem, since a minus sign appears before the expression $\left(4y^2 - 5y + 2\right)$, the sign of every term in this expression is changed.

$\left(4y^2 - 5y + 2\right)$ becomes $-4y^2 + 5y - 2$.

After the signs have been changed, the new expression $-4y^2 + 5y - 2$ can be added to the expression $7y^2 - 6$ (changing the signs of the expression changes the original problem to an addition problem). Therefore

$$\left(7y^2 - 6\right) - \left(4y^2 - 5y + 2\right) = 7y^2 - 6 - 4y^2 + 5y - 2.$$

Place the terms with similar powers together. Therefore:

$$\left(7y^2 - 6\right) - \left(4y^2 - 5y + 2\right) = 7y^2 - 6 - 4y^2 + 5y - 2$$

$$= 7y^2 - 4y^2 + 5y - 6 - 2. \qquad (1)$$

Since $7y^2 - 4y^2 = 3y^2$ and $-6 - 2 = -8$, equation (1) becomes:

$$\left(7y^2 - 6\right) - \left(4y^2 - 5y + 2\right) = 3y^2 + 5y - 8,$$

which is the final answer.

From the sum of $6x^2 + 4xy - 8y^2 - 11$ and $3x^2 - 4y^2 + 8 + 5xy$ subtract $xy - 10 - 5x^2 + 7y^2$.

<u>Solution:</u> First find the sum of $6x^2 + 4xy - 8y^2 - 11$ and $3x^2 - 4y^2 + 8 + 5xy$. Adding these two polynomials together:

$$\left(6x^2 + 4xy - 8y^2 - 11\right) + \left(3x^2 - 4y^2 + 8 + 5xy\right)$$

$$= 6x^2 + 4xy - 8y^2 - 11 + 3x^2 - 4y^2 + 8 + 5xy$$

Grouping like terms together,

$$\left(6x^2 + 4xy - 8y^2 - 11\right) + \left(3x^2 - 4y^2 + 8 + 5xy\right)$$

$$= \left(6x^2 + 3x^2\right) + \left(4xy + 5xy\right) + \left(-8y^2 - 4y^2\right) + \left(-11 + 8\right)$$

$$= 9x^2 + 9xy + \left(-12y^2\right) + (-3) = 9x^2 + 9xy - 12y^2 - 3 \qquad (1)$$

Now subtract $xy - 10 - 5x^2 + 7y^2$ from the resultant sum, which is the right side of equation (1), or $9x^2 + 9xy - 12y^2 - 3$. Then,

$$\left(9x^2 + 9xy - 12y^2 - 3\right) - \left(xy - 10 - 5x^2 + 7y^2\right) =$$

$$= 9x^2 + 9xy - 12y^2 - 3 - xy + 10 + 6x^2 - 7y^2 \qquad (2)$$

Grouping like terms together, equation (2) becomes:

$$\left(9x^2 + 9xy - 12y^2 - 3\right) - \left(xy - 10 - 5x^2 + 7y^2\right)$$

$$= \left(9x^2 + 5x^2\right) + \left(9xy - xy\right) + \left(-12y^2 - 7y^2\right) + (-3 + 10)$$

$$= 14x^2 + 8xy + \left(-19y^2\right) + 7$$

$$= 14x^2 + 8xy - 19y^2 + 7,$$

which is the final answer.

● **PROBLEM 141**

Subtract $3x^4y^3 + 5x^2y - 4xy + 5x - 3$ from the polynomial $5x^4y^3 - 3x^2y + 7$.

Solution: $\left(5x^4y^3 - 3x^2y + 7\right) - \left(3x^4y^3 + 5x^2y - 4xy + 5x - 3\right)$

$$= \left(5x^4y^3 - 3x^2y + 7\right) + \left(-3x^4y^3 - 5x^2y + 4xy - 5x + 3\right)$$

$$= 5x^4y^3 + \left(-3x^4y^3\right) - 3x^2y + \left(-5x^2y\right) + 4xy - 5x + 7 + 3$$

$$= 2x^4y^3 - 8x^2y + 4xy - 5x + 10$$

The column form may also be used for subtraction. Here we align the like terms and subtract the coefficients.

$$
\begin{array}{l}
5x^4y^3 - 3x^2y \qquad\qquad\quad + 7 \\
- \left[3x^4y^3 + 5x^2y - 4xy + 5x - 3\right] \\
\hline
2x^4y^3 - 8x^2y + 4xy - 5x + 10
\end{array}
$$

● **PROBLEM 142**

Simplify $3ax(ax^2 - 5bx + c)$.

Solution: Using the distributive property,

$$3ax(ax^2 - 5bx + c) = 3ax(ax^2) + 3ax(-5bx) + 3ax(c)$$
$$= 3a^2x^3 - 15abx^2 + 3acx .$$

● **PROBLEM** 143

Expand $(x + 5)(x - 4)$.

Solution: Distributing the second term:

$$(x + 5)(x - 4) = (x + 5)x + (x + 5)(-4) \qquad (1)$$

Distributing twice on the right side of equation (1):

$$(x + 5)(x - 4) = (x + 5)x + (x + 5)(-4)$$
$$= (x^2 + 5x) + (-4x - 20)$$
$$= x^2 + 5x - 4x - 20$$
$$= x^2 + x - 20.$$

● **PROBLEM** 144

Find the product $(2x - 5y)(x + 2y)$.

Solution: We use the distributive property and simplify.

$(2x - 5y)(x + 2y) = (2x - 5y)x + (2x - 5y)2y$ Distributive property

$= [x(2x - 5y)]+[2y(2x - 5y)]$ Commutative property of multiplication

$= [x \cdot 2x + x \cdot (-5y)]+[2y \cdot 2x + 2y \cdot (-5y)]$ Distributive property

$= 2x^2 - 5xy + 4xy - 10y^2$ Simplifying

$= 2x^2 - xy - 10y^2$ Combining like terms

$(2x - 5y)(x + 2y) = 2x^2 - xy - 10y^2$

We can use the properties of a field because the algebraic expressions represent members of the field of real numbers for any replacement of the variables.

● **PROBLEM** 145

Multiply $(4x - 5)(6x - 7)$.

Solution: We can apply the FOIL method. The letters indicate the order in which the terms are to be multiplied.

     F = first terms
     O = outer terms
     I = inner terms
     L = last terms

Thus,

$$(4x - 5)(6x - 7) = (4x)(6x) + (4x)(-7) + (-5)(6x) + (-5)(-7)$$
$$= 24x^2 - 28x - 30x + 35 = 24x^2 - 58x + 35.$$

Another way to multiply algebraic expressions is to apply the distributive law of multiplication with respect to addition. If a,b, and

83

c are real numbers, then a(b+c) = ab + ac. In this case let
a = (4x-5) and b + c = 6x - 7.

$$(4x - 5)(6x - 7) = (4x - 5)(6x) + (4x - 5)(-7)$$

Then, apply the law again.

$$(4x - 5)(6x - 7) = (4x)(6x) - (5)(6x) + (4x)(-7) + (-5)(-7)$$
$$(4x - 5)(6x - 7) = 24x^2 - 30x - 28x + 35$$

Add like terms.

$$(4x - 5)(6x - 7) = 24x^2 - 58x + 35.$$

● **PROBLEM 146**

Show that $(3x - 2)(x + 5) + 15 = 3x^2 + 13x + 5$ is an
identity.

Solution: An equation in x is an identity if it holds for
all real values of x. Thus, the given equation is an
identity since for each x ε R,

$$(3x - 2)(x + 5) + 15 = 3x^2 + 13x - 10 + 15$$

$$= 3x^2 + 13x + 5$$

● **PROBLEM 147**

Multiply $(2x + 3y)(4x - 5y)$.

Solution: Instead of writing one factor beneath the other, we shall
use the following process to find the product mentally. Multiply the
two first terms, 2x and 4x, to obtain the first term in the product;
also, multiply the two last terms, 3y and -5y, to obtain the last
term in the product. Thus, we have $8x^2$ and $-15y^2$ as the first and
last terms, respectively. We must now determine the middle term in the
desired product. This is done by multiplying the two inner terms, 3y
and 4x, and the two outer terms, 2x and -5y, and adding these. This
is shown below, and the middle term is 12xy - 10xy = 2xy.

$$(2x + 3y)(4x - 5y) = 8x^2 + 2xy - 15y^2$$

12xy

-10xy

It is to be noted that the final result is written immediately with
only one intermediate step: the two cross products, 12xy and -10xy,
are kept in mind and added mentally to produce the middle term 2xy.

● **PROBLEM 148**

Simplify $(5ax + by)(2ax - 3by)$.

Solution: The following formula can be used to simplify the given expression:

$$\left(N_1 + N_2\right)N_3 = N_1N_3 + N_2N_3$$

where N_1, N_2 and N_3 are any three numbers. Note that the distributive

84

property is used in this formula. N_1 is replaced by 5ax and by replaces N_2. Also, N_3 is replaced by (2ax - 3by). Therefore:

$$(5ax + by)(2ax - 3by) = 5ax(2ax - 3by) + by(2ax - 3by)$$

Use the distributive property to simplify the right side of the above equation:

$$(5ax + by)(2ax - 3by) = 5ax(2ax) + 5ax(-3by) + by(2ax) + by(-3by)$$
$$= 10a^2x^2 - 15abxy + 2abxy - 3b^2y^2$$
$$= 10a^2x^2 - 13abxy - 3b^2y^2$$

It is often convenient to arrange the two factors vertically as we do in ordinary arithmetic. Hence, the problem is an ordinary multiplication problem.

$$
\begin{array}{r}
5ax + by \\
\times \quad 2ax - 3by \\
\hline
-15abxy - 3b^2y^2 \\
+ \quad 10a^2x^2 + 2abxy \\
\hline
10a^2x^2 - 13abxy - 3b^2y^2
\end{array}
$$

Note that this answer (i.e., product) is the same as the answer obtained above.

● PROBLEM 149

Expand $(a + b - 2)^2$.

Solution: When we enclose $a + b$ in parentheses we may write
$$(a + b - 2)^2 = [(a + b) - 2]^2$$
$$= [(a + b) - 2][(a + b) - 2]$$

Employing our method of polynomial multiplication
$$(x + y)(u + v) = xu + xv + yu + yv$$

Substituting $(a + b)$ for x and u, and (-2) for y and v we obtain
$$[(a + b) - 2]^2 = (a + b)^2 - 4(a + b) + 4$$

Once again employ our method of polynomial multiplication on
$$(a + b)^2 = (a + b)(a + b).$$

Thus
$$(a + b - 2)^2 = a^2 + 2ab + b^2 - 4a - 4b + 4$$

● PROBLEM 150

Expand $(x + 3y - 5t)^2$.

Solution: It is sometimes convenient to group two or more terms and treat them as a single term. $x + 3y$ will be considered as one term. Hence,
$$(x + 3y - 5t)^2 = [(x + 3y) - 5t]^2$$
$$= [(x + 3y) - 5t][(x + 3y) - 5t]$$

Apply the FOIL method:
$$(x + 3y - 5t)^2 = (x + 3y)^2 - 5t(x + 3y) - 5t(x + 3y) + (-5t)^2$$
$$= (x + 3y)^2 + 2(x + 3y)(-5t) + (-5t)^2$$
$$= (x + 3y)(x + 3y) + 2(x + 3y)(-5t) + (-5t)^2$$

$$= x^2 + 3xy + 3xy + 9y^2 + (2x + 6y)(-5t) + 25t^2$$
$$= x^2 + 6xy + 9y^2 - 10tx - 30ty + 25t^2 .$$

● **PROBLEM** 151

Simplify $(x - y)(x^2 + xy + y^2)$.

<u>Solution:</u> By distributing,

$$(x - y)(x^2 + xy + y^2) = x(x^2 + xy + y^2) - y(x^2 + xy + y^2) \qquad (1)$$

Now, distribute the right side of equation (1):

$$(x - y)(x^2 + xy + y^2) = (x^3 + x^2y + xy^2) - (x^2y + xy^2 + y^3).$$

Combining terms:

$$(x - y)(x^2 + xy + y^2) = x^3 + \cancel{x^2y} + \cancel{xy^2} - \cancel{x^2y} - \cancel{xy^2} - y^3$$
$$= x^3 - y^3 .$$

● **PROBLEM** 152

Multiply $3x^2 - 5y^2 - 4xy$ by $2x - 7y$.

<u>Solution:</u> Write one algebraic expression under the other, and multiply the first expression by each term of the second expression, placing similar product terms in the same vertical column.

$$3x^2 - 5y^2 - 4xy$$
$$2x - 7y$$
$$\overline{}$$

$6x^3 - 10xy^2 - 8x^2y$	multiplying $3x^2 - 5y^2 - 4xy$ by $2x$
$28xy^2 -21x^2y + 35y^3$	multiplying $3x^2 - 5y^2 - 4xy$ by $-7y$
$6x^3 + 18xy^2 - 29x^2y + 35y^3$	adding the partial products

● **PROBLEM** 153

Find the product
$$(2x^2 - 3xy + y^2)(2x - y)$$

<u>Solution:</u> Multiplication of polynomials can be carried out very much the same way we multiply numbers. One polynomial is written under the other, and then multiplied term by term. Like terms in the product are arranged in columns and added.

$$2x^2 - 3xy + y^2$$
$$2x - y$$
$$\overline{4x^3 - 6x^2y + 2xy^2}$$
$$\underline{- 2x^2y + 3xy^2 - y^3}$$
$$4x^3 - 8x^2y + 5xy^2 - y^3$$

We are applying the Distributive Law in the following way:

$$(2x^2 - 3xy + y^2)(2x - y) = (2x^2 - 3xy + y^2)(2x) = 4x^3 - 6x^2y + 2xy^2$$
$$+ (2x^2 - 3xy + y^2)(-y) = \underline{-2x^2y + 3xy^2 - y^3}$$
$$(2x^2 - 3xy + y^2)(2x-y) = 4x^3 - 8x^2y + 5xy^2 - y^3$$

Divide $(37 + 8x^3 - 4x)$ by $(2x + 3)$.

Solution: Arrange both polynomials in descending powers of the variable. The first polynomial becomes: $8x^3 - 4x + 37$. The second polynomial stays the same: $2x + 3$. The problem is: $2x + 3\sqrt{8x^3 - 4x + 37}$. In the dividend, $8x^3 - 4x + 37$, all powers of x must be included. The only missing power of x is x^2. To include this power of x, a coefficient of 0 is used; that is, $0x^2$. This term, $0x^2$, can be added to the dividend without changing the dividend because $0x^2 = 0$ (anything multiplied by 0 is 0).

Now to accomplish the division we proceed as follows: divide the first term of the divisor into the first term of the dividend. Multiply the quotient from this division by each term of the divisor and subtract the products of each term from the dividend. We then obtain a new dividend. Use this dividend, and again divide by the first term of the divisor, and repeat all steps again until we obtain a remainder which is of degree lower than that of the divisor or zero. Following this procedure we obtain :

$$
\begin{array}{r}
4x^2 - 6x + 7 \\
2x + 3 \overline{\smash{\big)}\, 8x^3 + 0x^2 - 4x + 37} \\
\underline{8x^3 + 12x^2} \\
-12x^2 - 4x + 37 \\
\underline{-12x^2 - 18x} \\
14x + 37 \\
\underline{14x + 21} \\
16
\end{array}
$$

The degree of a polynomial is the highest power of the variable in the polynomial.

The degree of the divisor is 1. The number 16 can be written as $16x^0$ where $x^0 = 1$. Therefore, the number 16 has degree 0. When the degree of the divisor is greater than the degree of the dividend, we stop dividing.

Since the degree of the divisor in this problem is 1 and the degree of the dividend (16) is 0, the degree of the divisor is greater than the degree of the dividend. Therefore, dividing is stopped and the remainder is 16. Therefore, the quotient is $4x^2 - 6x + 7$ and the remainder is 16.

In order to verify this, multiply the quotient, $4x^2 - 6x + 7$, by the divisor, $2x + 3$, and then add 16. These two operations should total up to the dividend $8x^3 - 4x + 37$. Thus,

$$\left(4x^2 - 6x + 7\right)(2x + 3) + 16 =$$

$$8x^3 - 12x^2 + 14x + 12x^2 - 18x + 21 + 16 =$$

$$8x^3 - 4x + 37,$$

which is the desired result.

Divide $3x^5 - 8x^4 - 5x^3 + 26x^2 - 33x + 26$ by
$x^3 - 2x^2 - 4x + 8.$

<u>Solution:</u> To divide a polynomial by another poly-
nomial we set up the divisor and the dividend as
shown below. Then we divide the first term of the
divisor into the first term of the dividend. We
multiply the quotient from this division by each term
of the divisor, and subtract the products of each term
from the dividend. We then obtain a new dividend. Use
this dividend, and again divide by the first term of
the divisor, and repeat all steps again until we
obtain a remainder which is of degree lower than that
of the divisor or = zero. Following this procedure we
obtain:

$$
\require{enclose}
\begin{array}{r}
3x^2 - 2x + 3 \\[-2pt]
x^3-2x^2-4x+8 \enclose{longdiv}{3x^5 - 8x^4 - 5x^3 + 26x^2 - 33x + 26} \\
\end{array}
$$

$$3x^5 - 6x^4 - 12x^3 + 24x^2$$

$$- 2x^4 + 7x^3 + 2x^2 - 33x + 26$$

$$- 2x^4 + 4x^3 + 8x^2 - 16x$$

$$3x^3 - 6x^2 - 17x + 26$$

$$3x^3 - 6x^2 - 12x + 24$$

$$- 5x + 2$$

Thus, the quotient is $3x^2 - 2x + 3$ and the remainder
is $- 5x + 2$.

Find the quotient and remainder when $3x^7 - x^6 + 31x^4$
$+ 21x + 5$ is divided by $x + 2$.

<u>Solution:</u> To divide a polynomial by another polynomial
we set up the divisor and the dividend as shown below.
Then we divide the first term of the divisor into the
first term of the dividend. We multiply the quotient
from this division by each term of the divisor, and
subtract the products of each term from the dividend.
We then obtain a new dividend. Use this dividend, and
again divide by the first term of the divisor, and repeat all steps again until we obtain a remainder which
is of degree lower than that of the divisor, or which
is zero. Following this procedure we obtain:

$$
\begin{array}{l}
\phantom{x+2\sqrt{\,}}\;3x^6-7x^5+14x^4+3x^3-6x^2+12x-3 \\
x\;+\;2\;\overline{\smash{)}\;3x^7-\;x^6\quad\quad+31x^4\quad\quad\quad\quad\;+21x\,.+\;5}
\end{array}
$$

$$3x^7+6x^6$$
$$\overline{-7x^6\quad\quad+31x^4\quad\quad\quad\quad+21x\;+\;5}$$
$$-7x^6-14x^5$$
$$\overline{14x^5+31x^4\quad\quad\quad\quad+21x\;+\;5}$$
$$14x^5+28x^4$$
$$\overline{3x^4\quad\quad\quad\quad+21x\;+\;5}$$
$$3x^4+6x^3$$
$$\overline{-6x^3\quad\quad+21x\;+\;5}$$
$$-6x^3-12x^2$$
$$\overline{12x^2+21x\;+\;5}$$
$$12x^2+24x$$
$$\overline{-\;3x\;+\;5}$$
$$-\;3x\;-\;6$$
$$\overline{11}$$

Thus the quotient is $3x^6 - 7x^5 + 14x^4 + 3x^3 - 6x^2 +12x - 3$, and the remainder is 11.

CHAPTER 9

FUNCTIONS AND RELATIONS

> **Basic Attacks and Strategies for Solving Problems in this Chapter. See pages 90 to 104 for step-by-step solutions to problems.**

A relation is any set of ordered pairs, for example, the set

$A = \{(2, 3), (5, 2), (6, 3)\}$

is a relation. The set of all first coordinates of set A is called the domain of the relation, and the set of all second coordinates is said to be the range of the relation. Thus, for relation A, the sets $\{2, 5, 6\}$ and $\{3, 2\}$ are the domain and range, respectively.

There are two ways to specify the ordered pairs in a relation. One method is simply to list them. The other method is to give the rule (equation) for obtaining them.

One of the most fundamental ideas of mathematics is that of a function. There are two essentially equivalent ways of defining the concept. First, a function is a relation in which no two different ordered pairs have the same first coordinates. The domain and range of a function are the sets of first and second coordinates, respectively. For example, the above relation A is a function. On the other hand, the relation

$B = \{(1, 5), (3, 9), (1, 9)\}$

is not a function because the ordered pairs $(1, 5)$ and $(1, 9)$ have the same first coordinates.

The second definition of a function emphasizes that a function is a rule (equation) that pairs every element x in a set D with a unique element y in a set R, where set D is the domain and set R is the range or image set of x. When a function (or relation) is given in terms of a rule (equation), the domain is the set of all possible replacements for the variable x. If the domain of a function (or relation) is not specified, it is assumed to be all real numbers that do not yield undefined terms in the equation. That is, we cannot use values of x in the domain that will produce 0 in a denominator or the square root of a negative number.

The graph of a function (or relation) is sometimes helpful in determining the domain and range. Also, the notation,

$$y = f(x),$$

defined to be the value of the function f at x or the value of y associated with a given value of x, is useful in determining the domain and range of the function.

Step-by-Step Solutions to Problems in this Chapter, "Functions and Relations"

● **PROBLEM** 157

If $X = \{1,2,3,4,5,6,7,8\}$ and $Y = \{2,4,6,8\}$, use $f(x)$ notation to indicate the image of each element of X in the following mapping.

Set X   1   2   3   4   5   6   7   8

Set Y   2   4   6   8

<u>Solution:</u> The mapping of an element x in the set X to an element y in the set Y may be written in $f(x)$ notation as $f(x) = y$ when f is a function mapping x to y. That is, $f: x \to y$. Therefore,

$f(1) = 2$, $f(2) = 2$, $f(3) = 4$, $f(4) = 4$, $f(5) = 6$, $f(6) = 6$,

$f(7) = 8$, $f(8) = 8$

● **PROBLEM** 158

Find the image of each element in

$$A = \{1,2,3,4,5,6,7,8,9\}$$

under the following mapping:
$$f(x) = \begin{cases} 2x, & \text{if } x < 5 \\ 8, & \text{if } x \geq 5 \end{cases}$$

<u>Solution:</u> The image of each element in $A = \{1,2,3,4,5,6,7,8,9\}$ under the mapping $f(x)$, is $f(1),f(2),f(3),f(4),f(5),f(6),f(7),f(8),$ $f(9)$. $f(x)$ has two corresponding values, depending on the value of x. If $x < 5$, $f(x) = 2x$, thus for

$$x = 1, \ f(1) = 2(1) = 2$$
$$x = 2, \ f(2) = 2(2) = 4$$
$$x = 3, \ f(3) = 2(3) = 6$$
$$x = 4, \ f(4) = 2(4) = 8$$

and if $x \geq 5, \ f(x) = 8$,   thus for

$$x = 5, \ f(5) = 8$$
$$x = 6, \ f(6) = 8$$
$$x = 7, \ f(7) = 8$$
$$x = 8, \ f(8) = 8$$
$$x = 9, \ f(9) = 8$$

● **PROBLEM** 159

Find the relation defined by $y^2 = 25 - x^2$, where x belongs to

90

$D = \{0,3,4,5\}$.

Solution: x takes on the values 0,3,4, and 5. Replacing x by these values in the equation $y^2 = 25 - x^2$ we obtain the corresponding values of y:

x	$y^2 = 25 - x^2$	y
0	$y^2 = 25 - 0$ $y^2 = 25$ $y = \sqrt{25}$ $y = \pm 5$	± 5
3	$y^2 = 25 - 3^2$ $y^2 = 25 - 9$ $y^2 = 16$ $y = \sqrt{16}$ $y = \pm 4$	± 4
4	$y^2 = 25 - 4^2$ $y^2 = 25 - 16$ $y^2 = 9$ $y = \sqrt{9}$ $y = \pm 3$	± 3
5	$y^2 = 25 - 5^2$ $y^2 = 25 - 25$ $y^2 = 0$ $y = 0$	0

Hence the relation defined by $y^2 = 25 - x^2$ where x belongs to $D = \{0,3,4,5\}$ is

$$\{(0,5),(0,-5),(3,4),(3,-4),(4,3),(4,-3),(5,0)\} .$$

● **PROBLEM** 160

Find the relation M over set $S = \{1,2,3\}$ if
$$M = \{ (x,r(x)) : r(x) = 2x - 1 \}$$

Solution: x takes on the values 1,2, and 3. Replacing x by these values in the equation $r(x) = 2x - 1$ we obtain the corresponding values of r(x):

x	$r(x) = 2x - 1$	r(x)
1	$r(x) = 2(1)-1$ $= 2 - 1$ $= 1$	1
2	$r(x) = 2(2)-1$ $= 4 - 1$ $= 3$	3
3	$r(x) = 2(3)-1$ $= 6 - 1$ $= 5$	5

Thus the rule of correspondence $r(x) = 2x - 1$ determines the set of ordered pairs
$$\{(1,1),(2,3),(3,5)\}$$

But the relation must be a subset of S X S. Since (3,5) is not a subset of S X S (5 is not a member of S) we eliminate this pair. Hence,
$$M = \{(1,1),(2,3)\}$$

Find the relation Q over S × T if S = { 1,2,3}, T = { 4,5},
and the rule of correspondence is

$$r(x) = x + 2.$$

Solution: We first find the image of each element in S by
substituting each element for x in the rule of correspondence
$r(x) = x + 2$.

$$r(1) = 1 + 2 = 3 \qquad r(2) = 2 + 2 = 4$$

$$r(3) = 3 + 2 = 5.$$

Thus, the rule of correspondence determines the following
set of ordered pairs:

$$\{ (1,3), (2,4), (3,5)\}.$$

However, the relation Q must be a subset of S × T, which
equals { (1,4), (1,5), (2,4), (2,5), (3,4), (3,5)}. Therefore
the point (1,3) won't appear in Q because (1,3) doesn't appear in S × T. Therefore, the relation over S × T determined by $r(x) = x + 2$ is

$$Q = \{ (2,4), (3,5)\}.$$

We can use set-builder notation to describe the relation discussed in the above example. In the example, a set
of ordered pairs was determined by a rule of correspondence.
The first component, x, was chosen from the domain, S. The
second component, r(x), was the corresponding image from
the range, T. Thus, we can describe the relation Q in the
following manner:

$$Q = \{ \left(x, r(x)\right): r(x) = x + 2\}.$$

This notation refers to all ordered pairs [x, r(x)], such
that $r(x) = x + 2$.

Which of the following sets are functions of x?

A = { (5,1),(4,2),(4,3),(6,4)},

B = { (x,y) |y = |x|},

C = { (x,y) |x = |y|}?

Solution: A function is a relation having the property
that each member of its domain is paired with exactly
one member of its range. Thus, set A is not a function,
for it contains the pairs (4,2) and (4,3) - that is, one
member of its domain, 4, is paired with more than one
member of its range, 2 and 3. If each x value has only
one corresponding y value, any vertical line only intersects the graph of a function at one point. Thus, from

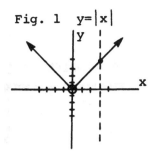

Fig. 1   y=|x|

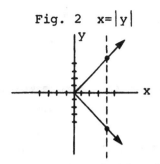

Fig. 2   x=|y|

figure 1 we note that set B is a function. Notice that a function may contain two pairs with the same second member; for example, our function B contains the pairs (1,1) and (- 1,1).

From figure 2 we note that a vertical line inter-sects the graph of C in two places, thus there are x values of C which have more than one corresponding y value, and C is not a function.

● PROBLEM 163

Let the domain of $M = \{(x,y): y = x\}$ be the set of real numbers. Is M a function?

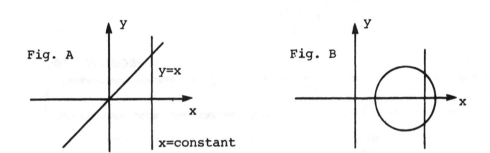

Fig. A

y=x

x=constant

Fig. B

Solution: The range is also the set of real numbers since $y = \{y | y = x\}$. The graph of y = x is the graph of a line (y = mx + b where m = 1 and b = 0). See fig. A. If for every value of x in the domain, there corresponds only one y value then y is said to be a function of x. Since each element in the domain of M has exactly one element for its image, M is a function. Also notice that a vertical line (x = constant) crosses the graph y = x only once. Whenever this is true the graph defines a function. Consult figure B.
The vertical line (x = constant) crosses the graph of the circle twice; i.e., for each x,y is not unique, therefore the graph does not de-fine a function.

● PROBLEM 164

If $g(x) = x^2 + 5x - 3$, find g(- 7).

Solution: Substitute - 7 for x everywhere in the equation:

$$g(-7) = (-7)^2 + 5(-7) - 3$$
$$= 49 - 35 - 3 = 11.$$

Let f be a mapping with the rule of correspondence
$$f(x) = 3x^2 - 2x + 1.$$
Find f(1), f(-3), f(-b).

Solution: In order to find f(1), f(-3), f(-b), we replace x by 1, (-3), and (-b) respectively in our equation for f(x), $f(x) = 3x^2 - 2x + 1$. Thus

$$f(1) = 3(1)^2 - 2(1) + 1$$
$$= 3 - 2 + 1$$
$$= 1 + 1$$
$$= 2$$
$$f(-3) = 3(-3)^2 - 2(-3) + 1$$
$$= 3(9) + 6 + 1$$
$$= 27 + 7$$
$$= 34$$
$$f(-b) = 3(-b)^2 - 2(-b) + 1$$
$$= 3(b^2) + 2b + 1$$
$$= 3b^2 + 2b + 1$$

If f(t) = 6t + 13, find f(5) - f(4).

Solution: To find f(5) we substitute 5 for t everywhere in the equation, that is:

f(5) = 6(5) + 13 = 43

Similarly, f(4) = 6(4) + 13 = 37

and now subtract f(4) from f(5). Therefore,

f(5) - f(4) = 43 - 37 = 6.

If f(x) = (x - 2)/(x + 1), find the function values f(2), f(½), and f(- 3/4).

Solution: To find f(2), we replace x by 2 in the given formula for f(x), f(x) = x - 2/x + 1; thus

$$f(2) = \frac{2 - 2}{2 + 1} = \frac{0}{3} = 0.$$

Similarly, $f(½) = \frac{½ - 2}{½ + 1}$.

Multiply numerator and denominator by 2,

$$= \frac{2(\frac{1}{2} - 2)}{2(\frac{1}{2} + 1)} \quad .$$

Distribute,

$$= \frac{2(\frac{1}{2}) - 2 \cdot 2}{2(\frac{1}{2}) + 2}$$

$$= \frac{1 - 4}{1 + 2}$$

$$= - \frac{3}{3} = - 1.$$

$$f(- 3/4) = \frac{- 3/4 - 2}{- 3/4 + 1} \quad .$$

Multiply numerator and denominator by 4,

$$= \frac{4(- 3/4 - 2)}{4(- 3/4 + 1)} \quad .$$

Distribute,

$$= \frac{4(- 3/4) - 4(2)}{4(- 3/4) + 4(1)}$$

$$= \frac{- 3 - 8}{- 3 + 4}$$

$$= \frac{- 11}{1}$$

$$= - 11.$$

● **PROBLEM   168**

If $g(x) = x^2 - 2x + 1$, find the given element in the range.

    a) $g(-2)$ b) $g(0)$ c) $g(a + 1)$ d) $g(a - 1)$

<u>Solution:</u> a) To find $g(-2)$, substitute -2 for x in the given equation.

$$g(x) = g(-2)$$

$$= (-2)^2 - 2(-2) + 1$$

$$= 4 + 4 + 1$$

$$= 8 + 1$$

$$= 9$$

Hence, $g(-2) = 9$

      b) To find $g(0)$, substitute 0 for x in the given equation.

$$g(x) = g(0)$$

$$= (0)^2 - 2(0) + 1$$

$$= 0 - 0 + 1$$

$$= 1$$

Hence, $g(0) = 1$

c) To find $g(a + 1)$, substitute $a + 1$ for x in given equation.

$$g(x) = g(a + 1)$$

$$= (a + 1)^2 - 2(a + 1) + 1$$

$$= (a^2 + 2a + 1) - 2a - 2 + 1$$

$$= a^2 + \cancel{2a} + 1 - \cancel{2a} - 2 + 1$$

$$= a^2 + 1 - 2 + 1$$

$$= a^2 + 0$$

$$= a^2$$

Hence, $g(a + 1) = a^2$.

d) To find $g(a - 1)$, substitute $a - 1$ for x in given equation.

$$g(x) = g(a - 1)$$

$$= (a - 1)^2 - 2(a - 1) + 1$$

$$= (a^2 - 2a + 1) - 2a + 2 + 1$$

$$= a^2 - 2a + 1 - 2a + 2 + 1$$

$$= a^2 - 4a + 4$$

Hence, $g(a - 1) = a^2 - 4a + 4$

● **PROBLEM** 169

Let f be the function whose domain is the set of all real numbers, whose range is the set of all numbers greater than or equal to 2, and whose rule of correspondence is given by the equation $f(x) = x^2 + 2$. Find $3f(0) + f(-1)f(2)$.

<u>Solution:</u> The rule of correspondence in this example is expressed by the equation $f(x) = x^2 + 2$. To find the number in the range that is associated with any particular number in the domain, we merely replace the letter x wherever it appears in the equation $f(x) = x^2 + 2$ by the given number. Thus

$$f(0) = 0^2 + 2 = 2 , \qquad f(-1) = (-1)^2 + 2 = 1 + 2 = 3 ,$$

$$f(2) = 2^2 + 2 = 4 + 2 = 6, \quad \text{and}$$

$$3f(0) + f(-1)f(2) = 3(2) + (3)(6) = 6 + 18 = 24.$$

● **PROBLEM** 170

Show that $f(a) = f(-a)$ if $f(x) = x^2 + 3$.

__Solution:__ If $f(a) = f(-a)$, one should obtain an identity when a and then - a are substituted into the equation. For the given equation we have:

$$f(-a) = (-a)^2 + 3 = a^2 + 3 = f(a).$$

● **PROBLEM** 171

Find the domain D and the range R of the function $\left(x, \dfrac{x}{|x|}\right)$.

Solution: Note that the y-value of any coordinate pair (x,y) is $\dfrac{x}{|x|}$.   We can replace x in the formula $\dfrac{x}{|x|}$ with any number except 0, since the denominator, $|x|$, can not equal 0, (i.e. $|x| \neq 0$) which is equivalent to $x \neq 0$.   This is because division by 0 is undefined.   Therefore, the domain D is the set of all real numbers except 0.   If x is negative, i.e. $x < 0$, then $|x| = -x$ by definition. Hence, if x is negative, then $\dfrac{x}{|x|} = \dfrac{x}{-x} = -1$. If x is positive, i.e. $x > 0$, then $|x| = x$ by definition.   Hence, if x is posi-tive, then $\dfrac{x}{|x|} = \dfrac{x}{x} = 1$.(The case where $x = 0$ has already been found to be undefined).   Thus, there are only two numbers -1 and 1 in the range R of the function; that is, $R = \{-1, 1\}$.

● **PROBLEM** 172

Describe the domain and range of the function
$f = (x,y) | y = \sqrt{9 - x^2}\}$ if x and y are real numbers.

__Solution:__ In determining the domain we are interested in the values of x which yield a real value for y. Since the square root of a negative number is not a real number, the domain is restricted to those values of x which make the radicand positive or zero. Therefore x^2 cannot exceed 9 which means that x cannot exceed 3 or be less than - 3. A convenient way to express this is to write $- 3 \leq x \leq 3$, which is read "x is greater than or equal to - 3 and less than or equal to 3." This is the domain of the function the range is the set of values that y can assume. To determine the range of the function we note that the largest value of y occurs when $x = 0$. Then $y = \sqrt{9 - 0} = 3$. Likewise, the smallest value of y occurs when $x = 3$ or $x = - 3$. Then $y = \sqrt{9 - 9} = 0$. Since this is an inclusive interval of the real axis, the range of y is $0 \leq y \leq 3$.

Find the set of ordered pairs $\{(x,y)\}$ if $y = x^2 - 2x - 3$ and $D = \{x \mid x$ is an integer and $1 \le x \le 4\}$.

Solution: We first note that $D = \{1,2,3,4\}$. Substituting these values of x in the equation

$$y = x^2 - 2x - 3,$$

we find the corresponding y values. Thus,

for $x = 1$, $y = 1^2 - 2(1) - 3 = 1 - 2 - 3 = -4$

for $x = 2$, $y = 2^2 - 2(2) - 3 = 4 - 4 - 3 = -3$,

for $x = 3$, $y = 3^2 - 2(3) - 3 = 9 - 6 - 3 = 0$,   and

for $x = 4$, $y = 4^2 - 2(4) - 3 = 16 - 8 - 3 = 5$.

Hence $\{(x,y)\} = \{(1,-4), (2,-3), (3,0), (4,5)\}$.

If $f(x) = 3x + 4$ and $D = \{x \mid -1 \le x \le 3\}$, find the range of $f(x)$.

Solution: We first prove that the value of $3x + 4$ increases when x increases. If $X > x$, then we may multiply both sides of the inequality by a positive number to obtain an equivalent inequality. Thus, $3X > 3x$. We may also add a number to both sides of the inequality to obtain an equivalent inequality. Thus

$$3X + 4 > 3x + 4.$$

Hence, if x belongs to D, the function value $f(x) = 3x + 4$ is least when $x = -1$ and greatest when $x = 3$. Consequently, since $f(-1) = -3 + 4 = 1$ and $f(3) = 9 + 4 = 13$, the range is all y from 1 to 13; that is,

$$R = \{y \mid 1 \le y \le 13\}.$$

Find the zeros of the function f if $f(x) = 3x - 5$.

Solution: The zeros of the function $f(x) = 3x - 5$ are those values of x for which $3x - 5 = 0$:

$$3x - 5 = 0$$
$$3x = 5$$
$$x = \frac{5}{3}$$

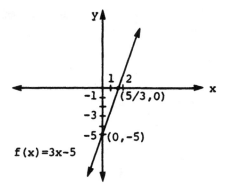

$$f(x) = 3x - 5$$

Thus $x = 5/3$ is a zero of $f(x) = 3x - 5$, which means that the graph of $f(x)$ crosses the x axis at the point $(5/3, 0)$ (see figure).

● **PROBLEM** 176

Find the zeros of the function

$$\frac{2x + 7}{5} + \frac{3x - 5}{4} + \frac{33}{10}.$$

Solution: Let the function $f(x)$ be equal to $\frac{2x + 7}{5} + \frac{3x - 5}{4} + \frac{33}{10}$. A number, a, is a zero of a function $f(x)$ if $f(a) = 0$. A zero of $f(x)$ is a root of the equation $f(x) = 0$. Thus, the zeros of the function are the roots of the equation

$$\frac{2x + 7}{5} + \frac{3x - 5}{4} + \frac{33}{10} = 0.$$

The least common denominator, LCD, of the denominators of 5, 4, and 10 is 20. This is a fractional equation which can be solved by multiplying both members of the equation by the LCD.

$$20\left(\frac{2x + 7}{5} + \frac{3x - 5}{4} + \frac{33}{10}\right) = (20)(0)$$

$$4(2x + 7) + 5(3x - 5) + (2 \cdot 33) = 0.$$

Distributing,

$$8x + 28 + 15x - 25 + 66 = 0.$$
$$23x + 69 = 0$$
$$23x = -69$$
$$x = -3$$

Hence $x = -3$ is the zero of the given function.

● **PROBLEM** 177

For each of the following functions, with domain equal to the set of all whole numbers, find: (a) $f(0)$; (b) $f(1)$; (c) $f(-1)$; (d) $f(2)$; (e) $f(-2)$.

(1) $f(x) = 2x^3 - 3x + 4$

(2) $f(x) = x^2 + 1$.

Solution: In order to find each f(x) value, we replace x by the given value in each equation. Thus:

(1) $\qquad f(x) = 2x^3 - 3x + 4$

   (a) $\qquad$ $f(0) = 2(0)^3 - 3(0) + 4$
$$= 2(0) - 0 + 4$$
$$= 0 + 4$$
$$= 4$$

   (b) $\qquad$ $f(1) = 2(1)^3 - 3(1) + 4$
$$= 2(1) - 3 + 4$$
$$= 2 - 3 + 4$$
$$= 3$$

   (c) $\qquad$ $f(-1) = 2(-1)^3 - 3(-1) + 4$
$$= 2(-1) + 3 + 4$$
$$= -2 + 3 + 4$$
$$= 5$$

   (d) $\qquad$ $f(2) = 2(2)^3 - 3(2) + 4$
$$= 2(8) - 6 + 4$$
$$= 16 - 6 + 4$$
$$= 14$$

   (e) $\qquad$ $f(-2) = 2(-2)^3 - 3(-2) + 4$
$$= 2(-8) + 6 + 4$$
$$= -16 + 10$$
$$= -6$$

(2) $\qquad f(x) = x^2 + 1$

   (a) $\qquad$ $f(0) = (0)^2 + 1$
$$= 0 + 1$$
$$= 1$$

   (b) $\qquad$ $f(1) = (1)^2 + 1$
$$= 1 + 1$$
$$= 2$$

   (c) $\qquad$ $f(-1) = (-1)^2 + 1$
$$= 1 + 1$$
$$= 2$$

   (d) $\qquad$ $f(2) = (2)^2 + 1$
$$= 4 + 1$$
$$= 5$$

   (e) $\qquad$ $f(-2) = (-2)^2 + 1$

$$= 4 + 1$$
$$= 5$$

Notice that the range is contained in the set of whole numbers.

If $y = f(x) = (x^2 - 2)/(x^2 + 4)$ and $x = t + 1$, express y as a function of t.

Solution: y is given as a function of x, $y = f(x) = \dfrac{x^2 - 2}{x^2 + 4}$.

To express y as a function of t, replace x by t + 1 (since x = t + 1) in the formula for y. Thus,

$$y = f(x) = f(t + 1) = \frac{(t + 1)^2 - 2}{(t + 1)^2 + 4}$$

$$= \frac{(t^2 + 2t + 1) - 2}{(t^2 + 2t + 1) + 4}$$

$$= \frac{t^2 + 2t - 1}{t^2 + 2t + 5}$$

$$= g(t).$$

Hence, y = g(t); that is, y is now a function of t since y has been expressed in terms of t.

If $f(x) = x^2 - x - 3$, $g(x) = (x^2 - 1)/(x + 2)$, and $h(x) = f(x) + g(x)$, find h(2).

Solution:

h(x) = f(x) + g(x), and we are told that f(x) = x^2 - x - 3 and g(x) = x^2 - 1/x + 2; thus h(x) = $(x^2 - x - 3) + (x^2 - 1)/(x + 2)$.

To find h(2), we replace x by 2 in the above formula for h(x),

$$h(2) = \left[(2)^2 - 2 - 3 \right] + \left(\frac{2^2 + 1}{2 + 2} \right)$$

$$= (4 - 2 - 3) + \left(\frac{4 - 1}{4} \right)$$

$$= (-1) + \left(\frac{3}{4} \right)$$

$$= -\frac{4}{4} + \frac{3}{4}$$

$$= -\frac{1}{4}.$$

Thus, $h(2) = -\frac{1}{4}$.

Let $f(x) = 2x^2$ with domain $D_f = R$ (or, alternatively, C) and $g(x) = x - 5$ with $D_g = R$ (or C). Find (a) $f + g$ (b) $f - g$ (c) fg (d) $\frac{f}{g}$.

Solution: (a) $f + g$ has domain R (or C) and

$$(f + g)(x) = f(x) + g(x) = 2x^2 + x - 5$$

for each number x. For example, $(f + g)(1) = f(1) + g(1)$ $= 2(1)^2 + 1 - 5 = 2 - 4 = -2$.

(b) $f - g$ has domain R (or C) and

$$(f - g)(x) = f(x) - g(x) = 2x^2 - (x - 5) = 2x^2 - x + 5$$

for each number x. For example, $(f - g)(1) = f(1) - g(1) =$ $2(1)^2 - 1 + 5 = 2 + 4 = 6$.

(c) fg has domain R (or C) and

$$(fg)(x) = f(x) \cdot g(x) = 2x^2 \cdot (x - 5) = 2x^3 - 10x^2$$

for each number x. In particular, $(fg)(1) = 2(1)^3 - 10(1)^2$ $= 2 - 10 = -8$.

(d) $\frac{f}{g}$ has domain R (or C) excluding the number x = 5 (when x = 5, $g(x) = 0$ and division by zero is undefined) and

$$\left(\frac{f}{g}\right)(x) = \frac{f(x)}{g(x)} = \frac{2x^2}{x-5}$$

for each number $x \neq 5$. In particular, $\left(\frac{f}{g}\right)(1) = \frac{2(1)^2}{1 - 5}$

$$= \frac{2}{-4} = -\frac{1}{2}.$$

If $D = \{x \mid x$ is an integer and $-2 \leq x \leq 1\}$, find the function $\{(x, f(x)) \mid f(x) = x^3 - 3$ and x belongs to D$\}$.

Solution: $D = \{-2, -1, 0, 1\}$. Substituting these values of x in the equation $f(x) = x^3 - 3$, we find the corresponding $f(x)$ values. Thus,

$$f(-2) = (-2)^3 - 3 = -8 - 3 = -11$$

$$f(-1) = (-1)^3 - 3 = -1 - 3 = -4$$

$$f(0) = 0^3 - 3 = 0 - 3 = -3$$

and $\quad f(1) = 1^3 - 3 = 1 - 3 = -2.$

Hence, $\quad f = \left\{ \left(x, f(x) \right) \middle| \, f(x) = x^3 - 3 \text{ and } x \text{ belongs to } D \right\}$

$\qquad = \left\{ (-2, -11), \ (-1, -4), \ (0, -3), \ (1, -2) \right\}$

● **PROBLEM** 182

Let f be the linear function that is defined by the equation
$f(x) = 3x + 2$. Find the equation that defines the inverse
function f^{-1}.

<u>Solution:</u> To find the inverse function f^{-1}, the given equation must be solved for x in terms of y. Let $x = f^{-1}(y)$.

Solving the given equation for x:

$$y = 3x + 2, \text{ where } y = f(x).$$

Subtract 2 from both sides of this equation:

$$y - 2 = 3x + \cancel{2} - \cancel{2}$$

$$y - 2 = 3x.$$

Divide both sides of this equation by 3:

$$\frac{y - 2}{3} = \frac{3x}{3}$$

$$\frac{y - 2}{3} = x$$

or $\qquad x = \frac{y}{3} - \frac{2}{3}$

Hence, the inverse function f^{-1} is given by::

$$x = f^{-1}(y) = \frac{y}{3} - \frac{2}{3}$$

$$\text{or } x = f^{-1}(y) = \frac{1}{3}y - \frac{2}{3}.$$

Of course, the letter that we use to denote a number in the domain of the inverse function is of no importance whatsoever, so this last equation can be rewritten $f^{-1}(u) = \frac{1}{3}u - \frac{2}{3}$, or $f^{-1}(s) = \frac{1}{3}s - \frac{2}{3}$, and it will still define the same function f^{-1}.

> Show that the inverse of the function $y = x^2 + 4x - 5$ is not a function.

Solution: Given the function f such that no two of its ordered pairs have the same second element, the inverse function f^{-1} is the set of ordered pairs obtained from f by interchanging in each ordered pair the first and second elements. Thus, the inverse of the function

$y = x^2 + 4x - 5$ is $x = y^2 + 4y - 5$.

The given function has more than one first component corresponding to a given second component. For example, if $y = 0$, then $x = -5$ or 1. If the elements $(-5,0)$ and $(1,0)$ are reversed, we have $(0,-5)$ and $(0,1)$ as elements of the inverse. Since the first component 0 has more than one second component, the inverse is not a function (a function can have only one y value corresponding to each x value).

CHAPTER 10

SOLVING LINEAR EQUATIONS

Basic Attacks and Strategies for Solving Problems in this Chapter. See pages 105 to 129 for step-by-step solutions to problems.

When solving a linear equation, the object is to simplify the equation using one or more of the four fundamental operations such that the equation has as its final form

variable = constant.

The constant is the solution.

TYPE I: Linear equations with the unknown in the numerator.

This category of linear equations consists of a number of general forms and variations. Three of these forms are the equations

$$ax + b = c, \quad ax + b = dx + c, \quad \text{and} \quad a(x + e) = f.$$

When solving an equation of form

$$ax + b = c,$$

where $a \neq 0$, the first step is to add the opposite of the constant b to each side of the equation and simplify. If coefficient $a = 1$, then the final equation is in the form *variable = constant* which is the solution. On the other hand, if coefficient $a \neq 1$ and $a \neq 0$, then the next step is to remove the coefficient from the variable by multiplying each side of the equation by the reciprocal of the coefficient or dividing each side of the equation by the coefficient and simplify. If coefficient a and other constant terms in the equation involve fractions, then it may be easier to multiply both sides of the equation by the LCD and simplify. In any case, the result is the form *variable = constant* which is the solution.

To solve an equation of form

$$ax + b = dx + c,$$

where a and d are not zero, the beginning strategy is to rewrite the equation so that only one variable term exists in the equation. This is done by adding the opposite of the variable term on the right-hand side to both sides of the equation and simplifying. With only one variable term, the next step is to add the opposite of the constant term on the left-hand side of the equation to both sides and simplify so that only one constant term remains. Finally, multiply each side of the equation by the reciprocal of the coefficient of the variable and simplify. The result is the equation *variable = constant* which is the solution.

When solving an equation involving one or more sets of parentheses, (e.g.,

$$a(x + e) = f,$$

where a and e are not zero), the first step in the solution procedure is to apply the distributive property to remove the parentheses. The remainder of the solution procedure is as indicated for the equation of the form

$$ax + b = c.$$

TYPE II: Linear equations with the unknown in the numerator and/or denominator.

A typical form of Type II linear equations is as follows:

$$\frac{ax + b}{cx + d} = \frac{e}{fx + g},$$

where a, c and constant terms are not zero. The beginning strategy for this and other variations of the equation is to remove or eliminate the denominators by multiplying both sides of the equation by the LCD and simplify. The resulting equation will conform to one of the forms in Type I linear equations. Thus, the remainder of the solution procedure follows what is indicated for Type I equations. However, a critical final step in the procedure must be observed. It is necessary to check the final constant value of the variable by substituting it in the original equation to determine if it really is the solution.

TYPE III: Linear equations with the unknown under a radical sign.

The property of squaring both sides of an equation is used as the first step in the solution procedure of this type of linear equation. When the result is simplified, this step should have eliminated all radical signs from the variable in the equation. However, if all of the radical signs are not removed from the variable, then usually it is necessary to rewrite the equation by transposing appropriately and again squaring both sides of the equation and simplify. Once the radical signs are removed, then the remainder of the solution procedure follows what is given above in Type I equations. It is necessary to check the final result in the original equation to determine if it is a solution.

Step-by-Step Solutions to Problems in this Chapter, "Solving Linear Equations"

UNKNOWN IN NUMERATOR

● **PROBLEM** 184

Solve 3x - 5 = 4 for x.

Solution: Since this equation is to be solved for x, place the term x on one side of the equation, 3x - 5 = 4. Add 5 to both sides of the equation.

$$3x - 5 + 5 = 4 + 5 \tag{1}$$

Since -5 + 5 = 0 and 4 + 5 = 9, Equation (1) reduces to:

$$3x = 9. \tag{2}$$

Since it is desired to get the term x on one side of the equation, divide both sides of Equation (2) by 3.

$$\frac{3x}{3} = \frac{9}{3}. \tag{3}$$

Since $\frac{3x}{3}$ reduces to 1x and since $\frac{9}{3}$ reduces to 3, Equation(3) becomes: 1x = 3. Since 1x = x, x = 3. Therefore, the equation has been solved for x.

Check: By substituting x = 3 into the original equation we have

$$3(3) - 5 \overset{?}{=} 4$$

$$9 - 5 \overset{?}{=} 4$$

$$4 = 4.$$

Note that, upon substitution of the solution into the original equation, the equation is reduced to the identity 4 = 4.

● **PROBLEM** 185

Solve the equation   6x - 3 = 7 + 5x.

Solution: To solve for x in the equation 6x - 3 = 7 + 5x, we wish to obtain an equivalent equation in which each term in one member involves x, and each term in the other member is a constant. If we add (- 5x) to both members, then only one side of the equation will have an x term:

$$6x - 3 + (- 5x) = 7 + 5x + (- 5x)$$

$$6x + (- 5x) - 3 = 7 + 0$$

$$x - 3 = 7$$

Now, adding 3 to both sides of the equation we obtain,

$$x - 3 + 3 = 7 + 3$$

$$x + 0 = 10$$

$$x = 10$$

Thus, our solution is $x = 10$. Now we check this value.

Check: Substitute 10 for x in the original equation:

$$6x - 3 = 7 + 5x$$

$$6(10) - 3 = 7 + 5(10)$$

$$60 - 3 = 7 + 50$$

$$57 = 57.$$

● **PROBLEM** 186

Solve the equation $4x - 5 = x + 7$.

Solution: The problem here is to list the elements in the set

$$S = \{x \mid 4x - 5 = x + 7\}$$

To find these elements, use the additive principle and the multiplicative principle. By using these principles, convert the description of S to the form $S = \{x \mid x = \ldots\}$.

The computation can be arranged in the following manner:

$4x - 5 = x + 7$	Original equation
$4x - 5 + 5 = x + 7 + 5$	Adding 5 to both sides
$4x = x + 12$	
$4x + (-x) = x + 12 + (-x)$	Adding $-x$ to both sides
$3x = 12$	
$\frac{1}{3}(3x) = \frac{1}{3}(12)$	Multiplying both sides by $\frac{1}{3}$
$x = 4$	

Hence the solution set is $S = \{x \mid x = 4\} = \{4\}$.

● **PROBLEM** 187

Solve, justifying each step. $3x - 8 = 7x + 8$.

Solution:	$3x - 8 = 7x + 8$
Adding 8 to both members,	$3x - 8 + 8 = 7x + 8 + 8$
Additive inverse property,	$3x + 0 = 7x + 16$
Additive identity property,	$3x = 7x + 16$
Adding $(-7x)$ to both members,	$3x - 7x = 7x + 16 - 7x$
Commuting,	$-4x = 7x - 7x + 16$
Additive inverse property,	$-4x = 0 + 16$
Additive identity property,	$-4x = 16$

106

Dividing both sides by -4, $x = \frac{16}{-4}$

$x = -4$

Check: Replacing x by -4 in the original equation:

$$3x \quad - 8 \quad = \quad 7x \quad + \quad 8$$
$$3(-4) - 8 \quad = \quad 7(-4) + \quad 8$$
$$- 12 - 8 \quad = \quad - 28 + \quad 8$$
$$-20 \quad = \quad - 20$$

● **PROBLEM 188**

Solve for x: 2x + 5 = 7 - x.

<u>Solution:</u> Add x to both sides: $2x + 5 + x = 7 - x + x$

Combine terms: $3x + 5 = 7$

Subtract 5 from both sides: $3x + 5 - 5 = 7 - 5$

Combine terms: $3x = 2$

Divide both sides by 3: $\frac{3x}{3} = \frac{2}{3}$

$$x = \frac{2}{3}$$

● **PROBLEM 189**

Solve for x: 7x - 3 = 2(x + 3).

<u>Solution:</u>

$$7x - 3 = 2(x + 3)$$

Distributing,

$$7x - 3 = 2x + 6$$

Adding (-2x) to both sides, $7x - 3 - 2x = 6$

Combining terms, $5x - 3 = 6$

Adding 3 to both sides, $5x = 6 + 3$

Combining terms, $5x = 9$

Dividing both sides by 5, $x = \frac{9}{5}$

Check: Replacing x by $\frac{9}{5}$ in the original equation,

$$7x \quad - \quad 3 \quad = \quad 2(x + 3)$$
$$7\left(\frac{9}{5}\right) - \quad 3 \quad = \quad 2\left(\frac{9}{5} + 3\right)$$
$$\frac{63}{5} - \frac{15}{5} \quad = \quad \frac{18}{5} + 6$$
$$\frac{48}{5} \quad = \quad \frac{18}{5} + \frac{30}{5}$$
$$\frac{48}{5} \quad = \quad \frac{48}{5}$$

107

> Solve each equation (find the solution set), and check each solution.
>
> (a) $4(6x + 5) - 3(x - 5) = 0$.   (b) $8 + 3x = -4(x - 2)$.

<u>Solution:</u>   (a) $4(6x + 5) - 3(x - 5) = 0$

distributing, $24x + 20 - 3x + 15 = 0$

combining like terms,
$$21x + 35 = 0$$

adding (-35) to both sides,
$$21x = -35$$

dividing both sides by 21,
$$x = \frac{-35}{21} = \frac{-5}{3}$$

Therefore the solution set to this equation is $\left\{\frac{-5}{3}\right\}$.

Check: Replace x by $\frac{-5}{3}$ in the equation,

$$4(6x + 5) - 3(x - 5) = 0$$
$$4\left[6\left(\frac{-5}{3}\right) + 5\right] - 3\left[\frac{-5}{3} - 5\right] = 0$$
$$4\left(\frac{-30}{3} + 5\right) - 3\left(\frac{-5}{3} - \frac{15}{3}\right) = 0$$
$$4(-10 + 5) - 3\left(\frac{-20}{3}\right) = 0$$
$$4(-5) + 20 = 0$$
$$-20 + 20 = 0$$
$$0 = 0$$

(b) $8 + 3x = -4(x - 2)$

distributing, $8 + 3x = -4x + 8$

adding 4x to both sides,
$$8 + 7x = 8$$

adding (-8) to both sides,
$$7x = 0$$

dividing both sides by 7, $x = 0$

Therefore the solution set to this equation is $\{0\}$, (not to be confused with the null set $\{\ \}$).

Check: Replace x by 0 in the equation,

$$8 + 3x = -4(x - 2)$$
$$8 + 3(0) = -4(0 - 2)$$
$$8 + 0 = -4(-2)$$
$$8 = 8$$

> Solve the equation $2(x + 3) = (3x + 5) - (x - 5)$.

<u>Solution:</u>   We transform the given equation to an equivalent equation where we can easily recognize the solution set.

$$2(x + 3) = 3x + 5 - (x - 5)$$

Distribute, $2x + 6 = 3x + 5 - x + 5$

Combine terms, $2x + 6 = 2x + 10$

Subtract 2x from both sides, $6 = 10$

Since $6 = 10$ is not a true statement, there is no real number which will make the original equation true. The equation is inconsistent and the solution set is ϕ, the empty set.

● **PROBLEM** 192

Solve the equation $\frac{3}{4}x + \frac{7}{8} + 1 = 0$.

<u>Solution:</u> There are several ways to proceed. First we observe that $\frac{3}{4}x + \frac{7}{8} + 1 = 0$ is equivalent to

$\frac{3}{4}x + \frac{7}{8} + \frac{8}{8} = 0$, where we have converted

1 into $\frac{8}{8}$. Now, combining fractions we obtain:

$$\frac{3}{4}x + \frac{15}{8} = 0$$

Subtract $\frac{15}{8}$ from both sides:

$$\frac{3}{4}x = \frac{-15}{8}$$

Multiplying both sides by $\frac{4}{3}$:

$$\left(\frac{4}{3}\right) \frac{3}{4}x = \left(\frac{4}{3}\right)\left(\frac{-15}{8}\right)$$

Cancelling like terms in numerator and denominator:

$$x = \frac{-5}{2}$$

A second method is to multiply both sides of the equation by the least common denominator, 8:

$$8\left[\frac{3}{4}x + \frac{7}{8} + 1\right] = 8(0)$$

Distributing: $\quad 8\left(\frac{3}{4}\right)x + 8\left(\frac{7}{8}\right) + 8 \cdot 1 = 0$

$$(2 \cdot 3)x + 7 + 8 = 0$$

$$6x + 15 = 0$$

Subtract 15 from both sides: $\quad 6x = -15$

Divide both sides by 6: $\quad x = \frac{-15}{6}$

Cancelling 3 from numerator and denominator: $x = \frac{-5}{2}$

Solve the equation $2\left(\frac{2}{3}y + 5\right) + 2(y + 5) = 130$.

<u>Solution:</u> The procedure for solving this equation is as follows:

$\frac{4}{3}y + 10 + 2y + 10 = 130$,   Distributive property

$\frac{4}{3}y + 2y + 20 = 130$,   Combining like terms

$\frac{4}{3}y + 2y = 110$,   Subtracting 20 from both sides

$\frac{4}{3}y + \frac{6}{3}y = 110$,   Converting 2y into a fraction with denominator 3

$\frac{10}{3}y = 110$,   Combining like terms

$y = 110 \cdot \frac{3}{10} = 33$,   Dividing by $\frac{10}{3}$

Check: Replace y by 33 in our original equation,

$2\left[\frac{2}{3}(33) + 5\right] + 2(33 + 5) = 130$

$2(22 + 5) + 2(38) = 130$

$2(27) + 76 = 130$

$54 + 76 = 130$

$130 = 130$

Therefore the solution to the given equation is y=33.

Solve the equation

$\frac{1}{2}x + \frac{2}{3} = \frac{1}{4}x - \frac{1}{6}$

<u>Solution:</u> Since 2, 3, 4, and 6, the denominators of the fractions, are all factors of 12, and there is no smaller number which contains 2, 3, 4, and 6 as factors, 12 is the least common multiple (LCM). We may therefore multiply both sides of the given equation by 12 to eliminate the fractions.

$\left(\frac{1}{2}x + \frac{2}{3}\right)12 = \left(\frac{1}{4}x - \frac{1}{6}\right)12$

Distribute, $\left(\frac{1}{2}x\right)12 + \left(\frac{2}{3}\right)12 = \left(\frac{1}{4}x\right)12 - \left(\frac{1}{6}\right)(12)$

$$6x + 8 = 3x - 2$$

Add (- 3x) to both sides,

$$6x + 8 + (- 3x) = 3x - 2 + (- 3x)$$

$$6x + (- 3x) + 8 = 3x + (- 3x) - 2$$

$$3x + 8 = - 2$$

Add (- 8) to both sides

$$3x + 8 + (- 8) = - 2 + (- 8)$$

$$3x = - 10$$

Divide both sides by 3, $\quad x = - \dfrac{10}{3}$

Thus, the solution is x = - 10/3, and we have

$$\left\{ x \left| \dfrac{1}{2}x + \dfrac{2}{3} = \dfrac{1}{4}x - \dfrac{1}{6} \right. \right\} = \left\{ \dfrac{-10}{3} \right\} \qquad \text{To verify this}$$

statement we perform the following check:

Check: Replace x by - 10/3 in the original equation,

$$\dfrac{1}{2}x + \dfrac{2}{3} = \dfrac{1}{4}x - \dfrac{1}{6}$$

$$\dfrac{1}{2} \left(\dfrac{- 10}{3} \right) + \dfrac{2}{3} = \dfrac{1}{4}\left(\dfrac{- 10}{3} \right) - \dfrac{1}{6}$$

$$\dfrac{- 10}{6} + \dfrac{2}{3} = \dfrac{- 10}{12} - \dfrac{1}{6}$$

Convert each fraction into a fraction whose denominator is 12. Here we are using the fact that 12 is the least common denominator (this is an alternative method to multiplying both members by the LCM 12). Thus

$$\dfrac{2}{2} \left(\dfrac{- 10}{6} \right) + \dfrac{4}{4}\left(\dfrac{2}{3} \right) = \dfrac{- 10}{12} - \dfrac{2}{2}\left(\dfrac{1}{6} \right)$$

$$\dfrac{- 20}{12} + \dfrac{8}{12} = \dfrac{- 10}{12} - \dfrac{2}{12}$$

$$\dfrac{- 12}{12} = \dfrac{- 12}{12}$$

$$- 1 = - 1$$

Since substitution of x by (- 10/3) results in this equivalent equation which is always true, - 10/3 is indeed a root of the equation.

● **PROBLEM** 195

Solve for x:

$$\dfrac{x}{2} + \dfrac{x}{3} = 12.$$

Solution: The Least Common Denominator is 6. Multiply both members of the equation by 6:

$$6\left(\frac{x}{2} + \frac{x}{3}\right) = 6(12).$$

Use distributive law: $3x + 2x = 72.$

Collect terms: $5x = 72.$

Divide by 5: Therefore, $x = 14\frac{2}{5}.$

Check: Substitute $14\frac{2}{5} = \frac{72}{5}$ for x in the given equation:

$$\frac{\frac{72}{5} + \frac{72}{5}}{2 \qquad 3} = 12$$

$$\left(\frac{72}{5} \cdot \frac{1}{2}\right) + \left(\frac{72}{5} \cdot \frac{1}{3}\right) = 12$$

$$\frac{36}{5} + \frac{24}{5} = 12$$

$$\frac{60}{5} = 12$$

$$12 = 12$$

● **PROBLEM** 196

Find the set indicated by
$$\left\{ x \mid \tfrac{1}{2}x - \frac{2}{3} = \frac{3}{4}x + \frac{1}{12} \right\}$$

Solution: The set indicated by $\left\{ x \mid \tfrac{1}{2}x - \frac{2}{3} = \frac{3}{4}x + \frac{1}{12} \right\}$ is the set of all x such that x makes the statement

$$\tfrac{1}{2}x - \frac{2}{3} = \frac{3}{4}x + \frac{1}{12}$$

true. Hence to obtain the required set, we must solve the equation

$$\tfrac{1}{2}x - \frac{2}{3} = \frac{3}{4}x + \frac{1}{12}.$$

Since 2, 3, 4, and 12, the denominators of the fractions are all factors of 12, we may multiply both sides of the equation by 12 to eliminate the fractions. Therefore, 12 is called the least common multiple (LCM). Thus,

$$12\left[\tfrac{1}{2}x - \frac{2}{3}\right] = 12\left[\frac{3}{4}x + \frac{1}{12}\right]$$

Distribute, $12(\tfrac{1}{2}x) - 12\left(\frac{2}{3}\right) = 12\left(\frac{3}{4}x\right) + 12\left(\frac{1}{12}\right)$

$$6x - 8 = 9x + 1$$

Add (-9x) to both sides, $6x - 8 + (-9x) = 9x + 1 + (-9x)$

commute, $6x + (-9x) - 8 = 9x + (-9x) + 1$

$$-3x - 8 = 1$$

Add 8 to both sides, $-3x - 8 + 8 = 1 + 8$

$$- 3x = 9$$

Divide both sides by - 3 to obtain,

$$x = - 3.$$

Thus, our solution is x = - 3, and the set indicated by $\left\{ x \mid \frac{1}{2}x - \frac{2}{3} = \frac{3}{4}x + \frac{1}{12} \right\}$ is $\{x \mid x = - 3\}$. Now, we check this solution.

Check: Substitute (- 3) for x in our original equation,

$$\tfrac{1}{2}x - \frac{2}{3} = \frac{3}{4}x + \frac{1}{12}$$

$$(\tfrac{1}{2})(- 3) - \frac{2}{3} = \frac{3}{4}(- 3) + \frac{1}{12}$$

$$\frac{- 3}{2} - \frac{2}{3} = \frac{- 9}{4} + \frac{1}{12}$$

Convert each fraction into a fraction whose denominator is 12. Here we are using the fact that 12 is the least common denominator (we could also multiply both members by the LCM 12 as before). Thus,

$$\frac{6}{6}\left(\frac{- 3}{2}\right) - \frac{4}{4}\left(\frac{2}{3}\right) = \frac{3}{3}\left(\frac{- 9}{4}\right) + \frac{1}{12}$$

$$\frac{- 18}{12} - \frac{8}{12} = \frac{- 27}{12} + \frac{1}{12}$$

$$\frac{- 26}{12} = \frac{- 26}{12}$$

Since substitution for x by (- 3) results in this equivalent equation, which is always true, (- 3) is indeed a root of the equation.

● **PROBLEM** 197

Solve the equation $\frac{3}{2}x - \frac{2}{3} = 2x + 1$.

<u>Solution:</u> Subtract $\frac{3}{2}x$ from both sides of the given equation:

$$\frac{3}{2}x - \frac{2}{3} - \frac{3}{2}x = 2x + 1 - \frac{3}{2}x$$

$$-\frac{2}{3} = \frac{4}{2}x - \frac{3}{2}x + 1$$

$$-\frac{2}{3} = \frac{1}{2}x + 1$$

$$-\frac{2}{3} = \frac{x}{2} + 1$$

Subtract 1 from both sides of this equation:

$$-\frac{2}{3} - 1 = \frac{x}{2} + \cancel{x} - \cancel{x}$$

$$-\frac{2}{3} - \frac{3}{3} = \frac{x}{2}$$

$$-\frac{5}{3} = \frac{x}{2}$$

Multiply both sides of this equation by 2:

$$2\left(-\frac{5}{3}\right) = \cancel{2}\left(\frac{x}{\cancel{2}}\right)$$

$$-\frac{10}{3} = x$$

Thus the solution set of our given equation is the set $\left\{-\frac{10}{3}\right\}$.

● **PROBLEM** 198

Solve $\quad \frac{1}{2x} - \frac{5}{16} = \frac{1}{x}$.

Solution: In order to rid an equation of fractions we multiply both sides by the least common multiple. In this case our L.C.M. is 16x:

$$16x \left[\frac{1}{2x} - \frac{5}{16}\right] = 16x \left[\frac{1}{x}\right]$$

Distributing, $\quad 16x \left[\frac{1}{2x}\right] - 16x \left[\frac{5}{16}\right] = \frac{16x}{x}$

Cancelling out like terms in numerator and denominator:

$$8 - 5x = 16$$

Subtracting 8 from both sides:

$$-5x = 16 - 8$$

$$-5x = 8$$

Dividing both sides by - 5:

$$x = -\frac{8}{5}$$

Check:

Substitute $\frac{-8}{5}$ for x in $\quad \frac{1}{2x} - \frac{5}{16} = \frac{1}{x}$:

$$\frac{1}{2\left(-\frac{8}{5}\right)} - \frac{5}{16} = \frac{1}{\left(-\frac{8}{5}\right)}$$

$$\frac{1}{\left(\frac{-16}{5}\right)} - \frac{5}{16} = \frac{1}{-\frac{8}{5}}$$

114

Since division by a fraction is equivalent to multiplication by the reciprocal

$$\frac{1}{\left(\frac{-\frac{16}{5}}{}\right)} = 1 \cdot \left(\frac{5}{-16}\right) = \frac{-5}{16}$$

and $\quad \dfrac{1}{\left(\frac{-\frac{8}{5}}{}\right)} = 1 \cdot \left(\frac{5}{-8}\right) = \frac{-5}{8}$

Hence, $\qquad \dfrac{-5}{16} - \dfrac{5}{16} = \dfrac{-5}{8}$

$$\frac{-10}{16} = \frac{-5}{8}$$

$$\frac{-5}{8} = \frac{-5}{8}$$

● **PROBLEM** 199

Solve $A = \dfrac{h}{2}(b + B)$ for h.

Solution: Since the given equation is to be solved for h, obtain h on one side of the equation. Multiply both sides of the equation $A = \dfrac{h}{2}(b + B)$ by 2. Then, we have:

$$2(A) = 2\left[\frac{h}{2}(b + B)\right].$$

Therefore: $\qquad 2(A) = \dfrac{\not2 h}{\not2}(b + B)$

$$2A = h(b + B). \tag{1}$$

Since it is desired to obtain h on one side of the equation, divide both sides of equation (1) by (b + B).

$$\frac{2A}{(b + B)} = \frac{h(b + \not B)}{(\not b + \not B)}.$$

Therefore: $\qquad \dfrac{2A}{b + B} = h.$

Thus, the given equation, $A = \dfrac{h}{2}(b + B)$, is solved for h.

This is the form of the formula used to determine values of h for a set of trapezoids, if the area and lengths of the bases are known.

● **PROBLEM** 200

Solve $\dfrac{1}{R} = \dfrac{1}{a} + \dfrac{1}{b}$ for a.

Solution: To solve for a we must obtain a alone on one side of the equation,

$$\frac{1}{R} = \frac{1}{a} + \frac{1}{b}. \tag{1}$$

To do this we proceed as follows: Multiply Equation (1) by Rab. Then,

$$Rab\left(\frac{1}{R}\right) = Rab\left(\frac{1}{a} + \frac{1}{b}\right).$$

Therefore: $\dfrac{Rab}{R} = \dfrac{Rab}{a} + \dfrac{Rab}{b}.$

Therefore: $ab = Rb + Ra.$ (2)

Subtracting Ra from both sides of Equation (2), we obtain:

$$ab - Ra = Rb + Ra - Ra.$$

Therefore: $ab - Ra = Rb.$ (3)

We can now factor a from both terms of the left side of Equation (3), obtaining:

$$a(b - R) = Rb.$$ (4)

Now, we divide both sides of Equation (4) by (b - R):

$$\frac{a(b - R)}{(b - R)} = \frac{Rb}{(b - R)}.$$

Thus, we find $a = \dfrac{Rb}{b - R}.$

● **PROBLEM** 201

Solve the equation $a(x + b) = bx + c$ for x if $a \neq b.$

Solution:

$ax + ab = bx + c$	Distributive property
$ax + ab + (-bx) = bx + c + (-bx)$	adding (-bx) to both sides
$ax + (-bx) + ab = bx + (-bx) + c$	commutative law of addition
$ax + (-bx) + ab = 0 + c$	additive inverse property
$ax + (-bx) + ab = c$	additive identity property
$ax + (-bx) + ab + (-ab) = c + (-ab)$	adding (-ab) to both sides
$ax + (-bx) + 0 = c + (-ab)$	additive inverse property
$ax - bx = c - ab$	additive identity property
$(a - b)x = c - ab$	factoring out x
$x = \dfrac{c - ab}{a - b}$ if $a \neq b$	Dividing by (a - b)

If a = b the denominator of this fraction is zero, and thus the fraction has no meaning.

● **PROBLEM** 202

Find a solution of the equation
$$3x + 4y + 5z = 13$$ (1)

Solution: The above equation is linear in x, y, and z. Any ordered triple (x, y, z) which satisfies it is a solution. If we chose x = 2, and y = 3, by substitution

$$6 + 12 + 5z = 13$$
$$z = -1$$

The one solution of Equation 1 is $x = 2$, $y = 3$, and $z = -1$.

Obviously, the number of solutions is unlimited, since any choice of values for two of the variables will determine the value of the third variable.

UNKNOWN IN NUMERATOR AND / OR DENOMINATOR

Find the solutions of the equation $\dfrac{4x - 7}{x - 2} = 3 + \dfrac{1}{x - 2}$.

Solution: Assume that there is a number x such that

$$\frac{4x - 7}{x - 2} = 3 + \frac{1}{x - 2}$$

In order to eliminate the fractions multiply both sides of the equation by $x - 2$ to obtain

$$(x - 2)\ \frac{4x - 7}{x - 2} = \left(3 + \frac{1}{x - 2}\right)(x - 2)$$

Thus

$$4x - 7 = 3(x - 2) + \frac{x - 2}{x - 2}$$

$$4x - 7 = 3(x - 2) + 1$$
$$4x - 7 = 3x - 6 + 1$$
$$4x - 7 = 3x - 5$$

Add $(-3x)$ to both sides, $\qquad 4x - 7 + (-3x) = -5$

$$x - 7 = -5$$

Add 7 to both sides, $\qquad x = -5 + 7$

and hence $x = 2$.

We have shown that if x is a solution of the equation

$$\frac{4x - 7}{x - 2} = 3 + \frac{1}{x - 2} ,$$

then $x = 2$. But if we substitute $x = 2$ in the right-hand member of the equation we obtain

$$3 + \frac{1}{0}$$

and we know that we cannot divide by zero. Hence 2 is not a solution.

Before we analyze the process which led to the conclusions that 2 was a possible solution to our equation, let us see exactly why our equation has no solution. To do this, we note that

$$3 + \frac{1}{x - 2} = 3 \cdot \frac{x - 2}{x - 2} + \frac{1}{x - 2} = \frac{3(x - 2) + 1}{x - 2} = \frac{3x - 6 + 1}{x - 2} = \frac{3x - 5}{x - 2}$$

and hence that the original equation is equivalent to

$$\frac{4x - 7}{x - 2} = \frac{3x - 5}{x - 2} \qquad (1)$$

Now we know that two fractions, $\frac{a}{b}$ and $\frac{c}{d}$ are equal if and only if $ad = bc$. Thus (1) holds, providing that $x \neq 2$, if and only if

$$(x - 2)(4x - 7) = (x - 2)(3x - 5) \qquad (2)$$

holds. But, since $x \neq 2$, $x - 2 \neq 0$, and we can divide both sides of (2) by $x - 2$ and have

$$4x - 7 = 3x - 5$$

which gives $x = 2$, a contradiction. In other words, the only possible solution is a number which we knew in advance could not be a solution, and hence there are no solutions to our given equation.

● **PROBLEM 204**

Solve the equation

$$\frac{x}{x + 1} + \frac{5}{8} = \frac{5}{2(x + 1)} + \frac{3}{4}$$

<u>Solution:</u> Since $(x + 1)$, 8, $2(x + 1)$, and 4, the denominators of the fractions, are all factors of $8(x + 1)$, and there is no smaller number which contains $(x + 1)$, 8, $2(x + 1)$, and 4 as factors, $8(x + 1)$ is the least common multiple (LCM). We may therefore multiply both sides of the given equation by $8(x + 1)$ to eliminate the fractions.

$$8(x + 1)\left[\frac{x}{x + 1} + \frac{5}{8}\right] = 8(x + 1)\left[\frac{5}{2(x + 1)} + \frac{3}{4}\right] .$$

Distribute,

$$8(x + 1)\left[\frac{x}{x + 1}\right] + 8(x + 1)\left(\frac{5}{8}\right)$$

$$= 8(x + 1)\left[\frac{5}{2(x + 1)}\right] + 8(x + 1)\left(\frac{3}{4}\right)$$

Cancel like terms, $\quad 8x + 5(x + 1) = 4(5) + 6(x + 1)$

Distribute, $\qquad\qquad 8x + 5x + 5 = 20 + 6x + 6$

Combine terms, $\qquad\qquad 13x + 5 = 26 + 6x$

Add $(- 6x)$ to both sides

$$13x + 5 + (-6x) = 26 + 6x + (- 6x)$$

$$7x + 5 = 26$$

Add $(- 5)$ to both sides

$$7x + 5 + (- 5) = 26 + (- 5)$$

$$7x = 21$$

Divide both sides by 7, $\qquad x = 3.$

Thus, our solution is $x = 3$, and we have

$$\left\{ x \middle| \frac{x}{x+1} + \frac{5}{8} = \frac{5}{2(x+1)} + \frac{3}{4} \right\} = \{3\}.$$

To verify this statement we perform the following check.

Check: Replace x by 3 in the original equation,

$$\frac{x}{x+1} + \frac{5}{8} = \frac{5}{2(x+1)} + \frac{3}{4}$$

$$\frac{3}{3+1} + \frac{5}{8} = \frac{5}{2(3+1)} + \frac{3}{4}$$

$$\frac{3}{4} + \frac{5}{8} = \frac{5}{2(4)} + \frac{3}{4}$$

$$\frac{3}{4} + \frac{5}{8} = \frac{5}{8} + \frac{3}{4}$$

$$\frac{3}{4} + \frac{5}{8} = \frac{3}{4} + \frac{5}{8}$$

Since substitution of x by 3 results in this equivalent equation, which is always true, 3 is indeed a root of the equation.

● **PROBLEM** 205

Solve the equation

$$\frac{5}{x-1} + \frac{1}{4-3x} = \frac{3}{6x-8} .$$

Solution: By factoring out a common factor of -2 from the denominator of the term on the right side of the given equation, the given equation becomes:

$$\frac{5}{x-1} + \frac{1}{4-3x} = \frac{3}{-2(-3x+4)} = \frac{3}{-2(4-3x)} = \frac{3}{2(4-3x)}$$

Hence,

$$\frac{5}{x-1} + \frac{1}{4-3x} = -\frac{3}{2(4-3x)}$$

Adding $\frac{3}{2(4-3x)}$ to both sides of this equation:

$$\frac{5}{x-1} + \frac{1}{4-3x} + \frac{3}{2(4-3x)} = 0. \qquad (1)$$

Now, in order to combine the fractions, the least common denominator (l.c.d.) must be found. The l.c.d. is found in the following way: list all the different factors of the denominators of the fractions. The exponent to be used for each factor in the l.c.d. is the greatest value of the exponent for each factor in any denominator. Therefore, the l.c.d. of the given fractions is:

$$2^1(x-1)^1(4-3x)^1 = 2(x-1)(4-3x)$$

Hence, equation (1) becomes:

119

$$\frac{(2)(4-3x)(5)}{(2)(4-3x)(x-1)} + \frac{(2)(x-1)(1)}{(2)(x-1)(4-3x)} + \frac{(x-1)(3)}{(x-1)(2)(4-3x)} = 0 \quad (2)$$

Simplifying equation (2):

$$\frac{10(4-3x) + 2(x-1) + 3(x-1)}{2(x-1)(4-3x)} = 0$$

$$\frac{40 - 30x + 2x - 2 + 3x - 3}{2(x-1)(4-3x)} = 0$$

$$\frac{-25x + 35}{2(x-1)(4-3x)} = 0$$

Multiplying both sides of this equation by $2(x-1)(4-3x)$:

$$2(x-1)(4-3x)\frac{-25x+35}{2(x-1)(4-3x)} = 2(x-1)(4-3x)(0)$$

$$-25x + 35 = 0$$

Adding 25x to both sides of this equation:

$$-25x + 35 + 25x = 0 + 25x$$

$$35 = 25x$$

Dividing both sides of this equation by 25:

$$\frac{35}{25} = \frac{25x}{25}$$

$$\frac{7}{5} = x$$

Therefore, the solution set to the equation $\frac{5}{x-1} + \frac{1}{4-3x} = \frac{3}{6x-8}$ is: $\left\{\frac{7}{5}\right\}$.

● **PROBLEM 206**

Find $\left\{x \middle| \frac{2}{x+1} - 3 = \frac{4x+6}{x+1}\right\}$.

<u>Solution:</u> The required set is the set of all x such that

$$\frac{2}{x+1} - 3 = \frac{4x+6}{x+1}$$

Multiplying each member by (x + 1) to eliminate the fractions, we obtain

$$\left(x+1\right)\left(\frac{2}{x+1} - 3\right) = \left(\frac{4x+6}{x+1}\right)\left(x+1\right)$$

Distributing, $\quad (x+1)\left(\frac{2}{x+1}\right) - (x+1)3 = 4x + 6$

$$2 - (3x + 3) = 4x + 6$$

$$2 - 3x - 3 = 4x + 6$$

$$-1 - 3x = 4x + 6$$

Adding (-4x) to both sides,

$$-1 - 3x - 4x = 6$$

$$-1 - 7x = 6$$

Adding 1 to both sides,

$$-7x = 7$$
$$x = -1$$

If we now substitute (-1) for x in our original equation,

$$\frac{2}{x + 1} - 3 = \frac{4x + 6}{x + 1}$$

$$\frac{2}{-1 + 1} - 3 = \frac{4(-1) + 6}{-1 + 1}$$

$$\frac{2}{0} - 3 = \frac{-4 + 6}{0}$$

Since division by zero is impossible the above equation is not defined for x = -1. Hence we conclude that the equation has no roots and

$$\left\{ x \mid \frac{2}{x + 1} - 3 = \frac{4x + 6}{x + 1} \right\} = \emptyset \text{ , where } \emptyset \text{ is the empty set.}$$

● **PROBLEM** 207

Solve

$$\frac{3}{x - 1} + \frac{1}{x - 2} = \frac{5}{(x - 1)(x - 2)} \text{ .}$$

Solution: First we will eliminate the fractions by finding the least common denominator, LCD. This is done by multiplying the denominators and taking the highest power of each factor which appears, only once.

$$(x - 1)(x - 2)(x - 1)(x - 2)$$

$$LCD = (x - 1)(x - 2)$$

Multiplying both sides of the equation by the LCD will remove the fractions and give:

$$(x - 1)(x - 2)\left[\frac{3}{x - 1} + \frac{1}{x - 2} \right]$$

$$= \left[\frac{5}{(x - 1)(x - 2)} \right] (x - 1)(x - 2)$$

$$3(x - 2) + (x - 1) = 5$$

$$3x - 6 + x - 1 = 5$$

$$4x - 7 = 5$$

$$4x = 12$$

$$x = 3$$

Substituting x = 3 into the original equation

121

$$\frac{3}{2} + 1 = \frac{5}{(2)(1)}$$

$$\frac{5}{2} = \frac{5}{2}$$

we find x = 3 satisfies the original equation.

● **PROBLEM** 208

Solve $\dfrac{3}{x - 1} + \dfrac{2}{x + 1} = \dfrac{6}{x^2 - 1}$.

<u>Solution:</u> First we obtain the Least Common Denominator, LCD, by multiplying the denominators,

$$(x - 1)(x + 1)(x^2 - 1), \text{ or}$$

$(x - 1)(x + 1)[(x - 1)(x + 1)]$, and taking each factor's highest power once.

$$\text{LCD} = (x - 1)(x + 1) = (x^2 - 1) \tag{1}$$

Then multiply both sides of the equation by the LCD to remove the fractions and obtain:

$$(x - 1)(x + 1)\left(\frac{3}{x - 1} + \frac{2}{x + 1}\right)$$

$$= \left(\frac{6}{x^2 - 1}\right)(x - 1)(x + 1)$$

$$3(x + 1) + 2(x - 1) = 6 \tag{2}$$

$$3x + 3 + 2x - 2 = 6$$

$$5x + 1 = 6$$

$$5x = 5$$

$$x = 1$$

Substituting x = 1 into **Equation** 2, we can readily see that it is a solution of that equation. However, x = 1 is not an admissible value of x for Equation 1, because division by 0 is undefined; therefore, x = 1 is not a solution of Equation 1. It is an extraneous root that was introduced by the multiplication by the LCD. The original equation does not have a solution.

● **PROBLEM** 209

Solve the equation

$$\frac{2x}{3 + x} + \frac{3 + x}{3} = 2 + \frac{x^2}{3(x - 3)} \ .$$

<u>Solution:</u> In order to eliminate the fractions in this equation we multiply both members of the equation by the Least Common Denominator (the LCD). Our denominators are $(3 + x), 3,$ and $3(x - 3)$. Thus the LCD is $3(3 + x)(x - 3)$. Therefore

$$[3(3 + x)(x - 3)]\left[\frac{2x}{3 + x} + \frac{3 + x}{3}\right] = [3(3 + x)(x - 3)]\left[2 + \frac{x^2}{3(x-3)}\right].$$

Distribute:

$$[3(3 + x)(x - 3)]\left[\frac{2x}{3 + x}\right] + [3(3 + x)(x - 3)]\left[\frac{3 + x}{3}\right]$$

$$= [3(3 + x)(x - 3)](2) + [3(3 + x)(x - 3)]\left[\frac{x^2}{3(x - 3)}\right].$$

Cancelling like terms in numerator and denominator,

$$3(x - 3)(2x) + (3 + x)(x - 3)(3 + x) = 3 \cdot 2(3 + x)(x - 3)$$
$$+ (3 + x)x^2$$

Factoring both sides of the equation,

$$6x(x - 3) + (9 + 6x + x^2)(x - 3) = 6(3x + x^2 - 9 - 3x) + (3x^2 + x^3)$$
$$6x(x - 3) + (9 + 6x + x^2)(x - 3) = 6(x^2 - 9) + (3x^2 + x^3)$$

Distributing the two terms on the left side and the one term on the right side of this equation,

$$(6x^2 - 18x) + (9x + 6x^2 + x^3) - (27 + 18x + 3x^2) = (6x^2 - 54) + (3x^2 + x^3)$$

Grouping terms and simplifying,

$$6x^2 - 18x + 9x + 6x^2 + x^3 - 27 - 18x - 3x^2 = 6x^2 - 54 + 3x^2 + x^3$$
$$x^3 + 9x^2 - 27x - 27 = x^3 + 9x^2 - 54$$

Subtract x^3 from both sides of this equation:

$$x^3 + 9x^2 - 27x - 27 - x^3 = x^3 + 9x^2 - 54 - x^3$$
$$9x^2 - 27x - 27 = 9x^2 - 54$$

Subtract $9x^2$ from both sides of this equation:

$$9x^2 - 27x - 27 - 9x^2 = 9x^2 - 54 - 9x^2$$

$$-27x - 27 = -54$$

Add   27 to both sides of this equation:

$$-27x - 27 + 27 = -54 + 27$$

$$27x = -27$$

Divide both sides of this equation by -27:

$$\frac{-27x}{-27} = \frac{-27}{-27}$$

Therefore, $x = 1$.

To verify that $x = 1$ is the solution to our problem, we perform the following check:

Check: Replace x by 1 in our original equation,

$$\frac{2x}{3 + x} + \frac{3 + x}{3} = 2 + \frac{x^2}{3(x - 3)}$$

$$\frac{2(1)}{3 + 1} + \frac{3 + 1}{3} = 2 + \frac{(1)^2}{3(1 - 3)}$$

$$\frac{2}{4} + \frac{4}{3} = 2 + \frac{1}{3(-2)}$$

$$\frac{2}{4} + \frac{4}{3} = 2 + \frac{1}{-6}$$

$$\frac{2}{4} + \frac{4}{3} = 2 - \frac{1}{6}$$

Multiplying both members by LCD, 12:

$$12\left(\frac{2}{4} + \frac{4}{3}\right) = 12\left(2 - \frac{1}{6}\right)$$

$$12\left(\frac{2}{4}\right) + 12\left(\frac{4}{3}\right) = 12\left(2\right) - 12\left(\frac{1}{6}\right)$$

$$6 + 16 = 24 - 2$$

$$22 = 22$$

Hence, x = 1 is our solution, and our solution set is {1}.

UNKNOWN UNDER RADICAL SIGN

● PROBLEM 210

Solve $\sqrt{x - 3} = 4$.

Solution: Square both sides of the given equation to obtain:

$$\left(\sqrt{x - 3}\right)^2 = 4^2$$

Note $\left(\sqrt{a}\right)^2 = \sqrt{a} \cdot \sqrt{a} = \sqrt{a \cdot a} = \sqrt{a^2} = a$; thus $\left(\sqrt{x - 3}\right)^2 = x - 3$, and we obtain:

$$x - 3 = 16$$

$$x = 19$$

Check: Substitute 19 for x in the original equation,

$$\sqrt{x - 3} = 4$$

$$\sqrt{19 - 3} = 4$$

$$\sqrt{16} = 4$$

$$4 = 4$$

● PROBLEM 211

Solve the equation $\sqrt{3x + 1} = 5$.

Solution: Square both members:

$$3x + 1 = 25.$$

Solve for x: x = 8.

Check: $\sqrt{3(8) + 1} = \sqrt{25} = 5$.

It should be recalled that $\sqrt{25} = +5$, and does not

equal +5; that is, when no sign precedes the radical the positive value of the root is to be taken. If both positive and negative roots are meant, we shall write both signs before the radical.

● **PROBLEM 212**

Solve $\sqrt{5x} + \sqrt{3} = \sqrt{3x} - \sqrt{5}$.

<u>Solution:</u>   Add $-\sqrt{3x}$ to both sides of the given equation,

$$\sqrt{5x} + \sqrt{3} - \sqrt{3x} = \sqrt{3x} - \sqrt{5} - \sqrt{3x} .$$

Commute terms and add $-\sqrt{3}$ to both sides,

$$\sqrt{5x} - \sqrt{3x} + \sqrt{3} - \sqrt{3} = \sqrt{3x} - \sqrt{3x} - \sqrt{5} - \sqrt{3}$$

$$\sqrt{5x} - \sqrt{3x} = -\sqrt{5} - \sqrt{3} .$$

Using the distributive law,

$$x\left(\sqrt{5} - \sqrt{3}\right) = -\sqrt{5} - \sqrt{3}$$

$$x = \frac{-\sqrt{5} - \sqrt{3}}{\sqrt{5} - \sqrt{3}} .$$

Factor (-1) from the numerator:

$$x = \frac{-\left(\sqrt{5} + \sqrt{3}\right)}{\sqrt{5} - \sqrt{3}}$$

To rationalize the denominator, we multiply numerator and denominator by the conjugate $\left(\sqrt{5} + \sqrt{3}\right)$ of the denominator; hence:

$$x = \frac{-\left(\sqrt{5} + \sqrt{3}\right)\left(\sqrt{5} + \sqrt{3}\right)}{\left(\sqrt{5} - \sqrt{3}\right)\left(\sqrt{5} + \sqrt{3}\right)}$$

$$x = \frac{-\left(\sqrt{5}\sqrt{5} + 2\sqrt{3}\sqrt{5} + \sqrt{3}\sqrt{3}\right)}{\sqrt{5}\cdot\sqrt{5} + \sqrt{5}\sqrt{3} - \sqrt{5}\sqrt{3} - \sqrt{3}\sqrt{3}}$$

Note $\sqrt{a}\cdot\sqrt{b} = \sqrt{ab}$, therefore:

$$x = \frac{-\left(\sqrt{25} + 2\sqrt{15} + \sqrt{9}\right)}{\sqrt{25} - \sqrt{9}}$$

$$x = \frac{-(5 + 2\sqrt{15} + 3)}{5 - 3}$$

$$x = \frac{-\left(8 + 2\sqrt{15}\right)}{2}$$

$$x = \frac{-2\left(4 + \sqrt{15}\right)}{2}$$

$$x = -\left(4 + \sqrt{15}\right)$$

$$x = -4 - \sqrt{15}$$

● **PROBLEM 213**

Solve the equation

$$\sqrt{x} = 7 + \sqrt{x - 7}$$

Solution: Squaring both sides of the given equation,

$$x = 49 + 14 \sqrt{x - 7} + x - 7$$

Simplifying

$$- 42 = 14 \sqrt{x - 7}$$

$$- 3 = \sqrt{x - 7} \qquad (1)$$

Squaring both sides of equation (1),

$$9 = x - 7$$

$$x = 16$$

Checking the root by substitution in the given equation:

$$\sqrt{16} \neq 7 + \sqrt{16 - 7}$$

$$4 \neq 7 + 3$$

Clearly x = 16 does not satisfy the given equation, and therefore the equation has no roots. The fact that the given equation has no roots could have been anticipated from equation (1), $- 3 = \sqrt{x - 7}$, since the positive root is indicated in the original equation.

● **PROBLEM** 214

Solve $\sqrt{2}x - 2 = 2x - \sqrt{2}$.

Solution: Add (-2x) to both sides of the given equation:

$$\sqrt{2}x - 2 - 2x = -\sqrt{2}$$

Now, add 2 to both sides:

$$\sqrt{2}x - 2x = 2 - \sqrt{2}$$

Use the distributive law:

$$x(\sqrt{2} - 2) = 2 - \sqrt{2}$$

$$x = \frac{2 - \sqrt{2}}{\sqrt{2} - 2}$$

Multiply both sides by (-1):

$$-x = \frac{-(2 - \sqrt{2})}{\sqrt{2} - 2}$$

$$-x = \frac{-2 + \sqrt{2}}{-2 + \sqrt{2}}$$

$$- x = 1$$

$$x = -1$$

Solve $\sqrt{4x + 5} + 2\sqrt{x - 3} = 17$.

Solution: Transpose:

$$\sqrt{4x + 5} - 17 = -2\sqrt{x - 3}.$$

Square: $4x + 5 - 34\sqrt{4x + 5} + 289 = 4(x - 3)$

$$4x + 5 - 34\sqrt{4x + 5} + 289 = 4x - 12.$$

Transpose: $-34\sqrt{4x + 5} = 4x - 12 - 4x - 5 - 289.$

Simplify: $-34\sqrt{4x + 5} = -306$

$$\sqrt{4x + 5} = 9.$$

Square: $4x + 5 = 81.$

Solve for x: $x = 19.$

Check: $\sqrt{4(19) + 5} + 2\sqrt{(19) - 3} \stackrel{?}{=} 17$

$$\sqrt{81} + \qquad 2\sqrt{16} \stackrel{?}{=} 17$$

$$9 + \qquad 2(4) \stackrel{?}{=} 17$$

$$17 = 17.$$

Sol.: $x = 19.$

Solve $\sqrt{x - 2} - \sqrt{x + 3} = 1$.

Solution: $\sqrt{x - 2} - \sqrt{x + 3} = 1$

$\sqrt{x - 2} = \sqrt{x + 3} + 1.$ $\qquad$ Transpose $\sqrt{x + 3}$,

$x - 2 = x + 3 + 2\sqrt{x + 3} + 1.$ Square both sides

of equation.

$x - 2 - x - 3 - 1 = 2\sqrt{x + 3}.$ Transpose and combine terms.

$$-6 = 2\sqrt{x + 3}$$

$$-3 = \sqrt{x + 3}$$

$$9 = x + 3. \quad \text{Square both sides of equation.}$$

Solving gives $x = 6.$

Check: $\sqrt{6 - 2} - \sqrt{6 + 3} \stackrel{?}{=} 1$

$$\sqrt{4} - \sqrt{9} \quad \stackrel{?}{=} 1$$

$$2 - 3 \neq 1$$
$$-1 \neq 1$$

Therefore x = 6 is not a solution. x = 6 is an extraneous root. The two expressions are not equal for any value of the unknown.

● **PROBLEM** 217

Solve the equation

$$\frac{\sqrt{x + 1} + \sqrt{x - 1}}{\sqrt{x + 1} - \sqrt{x - 1}} = \frac{4x - 1}{2} .$$

Solution: We can use the following law to rewrite the given proportion: If $\frac{a}{b} = \frac{c}{d}$, then $\frac{a + b}{a - b} = \frac{c + d}{c - d}$. Applying this law we have:

$$\frac{\sqrt{x + 1} + \sqrt{x - 1} + (\sqrt{x + 1} - \sqrt{x - 1})}{\sqrt{x + 1} + \sqrt{x - 1} - (\sqrt{x + 1} - \sqrt{x - 1})} = \frac{4x - 1 + (2)}{4x - 1 - (2)} \quad \text{or,}$$

$$\frac{2\sqrt{x + 1}}{2\sqrt{x - 1}} = \frac{4x + 1}{4x - 3} . \text{ Eliminating } \frac{2}{2} \text{ we have:}$$

$$\frac{\sqrt{x + 1}}{\sqrt{x - 1}} = \frac{4x + 1}{4x - 3} .$$

Squaring both sides of the equation gives us,

$$\left(\frac{\sqrt{x + 1}}{\sqrt{x - 1}}\right)^2 = \left(\frac{4x + 1}{4x - 3}\right)^2 \quad \text{or} \quad \frac{(\sqrt{x + 1})^2}{(\sqrt{x - 1})^2} = \frac{(4x + 1)^2}{(4x - 3)^2} .$$

Finding the above squares we obtain:

$$\frac{x + 1}{x - 1} = \frac{16x^2 + 8x + 1}{16x^2 - 24x + 9} .$$

We can again rewrite this new proportion as:

$$\frac{x + 1 + (x - 1)}{x + 1 - (x - 1)} =$$

$$\frac{16x^2 + 8x + 1 + (16x^2 - 24x + 9)}{16x^2 + 8x + 1 - (16x^2 - 24x + 9)} \quad \text{or,}$$

$$\frac{2x}{2} = \frac{32x^2 - 16x + 10}{32x - 8} ;$$

therefore, $x = \frac{32x^2 - 16x + 10}{32x - 8} = \frac{2(16x^2 - 8x + 5)}{2(16x - 4)} ;$

thus, $x = \frac{16x^2 - 8x + 5}{16x - 4} ;$

and multiplying both sides of this equation by (16x - 4) we have,

$$x(16x - 4) = 16x^2 - 8x + 5 \qquad \text{or,}$$

$$16x^2 - 4x = 16x^2 - 8x + 5.$$

Now, combining similar terms we obtain:

$$16x^2 - 16x^2 - 4x + 8x = 5 \qquad \text{or}$$

$$4x = 5.$$

Therefore, $x = \dfrac{5}{4}$.

CHAPTER 11

PROPERTIES OF STRAIGHT LINES

Basic Attacks and Strategies for Solving Problems in this Chapter. See pages 130 to 152 for step-by-step solutions to problems.

Any equation that can be put in the standard form

$ax + by = c,$

where a, b, and c are real numbers and a and b are not both 0, is defined to be a linear equation in two variables. The graph of this form is a straight line. In order to graph a linear equation in two variables, we simply graph its solution set. This means that we draw a straight line through all the points whose coordinates satisfy the equation.

Two important points on the graph of a straight line, if they exist, are the x- and y-intercepts or the points where the graph crosses the axes. The x-intercept of the line is the x-coordinate of the point that the line has in common with the x-axis, and the y-intercept is the y-coordinate of the point that the line has in common with the y-axis. Since any point on the x-axis has a y-coordinate of 0, we can find the x-intercept by letting $y = 0$ and solving the equation for x. Similarly, we can find the y-intercept by letting $x = 0$ and solving for y. For example, the x-intercept and y-intercept of the equation

$3x + 4y = 12$

are given by

$3x + 4(0) = 12$

or $x = 4$ and

$3(0) + 4y = 12$

or $y = 3$, respectively.

Graphing straight lines by finding the intercepts works best when the coefficients of x and y are factors of the constant term in the equation.

A linear equation or linear function, whose graph is a nonvertical straight line, has a slope. The slope, m, of a line is a measure of the steepness of the line in that it is the ratio of the vertical change (difference of ordinates) between any two points, (x_1, y_1) and (x_2, y_2), on the line to the horizontal change (difference of abscissas) between the two points. Thus, the most popular procedure for finding the slope of a line is given by the formula

$$m = \frac{y_2 - y_1}{x_2 - x_1},$$

where x_1 is not equal to x_2.

When calculating the slope using this formula, it does not matter which point is designated as the "first" point and which is designated as the "second" point. The only warning is that once the points are designated then they must not be changed throughout the problem. Geometrically, a straight line that gets higher as it goes from left to right has a positive slope; a line that gets lower as it goes from left to right has a negative slope; a horizontal line has a slope of 0; and a vertical line has an undefined slope.

If a linear equation can be written in the form

$$y = mx + b,$$

then the m represents the slope of the graph of the equation and the b is the y-intercept.

Some techniques for determining the equation of a line, when given certain facts about the line, are as follows:

(1) A procedure for finding the equation of a line having a slope of r and containing a point (c, d) includes first using the slope formula,

$$\frac{y - y_1}{x - x_1} = m \quad \text{or} \quad y - y_1 = m(x - x_1),$$

and substituting r for m, d for y_1, and c for x_1. Then write the resulting equation in the form

$$ax + by = c.$$

(2) The procedure for finding the equation of the line having a slope of r and y-intercept of c includes the use of the slope-intercept form of the equation,

$$y = mx + b.$$

Then, substitute r for m and c for b in the equation.

Of the various graphing techniques for a linear equation, the most obvious way is to calculate any two ordered pairs that satisfy the equation, plot the points, and draw a straight line through them. Depending on how the equation is arranged, there may be faster and more convenient ways to obtain the graph.

If the graph of a line is in slope-intercept form, then the graph can be obtained by marking the y-intercept and using the slope to move from that point to the second point on the line. One must remember that the slope is the ratio of how much the line moves up or down compared to how much the line moves right or left. If the equation of a line is in intercept form, then it is easy to mark the two intercepts and draw the straight line through them.

Step-by-Step Solutions to Problems in this Chapter, "Properties of Straight Lines"

SLOPES, INTERCEPTS AND POINTS ON GIVEN LINES

● **PROBLEM** 218

Find the slope of $f(x) = 3x + 4$.

Solution: Two points on the line determined by $f(x) = 3x + 4$ are A(0,4) and B(1,7).

$$\frac{\text{difference of ordinates}}{\text{difference of abscissas}} = \frac{7-4}{1-0} = 3$$

Note that the ordinates are the y-coordinates and the abscissas are the x-coordinates. The slope determined by points A and B is 3. Hence, the slope of $f(x) = 3x + 4$ is 3. In general, the slope of a linear function of the form $f(x) = mx + b$ is m.

● **PROBLEM** 219

Show that the slope of the segment joining (1,2) and (2,6) is equal to the slope of the segment joining (5,15) and (10,35).

Solution: The slope of the line segment, m, joining the points $\left(x_1, y_1\right)$ and $\left(x_2, y_2\right)$ is given by the formula

$$m = \frac{\left(y_2 - y_1\right)}{\left(x_2 - x_1\right)}$$

Therefore, the slope of the segment joining (1,2) and (2,6) is

$$\frac{6-2}{2-1} = \frac{4}{1} = 4.$$

The slope of the segment joining (5,15) and (10,35) is

$$\frac{35-15}{10-5} = \frac{20}{5} = 4.$$

Thus, the slopes of the two segments are equal.

● **PROBLEM** 220

Determine the constant A so that the lines $3x - 4y = 12$ and $Ax + 6y = -9$ are parallel.

Solution: If two non-vertical lines are parallel, their slopes are equal. Thus the lines $Ax + By + C = 0$ and $Ax + By + D = 0$ are parallel (since both have slope $= -A/B$). We are given two lines:

$$3x - 4y = 12 \qquad\qquad (1)$$

$$Ax + 6y = -9 \qquad\qquad (2)$$

We must make the coefficients of y the same for both equations in order to equate the coefficients of x. Multiply (1) by $-3/2$ to obtain

$$\frac{-3}{2}(3x - 4y) = -\frac{3}{2}(12)$$

$$-\frac{9}{2}x + 6y = -18 \qquad\qquad (3)$$

$$Ax + 6y = -9 \qquad\qquad (2)$$

Transpose the constant terms of (3) and (2) to the other side.

Adding 18 to both sides, $\qquad -\frac{9}{2}x + 6y + 18 = 0 \qquad (4)$

Adding 9 to both sides. $\qquad\qquad Ax + 6y + 9 = 0 \qquad (5)$

             (4) and (5) will now be parallel if the co-efficients of the x-terms are the same. Thus the constant A is -9/2 . Then equation (5) becomes -9/2x + 6y + 9 = 0. We can also express (5) in its given form, Ax + 6y = -9 or - 9/2 x + 6y = -9.
       We also can write it in a form that has the same coefficient of x as (1), which clearly shows that they have equal slopes.

$$3x - 4y = 12 \qquad\qquad\qquad (1)$$

$$-\frac{9}{2}x + 6y = -9$$

Multiply the second equation by - 2/3 to obtain a coefficient of x equal to 3.

$$-\frac{2}{3}\left(-\frac{9}{2}x + 6y\right) = -\frac{2}{3}(-9)$$

$$3x - 4y = 6$$

Now equations (1), 3x - 4y = 12, and the equation 3x - 4y = 6 are parallel since the coefficients of x and y are identical.

● **PROBLEM 221**

Find the slope and Y-intercept of the following lines.

(a) y = 3x - 1   (b) y = 1 - 4x   (c) 2y = 4x + 7

Solution: a) The equation of a line is: y = mx + b, where m is the slope of the line and b is the y-intercept of the line. Hence, the line y = 3x - 1 has slope = 3 and y-intercept = -1.

b) The line y = 1 - 4x can be rewritten, using the commutative law, as y = -4x + 1. Hence, the slope of this line = -4 and the y-intercept = 1.

c) The line 2y = 4x + 7, after dividing both sides by 2, can be rewritten as:

$$\frac{2y}{2} = \frac{4x + 7}{2}$$

$$y = 2x + \frac{7}{2}$$

Hence, the slope = 2 and the y-intercept = $\frac{7}{2}$.

● **PROBLEM 222**

Find the slope, the y-intercept, and the x-intercept of the equation 2x - 3y - 18 = 0.

Solution: The equation 2x - 3y - 18 = 0 can be written in the form of the general linear equation, ax + by = c.

$$2x - 3y - 18 = 0$$

$$2x - 3y = 18$$

To find the slope and y-intercept we derive them from the formula of the general linear equation $ax + by = c$. Dividing by b and solving for y we obtain:

$$\frac{a}{b}x + y = \frac{c}{b}$$

$$y = \frac{c}{b} - \frac{a}{b}x$$

where $-\frac{a}{b}$ = slope and $\frac{c}{b}$ = y-intercept.

To find the x-intercept, solve for x and let $y = 0$:

$$x = \frac{c}{b} - \frac{b}{a}y$$

$$x = \frac{c}{a}$$

In this form we have $a = 2$, $b = -3$, and $c = 18$. Thus,

$$\text{slope} = -\frac{a}{b} = -\frac{2}{-3} = \frac{2}{3}$$

$$\text{y-intercept} = \frac{c}{b} = \frac{18}{-3} = -6$$

$$\text{x-intercept} = \frac{c}{a} = \frac{18}{2} = 9$$

● **PROBLEM ·223**

The equation $F = \frac{9}{5}C + 32$ relates the Fahrenheit and centigrade temperature scales. What do the numbers $\frac{9}{5}$ and 32 represent?

Solution: An equation in the form $y = mx + b$ is a linear equation with slope m and y-intercept b. Thus, with $F = 9/5\ C + 32$, 32 is the y-intercept and 9/5 is the slope. That is, the number 32 tells us that when the centigrade thermometer reads 0, the Fahrenheit thermometer reads 32. The number 9/5 is the slope of the line we would obtain if we graphed our equation in an axis system in which centigrade temperatures are measured on the horizontal axis and Fahrenheit temperatures are measured on the vertical axis; that is, the number 9/5 is the number of units of Fahrenheit temperature rise per unit of centigrade temperature rise. If a body's temperature increases $1^0 C$, then it increases $9/5^0 F$. If a body's temperature increases -10^0 (decreases 10^0) C, then it increases $9/5(-10)^0 = -18^0 F$.

● **PROBLEM 224**

The slope and one point of a line are given. Is the Y-intercept positive or negative?

(a) $m = \frac{22}{7}$, $(1, \pi)$ (b) $m = \sqrt{2}$, $(1, 1.414)$

Solution: a) The equation of a line is: $y = mx + b$, where m is the slope and b is the y-intercept. Given the slope m and one point of the line, the y-intercept b can be found. Thus it can be determined

whether the y-intercept b is positive or negative. For the line with
slope m = 22/7 and which contains the point $(1,\pi)$:

$$y = mx + b$$
$$\pi = \frac{22}{7}(1) + b$$
$$\pi = \frac{22}{7} + b \qquad\qquad\qquad (1)$$

Since π is approximately $\frac{22}{7}$, equation (1) becomes:

$$\frac{22}{7} = \frac{22}{7} + b$$

Subtract 22/7 from both sides to obtain:

$$\frac{22}{7} - \frac{22}{7} = \frac{22}{7} + b - \frac{22}{7}$$
$$0 = b$$

Hence, the y-intercept b is neither positive nor negative, since the
y-intercept b = 0.

b) For the line with slope $m = \sqrt{2}$ and which contains the point
$(1,1.414)$:

$$y = mx + b$$
$$1.414 = \sqrt{2}(1) + b$$
$$1.414 = \sqrt{2} + b \qquad\qquad\qquad (2)$$

Since $\sqrt{2}$ is approximately 1.414, equation (2) becomes:

$$1.414 = 1.414 + b$$

Subtract 1.414 from both sides to obtain:

$$1.414 - 1.414 = 1.414 + b - 1.414$$
$$0 = b$$

Again, the y-intercept b is neither positive nor negative, since b = 0.

● **PROBLEM** 225

Show that the slope of the segment joining (1,2) and (3,8) is
equal to the slope of the segment joining (4,11) and (8,23)

<u>Solution:</u> The slope of the segment joining (1,2) and (3,8) is

$$\frac{8 - 2}{3 - 1} = \frac{6}{2} = 3 .$$

The slope of the segment joining (4,11) and (8,23) is

$$\frac{23 - 11}{8 - 4} = \frac{12}{4} = 3 .$$

Therefore, the slopes of the two segments are equal.

FINDING EQUATIONS OF LINES

● **PROBLEM** 226

The two points $P_1(1,-2)$ and $P_2(4,1)$ determine a line. What
is the equation of the line?

The slope of the line segment connecting the two points is

$$m = \frac{\Delta y}{\Delta x} = \frac{y_2 - y_1}{x_2 - x_1} = \frac{1 - (-2)}{4 - 1} = \frac{3}{3} = 1.$$

Now we know the slope and at least one point on the line. Therefore, let $P(x,y)$ be any point on the line. Then the slope between the points P and P_1 (or, alternatively, between P and P_2) must be 1. Therefore,

$$m = \frac{\Delta y}{\Delta x} = 1 = \frac{y - (-2)}{x - 1}$$

$$x - 1 = y + 2 \quad \text{by cross multiplying}$$

$$y = x - 3 \quad \text{by solving for y.}$$

The required equation is $y = x - 3$. Note that both of the given points satisfy this equation.

for $P_1(1,-2)$:   for $P_2(4,1)$:

$y = x - 3$ $y = x - 3$

$-2 \overset{?}{=} 1 - 3$ $1 \overset{?}{=} 4 - 3$

$-2 = -2$ $1 = 1.$

● **PROBLEM 227**

Find the equation for the line passing through (3,5) and (-1,2).

Solution A: We use the two-point form with $(x_1,y_1) = (3,5)$ and $(x_2,y_2) = (-1,2)$. Then

$$\frac{y - y_1}{x - x_1} = m = \frac{y_2 - y_1}{x_2 - x_1}.$$

$$\frac{y_2 - y_1}{x_2 - x_1} = \frac{2 - 5}{-1 - 3} \quad \text{thus} \quad \frac{y - 5}{x - 3} = \frac{-3}{-4}.$$

Cross multiply, $-4(y - 5) = -3(x - 3)$.

Distributing, $-4y + 20 = -3x + 9$

$$3x - 4y = -11.$$

Solution B: Does the same equation result if we let $(x_1,y_1) = (-1,2)$ and $(x_2,y_2) = (3,5)$?

$$\frac{y_2 - y_1}{x_2 - x_1} = \frac{5 - 2}{3 - (-1)} \quad \text{thus} \quad \frac{y - 2}{x + 1} = \frac{3}{4}.$$

Cross multiply, $4(y - 2) = 3(x + 1)$

$$3x - 4y = -11.$$

Hence, either replacement results in the same equation.

● **PROBLEM** 228

Write the equation of the lines that contain the following points:
(a) (1,2) and (3,4)   (b) (-2,1) and (2,3)

Solution: a) The equation of a line is: $y = mx + b$, where m = slope and b = y-intercept. The slope, m, of any line can be found by using the equation:

$$m = \frac{y_2 - y_1}{x_2 - x_1},$$

where (x_1, y_1) and (x_2, y_2) are two points. After the slope m is found, the y-intercept b can be found by substituting one point of the line and the value of the slope m into the equation of the line. Hence, for the line that contains the points (1,2) and (3,4) where (x_1, y_1) = (1,2) and (3,4) = (x_2, y_2):

$$\text{slope: } m = \frac{4 - 2}{3 - 1}$$
$$= 2/2$$
$$= 1$$

Using $m = 1$ and the point (1,2) to find the y-intercept b we have:

$$y = mx + b$$
$$2 = 1(1) + b$$
$$2 = 1 + b$$
$$1 = b$$

Therefore, the equation of the line that contains the points (1,2) and (3,4) is:

$$y = 1x + 1$$

or

$$y = x + 1 .$$

b) For the line that contains the points (-2,1) and (2,3), where (x_1, y_1) = (-2,1) and (x_2, y_2) = (2,3):

$$\text{slope} = m = \frac{3 - 1}{2 - (-2)}$$
$$= \frac{2}{4} = \frac{1}{2}$$

Using $m = 1/2$ and the point (2,3) to find the y-intercept b we have:

$$y = mx + b$$
$$3 = \frac{1}{2}(2) + b$$
$$3 = \frac{2}{2} + b$$
$$3 = 1 + b$$
$$b = 2$$

Therefore, the equation of the line that contains the points (-2,1) and (2,3) is:

$$y = \frac{1}{2} x + 2 .$$

Find the equation of the line which passes through the points (-3,1) and (7,11).

Solution: The general equation for a line is y = mx + b, where m is the slope of the line and b is the y-intercept. Replacing (-3,1) and (7,11) for x and y in this equation, we obtain the equations 1 = m(-3) + b, and 11 = m(7) + b; or:

$$1 = -3m + b \qquad\qquad (1)$$

and

$$11 = 7m + b \qquad\qquad (2)$$

Thus, we solve equations (1) and (2) for m and b. Subtracting equation (2) from (1):

$$\begin{array}{r} 1 = -3m + b \\ - \underline{(11 = 7m + b\ )} \\ -10 = -10m \\ m = 1 \end{array}$$

Replacing m by 1 in equation (1) we solve for b:

$$1 = (-3)(1) + b$$
$$1 = -3 + b$$
$$b = 4$$

Hence the equation of the line passing through (-3,1) and (7,11) is y = (1)x + 4 or y = x + 4.

(a)   Find the equation of the line passing through (2,5) with slope 3.

(b)   Suppose a line passes through the y-axis at (0,b). How can we write the equation if the point-slope form is used?

Solution: (a) In the point-slope form, let x_1 = 2, y_1 = 5, and m = 3. The point-slope form of a line is:

$$y - y_1 = m(x - x_1)$$
$$y - 5 = 3(x - 2)$$
$$y - 5 = 3x - 6 \qquad \text{Distributive property}$$
$$y = 3x - 1 \qquad \text{Transposition}$$

(b) $y - b = m (x - 0)$

$y = mx + b$.

Find the equation of the line through (-1,2) and (3,1).

Solution: The equation of a line is in the form y = mx + b, where m is the slope of the line, and b is the y-intercept. Given 2 points on a line, (x_1,y_1) and (x_2,y_2), we can determine the slope of the line by means of the formula

$$m = \frac{(y_2 - y_1)}{(x_2 - x_1)}$$

Therefore

$$m = \frac{1 - 2}{3 - (-1)} = \frac{-1}{4}$$

[Observe that it makes no difference which point is considered as (x_1, y_1).]

To determine the y-intercept, we replace (x,y) by either of the given points, and m by $-1/4$ in the equation $y = mx + b$. Thus, using $(-1,2)$ and solving for b:

$$2 = -\frac{1}{4}(-1) + b$$

$$2 = \frac{1}{4} + b$$

$$b = 2 - \frac{1}{4} = \frac{8}{4} - \frac{1}{4} = \frac{7}{4}$$

Therefore, the equation of the line through $(-1,2)$ and $(3,1)$ is

$$y = -\frac{1}{4}x + \frac{7}{4} . \tag{1}$$

We may multiply both members by 4 to obtain an equivalent equation to equation (1):

$$4(y) = 4\left(\frac{-1}{4}x + \frac{7}{4}\right)$$

Distributing,

$$4y = -1x + 7$$

$$4y = -x + 7$$

Adding x to both sides of this equation:

$$4y + x = -x + 7 + x \quad \text{or} \quad x + 4y = 7.$$

● **PROBLEM 232**

What is the equation of the line through the point $(3,5)$ whose slope is 2?

Solution: Let $P(x,y)$ be any point on this line other than $(3,5)$. Using the definition of slope we have

$$m = \frac{\Delta y}{\Delta x} \qquad 2 = \frac{y - 5}{x - 3}$$

$$2x - 6 = y - 5 \qquad \text{by cross-multiplying}$$

$$y = 2x - 1 \qquad \text{by solving for y.}$$

This equation is satisfied by the coordinates $(3,5)$ and represents a line with a slope of 2.

GRAPHING TECHNIQUES

● **PROBLEM 233**

Construct the graph of the function defined by $y = 3x - 9$.

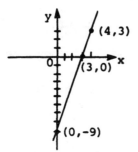

Solution: An equation of the form y = mx + b is a linear equation; that is, the equation of a straight line.

A straight line can be determined by two points. Let us choose the intercepts. The x-intercept lies on the x-axis and the y-intercept is on the y-axis.

We find the intercepts by assigning 0 to x and solving for y and by assigning 0 to y and solving for x. It is helpful to have a third point. We find a third point by assigning 4 to x and solving for y. Thus we get the following table of corresponding numbers:

x	y = 3x − 9	y
0	y = 3(0) − 9 = 0 − 9 =	−9
4	y = 3(4) − 9 = 12 − 9 =	3

Solving for x to get the x-intercept:

y = 3x − 9

y + 9 = 3x

$$x = \frac{y + 9}{3}$$

When y = 0, $x = \frac{9}{3} = 3$. The three points are (0,−9), (4,3), and (3,0). Draw a line through them (see sketch).

● **PROBLEM 234**

Graph the constant function 2y = 4.

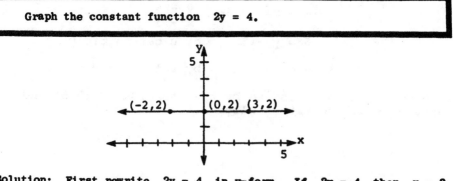

Solution: First rewrite 2y = 4 in y-form. If 2y = 4, then y = 2. Hence, g = {(x,y): y = 2}.

x	−2	0	3
y	2	2	2

For all values of x, y is equal to 2. The graph of g is a straight line with slope 0 and y-intercept (0,2).

● PROBLEM 235

Graph the function 3x - 5.

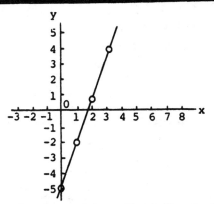

Solution: Let y = 3x - 5; then assign values to x and compute the corresponding values of y, the results being conveniently arranged in a table.

x	y = 3x - 5	y
0	y = 3(0) - 5 = 0 - 5 = -5	-5
1	y = 3(1) - 5 = 3 - 5 = -2	-2
2	y = 3(2) - 5 = 6 - 5 = 1	1
3	y = 3(3) - 5 = 9 - 5 = 4	4

The various points (x,y) are then plotted and joined by a smooth curve, which turns out to be a straight line. See the accompanying figure.

● PROBLEM 236

Find the equation of the line passing through the points (2,5) and (6,2). Check the results graphically.

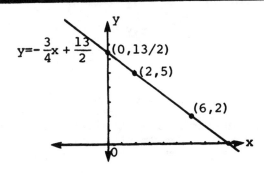

$y = -\frac{3}{4}x + \frac{13}{2}$ (0,13/2)

(2,5)

(6,2)

<u>Solution:</u> The general equation for a line is $y = mx + b$, where m is the slope of the line and b the y-intercept. Replacing (2,5) and (6,2) for x and y in this equation we obtain

$$5 = m(2) + b \qquad\qquad (1)$$

and

$$2 = m(6) + b \qquad\qquad (2)$$

Thus, we solve $5 = 2m + b$ and $2 = 6m + b$ for m and b:

subtracting equation (2) from (1):

$$
\begin{array}{r}
5 = 2m + b \\
- \ (2 = 6m + b) \\
\hline
3 = -4m
\end{array}
$$

$$m = \frac{-3}{4}$$

Replacing m by $\frac{-3}{4}$ in equation (1) we solve for b:

$$5 = \left(\frac{-3}{4}\right)(2) + b$$

$$5 = \frac{-6}{4} + b$$

$$b = 5 + \frac{6}{4} = \frac{20}{4} + \frac{6}{4} = \frac{26}{4} = \frac{13}{2}$$

Hence, the equation of the line passing through (2,5) and (6,2) is

$$y = \frac{-3}{4} x + \frac{13}{2} .$$

This result is shown graphically in the accompanying figure.

● **PROBLEM   237**

Graph the function defined by $3x - 4y = 12$.

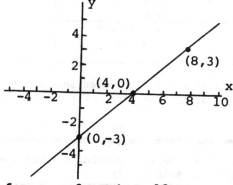

<u>Solution:</u> Solve for y: $3x - 4y = 12$

$$-4y = 12 - 3x$$

$$y = -3 + \frac{3}{4}x$$

$$y = \frac{3}{4}x - 3.$$

The graph of this function is a straight line since it is of the form $y = mx + b$. The y-intercept is the point $(0, -3)$ since for $x = 0$, $y = b = -3$. The x-intercept is

140

the point (4, 0) since for y = 0,

x = (y + 3) · $\frac{4}{3}$ = (0 + 3) · $\frac{4}{3}$ = 4. These two points, (0,-3) and (4,0) are sufficient to determine the graph (see the figure). A third point, (8,3), satisfying the equation of the function is plotted as a partial check of the inter-cepts. Note that the slope of the line is m = $\frac{3}{4}$. This means that y increases 3 units as x increases 4 units any-where along the line.

● **PROBLEM 238**

If f(x) = -2x - 5, find the (a) slope, (b) x-intercept, and (c) y-intercept. (d) Graph the function.

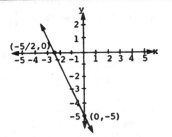

Solution: f(x) = mx + b is called a linear function where m and b are constants. m is the slope of the line and b is the y-intercept of the line. In this case, f(x) = -2x - 5, m = -2 and b = -5. Therefore,

(a) slope: m = -2

The x-intercept is located on the x-axis where f(x) = 0. Then we solve for x.

f(x) = mx + b = 0

$$mx = -b$$
$$x = \frac{-b}{m} = \text{x-intercept}$$

Hence,

(b) x-intercept: $\frac{-b}{m} = \frac{-(-5)}{-2} = \frac{5}{-2} = \frac{-5}{2}$

(c) y-intercept: b = -5

(d) We can graph the function by locating the points where the graph crosses the y-axis, (0, -5), and the x-axis, $\left(-\frac{5}{2}, 0\right)$. Recall again that (0,b) is the y-intercept and that $\left(\frac{-b}{m}, 0\right)$ is the x-intercept.

● **PROBLEM 239**

Graph the function y = 9 - $\frac{x}{2}$.

Solution: Writing the function as y = -$\frac{1}{2}$x + 9, we see that

141

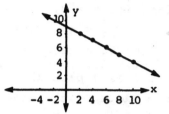

it is a straight line since it is of the form y = mx + b.
The y-intercept of the graph is (0,b), i.e., (0,9). The
slope of the line is m = -$\frac{1}{2}$. This means that y decreases
1 unit as x increases 2 units anywhere along the line. To
see this, choose even values of x in the interval
0 ≤ x ≤ 10. Determine the ordered pairs listed in the following table using the equation y = 9 - $\frac{x}{2}$.

x	0	2	4	6	8	10
y	9	8	7	6	5	4

Plotting these points, as illustrated in the figure, the
graph is a straight line passing through the first, second
and fourth quadrants. Although we have plotted a limited
number of points, the coordinates of each point on the line
will satisfy the equation y = 9 - $\frac{x}{2}$ and, conversely, each
ordered pair which satisfies the equation will determine
a point on the line.

● **PROBLEM 240**

Write the equation in slope-intercept form; specify the
slope and the y-intercept of the line. Sketch the graph
of the equation.

$$2x - 3y = 5$$

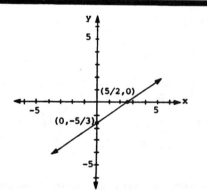

Solution: The equation y = mx + b is the slope-intercept
form of a line in which m is the slope of the line and b is
the y-intercept of the line. If the given equation is
solved for y, then the slope-intercept form of a line will
be obtained. Thus, adding 3y to both sides of the given
equation:

$$2x - \cancel{3y} + \cancel{3y} = 5 + 3y$$

$$2x = 5 + 3y$$

Subtracting 5 from both sides of this equation:

$$2x - 5 = \cancel{5} + 3y - \cancel{5}$$

$$2x - 5 = 3y$$

Dividing both sides by 3:

$$\frac{2x - 5}{3} = \frac{3y}{3}$$

$$\frac{2x - 5}{3} = y$$

$$\frac{2}{3}x - \frac{5}{3} = y \qquad\qquad y = \frac{2}{3}x - \frac{5}{3} \qquad (1)$$

or

Equation (1) is an equation in the slope-intercept form of a line. Comparing equation (1) with the slope-intercept form, $y = mx + b$, slope $= m = \frac{2}{3}$ and y - intercept $= b = -\frac{5}{3}$. To graph this equation it is sufficient to find one more point besides the y - intercept $\left(0, -\frac{5}{3}\right)$ which lies on the line (since two points determine a line). We can find the x - intercept by solving for x in the given equation, and use this as our second point on the line. Thus,

$$2x - 3y = 5$$

$$2x = 3y + 5$$

$$x = \frac{3}{2}y + \frac{5}{2},$$

and $\frac{5}{2}$ is the x - intercept. Thus, the point $\left(\frac{5}{2}, 0\right)$ is on the line (see figure).

● **PROBLEM** 241

The following table was constructed by reading the coordinates of selected points on a graphed line. Determine the equation of the line.

x	-2	-1	0	1	2	3
y	5	3	1	-1	-3	-5

<u>Solution:</u> The equation of the line is of the form $y = mx + b$ where m represents the slope and b represents the y-intercept.

For each interval in the table, as x increases 1 unit, y decreases 2 units. Therefore, the slope of the line connecting these points is m = $\frac{\Delta y}{\Delta x}$ = $\frac{-2}{1}$ = -2. The y-intercept is given as (0,1) since for x = 0, y = 1 as given in the table. The required equation is one which represents a straight line with a slope of -2 and y-intercept of 1, that is, y = -2x + 1. It can be verified that each listed ordered pair will satisfy this equation.

● **PROBLEM 242**

Use the intercept form to graph 3x + 4y = 12.

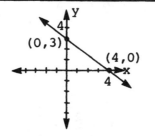

<u>Solution:</u> To find the x intercept substitute y = 0 in the expression and solve for x.

$$3x + 0 = 12$$

$$3x = 12$$

$$x = 4$$

The x intercept is (4,0).

Find the y intercept by substituting x = 0 in the expression and solve for y.

$$0 + 4y = 12$$

$$4y = 12$$

$$y = 3$$

The y intercept is (0,3).

Locate the points (4,0) and (0,3) and join them with the required straight line. See the figure.

● **PROBLEM 243**

a) Find the zeros of the function f if f(x) = 3x - 5.
b) Sketch the graph of the equation y = 3x - 5.

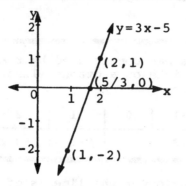

Solution: a) The zeros of a function are the numbers for which the value of the function is 0. Therefore, let $f(x) = 3x - 5 = 0$. Solving this equation:

$$3x - 5 = 0$$

Add 5 to both sides of this equation.

$$3x - 5 + 5 = 0 + 5$$

$$3x = 5$$

Divide both sides of this equation by 3.

$$\frac{3x}{3} = \frac{5}{3}$$

$$x = \frac{5}{3}$$

This number is the only zero of f (see the graph of f in the figure).

b) Note that the equation of a line is: $y = mx + b$ where m is the slope of the line and b is the y-intercept. Since the given equation is in this form, the graph is a line. It is only necessary to find two points of the graph in order to draw it. Let $x = 1$. Then $f(x) = f(1) = = 3(1) - 5 = 3 - 5 = -2$. Hence, one point is $(1,-2)$. Let $x = 2$. Then $f(x) = f(2) = 3(2) - 5 = 6 - 5 = 1$. Therefore, $(2,1)$ is the other point. These two points determine the straight line shown in the figure.

● **PROBLEM 244**

Discuss the graph of the function $y = -3x + 4$.

Solution: The graph is a straight line since it is of the form $y = mx + b$. The line intersects the y-axis at the point $(0,4)$. That is, when $x = 0$ then $y = 4$. The y-intercept in this example corresponds to $b = 4$. The slope of the line is $m = -3$. This means that y decreases 3 units as x increases 1 unit, anywhere along the line.

● **PROBLEM 245**

Are the following points on the graph of the equation $3x-2y=0$? (a) point $(2,3)$? (b) point $(3,2)$? (c) point $(4,6)$?

Solution: The point (a,b) lies on the graph of the equation $3x - 2y = 0$ if replacement of x and y by a and b, respectively, in the given equation results in an equation which is true.

(a) Replacing (x,y) by $(2,3)$:

$$3x - 2y = 0$$
$$3(2) - 2(3) = 0$$
$$6 - 6 = 0$$
$$0 = 0 , \quad \text{which is true.}$$

Therefore $(2,3)$ is a point on the graph.

(b) Replacing (x,y) by $(3,2)$:

$$3x - 2y = 0$$
$$3(3) - 2(2) = 0$$
$$9 - 4 = 0$$
$$5 = 0 , \quad \text{which is not true.}$$

Therefore $(3,2)$ is not a point on the graph.

(c) Replacing (x,y) by $(4,6)$:

$$3x - 2y = 0$$
$$3(4) - 2(6) = 0$$
$$12 - 12 = 0$$
$$0 = 0 , \quad \text{which is true.}$$

Therefore $(4,6)$ is a point on the graph.

This problem may also be solved geometrically as follows: draw the graph of the line $3x - 2y = 0$ on the coordinate axes. This can be done by solving for y:

$$3x - 2y = 0$$
$$- 2y = - 3x$$
$$y = \frac{-3}{-2} x = \frac{3}{2} x ,$$

and plotting the points shown in the following table:

x	$y = \frac{3}{2} x$
0	0
1	$3/2 = 1\frac{1}{2}$
2	3
-2	-3

(See accompanying figure.)

Observe that we obtain the same result as in our algebraic solution. The points $(2,3)$ and $(4,6)$ lie on the line $3x - 2y = 0$, whereas $(3,2)$ does not.

● **PROBLEM 246**

Show that the graphs of $3x - y = 9$ and $6x - 2y + 9 = 0$ are parallel lines.

<u>Solution:</u> If the slopes of two lines are equal, the lines

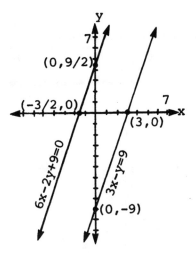

are parallel. Thus we must show that the two slopes are
equal. In standard form the equation of a line is y = mx + b,
where m is the slope.

Putting 3x - y = 9 in standard form,

$$-y = 9 - 3x$$

$$y = -9 + 3x$$

$$y = 3x - 9.$$

Thus the slope of the first line is 3. Putting
6x - 2y + 9 = 0 in standard form,

$$-2y + 9 = -6x$$

$$-2y = -6x - 9$$

$$y = 3x + \frac{9}{2}.$$

Thus the slope of this line is also 3. The slopes are equal.
Hence, the lines are parallel.

To graph these equations pick values of x and substitute them into the equation to determine the corresponding values of y. Thus we obtain the following tables of
values. Notice we need only <u>two</u> points to plot a line (2
points determine a line).

6x - 2y + 9 = 0   3x - y = 9

$$y = 3x + \frac{9}{2} \qquad\qquad y = 3x - 9$$

x	0	$-\frac{3}{2}$
y	$\frac{9}{2}$	0

x	0	3
y	-9	0

(See accompanying figure)

Determine whether there is a point of intersection of the graphs of 2x - 3y = 5 and 6x - 9y = 10.

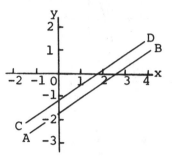

<u>Solution:</u> Geometric discussion. Rewriting the given linear equations in standard form, y = mx + b, the slope, m, can be read directly.

$$2x - 3y = 5 \qquad\qquad 6x - 9y = 10$$

$$- 3y = 5 - 2x \qquad\qquad - 9y = 10 - 6x$$

$$y = \frac{2}{3}x - \frac{5}{3} \qquad\qquad y = \frac{6}{9}x - \frac{10}{9}$$

$$m = \frac{2}{3} \qquad\qquad m = \frac{6}{9} = \frac{2}{3}$$

Recall that if the slope, m, of two lines are equal, the lines are parallel. This can be seen in the figure.

The graph of the first equation is the line AB through the point (1, -1) with slope $\frac{2}{3}$, and the graph of the second equation is the line CD through the point $\left(\frac{2}{3}, -\frac{2}{3}\right)$, with slope $\frac{2}{3}$. The lines are parallel, hence there is no point of intersection and the equations are inconsistent.

Algebraic discussion. If the members of the first equation are multiplied by 3, and if the members of the resulting equation are subtracted from the corresponding members of the second equation, we obtain

$$3(2x - 3y) = 3(5)$$

$$6x - 9y = 15$$

Then, $\quad 6x - 9y = 10$

$\qquad\quad -(6x - 9y = 15)$

$$\overline{\qquad\qquad\qquad 0 = -5, \text{ which is impossible.}}$$

The steps that were taken were based on the assumption that the given equations had a solution. The fact that an impossible conclusion results proves that the assumption was false. In other words, the two equations have no common solution, and are therefore inconsistent.

> Find the point of intersection of the graphs of $3x - y = 5$ and $9x - 3y = 15$.

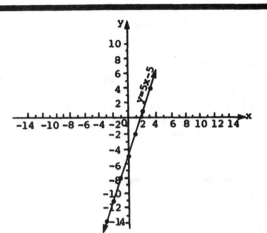

Solution: (1) $3x - y = 5$
(2) $9x - 3y = 15$

Divide the second equation by 3 to obtain:

(1) $3x - y = 5$

Thus any pair of values (x,y) which satisfies the first equation also satisfies the second equation. Hence the same straight line is the graph of both equations. It follows that there is no unique solution, but rather that every point on the common line is a solution. The two equations are dependent.

To solve the pair of dependent equations algebraically it is sufficient to assign an arbitrary value to x (or y), and then to solve for y (or x) in either equation.

$$3x - y = 5$$
$$-y = 5 - 3x \qquad \text{Add } -3x \text{ to both sides.}$$
$$y = 3x - 5 \qquad \text{Multiply by } -1.$$

To graph this equation, choose values of x and obtain their corresponding values of y from $y = 3x - 5$. The following table is constructed.

x	-3	-2	-1	0	1	2	3
y	-14	-11	-8	-5	-2	1	4

We then plot the points found in the table and draw a smooth curve (which turns out to be a straight line) through them.

> Sketch the graph of the function $H = \{(x, H(x)) | H(x) = 6\}$.

Solution: $H(x) = 6$ can be expressed in the form $H(x) = mx + b$, for in this particular example $m = 0$; hence $H(x) = 6$ can be written as $H(x) = 0 \cdot x + 6$. From this, regardless of the choice of a value for x,

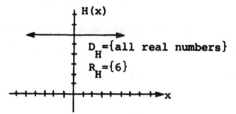

D$_H$={all real numbers}

R$_H$={6}

the corresponding value for H(x) will be 6. When there is no domain
set given, it is taken to be the largest subset of the real numbers for
which the corresponding H(x) value is also real. Hence the domain of
H is {all real numbers} and the range is {6}. The graph of H is a
horizontal line, i.e., has slope = m = 0 and H-intercept = b = 6.
The graph is sketched in the figure.

● **PROBLEM 250**

Given the function f defined by the equation

$$y = f(x) = \frac{3x + 4}{5} ,\qquad (1)$$

where the domain (and the range) of f is the set R of all real
numbers.

(a) Find the equation $x = g(y) = f^{-1}(x)$ that defines f^{-1} .

(b) Show that $f^{-1}(f(x)) = x$.

(c) Show that $f(f^{-1}(x)) = y$.

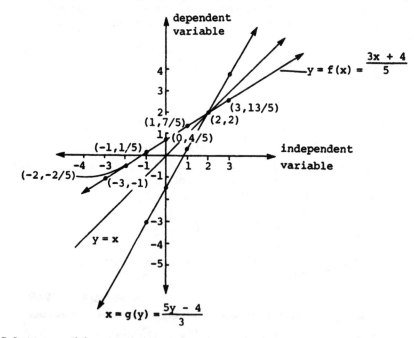

<u>Solution:</u> (a) The definition of a function f is a set of ordered
pairs (x,y) where

1) x is an element of a set X
2) y is an element of a set Y, and
3) no two pairs in f have the same first element.

By definition, f is the infinite set of ordered pairs

$$\left\{\left(x,\ \frac{3x + 4}{5}\right) \middle| x \in R\right\},$$

which includes $(0,4/5)$, $(2,2)$, $(7,5)$, $(12,8)$, $(-3,-1)$, etc. Furthermore, no two ordered pairs have the same first element . That is, for each element of X a unique value of Y is assigned. For example, if $x = 0$, we obtain only one y value, $4/5$.

We construct the following table to calculate the x and corresponding y values. Note that x is the independent variable and y is the dependent variable.

x	$\dfrac{3x+4}{5}$ =	y	(x,y)
-3	$\dfrac{3(-3)+4}{5} = \dfrac{-5}{5}$	-1	$(-3,-1)$
-2	$\dfrac{3(-2)+4}{5} = \dfrac{-2}{5}$	$\dfrac{-2}{5}$	$\left(-2, \dfrac{-2}{5}\right)$
-1	$\dfrac{3(-1)+4}{5} = \dfrac{1}{5}$	$\dfrac{1}{5}$	$\left(-1, \dfrac{1}{5}\right)$
0	$\dfrac{3(0)+4}{5} = \dfrac{4}{5}$	$\dfrac{4}{5}$	$\left(0, \dfrac{4}{5}\right)$
1	$\dfrac{3(1)+4}{5} = \dfrac{7}{5}$	$\dfrac{7}{5}$	$\left(1, \dfrac{7}{5}\right)$
2	$\dfrac{3(2)+4}{5} = \dfrac{10}{5}$	2	$(2,2)$
3	$\dfrac{3(3)+4}{5} = \dfrac{13}{5}$	$\dfrac{13}{5}$	$\left(3, \dfrac{13}{5}\right)$

See the accompanying figure, which shows the graph of the function f $\left(\text{which is also the graph of the equation } y = \frac{3x+4}{5}\right)$. We can say that f carries (or maps) any real number x into the number $\dfrac{3x+4}{5}$:

$$f:\ x \to \frac{3x + 4}{5}\ .$$

Now to find the inverse function, we must find a function which takes each element of the original set Y and relates it to a unique value of X. There cannot be two values of X for a given value of Y in order for the inverse function to exist. That is, if this is true: $\left(x_1, y\right)$ and $\left(x_2, y\right)$, then there is no f^{-1}.

To find $x = g(y)$, we solve for x in terms of y.

Given: $\qquad y = \dfrac{3x + 4}{5}$

Multiply both sides by 5 ,

$$5y = 3x + 4$$

Subtract 4 from both sides,

$$5y - 4 = 3x$$

Divide by 3 and solve for x,

$$x = \frac{5y - 4}{3} = f^{-1}(x) = g(y)\ .$$

Choose y values and find their corresponding x-values, as shown in the following table. Note that y is the independent variable and x is the dependent variable.

y	$g(y)=\dfrac{5y-4}{3}=$	x	(y,x)
-3	$\dfrac{5(-3)-4}{3}$	$\dfrac{-19}{3}=-6\dfrac{1}{3}$	$\left(-3,-6\dfrac{1}{3}\right)$
-2	$\dfrac{5(-2)-4}{3}$	$\dfrac{-14}{3}=-4\dfrac{2}{3}$	$\left(-2,-4\dfrac{2}{3}\right)$
-1	$\dfrac{5(-1)-4}{3}$	$\dfrac{-9}{3}=-3$	$(-1,-3)$
0	$\dfrac{5(0)-4}{3}$	$\dfrac{-4}{3}=-1\dfrac{1}{3}$	$\left(0,-1\dfrac{1}{3}\right)$
1	$\dfrac{5(1)-4}{3}$	$\dfrac{1}{3}$	$\left(1,\dfrac{1}{3}\right)$
2	$\dfrac{5(2)-4}{3}$	2	$(2,2)$
3	$\dfrac{5(3)-4}{3}$	$\dfrac{11}{3}=3\dfrac{2}{3}$	$\left(3,3\dfrac{2}{3}\right)$

See graph. Since there is only one value of y for each value of x, this equation defines the inverse function f^{-1}. The graph of f^{-1} is the image of the graph of f in the mirror y = x.

(b) Given $f(x) = \dfrac{3x + 4}{5} = y.$

Then perform the operation of f^{-1} on y = f(x) where $f^{-1}(x) = \dfrac{5y - 4}{3}$. That is, substitute for y: $\dfrac{3x + 4}{5}$.

$$f^{-1}\big(f(x)\big) = f^{-1}\left(\dfrac{3x + 4}{5}\right) = \dfrac{5\left(\dfrac{3x + 4}{5}\right)-4}{3} = \dfrac{\dfrac{15x + 20}{5}-4}{3}$$

$$= \dfrac{\dfrac{15x + 20 - 20}{5}}{3} = \dfrac{3x}{3} = x , \qquad \text{or}$$

$$f^{-1}\big(f(x)\big) = \dfrac{5f(x) - 4}{3} = \dfrac{5\left(\dfrac{3x + 4}{5}\right) - 4}{3} = x .$$

(c) We now perform the operation of f on $f^{-1}(x)$. Substitute for $f^{-1}(x)$: $\dfrac{5y - 4}{3} = x.$ Note $f(x) = \dfrac{3x + 4}{5}$

$$f\big(f^{-1}(x)\big) = f\left(\dfrac{5y - 4}{3}\right) = \dfrac{3\left(\dfrac{5y - 4}{3}\right)+ 4}{5}$$

$$= \dfrac{5y - 4 + 4}{5} = \dfrac{5y}{5} = y = f(x)$$

Comment. Since $f = \left\{\left(x, \dfrac{3x + 4}{5}\right)\right\}$, this function f may be thought of as a sequence of directions listing the operations that must be per- formed on x to get $\dfrac{3x + 4}{5}$. These operations are, in order: take any number x, multiply it by 3, add 4, and then divide by 5. The inverse function

$$f^{-1} = \left\{\left(y, \dfrac{5y - 4}{3}\right)\right\}$$

tells us to multiply by 5, subtract 4, and then divide by 3. This "undoes," in reverse order, the operations performed by f. The function f^{-1} could be called "the undoing function" because it undoes what the function f has done.

CHAPTER 12

LINEAR INEQUALITIES

> **Basic Attacks and Strategies for Solving Problems in this Chapter. See pages 153 to 176 for step-by-step solutions to problems.**

There is a wide variety of techniques for solving different kinds of inequalities. The most basic technique is the use of the properties of inequality. The properties can be summarized by saying that, except in two cases, if a certain inequality originally exists between two quantities and the both quantities are changed in the same way (by one of the four fundamental operations), then the same kind of inequality exists between the resulting quantities. The two exceptions are that whenever you multiply or divide both sides of an inequality by the same negative number, the direction of the inequality is reversed. An example of applying the basic technique is as follows:

$$-3x + 5 > -7$$
$$-3x + 5 + (-5) > -7 + (-5)$$
$$-3x > -12$$
$$-3x/(-3) < -12/(-3)$$
$$x < 4$$

The procedure for graphing inequalities involving one variable includes the following steps:

(1) Solve the inequality as shown in the above example.

(2) If the equation is of the form $x < a$, draw a number line and draw the endpoint a as an open circle to show that it is not a part of the graph.

(3) Draw a solid line from the open circle to the left to show the numbers included in the graph.

If the form of the solution is $x > a$, then a solid line is drawn from the open circle to the right. If the form of the solution of the inequality is $x \leq a$ or $x \geq a$, then the only change from the above procedure is a closed or solid circle in the respective graphs.

The procedure for graphing a compound inequality, that is, two inequalities connected by "and" and "or," begins by graphing each inequality separately. If the two inequalities are connected by the word "or," then we graph the union of all points on either graph on a real number line. If the two inequalities are connected by the word "and," then we graph their intersection, or the parts they have in common, on a real number line.

The procedure for graphing a linear inequality in two variables involves the following steps:

(1) Replace the inequality symbol with an equal sign. The resulting equality represents the boundary for the solution set.

(2) Graph the boundary found in Step 1 using a solid line if the boundary is included in the solution set (that is, if the original inequality symbol was either $\leq$ or $\geq$). Use a broken line to graph the boundary if it is not included in the solution set. (It is not included if the original inequality was either $<$ or $>$.)

(3) Choose any convenient point not on the boundary and substitute the coordinates into the original inequality. If the resulting statement is true, the graph (shaded) lies on the same side of the boundary as the chosen point. If the resulting statement is false, the graph (shaded) lies on the opposite side of the boundary.

To solve an inequality that involves absolute value, one first isolates the absolute value on the left side of the inequality symbol. Then, rewrites the absolute value inequality as an equivalent continued or compound inequality that does not contain absolute value symbols. In general, if a is a positive number, then $|x| < a$ is equivalent to $-a < x < a$ and $|x| > a$ is equivalent to $a < -x$ or $x > a$.

Step-by-Step Solutions to Problems in this Chapter, "Linear Inequalities"

SOLVING INEQUALITIES AND GRAPHING

● PROBLEM 251

Solve the inequality $2x + 5 > 9$.

Solution: $2x + 5 + (-5) > 9 + (-5)$. Adding -5 to both sides.

$2x + 0 > 9 + (-5)$ Additive inverse property

$2x > 9 + (-5)$ Additive identity property

$2x > 4$ Combining terms

$\frac{1}{2}(2x) > \frac{1}{2} \cdot 4$ Multiplying both sides by $\frac{1}{2}$.

$x > 2$

The solution set is

$$X = \{x \mid 2x + 5 > 9\}$$
$$= \{x \mid x > 2\}$$

(that is all x, such that x is greater than 2).

● PROBLEM 252

Determine the values of x for which $3x + 2 < 0$.

Solution: We may add - 2 to both members, to give $3x < - 2$. We may then multply both sides of the inequality by 1/3; hence $x < - 2/3$, which is the solution. In other words, the solution consists of the set of all numbers which are less than - 2/3, and can be expressed

in solution set notation: $\{x : x < - 2/3\}$, (meaning the set of all x such that x is less than - 2/3).

● PROBLEM 253

Solve the inequality $2x - 5 > 3$.

Solution: $2x - 5 > 3$ **Given**

Add 5 to both sides of the given inequality.

$2x - 5 + 5 > 3 + 5$

Therefore: $2x > 8$ (1).

Divide both sides of inequality (1) by 2. Dividing both sides of an inequality by a positive number does not change the direction of the inequality.

$$\frac{2x}{2} > \frac{8}{2}$$

Therefore: x > 4, and x is any real number greater than 4.

● PROBLEM 254

Solve 4 - 5x < - 3.

Solution: 4 - 5x < - 3   (1)

Subtract 4 from both sides of inequality (1).

4 - 5x - 4 < - 3 - 4

Therefore: - 5x < - 7   (2)

Divide both sides of inequality (2) by - 5. Dividing both sides of an inequality by a negative number changes the direction of the inequality.

$$\frac{-5x}{-5} > \frac{-7}{-5}$$

Therefore: x > $\frac{7}{5}$.

Hence, x is any real number greater than $\frac{7}{5}$.

● PROBLEM 255

Solve the inequality $\frac{1}{3}$x + 6 < 2.

Solution: $\frac{1}{3}$x + 6 < 2   (1)

Subtract 6 from both sides of inequality (1).

$\frac{1}{3}$x + 6 - 6 < 2 - 6

Therefore: $\frac{1}{3}$x < - 4   (2).

Multiply both sides of inequality (2) by 3. Multiplying both sides of an inequality by a positive number does not change the direction of the inequality.

$$3\left(\frac{1}{3}x\right) < 3(-4)$$

Therefore: x < - 12 and x is any real number less than - 12.

● PROBLEM 256

Illustrate one (a) conditional inequality, (b) identity, and (c) inconsistent inequality.

Solution: (a) A conditional inequality is an inequality whose validity depends on the values of the variables in the sentence. That is, certain values of the variables will make the sentence true, and others will make it false. $3 - y > 3 + y$ is a conditional inequality for the set of real numbers, since it is true for any replacement less than zero and false for all others.

(b) $x + 5 > x + 2$ is an identity for the set of real numbers, since for any real valued x, the expression on the left is greater than the expression on the right.

(c) $5y < 2y + y$ is inconsistent for the set of non-negative real numbers. For any x greater than 0 the sentence is always false. A sentence is inconsistent if it is always false when its variables assume allowable values.

● PROBLEM 257

Solve the inequality $4x + 3 < 6x + 8$.

Solution: In order to solve the inequality $4x + 3 < 6x + 8$, we must find all values of x which make it true. Thus, we wish to obtain x alone on one side of the inequality.

Add − 3 to both sides:

$$\begin{array}{r} 4x + 3 < 6x + 8 \\ -\ 3 \qquad\quad -\ 3 \\ \hline 4x < 6x + 5 \end{array}$$

Add − 6x to both sides:

$$\begin{array}{r} 4x < \quad 6x + 5 \\ -\ 6x \qquad -\ 6x \\ \hline -\ 2x < \quad 5 \end{array}$$

In order to obtain x alone we must divide both sides by (− 2). Recall that dividing an inequality by a negative number reverses the inequality sign, hence

$$\frac{-\ 2x}{-\ 2} > \frac{5}{-\ 2}$$

Cancelling $\frac{-\ 2}{-\ 2}$ we obtain, $x > -\frac{5}{2}$

Thus, our solution is $\left(x : x > -\frac{5}{2}\right)$ (the set of all x such that x is greater than $-\frac{5}{2}$).

● PROBLEM 258

Solve the inequality $4x - 5 \geq - 6x + 5$.

$$\{x \mid x \geq 1\} = [1, \infty)$$

155

Solution: To solve this compound statement we solve for x as follows:

Adding 5 to both sides of the given inequality we have:

$$4x \geq -6x + 10$$

Adding 6x to both sides: $10x \geq 10$

Multiplying both sides by $\frac{1}{10}$: $x \geq 1$

Therefore $S = \{x \mid x \geq 1\}$

The region representing this set on the number line is shown in the diagram.

You should note that the bracket in the graph includes the point 1.

● PROBLEM 259

Find the solution set of inequality $5x - 9 > 2x + 3$.

Solution: To find the solution set of the inequality $5x - 9 > 2x + 3$, we wish to obtain an equivalent inequality in which each term in one member involves x, and each term in the other member is a constant. Thus, if we add $(- 2x)$ to both members, only one side of the inequality will have an x term:

$$5x - 9 + (- 2x) > 2x + 3 + (- 2x)$$

$$5x + (-2x) - 9 > 2x + (- 2x) + 3$$

$$3x - 9 > 3$$

Now, adding 9 to both sides of the inequality we obtain,

$$3x - 9 + 9 > 3 + 9$$

$$3x > 12$$

Dividing both sides by 3, we arrive at $x > 4$.

Hence the solution set is $\{x \mid x > 4\}$, and is pictured in the accompanying figure.

● PROBLEM 260

Solve $3(x + 2) < 5x$.

Solution:

$3(x + 2) < 5x$	Given
$3x + 6 < 5x$	Distributive property
$-2x < -6$	Transposition (adding (-6) and (-5x) to both sides of the inequality and simplifying)

$$x > 3$$

Multiplicative property (multi-ply both members of the previous inequality by $-\frac{1}{2}$). Note that multiplying both sides of an inequality by a negative number changes the sense of the in-equality.

The solution set is $\{x: x > 3\}$ (see figure).
The unshaded circle above 3 on the number line indicates that $x = 3$ is not included in the solution set.

● **PROBLEM** 261

Solve $2(x + 1) < 4$.

The solution set of $2(x + 1) < 4$ is $\{x: x < 1\}$. The graph of the solution set can be seen in the accompanying figure.

Solution:		
$2(x + 1) < 4$		Given
$2x + 2 < 4$		Distributive property
$2x < 2$		Additive property (with -2)
$x < 1$		Multiplicative property (with $\frac{1}{2}$)

The solution set of $2(x + 1) < 4$ is $\{x: x < 1\}$. The graph of the solution set can be seen in the accompanying figure.

The solution set of $x < 1$ is equal to the solution set of $2(x + 1) < 4$ because the inequalities are equivalent. We have solved an inequality when we know its solution set.

● **PROBLEM** 262

Solve $-3(x - 5) > x + 7$.

Solution:

$-3(x - 5) > x + 7$	Given
$-3x + 15 > x + 7$	$a(b - c) = ab - ac$, Distributive Law
$-4x > -8$	Subtracting 15 and x from both members of the inequality, Transposition
$x < 2$	Multiplying both members of the

inequality by $-\frac{1}{4}$, Multiplicative Property. Note that multiplying both members of an inequality by a negative number changes the sense of the inequality.

The solution set is $\{x: x < 2\}$ and the graph can be seen in the accompanying figure.

The unshaded circle above 2 on the number line indicates that $x = 2$ is not included in the solution set.

● **PROBLEM** 263

Solve the inequality $3x - 4 < 5x + 7$.

Solution: By subtracting 3x from both sides of the given inequality, we obtain the equivalent inequality

157

$-4 < 2x + 7$. Now we subtract 7 from both sides, and we obtain the equivalent inequality $-11 < 2x$. Finally, we divide by the positive number 2, and we have $-11/2 < x$. Thus, our solution set consists of all points

greater than $\frac{-11}{2}$, $\left\{ x : x > \frac{-11}{2} \right\}$, as pictured in the

accompanying figure.

● **PROBLEM** 264

Solve the inequality $-5(x - 1) \geq 3(x - 3)$.

<u>Solution:</u> $-5x + 5 \geq 3x - 9$ Distributing

$-5x + 5 + (-5) \geq 3x - 9 + (-5)$ Adding (-5) to both sides

$-5x \geq 3x - 9 + (-5)$ Additive inverse property

$-5x \geq 3x - 14$ Combining terms

$-5x + (-3x) \geq 3x - 14 + (-3x)$ Adding $(-3x)$ to both sides

$-5x + (-3x) \geq -14 + 3x + (-3x)$ Commuting

$-5x + (-3x) \geq -14$ Additive inverse property

$-8x \geq -14$ Simplifying

$\left(-\frac{1}{8}\right)(-8x) \leq (-14)\left(-\frac{1}{8}\right)$ Multiplying both sides by $-\frac{1}{8}$

Notice that multiplying by a negative number reverses the inequality, that is it goes from greater than and equal to, to less than and equal to.

$$x \leq \frac{14}{8}$$

Reducing to lowest terms $x \leq \frac{7}{4}$

The solution set is

$$X = \{x \mid -5(x - 1) \geq 3(x - 3)\}$$
$$= \left\{ x \mid x \leq \frac{7}{4} \right\},$$

that is, all x such that X is less than or equal to $\frac{7}{4}$. The solution is pictured above.

● **PROBLEM** 265

Solve $\frac{1}{6}x - 3 < \frac{3}{4}x + \frac{1}{2}$.

Solution: We can eliminate the fractional coefficients by multiplying both members of the inequality by the least common denominator of the fractions. Multiplying by 12, we have

$$2x - 36 < 9x + 6.$$

Isolating the constant terms on the left side of the inequality sign and the x-terms on the right side by transposition, we have:

$$-36 - 6 < 9x - 2x .$$

Then simplifying:

$$-42 < 7x.$$

Dividing both members of this inequality by 7 yields:

$$-6 < x$$

Hence, the solution set is $\{x: x > -6\}$ and the graph is shown in the accompanying figure.

The unshaded circle above -6 on the number line indicates that $x = -6$ is not included in the solution set.

● **PROBLEM 266**

Solve $7\left(\frac{2}{3}x - 1\right) > 2(x - 6)$

Solution: $\qquad 7\left(\frac{2}{3}x - 1\right) > 2(x - 6) \qquad (1)$

$$-\frac{14}{3}x - 7 > 2x - 12 \qquad (2) \text{ Distributive Property}$$

Subtract 2x from both sides of inequality (2).

$$\frac{14}{3}x - 7 - 2x > 2x - 12 - 2x$$

Therefore: $\frac{8}{3}x - 7 > -12 \qquad (3),$

Add 7 to both sides of inequality (3),

$$\frac{8}{3}x - 7 + 7 > -12 + 7$$

Therefore: $\frac{8}{3}x > -5 \qquad (4),$

Multiply both sides of inequality (4) by $\frac{3}{8}$.

Multiplying both sides of an inequality by a positive number does not change the direction of the inequality. Therefore:

$$\frac{3}{8}\left(\frac{8}{3}x\right) > \frac{3}{8}(-5) \quad \text{and} \quad x > -\frac{15}{8}.$$

Hence, x is any real number greater than $-\frac{15}{8}$.

● **PROBLEM 267**

Solve the inequality

159

$$\frac{1}{x-1} > \frac{1}{3} .$$

Key

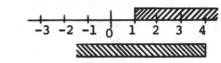

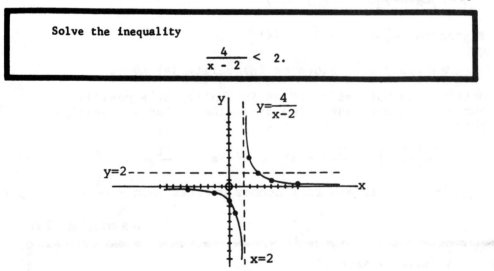

///// $x > 1$

\\\\\ $x < 4$

Solution: Since the fraction $\frac{1}{x-1} > \frac{1}{3}$; that is, since the fraction $\frac{1}{x-1}$ is greater than 0, $x - 1$ must be positive. Hence, $x - 1 > 0$. If both sides of the given equation are multiplied by $3(x - 1)$, then:

$$3(x - 1) \frac{1}{x-1} > 3(x - 1)\left(\frac{1}{3}\right)$$

$$3 > x - 1 .$$

Note that multiplying both sides of an inequality by a positive number (in this case, $3(x - 1)$) does not change the sign of the inequality.

Now we have the double restrictions

$$x - 1 > 0 \quad \text{and} \quad 3 > x - 1$$

and the solution set is the intersection of the solution sets of these two inequalities. Solving each of the two inequalities, we find that

$$x > 1 \quad \text{and} \quad x < 4$$

The solution set is the intersection of the two inequalities, as can be seen on a number line (see diagrams). The intersection of these two inequalities is the set $1 < x < 4$. Hence, the solution set is:

$$X = \{x \mid 1 < x < 4\}$$

(The endpoints $X = 1$ and $X = 4$ are not included in the solution set).

● **PROBLEM** 268

Solve the inequality

$$\frac{4}{x-2} < 2.$$

<u>Solution:</u> The inequality is meaningless for x = 2 because when x = 2 the denominator of the left member is 0, making the fraction undefined.

If x > 2, x - 2 is positive (since x > 2 is equivalent to x - 2 > 0), and multiplication of the given inequality by x - 2 yields

$$4 < 2(x - 2)$$
$$4 < 2x - 4$$
$$8 < 2x$$
$$4 < x$$
$$x > 4.$$

Thus, the solution is the intersection of x > 2 and x > 4, x > 2 ∩ x > 4, which is {x|x > 4}.

If x < 2, x - 2 is negative (since x < 2 is equivalent to x - 2 < 0), and multiplication by x - 2 yields

$$4 > 2(x - 2)$$

because multiplication by a negative number reverses an inequality. Distributing, $4 > 2x - 4$

Adding 4 to both sides,
$$8 > 2x$$

Dividing both sides by 2,

$$4 > x, \text{ or } x < 4.$$

Thus the solution is the intersection of x < 2 and x < 4, x < 2 ∩ x < 4, which is
{x|x < 2}. Hence
$$\frac{4}{x - 2} < 2$$

if x < 2 or if x > 4.

A graphical solution of the problem (see diagram) can be obtained by sketching the equilateral hyperbola y = 4/(x - 2) and the line y = 2. The hyperbola may be sketched from its vertical asymptote x = 2, its horizontal asymptote y = 0, its intercepts x = 0, y = -2, and a few other points obtained by substitution and symmetry. It is then possible to observe the values of x for which the hyperbola is below the line, namely,
$$x < 2 \text{ and } x > 4.$$

The same diagram also shows that [4/(x - 2)] > 2 for 2 < x < 4.

● PROBLEM 269

Find the solution set of the disjunction

$$2 - 3x > 5 \text{ or } 2x - 1 > 5.$$

<u>Solution:</u> A disjunction is a compound sentence using the connective 'or'. The union of the solution sets of the two sentences comprising the compound sentence is the solution set of the disjunction. For this problem the solution set is

$${x: 2 - 3x > 5} \cup {x: 2x - 1 > 5}$$

We solve each inequality independently and find the union of their solution set:

$$2 - 3x > 5 \text{ or } 2x - 1 > 5$$
$$-3x > 3 \text{ or } 2x > 6$$

Solving for x, we divide both members of the inequality by a negative
number (-3). Therefore the direction of the inequality is reversed.

$$x < -1 \quad \text{or} \quad x > 3$$

The solution set of $2 - 3x > 5$ or $2x - 1 > 5$ is shown on the graph.
The unshaded circles above -1 and 3 on the number line
indicate that these values are not included in the solution set.

Compound sentences can also be formed by connecting two sentences with
the word 'and.' A compound sentence using the connective and is
called a conjunction. The solution set of a conjunction is the set of
replacements that are common to the solution sets of the sentences
making-up the conjunction (their intersection). We may write the
solution set as:

$$\{x: \ x > a\} \cap \{x: \ x < b\} = \{x: \ a < x < b\}.$$

● **PROBLEM 270**

Find the solution set of the conjunction

$$\tfrac{1}{2}x + 1 > 3 \quad \text{and} \quad x > 2x - 6$$

Solution: The solution set must be such that x satisfies both inequalities simultaneously. The solution set of the conjunction is

$$\{x: \ \tfrac{1}{2}x + 1 > 3\} \cap \{x: \ x > 2x - 6\}$$

$$\tfrac{1}{2}x + 1 > 3 \quad \text{and} \quad x > 2x - 6$$

$$\tfrac{1}{2}x > 2 \quad \text{and} \quad -x > -6$$

$$x > 4 \quad \text{and} \quad x < 6$$

(multiplying by negative 1 reverses the inequality).

Note: if $x = 4$ the sentence $\tfrac{1}{2}x + 1 > 3$ becomes false, i.e.,

$$\tfrac{1}{2}(4) + 1 = 3 > 3 \quad \text{false}$$

and if $x = 6$ the sentence $x > 2x - 6$ becomes false

also $\quad (6 > 2(6) - 6 = 6 \quad \text{false}).$

Therefore the solution set cannot include these two points, and
the values of x which make both sentences true simultaneously are
the inequalities x greater than but not equal to 4 and less
than but not equal to 6. That is $\{x: \ 4 < x < 6\}$. (See the number
line).

● **PROBLEM 271**

If $1 < a$, show that $a < a^2$.

Solution: We are given a>1, and we know 1>0, thus using
the transitive property (if x>y and y>z, then x>z) we conclude a>o. Since a is positive we may multiply both sides
of the inequality 1<a by a to obtain an equivalent inequality,

$$1 \cdot a < a \cdot a$$

$$a < a^2$$

Thus we have shown if,
$$1 < a, a < a^2.$$

● PROBLEM 272

Show that if $0 < a < 1$, then $a^2 < a$.

Solution: The relation $0 < a < 1$ means that $a > 0$ and $a < 1$. Thus a is positive and less than one. Since a is positive we may multiply both sides of the inequality, $a < 1$, by a.

$$a(a) < a(1)$$
hence
$$a^2 < a .$$

(To visualize this concept pick a number between 0 and 1. If we choose $\frac{1}{2}$, then
$$0 < \tfrac{1}{2} < 1, \text{ and } (\tfrac{1}{2})^2 < \tfrac{1}{2} \text{ or } \tfrac{1}{4} < \tfrac{1}{2}).$$

● PROBLEM 273

Prove that if $a > b > 0$, then
$$\frac{1}{a} < \frac{1}{b}.$$

Solution: Since a and b are both positive (given), ab is positive because the product of two positive numbers is always positive. Now, since $ab > 0$, we may divide both sides of $a > b$ by ab to obtain

$$\frac{a}{ab} > \frac{b}{ab}.$$

Cancelling like terms in numerator and denominator,

$$\frac{1}{b} > \frac{1}{a},$$

which is equivalent to $\frac{1}{a} < \frac{1}{b}$. To complete the proof, the student should check that the steps are reversible, as follows:

Check. $\frac{1}{b} > \frac{1}{a}$. Multiply both sides of the inequality by the least common denominator obtained by multiplying the two denominators together.

$$\frac{ab}{b} > \frac{ab}{a} = a > b.$$

● PROBLEM 274

Under what conditions does the inequality $1/a \le 1/b$ imply that $b \le a$?

Solution: If $ab > 0$, we may multiply both sides of the first inequality by ab to obtain
$$ab\left(\frac{1}{a}\right) \le ab\left(\frac{1}{b}\right) \quad \text{or} \quad b \le a ,$$

163

whereas if ab < 0, such multiplication produces the reverse inequality a ≤ b. Therefore, the inequality 1/a ≤ 1/b implies that b ≤ a provided that ab > 0. We don't have to consider the case ab = 0 because if ab = 0, either a = 0 or b = 0, which would mean that one side of our original inequality 1/a ≤ 1/b would be undefined (as it would be in the form 1/0).

● **PROBLEM** 275

Solve the inequality $\sqrt{x-3} \leq 2 - \sqrt{x+1}$.

<u>Solution:</u>

$\sqrt{x-3} \leq 2 - \sqrt{x+1}$	Given
$x - 3 \leq 4 - 4\sqrt{x+1} + x + 1$	Squaring
$-8 \leq -4\sqrt{x+1}$	Transposing and simplifying
$2 \geq \sqrt{x+1}$	Dividing by −4.

Note that dividing both sides of an inequality by a negative number changes the sense of the inequality.

$4 \geq x + 1$	Squaring
$3 \geq x$ or $x \leq 3$	Solving for x

Check: If x = 3, $\sqrt{3-3} = 2 - \sqrt{3+1}$.

If x > 3, the left member, $\sqrt{x-3}$, is positive, but the right member, $2 - \sqrt{x+1}$, is negative. Hence, the inequality is not satisfied. Nor is the inequality satisfied if x < 3, for in this case the left member is not a real number. Hence, the solution set is {3}.

If the inequality were a strict inequality, the solution set would be the null set. If the left member is to be a real number, it must be positive and x must be greater than 3, but in this case the right member must be negative. However, it is impossible for a positive number to be less than a negative number.

● **PROBLEM** 276

What is the set {x < -2} ∩ {x > 3}?

-5 -4 -3 -2 -1 0 1 2 3 4 5

<u>Solution:</u> An element belongs to the intersection of two sets,if, and only if, it belongs to both of them. Thus, in order for a number to belong to our intersection, it would have to be both less than -2 and greater than 3. There is no such number, so the intersection is the empty set; that is,

$$\{x < -2\} \cap \{x > 3\} = \emptyset .$$

This can be seen from the accompanying number line representation, which illustrates that the two graphs have no points in common.

INEQUALITIES WITH TWO VARIABLES

● **PROBLEM** 277

Solve 2x - 3y ≥ 6

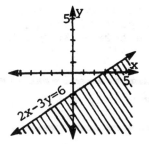

Solution: The statement $2x - 3y \geq 6$ means $2x - 3y$ is greater than or equal to 6. Symbolically, we have $2x - 3y > 6$ or $2x - 3y = 6$. Consider the corresponding equality and graph $2x - 3y = 6$. To find the x-intercept, set $y = 0$

$$2x - 3y = 6$$
$$2x - 3(0) = 6$$
$$2x = 6$$
$$x = 3$$

{3,0} is the x-intercept.

To find the y-intercept, set x=0

$$2x - 3y = 6$$
$$2(0) - 3y = 6$$
$$-3y = 6$$
$$y = -2$$

{0,-2} is the y-intercept.

A line is determined by two points. Therefore draw a straight line through the two intercepts {3,0} and {0,-2}. Since the inequality is mixed, a solid line is drawn through the intercepts. This line represents the part of the statement $2x - 3y = 6$.

We must now determine the region for which the inequality $2x - 3y > 6$ holds.

Choose two points to decide on which side of the line the region $x - 3y > 6$ lies. We shall try the points (0,0) and (5,1).

For (0,0)	For (5,1)
$2x - 3y > 6$	$2x - 3y > 6$
$2(0) - 3(0) > 6$	$2(5) - 3(1) > 6$
$0 - 0 > 6$	$10 - 3 > 6$
$0 > 6$	$7 > 6$
False	True

The inequality, $2x - 3y > 6$, holds true for the point (5,1). We shade this region of the xy-plane. That is, the area lying below the line $2x - 3y = 6$ and containing (5,1).

165

Therefore, the solution contains the solid line, 2x - 3y = 6, and the part of the plane below this line for which the statement 2x - 3y > 6 holds.

Solve the inequality x + 2y ≥ 6 for y in terms of x and draw its graph.

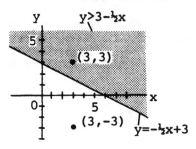

Solution: To solve for y in terms of x, obtain y alone on one side of the inequality and x and any constants on the other. Subtracting x from both sides of

x + 2y ≥ 6 gives 2y ≥ 6 - x

Divide the equation by 2

$y \geq 3 - \frac{1}{2} x$

The points in the x - y plane which will satisfy this equation are those satisfying

$y > 3 - \frac{1}{2} x$ and $y = 3 - \frac{1}{2} x.$

Consider the case,

$y = 3 - \frac{1}{2} x$

which is a graph of the solid straight line with y -intercept 3 and slope - ½. (See diagram.)

We must also find those points which satisfy

$y > 3 - \frac{1}{2} x.$

Choose two points which lie on either side of the line

$y = 3 - \frac{1}{2} x$ to find the region where

$y > 3 - \frac{1}{2} x.$

We shall choose (3, 3) and (3, - 3) (see diagram).

For (3, 3)   For (3, - 3)

$y > 3 - \frac{1}{2} x$ $y > 3 - \frac{1}{2} x$

$3 > 3 - \frac{1}{2}$ (3) $-3 > 3 - \frac{1}{2}$ (3)

$3 > \frac{3}{2}$ $-3 \nmid \frac{3}{2}$

(3, 3) satisfies the   (3, - 3) does not
inequality.   satisfy the inequality.

    Thus, all the points in the region where (3, 3)
lies satisfy $y > 3 - \frac{1}{2} x$. That is all those points above
the line $y = 3 - \frac{1}{2} x$ satisfy $y > 3 - \frac{1}{2} x$ and lie in the
shaded area.

    Consequently, the graphical solution of
$y \geq - \frac{1}{2} x + 3$ are those points which lie on the solid

line $y = - \frac{1}{2} + 3$ and those points in the shaded area
$y > - \frac{1}{2} x + 3$.

● **PROBLEM 279**

Solve the inequality 2x - y > 4 for y in terms of x, and
draw its graph.

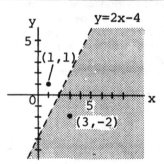

<u>Solution:</u> To solve for y in terms of x obtain y on one
side of the inequality and x on the other. Given

        2x - y > 4   (1)

    Add - 2x to both sides of

        2x - y > 4   (1)

We obtain - y > 4 - 2x   (2)

    Multiply (2) by - 1 and reverse the inequality
sign since we are multiplying by a negative number. We
obtain y in terms of x

        y < - 4 + 2x   (3)

    Rewriting (3)

        y < 2x - 4   (3)

Graphing (3) consider the equation first as an equality

$$y = 2x - 4 \qquad\qquad (4)$$

We draw the graph of (4) as a dotted line since the points of the given inequality $y < 2x - 4$ do not satisfy $y = 2x - 4$. To draw $y = 2x - 4$, we note the slope is 2 and the y-intercept is -4.

To determine what region of the $x - y$ plane satisfies $y < 2x - 4$ choose a point on either side of the dotted line. Let us take the points $(3, -2)$ and $(1, 1)$ (see diagram). Substitute these points into the given inequality and see which point will satisfy it.

For $(3, -2)$ For $(1, 1)$

$\quad y < 2x - 4$ $\quad y < 2x - 4$

$\quad -2 < 2(3) - 4$ $\quad 1 < 2 - 4$

$\quad -2 < 2$ $\quad 1 \nless -2$

Now hatch in that portion of the plane containing $(3, -2)$. All the points to the right of the dotted line $y = 2x - 4$ will satisfy the given inequality.

● **PROBLEM** 280

A livestock farmer has 500 acres to devote to grazing. He estimates that cattle require 5 acres per head and sheep require 3 acres per head. He has winter shelter facilities for 40 head of cattle and for 125 sheep. What constraints are imposed on the number of cattle and sheep he can raise?

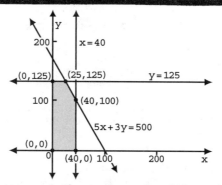

Solution: Let x represent the number of cattle raised and y the number of sheep. Since he cannot raise a negative number of either cattle or sheep, we have the constraints

$$x \geq 0 \qquad\qquad (1)$$
$$y \geq 0 \qquad\qquad (2)$$

Since $5x$ acres are required for the cattle and $3y$ acres for the sheep and there are only 500 acres available, we have

$$5x + 3y \leq 500 \qquad\qquad (3)$$

Since he can winter only 40 cattle,

$$x \leq 40 \qquad\qquad (4)$$

Since he can winter only 125 sheep,

$$y \leq 125 \qquad\qquad (5)$$

Relations (1) through (5) are the constraints.

The graph of the constraints in this example is a convex set of points. The corner points of the shaded polygon are (0,0), (40,0), (40,100), (25,125), and (0,125).

INEQUALITIES COMBINED WITH ABSOLUTE VALUES

● **PROBLEM** 281

Express the inequality $|x| < 3$ without using absolute value signs.

Solution: According to the law of absolute values which states that $|a| < b$ is equivalent to $-b < a < b$, where b is any positive number, $|x| < 3$ is equivalent to $-3 < x < 3$.

● **PROBLEM** 282

Solve the inequality $|5 - 2x| > 3$.

Solution: The property of absolute values states that $|a| = +a$ or $|a| = -a$. Therefore: $|5 - 2x| = 5 - 2x$ or $-(5 - 2x)$. Thus, the given inequality becomes two new inequalities:

$$5 - 2x > 3, \quad -(5 - 2x) > 3.$$

Now, we must solve for x in both inequalities. For the first, we subtract 5 from both sides of the inequality, and then divide by -2. We must keep in mind that division or multiplication by a negative number reverses the inequality sign. Thus, for $5 - 2x > 3$ we have:

$$5 - 5 - 2x > 3 - 5$$

$$-2x > -2$$

$$\frac{-2x}{-2} > \frac{-2}{-2}$$

$$x < 1.$$

For the second inequality, we first take the negative of all the terms inside the parentheses. Thus, for $-(5 - 2x) > 3$ we have:

$$-5 + 2x > 3.$$

Now, we add 5 to both sides of the inequality, and then divide by 2. Thus, we obtain:

$$-5 + 5 + 2x > 3 + 5$$

$$2x > 8$$

$$\frac{2x}{2} > \frac{8}{2}$$

$$x > 4.$$

Therefore, the above inequality holds when x < 1, and when x > 4.

● **PROBLEM** 283

Solve

$$\left|\frac{4x}{5} - 1\right| > 3$$

$|a| > b$

-b 0 b

Solution: We note the following about absolute values. If b is a nonnegative real number, then a is a real number for which $|a| > b$ if and only if a > b or a < - b. See the figure.

In the given example, this inequality is satisfied if either

$$\frac{4x}{5} - 1 > 3 \quad \text{or} \quad \frac{4x}{5} - 1 < - 3$$

is satisfied. By adding 1 to each member of these inequalities, we get

$$\frac{4x}{5} > 4 \quad \text{and} \quad \frac{4x}{5} < - 2$$

Hence, by multiplying by 5/4 in each case, we note that the original inequality is satisfied by values of x that are greater than 5 and by values of x that are less than - 5/2, that is, by x > 5 and by x < - 5/2.

The solution set is therefore

$$\{x|\ x > 5\} \cup \{x|x < - 5/2\}.$$

● **PROBLEM** 284

Find the solution set of the inequality

$$|- 2x + 6| > 8$$

Solution: By a property of inequalities involving absolute values, the solution set of the given inequality is the union of the solution sets of

- 2x + 6 > 8 and	- 2x + 6 < - 8
- 2x > 2	- 2x < - 14
x < - 1	x > 7

Hence the solution set is

$$\{x|x < - 1\} \cup \{x|x > 7\}.$$

Graph $\{x: |3x - 4| \geq 2\}$.

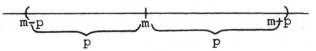

Solution: In general, the required graph of $\{x: |ax + b| \geq c\}$ is the union of two sets: $\{x: ax + b \geq c\} \cup \{x: ax + b \leq -c\}$. Therefore, the required graph of $\{x: |3x - 4| \geq 2\}$ is the union of two sets:

$$\{x: 3x - 4 \geq 2\} \cup \{x: 3x - 4 \leq -2\}$$

$$3x - 4 \geq 2 \quad \text{or} \quad 3x - 4 \leq -2$$

$$3x \geq 6 \quad \text{or} \quad 3x \leq 2$$

$$x \geq 2 \qquad\qquad x \leq 2/3$$

The solution set of $\{x: |3x - 4| \geq 2\}$ is

$$\{x: x \geq 2\} \cup \{x: x \leq 2/3\}$$

The graph of the solution set is the union of two rays. Notice that the shaded circles above 2/3 and 2 on the number line indicate that 2/3 and 2 are included in the solution set.

Express the inequality $|2x - 1| < 5$ without using absolute value signs.

Solution: If x is a number such that $|x - m| < p$, then x must lie in the interval between the points m - p and m + p; that is, if $|x - m| < p$, then $-p < x - m < p$. Then add +m to all the members of the inequality to obtain m - p < x < m + p. Observe that the point m is the midpoint of this interval and that the length of the interval is 2p. Note the number line.

$$\overset{\frown}{m-p} \qquad \overset{|}{m} \qquad \overset{\frown}{m+p}$$

$$\underbrace{\qquad}_{p} \underbrace{\qquad}_{p}$$

$$\text{length} = p + p = 2p$$

Therefore, the inequalities m - p < x < m + p and $|x - m| < p$ are equivalent. Hence, the given inequality $|2x - 1| < 5$ reduces to:

$$1 - 5 < 2x < 1 + 5 \quad \text{where x is replaced by 2x, m is}$$

replaced by 1, and p is replaced by 5.

Therefore,

$$-4 < 2x < 6$$

Now, divide each term of these inequalities by 2:

$$\frac{-4}{2} < \frac{2x}{2} < \frac{6}{2}$$

$$-2 < x < 3.$$

Solve the inequality $|2x - 3| \leq 4$.

$$x = [-1/2, 7/2]$$

-1/2 0   7/2

Solution: In general, $|ax + b| \leq c$ is equivalent to $-c \leq ax + b \leq c$. Therefore $|2x - 3| \leq 4$ implies that $-4 \leq 2x - 3 \leq 4$. This statement is the conjunction of the statement $-4 \leq 2x - 3$ and the statement $2x - 3 \leq 4$. Hence the solution set of the conjunction is the intersection of the solution sets of the two component propositions.

The computation may be arranged in the following manner:

$-4 \leq 2x - 3 \leq 4$

$-1 \leq 2x \leq 7$ Adding 3 to both sides of both inequalities

$-\frac{1}{2} \leq x \leq \frac{7}{2}$ Multiplying both sides of both inequalities by $\frac{1}{2}$

The solution set is represented on the number line as in the figure.

Solve $|3x - 1| \leq 8$.

-4 -3 -2 -1 0 1 2 3 4 5 6

Solution: Since $|a| = a$ if $a \geq 0$ and $|a| = -a$ if $a \leq 0$. We must solve two equations

$$3x - 1 \leq 8$$

$$-(3x - 1) \leq 8 \text{ or } 3x - 1 \geq -8.$$

(Note that multiplying an inequality by a negative number, i.e., -1, reverses the inequality.)
     The solution set will be the conjunction of the solution sets of each equation; that is,

$$\{x: 3x - 1 \leq 8\} \text{ and } \{x: 3x - 1 \geq -8\}.$$

We must find

$$\{x: 3x - 1 \leq 8\} \cap \{x: 3x - 1 \geq -8\}$$

$$3x - 1 \leq 8 \text{ and } 3x - 1 \geq -8$$

$$3x \leq 9 \text{ and } \quad 3x \geq -7$$

$$x \leq 3 \text{ and } \quad x \geq -2\frac{1}{3}.$$

The solution set is $\left\{x: -2\frac{1}{3} \leq x \leq 3\right\}$. See the figure.

Find the values of x satisfying the statement $\left|\frac{x}{3} - 7\right| \geq 5$.

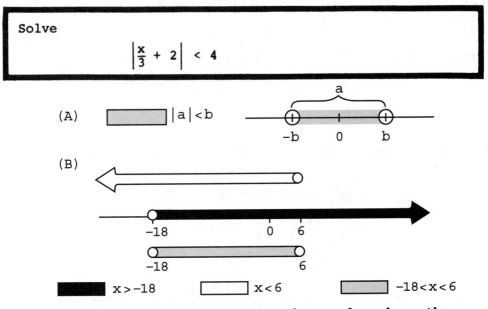

$(-\infty, 6] \cup [36, \infty)$

Solution: In general, $|ax + b| \geq c$ implies that $ax + b \geq c$ or $ax + b \leq -c$. Therefore, for

$$\left|\frac{x}{3} - 7\right| \geq 5 :$$

(1) $\frac{x}{3} - 7 \geq 5$ or   (2) $\frac{x}{3} - 7 \leq -5$

Case (1) $\frac{x}{3} - 7 \geq 5$ Case (2) $\frac{x}{3} - 7 \leq -5$

$\frac{x}{3} \geq 12$ $\frac{x}{3} \leq 2$

$x \geq 36$ $x \leq 6$

We may consider the solution set of the inequality in this example as the union of the two disjoint sets $\{x | x \geq 36\}$ and $\{x | x \leq 6\}$. This union may be represented on the number line as in the accompanying figure.

Solve

$$\left|\frac{x}{3} + 2\right| < 4$$

(A) $|a| < b$

(B)

$x > -18$ $x < 6$ $-18 < x < 6$

Solution: Now, if b is a nonnegative real number, then a is a real number for which $|a| < b$ if and only if $-b < a < b$. See number line (A).

We apply this rule to the given problem. Therefore,

$\left|\frac{x}{3} + 2\right| < 4$ is equivalent to

$-4 < \frac{x}{3} + 2 < 4$. In other words, this inequality is

satisfied if and only if both

$$\frac{x}{3} + 2 < 4 \qquad \text{and} \qquad \frac{x}{3} + 2 > -4$$

are satisfied. By adding -2 to each member of these inequalities, we get

$$\frac{x}{3} < 2 \qquad \text{and} \qquad \frac{x}{3} > -6$$

Hence, multiplying by 3 in each case, we note that the original inequality is satisfied by values of x that satisfy both $x < 6$ and $x > -18$. We can observe the solution from diagram (B). Therefore, the solution set is

$$\{x| \quad -18 < x < 6\} \qquad \text{or} \quad \{x|x < 6\} \cap \{x|x > -18\}.$$

● **PROBLEM** 291

Find all x for which

$$\left|\frac{4}{3} + x\right| \leq \frac{2}{5}.$$

$$\cdots -2 \ -26/15 \ -1 \ -14/15 \quad 0 \qquad\qquad 1\cdots$$

Solution: Note the following rule for absolute values: for

$|a + b| \leq c$ where a,b,c are any real numbers, $-c \leq a + b \leq c$. Therefore, the given inequality, involving the absolute value, reduces to:

$$-\frac{2}{5} \leq \frac{4}{3} + x \leq \frac{2}{5}.$$

Subtract $\frac{4}{3}$ from the three parts of the inequality above,

$$-\frac{2}{5} - \frac{4}{3} \leq \frac{4}{3} + x - \frac{4}{3} \leq \frac{2}{5} - \frac{4}{3}$$

$$-\frac{2}{5} - \frac{4}{3} \leq x \leq \frac{2}{5} - \frac{4}{3}.$$

Getting a common denominator of 15 for the fractions involved in the above inequality,

$$-\frac{3(2)}{3(5)} - \frac{5(4)}{5(3)} \leq x \leq \frac{3(2)}{3(5)} - \frac{5(4)}{5(3)}$$

$$-\frac{6}{15} - \frac{20}{15} \leq x \leq \frac{6}{15} - \frac{20}{15}$$

$$-\frac{26}{15} \leq x \leq -\frac{14}{15}$$

Thus, all rational numbers between $\frac{-26}{15}$ and $\frac{-14}{15}$ are solutions. The figure gives the solution set on the rational line.

● **PROBLEM** 292

Find the solution set of $|2x + 5| \leq x + 3$.

174

Fig. A

Key $x \geq -5/2$ $x \leq -2$

Fig. B

Key $x \geq -8/3$ $x < -5/2$

Solution: Case 1. If $2x + 5 \geq 0$, then $|2x + 5| = 2x + 5$ and the inequality becomes

$$2x + 5 \leq x + 3$$

$$x \leq -2$$

For Case 1 we have the simultaneous restrictions

$$2x + 5 \geq 0 \quad \text{and} \quad x \leq -2$$

or

$$x \geq -\frac{5}{2} \quad \text{and} \quad x \leq -2$$

The solution set for Case 1 is

$$X_1 = \left(x \,\middle|\, x \geq -\frac{5}{2} \text{ and } x \leq -2\right)$$

$$= \left(x \,\middle|\, -\frac{5}{2} \leq x \leq -2\right)$$

This inequality holds by noting Figure A. The solution set for Case 1 is the intersection of the two inequalities on the number line. This intersection is the set

$$-2\tfrac{1}{2} \leq x \leq -2.$$

Case 2. If $2x + 5 < 0$, then $|2x + 5| = -(2x + 5)$ and the inequality becomes

$$-(2x + 5) \leq x + 3$$

Multiplying an inequality by -1 reverses the direction of the inequality.

$$2x + 5 \geq -(x + 3) = -x - 3$$
$$3x \geq -8$$
$$x \geq -\frac{8}{3}$$

For Case 2 we have

$$2x + 5 < 0 \quad \text{and} \quad x \geq -\frac{8}{3}$$

or

$$x < -\frac{5}{2} \quad \text{and} \quad x \geq -\frac{8}{3}$$

The solution set is

$$X_2 = \left(x \,\middle|\, x < -\frac{5}{2} \text{ and } x \geq -\frac{8}{3}\right)$$

$$= \left(x \,\middle|\, -\frac{8}{3} \leq x < -\frac{5}{2}\right)$$

This inequality holds by noting Figure B. The solution set for Case 2
is the intersection of the two inequalities on the number line. This
intersection is the set

$$-\frac{8}{3} \le x < -\frac{5}{2} .$$

Finally, the solution set, X, of the given inequality is the union of
X_1 and X_2 .

$$X = X_1 \cup X_2$$

$$= \left\{ x \left| -\frac{8}{3} \le x < -\frac{5}{2} \right. \right\} \cup \left\{ x \left| -\frac{5}{2} \le x \le -2 \right. \right\}$$

$$= \left\{ x \left| -\frac{8}{3} \le x \le -2 \right. \right\}$$

● **PROBLEM 293**

Replace the inequality $1 < x < 3$ by a single inequality
involving an absolute value.

Solution: Recall:

$$a - b < x < a + b$$

$$= - b < x - a < b , \quad \text{subtracting} \quad a$$

$$= |x - a| < b , \quad \text{definition of absolute value}$$

Replacing a by 2 and b by 1 we obtain:

$$2 - 1 < x < 2 + 1$$

$$- 1 < x - 2 < 1$$

$$|x - 2| < 1.$$

● **PROBLEM 294**

Let $A = \{x | x > -2\}$ and $B = \{x | x < 3\}$. Describe these sets as
collections of points of the number scale. What is $A \cap B$? $A \cup B$?

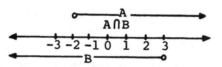

Solution: The set A consists of all numbers that are greater than
-2, so a point belongs to A if, and only if, it lies to the right
of the point -2 of the number scale. Similarly, we think of B
graphically as the set of points to the left of 3. The intersection
A ∩ B, that is, the set of points in both A and B, is illustrated
in the diagram. It consists of the points between -2 and 3, so we
have the set equation

$$A \cap B = \{x | x > -2\} \cap \{x | x < 3\} = \{x | -2 < x < 3\} .$$

Every real number belongs to at least one of the sets A or B, since
every real number is either greater than -2 or less than 3 (and
some numbers are both). In other words, the union A ∪ B = R, the set
of all real numbers.

CHAPTER 13

SYSTEMS OF LINEAR EQUATIONS AND INEQUALITIES

Basic Attacks and Strategies for Solving Problems in this Chapter. See pages 177 to 198 for step-by-step solutions to problems.

Algebraic methods of solving systems of linear equations in two variables involve applying certain laws of algebra to individual equations or combinations of equations. One of the algebraic solution procedures for a 2 by 2 system of linear equations is called the substitution method. The procedure is:

(1) Choose one of the variables and one of the equations. Rearrange the chosen equation so that it is solved for the chosen variable in terms of the other variable.

(2) Substitute the expression for the chosen variable obtained in Step 1 into the remaining equation. The result will be an equation that involves only one variable.

(3) Solve the equation obtained in Step 2. The number obtained is one of the coordinates of the solution for the system.

(4) Substitute the number obtained in Step 3 into the expression for the variable chosen in Step 1 and simplify. The result is the other coordinate of the ordered pair solution.

When the system of equations is inconsistent, the substitution method produces a false mathematical statement which means that the system has no solution. On the other hand, when the system of equations is dependent, the substitution method produces a true statement in which no variable appears. An infinite number of ordered pairs represents the solution set.

A second algebraic solution procedure for a 2 by 2 system of linear equations is called the addition-subtraction or elimination method. Its strategy is to add or subtract multiples of the two given equations so that one of the variables is eliminated. The resulting equation involves only one variable whose value can be easily solved for. Substitute the value of this variable in either of the original equations and solve for the other variable. The values of the variables represent the coordinates of the solution. As with the previous solution method, the addition-subtraction method can be used to alert the problem solver to inconsistent and dependent systems of equations.

An algebraic solution of a system of three linear equations in three variables can be achieved by using the substitution and/or addition-subtraction method(s). The primary step is to use one of the methods to reduce the original system to a 2 by 2 system and then solve the reduced system using an algebraic method. Once a solution is found for the reduced system, choose one of the original equations, substitute the values already obtained, and solve for the remaining unknown. The result is the third coordinate of the ordered triplet which is the solution for the original system.

GRAPHING METHOD

The graphing method of solving a 2 × 2 system of linear equations involves graphing each equation and determining a point of intersection of the graphs. If the graphs of the equations are non-parallel, non-coincident lines, there is exactly one point of intersection which is the solution. Parallel lines indicate no solution and coincident lines indicate an infinite set of solutions.

SOLVING SYSTEMS OF INEQUALITIES AND GRAPHING

When solving by graphing a system of inequalities in two variables (where the inequality symbols are all strictly > and/or <), the first step is to rewrite each inequality in the system in the form of

$y > mx + b$ or $y < mx + b$.

The second step is to graph the linear equation,

$y = mx + b,$

for each inequality as a straight dotted line. Determine in what region of the x-y plane each inequality holds true by selecting points on both sides of the corresponding dotted line and substitute them into the variable statement of the inequality. Shade in the side of the line whose points make the inequality a true statement. Represent each shaded area with a unique pattern (e.g., diagonal shading, vertical shading, etc.). The solution is the intersection of all the shaded areas representing points whose ordered pairs satisfy all conditions in the original system of inequalities.

If the system of inequalities contains the ≥ and/or ≤ symbols, then the only change in the graphing solution procedure above is to graph the linear equation,

$y = mx + b,$

for each inequality as a straight non-dotted (solid) line which will be a part of the solution. If unique pairs of these equations are formed into 2 by 2 linear systems, then they can be solved algebraically as indicated above. The result of the solution of each system yields the coordinates of one of the vertices of the shaded area that represents the solution of the original system.

Step-by-Step Solutions to Problems in this Chapter, "Systems of Linear Equations and Inequalities"

SOLVING EQUATIONS IN TWO VARIABLES AND GRAPHING

Solve the simultaneous equations $2x + 4y = 11$, $-5x + 3y = 5$ by the method of substitution and by the method of elimination by addition.

<u>Solution:</u> The method of substitution involves solving for one variable in terms of the other and then substituting the obtained value into the second equation. Thus, we solve the first equation for x and substitute in the second:

$$2x + 4y = 11$$
$$2x = 11 - 4y$$
$$x = \frac{11 - 4y}{2}$$

Replacing x by $\left(\frac{11 - 4y}{2}\right)$ in the second equation,

$$-5\left(\frac{11 - 4y}{2}\right) + 3y = 5$$

$$\frac{-55 + 20y}{2} + 3y = 5$$

$$\frac{-55}{2} + 10y + 3y = 5$$

Multiply both sides by 2,

$$-55 + 20y + 6y = 10$$
$$26y = 65$$

$$y = \frac{65}{26} = \frac{5}{2} .$$

Substituting this value for y into the first equation:

$$2x + 4\left(\frac{5}{2}\right) = 11$$

$$2x + 10 = 11$$
$$2x = 1$$
$$x = \frac{1}{2} .$$

We obtain the same result by the method of elimination by addition.

$$2x + 4y = 11 \qquad\qquad\qquad (1)$$
$$-5x + 3y = 5 \qquad\qquad\qquad (2)$$

Multiplying equation (1) by 5 and equation (2) by 2 and adding the result we obtain:

$$10x + 20y = 55$$
$$-\ 10x + 6y = \ \ 10$$
$$\overline{\qquad\qquad 26y = 65}$$
$$y = \frac{65}{26} = \frac{5}{2}$$

Once again, replacing y by $\frac{5}{2}$ in equation (1):

$$2x + 4\left(\frac{5}{2}\right) = 11$$

$$2x + 10 = 11$$
$$2x = 1$$
$$x = \frac{1}{2}$$

Thus $\left\{\left(\frac{1}{2}, \frac{5}{2}\right)\right\}$ is the solution to the given system of equations.

Solve the equations $3x + 2y = 1$ and $5x - 3y = 8$ simultaneously.

Solution: We have 2 equations in 2 unknowns,

$$3x + 2y = 1 \qquad\qquad (1)$$

and

$$5x - 3y = 8 \qquad\qquad (2)$$

There are several methods to solve this problem. We have chosen to multiply each equation by a different number so that when the two equations are added, one of the variables drops out. Thus

multiplying the first by 3: $\qquad 9x + 6y = 3$

and the second by 2: $\qquad 10x - 6y = 16$

and adding: $\qquad\qquad\qquad 19x = 19$

$$x = 1$$

Substituting $x = 1$ in the first equation:
$$3(1) + 2y = 3 + 2y = 1$$
$$2y = -2$$
$$y = -1$$

(Alternatively, y might have been found by multiplying the first equation by 5, the second by -3, and adding.)

In this case, then, there is a unique solution: $x = 1$, $y = -1$. This may be checked by replacing x by 1 and y by (-1) in each equation. In equation (1):
$$3x + 2y = 1$$
$$3(1) + 2(-1) = 1$$
$$3 - 2 = 1$$
$$1 = 1$$

In equation (2):
$$5x - 3y = 8$$
$$5(1) - 3(-1) = 8$$
$$5 - (-3) = 8$$
$$5 + 3 = 8$$
$$8 = 8$$

In other words, the lines whose equations are 3x+2y=1 and
5x-3y=8 meet in one and only one point: (1,-1). This,
again, may be checked graphically, as seen in the diagram.

● **PROBLEM 297**

Solve the equations 2x + 3y = 6 and 4x + 6y = 7 simultaneously.

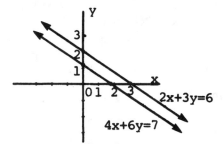

Solution: We have 2 equations in 2 unknowns,

$$2x + 3y = 6 \tag{1}$$

and

$$4x + 6y = 7 \tag{2}$$

There are several methods to solve this problem. We have chosen to
multiply each equation by a different number so that when the two
equations are added, one of the variables drops out. Thus

multiplying equation (1) by 2: $4x + 6y = 12$ (3)

multiplying equation (2) by -1: $\underline{-4x - 6y = -7}$ (4)

adding equations (3) and (4): $0 = 5$

We obtain a peculiar result!

Actually, what we have shown in this case is that if there were a
simultaneous solution to the given equations, then 0 would equal 5.
But the conclusion is impossible; therefore there can be no simultaneous
solution to these two equations, hence no point satisfying both.

The straight lines which are the graphs of these equations must be
parallel if they never intersect, but not identical, which can be seen
from the graph of these equations (see the accompanying diagram).

● **PROBLEM 298**

Solve for x and y .

$$x + 2y = 8 \tag{1}$$
$$3x + 4y = 20 \tag{2}$$

Solution: Solve equation (1) for x in terms of y:

$$x = 8 - 2y \tag{3}$$

Substitute (8 - 2y) for x in (2):

$$3(8 - 2y) + 4y = 20 \tag{4}$$

Solve (4) for y as follows:

Distribute: $24 - 6y + 4y = 20$

Combine like terms and then subtract 24 from both sides:

$$24 - 2y = 20$$

$$24 - 24 - 2y = 20 - 24$$

$$-2y = -4 \qquad \text{Divide both sides by } -2: \quad y = 2$$

Substitute 2 for y in equation (1):

$$x + 2(2) = 8$$

$$x = 4$$

Thus, our solution is x = 4, y = 2.

Check: Substitute   x = 4, y = 2   in equations (1) and (2):

$$4 + 2(2) = 8$$

$$8 = 8$$

$$3(4) + 4(2) = 20$$

$$20 = 20$$

● **PROBLEM   299**

Solve the equations $2x + 3y = 6$ and $y = -(2x/3) + 2$ simultaneously.

Solution: We have 2 equations in 2 unknowns,

$$2x + 3y = 6 \qquad\qquad (1)$$

$$y = -(2x/3) + 2 \qquad\qquad (2)$$

There are several methods of solution for this problem. Since equation (2) already gives us an expression for y, we use the method of substitution. Substituting $-(2x/3) + 2$ for y in the first equation:

$$2x + 3\left(-\frac{2x}{3} + 2\right) = 6$$

Distributing, $2x - 2x + 6 = 6$

$$6 = 6$$

Apparently we have gotten nowhere! The result 6 = 6 is true, but indicates no solution. Actually, our work shows that no matter what real number x is, if y is determined by the second equation, then the first equation will always be satisfied.

The reason for this peculiarity may be seen if we take a closer look at the equation $y = -(2x/3) + 2$. It is equivalent to $3y = -2x + 6$, or $2x + 3y = 6$.

In other words, the two equations are equivalent. Any pair of values of x and y which satisfies one satisfies the other.

It is hardly necessary to verify that in this case the graphs of the given equations are identical lines, and that there are an infinite number of simultaneous solutions of these equations.

● **PROBLEM   300**

Find the solution set for the system:

$$3x + 5y = -9$$

$$x - 5y = 17$$

180

Solution: Upon examination we see that if the left members of the two equations are added and the right members of the two equations are added, we obtain the equation $4x = 8$. (This is justified by the additive principle of equations; we are simply adding equal quantities to both sides of an equation.) Hence the y terms have been eliminated, since the coefficients were additive inverses. This new equation, $4x = 8$, can be easily seen to have $\{2\}$ for its solution set. Now, if we use this value for x, we see that upon substituting it into either of the two equations in our system, for example, $3x + 5y = -9$, we obtain $3(2) + 5y = -9$. Upon simplifying, this becomes $y = -3$. (You should convince yourself that had the other equation in the system been selected, the value of y would have been found to be -3.)

Therefore the single solution for our system is the ordered pair $(2,-3)$.

● **PROBLEM** 301

Solve algebraically:
$$\begin{cases} 4x + 2y = -1 & \quad\text{(1)} \\ 5x - 3y = 7 & \quad\text{(2)} \end{cases}$$

Solution: We arbitrarily choose to eliminate x   first.

Multiply (1) by 5: $20x + 10y = -5$ (3)

Multiply (2) by 4: $20x - 12y = 28$ (4)

Subtract, (3) - (4): $22y = -33$ (5)

Divide (5) by 22: $y = -\dfrac{33}{22} = -\dfrac{3}{2}$.

To find x, substitute $y = -\dfrac{3}{2}$ in either of the original equations. If we use Eq. (1), we obtain $4x + 2(-3/2) = -1$, $4x - 3 = -1$, $4x = 2$, $x = \dfrac{1}{2}$.

The solution $\left(\dfrac{1}{2}, -\dfrac{3}{2}\right)$ should be checked in both equations of the given system.

Replacing $\left(\dfrac{1}{2}, -\dfrac{3}{2}\right)$ in Eq. (1):

$$4x + 2y = -1$$
$$4\left(\tfrac{1}{2}\right) + 2\left(-\tfrac{3}{2}\right) = -1$$
$$\tfrac{4}{2} - 3 = -1$$
$$2 - 3 = -1$$
$$-1 = -1$$

Replacing $\left(\dfrac{1}{2}, -\dfrac{3}{2}\right)$ in Eq. (2):

$$5x - 3y = 7$$
$$5\left(\tfrac{1}{2}\right) - 3\left(-\tfrac{3}{2}\right) = 7$$
$$\tfrac{5}{2} + \tfrac{9}{2} = 7$$
$$\tfrac{14}{2} = 7$$
$$7 = 7$$

(Instead of eliminating x from the two given equations, we could have eliminated y by multiplying Eq. (1) by 3, multiplying Eq. (2) by 2, and then adding the two derived equations.)

Solve one equation for one unknown and substitute in the other equation to find the solutions of the following systems.

$$xy = 1 \tag{1}$$

$$x + 2y = 3 \tag{2}$$

Solution: We can solve equation (2) for x by adding (-2y) to both sides:

$$x = 3 - 2y \tag{3}$$

Substituting (3 - 2y) for x in equation (1):

$$(3-2y)y = 1$$

Distributing,

$$3y - 2y^2 = 1$$

$$3y - 2y^2 - 1 = 0$$

$$-2y^2 + 3y - 1 = 0$$

Factoring,

$$(-2y + 1)(y-1) = 0$$

Whenever the product of two numbers ab = 0, either a = 0 or b = 0. Thus, either

$$-2y + 1 = 0 \quad \text{or} \quad y - 1 = 0$$

$$-2y = -1$$

and

$$y = \frac{-1}{-2} = \frac{1}{2} \text{ or } y = 1$$

Replacing y by $\frac{1}{2}$ in equation (3), we obtain the corresponding x value:

$$x = 3 - 2\left(\frac{1}{2}\right)$$

$$x = 3 - 1$$

$$x = 2$$

Replacing y by 1 in equation (3), we obtain

$$x = 3 - 2(1)$$

$$x = 3 - 2$$

$$x = 1$$

Thus, the two solutions to this system appear to be $\left(2, \frac{1}{2}\right)$ and (1,1). These can be verified by the following check: Replace (x,y) by (2,1/2) in equations (1) and (2):

$$xy = 1 \tag{1}$$

$$(2)\left(\frac{1}{2}\right) = 1$$

$$\frac{2}{2} = 1$$

$$1 = 1$$

$$x + 2y = 3 \tag{2}$$

$$2 + 2\left(\frac{1}{2}\right) = 3$$

$$2 + \frac{2}{2} = 3$$

$$2 + 1 = 3$$
$$3 = 3$$

Replace (x,y) by $(1,1)$ in equations (1) and (2):

$$xy = 1 \qquad\qquad (1)$$
$$1 \cdot 1 = 1$$
$$1 = 1$$

$$x + 2y = 3 \qquad\qquad (2)$$
$$1 + 2(1) = 3$$
$$1 + 2 = 3$$
$$3 = 3$$

Thus, the solutions to this system are indeed $\left(2, \frac{1}{2}\right)$ and $(1,1)$.

● **PROBLEM 303**

Determine the nature of the system of linear equations

$$2x + y = 6 \qquad\qquad (1)$$
$$4x + 2y = 8 \qquad\qquad (2)$$

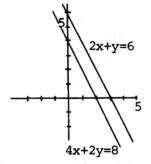

$2x+y=6$

$4x+2y=8$

Solution: These linear equations may be written in the standard form $y = mx + b$:

$$y = -2x + 6 \qquad\qquad (3\text{-}1)$$
$$\text{and } y = -2x + 4 \qquad\qquad (4\text{-}2)$$

Observe that the slope of each line is $m = -2$, but the y-intercepts are different, that is, $b = 6$ for equation (3-1) and $b = 4$ for equation (4-2). The lines are therefore parallel and distinct. The graph below also indicates that the lines are parallel. The system is therefore inconsistent, and there is no solution.

● **PROBLEM 304**

Determine the nature of the system of linear equations

$$x + 2y = 8$$

$$x - 2y = 2.$$

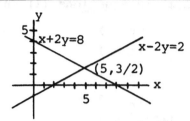

Solution: Add the two equations, eliminating the y-terms, to obtain a single equation in terms of x. Values of x satisfying this equation will yield solutions of the system.

$$x + 2y = 8 \qquad\qquad (1)$$

$$\underline{+\ x - 2y = 2} \qquad\qquad (2)$$

$$2x \qquad = 10$$

$$x \qquad = 5$$

Substituting x = 5 into Equation (1) yields
$y = (8 - x)/2 = (8 - 5)/2 = \frac{3}{2}$ or into Equation (2) yields
$y = (2 - x)/(-2) = (2 - 5)/(-2) = \frac{3}{2}$. Thus we have x = 5,
$y = \frac{3}{2}$ as the only solution of the system. Alternately, the figure indicates that the lines intersect in the point $\left(5, \frac{3}{2}\right)$. The system is therefore consistent and independent. Substitution of x = 5 and $y = \frac{3}{2}$ in both equations yields

$$5 + 2\left(\frac{3}{2}\right) = 8, \text{ or } 8 = 8$$

$$5 - 2\left(\frac{3}{2}\right) = 2, \text{ or } 2 = 2$$

so that x = 5, y = 3/2, is a solution, and the only solution of the system.

● PROBLEM 305

Show that the following pair of equations is dependent by showing that the two equations are equivalent.

$$3x - 2y = 9$$

$$4y - 6x = -18$$

Solution: We can derive 4y - 6x = -18 from 3x - 2y = 9 by applying field properties.

$$3x - 2y = 9$$

Multiplying both sides by (-2), $(-2)(3x - 2y) = (-2)9$

Distributing, $-6x + 4y = -18$

Commuting, $4y - 6x = -18$

Thus, since $3x - 2y = 9$ is equivalent to $4y - 6x = -18$ the two equations are dependent.

● **PROBLEM** 306

Determine the nature of the system of linear equations

$$x + 3y = 4 \qquad\qquad (1)$$

$$2x + 6y = 8. \qquad\qquad (2)$$

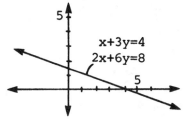

Solution: If the first equation is multiplied by 2, the solution of the system will not be altered. Note, however, that the two equations are then identical. The graph too, indicates that the lines coincide, and therefore the system is consistent and dependent. It can be verified by substitution that three of the solutions are $x = 1$, $y = 1$; $x = 7$, $y = -1$; and $x = -5$, $y = 3$.

● **PROBLEM** 307

Solve for x and y: $\begin{cases} 3x + 5y = 9, & (1) \\ 7x - 10y = 8 & (2) \end{cases}$

Solution: Multiply each member of the first equation by 2; thus $6x + 10y = 18$. Now add each member of the resulting equation to the corresponding member of the second equation.

$$\begin{array}{r} 6x + 10y = 18 \\ \underline{7x - 10y = 18} \\ 13x = 26 \\ x = 2 \end{array}$$

To solve for y, replace x by 2 in either equation. Using equation (1),

$$\begin{array}{r} 3x + 5y = 9 \\ 3(2) + 5y = 9 \\ 6 + 5y = 9 \\ 5y = 3 \\ y = 3/5 \end{array}$$

Therefore our solution is $x = 2$, $y = 3/5$.

Check: To verify our solutions, we substitute the values 2 and $3/5$ for x and y in equations (1) and (2):

$$3x + 5y = 9$$
$$3(2) + 5(3/5) = 9$$
$$6 + 3 = 9$$
$$9 = 9$$

$$7x - 10y = 8$$
$$7(2) - 10(3/5) = 8$$
$$14 - 6 = 8$$
$$8 = 8$$

● **PROBLEM** 308

Solve for x and y.

$$3x + 2y = 23 \qquad (1)$$
$$x + y = 9 \qquad (2)$$

Solution: Multiply equation (2) by -3:

$$-3x - 3y = -27 \qquad (3)$$

Add equations (1) and (3):

$$3x + 2y = 23$$
$$\underline{-3x - 3y = -27}$$
$$-y = -4$$
$$y = 4$$

Substitute 4 for y in equation (1):

$$3x + 2(4) = 23$$
$$3x + 8 \quad = 23$$

Subtract 8 from both sides: $3x = 15$

Divide each side by 3: $\qquad x = 5$

Hence our solution is, $x = 5$ and $y = 4$.

Check: Substitute 5 for x and 4 for y in equation (1):

$$3(5) + 2(4) = 23$$
$$23 = 23$$

Substitute 5 for x and 4 for y in equation (2):

$$5 + 4 = 9$$
$$9 = 9 .$$

● **PROBLEM** 309

Solve for x and y.

$$4x + 3y = 23 \qquad (1)$$
$$2x - 5y = 31 \qquad (2)$$

Solution: Multiply equation (2) by -2:

$$-4x + 10y = -62 \qquad (3)$$

Add equation (3) to equation (1):

$$4x + 3y = 23$$
$$\underline{+ (-4x + 10y = -62)}$$
$$13y = -39$$

$$y = -3$$

Substitute -3 for y in equation (1):

$$4x + 3(-3) = 23$$
$$4x - 9 = 23$$
$$4x = 32$$
$$x = 8$$

Hence the solution is x = 8, y = -3.

Check: Substitute 8 for x and -3 for y in (1),

$$4(8) + 3(-3) = 23$$
$$32 - 9 = 23$$
$$23 = 23.$$

Substitute 8 for x and -3 for y in (2),

$$2(8) - 5(-3) = 31$$
$$16 + 15 = 31$$
$$31 = 31 .$$

● **PROBLEM** 310

Find the point of intersection of the graphs of the equations:
$$\begin{cases} x + y = 3, \\ 3x - 2y = 14. \end{cases}$$

Solution: To solve these linear equations, solve for y in terms of x. The equations will be in the form y = mx + b, where m is the slope and b is the intercept on the y-axis.

$$x + y = 3$$
$$y = 3 - x \quad \text{subtract } x \text{ from both sides}$$

$$3x - 2y = 14 \quad \text{subtract } 3x \text{ from both sides}$$

$$-2y = 14 - 3x \quad \text{divide by } -2.$$

$$y = -7 + \frac{3}{2} x$$

The graphs of the linear functions, y = 3 - x and $y = -7 + \frac{3}{2} x$, can be determined by plotting only two points. For example, for y = 3 - x, let x = 0, then y = 3. Let x = 1, then y = 2. The two points on this first line are (0,3) and (1,2). For y = -7 + 3/2 x, let x = 0, then

187

$y = -7$. Let $x = 1$, then $y = -5\frac{1}{2}$. The two points on this second line are $(0,-7)$ and $(1, -5\frac{1}{2})$.

To find the point of intersection P of

$$x + y = 3$$

and

$$3x - 2y = 14,$$

solve them algebraically. Multiply the first equation by 2. Add these two equations to eliminate the variable y.

$$\begin{array}{r} 2x + 2y = 6 \\ \underline{3x - 2y = 14} \\ 5x \qquad = 20 \end{array}$$

Solve for x to obtain $x = 4$. Substitute this into $y = 3 - x$ to get $y = 3 - 4 = -1$. P is $(4,1)$. AB is the graph of the first equation, and CD is the graph of the second equation. The point of intersection P of the two graphs is the only point on both lines. The coordinates of P satisfy both equations and represent the desired solution of the problem. From the graph, P seems to be the point $(4,-1)$. These coordinates satisfy both equations, and hence are the exact coordinates of the point of intersection of the two lines.

To show that $(4,-1)$ satisfies both equations, substitute this point into both equations.

$$\begin{array}{ll}
x + y = 3 & \qquad 3x - 2y = 14 \\
4 + (-1) = 3 & \qquad 3(4) - 2(-1) = 14 \\
4 - 1 = 3 & \qquad 12 + 2 = 14 \\
3 = 3 & \qquad 14 = 14
\end{array}$$

● **PROBLEM 311**

Find the solution set of the system

$$2x - 12y = 3 \qquad (1)$$

$$3x + 9y = 4 \qquad (2)$$

<u>Solution:</u> This system can be solved using the multiplication-addition method: The least common multiple of the x coefficients is 6. Multiply equation (1) by 3 and equation (2) by -2 to obtain

$$3(2x - 12y) = 3 \cdot 3$$

$$6x - 36y = 9 \qquad (3)$$

$$-2(3x + 9y) = 4 \cdot (-2)$$

$$-6x - 18y = -8 \qquad (4)$$

Adding equations (3) and (4),

$$\begin{array}{r} 6x - 36y = 9 \\ \underline{-6x - 18y = -8} \\ -54y = 1 \\ y = -\frac{1}{54} \end{array}$$

To solve for x, we substitute this value of y in either of our given equations. Substituting $\left(\frac{-1}{54}\right)$ for y in (2):

$$3x + 9y = 4$$

$$3x + 9\left(-\frac{1}{54}\right) = 4$$

$$3x - \frac{1}{6} = 4$$

$$3x = 4 + \frac{1}{6}$$

$$3x = \frac{24}{6} + \frac{1}{6}$$

$$3x = \frac{25}{6}$$

$$x = \frac{25}{18}$$

Thus our solution set is $\left\{ \left(\frac{25}{18}, -\frac{1}{54} \right) \right\}$, and we perform the following check to verify this result.

Check: Replace x and y by $\left(\frac{25}{18}, -\frac{1}{54} \right)$ respectively in (1) and (2):

(1) $\qquad 2x - 12y = 3$

$$2\left(\frac{25}{18}\right) - 12\left(\frac{-1}{54}\right) = 3$$

$$\frac{50}{18} + \frac{12}{54} = 3$$

$$\frac{50}{18} + \frac{4}{18} = 3$$

$$\frac{54}{18} = 3$$

$$3 = 3$$

(2) $\qquad 3x + 9y = 4$

$$3\left(\frac{25}{18}\right) + 9\left(\frac{-1}{54}\right) = 4$$

$$\frac{75}{18} - \frac{9}{54} = 4$$

$$\frac{75}{18} - \frac{3}{18} = 4$$

$$\frac{72}{18} = 4$$

$$4 = 4$$

● **PROBLEM 312**

Obtain the simultaneous solution set of the equations

$$3x + 4y = -6 \qquad\qquad (1)$$

$$5x + 6y = -8 \qquad\qquad (2)$$

Solution: We eliminate one of the unknowns to obtain an equation in one unknown whose root is one of the numbers in a simultaneous solution pair. We arbitrarily choose to eliminate y. Notice that the lowest common multiple, LCM, of the coefficients of y in (1) and (2) is 12. This is because the coefficients of y are $4 = 2 \cdot 2$ and $6 = 2 \cdot 3$. Thus their LCM will be $2 \cdot 2 \cdot 3 = 12$.

To obtain a coefficient of y equal to 12 in equation (1) we multiply the equation by 3,

$$3(3x + 4y) = 3(-6)$$

$$9x + 12y = -18 \qquad\qquad (3)$$

To obtain a coefficient of y equal to 12 in equation (2) we multiply equation (2) by 2,

$$2(5x + 6y) = 2(-8)$$

$$10x + 12y = -16 \qquad (4)$$

Now subtract equation (4) from equation (3),

$$9x + 12y = -18$$
$$\underline{10x + 12y = -16}$$
$$\frac{-x \qquad\quad = -2}{x = 2}$$

Consequently the first number in (x,y) is 2. We obtain the second number by replacing x by 2 in either (1) or (2) and solving for y. We shall choose (1) and get

$$3(2) + 4y = -6$$

$$6 + 4y = -6$$

$$4y = -12$$
$$y = -3$$

Hence (x,y) = (2,-3).

Check: Replacing x by 2 and y by -3 in (1) and (2), we get

from equation (1): $3x + 4y = -6$

$$3(2) + 4(-3) = -6$$

$$6 + (-12) = -6$$

$$-6 = -6$$

from equation (2): $5x + 6y = -8$

$$5(2) + 6(-3) = -8$$

$$10 + (-18) = -8$$

$$-8 = -8$$

Therefore the simultaneous solution set is $\{(2,-3)\}$.

SOLVING EQUATIONS IN THREE VARIABLES

● **PROBLEM** 313

Solve the system of equations,

$2x - y - 4z = 3$ (1)

$-x + 3y + z = -10$ (2)

$3x + 2y - 2z = -2$ (3)

<u>Solution:</u> To solve a system of 3 equations in 3 unknowns, we first reduce it to a system of 2 equations in 2 unknowns, a process which can often be done many ways. Although various other algebraic manipulations may be used to arrive at the same result, we will employ the following method: Multiplying equation 1 by (-1) we obtain,

$-2x + y + 4z = -3$ (4)

Adding equations (4), (2), and (3) we obtain,

$$-2x + y + 4z = -3$$
$$-x + 3y + z = -10$$
$$\underline{3x + 2y - 2z = -2}$$
$$6y + 3z = -15 \qquad (5)$$

Multiplying equation (2) by 3 we obtain,

$$-3x + 9y + 3z = -30 \qquad (6)$$

Adding equations (6) and (3) we obtain,

$$-3x + 9y + 3z = -30$$
$$\underline{3x + 2y - 2z = -2}$$
$$11y + z = -32 \qquad (7)$$

Multiplying equation (7) by (-3) we obtain,

$$-33y - 3z = 96 \qquad (8)$$

Adding equations (8) and (5) we obtain,

$$-33y - 3z = 96$$
$$\underline{6y + 3z = -15}$$
$$-27y = 81$$
$$y = -3$$

Solving for z, we replace, y by (-3) in equation (5):

$$6y + 3z = -15$$
$$6(-3) + 3z = -15$$
$$-18 + 3z = -15$$
$$3z = 3$$
$$z = 1$$

Solving for x, we replace y by (-3) and z by 1 in equation (1):

$$2x - y - 4z = 3$$
$$2x - (-3) - 4(1) = 3$$
$$2x + 3 - 4 = 3$$
$$2x - 1 = 3$$
$$2x = 4$$
$$x = 2$$

Thus the solution to this system is x = 2, y = -3, and z = 1.

Check: Replace x,y, and z by 2, -3, and 1 in each equation.

$$2x - y - 4z = 3 \qquad (1)$$

$$2(2) - (-3) - 4(1) = 3$$

$$4 + 3 - 4 = 3$$

$$3 = 3$$

$$-x + 3y + z = -10 \qquad (2)$$

$$-(2) + 3(-3) + 1 = -10$$

$$-2 - 9 + 1 = -10$$

$$-10 = -10$$

$$3x + 2y - 2z = -2 \qquad (3)$$

$$3(2) + 2(-3) - 2(1) = -2$$

$$6 - 6 - 2 = -2$$

$$-2 = -2$$

● **PROBLEM 314**

Solve the system

$$2x - y + 4z = 1 \qquad (1)$$

$$x - y + z = 0 \qquad (2)$$

$$x + y + z = 1 \qquad (3)$$

<u>Solution:</u> It is easiest to eliminate the variable y since the equations (1), (2), (3) differ only by a factor of +1 or - 1 for the variable y. (For the other variables x and z, the equations differ by factors of ± 2 for x and ± 4 for z).

Multiplying equation (1) by - 1 we obtain:
$$- 2x + y - 4z = - 1 \qquad (4)$$

$$x - y + z = 0 \qquad (2)$$

$$x + y + z = 1 \qquad (3)$$

Add equations (4) and (2) to eliminate the variable y and we obtain a new equation (5) in x and z.
$$- 2x + y - 4z = - 1 \qquad (4)$$

$$\underline{x - y + z = 0} \qquad (2)$$

$$- x - 3z = - 1 \qquad (5)$$

192

Add (2) and (3) to obtain another equation (6) in the variables x and z.

$$x - y + z = 0 \qquad\qquad (2)$$

$$\underline{x + y + z = 1} \qquad\qquad (3)$$

$$2x \qquad + 2z = 1 \qquad\qquad (6)$$

Now we have a new system of 2 equations in 2 unknowns x and z:

$$- x - 3z = -1 \qquad\qquad (5)$$

$$2x + 2z = \;\; 1 \qquad\qquad (6)$$

We must solve for one variable. The simplest way is to eliminate x. Multiply equation (5) by 2 and we obtain:

$$- 2x - 6z = -2 \qquad\qquad (7)$$

$$2x + 2z = 1 \qquad\qquad (6)$$

Add equations (7) and (6) to obtain:

$$- 4z = -1 \qquad\qquad (8)$$

Divide equation (8) by - 4 to solve for z.

$$z = \frac{1}{4}$$

Substitute z into either (5) or (6) to find x. For equation (5) then we have:

$$- x - 3z = -1 \qquad\qquad (5)$$

$$z = \frac{1}{4}$$

$$- x - 3\left(\frac{1}{4}\right) = -1$$

$$- x - \frac{3}{4} = -1$$

$$x = \frac{1}{4}$$

Given x and z we can now solve for y by substituting x and z into any of the three original equations (1), (2) or (3). For equation (2)

$$x - y + z = 0 \qquad\qquad (2)$$

$$x = \frac{1}{4}$$

$$z = \frac{1}{4}$$

$$\frac{1}{4} - y + \frac{1}{4} = 0$$

$$- y + \frac{2}{4} = 0$$

$$y = \frac{1}{2}$$

The solution of the original system is then:
$x = \frac{1}{4}; \ y = \frac{1}{2}; \ z = \frac{1}{4}$.

To check, substitute the solution into each of the three original equations (1), (2), and (3).

For (1): $2x - y + 4z = 1$ (1)

$$2\left(\frac{1}{4}\right) - \frac{1}{2} + 4\left(\frac{1}{4}\right) = 1$$

$$\frac{1}{2} - \frac{1}{2} + \quad 1 = 1$$

$$1 = 1$$

For (2): $x - y + z = 0$ (2)

$$\frac{1}{4} - \frac{1}{2} + \frac{1}{4} = 0$$

$$\frac{2}{4} - \frac{1}{2} \quad = 0$$

$$\frac{2}{4} - \frac{2}{4} \quad = 0$$

$$0 = 0$$

For (3): $x + y + z = 1$ (3)

$$\frac{1}{4} + \frac{1}{2} + \frac{1}{4} = 1$$

$$\frac{2}{4} + \frac{2}{4} \quad = 1$$

$$1 = 1$$

● **PROBLEM** 315

Solve the system

$$2x + 3y - 4z = -8 \qquad (1)$$

$$x + y - 2z = -5 \qquad (2)$$

$$7x - 2y + 5z = 4 \qquad (3)$$

Solution: We cannot eliminate any variable from two pairs of equations by a single multiplication. However, both x and z may be eliminated from equations 1 and 2 by multiplying Equation 2 by - 2. Then

$$2x + 3y - 4z = -8 \qquad (1)$$

$$-2x - 2y + 4z = 10 \qquad (4)$$

By addition, we have y = 2. Although, we may now eliminate either x or z from another pair of equations, we can more conveniently substitute y = 2 in Equations 2 and 3 to get two equations in two variables. Thus, making the substitution y = 2 in Equations 2 and 3, we have

$$x - 2z = -7 \qquad\qquad (5)$$

$$7x + 5z = 8 \qquad\qquad (6)$$

Multiply (5) by 5 and multiply (6) by 2. Then add the two new equations. Then $x = -1$. Substitute x in either (5) or (6) to find z.

The solution of the system is $x = -1$, $y = 2$, and $z = 3$. Check by substitution.

● **PROBLEM** 316

Find the solution set for the system:

$$3x + 4y - z = -2$$
$$2x - 3y + z = 4$$
$$x - 6y + 2z = 5$$

Solution: Adding the first and second equations, we obtain another equation without a term involving z:

$$\begin{array}{rcl} 3x + 4y - z &=& -2 \\ 2x - 3y + z &=& 4 \\ \hline 5x + y &=& 2 \end{array}$$

Similarly, after multiplying through by -2 in the second equation, we can use this new equation and the third one to obtain another equation without a term involving z:

$$\begin{array}{rcl} -4x + 6y - 2z &=& -8 \\ x - 6y + 2z &=& 5 \\ \hline -3x &=& -3 \end{array}$$

Our problem has been somewhat simplified in that not only have we obtained an equation without a term involving z, but we have obtained one without a y term.

The solution set of $-3x = -3$ is $\{1\}$. Upon substituting this into the equation $5x + y = 2$, we find that $y = -3$. Finally, upon substituting these values for x and y in either of the three equations of the system, we can obtain a value for z. If we use the first equation, $3x + 4y - z = -2$, we find that $z = -7$.

Hence the solution set for this system is $\{(1, -3, -7)\}$.

● **PROBLEM** 317

Solve for x, y and z:

$$5x + y - z = 9 , \qquad\qquad (1)$$

$$3x + y + 2z = 17 , \qquad\qquad (2)$$

$$x + 2y + 3z = 20. \qquad\qquad (3)$$

Solution: Subtract (2) from (1):

$$\begin{array}{rcl} 5x + y - z &=& 9 \\ -(3x + y + 2z &=& 17) \\ \hline 2x - 3z &=& -8 \end{array} \qquad (4)$$

Multiply (2) by 2: $6x + 2y + 4z = 34.$ $\qquad\qquad (5)$

195

Subtract (3) from (5):

$$6x + 2y + 4z = 34$$
$$\underline{- (x + 2y + 3z = 20)}$$
$$5x + z = 14 \qquad\qquad (6)$$

Subtract 5x from both sides: $z = 14 - 5x$ $\qquad\qquad$ (7)

Substitute (14 - 5x) for z in equation (4):

$$2x - 3(14 - 5x) = -8 \ .$$

Distribute: $\qquad 2x - 42 + 15x = -8$

$$17x - 42 = -8$$

Add 42 to both sides: $\qquad 17x = 34$

$$x = 2$$

Substitute 2 for x in equation (7)

$$z = 14 - 5(2) = 14 - 10 = 4 \ .$$

Therefore, x = 2, and z = 4.

Substitute in (1): $5(2) + y - 4 = 9$

$$10 + y - 4 = 9$$
$$6 + y = 9$$

Subtract 6 from both sides: $y = 3$

Thus, x = 2, y = 3, z = 4.

Check: $\qquad\qquad 5(2) + 3 - 4 = 9,$

$$9 = 9.$$

$$3(2) + 3 + 2(4) = 17,$$

$$17 = 17.$$

$$2 + 2(3) + 3(4) = 20,$$

$$20 = 20.$$

● **PROBLEM** 318

Solve the system:
$$2a - 3b + c = 2 \qquad (1)$$
$$3a + 2b - c = 4 \qquad (2)$$
$$2a - 3b + c = 5 \qquad (3)$$

Solution: Observe that equations (1) and (3) are inconsistent.

$$2a - 3b + c = 2 \neq 5 = 2a - 3b + c$$

This is a contradiction; that is, 2a - 3b + c cannot equal both 2 and 5 at the same time. This implies that there are no values of a,b, and c which will solve this set of simultaneous equations, hence this system has no solution.

SOLVING SYSTEMS OF INEQUALITIES AND GRAPHING

Solve the following system graphically.

$$y - x > -3$$
$$y - 2x < 2$$
$$x + y - 3 < 0$$

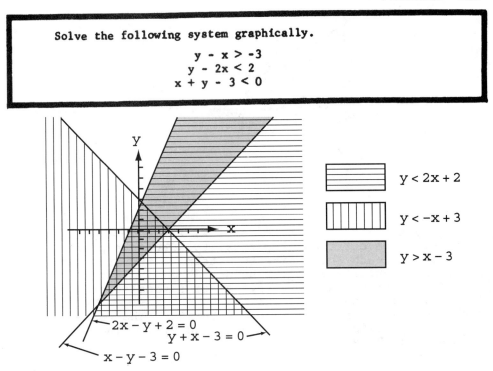

$y < 2x + 2$

$y < -x + 3$

$y > x - 3$

Solution: We may rewrite the system:

$$y > x - 3$$
$$y < 2x + 2$$
$$y < -x + 3$$

Graph the linear equation, $y = mx + b$, for each inequality as a straight dotted line. Thus, we graph

$$y = x - 3$$
$$y = 2x + 2$$
$$y = -x + 3$$

To determine in what region of the x - y plane the inequality holds, select points on both sides of the corresponding dotted line and substitute them into the variable statement. Shade in the side of the line whose point makes the inequality a true statement.

The graphs of the variable sentences are represented in the accompanying figure by diagonal, horizontal, and vertical shading, respectively.

The triple-shaded triangular region is the set of all points whose coordinate pairs satisfy all three conditions as defined by the three inequalities in the system.

Draw the graph of the given system of inequalities, and determine the coordinates of the vertices of the polygon which forms the boundary.

$$y < 3x - 3 \qquad (1)$$
$$3y \leq 24 - 2x \qquad (2)$$

$$2y \geq 3x - 10 \qquad\qquad (3)$$
$$y \geq -x + 5 \qquad\qquad (4)$$

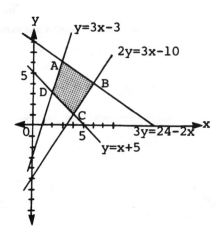

Solution: y is expressed in terms of x. For each inequality draw the corresponding equality. Choose a point on each side of each solid line to determine the area where the particular inequality holds. Shade in that region. The graph of the given system of inequalities consists of the hatched area and the four lines which form the boundary, that is, the polygon ABCD. The vertex A is found by solving the system obtained by writing Equations 1 and 2 as

$$y = 3x - 3 \qquad\qquad (5)$$

$$3y = 24 - 2x \qquad\qquad (6)$$

Solving the system of equations 5 and 6, we have

$$x = 3, \qquad y = 6$$

The coordinates of the vertex A are therefore (3, 6). In a similar manner, the coordinates of B, C, and D are found to be (6, 4) (4, 1) and (2, 3), respectively.

CHAPTER 14

DETERMINANTS AND MATRICES

> **Basic Attacks and Strategies for Solving Problems in this Chapter. See pages 199 to 231 for step-by-step solutions to problems.**

Any rectangular array of numbers, given by m rows and n columns, is a matrix. Associated with a square matrix A is a real number called a determinant of A, denoted by $|A|$. The procedure for finding the value of a second order determinant, that is, a determinant of a 2×2 matrix, is given by

$$\begin{bmatrix} a & c \\ b & d \end{bmatrix} = ad - bc,$$

where a, b, c, and d are the elements in the matrix.

Second order determinants are used in finding the solution of any system of two linear equations in two variables by using the well-known Cramer's Rule. The procedure for using this rule is clearly explained in Problem #324 in this chapter. Notice that the system of equations must be in standard form before Cramer's Rule can be applied.

Another method for finding the solution of a system of two linear equations in two variables (in standard form) involves writing the system in matrix format and solving for the variables. The procedure includes finding the multiplicative inverse of the matrix involving the coefficients of the variables in the system and then multiplying this matrix throughout the matrix equation. For example, given the system

$$5x - y = 7$$

$$2x + 3y = -1$$

we can write it in matrix form as follows:

$$\begin{bmatrix} 5 & -1 \\ 2 & 3 \end{bmatrix} \cdot \begin{bmatrix} x \\ y \end{bmatrix} = \begin{bmatrix} 7 \\ -1 \end{bmatrix}.$$

Then, the inverse matrix C^{-1} for the coefficients of the variables matrix

$$C = \begin{bmatrix} 5 & -1 \\ 2 & 3 \end{bmatrix}$$

is given by

$$C^{-1} = \left(\frac{1}{\det C}\right)\begin{bmatrix} 3 & 1 \\ -2 & 5 \end{bmatrix} = \left(\frac{1}{17}\right)\begin{bmatrix} 3 & 1 \\ -2 & 5 \end{bmatrix}.$$

Multiplying throughout the equation by C^{-1} we obtain

$$\begin{bmatrix} x \\ y \end{bmatrix} = \frac{1}{17}\cdot\begin{bmatrix} 3 & 1 \\ -2 & 5 \end{bmatrix}\cdot\begin{bmatrix} 7 \\ -1 \end{bmatrix} \quad \text{or}$$

$$\begin{bmatrix} x \\ y \end{bmatrix} = \frac{1}{17}\cdot\begin{bmatrix} 20 \\ -19 \end{bmatrix}.$$

Thus, $x = \dfrac{20}{17}$ and $y = \dfrac{19}{17}$.

A determinant of a 3 × 3 matrix can be found in more than one way. Two popular procedures are highlighted. The first procedure is the so-called "crossing pattern" which starts with rewriting the first two columns in the matrix to the right of the original matrix as follows:

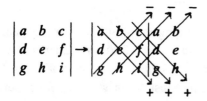

Then, lightly draw in six arrows (as shown above) and form products among the numbers along each arrow. Each arrow pointing upward yields a negative product, while each pointing downward yields a positive product. Finally, the sum of the products is the value of the determinant.

The second approach to finding the value of a determinant of a 3 × 3 matrix is to use the concept of cofactor. The procedure for this concept is formed by multiplying each of the three elements of the first row of a 3 × 3 matrix by its corresponding cofactor and then adding the three results to obtain the determinant. Problem #339 in this chapter clearly illustrates this procedure.

Step-by-Step Solutions to Problems in this Chapter, "Determinants and Matrices"

DETERMINANTS OF SECOND ORDER

● **PROBLEM** 321

Find the value of the determinant $\begin{vmatrix} 1 & 2 \\ 2 & 3 \end{vmatrix}$.

<u>Solution:</u> The value of a 2X2 determinant can be found by the following equation:

$$\begin{vmatrix} a & c \\ b & d \end{vmatrix} = ad - bc .$$

Hence, the value of $\begin{vmatrix} 1 & 2 \\ 2 & 3 \end{vmatrix}$ is :

$$\begin{vmatrix} 1 & 2 \\ 2 & 3 \end{vmatrix} = (1)(3) - (2)(2) = 3 - 4 = -1.$$

● **PROBLEM** 322

Evaluate the determinant

$$\begin{vmatrix} 3 & 5 \\ -2 & 3 \end{vmatrix}$$

<u>Solution:</u> The determinant of any 2 × 2 matrix is

$$\begin{vmatrix} a & b \\ c & d \end{vmatrix} = ad - bc.$$

Apply this rule to the given 2 × 2 matrix.

$$\begin{vmatrix} 3 & 5 \\ -2 & 3 \end{vmatrix} = (3)(3) - (-2)(5) = 9 + 10 = 19$$

● **PROBLEM** 323

Evaluate, or expand, the determinant

$$\begin{vmatrix} 2 & 3 \\ 3 & -1 \end{vmatrix}$$

<u>Solution:</u> The determinant of a 2 by 2 matrix is defined to be

$$\begin{vmatrix} a & b \\ c & d \end{vmatrix} = ad - bc. \text{ Thus the solution is:}$$

$$\begin{vmatrix} 2 & 3 \\ 3 & -1 \end{vmatrix} = (2)(-1) - (3)(3) = -2 - 9 = -11.$$

Solve $\qquad$ $x + y = 3$

$\qquad$ $2x + 3y = 1$

Solution: The values for x and y can be determined by use of Cramer's Rule and determinants. The value for x is the quotient of two determinants. The determinant in the denominator consists of vertical columns in which the numbers are the coefficients of the variables. The determinant in the numerator is the same as the determinant in the denominator, except that the first vertical column is replaced by the constant terms. Note that the first vertical column in the two given equations corresponds to the x term.

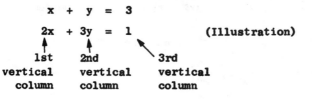

$$x + y = 3$$
$$2x + 3y = 1 \qquad \text{(Illustration)}$$

1st   2nd   3rd
vertical   vertical   vertical
column   column   column

The third vertical column consists of the constant terms.
Hence,

$$x = \frac{\begin{vmatrix} 3 & 1 \\ 1 & 3 \end{vmatrix}}{\begin{vmatrix} 1 & 1 \\ 2 & 3 \end{vmatrix}} = \frac{(3)(3) - (1)(1)}{(1)(3) - (2)(1)} = \frac{9-1}{3-2} = \frac{8}{1} = 8.$$

The value for y is also the quotient of two determinants. The determinant in the denominator is the same as the determinant in the denominator used for finding x. The determinant in the numerator is the same as the determinant in the denominator, except that the second vertical column is replaced by the constant terms. Note that the second vertical column in the two given equations corresponds to the y term. (See the Illustration). Hence,

$$y = \frac{\begin{vmatrix} 1 & 3 \\ 2 & 1 \end{vmatrix}}{\begin{vmatrix} 1 & 1 \\ 2 & 3 \end{vmatrix}} = \frac{(1)(1) - (2)(3)}{(1)(3) - (2)(1)} = \frac{1-6}{3-2} = \frac{-5}{1} = -5 .$$

Solve the equations $2x + 4y = 11$, $-5x + 3y = 5$ graphically and by means of determinants.

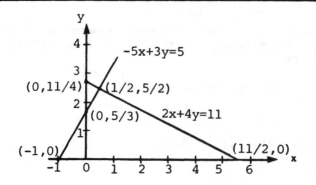

<u>Solution A:</u> To solve a set of linear equations graphically we find their point of intersection (which satisfies both equations simultaneously). Draw both lines by determining their y- and x-intercepts by setting x = 0 and y = 0 respectively. See the following tables. Note: We solve for y before finding the y-intercept and solve for x to find the x-intercept.

$$2x + 4y = 11 \quad ; \qquad 2x + 4y = 11 \qquad\qquad -5x + 3y = 5 \quad ; \quad -5x + 3y = 5$$

$$4y = 11 - 2x \qquad 2x = 11 - 4y \qquad\qquad 3y = 5 + 5x \quad ; \qquad 3y = 5 + 5x$$

$$y = \frac{11 - 2x}{4} \qquad x = \frac{11 - 4y}{2} \qquad\qquad y = \frac{5 + 5x}{3} \qquad 3y = 5 + 5x$$

$$3y - 5 = 5x$$

$$\frac{3y - 5}{5} = x$$

2x + 4y = 11

x	$\frac{11 - 2x}{4}$ =y	y
0	$\frac{11 - 2(0)}{4}$	$\frac{11}{4}$
y	$\frac{11 - 4y}{2}$ = x	x
0	$\frac{11 - 4(0)}{2}$	$\frac{11}{2}$

−5x + 3y = 5

x	$\frac{5 + 5x}{3}$ =y	y
0	$\frac{5 + 5(0)}{3}$	$\frac{5}{3}$
y	$\frac{3y - 5}{5}$ =x	x
0	$\frac{3(0) - 5}{5}$	−1

Now each line can be plotted from two points. We see from the graph that the point of intersection is $\left(\frac{1}{2}, \frac{5}{2}\right)$.

<u>Solution B:</u> To solve a system of two linear equations in two unknowns by determinants, we set up the following solutions in determinant form, derived from the linear equations in standard form.

$$a_1x + b_1y = c_1$$

$$a_2x + b_2y = c_2$$

$$x = \frac{\begin{vmatrix} c_1 & b_1 \\ c_2 & b_2 \end{vmatrix}}{\begin{vmatrix} a_1 & b_1 \\ a_2 & b_2 \end{vmatrix}} \qquad\qquad y = \frac{\begin{vmatrix} a_1 & c_1 \\ a_2 & c_2 \end{vmatrix}}{\begin{vmatrix} a_1 & b_1 \\ a_2 & b_2 \end{vmatrix}}$$

The denominator for both variables is formed by writing the coefficients of x and y in the linear equations. The numerators are formed from the denominator by replacing the column of coefficients of that unknown by the column of constants.

In this case the linear equations in standard form are:

$$2x + 4y = 11$$

$$-5x + 3y = 5$$

Then, the solution by determinants is

$$x = \frac{\begin{vmatrix} 11 & 4 \\ 5 & 3 \end{vmatrix}}{\begin{vmatrix} 2 & 4 \\ -5 & 3 \end{vmatrix}}.$$ The value of a 2X2 determinant is defined to be:

$$\begin{vmatrix} a & b \\ c & d \end{vmatrix} = ad - bc .$$ Therefore,

$$x = \frac{(11)(3) - (4)(5)}{(2)(3) - (4)(-5)} = \frac{33 - 20}{6 - (-20)} = \frac{13}{26} = \frac{1}{2} \ ;$$

$$y = \frac{\begin{vmatrix} 2 & 11 \\ -5 & 5 \end{vmatrix}}{26} = \frac{(2)(5) - (-5)(11)}{26} = \frac{10 - (-55)}{26} = \frac{65}{26} = \frac{5}{2} \ .$$

This agrees with the geometrical solution.

● **PROBLEM** 326

Find the simultaneous solution set of the equations

$$3x - 6y - 2 = 0 \qquad\qquad (9)$$
$$4x + 7y + 3 = 0 \qquad\qquad (10)$$

by use of Cramer's rule.

Solution: We first add 2 to each members of (9) and -3 to each member of (10) and get

$$3x - 6y = 2$$
$$4x + 7y = -3$$

We now obtain the solution set by the following steps:

1. We form the determinant D whose elements are the coefficients of the unknowns in the order in which they appear, and get

$$D = \begin{vmatrix} 3 & -6 \\ 4 & 7 \end{vmatrix}$$

Then recalling that a 2X2 determinant $\begin{vmatrix} a & b \\ c & d \end{vmatrix}$ may be evaluated as $\begin{vmatrix} a & b \\ c & d \end{vmatrix} = ad - bc$, we get

$$(3)(7) - (-6)(4) = 45$$

2. Replace the column of coefficients of x in D by the constant terms and get

$$D_x = \begin{vmatrix} 2 & -6 \\ -3 & 7 \end{vmatrix} = (2)(7) - (-6)(-3) = -4$$

3. Replace the column of coefficients of y in D by the constant terms and get

$$D_y = \begin{vmatrix} 3 & 2 \\ 4 & -3 \end{vmatrix} = (3)(-3) - (2)(4) = -17$$

4. By Cramer's rule

$$x = \frac{D_x}{D} = \frac{-4}{45} = -\frac{4}{45}$$

$$y = \frac{D_y}{D} = \frac{-17}{45} = -\frac{17}{45}$$

Hence the simultaneous solution set is $\left\{ \left(-\frac{4}{45} , -\frac{17}{45} \right) \right\}$.

Check: Replacing x and y in the given equations by the appropriate elements of the solution set, we have

$$3\left(-\frac{4}{45} \right) - 6\left(-\frac{17}{45} \right) - 2 = \frac{-12 + 102 - 90}{45} = 0 \quad \text{from (9)}$$

$$4\left(-\frac{4}{45} \right) + 7\left(-\frac{17}{45} \right) + 3 = \frac{-16 - 119 + 135}{45} = 0 \quad \text{from (10)}$$

● **PROBLEM** 327

Solve by determinants:

$$3x - 5y = 4$$
$$7x + 4y = 25.$$

<u>Solution:</u> The equations, as given, are in standard form for applying Cramer's rule. Therefore,

$$x = \frac{\begin{vmatrix} 4 & -5 \\ 25 & 4 \end{vmatrix}}{\begin{vmatrix} 3 & -5 \\ 7 & 4 \end{vmatrix}} = \frac{4 \cdot 4 - 25(-5)}{3 \cdot 4 - 7(-5)} = \frac{16 + 125}{12 + 35} = \frac{141}{47} = 3$$

$$y = \frac{\begin{vmatrix} 3 & 4 \\ 7 & 25 \end{vmatrix}}{\begin{vmatrix} 3 & -5 \\ 7 & 4 \end{vmatrix}} = \frac{3 \cdot 25 - 7 \cdot 4}{47} = \frac{75 - 28}{47} = \frac{47}{47} = 1$$

This process always yields a unique solution unless the denominator determinant is equal to zero.

● **PROBLEM** 328

Solve the system

$$2x + 3y = 4$$
$$3x - 2y = -2$$

<u>Solution:</u> This system can be solved by Cramer's rule.

The equations are in the standard form to apply the rule. That is, the constants are on one side of the equation while the unknowns are on the opposite side.

Each unknown is the quotient of two determinants. The denominator is the determinant of the coefficients. The numerator is derived from the denominator by substituting the constant terms for the coefficients of the unknown. Thus:

$$x = \frac{\begin{vmatrix} 4 & 3 \\ -2 & -2 \end{vmatrix}}{\begin{vmatrix} 2 & 3 \\ 3 & -2 \end{vmatrix}} = \frac{-8 + 6}{-4 - 9} = \frac{2}{13}$$

$$y = \frac{\begin{vmatrix} 2 & 4 \\ 3 & -2 \end{vmatrix}}{\begin{vmatrix} 2 & 3 \\ 3 & -2 \end{vmatrix}} = \frac{-4 - 12}{-4 - 9} = \frac{16}{13}$$

Check by substitution into the original equations:

$$x = \frac{2}{13} \qquad y = \frac{16}{13}$$

$$2x + 3y = 4 \qquad\qquad 3x - 2y = -2$$

$$2\left(\frac{2}{13}\right) + 3\left(\frac{16}{13}\right) = 4 \qquad 3\left(\frac{2}{13}\right) - 2\left(\frac{16}{13}\right) = -2$$

$$\frac{52}{13} = 4 \qquad\qquad\qquad - 2 = - 2$$

$$4 = 4$$

Solve the system

$$3x + 2y = 12$$

$$4x - 3y = - 1$$

<u>Solution:</u> This system can be solved algebraically by eliminating one variable from the pair of equations or by Cramer's rule of determinants. The equations are in the proper form to use determinants to find x and y. Hence:

$$x = \frac{\begin{vmatrix} 12 & 2 \\ -1 & -3 \end{vmatrix}}{\begin{vmatrix} 3 & 2 \\ 4 & -3 \end{vmatrix}} = \frac{12(-3) - 2(-1)}{3(-3) - 4(2)} = \frac{-34}{-17} = 2$$

$$y = \frac{\begin{vmatrix} 3 & 12 \\ 4 & -1 \end{vmatrix}}{\begin{vmatrix} 3 & 2 \\ 4 & -3 \end{vmatrix}} = \frac{3(-1) - 12(4)}{3(-3) - 4(2)} = \frac{-51}{-17} = 3$$

To verify if the solution x = 2 and y = 3 is correct, substitute both values into the original equations.

Solve the system

$$2x + 3y - 6 = 0$$
$$2y = 3x$$

<u>Solution:</u> We can solve this system of two equations in two unknowns by determinants or by adding and subtracting the equations to eliminate one variable.

For purposes of illustration, we shall use determinants to solve this system. Use Cramer's rule. We must have all the unknowns on one side of the equal sign and the constant terms on the other to apply this rule.

$$2x + 3y = 6$$
$$3x - 2y = 0$$

Each unknown, x and y, is the quotient of two determinants. Let D = denominator which is the determinant of the coefficients.

$$D = \begin{vmatrix} 2 & 3 \\ 3 & -2 \end{vmatrix} = 2(-2) - 3(3) = -4 - 9 = -13$$

The numerator, for each unknown, is obtained from the denominator by substituting the constant terms for the coefficients of the unknown. Thus the numerator for x is:

$$\begin{vmatrix} 6 & 3 \\ 0 & -2 \end{vmatrix} = 6(-2) - 3(0) = -12$$

The numerator for y is:

$$\begin{vmatrix} 2 & 6 \\ 3 & 0 \end{vmatrix} = 2(0) - 6(3) = -18$$

$$\text{and } x = \frac{\begin{vmatrix} 6 & 3 \\ 0 & -2 \end{vmatrix}}{\begin{vmatrix} 2 & 3 \\ 3 & -2 \end{vmatrix}} = -\frac{12}{13}$$

$$y = \frac{\begin{vmatrix} 2 & 6 \\ 3 & 0 \end{vmatrix}}{\begin{vmatrix} 2 & 3 \\ 3 & -2 \end{vmatrix}} = -\frac{18}{13}$$

To check the solution $x = \frac{12}{13}$ and $y = \frac{18}{13}$, substitute these values into the original equations.

$$2x + 3y - 6 = 0$$

$$2\left(\frac{12}{13}\right) + 3\left(\frac{18}{13}\right) - 6 = 0$$

$$\frac{24}{13} + \frac{54}{13} - 6 = 0$$

$$\frac{78}{13} - \frac{78}{13} = 0$$

$$0 = 0$$

$$2y = 3x$$

$$2\left(\frac{18}{13}\right) = 3\left(\frac{12}{13}\right)$$

$$\frac{36}{13} = \frac{36}{13}$$

● **PROBLEM** 331

Use determinants to show that the following system is inconsistent.

$$x + y = 3(x - 2y) + 5 \qquad (1)$$
$$14y - 4x = 11 \qquad (2)$$

Solution: The method of solving a system of equations by determinants is based upon Cramer's rule. Cramer's rule is stated:

205

In a system of n linear equations in n variables, if the determinant of the coefficients is not zero, the system has a unique solution. The value of each variable is a fraction whose denominator is the determinant of the coefficients and whose numerator is the same determinant, with the coefficients of that variable replaced by the corresponding constants.

Thus, if we have two equations arranged in standard form $ax + by = c$ and $dx + ey = f$, then

$$x = \frac{\begin{vmatrix} c & b \\ f & e \end{vmatrix}}{\begin{vmatrix} a & b \\ d & e \end{vmatrix}} = \frac{ce - fb}{ae - db}$$

$$y = \frac{\begin{vmatrix} a & c \\ d & f \end{vmatrix}}{\begin{vmatrix} a & b \\ d & e \end{vmatrix}} = \frac{af - dc}{ae - db}$$

If the numerator is not zero and the denominator is zero, the system is inconsistent. We can approach this particular problem as follows: The determinant can be obtained more readily if the terms are arranged in the standard form, $ax + by = c$. Equation (1) becomes $2x - 7y = -5$ and equation (2) becomes $-4 + 14y = 11$. Hence,

$$x = \frac{\begin{vmatrix} -5 & -7 \\ 11 & 14 \end{vmatrix}}{\begin{vmatrix} 2 & -7 \\ -4 & 14 \end{vmatrix}} = \frac{-5(14) - 11(-7)}{2(14) - (-4)(-7)} = \frac{-70 + 77}{28 - 28} = \frac{7}{0}$$

$$y = \frac{\begin{vmatrix} 2 & -5 \\ -4 & 11 \end{vmatrix}}{\begin{vmatrix} 2 & -7 \\ -4 & 14 \end{vmatrix}} = \frac{2(11) - (-4)(-5)}{0} = \frac{22 - 20}{0} = \frac{2}{0}$$

Since both x and y are of the form $\frac{a}{0}$, $a \ne 0$, the solution set is the empty set, and the system is inconsistent.

If both numerator and denominator are zero, the values of x and y are indeterminate; any (x,y) pair that satisfies one equation will satisfy the other also. Since the two equations have the same solution set, they are dependent equations.

● **PROBLEM 332**

Obtain the simultaneous solution set of the system of equations

$$3x - 4y = -6$$
$$2x + 5y = 19$$

by use of the multiplicative inverse of a matrix.

Solution: We first express the system in matrix notation as

$$\begin{bmatrix} 3 & -4 \\ 2 & 5 \end{bmatrix} \begin{bmatrix} x \\ y \end{bmatrix} = \begin{bmatrix} -6 \\ 19 \end{bmatrix} \qquad (1)$$

The determinant of the matrix $M = \begin{bmatrix} 3 & -4 \\ 2 & 5 \end{bmatrix}$ is $3 \cdot 5 - [(2)(-4)] = 15 + 8 = 23$. The multiplicative inverse of M is

$$m^{-1} = \frac{1}{\det(m)} \times \text{(matrix of the cofactors of each element of the original matrix)}$$

$$= \frac{1}{23}\begin{bmatrix} 5 & 4 \\ -2 & 3 \end{bmatrix}$$

since the cofactor of 3 is 5, the cofactor of −4 is −2, the cofactor of 2 is −(−4) = 4, and the cofactor of 5 is 3.

We now multiply each member of (1) by the inverse of M and get

$$M^{-1}M\begin{bmatrix} x \\ y \end{bmatrix} = M^{-1}\begin{bmatrix} -6 \\ 19 \end{bmatrix}$$

or

$$\frac{1}{23}\begin{bmatrix} 5 & 4 \\ -2 & 3 \end{bmatrix}\begin{bmatrix} 3 & -4 \\ 2 & 5 \end{bmatrix}\begin{bmatrix} x \\ y \end{bmatrix} = \frac{1}{23}\begin{bmatrix} 5 & 4 \\ -2 & 3 \end{bmatrix}\begin{bmatrix} -6 \\ 19 \end{bmatrix}$$

We complete the process as follows:

$$\frac{1}{23}\begin{bmatrix} 15+8 & -20+20 \\ -6+6 & 8+15 \end{bmatrix}\begin{bmatrix} x \\ y \end{bmatrix} = \frac{1}{23}\begin{bmatrix} -30+76 \\ 12+57 \end{bmatrix}$$

$$\frac{1}{23}\begin{bmatrix} 23 & 0 \\ 0 & 23 \end{bmatrix}\begin{bmatrix} x \\ y \end{bmatrix} = \frac{1}{23}\begin{bmatrix} 46 \\ 69 \end{bmatrix}$$

$$\begin{bmatrix} 1 & 0 \\ 0 & 1 \end{bmatrix}\begin{bmatrix} x \\ y \end{bmatrix} = \begin{bmatrix} 2 \\ 3 \end{bmatrix}$$

$$\begin{bmatrix} x \\ y \end{bmatrix} = \begin{bmatrix} 2 \\ 3 \end{bmatrix}$$

Therefore, the simultaneous solution set is $\{(2,3)\}$.

● **PROBLEM 333**

If
$$A = \begin{bmatrix} a_1 & a_2 & a_3 \\ b_1 & b_2 & b_3 \end{bmatrix} \text{ and } B = \begin{bmatrix} x \\ y \\ z \end{bmatrix}$$

find A × B.

Solution: A matrix is a set of numbers in a rectangular arrangement. The numbers which make up a matrix are its elements. Matrix A has 2 rows and 3 columns; it is called a 2×3 matrix, the number of rows being written first. Matrix B has 3 rows and 1 column; it is called a 3×1 matrix. The product of an m×n matrix A by an n×p matrix B is an m×p matrix whose element in the ith row and jth column is the single element in the product of the ith row vector of A by the jth column vector of B. An ith row vector is a 1×n matrix of the form

$$[a_{i1} \ a_{i2} \ \cdots \ a_{in}].$$

A jth column vector is n×1 like
$$\begin{bmatrix} b_{1j} \\ b_{2j} \\ \cdot \\ \cdot \\ \cdot \\ b_{nj} \end{bmatrix}.$$

$$i \times j = [a_{i1} \ a_{i2} \ \cdots \ a_{in}]\begin{bmatrix} b_{1j} \\ b_{2j} \\ \vdots \\ b_{nj} \end{bmatrix} = [a_{i1} \ b_{1j} \cdots \ a_{in} \ b_{nj}].$$

Since A is 2×3, and B is 3×1, A × B is 2×1.

$$A \times B = \begin{bmatrix} a_1 x + a_2 y + a_3 z \\ b_1 x + b_2 y + b_3 z \end{bmatrix}$$

The first row is $a_1 x + a_2 y + a_3 z$. The second row is $b_1 x + b_2 y + b_3 z$. The one column is

$$a_1 x + a_2 y + a_3 z$$
$$b_1 x + b_2 y + b_3 z \quad .$$

If the number of columns in A is not equal to the number of rows in B, the product $A \times B$ is not defined. Furthermore, if A and B are square matrices, (matrices which have the same number of rows as columns), $A \times B$ is usually not equal to $B \times A$.

● **PROBLEM** 334

Obtain the product of

$$A = \begin{bmatrix} 3 & -2 \\ 1 & 4 \end{bmatrix} \quad \text{and} \quad B = \begin{bmatrix} 5 & 1 \\ -2 & 3 \end{bmatrix}$$

Solution: A is a 2×2 matrix. B is a 2×2 matrix. Therefore $A \times B$ is a 2×2 matrix. (A, B, and $A \times B$ are square matrices.)

$$A \times B = \begin{bmatrix} (3)(5) + (-2)(-2) & (3)(1) + (-2)(3) \\ (1)(5) + (4)(-2) & (1)(1) + (4)(3) \end{bmatrix} = \begin{bmatrix} 19 & -3 \\ -3 & 13 \end{bmatrix}$$

Note that

$$B \times A = \begin{bmatrix} 5 & 1 \\ -2 & 3 \end{bmatrix} \times \begin{bmatrix} 3 & -2 \\ 1 & 4 \end{bmatrix} = \begin{bmatrix} (5)(3)+(1)(1) & (5)(-2)+(1)(4) \\ (-2)(3)+(3)(1) & (-2)(-2)+(3)(4) \end{bmatrix}$$

$$= \begin{bmatrix} 16 & -6 \\ -3 & 16 \end{bmatrix} \neq \begin{bmatrix} 19 & -3 \\ -3 & 13 \end{bmatrix} = A \times B$$

● **PROBLEM** 335

If $A = \begin{bmatrix} 3 & -5 \\ 7 & 0 \end{bmatrix}$ and $B = \begin{bmatrix} 2 & 4 \\ -8 & 9 \end{bmatrix}$, find AB and BA.

Solution: The product of two 2×2 matrices is the 2×2 matrix given by the following formula:

$$\begin{pmatrix} a & b \\ c & d \end{pmatrix}\begin{pmatrix} p & q \\ r & s \end{pmatrix} = \begin{pmatrix} ap + br & aq + bs \\ cp + dr & cq + ds \end{pmatrix}$$

We consider the first row of the first matrix and the first column of the second matrix. (See the dotted line between the two matrices). Multiply the number in the first row and first column of the first matrix by the number in the second matrix which is in the same position. Then we multiply the number in the first row and second column of the first matrix by the number in the first column and second row of the second matrix. Adding the two products, we obtain the term in the first row, first column of the product matrix. We perform the multiplication in a similar manner on the second row of the first matrix and the first column of the second matrix, to obtain the second row, first column of the product matrix.

$$\begin{pmatrix} a & b \\ c & d \end{pmatrix}\begin{pmatrix} p & q \\ r & s \end{pmatrix}$$

We do the same for the second column of the second matrix. Therefore,

$$AB = \begin{bmatrix} 3 & -5 \\ 7 & 0 \end{bmatrix} \begin{bmatrix} 2 & 4 \\ -8 & 9 \end{bmatrix} = \begin{bmatrix} 3(2)+(-5)(-8) & 3(4)+(-5)9 \\ 7(2)+(0)(-8) & 7(4)+(0)9 \end{bmatrix} = \begin{bmatrix} 46 & -33 \\ 14 & 28 \end{bmatrix}$$

$$BA = \begin{bmatrix} 2 & 4 \\ -8 & 9 \end{bmatrix} \begin{bmatrix} 3 & -5 \\ 7 & 0 \end{bmatrix} = \begin{bmatrix} 2(3)+4(7) & 2(-5)+4(0) \\ -8(3)+9(7) & (-8)(-5)+9(0) \end{bmatrix} = \begin{bmatrix} 34 & -10 \\ 39 & 40 \end{bmatrix}$$

In this case $AB \neq BA$. Hence, matrix multiplication is not commutative.

● **PROBLEM** 336

Write the solution of the system

$$2x + \pi y = \sqrt{17}$$

$$ix - 23y = 89$$

in terms of determinants.

Solution: The variables x and y can both be written as the quotient of two determinants, using Cramer's Rule. For the variable x, the determinant in the denominator has the coefficients of the x-terms as its first vertical column and the coefficients of the y-terms as its second vertical column. Also for the variable x, the determinant in the numerator is the same as the determinant in the denominator except that the first vertical column is replaced by the constant terms. Hence,

$$x = \frac{\begin{vmatrix} \sqrt{17} & \pi \\ 89 & -23 \end{vmatrix}}{\begin{vmatrix} 2 & \pi \\ i & -23 \end{vmatrix}}$$

For the variable y, the determinant in the denominator is the same as the determinant in the denominator used for the variable x. Also for the variable y, the determinant in the numerator is the same as the determinant in the denominator except that the second vertical column is replaced by the constant terms. Hence,

$$y = \frac{\begin{vmatrix} 2 & \sqrt{17} \\ i & 89 \end{vmatrix}}{\begin{vmatrix} 2 & \pi \\ i & -23 \end{vmatrix}}$$

Note that the column which is replaced by the constant terms in the numerator, is the column which contains the coefficients of the variable we are solving for.

● **PROBLEM** 337

Show, using determinants, that the equations of the following system are dependent.

$$5x - 3y + 7 = 0 \qquad (1)$$

$$x - 2y = 4\left(4x - \frac{11}{4}y + 5\right) + 1 \quad (2)$$

Solution: Rewrite each equation in standard form.

(1) $5x - 3y + 7 = 0$

$5x - 3y = -7$

(2) $x - 2y = 4\left(4x - \frac{11}{4}y + 5\right) + 1$

$$x - 2y = 16x - 11y + 20 + 1$$
$$15x - 9y = -21$$

Hence, we have

$$x = \frac{\begin{vmatrix} -7 & -3 \\ -21 & -9 \end{vmatrix}}{\begin{vmatrix} 5 & -3 \\ 15 & -9 \end{vmatrix}} = \frac{(-7)(-9)-(-21)(-3)}{5(-9) - 15(-3)} = \frac{63 - 63}{-45 + 45} = \frac{0}{0}$$

$$y = \frac{\begin{vmatrix} 5 & -7 \\ 15 & -21 \end{vmatrix}}{\begin{vmatrix} 5 & -3 \\ 15 & -9 \end{vmatrix}} = \frac{5(-21)-15(-7)}{0} = \frac{-105 + 105}{0} = \frac{0}{0}$$

Since both x and y are in the indeterminate form $\frac{0}{0}$, any value may be assigned to one of the variables, and the same corresponding value of the other will satisfy equations.

DETERMINANTS AND MATRICES OF THIRD AND HIGHER ORDERS

● **PROBLEM** 338

Evaluate
$$\begin{vmatrix} 2 & 3 & 5 \\ 1 & 4 & 6 \\ 7 & 2 & 8 \end{vmatrix}$$

Solution: We will use minors to evaluate this determinant. Choose the first row, and call its elements a_1, b_1, c_1 . Then their corresponding minors are A_1, B_1, C_1 . We form the products a_1A_1, b_1B_1, c_1C_1 . Since a_1 is in the first row and the first column, and $1 + 1 = 2$, which is even, the sign of a_1A_1 is positive. Similarly, the sign of b_1B_1 is negative, and that of c_3C_3 is positive. Thus, we have:

$$a_1A_1 - b_1B_1 + c_1C_1 ,$$

and substituting we obtain:

$$2A_1 - 3B_1 + 5C_1 .$$

We find the minors A_1, B_1, and C_1 by eliminating from the determinant the row and column that a_1, b_1, and c_1 are found in. Thus,

$$A_1 = \begin{vmatrix} 4 & 6 \\ 2 & 8 \end{vmatrix}, \quad B_1 = \begin{vmatrix} 1 & 6 \\ 7 & 8 \end{vmatrix}, \quad C_1 = \begin{vmatrix} 1 & 4 \\ 7 & 2 \end{vmatrix}; \text{ we obtain:}$$

$$2\begin{vmatrix} 4 & 6 \\ 2 & 8 \end{vmatrix} - 3\begin{vmatrix} 1 & 6 \\ 7 & 8 \end{vmatrix} + 5\begin{vmatrix} 1 & 4 \\ 7 & 2 \end{vmatrix} .$$

Now, since a determinant of the form $\begin{vmatrix} a & b \\ c & d \end{vmatrix}$ can be equivalently

written as: $ad - bc$, we have:

$$2[(4)(8) - (6)(2)] - 3[(1)(8) - (6)(7)] + 5[(1)(2) - (4)(7)]$$

$$= 2(32 - 12) - 3(8 - 42) + 5(2 - 28)$$

$$= 2(20) - 3(-34) + 5(-26)$$

$$= 40 + 102 - 130$$

$$= 142 - 130$$

$$= 12$$

Therefore, the value of our given determinant is 12.

● **PROBLEM** 339

Evaluate the determinant:
$$D = \begin{vmatrix} 2 & -1 & 2 \\ 3 & 3 & 6 \\ 5 & 0 & -1 \end{vmatrix}$$

Solution: D is a determinant of order three, which may be evaluated by making use of six lines each of which joins the three elements whose product is to be formed.

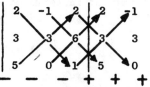

Here the first two columns are rewritten to the right of the determinant. The products formed by following the lines running down from left to right have a plus sign attached, and those formed by following the lines running up from left to right have a negative sign attached. The algebraic sum of the products thus formed is the value of the determinant:

$$D = (2)(3)(-1) + (-1)(6)(5) + (2)(3)(0) - (5)(3)(2) - (0)(6)(2)$$
$$- (-1)(3)(-1)$$

$$D = -6 - 30 + 0 - 30 - 0 - 3$$

$$D = -69$$

● **PROBLEM** 340

Expand the determinant
$$D = \begin{vmatrix} 3 & 2 & 4 \\ 0 & 2 & 0 \\ 1 & 3 & 2 \end{vmatrix}$$

Solution: Here the second row contains two zeros. Hence we shall use this row to get the expansion:

$$D = -0\begin{vmatrix} 2 & 4 \\ 3 & 2 \end{vmatrix} + 2\begin{vmatrix} 3 & 4 \\ 1 & 2 \end{vmatrix} - 0\begin{vmatrix} 3 & 2 \\ 1 & 3 \end{vmatrix}$$

$$= -0[(2)(2) - (4)(3)] + 2[(3)(2) - (4)(1)]$$
$$- 0[(3)(3) - (2)(1)]$$

$$= 0 + 2(6 - 4) - 0 = 0 + 4 - 0 = 4$$

211

Obtain the value of
$$\begin{vmatrix} 2 & 4 & 1 \\ 3 & 6 & 2 \\ 5 & 2 & 4 \end{vmatrix}$$

Solution: If to the elements of any row (any column) of a determinant there is added m times the corresponding elements of another row (another column), the value of the determinant is unchanged. Therefore since the first two elements of the second column are twice the corresponding elements of the first column, if we multiply each element of the first column by -2 and add the product to the corresponding element in the second column, we get

$$\begin{vmatrix} 2 & 4 & 1 \\ 3 & 6 & 2 \\ 5 & 2 & 4 \end{vmatrix} = \begin{vmatrix} 2 & 4+(-2)(2) & 1 \\ 3 & 6+(-2)(3) & 2 \\ 5 & 2+(-2)5 & 4 \end{vmatrix} = \begin{vmatrix} 2 & 0 & 1 \\ 3 & 0 & 2 \\ 5 & -8 & 4 \end{vmatrix}$$

Now if we expand this determinant in terms of the elements of the second column, using the signs -, +, - that appear in the second column of the sign diagram, we get

$$-0 + 0 - (-8)\begin{vmatrix} 2 & 1 \\ 3 & 2 \end{vmatrix} = 8(4 - 3) = 8$$

Find the value of
$$\begin{vmatrix} 29 & 26 & 22 \\ 25 & 31 & 27 \\ 63 & 54 & 46 \end{vmatrix}$$

Solution: We wish to rewrite the given determinant in a simpler form so as to make evaluation less complicated. Adding a multiple of each element in one column to the corresponding element in another column does not change the value of the determinant. Therefore, adding - 1 times the elements of column two to the corresponding elements of column one gives us:

$$\begin{vmatrix} 29 & 26 & 22 \\ 25 & 31 & 27 \\ 63 & 54 & 46 \end{vmatrix} = \begin{vmatrix} 29 + (-1)(26) & 26 & 22 \\ 25 + (-1)(31) & 31 & 27 \\ 63 + (-1)(54) & 54 & 46 \end{vmatrix} = \begin{vmatrix} 3 & 26 & 22 \\ -6 & 31 & 27 \\ 9 & 54 & 46 \end{vmatrix}.$$

Now, adding - 1 times the elements of column two to the corresponding elements of column three, we obtain:

$$\begin{vmatrix} 3 & 26 & 22 + (-1)(26) \\ -6 & 31 & 27 + (-1)(31) \\ 9 & 54 & 46 + (-1)(54) \end{vmatrix} = \begin{vmatrix} 3 & 26 & -4 \\ -6 & 31 & -4 \\ 9 & 54 & -8 \end{vmatrix}.$$

We can again rewrite this determinant as

$$\begin{vmatrix} 3(\ 1) & 26 & -4 \\ 3(-2) & 31 & -4 \\ 3(\ 3) & 54 & -8 \end{vmatrix}$$, and since multiplying each

element of a column of a determinant by a number is
equivalent to multiplying the determinant by that
number, we can write:

$$\begin{vmatrix} 3(\ 1) & 26 & -4 \\ 3(-2) & 31 & -4 \\ 3(\ 3) & 54 & -8 \end{vmatrix} = 3 \begin{vmatrix} 1 & 26 & -4 \\ -2 & 31 & -4 \\ 3 & 54 & -8 \end{vmatrix}$$. Now,

since each element in the last column of the determinant
can be written as a multiple of - 4, we obtain:

$$(3)\,(-4) \begin{vmatrix} 1 & 26 & 1 \\ -2 & 31 & 1 \\ 3 & 54 & 2 \end{vmatrix}$$. We now add

- 1 times the first row to the second row. This gives us:

$$- 12 \begin{vmatrix} 1 & 26 & 1 \\ -2-1 & 31-26 & 1-1 \\ 3 & 54 & 2 \end{vmatrix} = - 12 \begin{vmatrix} 1 & 26 & 1 \\ -3 & 5 & 0 \\ 2 & 54 & 2 \end{vmatrix}$$.

Again, adding - 2 times the first row to the third, we
obtain: $- 12 \begin{vmatrix} 1 & 26 & 1 \\ -3 & 5 & 0 \\ 1 & 2 & 0 \end{vmatrix}$.

We can now use minors to determine the value of
the determinant. Let us choose column three, and call
its elements c_1, c_2, c_3. Then their corresponding minors
are C_1, C_2, C_3. We form the products $c_1 C_1$, $c_2 C_2$, $c_3 C_3$.
Since c_1 is in the first row and the third column, and
$1 + 3 = 4$, which is even, the sign of $c_1 C_1$ is positive.
Similarly, that of $c_2 C_2$ is negative, and that of $c_3 C_3$
is positive. Thus, we have: $c_1 C_1 - c_2 C_2 + c_3 C_3$. Substituting we obtain: $1C_1 - 0C_2 + 0C_3$. The last two
terms vanish. We find the minor C_1 by eliminating from
the determinant the row and column that c_1 is found in.
Thus, $C_1 = \begin{vmatrix} - 3 & 5 \\ 1 & 2 \end{vmatrix}$, and since $c_1 = 1$,

$$c_1 C_1 \text{ also} = \begin{vmatrix} -3 & 5 \\ 1 & 2 \end{vmatrix}$$. Thus, our given determinant

equals: $-12 \begin{vmatrix} -3 & 5 \\ 1 & 2 \end{vmatrix} = -12[(-3)(2) - (5)(1)]$

$$= -12 \ (-6-5)$$

$$= -12 \ (-11) = 132.$$

● **PROBLEM** 343

Find the value of $\begin{vmatrix} 67 & 19 & 21 \\ 39 & 13 & 14 \\ 81 & 24 & 26 \end{vmatrix}$

<u>Solution:</u> Our aim in this problem is to break down the given determinant into one that is easier to evaluate. We can therefore rewrite our determinant as:

$$\begin{vmatrix} 67 & 19 & 21 \\ 39 & 13 & 14 \\ 81 & 24 & 26 \end{vmatrix} = \begin{vmatrix} 10 + 57 & 19 & 21 \\ 0 + 39 & 13 & 14 \\ 9 + 72 & 24 & 26 \end{vmatrix} .$$

Now we can make use of one of the well-known properties of determinants; that is, if each element of a column of a determinant is expressed as the sum of two terms, the determinant can be expressed as the sum of two determinants. Thus,

$$\begin{vmatrix} 10 + 57 & 19 & 21 \\ 0 + 39 & 13 & 14 \\ 9 + 72 & 24 & 26 \end{vmatrix} = \begin{vmatrix} 10 & 19 & 21 \\ 0 & 13 & 14 \\ 9 & 24 & 26 \end{vmatrix} + \begin{vmatrix} 57 & 19 & 21 \\ 39 & 13 & 14 \\ 72 & 24 & 26 \end{vmatrix} .$$

The determinant can again be simplified further. Let us examine the second determinant in the above sum. Remember that multiplying each element in a column of a determinant by a number and adding that product to the corresponding elements in another column does not change the value of the determinant. Therefore, we can perform this on the determinant using - 3 as the number, and adding the product of - 3 and the elements of column two to the corresponding elements of column one. Thus, we obtain:

$$\begin{vmatrix} 57 & 19 & 21 \\ 39 & 13 & 14 \\ 72 & 24 & 26 \end{vmatrix} = \begin{vmatrix} 57 + (-3)(19) & 19 & 21 \\ 39 + (-3)(13) & 13 & 14 \\ 72 + (-3)(24) & 24 & 26 \end{vmatrix} .$$

$$= \begin{vmatrix} 0 & 19 & 21 \\ 0 & 13 & 14 \\ 0 & 24 & 26 \end{vmatrix}.$$

Now, since each element in a column of a determinant is zero, the value of the determinant is zero. Thus, the value of the second determinant in the above sum is zero, and we have:

$$\begin{vmatrix} 67 & 19 & 21 \\ 39 & 13 & 14 \\ 81 & 24 & 26 \end{vmatrix} = \begin{vmatrix} 10 & 19 & 21 \\ 0 & 13 & 14 \\ 9 & 24 & 26 \end{vmatrix}.$$

But, this can be rewritten as:

$$\begin{vmatrix} 10 & 19 & 19+2 \\ 0 & 13 & 13+1 \\ 9 & 24 & 24+2 \end{vmatrix} = \begin{vmatrix} 10 & 19 & 19 \\ 0 & 13 & 13 \\ 9 & 24 & 24 \end{vmatrix} + \begin{vmatrix} 10 & 19 & 2 \\ 0 & 13 & 1 \\ 9 & 24 & 2 \end{vmatrix}.$$

If two columns of a determinant have the same elements, then its value is zero. Thus the first determinant in the above sum is zero, and we are left with:

$$\begin{vmatrix} 10 & 19 & 2 \\ 0 & 13 & 1 \\ 9 & 24 & 2 \end{vmatrix}.$$

We now use minors to determine the value of the determinant. Let us choose column one, and call its elements a_1, a_2, a_3. Then their corresponding minors are A_1, A_2, A_3. We form the products a_1A_1, a_2A_2, a_3A_3. Since a_1 is in the first row and the first column, and $1 + 1 = 2$, which is even, the sign of a_1A_1 is positive. Similarly, the sign of a_2A_2 is negative, and that of a_3A_3 is positive. Thus, we have:

$a_1A_1 - a_2A_2 + a_3A_3$, and substituting we obtain:

$10A_1 - 0A_2 + 9A_3$. The second term vanishes. We find the minors A_1 and A_3 by eliminating from the determinant the row and column that a_1 and a_3 are found in. Thus,

$$A_1 = \begin{vmatrix} 13 & 1 \\ 24 & 2 \end{vmatrix}, \qquad A_3 = \begin{vmatrix} 19 & 2 \\ 13 & 1 \end{vmatrix}$$

and
$$\begin{vmatrix} 67 & 19 & 21 \\ 39 & 13 & 14 \\ 81 & 24 & 26 \end{vmatrix} = \begin{vmatrix} 10 & 19 & 2 \\ 0 & 13 & 1 \\ 9 & 24 & 2 \end{vmatrix}$$

$$= 10 \begin{vmatrix} 13 & 1 \\ 24 & 2 \end{vmatrix} + 9 \begin{vmatrix} 19 & 2 \\ 13 & 1 \end{vmatrix}.$$

Now, these two determinants are easily evaluated. The first,

$$\begin{vmatrix} 13 & 1 \\ 24 & 2 \end{vmatrix} = (13)(2) - (1)(24) = 26 - 24 = 2,$$ and the second

$$\begin{vmatrix} 19 & 2 \\ 13 & 1 \end{vmatrix} = (19)(1) - (2)(13) = 19 - 26 = -7.$$

Thus,
$$\begin{vmatrix} 67 & 19 & 21 \\ 39 & 13 & 14 \\ 81 & 24 & 26 \end{vmatrix} = 10(2) + 9(-7) = 20 - 63 = -43.$$

Another way to approach this problem is to use the expansion scheme for determinants of third order. Using this method we rewrite the given determinant as follows:

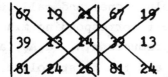

We multiply the elements falling on the same diagonal, thus obtaining six terms. The three terms on the lines sloping downward from left to right have a positive value, and the three on the lines sloping downward from right to left have a negative value. Upon expanding we obtain:

$$(67)(13)(26) + (19)(14)(81) + (21)(39)(24) - (21)(13)(81)$$
$$- (67)(14)(24) - (19)(39)(26).$$

Performing the indicated operations should give us the same value obtained using our previous method, that is -43.

The advantage of the first method is that it does not involve a long multiplication process.

Find the value of

$$\begin{vmatrix} 30 & 11 & 20 & 38 \\ 6 & 3 & 0 & 9 \\ 11 & -2 & 36 & 3 \\ 19 & 6 & 17 & 22 \end{vmatrix}$$

Solution: We can simplify this determinant by multiplying each element of the second column by - 2, and adding this value to the corresponding element in the first column. This does not change the value of the determinant. Doing this we obtain:

$$\begin{vmatrix} 30 + (-2)(11) & 11 & 20 & 38 \\ 6 + (-2)(3) & 3 & 0 & 9 \\ 11 + (-2)(-2) & -2 & 36 & 3 \\ 19 + (-2)(6) & 6 & 17 & 22 \end{vmatrix} \quad \text{or,}$$

$$\begin{vmatrix} 8 & 11 & 20 & 38 \\ 0 & 3 & 0 & 9 \\ 15 & -2 & 36 & 3 \\ 7 & 6 & 17 & 22 \end{vmatrix} .$$

Now, we multiply each element of the second column by - 3, and add this value to the corresponding element in the fourth column. We thus obtain:

$$\begin{vmatrix} 8 & 11 & 20 & 38 + (-3)(11) \\ 0 & 3 & 0 & 9 + (-3)(3) \\ 15 & -2 & 36 & 3 + (-3)(-2) \\ 7 & 6 & 17 & 22 + (-3)(6) \end{vmatrix} \quad \text{or,}$$

$$\begin{vmatrix} 8 & 11 & 20 & 5 \\ 0 & 3 & 0 & 0 \\ 15 & -2 & 36 & 9 \\ 7 & 6 & 17 & 4 \end{vmatrix}$$

We now find the value of the determinant in terms of minors. Let us choose the second row of the determinant. We will call the elements of this row a_2, b_2, c_2, d_2, respectively. Their corresponding minors are A_2, B_2, C_2, D_2. We obtain the value of the determinant by multiplying each element in the chosen row by its corresponding minor as follows:

a_2A_2, b_2B_2, c_2C_2, d_2D_2. We now add each of these.

The signs are determined by the row and column of each element. Since a_2 is the term in the second row and the first column, and $2 + 1 = 3$, an odd number, the sign is negative. Since b_2 is the term in the second row, second column, and $2 + 2 = 4$, an even number, the sign is positive. Similarly, the sign for c_2C_2 is negative, and that for d_2D_2 is positive. Thus the value of the determinant is:

$- a_2A_2 + b_2B_2 - c_2C_2 + d_2D_2$.   Substituting we obtain:

$- 0A_2 + 3B_2 - 0C_2 + 0D_2$.   All terms vanish except the second, $3B_2$. We must now find the minor, B_2. This is done by eliminating the row and column in the determinant which contains $b_2 = 3$. Thus,

$$B_2 = \begin{vmatrix} 8 & 20 & 5 \\ 15 & 36 & 9 \\ 7 & 17 & 4 \end{vmatrix}, \text{ and}$$

the value of the given determinant up to this points is:

$$3 \begin{vmatrix} 8 & 20 & 5 \\ 15 & 36 & 9 \\ 7 & 17 & 4 \end{vmatrix}.$$

We can simplify the above determinant in the following manner; multiply each element of row three by -1 and add this value to the corresponding element in row two. We obtain:

$$3 \begin{vmatrix} 8 & 20 & 5 \\ 15 + (-1)(7) & 36 + (-1)(17) & 9 + (-1)(4) \\ 7 & 17 & 4 \end{vmatrix} =$$

$$3 \begin{vmatrix} 8 & 20 & 5 \\ 8 & 19 & 5 \\ 7 & 17 & 4 \end{vmatrix}$$

Now, multiplying each element of row 2 by -1 and adding this value to the corresponding element in row 1 we have:

218

$$3 \begin{vmatrix} 8 + (-1)(8) & 20 + (-1)(19) & 5 + (-1)(5) \\ 8 & 19 & 5 \\ 7 & 17 & 4 \end{vmatrix} =$$

$$3 \begin{vmatrix} 0 & 1 & 0 \\ 8 & 19 & 5 \\ 7 & 17 & 4 \end{vmatrix} .$$

We can now obtain the value of this determinant in terms of minors. Choose row 1. Then,

$$\begin{vmatrix} 0 & 1 & 0 \\ 8 & 19 & 5 \\ 7 & 17 & 4 \end{vmatrix} = 0A_1 - 1B_1 + 0C_1. \quad \text{But,}$$

$B_1 = \begin{vmatrix} 8 & 5 \\ 7 & 4 \end{vmatrix}$. Thus the value of our given determinant is

$$(3)(-1) \begin{vmatrix} 8 & 5 \\ 7 & 4 \end{vmatrix} = -3 \begin{vmatrix} 8 & 5 \\ 7 & 4 \end{vmatrix} .$$

Now, $\begin{vmatrix} 8 & 5 \\ 7 & 4 \end{vmatrix} = (8)(4) - (5)(7) = 32 - 35 = -3,$

and $(-3)(-3) = 9$. Therefore,

$$\begin{vmatrix} 30 & 11 & 20 & 38 \\ 6 & 3 & 0 & 9 \\ 11 & -2 & 36 & 3 \\ 19 & 6 & 17 & 22 \end{vmatrix} = 9.$$

● PROBLEM 345

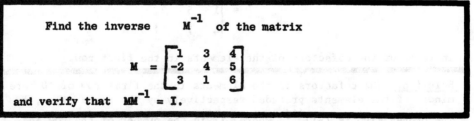

Find the inverse M^{-1} of the matrix

$$M = \begin{bmatrix} 1 & 3 & 4 \\ -2 & 4 & 5 \\ 3 & 1 & 6 \end{bmatrix}$$

and verify that $MM^{-1} = I$.

<u>Solution:</u> We find the determinant of the matrix **M** using the method of minors and cofactors, and expanding in the first row, with the following scheme giving the sign of each minor (in this case each element of the first row).

$$\begin{bmatrix} + & - & + \\ - & + & - \\ + & - & + \end{bmatrix}$$

$$D = \quad 1 \begin{vmatrix} 4 & 5 \\ 1 & 6 \end{vmatrix} -3 \begin{vmatrix} -2 & 5 \\ 3 & 6 \end{vmatrix} +4 \begin{vmatrix} -2 & 4 \\ 3 & 1 \end{vmatrix}$$

$$= 1(24-5) -3(-12-15) +4(-2-12) = 19 + 81 - 56 = 44$$

Furthermore, each entry A_{ij} consists of the cofactor (the determinant of the two by two matrix of entries resulting when any combination of row i and column j is excluded) of each element of the original matrix.

$$A_{11} = \begin{vmatrix} 4 & 5 \\ 1 & 6 \end{vmatrix} = 19 \quad A_{12} = - \begin{vmatrix} -2 & 5 \\ 3 & 6 \end{vmatrix} = 27 \quad \text{and} \quad A_{13} = \begin{vmatrix} -2 & 4 \\ 3 & 1 \end{vmatrix} = -14$$

Similarly, $A_{21} = -14$, $A_{22} = -6$, $A_{23} = 8$, $A_{31} = -1$, $A_{32} = -13$, and $A_{33} = 10$.

Hence, by the definition of the multiplicative inverse, the inverse matrix M^{-1} becomes

$$\frac{1}{\text{determinant of } M} [\text{matrix A}]$$

where A is the matrix of cofactors.

$$M^{-1} = \frac{1}{44} \begin{bmatrix} 19 & -14 & -1 \\ 27 & -6 & -13 \\ -14 & 8 & 10 \end{bmatrix}$$

Therefore,

$$M^{-1}M = \frac{1}{44} \begin{bmatrix} 19 & -14 & -1 \\ 27 & -6 & -13 \\ -14 & 8 & 10 \end{bmatrix} \begin{bmatrix} 1 & 3 & 4 \\ -2 & 4 & 5 \\ 3 & 1 & 6 \end{bmatrix}$$

$$= \frac{1}{44} \begin{bmatrix} 19 + 28 - 3 & 57 - 56 - 1 & 76 - 70 - 6 \\ 27 + 12 - 39 & 81 - 24 - 13 & 108 - 30 - 78 \\ -14 - 16 + 30 & -42 + 32 + 10 & -56 + 40 + 60 \end{bmatrix}$$

$$= \frac{1}{44} \begin{bmatrix} 44 & 0 & 0 \\ 0 & 44 & 0 \\ 0 & 0 & 44 \end{bmatrix} = \begin{bmatrix} 1 & 0 & 0 \\ 0 & 1 & 0 \\ 0 & 0 & 1 \end{bmatrix}$$

● **PROBLEM** 346

Expand the determinant

$$D = \begin{vmatrix} 3 & 2 & 4 \\ 1 & 5 & 2 \\ 4 & 7 & 6 \end{vmatrix}$$

in terms of the cofactors of the elements in the first row.

Solution: The cofactors of the elements in the first row of D are the minors of the elements preceded respectively by the signs $+$, $-$, $+$.

The proper sign is $(-1)^{i+j}$ where i is the number of the row and j is the number of the column in which the element stands. The minor of a given element of a determinant is the determinant of the elements which remain after deleting the row and the column in which the given element is found. The minor of an element a may be denoted by $m(a)$. Then since a = 3, b = 2, and c = 4, we have the corresponding cofactors

$$A = + m(3) = \begin{vmatrix} 5 & 2 \\ 7 & 6 \end{vmatrix} \quad B = -m(2) = -\begin{vmatrix} 1 & 2 \\ 4 & 6 \end{vmatrix} \quad C = +m(4) = \begin{vmatrix} 1 & 5 \\ 4 & 7 \end{vmatrix}$$

Since a determinant may be expressed as the sum of the products formed by multiplying each element of any chosen row (column) by its cofactor, the expansion of D in terms of the cofactors of the elements in the first row is

$$D = 3A - 2B + 4c \ .$$

Hence we have

$$D = 3\begin{vmatrix} 5 & 2 \\ 7 & 6 \end{vmatrix} - 2\begin{vmatrix} 1 & 2 \\ 4 & 6 \end{vmatrix} + 4\begin{vmatrix} 1 & 5 \\ 4 & 7 \end{vmatrix}$$

Evaluating a 2X2 determinant $\begin{vmatrix} a & b \\ c & d \end{vmatrix}$ as ad - bc, D becomes

$$D = 3[(5)(6) - (2)(7)] - 2[(1)(6) - (2)(4)]$$
$$+ 4[(1)(7) - (5)(4)]$$
$$= 3(30 - 14) - 2(6 - 8) + 4(7 - 20)$$
$$= 3(16) - 2(-2) + 4(-13) = 48 + 4 - 52 = 0$$

● PROBLEM 347

Expand the determinant

$$D = \begin{vmatrix} -2 & 4 & 3 \\ 1 & -5 & -6 \\ 3 & 1 & -2 \end{vmatrix}$$

in terms of the cofactors of the elements of the third column and check the result by expanding in terms of the cofactors of the second row.

Solution: Since in the third column $c_1 = 3$, $c_2 = -6$, and $c_3 = -2$ and the signs in the third column of the sign diagram for a 3X3 determinant,

$$\begin{vmatrix} + & - & + \\ - & + & - \\ + & - & + \end{vmatrix},$$

are +, -, +, we have the corresponding cofactors

$$C_1 = +m(3) = \begin{vmatrix} 1 & -5 \\ 3 & 1 \end{vmatrix}, \ C_2 = -m(-6) = -\begin{vmatrix} -2 & 4 \\ 3 & 1 \end{vmatrix},$$

$$C_3 = +m(-2) = \begin{vmatrix} -2 & 4 \\ 1 & -5 \end{vmatrix}$$

$$D = 3C_1 - (-6)C_2 + (-2)C_3 \ .$$

$$D = 3\begin{vmatrix} 1 & -5 \\ 3 & 1 \end{vmatrix} - (-6)\begin{vmatrix} -2 & 4 \\ 3 & 1 \end{vmatrix} + (-2)\begin{vmatrix} -2 & 4 \\ 1 & -5 \end{vmatrix}$$

$$= 3[(1)(1) - (-5)(3)] + 6[(-2)(1) - (4)(3)]$$
$$- 2[(-2)(-5) - (4)(1)]$$
$$= 3(1 + 15) + 6(-2-12) - 2(10 - 4) = 48 - 84 - 12 = -48$$

The signs in the second row of the sign diagram are -, +, -, and the elements are 1, -5, and -6. Hence

221

$$D = -1 \begin{vmatrix} 4 & 3 \\ 1 & -2 \end{vmatrix} + (-5) \begin{vmatrix} -2 & 3 \\ 3 & -2 \end{vmatrix} - (-6) \begin{vmatrix} -2 & 4 \\ 3 & 1 \end{vmatrix}$$

$$= -1[(4)(-2) - (3)(1)] - 5[(-2)(-2) - (3)(3)]$$
$$+ 6[(-2)(1) - (4)(3)]$$
$$= -1(-8-3) + (-5)(4-9) + 6(-2-12)$$
$$= 11 + 25 - 84 = -48$$

If one or more of the elements of a determinant are zero, it is advisable to expand the determinant in terms of the cofactors of the elements of the row or column that contains the greatest number of zeros.

• PROBLEM 348

Use Cramer's rule to solve the system of equations
$$3x + y - 2z = -3$$
$$2x + 7y + 3z = 9$$
$$4x - 3y - z = 7$$

<u>Solution:</u> The terms in the left members are arranged in the proper order and only the constant terms appear in the right members. Hence, we proceed as follows:

$$Ax = B \text{ where } A = \begin{bmatrix} 3 & 1 & -2 \\ 2 & 7 & 3 \\ 1 & -3 & -1 \end{bmatrix} ; \ x = \begin{bmatrix} x \\ y \\ z \end{bmatrix} ; \ B = \begin{bmatrix} -3 \\ 9 \\ 7 \end{bmatrix}$$

Step 1:
The determinant of A is:

$$D = \begin{vmatrix} 3 & 1 & -2 \\ 2 & 7 & 3 \\ 4 & -3 & -1 \end{vmatrix}$$

Compute the determinant D by the method of minors and cofactors. That is, compute the sum of the products of a number, the minor, and the determinant of the two by two matrix resulting when the row and column containing the minor is crossed out. The summation can be along any row or column and the following scheme provides the sign for each minor.

$$\begin{bmatrix} + & - & + \\ - & + & - \\ + & - & + \end{bmatrix}$$

The minors may be the elements along any chosen row or column of the matrix. The cofactor is the determinant of the four terms remaining when the row and column of the chosen minor is eliminated.

$$= 3 \begin{vmatrix} 7 & 3 \\ -3 & -1 \end{vmatrix} -1 \begin{vmatrix} 2 & 3 \\ 4 & -1 \end{vmatrix} -2 \begin{vmatrix} 2 & 7 \\ 4 & -3 \end{vmatrix}$$

The determinant of a two by two matrix is computed as follows
$$\det \begin{bmatrix} a & b \\ c & d \end{bmatrix} = ad - bc.$$

$$= 3(-7+9) - 1(-2-12) - 2(-6-28)$$

$$D = 6 + 14 + 68 = 88$$

Step 2: Now compute D_x, D_y, D_z i.e., the determinants of the matrices resulting when you replace the column containing the co-efficients of the variable under consideration by the constant terms in matrix B while keeping the other two columns the same.

$$D_x = \begin{vmatrix} -3 & 1 & -2 \\ 9 & 7 & 3 \\ 7 & -3 & -1 \end{vmatrix}$$

$$= -3 \begin{vmatrix} 7 & 3 \\ -3 & -1 \end{vmatrix} -1 \begin{vmatrix} 9 & 3 \\ 7 & -1 \end{vmatrix} -2 \begin{vmatrix} 9 & 7 \\ 7 & -3 \end{vmatrix}$$

$$= -3(-7+9) -1(-9-21) -2(-27-49)$$
$$= -6 + 30 + 152 = 176$$

Step 3:

$$D_y = \begin{vmatrix} 3 & -3 & -2 \\ 2 & 9 & 3 \\ 4 & 7 & -1 \end{vmatrix}$$

$$= 3 \begin{vmatrix} 9 & 3 \\ 7 & -1 \end{vmatrix} +3 \begin{vmatrix} 2 & 3 \\ 4 & -1 \end{vmatrix} -2 \begin{vmatrix} 2 & 9 \\ 4 & 7 \end{vmatrix}$$

$$=3(-9-21) +3(-2-12) -2(14-36)$$
$$= -90 - 42 + 44 = -88$$

Step 4:

$$D_z = \begin{vmatrix} 3 & 1 & -3 \\ 2 & 7 & 9 \\ 4 & -3 & 7 \end{vmatrix}$$

$$= 3 \begin{vmatrix} 7 & 9 \\ -3 & 7 \end{vmatrix} -1 \begin{vmatrix} 2 & 9 \\ 4 & 7 \end{vmatrix} -3 \begin{vmatrix} 2 & 7 \\ 4 & -3 \end{vmatrix}$$

$$= 3(49+27) -1(14-36) -3(-6-28)$$

$$= 228 + 22 + 102 = 352$$

Step 5:

$$x = \frac{D_x}{D} = \frac{176}{88} = 2$$

$$y = \frac{D_y}{D} = \frac{-88}{88} = -1$$

$$z = \frac{D_z}{D} = \frac{352}{88} = 4$$

by Cramer's rule. Hence the solution set is $\{(2,-1,4)\}$; it can be checked by the method of substitution.

● **PROBLEM** 349

Using determinants, solve the system

$$2x - y - 2z = 4,$$
$$x + 3y - z = -1,$$
$$x + 2y + 3z = 5.$$

Solution: Use Cramer's rule $\dfrac{\Delta_{i3}}{\Delta_3}$ to solve for the variables x,y,z, i = x,y,z. Δ_{i3} is the determinant of the system, Δ_3, with the ith column replaced by the elements to the right of the equal signs in the system's equations. The determinant of the system is

$$\Delta_3 = \begin{vmatrix} 2 & -1 & -2 \\ 1 & 3 & -1 \\ 1 & 2 & 3 \end{vmatrix}$$

Compute the determinant by expansion in minors and cofactors. For simplicity allow the minors to be the elements of the first column. The following scheme gives the means for determining the sign of the minors.

$$\begin{bmatrix} + & - & + \\ - & + & - \\ + & - & + \end{bmatrix}$$

Each minor is multiplied by its cofactor, i.e., the determinant of the two by two matrix resulting when the row and column containing the minor are crossed out.

$\Delta_3 = 2[3(3)-2(-1)] -1[(-1)(3) -2(-2)] +1[(-1)(-1) -3(-2)] = 2(11)-1(1)$

$+ 1(7) = 28$

Since $\Delta_3 \neq 0$, the system possesses a unique solution. Also allow the minors of Δ_{i3} to be the elements of the first column. Then

we get, by application of the rule stated above,

$$\Delta_3 x = \begin{vmatrix} 4 & -1 & -2 \\ -1 & 3 & -1 \\ 5 & 2 & 3 \end{vmatrix} =$$

$\Delta_{x3} = 4[3(3)-2(-1)] - (-1)[-1(3)-2(-2)]+5[-1(-1)-3(-2)]$

$= 4(11)+1(1)+5(7) = 80,$

$$\Delta_3 y = \begin{vmatrix} 2 & 4 & -2 \\ 1 & -1 & -1 \\ 1 & 5 & 3 \end{vmatrix} =$$

$\Delta_{y3} = 2[-1(3)-5(-1)]-1[4(3) - 5(-2)] +1[4(-1)-(-1)(-2)]$

$= 2(2)-1(22)+1(-6) = -24,$

$$\Delta_3 z = \begin{vmatrix} 2 & -1 & 4 \\ 1 & 3 & -1 \\ 1 & 2 & 5 \end{vmatrix} =$$

$\Delta_{z3} = 2[3(5)-2(-1)]-1[-1(5)-2(4)]+1[-1(-1)-3(4)] = 2(17)-1(-13)+1(-11)= 36,$

whence, since

$$x = \frac{\Delta_{x3}}{\Delta_3} \quad \text{or} \quad \Delta_3 X = \Delta_{x3}$$

$$y = \frac{\Delta_{y3}}{\Delta_3} \quad \text{or} \quad \Delta_3 Y = \Delta_{y3}$$

$$z = \frac{\Delta_{z3}}{\Delta_3} \quad \text{or} \quad \Delta_3 Z = \Delta_{z3},$$

then

$$x = \frac{80}{28} = \frac{20}{7}, \quad y = -\frac{24}{28} = -\frac{6}{7}, \quad z = \frac{36}{28} = \frac{9}{7}.$$

That this solution satisfies the given system of equations is readily verified.

● **PROBLEM** 350

Show that the following equations are not independent:

$$5x + 4y + 11z = 3$$
$$6x - 4y + 2z = 1$$
$$x + 3y + 5z = 2$$

Solution: To show that a system of linear equations is independent first compute the determinant of the matrix of the coefficients. If this result is equal to zero the equations are dependent and if the result is not equal to zero the equations are independent or consistent. The matrix of coefficients is:

$$m = \begin{bmatrix} 5 & 4 & 11 \\ 6 & -4 & 2 \\ 1 & 3 & 5 \end{bmatrix}$$

Call the determinant of (m), D. Thus

$$D = \begin{vmatrix} 5 & 4 & 11 \\ 6 & -4 & 2 \\ 1 & 3 & 5 \end{vmatrix} = 5 \begin{vmatrix} -4 & 2 \\ 3 & 5 \end{vmatrix} -4 \begin{vmatrix} 6 & 2 \\ 1 & 5 \end{vmatrix} +11 \begin{vmatrix} 6 & -4 \\ 1 & 3 \end{vmatrix}$$

$$= 5(-20 - 6) - 4(30 - 2) + 11(18 + 4)$$

$$D = -130 - 112 + 242 = 0$$

Hence, since D = 0, the equations are not independent, and no unique solution set exists.

● **PROBLEM** 351

Obtain the simultaneous solution set of
$$x + 3y + 4z = 15$$
$$-2x + 4y + 5z = 12$$
$$3x + y + 6z = 29$$

<u>Solution:</u> The matrix notation for the system of equations is

$$MX = B \text{ where } M = \begin{bmatrix} 1 & 3 & 4 \\ -2 & 4 & 5 \\ 3 & 1 & 6 \end{bmatrix} ; \quad X = \begin{bmatrix} x \\ y \\ z \end{bmatrix} ; \quad B = \begin{bmatrix} 15 \\ 12 \\ 29 \end{bmatrix}$$

or

$$\begin{bmatrix} 1 & 3 & 4 \\ -2 & 4 & 5 \\ 3 & 1 & 6 \end{bmatrix} \begin{bmatrix} x \\ y \\ z \end{bmatrix} = \begin{bmatrix} 15 \\ 12 \\ 29 \end{bmatrix} \qquad (2)$$

The matrix of the coefficients is the matrix M discussed in the previous problem, and its inverse M^{-1} was found to be

$$\frac{1}{44} \begin{bmatrix} 19 & -14 & -1 \\ 27 & -6 & -13 \\ -14 & 8 & 10 \end{bmatrix}$$

Therefore, if we multiply each member of (2) by M^{-1}, we get $M^{-1}MX = M^{-1}B$ or

$$\begin{bmatrix} 1 & 0 & 0 \\ 0 & 1 & 0 \\ 0 & 0 & 1 \end{bmatrix} \begin{bmatrix} x \\ y \\ z \end{bmatrix} = \frac{1}{44} \begin{bmatrix} 19 & -14 & -1 \\ 27 & -6 & -13 \\ -14 & 8 & 10 \end{bmatrix} \begin{bmatrix} 15 \\ 12 \\ 29 \end{bmatrix}$$

which is IX = M^{-1}B. Then, performing the indicated operations, we have X = M^{-1}B or

$$\begin{bmatrix} x \\ y \\ z \end{bmatrix} = \frac{1}{44} \begin{bmatrix} 285 - 168 - 29 \\ 405 - 72 - 377 \\ -210 + 96 + 290 \end{bmatrix} = \frac{1}{44} \begin{bmatrix} 88 \\ -44 \\ 176 \end{bmatrix} = \begin{bmatrix} 2 \\ -1 \\ 4 \end{bmatrix}$$

Therefore, the simultaneous solution set is $\{(2,-1,4)\}$.

● **PROBLEM** 352

Solve by determinants
$$2x - y - z = -3$$

$$x + y + z = 6$$

$$x - 2y + 3z = 6$$

<u>Solution:</u> Three equations in three unknowns can be solved algebrically by eliminating the same variable from two different pairs of equations.

We can also solve this system using Cramer's rule as in the system with only two unknowns.

The given equations are in the appropriate form to set up the needed determinants. Thus:

$$x = \frac{\begin{vmatrix} -3 & -1 & -1 \\ 6 & 1 & 1 \\ 6 & -2 & 3 \end{vmatrix}}{\begin{vmatrix} 2 & -1 & -1 \\ 1 & 1 & 1 \\ 1 & -2 & 3 \end{vmatrix}} = \frac{15}{15} = 1$$

$$y = \frac{\begin{vmatrix} 2 & -3 & -1 \\ 1 & 6 & 1 \\ 1 & 6 & 3 \end{vmatrix}}{\begin{vmatrix} 2 & -1 & -1 \\ 1 & 1 & 1 \\ 1 & -2 & 3 \end{vmatrix}} = \frac{30}{15} = 2$$

$$z = \frac{\begin{vmatrix} 2 & -1 & -3 \\ 1 & 1 & 6 \\ 1 & -2 & 6 \end{vmatrix}}{\begin{vmatrix} 2 & -1 & -1 \\ 1 & 1 & 1 \\ 1 & -2 & 3 \end{vmatrix}} = \frac{45}{15} = 3$$

If 1, 2, and 3, are substituted for x, y, and z, respectively, in each of the three equations, it will be found that they satisfy each equation, and therefore x = 1, y = 2, and z = 3, is the correct solution of the given system of equations.

Check:

$$2x - y - z = -3 \qquad x + y + z = 6 \qquad x - 2y + 3z = 6$$

$$2(1) - 2 - 3 = -3 \qquad 1 + 2 + 3 = 6 \qquad 1 - 2(2) + 3(3) = 6$$

$$-3 = -3 \qquad\qquad 6 = 6 \qquad\qquad 6 = 6$$

Without expanding, prove that

$$D = \begin{vmatrix} x & y & 2x \\ z & w & 2z \\ u & v & 2u \end{vmatrix} = 0$$

__Solution:__ If the elements of a column (row) of a determinant are multiplied by any number m, the determinant is multiplied by m. Therefore,

$$D = \begin{vmatrix} x & y & 2x \\ z & w & 2z \\ u & v & 2u \end{vmatrix} = 2 \begin{vmatrix} x & y & x \\ z & w & z \\ u & v & u \end{vmatrix}$$

since the elements of the third column of D are multiples of 2 .
Since two rows (two columns) of this determinant are identical, the value of

$$D = \begin{vmatrix} x & y & 2x \\ z & w & 2z \\ u & v & 2u \end{vmatrix} = 2 \begin{vmatrix} x & y & x \\ z & w & z \\ u & v & u \end{vmatrix} = 0$$

(the first and third columns are identical)

If $D_1 = \begin{vmatrix} a & b & c \\ d & e & f \\ g & h & k \end{vmatrix}$, $D_2 = \begin{vmatrix} a & g & x \\ b & h & y \\ c & k & z \end{vmatrix}$

and $d = tx$, $e = ty$, $f = tz$, prove without expanding that
$D_1 = -tD_2$.

__Solution:__ In D_1 if we replace d by tx, e by ty and f by
tz, we have

$$D_1 = \begin{vmatrix} a & b & c \\ tx & ty & tz \\ g & h & k \end{vmatrix}$$

If the elements of a row (a column) of a determinant are
multiplied by any number t, the determinant is multiplied
by t. Therefore

$$D_1 = \begin{vmatrix} a & b & c \\ tx & ty & tz \\ g & h & k \end{vmatrix} = \begin{vmatrix} a & b & c \\ t(x) & t(y) & t(z) \\ g & h & k \end{vmatrix}$$

$$= t \begin{vmatrix} a & b & c \\ x & y & z \\ g & h & k \end{vmatrix}$$

$$= t \begin{vmatrix} a & x & g \\ b & y & h \\ c & z & k \end{vmatrix}$$

since the rows of this determinant are the columns of the one just above.

Now, since interchanging two columns (rows) of a determinant changes the sign of the determinant, interchanging columns 2 and 3 gives us:

$$= -t \begin{vmatrix} a & g & x \\ b & h & y \\ c & k & z \end{vmatrix} \quad \text{But} \quad \begin{vmatrix} a & g & x \\ b & h & y \\ c & k & z \end{vmatrix} = D_2;$$

hence, $D_1 = -tD_2$.

● **PROBLEM** 355

Show that $\begin{vmatrix} b + c & a - b & a \\ c + a & b - c & b \\ a + b & c - a & c \end{vmatrix} = 3abc - a^3 - b^3 - c^3$.

Solution: The following is a known property of determinants:

$$\begin{vmatrix} a_1 + \overline{a}_1 & b_1 & c_1 \\ a_2 + \overline{a}_2 & b_2 & c_2 \\ a_3 + \overline{a}_3 & b_3 & c_3 \end{vmatrix} =$$

$$\begin{vmatrix} a_1 & b_1 & c_1 \\ a_2 & b_2 & c_2 \\ a_3 & b_3 & c_3 \end{vmatrix} + \begin{vmatrix} \overline{a}_1 & b_1 & c_1 \\ \overline{a}_2 & b_2 & b_2 \\ \overline{a}_3 & b_3 & c_3 \end{vmatrix}.$$ Notice that in our given

determinant there are two columns with their elements expressed as the sum of two terms. We will first apply the above property to the first column. We thus obtain:

$$\begin{vmatrix} b + c & a - b & a \\ c + a & b - c & b \\ a + b & c - a & c \end{vmatrix} = \begin{vmatrix} b & a - b & a \\ c & b - c & b \\ a & c - a & c \end{vmatrix} + \begin{vmatrix} c & a - b & a \\ a & b - c & b \\ b & c - a & c \end{vmatrix}.$$

Now, applying this property to the second column of both determinants on the right side of the equal sign we obtain:

$$\begin{vmatrix} b & a & a \\ c & b & b \\ a & c & c \end{vmatrix} + \begin{vmatrix} b & -b & a \\ c & -c & b \\ a & -a & c \end{vmatrix} + \begin{vmatrix} c & a & a \\ a & b & b \\ b & c & c \end{vmatrix} + \begin{vmatrix} c & -b & a \\ a & -c & b \\ b & -a & c \end{vmatrix}.$$

228

But, if each element in a column of a determinant is multiplied by a number p, in this case p = - 1, then the value of the determinant is multiplied by p. That is,

$$\begin{vmatrix} a_1 & pb_1 & c_1 \\ a_2 & pb_2 & c_2 \\ a_3 & pb_3 & c_3 \end{vmatrix} = p \begin{vmatrix} a_1 & b_1 & c_1 \\ a_2 & b_2 & c_2 \\ a_3 & b_3 & c_3 \end{vmatrix} .$$

Thus, our above determinants become,

$$= \begin{vmatrix} b & a & a \\ c & b & b \\ a & c & c \end{vmatrix} - \begin{vmatrix} b & b & a \\ c & c & b \\ a & a & c \end{vmatrix} + \begin{vmatrix} c & a & a \\ a & b & b \\ b & c & c \end{vmatrix} - \begin{vmatrix} c & b & a \\ a & c & b \\ b & a & c \end{vmatrix} .$$

Recall that when two columns of a determinant are identical, the value of the determinant is zero. Thus, the first three determinants vanish and we are left with

$$- \begin{vmatrix} c & b & a \\ a & c & b \\ b & a & c \end{vmatrix} .$$

To evaluate this third order determinant we employ the following method: rewrite the first two columns of the determinant next to the third column, obtaining:

$$- \begin{vmatrix} c & b & a \\ a & c & b \\ b & a & c \end{vmatrix} \begin{matrix} c & b \\ a & c \\ b & a \end{matrix} .$$

Draw three diagonal lines sloping downward from left to right, each of which encompasses three elements of the determinant. Do this also from right to left.

The diagram now looks like: -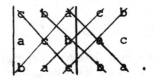

We now form the products of the elements in each of the six diagonals, preceding each of the terms in the left to right diagonals by a positive sign, and each of the terms in the right to left diagonals by a negative sign. The sum of the six products is the required expansion of the determinant. Thus, we obtain:

$$- (c \cdot c \cdot c + b \cdot b \cdot b + a \cdot a \cdot a - acb - cba - bac) =$$

$$- c^3 - b^3 - a^3 + 3abc = 3abc - a^3 - b^3 - c^3 .$$

Show that
$$\begin{vmatrix} a-b-c & 2a & 2a \\ 2b & b-c-a & 2b \\ 2c & 2c & c-a-b \end{vmatrix} = (a+b+c)^3.$$

<u>Solution:</u> We can rewrite the given determinant by adding each element in the third row to its corresponding element in the second row, and then adding each sum to the corresponding element in the first row. Thus,

$$\begin{vmatrix} a-b-c & 2a & 2a \\ 2b & b-c-a & 2b \\ 2c & 2c & c-a-b \end{vmatrix} =$$

$$\begin{vmatrix} a-b-c+(2b+2c) & 2a+(b-c-a+2c) & 2a+(2b+c-a-b) \\ 2b & b-c-a & 2b \\ 2c & 2c & c-a-b \end{vmatrix}$$

$$= \begin{vmatrix} a+b+c & a+b+c & a+b+c \\ 2b & b-c-a & 2b \\ 2c & 2c & c-a-b \end{vmatrix}.$$

Now, since multiplying each element of a row of a determinant by a number is equivalent to multiplying the determinant by that number, we can factor out $(a+b+c)$ from the first row obtaining:

$$(a+b+c) \times \begin{vmatrix} 1 & 1 & 1 \\ 2b & b-c-a & 2b \\ 2c & 2c & c-a-b \end{vmatrix}.$$

We now leave the first column unaltered, and subtract each element of the first column from the corresponding element in the second and third columns. Doing this we obtain:

$$(a+b+c) \times \begin{vmatrix} 1 & 1-1 & 1-1 \\ 2b & b-c-a-2b & 2b-2b \\ 2c & 2c-2c & c-a-b-2c \end{vmatrix} =$$

$$(a+b+c) \times \begin{vmatrix} 1 & 0 & 0 \\ 2b & -b-c-a & 0 \\ 2c & 0 & -c-a-b \end{vmatrix}.$$

230

We can now use minors to determine the value of the determinant. Let us choose row one, and call its elements a_1, b_1, c_1. Then their corresponding minors are A_1, B_1, C_1. We form the products a_1A_1, b_1B_1, c_1C_1. Since a_1 is in the first row and the first column, and $1 + 1 = 2$, which is even, the sign of a_1A_1 is positive. Similarly, that of b,B is negative, and that of $c_1 C_1$ is positive. Thus, we have $a_1A_1 - b_1B_1 + c_1C_1$. Substituting we obtain: $1A_1 - 0B_1 + 0C_1$. The last two terms vanish. We find the minor A_1 by eliminating from the determinant the row and column that a_1 is found in. Thus,

$$A_1 = \begin{vmatrix} -b-c-a & 0 \\ 0 & -c-a-b \end{vmatrix}, \text{ and since}$$

$a_1 = 1$, a_1A_1 also equals $\begin{vmatrix} -b-c-a & 0 \\ 0 & -c-a-b \end{vmatrix}$.

Thus, our given determinant equals:

$$= (a+b+c) \times \begin{vmatrix} -b-c-a & 0 \\ 0 & -c-a-b \end{vmatrix}.$$

But $\begin{vmatrix} -b-c-a & 0 \\ 0 & -c-a-b \end{vmatrix} = (-b-c-a)(-c-a-b) - (0)(0)$

$$= bc + c^2 + ac + ab + ac + a^2 + b^2 + bc + ab$$

$$= (a+b+c)^2,$$

and $(a+b+c) \times \begin{vmatrix} -b-c-a & 0 \\ 0 & -c-a-b \end{vmatrix} = (a+b+c)(a+b+c)^2$

$$= (a+b+c)^3.$$

CHAPTER 15

FACTORING EXPRESSIONS AND FUNCTIONS

> ## Basic Attacks and Strategies for Solving Problems in this Chapter. See pages 232 to 264 for step-by-step solutions to problems.

In carrying out the factorization of a non-fractional polynomial expression, the following general steps are applicable:

(1) Determine if there is a greatest common factor in the polynomial and then factor it out. For example, 3 is the greatest common monomial factor of $12x - 3$ since it can be factored as $3(4x - 1)$.

(2) If the polynomial has two terms (a binomial), then see if it is the difference of two squares, or the sum or difference of two cubes, and then factor accordingly. Remember, if it is the sum of two squares then it will not be factorable.

(3) If the polynomial has three terms (a trinomial), then it is either a perfect square trinomial which will factor into the square of a binomial, or it is not a perfect square trinomial, in which case you use a trial and error method. For example, the factorization of

$$6x^2 - x - 2 \quad \text{is} \quad (2x + 1)(3x - 2),$$

found by trial and error after finding the possible factors of the first and last terms in the polynomial.

(4) If the polynomial has more than three terms, then try to factor it by grouping. For example, the factorization of the expression

$$ax + bx + 2a + 2b$$

is found by grouping and factoring as follows:

$$x(a + b) + 2(a + b) = (x + 2)(a + b).$$

(5) As a final check, see if any of the factors you have written can be factored further. If you have overlooked a common factor, you can catch it as this point.

Factoring a fractional or rational expression is done by first factoring the numerator and denominator and then dividing the numerator and denominator by any factors they have in common. This process is called reducing the expression to the lowest terms.

The procedure for combining two or more rational expressions depends on the operation(s) involved and whether the denominators are alike or different. When adding or subtracting fractional expressions with different denominators, find the LCD for all denominators and change each rational expression to an equivalent expression that has the LCD. Then, add or subtract the numerators of the expressions and apply the common denominator. If the denominators are alike, simply add or subtract the numerators and apply the common denominator.

To multiply two or more fractional expressions, multiply numerators and multiply denominators to obtain the product. Reduce this product to the lowest terms.

Finally, to divide one fractional expression by another fractional expression, simply multiply the first expression by the reciprocal of the second expression. Then, reduce the result (quotient) to the lowest terms.

Step-by-Step Solutions to Problems in this Chapter, "Factoring Expressions and Functions"

NONFRACTIONAL

● **PROBLEM** 357

Factor $x(y + z) + u(y + z)$.

Solution: By the distributive law,

$$x(y + z) + u(y + z) = (x + u)(y + z) \qquad (1)$$

To check if this factoring is correct, distribute the two products on the left side of equation (1).

$$x(y + z) + u(y + z) = (xy + xz) + (uy + uz)$$

$$\text{or } x(y + z) + u(y + z) = xy + xz + uy + uz \qquad (2)$$

Calculating the product on the right side of equation (1):

$$(x + u)(y + z) = xy + uy + xz + uz$$

$$\text{or } (x + u)(y + z) = xy + xz + uy + uz \qquad (3)$$

Since the right side of equations (2) and (3) are equal, equation (1) is true.

● **PROBLEM** 358

Factor

A) $4a^2b - 2ab$

B) $9ab^2c^3 - 6a^2c + 12ac$

C) $ac + bc + ad + bd$

Solution: Find the highest common factor of each polynomial.

A) $4a^2b = 2 \cdot 2 \cdot a \cdot a \cdot b$

$2ab = 2 \cdot a \cdot b$

The highest common factor of the two terms is therefore 2ab. Hence,

$$4a^2b - 2ab = 2ab(2a - 1)$$

B) $9ab^2c^3 = 3 \cdot 3 \cdot a \cdot b \cdot b \cdot c \cdot c \cdot c$

$6a^2c = 3 \cdot 2 \cdot a \cdot a \cdot c$

$12ac = 3 \cdot 2 \cdot 2 \cdot a \cdot c$

The highest common factor of the three terms is 3ac. Then,

$$9ab^2c^3 - 6a^2c + 12ac = 3ac\left(3b^2c^2 - 2a + 4\right)$$

C) An expression may sometimes be factored by grouping terms having a common factor and thus getting new terms containing a common factor. The type form for this case is ac+bc+ ad+bd, because the terms ac and bc have the common factor c, and ad and bd have the common factor d. Then,

$$ac + bc + ad + bd = c(a + b) + d(a + b)$$

Factoring out $(a + b)$, we obtain:

$$= (a + b)(c + d).$$

● **PROBLEM** 359

Factor the following polynomials:

(a) $15\,ac + 6bc - 10ad - 4bd$

(b) $3a^2c + 3a^2d^2 + 2b^2c + 2b^2d^2$.

Solution: (a) Group terms which have a common factor. Here, they are already grouped. Then factor.

$$15\,ac + 6\,bc - 10ad - 4bd = 3c(5a + 2b) - 2d(5a + 2b)$$

Factoring out $(5a + 2b)$

$$3c(5a + 2b) - 2d(5a + 2b) = (5a + 2b)(3c - 2d),$$

(b) Apply the same method as in (a),

$$3a^2c + 3a^2d^2 + 2b^2c + 2b^2d^2 = 3a^2\left(c + d^2\right) + 2b^2\left(c + d^2\right)$$
$$= \left(c + d^2\right)\left(3a^2 + 2b^2\right).$$

● **PROBLEM** 360

Factor $ax + by + ay + bx$ completely.

Solution: Group the terms which have a common factor. The first and last terms have the factor x in common, while the second and third terms have the factor y in common. Hence we may rewrite the expression, and we have

$$ax + bx + ay + by = x(a + b) + y(a + b)$$

Now factor out $(a + b)$

$$= (a + b)(x + y).$$

● **PROBLEM** 361

Factor $x^2 + 7x + 12.$

Solution: Since $7x = 4x + 3x$, $x^2 + 7x + 12 = x^2 + 4x + 3x + 12$. Factor out the common factor of x from the first two terms. Also, factor out the common factor of 3 from the last two terms. Therefore,

$$x^2 + 7x + 12 = x(x + 4) + 3(x + 4).$$

Now factor out the common factor of $(x + 4)$ from the right side to obtain:

$$x^2 + 7x + 12 = (x + 4)(x + 3).$$

Factor $xy - 3y + y^2 - 3x$ completely.

<u>Solution:</u> Note that the first and last terms have a common factor of x. Also note that the second and third factors have a common factor of y. Hence, group the x and y terms together and factor out the x and y from their respective two terms. Therefore,

$$xy - 3y + y^2 - 3x = (xy - 3x) + \left(-3y + y^2\right)$$

Since $\left(-3y + y^2\right) = \left(y^2 - 3y\right),$

$$xy - 3y + y^2 - 3x = (xy - 3x) + \left(y^2 - 3y\right)$$
$$= x(y - 3) + y(y - 3)$$

Now factor out the common factor $(y - 3)$ from both terms:

$$xy - 3y + y^2 - 3x = (x + y)(y - 3).$$

Factor $2ax - 3by - 2ay + 3bx$ completely.

<u>Solution:</u> Note that the first and the last terms have a common factor of x. Also note that the second and third factors have a common factor of y. Hence, group the x and y terms and factor out the x and y from their respective two terms. Therefore:

$$2ax - 3by - 2ay + 3bx = (2ax + 3bx) + (-3by - 2ay)$$
$$= x(2a + 3b) + y(-3b - 2a)$$

Since $-3b - 2a = -2a - 3b,$

$$2ax - 3by - 2ay + 3bx = x(2a + 3b) + y(-2a - 3b)$$

Factor out -1 from the term $(-2a - 3b)$:

$$(-2a - 3b) = -1(2a + 3b).$$

Therefore,

$$2ax - 3by - 2ay + 3bx = x(2a + 3b) + y\left((-1)(2a + 3b)\right)$$
$$= x(2a + 3b) - y(2a + 3b)$$

Factoring out the common factor $(2a + 3b)$ from both terms:

$$2ax - 3by - 2ay + 3bx = (x - y)(2a + 3b).$$

Factor $x^2 - x - 12$ over the integers.

<u>Solution:</u> A quadratic equation whose roots are a and b may be written in the form: $(x - a)(x - b) = x^2 - (a + b)x + ab = 0$. Considering $x^2 - x - 12$, the coefficient of x is -1 and the constant is -12; thus, we want to find the values for a and b such that

$$a + b = 1 \quad \text{and} \quad a \cdot b = -12$$

One of the numbers must be negative and the other one positive, since only a negative multiplied by a positive will give us a negative quantity (-12) for $a \cdot b$.

After examining the possible factors of -12, we find that 4 and -3 are the desired ones since $4 + (-3) = 1$. Thus, let $a = 4$, $b = -3$, and

$$x^2 - x - 12 = x^2 - (4 - 3)x + (4)(-3)$$
$$= (x - 4)(x + 3).$$

● **PROBLEM 365**

Factor $x^2 + 7x + 10$.

<u>Solution:</u> We are given $x^2 + 7x + 10$. We may use the formula

$$x^2 + (b+c)x + bc = (x+b)(x+c) \qquad (1)$$

to factor this polynomial. That is, we set

$$x^2 + 7x + 10 = x^2 + (b+c)x + bc .$$

Thus the coefficient of the x term, $7 = b + c$ and $10 = bc$. We now must find the two numbers b and c whose sum is seven and whose product is 10. We first check all **pairs** of numbers whose product is ten:

    (a) $1 \times 10 = 10$; hence $b = 1$, $c = 10$.

We reject these values because $b + c$ must equal 7, but $1 + 10 = 11$.

    (b) $2 \times 5 = 10$; hence $b = 2$, $c = 5$.

We check the sum of these values, and note $b + c = 2 + 5 = 7$.

Thus $b = 2$ and $c = 5$ are the correct values. Now we go back to equation (1)

$$x^2 + (b+c)x + bc = (x+b)(x+c),$$ and substituting our values of b

and c we obtain:

$$x^2 + (2+5)x + (2 \cdot 5) = (x+2)(x+5)$$
$$x^2 + 7x + 10 = (x+2)(x+5) .$$

● **PROBLEM 366**

Factor $(z + 1)^2 - b^2$.

<u>Solution:</u> Since $(z + 1)^2 - b^2$ is the difference of two squares we apply the formula for the difference of two squares, $x^2 - y^2 = (x + y)(x - y)$, replacing x by $(z + 1)$ and y by b to obtain:

$$(z + 1)^2 - b^2 = [(z + 1) + b][(z + 1) - b].$$

● **PROBLEM 367**

Factor $2x^2 - 3y^2$.

Solution: Since a = $(\sqrt{a})^2$, 2 = $(\sqrt{2})^2$ and 3 = $(\sqrt{3})^2$.

Therefore, $2x^2 - 3y^2 = (\sqrt{2}x)^2 - (\sqrt{3}y)^2$

$(\sqrt{2}x)^2 - (\sqrt{3}y)^2$ is the difference of two squares, hence we apply the formula for the difference of two squares,

$u^2 - v^2 = (u + v)(u - v)$, letting u = $\sqrt{2}x$ and v = $\sqrt{3}y$.

Thus, we obtain: $(\sqrt{2}x + \sqrt{3}y)(\sqrt{2}x - \sqrt{3}y)$

● **PROBLEM** 368

Factor the expression $16a^2 - 4(b - c)^2$.

Solution: 16 = 4^2, thus $16a^2 = 4^2a^2$. Since $a^x b^x = (ab)^x$, $16a^2 = 4^2a^2 = (4a)^2$. Similarly 4 = 2^2, thus $4(b-c)^2 = 2^2(b-c)^2 = [2(b-c)]^2$. Hence $16a^2 - 4(b-c)^2 = (4a)^2 - [2(b-c)]^2$. We are now dealing with the difference of two squares. Applying the formula for the difference of two squares, $x^2 - y^2 = (x+y)(x-y)$, and replacing x by 4a and y by 2(b-c) we obtain:

$(4a)^2 - [2(b-c)]^2 = [4a + 2(b-c)][4a - 2(b-c)]$

$= (4a + 2b - 2c)(4a - 2b + 2c).$

Therefore, $16a^2 - 4(b-c)^2 = (4a + 2b - 2c)(4a - 2b + 2c)$.

● **PROBLEM** 369

Factor $4x^2y^2 - 36x^2z^2$ completely.

Solution: We observe,

$4x^2y^2 = 2^2x^2y^2$ and $36x^2z^2 = 6^2x^2z^2$.

Since,

$x^a y^a z^a = (xyz)^a$, $2^2x^2y^2 = (2xy)^2$

and,

$6^2x^2z^2 = (6xz)^2$.

Thus,

$4x^2y^2 - 36x^2z^2 = (2xy)^2 - (6xz)^2$,

the difference of two squares. We apply the formula for the difference of two squares,

$a^2 - b^2 = (a+b)(a-b)$,

replacing a by 2 xy and b by 6xz:

$4x^2y^2 - 36x^2z^2 = (2xy + 6xz)(2xy - 6xz)$.

We may now factor 2x from each of the above factors since,

and,

$$2xy + 6xz = 2x(y + 3z)$$

Thus,

$$2xy - 6xz = 2x(y - 3z).$$

$$4x^2y^2 - 36x^2z^2 = (2x)(y + 3z)(2x)(y - 3z)$$
$$= (2x)(2x)(y + 3z)(y - 3z)$$
$$= 4x^2(y + 3z)(y - 3z)$$

● **PROBLEM** 370

Factor $x^2 - y^2$.

Solution: Add and subtract xy from the given expression. Note that this procedure doesn't change the value of the expression because $x^2 - y^2 + xy - xy = x^2 - y^2 + (xy - xy) = x^2 - y^2 + 0 = x^2 - y^2$.

Therefore,

$$x^2 - y^2 = x^2 - y^2 + xy - xy.$$

Also, by the commutative law of addition:

$$x^2 - y^2 = x^2 - y^2 + xy - xy$$
$$= x^2 - y^2 + (xy - xy)$$
$$= x^2 + (xy - xy) - y^2$$
$$\text{or } x^2 - y^2 = x^2 + xy - xy - y^2 \qquad (1)$$

Again, applying the commutative law of addition to the second and third terms of equation (1):

$$x^2 - y^2 = x^2 - xy + xy - y^2 \qquad (2)$$

Factoring x from the first two terms of equation (2) and also factoring y from the last two terms of equation (2):

$$x^2 - y^2 = x(x - y) + y(x - y)$$

By the distributive property:

$$x^2 - y^2 = x(x - y) + y(x - y)$$
$$= (x + y)(x - y)$$

● **PROBLEM** 371

Simplify $(x + 2y)(x - 2y)(x^2 + 4y^2)$.

Solution: Here we use the factoring formula $a^2 - b^2 = (a - b)(a + b)$ to rewrite the product $(x + 2y)(x - 2y)$:

$$(x + 2y)(x - 2y) = (x)^2 - (2y)^2 \quad \text{difference of two squares.}$$
$$= x^2 - 4y^2.$$

Hence, $(x + 2y)(x - 2y)(x^2 + 4y^2) = (x^2 - 4y^2)(x^2 + 4y^2) \qquad (1)$

237

Now, again use the factoring formula given above to rewrite the right side of equation (1) in which $x^2 = a$ and $4y^2 = b$. Hence,

$$\left(x^2 - 4y^2\right)\left(x^2 + 4y^2\right) = \left(x^2\right)^2 - \left(4y^2\right)^2$$
$$= x^4 - 4^2y^4 \quad \text{since} \quad \left(a^x\right)^y = a^{x \cdot y}$$
$$= x^4 - 16y^4$$

Hence, equation (1) becomes:

$$(x + 2y)(x - 2y)\left(x^2 + 4y^2\right) = x^4 - 16y^4 \ .$$

● **PROBLEM** 372

Factor $25a^2 + 30ab + 9b^2$.

Solution: Note that $25a^2 = 5^2 a^2 = (5a)^2$, $9b^2 = 3^2b^2$ $= (3b)^2$ and $30ab = 2(5a)(3b)$. The given expression can be rewritten as:

$$25a^2 + 30ab + 9b^2 = (5a)^2 + 2(5a)(3b) + (3b)^2. \quad (1)$$

Also, the formula for the square of a binomial sum is:

$$(x + y)^2 = (x + y)(x + y) = x^2 + xy + xy + y^2$$
$$= x^2 + 2xy + y^2$$

thus,

$$(x + y)^2 = x^2 + 2xy + y^2 \quad (2)$$

The right side of equation (1) corresponds to the right side of equation (2), where $x = 5a$ and $y = 3b$. Hence, using the left side of equation (2), equation (1) can be rewritten as:

$$25a^2 + 30ab + 9b^2 = (5a)^2 + 2(5a)(3b) + (3b)^2$$
$$= (5a + 3b)^2$$

● **PROBLEM** 373

Factor $x^3 - 8$.

Solution: Since $8 = 2^3$, $x^3 - 8 = (x)^3 - (2)^3$. Therefore, $x^3 - 8$ is the difference of two cubes. The formula for the difference of two cubes is:

$$a^3 - b^3 = (a - b)\left(a^2 + ab + b^2\right)$$

Replacing a by x and b by 2;

$$x^3 - 8 = (x)^3 - (2)^3$$
$$= (x - 2)\left[x^2 + (x)(2) + (2)^2\right]$$
$$x^3 - 8 = (x - 2)\left(x^2 + 2x + 4\right)$$

238

Factor $8a^3 - 27$.

Solution: Note that $8 = (2)^3$, hence $8a^3 = (2^3)(a^3) = (2a)^3$

and $27 = (3)^3$

So, $8a^3 - 27 = (2a)^3 - (3)^3$.

Recall the formula for the difference of two cubes:

$$x^3 - y^3 = (x - y)(x^2 + xy + y^2)$$

Substituting 2a for x, and 3 for y:

$$(2a)^3 - (3)^3 = (2a - 3)\left[(2a)^2 + 3(2a) + 3^2\right]$$

$$= (2a - 3)(4a^2 + 6a + 9)$$

Hence, $8a^3 - 27 = (2a - 3)(4a^2 + 6a + 9)$.

Factor $27x^3 - 8y^3$.

Solution: Note that $27 = 3^3$; thus $27x^3 = (3^3)(x^3)$.

Since $a^x b^x = (ab)^x$, $(3^3)(x^3) = (3x)^3$.

Similarly, $8 = 2^3$; thus $8y^3 = (2^3)(y^3) = (2y)^3$.

Therefore $27x^3 - 8y^3 = (3x)^3 - (2y)^3$, the difference of two cubes. We apply the formula for the difference of two cubes, $(a^3 - b^3) = (a - b)(a^2 + ab + b^2)$, replacing a by 3x and b by 2y. Thus, we obtain:

$$27x^3 - 8y^3 = (3x)^3 - (2y)^3$$

$$= (3x - 2y)\left[(3x)^2 + (3x)(2y) + (2y)^2\right]$$

$$= (3x - 2y)(9x^2 + 6xy + 4y^2).$$

Find the factors of $125m^3n^6 - 8a^3$.

Solution: Note that $125 = 5 \cdot 5 \cdot 5 = 5^3$. Also since $a^{xy} = (a^x)^y$, $n^6 = n^{2 \cdot 3} = (n^2)^3$. Thus $125m^3n^6 = 5^3m^3(n^2)^3$. Since $a^x b^x c^x = (abc)^x$, $5^3m^3(n^2)^3 = (5mn^2)^3$. $8 = 2 \cdot 2 \cdot 2 = 2^3$, thus $8a^3 = 2^3a^3 = (2a)^3$. Now $125m^3n^6 - 8a^3 = (5mn^2)^3 - (2a)^3$, which is the difference of two cubes. Apply the formula for the difference of two cubes $x^3 - y^3 = (x - y)(x^2 + xy + y^2)$, replacing x by $5mn^2$ and y by 2a. Hence,

$$125m^3n^6 - 8a^3 = (5mn^2 - 2a)\left[(5mn^2)^2 + 5mn^2(2a) + (2a)^2\right].$$

$$= (5mn^2 - 2a)(25m^2n^4 + 10amn^2 + 4a^2)$$

Factor $(x + y)^3 + z^3$.

Solution: The given expression is the sum of two cubes. The formula for the sum of two cubes can be used to factor the given expression. This formula is:

$$a^3 + b^3 = (a + b)(a^2 - ab + b^2).$$

Using this formula and replacing a by x + y and b by z:

$$(x + y)^3 + z^3 = \left[(x+y) + z\right]\left[(x+y)^2 - (x+y)z + z^2\right]$$

Factor $125x^3 + 64y^3$.

Solution: Noting that $125x^3 = 5^3x^3 = (5x)^3$ and $64y^3 = 4^3y^3 = (4y)^3$, we obtain

$$125x^3 + 64y^3 = (5x)^3 + (4y)^3.$$

Thus we have the sum of two cubes. Applying the formula for the sum of two cubes,

$$a^3 + b^3 = (a+b)(a^2 - ab + b^2),$$

and replacing a by (5x) and b by (4y) we obtain $(5x)^3 + (4y)^3$

$$= (5x + 4y)\left[(5x)^2 - (5x)(4y) + (4y)^2\right]$$

$$= (5x + 4y)(5^2x^2 - 20xy + 4^2y^2)$$

$$= (5x + 4y)(25x^2 - 20xy + 16y^2).$$

Factor $5x^3 + 8y^3$.

Solution: Recall the formula for the sum of two cubes:
$$a^3 + b^3 = (a + b)(a^2 - ab + b^2) .$$

To obtain $5x^3 + 8y^3$ as the sum of two cubes, we must express $5x^3$ and $8y^3$ as perfect cubes: Note that

$$\left(\sqrt[3]{a}\right)^3 = \left(a^{\frac{1}{3}}\right)^3 = a^1 = a ;$$

thus,

$$\left(\sqrt[3]{5}\right)^3 = \left(5^{\frac{1}{3}}\right)^3 = 5^1 = 5 .$$

So, we can write

$$5x^3 = \left(\sqrt[3]{5}\right)^3 x^3 = \left[\sqrt[3]{5}x\right]^3 ;$$

and $8 = 2^3$, so

$$8y^3 = 2^3y^3 = (2y)^3 .$$

Thus,

$$5x^3 + 8y^3 = \left(\sqrt[3]{5}x\right)^3 + (2y)^3 .$$

Substituting $\sqrt[3]{5}x$ for a and 2y for b in the above general formula we obtain:

$$\left(\sqrt[3]{5x}\right)^3 + (2y)^3 = \left(\sqrt[3]{5x} + 2y\right)\left[\left(\sqrt[3]{5x}\right)^2 - \sqrt[3]{5x}\cdot 2y + (2y)^2\right],$$

and since
$$\left(\sqrt[3]{5x}\right)^2 = \left(5^{\frac{1}{3}}\right)^2 x^2 = 5^{\frac{2}{3}}x^2 = \sqrt[3]{5^2}x^2 = \sqrt[3]{25}x^2 ,$$

$$= \left(\sqrt[3]{5x} + 2y\right)\left(\sqrt[3]{25}x^2 - 2\sqrt[3]{5}xy + 4y^2\right).$$

● **PROBLEM 380**

Factor: (a) $2x^2 + 2y^2$ (b) $a^3x + b^3x.$

Solution: (a) First we factor out a 2 from this expression.
Thus $2x^2 + 2y^2 = 2\left(x^2 + y^2\right)$. $x^2 + y^2$, the sum of two like
even powers, cannot be factored. Thus, it is a prime expression.
Hence $2x^2 + 2y^2 = 2\left(x^2 + y^2\right)$.

(b) First we factor out an x from this expression.
Thus, $a^3x + b^3x = \left(a^3 + b^3\right)x$. $\left(a^3 + b^3\right)$ is the sum of two
cubes. Applying the formula for the sum of two cubes:
$c^3 + d^3 = (c + d)\left(c^2 - cd + d^2\right)$, replacing c by a and d by b
we obtain
$$a^3 + b^3 = (a + b)\left(a^2 - ab + b^2\right).$$

Hence, $a^3x + b^3x = (a + b)\left(a^2 - ab + b^2\right)x.$

● **PROBLEM 381**

Find the following special products:
(a) $3xy^2\left(2x^2 - 3x + 6\right)$

(b) $\left(2x^2 - 3y\right)\left(2x^2 + 3y\right)$

(c) $(3a + 2)(4a - 5)$

(d) $\left(9x^2 + 3xy^2 + y^4\right)\left(3x - y^2\right)$.

Solution:

(a) Use the distributive property.
$$3xy^2\left(2x^2 - 3x + 6\right) = 3xy^2\left(2x^2\right) - 3xy^2(3x) + 3xy^2(6)$$
$$= (3\cdot 2)\left(x\cdot x^2\right)y^2 - (3\cdot 3)(x\cdot x)y^2 + (3\cdot 6)xy^2$$
$$= 6x^3y^2 - 9x^2y^2 + 18xy^2$$

(b) Use the difference of two squares, $(a-b)(a+b) = a^2 - b^2$,
replacing a by $2x^2$ and b by 3y,
$$\left(2x^2-3y\right)\left(2x^2+3y\right) = \left(2x^2\right)^2 - (3y)^2 = 2^2\left(x^2\right)^2 - 3^2y^2$$
$$= 4x^{2\cdot 2} - 9y^2$$
$$= 4x^4 - 9y^2$$

(c) Use the distributive property.
$$(3a+2)(4a-5) = 3a(4a-5) + 2(4a-5)$$
$$= 12a^2 - 15a + 8a - 10$$
$$= 12a^2 - 7a - 10$$

(d) Use the difference of two cubes, $\left(a^2+ab+b^2\right)(a-b) = a^3 -b^3$, replacing a by 3x and b by y^2. Thus

$$\left(9x^2+3xy^2+y^4\right)\left(3x-y^2\right) = \left[(3x)^2+3xy^2+\left(y^2\right)^2\right]\left(3x-y^2\right) = (3x)^3 - \left(y^2\right)^3$$
$$= 27x^3 - y^6$$

● PROBLEM 382

Factor $a^4 - b^4$.

Solution: Note that $a^4 = (a^2)^2$ and $b^4 = (b^2)^2$; thus

$a^4 - b^4 = (a^2)^2 - (b^2)^2$, the difference of two squares. Thus we apply the formula for the difference of two squares, $x^2 - y^2 = (x + y)(x - y)$, replacing x by a^2 and y by b^2 to obtain:

$$a^4 - b^4 = (a^2)^2 - (b^2)^2 = (a^2 + b^2)(a^2 - b^2).$$

Since $a^2 - b^2$ is also the difference of two squares, we once again apply the above formula to obtain:

$$a^2 - b^2 = (a + b)(a - b).$$

Therefore, $a^4 - b^4 = (a^2 + b^2)(a + b)(a - b)$.

● PROBLEM 383

Evaluate $2x^2y - 8y^3$.

Solution: First note that there is a common factor of y in both terms. Thus

$$2x^2y - 8y^3 = y(2x^2 - 8y^2)$$

There is also a common factor of 2 which can be factored out,

$$= 2y(x^2 - 4y^2)$$

Observe $4y^2 = 2^2y^2 = (2y)^2$, thus $x^2 - 4y^2 = x^2 - (2y)^2$, the difference of two squares. Applying the formula for the difference of two squares, $a^2 - b^2 = (a+b)(a-b)$. Replacing a by x and b by 2y,

$$x^2 - (2y)^2 = (x + 2y)(x - 2y).$$

Thus

$$2y(x^2 - 4y^2) = 2y(x + 2y)(x - 2y)$$

and

$$2x^2 - 8y^3 = 2y(x + 2y)(x - 2y).$$

● PROBLEM 384

Factor $16x^4y^2 - 250xy^5$.

Solution: Factoring out the common monomial factor $2xy^2$, we have

$$16x^4y^2 - 250xy^5 = 2xy^2(8x^3 - 125y^3).$$

$8 - 2 \cdot 2 \cdot 2 = 2^3$, thus $8x^3 = 2^3x^3 = (2x)^3$, and
$125 = 5 \cdot 5 \cdot 5 = 5^3$, thus $125y^3 = 5^3y^3 = (5y)^3$.
Hence, $2xy^2(8x^3 - 125y^3) = 2xy^2[(2x)^3 - (5y)^3]$. Since
$(2x)^3 - (5y)^3$ is the difference of two cubes, we apply the formula for the difference of two cubes.

$$a^3 - b^3 = (a - b)(a^2 + ab + b^2),$$

replacing a by 2x and b by 5y. Thus

$$2xy^2(8x^3 - 125y^3) = 2xy^2(2x - 5y)\left[(2x)^2 + (2x)(5y) + (5y)^2\right]$$

$$= 2xy^2(2x - 5y)(4x^2 + 10xy + 25y^2).$$

Therefore $16x^4y^2 - 250xy^5 = 2xy^2(2x - 5y)(4x^2 + 10xy + 25y^2).$

● **PROBLEM** 385

Factor A) $\quad a^4 + 4a^2 + 4$

B) $\quad 9a^2 - 6ab^2 + b^4$.

Solution: The first example is a trinomial which is a perfect square, in the form:

$$x^2 + 2xy + y^2 = x^2 + xy + xy + y^2 = (x+y)(x+y) = (x+y)^2 .$$

For example A), replace x by a^2 and y by 2 to obtain

$$a^2 + 4a^2 + 4 = \left(a^2\right)^2 + 2 \cdot a^2 \cdot 2 + 2^2 = \left(a^2 + 2\right)^2 ,$$

The second example is a trinomial perfect square whose form is:

$$x^2 - 2xy + y^2 = x^2 - xy - xy + y^2 = (x-y)(x-y) = (x-y)^2 .$$

For example B) replace x by 3a and y by b^2 to obtain:

$$9a^2 - 6ab^2 + b^4 = (3a)^2 - 2(3a)\left(b^2\right) + \left(b^2\right)^2$$

$$= \left(3a - b^2\right)\left(3a - b^2\right)$$

$$= \left(3a - b^2\right)^2 .$$

● **PROBLEM** 386

Factor $a^4 - a^2 - 12$ completely.

Solution: We factor a trinomial of degree two in this manner:

$$x^2 + (c + d)x + cd = (x + c)(x + d).$$

In this case $x = \left(a^2\right)$

$$a^4 - a^2 - 12 = \left(a^2\right)^2 - \left(a^2\right) - 12.$$

$$c + d = -1 \qquad c \cdot d = -12.$$

We must find two numbers whose sum is -1 and whose product is -12. The two numbers which satisfy these two conditions are -4 and 3. Thus,

$$a^4 - a^2 - 12 = \left(a^2\right)^2 - a^2 - 12$$

$$= \left(a^2\right)^2 + (-4 + 3)\left(a^2\right) + (-4)(3)$$

243

$$= a^4 - a^2 - 12.$$

Therefore,

$$a^4 - a^2 - 12 = \left(a^2 + 3\right)\left(a^2 - 4\right).$$

The first factor on the right does not factor further, but the second factor; $a^2 - 4$, is a difference of two squares. Completion of the factorization gives

$$a^4 - a^2 - 12 = \left(a^2 + 3\right)(a + 2)(a - 2).$$

● **PROBLEM 387**

Factor the expression $49x^6 - 25y^6$.

Solution: Noting that $49x^6 = 7^2\left(x^3\right)^2 = \left(7x^3\right)^2$ and $25y^6 = 5^2\left(y^3\right)^2 = \left(5y^3\right)^2$, we obtain

$$49x^6 - 25y^6 = \left(7x^3\right)^2 - \left(5y^3\right)^2.$$

Thus we have the difference of two squares. Applying the formula for the difference of two squares $a^2 - b^2 = (a+b)(a-b)$, and replacing a by $\left(7x^3\right)$ and b by $\left(5y^3\right)$, we obtain $\left(7x^3\right)^2 - \left(5y^3\right)^2 = \left[\left(7x^3\right) + \left(5y^3\right)\right]\left[\left(7x^3\right) - \left(5y^3\right)\right] = \left(7x^3 + 5y^3\right)\left(7x^3 - 5y^3\right)$.

● **PROBLEM 388**

Factor $128x^6 - 2y^6$.

Solution: We first observe that 2 may be factored from this expression. Thus

$$128x^6 - 2y^6 = 2\left(64x^6 - y^6\right).$$

Now, since $a^{b \cdot c} = \left(a^b\right)^c$, $x^6 = x^{3 \cdot 2} = \left(x^3\right)^2$ and

$$y^6 = y^{3 \cdot 2} = \left(y^3\right)^2.$$

Therefore $2\left(64x^6 - y^6\right) = 2\left[64\left(x^3\right)^2 - \left(y^3\right)^2\right]$

$$= 2\left[8^2\left(x^3\right)^2 - \left(y^3\right)^2\right]$$

$$= \left[\left(8x^3\right)^2 - \left(y^3\right)^2\right].$$

Thus, we have the difference of two squares. Applying the formula for the difference of two squares, $a^2 - b^2 = (a+b)(a-b)$, and replacing a by $8x^3$ and b by y^3, we obtain

244

$$2[(8x^3)^2 - (y^3)^2] = 2\left[(8x^3 + y^3)(8x^3 - y^3)\right]$$

$$= 2\left[(2^3x^3 + y^3)(2^3x^3 - y^3)\right]$$

$$= 2\left[(2x)^3 + y^3\right][(2x)^3 - y^3].$$

Now since the expressions in brackets are the sum and difference of two cubes, respectively, we apply the formulas for the sum and difference of two cubes:

$$a^3 + b^3 = (a+b)\left(a^2 - ab + b^2\right)$$

$$a^3 - b^3 = (a-b)\left(a^2 + ab + b^2\right).$$

Replacing a by 2x and b by y we have

$$2\left[(2x)^3 + y^3\right]\left[(2x)^3 - y^3\right] = 2(2x+y)\left(4x^2-2xy+y^2\right)(2x-y)\left(4x^2+2xy+y^2\right)$$

Therefore, $128x^6 - 2y^6 = 2(2x+y)\left(4x^2-2xy+y^2\right)(2x-y)\left(4x^2+2xy+y^2\right).$

● **PROBLEM** 389

Factor: (a) $4a^4 - b^6$

(b) $a^2 - b^2 + 2bc - c^2$

Solution: (a) First note that $4a^4 - b^6$ may be expressed as the difference of two squares: $4a^4 - b^6 = \left(2a^2\right)^2 - \left(b^3\right)^2$. Recall the formula for the difference of two squares: $x^2 - y^2 = (x-y)(x+y)$. Replacing x by $2a^2$ and y by b^3 in this formula we obtain:

$$4a^4 - b^6 = \left(2a^2\right)^2 - \left(b^3\right)^2 = \left(2a^2 - b^3\right)\left(2a^2 + b^3\right)$$

(b) Observe that $b^2 - 2bc + c^2$ is a perfect square, since the first and last terms are perfect squares and positive, and the middle term is twice the product of the square roots of the end terms. Then, $b^2 - 2bc + c^2 = (b - c)^2$.

Thus, express the given algebraic expression as the difference of two squares.

$$a^2 - b^2 + 2bc - c^2 = a^2 - \left(b^2 - 2bc + c^2\right) = a^2 - (b - c)^2$$

Then apply the formula for the difference of two squares

$$\left(x^2 - y^2\right) = (x-y)(x+y),$$

replacing x by a and y by (b-c):

$$a^2 - (b-c)^2 = [a - (b-c)][a + (b-c)]$$

$$= (a - b + c)(a + b - c)$$

Thus,

$$a^2 - b^2 + 2bc - c^2 = (a - b + c)(a + b - c)$$

● **PROBLEM** 390

Factor $x^6 - 64$ completely.

Solution: We observe that,

$$x^6 = x^{3 \cdot 2} = \left(x^3\right)^2 \quad \text{and} \quad 64 = 8^2.$$

Thus,
$$x^6 - 64 = \left(x^3\right)^2 - 8^2,$$

the difference of two squares. Applying the formula for the difference of two squares,

$$a^2 - b^2 = (a + b)(a - b)$$

and replacing a by x^3 and b by 8, we obtain,

$$x^6 - 64 = \left(x^3 + 8\right)\left(x^3 - 8\right).$$

Since ,

$$8 = 2^3, \quad x^6 - 64 = \left(x^3 + 8\right)\left(x^3 - 8\right)$$
$$= \left(x^3 + 2^3\right)\left(x^3 - 2^3\right).$$

Thus each resulting factor, factors further, as a sum and as a difference of two cubes, respectively. Applying the formulas for the sum and difference of two cubes,

$$\left(a^3 + b^3\right) = (a + b)\left(a^2 - ab + b^2\right)$$

$$\left(a^3 - b^3\right) = (a - b)\left(a^2 + ab + b^2\right),$$

and replacing a by x and b by 2, we obtain,

$$x^6 - 64 = (x + 2)\left(x^2 - 2x + 4\right)(x - 2)\left(x^2 + 2x + 4\right).$$

It is of interest to see what results on factoring $x^6 - 64$ as a difference of two cubes. We get, in this case,

$$x^6 - 64 = \left(x^2\right)^3 - 4^3$$
$$= \left(x^2 - 4\right)\left(x^4 + 4x^2 + 16\right),$$

in which

$$x^2 - 4 = x^2 - 2^2$$

factors as $(x + 2)(x - 2)$. A comparison of the two results shows that $x^4 + 4x^2 + 16$ must be factorable as,

$$x^4 + 4x^2 + 16 = \left(x^2 - 2x + 4\right)\left(x^2 + 2x + 4\right).$$

This may be verified by multiplication, but it is not easy to see directly. Treating,

$$x^6 - 64$$

as the difference of two squares is much simpler than thinking of it as the difference of two cubes.

246

Find the LCM of: $x^2 + 2x + 1$, $x^2 - 1$, and $x^2 - 3x + 2$.

Solution: Factor each term.

$$x^2 + 2x + 1 = (x + 1)^2, \quad x^2 - 1 = (x - 1)(x + 1),$$

$$x^2 - 3x + 2 = (x - 1)(x - 2)$$

As with integers, the LCM must contain all of the prime factors as a product, and each prime factor must contain the largest exponent it has in any of the factored forms. Each factor is used only once, regardless of the number of times it appears. Therefore,

$$LCM = (x + 1)^2 (x - 1)(x - 2)$$

Find the LCM of: $6x^2 + 24x + 24$, $4x^2 - 8x - 12$, and $3x^2 + 9x + 6$.

Solution: Factor each expression completely. Constant factors should be written as a product of prime numbers.

$$6x^2 + 24x + 24 = 6\left(x^2 + 4x + 4\right) = 6(x + 2)^2 = (3)(2)(x + 2)^2$$

$$4x^2 - 8x - 12 = 4\left(x^2 - 2x - 3\right) = 4(x + 1)(x - 3)$$

$$= (2)^2(x + 1)(x - 3)$$

$$3x^2 + 9x + 6 = 3\left(x^2 + 3x + 2\right) = 3(x + 1)(x + 2).$$

Each of the factors of these expressions appears in the product known as the LCM. Each factor is raised to the highest power to which it appears in any one of the expressions. Therefore,

$$LCM = (2)^2(3)(x + 2)^2(x + 1)(x - 3).$$

Find the LCM of: $(x - 1)^2$, $(1 - x)^3$, $1 - x^3$.

Solution: Factor each polynomial completely. Notice in the factoring of the second and the third polynomials that -1 may be factored from the expressions first so that the terms of highest degree in the factors will have positive coefficients.

$$(x - 1)^2 = (x - 1)^2$$

$$(1 - x)^3 = [-1(x - 1)]^3 = (-1)^3(x - 1)^3 = -(x - 1)^3$$

$$1 - x^3 = (-1)\left(x^3 - 1\right) = -(x - 1)\left(x^2 + x + 1\right).$$

$\left(x^3 - 1\right.$ is the difference of two cubes.$\left.\right)$

Each of the factors of these expressions appears in the product known as the LCM. Each factor is raised to the highest power to which it appears in any one of the expressions. Therefore the

$$\text{LCM} = \left(x - 1\right)^3\left(x^2 + x + 1\right).$$

FRACTIONAL

● **PROBLEM** 394

Simplify $\dfrac{x^2 - y^2}{x + y}$.

Solution: $x^2 - y^2$ is the difference of two squares. Applying the formula for the difference of two squares, $a^2 - b^2 = (a + b)(a - b)$, with $x = a$ and $y = b$,

$$\frac{x^2 - y^2}{x + y} = \frac{(x + y)(x - y)}{x + y} = x - y .$$

● **PROBLEM** 395

Perform the indicated operation

$$\frac{x^2 - y^2}{2x} \cdot \frac{4x^2}{x + y}$$

Solution: $x^2 - y^2$ is the difference of two squares. Applying the formula for the difference of two squares

$$a^2 - b^2 = (a+b)(a-b)$$
$$x^2 - y^2 = (x+y)(x-y)$$

Thus
$$\frac{x^2 - y^2}{2x} \cdot \frac{4x^2}{x + y} = \frac{(x + y)(x - y)}{2x} \cdot \frac{4x^2}{x + y}$$

$$= \frac{(x+y)(x-y)\left(4x^2\right)}{(2x)(x+y)}$$

Since $4x^2 = (2x)(2x)$,

$$= \frac{(2x)(2x)(x+y)(x-y)}{(2x)(x+y)}$$

$$= 2x(x - y)$$

● **PROBLEM** 396

Combine $\dfrac{3x + y}{x^2 - y^2} - \dfrac{2y}{x(x - y)} - \dfrac{1}{x + y}$ into a single fraction.

<u>Solution:</u> Fractions which have unlike denominators must be transformed into fractions with the same denominator before they may be combined. This identical denominator is the least common denominator (L.C.D.), the least common multiple of the denominators of the fractions to be added. In the process of transforming the fractions to a common denominator we make use of the fact that the numerator and denominator of a fraction may be multiplied by the same non-zero number without changing the value of the fraction. In our case the denominators are $x^2 - y^2 = (x + y)(x - y)$, $x(x - y)$, and $x + y$. Therefore the LCD is $x(x + y)(x - y)$, and we proceed as follows:

$$\frac{3x + y}{x^2 - y^2} - \frac{2y}{x(x-y)} - \frac{1}{x+y} = \frac{3x + y}{(x+y)(x-y)} - \frac{2y}{x(x-y)} - \frac{1}{x+y}$$

$$= \frac{x(3x + y)}{x(x+y)(x-y)} - \frac{(x + y)2y}{(x+y)(x)(x-y)}$$

$$- \frac{x(x - y)}{x(x-y)(x+y)}$$

$$= \frac{3x^2 + xy}{x(x+y)(x-y)} - \frac{2xy + 2y^2}{x(x+y)(x-y)}$$

$$- \frac{x^2 - xy}{x(x+y)(x-y)}$$

$$= \frac{3x^2 + xy - \left(2xy + 2y^2\right) - \left(x^2 - xy\right)}{x(x+y)(x-y)}$$

$$= \frac{3x^2 + xy - 2xy - 2y^2 - x^2 + xy}{x(x+y)(x-y)}$$

$$= \frac{3x^2 - x^2 + xy + xy - 2xy - 2y^2}{x(x+y)(x-y)}$$

$$= \frac{2x^2 - 2y^2}{x(x+y)(x-y)}$$

$$= \frac{2\left(x^2 - y^2\right)}{x(x+y)(x-y)}$$

$$= \frac{2(x+y)(x-y)}{x(x+y)(x-y)}$$

$$= \frac{2}{x}$$

● **PROBLEM** 397

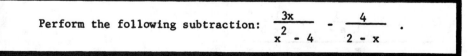

Perform the following subtraction: $\dfrac{3x}{x^2 - 4} - \dfrac{4}{2 - x}$.

<u>Solution:</u> Note that $x^2 - 4 = x^2 - 2^2$, the difference of two squares. Using the formula for the difference of two squares, $a^2 - b^2 = (a-b)(a+b)$,

249

we have:
$$x^2 - 4 = x^2 - 2^2 = (x-2)(x+2).$$

Thus,

$$\frac{3x}{x^2 - 4} - \frac{4}{2 - x} = \frac{3x}{(x-2)(x+2)} - \frac{4}{2 - x}$$

$$= \frac{3x}{(x-2)(x+2)} + \frac{-4}{2 - x} \ .$$

Multiplying numerator and denominator of $\frac{-4}{2 - x}$ by (-1):

$$\frac{-4(-1)}{(2-x)(-1)} = \frac{4}{-2 + x} = \frac{4}{x - 2} \ .$$

Thus,

$$\frac{3x}{x^2 - 4} - \frac{4}{2 - x} = \frac{3x}{(x-2)(x+2)} + \frac{4}{x - 2}$$

$$= \frac{3x}{(x-2)(x+2)} + \frac{4(x+2)}{(x-2)(x+2)}$$

(Notice that $\frac{x+2}{x+2} = 1$, therefore multiplication by this fraction does not alter the value of the term)

$$= \frac{3x + 4(x+2)}{(x-2)(x+2)} = \frac{3x + 4x + 8}{(x-2)(x+2)}$$

$$= \frac{7x + 8}{(x-2)(x+2)} \ .$$

● **PROBLEM** 398

Simplify:
$$\frac{\dfrac{1}{a - b} + \dfrac{1}{a + b}}{1 + \dfrac{b^2}{a^2 - b^2}}$$

Solution: This is a complex fraction, a fraction whose numerator and denominator both contain fractions. To simplify it, multiply the numerator and denominator by the least common denominator, LCD. To find the LCD of several fractions, first factor each denominator into its prime factors.

$$a - b = (a - b)$$
$$a + b = (a + b)$$
$$a^2 - b^2 = (a - b)(a + b)$$

The LCD of the fractions is the product of the highest power of the different prime factors, with each prime factor being used only once. Hence $(a - b)(a + b)$ is our LCD. Multiplying, we obtain:

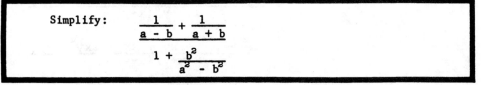

$$\frac{(a-b)(a+b)\left[\dfrac{1}{a - b} + \dfrac{1}{a + b}\right]}{(a-b)(a+b)\left[1 + \dfrac{b^2}{a^2 - b^2}\right]}$$

Distributing in the numerator and denominator, and recalling that

250

$a^2 - b^2 = (a - b)(a + b)$ we have:

$$\frac{\dfrac{\cancel{(a-b)}(a+b)}{\cancel{(a-b)}} + \dfrac{(a-b)\cancel{(a+b)}}{\cancel{(a+b)}}}{(a-b)(a+b) + \dfrac{b^2\cancel{(a-b)}\cancel{(a+b)}}{\cancel{(a-b)}\cancel{(a+b)}}} = \frac{(a+b) + (a-b)}{(a-b)(a+b) + b^2}$$

$$= \frac{a + b + a - b}{a^2 - b^2 + b^2} = \frac{2a}{a^2} = \frac{2}{a} \quad .$$

● **PROBLEM** 399

Combine the following fractions $\dfrac{x}{x^2 - y^2} + \dfrac{2}{y - x} - 5$.

<u>Solution:</u> Note that the denominator $x^2 - y^2$ is the difference of two squares. Using the formula for the difference of two squares

$$a^2 - b^2 = (a + b)(a - b) ,$$

factor $x^2 - y^2$ into $(x + y)(x - y)$. Next observe that $-(y-x) = -y+x = x-y$. Then multiplying numerator and denominator of

$$\left(\frac{2}{y - x}\right) \text{ by } -1,$$

$$\frac{-1}{-1}\left(\frac{2}{y - x}\right) = \frac{-2}{x - y} \quad .$$

Therefore,

$$\frac{x}{x^2 - y^2} + \frac{2}{y - x} - 5 = \frac{x}{(x+y)(x-y)} + \frac{-2}{x - y} - \frac{5}{1}$$

Thus the terms $(x+y),(x-y)$, and 1 appear in our denominators. In order to combine fractions, transform them into equivalent fractions with a common denominator. Using $(x+y)(x-y)$ as our least common denominator, LCD, multiply each term by the necessary factor to yield a denominator of $(x+y)(x-y)$:

$$\frac{x}{x^2 - y^2} + \frac{2}{y - x} - 5 = \frac{x}{(x+y)(x-y)} + \left(\frac{x+y}{x+y}\right)\left(\frac{-2}{x-y}\right) + \frac{(x+y)(x-y)}{(x+y)(x-y)}\left(\frac{-5}{1}\right)$$

$$= \frac{x}{(x+y)(x-y)} + \frac{-2x - 2y}{(x+y)(x-y)} + \frac{-5(x^2 - y^2)}{(x+y)(x-y)}$$

$$= \frac{x - 2x - 2y - 5x^2 + 5y^2}{(x+y)(x-y)}$$

$$= \frac{-x - 2y - 5x^2 + 5y^2}{(x+y)(x-y)}$$

$$= \frac{-x - 2y - 5x^2 + 5y^2}{x^2 - y^2}$$

● **PROBLEM** 400

Reduce $\dfrac{3x - 6}{x^2 - 4}$ to lowest terms.

<u>Solution:</u> Factor the expression in both the numerator and denominator. In the numerator we factor 3 from both terms, and observing that the denominator is the difference of two squares, $x^2 - 2^2$, we obtain:

$$\frac{3x - 6}{x^2 - 4} = \frac{3(x - 2)}{(x - 2)(x + 2)}$$

$$= \frac{3}{x + 2}$$

The numerator and denominator were divided by $x - 2$.

● **PROBLEM** 401

Combine the following fractions:

$$\frac{12b - 16a}{3a^2 - 3b^2} + \frac{5}{a + b} - \frac{1}{b - a} .$$

<u>Solution:</u> In order to add fractions with unlike denominators, we must find a least common denominator, L.C.D., which is the least common multiple of the denominators of the fractions to be added. First, factor the different denominators

$$3a^2 - 3b^2 = 3\left(a^2 - b^2\right) = 3(a+b)(a-b)$$

$$a + b = (a + b)$$

$$b - a = -1(a - b)$$

Notice that the last denominator differs from the 3rd factor $(a - b)$ of the first denominator by a factor of -1. Factor out this minus one. Hence

$$\frac{12b - 16a}{3a^2 - 3b^2} + \frac{5}{a + b} - \frac{1}{(-1)(a-b)} = \frac{12b - 16a}{3(a+b)(a-b)} + \frac{5}{a + b} + \frac{1}{a - b} .$$

Now, the least common denominator is $3(a-b)(a+b)$. Transform the fractions to equivalent fractions all having the same denominator. Then, add the numerators.

$$\frac{12b - 16a}{3(a+b)(a-b)} + \frac{5}{(a + b)} + \frac{1}{(a - b)} = \frac{12b - 16a}{3(a+b)(a-b)}$$

$$+ \frac{5(3)(a-b)}{3(a+b)(a-b)} + \frac{(1)3(a+b)}{3(a-b)(a+b)} =$$

$$\frac{12b - 16a + 15a - 15b + 3a + 3b}{3(a+b)(a-b)} = \frac{2a}{3\left(a^2 - b^2\right)} .$$

● **PROBLEM** 402

Simplify $\dfrac{4x + 10}{4x^2 + 20x + 25}$.

<u>Solution:</u> First we factor a 2 from the numerator, thus

252

$$\frac{4x + 10}{4x^2 + 20x + 25} \quad \frac{2(2x + 5)}{4x^2 + 20x + 25}.$$

Factoring the denominator, $= \dfrac{2(2x + 5)}{(2x + 5)(2x + 5)} = \dfrac{2}{2x + 5}.$

Since the factor $(2x + 5)$ appears in the denominator it may not equal zero, as division by zero is not defined. Thus

$$2x + 5 \neq 0$$

$$2x \neq -5$$

$$x \neq \frac{-5}{2}.$$

Therefore, $\dfrac{4x + 10}{4x^2 + 20x + 25} = \dfrac{2}{2x + 5}, \; x \neq \dfrac{-5}{2}.$

● **PROBLEM 403**

Reduce $\dfrac{4x - 20}{50 - 2x^2}$ to lowest terms.

Solution: Factor the numerator and the denominator:

$$\frac{4x - 20}{50 - 2x^2} = \frac{4(x - 5)}{2(25 - x^2)} = \frac{4(x - 5)}{2(5 - x)(5 + x)}$$

Multiply the numerator and denominator by (-1) to reverse the sign of the factor $(5 - x)$ in the denominator. Then divide both the numerator and denominator by $2(x - 5)$.

$$\frac{(-1)[4(x - 5)]}{(-1)[2(5 - x)(5 + x)]} = \frac{-4(x - 5)}{2(x - 5)(5 + x)}$$

Dividing, we obtain:

$$-\frac{2}{x + 5} .$$

● **PROBLEM 404**

Simplify $\dfrac{a^2 - 3ab + 2b^2}{2b^2 + ab - a^2}.$

Solution: Factoring the numerator and denominator of the given fraction we obtain:

$$\frac{a^2 - 3ab + 2b^2}{2b^2 + ab - a^2} = \frac{(a - 2b)(a - b)}{(2b - a)(b + a)}$$

If we negate both factors in the numerator we do not change the fraction's value(because negating both factors gives us -1 multiplied by -1, which equals 1; and multiplication by 1 does not change the expression's value). Thus, we have:

253

$$\frac{[-(a - 2b)][-(a - b)]}{(2b - a)(b + a)} = \frac{(2b - a)(b - a)}{(2b - a)(b + a)} = \frac{b - a}{b + a} ,$$

since $\frac{2b - a}{2b - a} = 1.$

Perform the following addition: $\frac{2x}{x^2 - 4} + \frac{3}{x^2 - 5x + 6}$.

<u>Solution:</u> Factor the denominators into polynomial factors. Hence,

$$x^2 - 4 = (x + 2)(x - 2)$$
$$x^2 - 5x + 6 = (x - 3)(x - 2).$$

Therefore:

$$\frac{2x}{x^2 - 4} + \frac{3}{x^2 - 5x + 6} = \frac{2x}{(x + 2)(x - 2)} + \frac{3}{(x - 3)(x - 2)} \quad (1)$$

Now, find the least common denominator (l.c.d) of the two fractions on the right side of equation (1). This is done by writing down all the different factors that appear in the two denominators. The exponent to be used for each factor is the smallest number of times that the factor appears in either denominator. Hence,

$$l.c.d. = (x + 2)^1(x - 2)^1(x - 3)^1 = (x + 2)(x - 2)(x - 3).$$

Multiplying each fraction by a fraction of the appropriate form, and with unit value, produces an equivalent fraction whose denominator is the l.c.d. Therefore:

$$\frac{2x}{x^2 - 4} + \frac{3}{x^2 - 5x + 6} = \frac{(x - 3)(2x)}{(x - 3)(x + 2)(x - 2)} + \frac{(x + 2)(3)}{(x + 2)(x - 3)(x-2)}$$

$$= \frac{(x - 3)(2x) + 3(x + 2)}{(x + 2)(x - 2)(x - 3)}$$

distribute, $= \frac{2x^2 - 6x + 3x + 6}{(x + 2)(x - 2)(x - 3)}$

combine terms $= \frac{2x^2 - 3x + 6}{(x + 2)(x - 2)(x - 3)}$.

Divide $\frac{2x - 8}{x + 1}$ by $\frac{3x^2 - 12x}{x^2 - 1}$.

<u>Solution:</u> The problem can be written as:

$$\frac{\dfrac{2x - 8}{x + 1}}{\dfrac{3x^2 - 12x}{x^2 - 1}}$$

To divide fractions invert the denominator and

multiply the inverted fraction by the numerator. Thus,

$$\frac{\dfrac{2x - 8}{x + 1}}{\dfrac{3x^2 - 12x}{x^2 - 1}} = \frac{2x - 8}{x + 1} \cdot \frac{x^2 - 1}{3x^2 - 12x}$$

$$= \frac{2(x - 4)}{x + 1} \cdot \frac{(x + 1)(x - 1)}{3x(x - 4)}$$

factoring and dividing out common factors

$$= \frac{2(x - 1)}{3x} \quad .$$

Subtract $\dfrac{2x - 3}{x^2 - 3x + 2}$ from $\dfrac{2 - x}{x^2 - 2x + 1}$.

<u>Solution:</u> To subtract $\dfrac{2x - 3}{x^2 - 3x + 2}$ from $\dfrac{2 - x}{x^2 - 2x + 1}$ we must change these fractions to equivalent fractions with a common denominator.

$$\frac{2 - x}{x^2 - 2x + 1} - \frac{2x - 3}{x^2 - 3x + 2} = \frac{2 - x}{(x - 1)(x - 1)} - \frac{2x - 3}{(x - 1)(x - 2)}$$

Denominators were factored for convenience. Multiplying the numerator and denominator of $\dfrac{2 - x}{(x - 1)(x - 1)}$ by the denominator of $\dfrac{2x - 3}{(x - 1)(x - 2)}$ does not change the value of the original fraction. Likewise, we can multiply the numerator and denominator of $\dfrac{2x - 3}{(x - 1)(x - 2)}$ by the denominator of $\dfrac{2 - x}{(x - 1)(x - 1)}$

$$\frac{2 - x}{x^2 - 2x + 1} - \frac{2x - 3}{x^2 - 3x + 2} = \frac{2 - x}{(x - 1)(x - 1)} - \frac{2x - 3}{(x - 1)(x - 2)}$$

$$= \frac{(2x - x)(x - 1)(x - 2)}{(x - 1)(x - 2)(x - 1)(x - 1)}$$

$$- \frac{(2x - 3)(x - 1)(x - 1)}{(x - 1)(x - 2)(x - 1)(x - 1)}$$

$$= \frac{(2 - x)(x - 1)(x - 2) - (2x - 3)(x - 1)(x - 1)}{(x - 1)(x - 2)(x - 1)(x - 1)}$$

Rule for subtracting fractions with common denominators

$$= \frac{(x - 1)[(2 - x)(x - 2) - (2x - 3)(x - 1)]}{(x - 1)[(x - 2)(x - 1)(x - 1)]}$$

$$= \frac{(2 - x)(x - 2) - (2x - 3)(x - 1)}{(x - 2)(x - 1)(x - 1)}$$

factor of (x - 1) cancels

$$= \frac{-x^2 + 4x - 4 - \left(2x^2 - 5x + 3\right)}{(x - 1)^2(x - 2)}$$

$$= \frac{-3x^2 + 9x - 7}{(x - 1)^2(x - 2)}$$

$$= -\frac{3x^2 - 9x + 7}{(x - 1)^2(x - 2)} .$$

The numerator is not factorable so the fraction can not be reduced. Either of the last two fractions could be given as the answer.

● **PROBLEM** 408

Combine

$$\frac{1}{x^2 + x} - \frac{4}{x^2 - 1} + \frac{1}{x^2 - x} , \quad (x \neq 0, 1, -1).$$

<u>Solution:</u> Note that the denominators are $x^2 + x = x(x+1)$, $x^2 - 1$, the difference of two squares, $= (x+1)(x-1)$, and $x^2 - x = x(x-1)$. Before fractions can be combined they must be transformed so that all will have the same denominator. This identical denominator is called the Least Common Denominator (L.C.D.). The L.C.D. is the least common multiple of the denominators of the fractions to be added. Thus, in our example the least common denominator is $x(x-1)(x+1)$. Multiply the numerator and denominator of the first fraction by $x-1$, those of the second fraction by x, and those of the third by $(x+1)$. Then the sum is given by

$$\left(\frac{x-1}{x-1}\right)\left(\frac{1}{x(x+1)}\right) - \left(\frac{x}{x}\right)\left(\frac{4}{(x+1)(x-1)}\right) + \left(\frac{x+1}{x+1}\right)\left(\frac{1}{x(x-1)}\right)$$

$$= \frac{x-1}{x(x-1)(x+1)} - \frac{4x}{x(x-1)(x+1)} + \frac{x+1}{x(x-1)(x+1)}$$

$$= \frac{(x-1) - 4x + x+1}{x(x-1)(x+1)} = \frac{-2x}{x(x-1)(x+1)} = \frac{-2}{(x-1)(x+1)} .$$

Thus,

$$\frac{1}{x^2 + x} - \frac{4}{x^2 - 1} + \frac{1}{x^2 - x} = \frac{-2}{(x-1)(x+1)} .$$

To check that these equations are equivalent, substitute any value of x (we use x = 2) in both sides of the equation. From the left side we obtain:

$$\frac{1}{2^2 + 2} - \frac{4}{2^2 - 1} + \frac{1}{2^2 - 2} = \frac{1}{4 + 2} - \frac{4}{4 - 1} + \frac{1}{4 - 2}$$

$$= \frac{1}{6} - \frac{4}{3} + \frac{1}{2}$$

$$= \frac{1}{6} - \frac{8}{6} + \frac{3}{6}$$

$$= \frac{-4}{6}$$

$$= \frac{-2}{3}$$

From the right side we obtain:

$$\frac{-2}{(2-1)(2+1)} = \frac{-2}{(1)(3)} = \frac{-2}{3}$$

Thus, both members yield the same result.

Find the product of

$$\frac{x^2 - x}{x^2 - x - 2} \quad \text{and} \quad \frac{x - 2}{x^2} .$$

Solution:

$$\frac{x^2 - x}{x^2 - x - 2} \cdot \frac{x - 2}{x^2} = \frac{x(x - 1)(x - 2)}{(x - 2)(x + 1)x^2} \quad \text{by}$$

factoring

$$= \frac{x - 1}{(x + 1)x} \quad \text{by dividing}$$

out common factors x and x - 2

$$= \frac{x - 1}{x^2 + x}$$

Combine into a single fraction

$$\frac{2x}{x^2 - 6x + 9} - \frac{8}{x^2 - 2x - 3} - \frac{1}{x+1} .$$

Solution: Fractions which have unlike denominators cannot be combined directly. First they must be transformed into fractions with the same denominator. This identical denominator is called the Least Common Denominator (L.C.D.).

Before we can obtain the L.C.D., we factor each individual denominator.

$$x^2 - 6x + 9 = (x-3)(x-3) = (x-3)^2$$
$$x^2 - 2x - 3 = (x-3)(x+1)$$
$$x + 1 = (x + 1)$$

To find the L.C.D., we consider all the different factors. Take the highest value of the exponent of each factor. Thus, factoring the denominators, we obtain

$$\frac{2x}{(x-3)^2} - \frac{8}{(x-3)(x+1)} - \frac{1}{x+1}$$

and the L.C.D. is $(x-3)^2(x+1)$. We shall now rewrite the three given fractions as equivalent fractions, each having the denominator $(x-3)^2(x+1)$. To this end, multiply numerator and denominator of the first fraction by x + 1, of the second fraction by x - 3, and of the third fraction by $(x-3)^2$. This gives

$$\frac{2x(x+1)}{(x-3)^2(x+1)} - \frac{8(x-3)}{(x-3)^2(x+1)} - \frac{(x-3)^2}{(x+1)(x-3)^2}$$

$$= \frac{2x^2+2x-(8x-24)-(x^2-6x+9)}{(x-3)^2(x+1)}$$

$$= \frac{2x^2+2x-8x+24-x^2+6x-9}{(x-3)^2(x+1)}$$

$$= \frac{2x^2-x^2+2x-8x+6x+24-9}{(x-3)^2(x+1)}$$

$$= \frac{x^2+15}{(x-3)^2(x+1)}$$

Check. The given expression should be equal to the resulting fraction for all permissible values of x, that is, for all values of x except 3 and -1 (x = 3 and x = -1 give us zero in the denominator of the fraction, which is undefined). Replacing x arbitrarily by 2,

$$\left. \frac{2x}{x^2-6x+9} - \frac{8}{x^2-2x-3} - \frac{1}{x+1} \right]_{x=2}$$

$$= \frac{2(2)}{(2)^2-6(2)+9} - \frac{8}{(2)^2-2(2)-3} - \frac{1}{2+1}$$

$$= \frac{4}{4-12+9} - \frac{8}{4-4-3} - \frac{1}{3}$$

$$= \frac{4}{1} - \frac{8}{-3} - \frac{1}{3}$$

$$= \frac{12}{3} + \frac{8}{3} - \frac{1}{3} = \frac{19}{3}$$

and

$$\left. \frac{x^2+15}{(x-3)^2(x+1)} \right]_{x=2} = \frac{(2)^2+15}{(2-3)^2(2+1)} = \frac{4+15}{(-1)^2(3)} = \frac{19}{1\cdot3} = \frac{19}{3}$$

Hence, we have shown that the given expression holds true for x = 2 in the uncombined and combined forms.

● PROBLEM 411

Reduce $\dfrac{x^2 - 5x + 4}{x^2 - 7x + 12}$ to lowest terms.

Solution: Factor the expressions in both the numerator and denominator and cancel like terms.

$$\frac{x^2 - 5x + 4}{x^2 - 7x + 12} = \frac{(x - 1)(x - 4)}{(x - 3)(x - 4)}$$

$$= \frac{x - 1}{x - 3}$$

The numerator and the denominator were both divided by x - 4.

● PROBLEM 412

Reduce to lowest form: $\dfrac{a^3 - 8b^3}{2a^2 - 8b^2}$.

258

Solution: Factor the numerator and denominator as completely as possible.

The numerator, $a^3 - 8b^3 = a^3 - (2b)^3$, is the difference of two cubes. Apply the following formula:

$$\left(x^3 - y^3\right) = (x - y)\left(x^2 + xy + y^3\right)$$

Replacing x by a and y by $2b$, we obtain:

$$\left(a^3 - 8b^3\right) = \left[a^3 - (2b)^3\right] = \left(a - (2b)\right)\left(a^2 + 2ab + 4b^2\right)$$

For the denominator, factor out the highest common factor 2.

$$2a^2 - 8b^2 = 2\left(a^2 - 4b^2\right)$$

where $a^2 - 4b^2$ is the difference of two squares. Recall the formula for the difference of two squares: $x^2 - y^2 = (x-y)(x+y)$. Substitute a for x and $2b$ for y to obtain:

$$\left(2a^2 - 8b^2\right) = 2\left(a^2 - 4b^2\right) = 2(a - 2b)(a + 2b)$$

Then, writing the factored forms and cancelling:

$$\frac{a^3 - 8b^3}{2a^2 - 8b^2} = \frac{\cancel{(a - 2b)}\left(a^2 + 2ab + 4b^2\right)}{2(a + 2b)\cancel{(a - 2b)}} = \frac{a^2 + 2ab + 4b^2}{2(a + 2b)}$$

● **PROBLEM 413**

Divide $\dfrac{y^2 + y - 20}{y - 3}$ by $\dfrac{y^2 - 16}{y^2 + y - 12}$.

Solution: Dividing by a nonzero polynomial is the same as multiplying by its reciprocal. That is,

$$\frac{y^2 + y - 20}{y - 3} \div \frac{y^2 - 16}{y^2 + y - 12} = \frac{y^2 + y - 20}{y - 3} \cdot \frac{y^2 + y - 12}{y^2 - 16}$$

Factor each numerator and denominator, where possible. Note that $y^2 + y - 20 = (y + 5)(y - 4)$

$$y^2 + y - 12 = (y + 4)(y - 3),$$

and $y^2 - 16 = y^2 - 4^2$, the difference of two squares. Using the formula for the difference of two squares, $\left(a^2 - b^2\right) = (a - b)(a + b)$, replace a by **y** and b by 4 to obtain, $\left(y^2 - 16\right) = (y - 4)(y + 4)$.

Thus, $\dfrac{y^2 + y - 20}{y - 3} \cdot \dfrac{y^2 + y - 12}{y^2 - 16}$

$$= \frac{(y + 5)(y - 4)}{y - 3} \cdot \frac{(y + 4)(y - 3)}{(y - 4)(y + 4)} \qquad (1)$$

$$= \frac{(y + 5)(y - 4)(y + 4)(y - 3)}{(y - 3)(y - 4)(y + 4)} \qquad (2)$$

$$= \frac{(y + 5)(y - 4)(y + 4)(y - 3)}{(y - 4)(y + 4)(y - 3)} \qquad (3)$$

$$= y + 5.$$

Note that in equation (2) we are dividing by $(y - 3)(y - 4)(y + 4)$. If any of these factors equal 0, then we are dividing by zero, making our fraction in-

259

valid. Thus, in order to be certain we are proceeding correctly, we must establish the following restrictions:

$$(y - 3) \neq 0, \quad (y - 4) \neq 0, \quad (y + 4) \neq 0;$$

thus, $\quad y \neq 3, \qquad\qquad y \neq 4, \qquad\qquad y \neq -4.$

Therefore, $\dfrac{y^2 + y - 20}{y - 3} \div \dfrac{y^2 - 16}{y^2 + y - 12} = y + 5,$

and $y \neq 3, 4, -4.$

● **PROBLEM 414**

Combine the following fractions:

$$\frac{3}{x^2 - 3x + 2} + \frac{1}{x^2 - 5x + 6} - \frac{2}{x^2 - 4x + 3}$$

<u>Solution:</u> Factor each denominator. Multiply together the highest power of each factor only once regardless of the number of times each factor appears to obtain the Least Common Denominator, LCD.

$$x^2 - 3x + 2 = (x - 1)(x - 2),$$

$$x^2 - 5x + 6 = (x - 2)(x - 3),$$

$$x^2 - 4x + 3 = (x - 1)(x - 3);$$

the LCD can be seen to be $(x - 1)(x - 2)(x - 3)$.

In factored form we have:

$$\frac{3}{(x - 1)(x - 2)} + \frac{1}{(x - 2)(x - 3)} - \frac{2}{(x - 1)(x - 3)}$$

Now multiply each fraction by the LCD to obtain:

$$\frac{3(x - 1)(x - 2)(x - 3)}{(x - 1)(x - 2)} + \frac{1(x - 1)(x - 2)(x - 3)}{(x - 2)(x - 3)}$$

$$- \frac{2(x - 1)(x - 2)(x - 3)}{(x - 1)(x - 3)};$$

and cancel like terms to obtain:

$$3(x - 3) + (x - 1) - 2(x - 2)$$

$$= 3x - 9 + (x - 1) - (2x - 4)$$

$$= 3x - 9 + x - 1 - 2x + 4 = 2x - 6.$$

Now, divide by the LCD, so as not to change the value of the given fraction. Thus, we have:

260

$$\frac{2x - 6}{(x - 1)(x - 2)(x - 3)}$$

and simplified to give:

$$\frac{2(x - 3)}{(x - 1)(x - 2)(x - 3)} = \frac{2}{(x - 1)(x - 2)} .$$

$$= \frac{2x - 6}{\text{LCD}}$$

$$= \frac{2(x - 3)}{(x - 1)(x - 2)(x - 3)}$$

$$= \frac{2}{(x - 1)(x - 2)} .$$

● **PROBLEM 415**

Multiply

$$\frac{y^2 + 3y + 2}{y - 3} \quad \text{by} \quad \frac{y^2 - 7y + 12}{y^2 + y - 2} .$$

Solution: Factor the terms in the numerators and denominators of both fractions and cancel like terms.

$$\left(\frac{y^2 + 3y + 2}{y - 3}\right)\left(\frac{y^2 - 7y + 12}{y^2 + y - 2}\right)$$

$$= \frac{(y + 1)(y + 2)(y - 3)(y - 4)}{(y - 3)(y + 2)(y - 1)}$$

$$= \frac{(y + 1)(y - 4)}{y - 1} \qquad \text{dividing out common factors} \\ \qquad\qquad\qquad (y + 2) \text{ and } (y - 3)$$

$$= \frac{y^2 - 3y - 4}{y - 1}$$

Either of the last two fractions may be accepted as correct results.

● **PROBLEM 416**

Divide $\dfrac{x - 3}{x^2 - 7x + 12}$ by $\dfrac{x^2 - 6x + 9}{x^2 - 8x + 16}$.

Solution: Dividing by a fraction is equivalent to multiplying by its reciprocal; hence:

$$\frac{x - 3}{x^2 - 7x + 12} \div \frac{x^2 - 6x + 9}{x^2 - 8x + 16} = \frac{x - 3}{x^2 - 7x + 12} \times \frac{x^2 - 8x + 16}{x^2 - 6x + 9}$$

Factoring the numerators and denominators where possible we obtain:

$$\frac{x - 3}{x^2 - 7x + 12} \times \frac{x^2 - 8x + 16}{x^2 - 6x + 9} = \frac{x - 3}{(x - 4)(x - 3)} \cdot \frac{(x - 4)(x - 4)}{(x - 3)(x - 3)}$$

$$= \frac{(x - 3)(x - 4)(x - 4)}{(x - 4)(x - 3)(x - 3)(x - 3)}$$

Cancelling out the like terms which appear in both the numerators and denominators (since $\frac{x - 3}{x - 3} = 1$, and $\frac{x - 4}{x - 4} = 1$) gives us:

$$\frac{(x - 3)(x - 4)(x - 4)}{(x - 4)(x - 3)(x - 3)(x - 3)} = \frac{x - 4}{(x - 3)(x - 3)}$$

Therefore,

$$\frac{x - 3}{x^2 - 7x + 12} \div \frac{x^2 - 6x + 9}{x^2 - 8x + 16} = \frac{x - 4}{(x - 3)(x - 3)}.$$

● **PROBLEM 417**

Divide

$$\frac{2y^2 - 11y + 12}{6y^2 - 6y - 12} \quad \text{by} \quad \frac{3y^2 - 14y + 8}{2y^2 - 6y + 4}.$$

Solution: Divide the two fractions by inverting the fraction to the right of the division sign and multiplying the inverted fraction by the fraction on the left of the division sign. Thus,

$$\frac{2y^2 - 11y + 12}{6y^2 - 6y - 12} \div \frac{3y^2 - 14y + 8}{2y^2 - 6y + 4}$$

$$= \frac{2y^2 - 11y + 12}{6y^2 - 6y - 12} \cdot \frac{2y^2 - 6y + 4}{3y^2 - 14y + 8}$$

Now factor the expressions in the numerator and denominator of each fraction:

$$\frac{(2y - 3)(y - 4)\ 2(y - 2)(y - 1)}{6(y - 2)(y + 1)(3y - 2)(y - 4)}$$

Cancel common terms in the numerator and denominator to obtain:

$$\frac{(2y - 3)(y - 1)}{3(y + 1)(3y - 2)}.$$

● **PROBLEM 418**

Find $\dfrac{4a^2 + 4ab + b^2}{2a^3 + 16b^3} \div \dfrac{4a^2 - b^2}{6a + 12b}$

Solution: Division by a fraction is equivalent to multiplication by its reciprocal, hence:

$$\frac{4a^2 + 4ab + b^2}{2a^3 + 16b^3} \div \frac{4a^2 - b^2}{6a + 12b} = \frac{4a^2 + 4ab + b^2}{2a^3 + 16b^3} \times \frac{6a + 12b}{4a^2 - b^2}$$

Factor the numerators and denominators as completely as possible.

$4a^2 + 4ab + b^2$ is called a trinomial perfect square for it is in the form $(2a)^2 + 2(2ab) + b^2$. The formula for factoring a trinomial perfect square is given by $x^2 + 2xy + y^2 = (x+y)(x+y) = (x+y)^2$.

Replacing x by $2a$ and y by b we obtain:

$$4a^2 + 4ab + b^2 = (2a + b)(2a + b) = (2a + b)^2$$
$$2a^3 + 16b^3 = 2\left(a^3 + 8b^3\right) = 2\left(a^3 + (2b)^3\right) \text{ where } a^3 + (2b)^3$$

us the sum of two cubes. The formula for the sum of two cubes is

$$x^3 + y^3 = (x + y)\left(x^2 - xy + y^2\right)$$

Replacing x by a and y by $2b$ we obtain:

$$a^3 + (2b)^3 = (a + 2b)\left(a^2 - 2ab + (2b)^2\right)$$
$$= (a + 2b)\left(a^2 - 2ab + 4b^2\right)$$

thus,

$$2a^3 + 16b^3 = 2(a + 2b)\left(a^2 - 2ab + 4b^2\right)$$

Remove the highest common factor from $6a + 12b$ which is 6. Hence,

$$6a + 12b = 6(a + 2b)$$

$4a^2 - b^2$ is the difference of two squares, $(2a)^2 - b^2$.

Applying the formula for the difference of two squares $x^2 - y^2 = (x-y)(x+y)$:

$$4a^2 - b^2 = (2a)^2 - b^2 = (2a - b)(2a + b).$$

Now, express all the denominators and numerators in their factored form, and cancel:

$$\frac{4a^2 + 4ab + b^2}{2a^3 + 16b^3} \times \frac{6a + 12b}{4a^2 - b^2} = \frac{(2a + b)(2a + b)}{2(a+2b)(a^2 - 2ab + 4b^2)} \times \frac{\overset{3}{\cancel{6}}(a + 2b)}{(2a - b)(2a + b)}$$
$$= \frac{3(2a + b)}{\left(a^2 - 2ab + 4b^2\right)(2a - b)}$$

● **PROBLEM** 419

Perform the indicated operation,

$$\frac{x^3 - y^3}{x^2 - 5x + 6} \cdot \frac{x^2 - 4}{x^2 - 2xy + y^2}$$

Solution: We factor numerators and denominators to enable us to cancel terms.

$$x^3 - y^3$$

is the difference of two cubes. Thus we factor it applying the formula for the difference of two cubes,

$$a^3 - b^3 = (a - b)\left(a^2 + ab + b^2\right),$$

replacing a by x and b by y. Thus,

$$x^3 - y^3 = (x - y)(x^2 + xy + y^2).$$
$$x^2 - 5x+6 \text{ is factored as } (x-2)(x-3).$$
$$x^2 - 4 = x^2 - 2^2,$$

the difference of two squares. Applying the formula for the difference of two squares,

$$a^2 - b^2 = (a + b)(a - b),$$

and replacing a by x and b by 2 we obtain,

$$x^2 - 4 = (x + 2)(x - 2).$$

$$x^2 - 2xy + y^2 = (x - y)(x - y).$$

Thus,

$$\frac{x^3-y^3}{x^2 - 5x + 6} \cdot \frac{x^2 - 4}{x^2 - 2xy + y^2} = \frac{(x - y)(x^2 + xy + y^2)}{(x - 2)(x - 3)}$$

$$\cdot \frac{(x + 2)(x - 2)}{(x - y)(x - y)}$$

$$= \frac{(x^2 + xy + y^2)(x - y)(x - 2)(x + 2)}{(x - 3)(x - 2)(x - y)(x - y)}$$

$$= \frac{(x^2 + xy + y^2)(x + 2)}{(x - 3)(x -y)}$$

CHAPTER 16

SOLVING QUADRATIC EQUATIONS BY FACTORING

> **Basic Attacks and Strategies for Solving Problems in this Chapter. See pages 265 to 310 for step-by-step solutions to problems.**

A quadratic equation in standard form is written as

$$ax^2 + bx + c = 0,$$

where a, b, and c are constants and $a \neq 0$. To solve a quadratic equation by factoring involves the following steps:

(1) Eliminate any fractions from the equation.

(2) Write the equation in standard form.

(3) Factor the left side or non-zero side of the equation.

(4) Use the zero-property to set each factor equal to 0 which yields linear equations. (The zero-property states that for real number x and y, $xy = 0$ if and only if $x = 0$ or $y = 0$ or both equal 0).

(5) Solve the resulting linear equations.

For example, the solution of the quadratic equation

$$x^2 - 8x + 15 = 0$$

by factoring is given as follows:

$$(x - 3)(x - 5) = 0$$

$$x - 3 = 0 \quad \text{and} \quad x - 5 = 0$$

$$x = 3 \qquad\qquad x = 5$$

Thus, the solution set is $\{3, 5\}$.

If the quadratic equation contains one or more radicals then the first step in the procedure for solving the equation is to eliminate the radical(s) from the equation. The standard procedure for doing this is to use the **squaring property of equality**, that is, square both sides of the original equation and simplify. Then, the solution procedure proceeds with the steps outlined above. Notice that if a radical expression in the equation equals a negative real number, then the solution is an extraneous solution. This means that it satisfies the equation obtained after removing the radical(s), but does not satisfy the original equation.

A quadratic equation that is not factorable in standard form can be solved by completing the square. This technique involves the following steps:

(1) Write the given standard quadratic equation in the form

$$x^2 + (b/a)x = -c/a,$$

where $a \neq 0$.

(2) Then, take $1/2$ of the coefficient of x, square it, and add the result to both sides of the equation in Step 1 to obtain a perfect square trinomial on the left-hand side of the equation as follows:

$$x^2 + (b/a)x + (b/2a)^2 = -c/a + (b/2a)^2$$

(3) Factor the left-hand side of the equation in Step 2 to get:

$$(x + b/2a)^2 = -c/a + (b/2a)^2.$$

(4) Find the values of x by taking the square root of both sides of the equation in Step 3.

For example, the solution of

$$x^2 + 4x - 9 = 0$$

by completing the square as follows:

$$x^2 + 4x = 9 \quad \text{or} \quad x^2 + 4x + (2)^2 = 9 + (2)^2$$

$$(x + 2)^2 = 13$$

$$x + 2 = \pm\sqrt{13}$$

$$x = -2 \pm\sqrt{13}.$$

Step-by-Step Solutions to Problems in this Chapter, "Solving Quadratic Equations by Factoring"

EQUATIONS WITHOUT RADICALS

● PROBLEM 420

Show that $x^2 + 2x + 5 = 20$ is a conditional equation.

Solution: A conditional equation is an equation for which there exists at least one value which may be substituted for the variable that makes the equation false, but is true for other values. It is sufficient to exhibit one replacement for x that makes the equation true and one that makes it false.

$$\text{Let } x = 3: \qquad (3)^2 + 2(3) + 5 \overset{?}{=} 20$$

$$9 + 6 + 5 \overset{?}{=} 20$$

$$20 = 20$$

$$\text{Let } x = -4: \qquad (-4)^2 + (-4) + 5 \overset{?}{=} 20$$

$$16 - 8 + 5 \overset{?}{=} 20$$

$$13 \neq 20$$

When $x = -4$, this value of x makes the equation false. For $x = 3$, the equation is true. Therefore, $x^2 + 2x + 5 = 20$ is a conditional equation.

Notice that we have not solved the equation in this example . An equation is solved when its solution set is completely known.

● PROBLEM 421

Solve the equation $(3x - 7)(x + 2) = 0$.

Solution: When a given product of two numbers that are equal to zero, $ab = 0$, either a must equal zero or b must equal zero (or both equal zero). So if $(3x - 7)(x + 2) = 0$, then $(3x - 7) = 0$ or $(x + 2) = 0$.

$3x - 7 = 0$	$x + 2 = 0$
Add 7 to both sides:	Subtract 2 from both sides:
$3x = 7$	
Divide both sides by 3:	
$x = \dfrac{7}{3}$	$x = -2$

Hence $x = \frac{7}{3}$ or $x = -2$, and our solution set is $\left\{ \frac{7}{3}, -2 \right\}$.

● PROBLEM 422

Solve the equation $3x^2 + 5x = 0$.

Solution: Because division by zero is impossible, we must not divide by x, since x might be equal to zero. Instead of dividing by x we factor x from the left side of the equation to obtain:

$$x(3x + 5) = 0.$$

Whenever we have a situation where ab = 0 (the product of two or more numbers equal to zero) either a = 0 or b = 0. Therefore x = 0, or 3x + 5 = 0. Subtract 5 from each side of the second equation to obtain:

$$3x = -5$$

Divide both sides by 3 to obtain $x = -\frac{5}{3}$. The two solutions of the given equation are x = 0 and $x = -\frac{5}{3}$.

To check the validity of the two solutions we substitute them into the given equation. Thus,

when x = 0

$$3x^2 + 5x = 0$$

$$3(0)^2 + 5(0) = 0$$

$$0 = 0$$

when $x = -\frac{5}{3}$

$$3x^2 + 5x = 0$$

$$3\left(-\frac{5}{3}\right)^2 + 5\left(-\frac{5}{3}\right) = 0$$

$$3\left(\frac{25}{9}\right) + 5\left(-\frac{5}{3}\right) = 0$$

$$\frac{25}{3} - \frac{25}{3} = 0$$

$$0 = 0$$

● PROBLEM 423

Solve the equation $x^2 + 8x + 15 = 0$.

Solution: Since $(x + a)(x + b) = x^2 + bx + ax + ab$

$= x^2 + (a + b) x + ab$, we may factor the given equation, $0 = x^2 + 8x + 15$, replacing $a + b$ by 8 and ab by 15. Thus,

$$a + b = 8, \quad \text{and}$$

$$ab = 15.$$

We want the two numbers a and b whose sum is 8 and whose product is 15. We check all pairs of numbers whose product is 15:

(a) $1 \cdot 15 = 15$; thus $a = 1$, $b = 15$ and $ab = 15$.

$1 + 15 = 16$, therefore we reject these values because $a + b \neq 8$.

(b) $3 \cdot 5 = 15$; thus $a = 3$, $b = 5$, and $ab = 15$.

$3 + 5 = 8$. Therefore $a + b = 8$, and we accept these values.

Hence $\quad x^2 + 8x + 15 = 0 \qquad$ is equivalent to

$$0 = x^2 + (3 + 5)x + 3 \cdot 5 = (x + 3)(x + 5)$$

Hence, $\quad x + 5 = 0 \qquad$ or $x + 3 = 0$

since the product of these two numbers is zero, one of the numbers must be zero. Hence, $x = -5$, or $x = -3$, and the solution set is $X = \{-5, -3\}$.

The student should note that $x = -5$ or $x = -3$. We are certainly not making the statement, that $x = -5$, and $x = -3$. Also, the student should check that both these numbers do actually satisfy the given equations and hence are solutions.

Check: Replacing x by (-5) in the original equation:

$$x^2 + 8x + 15 = 0$$

$$(-5)^2 + 8(-5) + 15 = 0$$

$$25 - 40 + 15 = 0$$

$$-15 + 15 = 0$$

$$0 = 0$$

Replacing x by (-3) in the original equation:

$$x^2 + 8x + 15 = 0$$

$$(-3)^2 + 8(-3) + 15 = 0$$

$$9 - 24 + 15 = 0$$

$$-15 + 15 = 0$$

$$0 = 0.$$

Find the roots of the equation $x^2 + 6x + 8 = 0$.

<u>Solution:</u>   In order to obtain the roots of this equation we must factor it, that is, put it in the form $(x + a) (x + b) = 0$. Using our method for multiplying polynomials we find $(x + a) (x + b) = x^2 + ax + bx + ab = x^2 + (a + b) x + ab$. We are given $x^2 + 6x + 8 = 0$. Thus in our case

$$a + b = 6 \quad (1)$$

and $ab = 8 \qquad (2)$

That is, we want to find 2 numbers a and b whose sum is 6 and whose product is 8.

Checking all pairs of numbers whose product is 8:

(a) $8 \times 1 = 8$; hence 8 and 1 satisfy equation 2, however

$8 + 1 = 9 \neq 6$; thus we reject these values, for they fail to satisfy equation (1)

(b) $4 \times 2 = 8$; hence 4 and 2 satisfy equation 2, and

$4 + 2 = 6$ satisfying equation (1).

Thus, we conclude $a = 4$, $b = 2$, and we may write

$x^2 + 6x + 8 = \left(x^2 + (4 + 2) x + 4 \cdot 2 \right)$ as $(x + 4) (x + 2)$

Since $x^2 + 6x + 8 = 0$ and $x^2 + 6x + 8 = (x + 4) (x + 2)$,

$(x + 4) (x + 2) = 0$

Recall that if the product of 2 numbers, $ab = 0$, either $a = 0$, or $b = 0$.

Hence $x + 4 = 0$ or $x + 2 = 0$

and $x = -4$ or $x = -2$

Check: Replace x by (-4) in our given equation,

$$(-4)^2 + 6 (-4) + 8 = 0$$

$$16 - 24 + 8 = 0$$

$$-8 + 8 = 0$$

$$0 = 0$$

Now, replace x by (-2) in our given equation,

$$(-2)^2 + 6 (-2) + 8 = 0$$

$$4 - 12 + 8 = 0$$

$$-8 + 8 = 0$$

$$0 = 0$$

Therefore, the roots of the given equation are $x = -4$, $x = -2$.

Solve: $x^2 - 5x - 14 = 0$.

Solution: To find the roots of this quadratic, we factor it (put it in the form $(x + a)(x + b) = 0$).

Note that $(x + a)(x + b) = x^2 + (a + b)x + ab$

Thus, in our quadratic, $x^2 + (- 5)x + (- 14)$,

$$a + b = - 5 \tag{1}$$

and $\qquad ab = - 14 \tag{2}$

That is, we want the two numbers, a and b, whose sum is $(- 5)$, and whose product is $(- 14)$.

To find these numbers, we can check the set of numbers whose product is $(- 14)$:

(a) $(- 14) \times (1) = - 14$, therefore equation (2) is satisfied, now check these values in equation (1):

$(- 14) + (1) = - 13 \neq - 5$ therefore we reject these values.

(b) $(- 7) \times (2) = - 14$, therefore equation (2) is satisfied, now check these values in equation (1):

$(- 7) + 2 = - 5$ hence both equations are satisfied and we conclude

$$a = - 7 \quad \text{and } b = 2.$$

Thus $\quad x^2 - 5x - 14 = x^2 + (- 7 + 2)x + (- 7)(2)$

$$= \left[x + (- 7)\right]\left[x + 2\right]$$

$$= (x - 7)(x + 2) = 0$$

By the fundamental principle, if the product of two numbers $yz = 0$, then either $y = 0$ or $z = 0$; hence if

$(x - 7)(x + 2) = 0$

either $x - 7 = 0 \qquad$ or $\quad x + 2 = 0$

add 7 to both sides $\qquad$ subtract 2 from both sides

$\qquad x = 7 \qquad\qquad$ or $\qquad x = - 2$

This proves that if the equation has roots, they must be either 7 or - 2. We check these values by substituting in the given equation:

If x = 7, then $x^2 - 5x - 14 = (7)^2 - 5(7) - 14$

$$= 49 - 35 - 14$$

$$= 49 - 49$$

$$= 0$$

If x = - 2, then $x^2 - 5x - 14 = (-2)^2 - 5(-2) - 14$

$$= 4 + 10 - 14$$

$$= 14 - 14$$

$$= 0$$

We may now conclude that the solution to our equation is
x = 7 or x = - 2.

● **PROBLEM   426**

Find the roots of $x^2 - 3x - 10 = 0$.

<u>Solution</u>:   To find the roots of this quadratic, we factor
it (put it in the form (x + a)(x + b) = 0).

Note that $(x + a)(x + b) = x^2 + (a + b)x + ab$
Thus in our quadratic, $x^2 + (-3)x + (-10)$,

$$a + b = -3 \qquad\qquad\qquad (1)$$

and   ab = - 10.   (2)

That is, we want the two numbers a and b whose sum is
(- 3), and whose product is (- 10).

To find these numbers, we can check the set of
numbers whose product is (- 10):

(a) (-10) × (1) = - 10, therefore equation (2) is satisfied,
now check these values in equation (1):
     $(-10) + (1) = -9 \neq -3$ therefore we reject these
values.

(b) (- 5) × (2) = - 10, therefore equation (2) is
satisfied, now checking these values in equation (1):
(- 5) + 2 = - 3.

Hence both equations are satisfied and we conclude

$$a = -5 \qquad \text{and} \qquad b = 2.$$

Thus, $x^2 - 3x - 10 = x^2 + (-5 + 2)x + (-5)(2)$

$$= \left[x + (-5)\right]\left[x + 2\right]$$

$$= (x - 5)(x + 2) = 0.$$

Hence, by the fundamental property which states that if
ab = 0, either a = 0 or b = 0, x - 5 = 0 or x + 2 = 0 and

$$x = 5 \qquad \text{or} \qquad x = -2.$$

This proves that if the equation has roots, they must be either 5 or - 2. So far we have not proved that these are roots. We can check this by substituting in the given equation. If x = 5, then

$$x^2 - 3x - 10 = (5)^2 - 3(5) - 10$$
$$= 25 - 15 - 10$$
$$= 25 - 25$$
$$= 0$$

Thus 5 is indeed a root of the equation.

If x = - 2, then

$$x^2 - 3x - 10 = (-2)^2 - 3(-2) - 10$$
$$= 4 + 6 - 10$$
$$= 10 - 10$$
$$= 0$$

Thus - 2 is also a root.

Such a check not only has a logical purpose, but it also assures us that we have not made a mistake in arithmetic. We may now conclude that the solution to our equation is x = 5 or x = - 2.

● **PROBLEM** 427

Find the roots of the function G defined by the rule
$$G(x) = x^2 + 5x + 6.$$

Solution: To find the roots of the function G, find those values of x which satisfy G(x) = 0. Let $x^2 + 5x + 6 = 0$. In factored form this may be written (x + 3)(x + 2) = 0. Values of x which make this product = 0 satisfy x + 3 = 0 or x + 2 = 0. Hence x = -3 or x = 2.

Check:

for x = -3   for x = -2

$(-3)^2 + 5(-3) + 6 = 9 - 15 + 6 = 0$ $(-2)^2 + 5(-2) + 6 = 4-10+6 = 0$

● **PROBLEM** 428

Solve and check the roots of the equation
$$2x^2 - 3x + 1 = 0$$

Solution: Factor the given equation into a product of two polynomials: therefore: (2x - 1)(x - 1) = 0.

When ab = 0, either a = 0 or b = 0 where a and b

271

are real numbers. Therefore, either $2x - 1 = 0$ or $x - 1 = 0$.

Therefore, either $x = \frac{1}{2}$ or $x = 1$.

Hence, the roots to the given equation are:

$x = \frac{1}{2}$ and $x = 1$.

To check these roots, substitute $x = \frac{1}{2}$ in the given equation:

$$2\left(\frac{1}{2}\right)^2 - 3\left(\frac{1}{2}\right) + 1 = 0$$

$$2\left(\frac{1}{4}\right) - \frac{3}{2} + 1 = 0$$

$$\frac{1}{2} - \frac{3}{2} + 1 = 0$$

$$-1 + 1 = 0$$

$$0 = 0$$

Now, substitute $x = 1$ in the given equation:

$$2(1)^2 - 3(1) + 1 = 0$$

$$2 - 3 + 1 = 0$$

$$-1 + 1 = 0$$

$$0 = 0$$

● **PROBLEM 429**

Solve the equation $2x^2 - 7x + 6 = 0$.

<u>Solution:</u> To solve this equation we must find the values of x. We find these values by factoring the given equation into a product of two polynomials. To do this, factors of 6 must be found which give a coefficient of -7 for x when the two polynomials are multiplied together. The factors $(2x - 3)(x - 2)$ accomplish this. When $ab = 0$ where a and b are any numbers, either $a = 0$ or $b = 0$. Therefore, either $2x - 3 = 0$ or $x - 2 = 0$. To solve for x in the first equation we add 3 to both sides of the equation, $2x - 3 = 0$, and then divide by 2. Thus, we have:

$$2x - 3 + 3 = 0 + 3$$

$$2x = 3$$

$$\frac{2x}{2} = \frac{3}{2}$$

$$x = \frac{3}{2} \ .$$

272

To solve for x in the second equation we add 2 to both sides of the equation, $x - 2 = 0$. Thus, we have:

$$x - 2 + 2 = 0 + 2$$

$$x = 2.$$

Therefore, the roots of the given equation are:

$$x = \frac{3}{2}, \; x = 2.$$

To check these roots, do the following: Substituting $x = \frac{3}{2}$ in the given equation, we find:

$$2\left(\frac{3}{2}\right)^2 - 7\left(\frac{3}{2}\right) + 6 = 0$$

$$\frac{9}{2} - \frac{21}{2} + \frac{12}{2} = 0$$

$$0 = 0.$$

Substituting $x = 2$ in the given equation, we find:

$$2(2)^2 - 7(2) + 6 = 0$$

$$8 - 14 + 6 = 0$$

$$0 = 0.$$

Thus, the two roots are valid.

● **PROBLEM** 430

Solve the following for x:

(a) $x^2 - 3x = 0$

(b) $6x^2 + 5x - 4 = 0$

Solution: (a) Factor the common factor x from the left side of the given equation:

$$x^2 - 3x = x(x - 3)$$

Since $x^2 - 3x = 0$,

$$x(x - 3) = 0. \tag{1}$$

Whenever a product $ab = 0$, where a and b are any two numbers, either $a = 0$ or $b = 0$. Then, equation (1) becomes,

$$x = 0 \text{ or } x - 3 = 0$$

$$x = 3$$

Hence, the solution set is: $\{0, 3\}$.

(b) Factor the left side of the given equation into a product of two polynomials:

$$6x^2 + 5x - 4 = (3x + 4)(2x - 1)$$

Since $6x^2 + 5x - 4 = 0$,

$$(3x + 4)(2x - 1) = 0$$

Thus,

$3x + 4 = 0$	or	$2x - 1 = 0$
$3x = -4$	or	$2x = 1$
$x = \dfrac{-4}{3}$	or	$x = \dfrac{1}{2}$

Hence, the solution set is: $\left\{ -\dfrac{4}{3}, \dfrac{1}{2} \right\}$.

● **PROBLEM 431**

Solve the equation $5y^2 = 6y$ by the factoring method.

<u>Solution:</u> Add $(- 6y)$ to both members of the given equation
$$5y^2 - 6y = 0$$

Factor y from the left member, $y(5y - 6) = 0$.

When the product of two numbers $ab = 0$ either $a = 0$, or $b = 0$. Thus, either $y = 0$ or $(5y - 6) = 0$.

Solving for y in the second equation, $5y - 6 = 0$:

add $- 6$ to both sides, $\qquad\qquad\qquad 5y = 6$

divide by 5, $\qquad\qquad\qquad\qquad\qquad y = 6/5.$

Therefore, the solution set is $\{0, 6/5\}$.

Check: To check these values we replace y by 0 and then by 6/5 in the original equation:

(a) when $y = 0$ $\qquad 5y^2 = 6y$

$$5(0)^2 = 6(0)$$

$$0 = 0$$

(b) when $y = \dfrac{6}{5}$ $\qquad 5\left(\dfrac{6}{5}\right)^2 = 6\left(\dfrac{6}{5}\right)$

$$5\left(\dfrac{36}{25}\right) = \dfrac{36}{5}$$

$$\dfrac{36}{5} = \dfrac{36}{5}$$

Thus, the solution set of the given equation is indeed $\{0, 6/5\}$.

Solve $4x^2 = 8x$.

Solution: The temptation to divide both sides by $4x$ to arrive at: $x = 2$, should be avoided, for if $x = 0$ we are performing an operation which is undefined. Although 2 actually is a root, there happens to be another root, which is lost in this process.

When solving equations, avoid multiplying or dividing by anything but nonzero numbers. In this case, there is no harm in dividing both sides by the number 4:

$$x^2 = 2x$$

We then add $-2x$ to both sides, to arrive at:

$$x^2 - 2x = 0$$

Factoring: $\qquad\qquad x(x-2) = 0$

Whenever the product of two numbers $ab = 0$, either $a = 0$ or $b = 0$. Therefore,

$$x = 0 \quad \text{or} \quad x - 2 = 0,$$

and

$$x = 0 \quad \text{or} \quad x = 2.$$

Check: To verify that the roots of this equation are $x = 0$ and $x = 2$, we replace x by each value in the original equation. Replacing x by 0 in $4x^2 = 8x$:

$$4(0)^2 = 8(0)$$
$$0 = 0$$

Replacing x by 2:

$$4(2)^2 = 8(2)$$
$$4(4) = 16$$
$$16 = 16$$

Thus, the roots of the equation are 0 and 2, and the solution set is

$$\{0,2\} .$$

Solve the equation $4x^2 = 9x$.

Solution: To solve the given equation we must find the values of x which satisfy the equation. To do this we proceed as follows: Subtract 9x from both sides of the given equation. Thus, we have:

$$4x^2 - 9x = 0. \qquad\qquad (1)$$

Now, factor x from the terms on the left side of Equation (1): $\qquad x(4x - 9) = 0.$

When ab = 0 where a and b are any numbers, then either a = 0 or b = 0. Therefore: either x = 0, or 4x - 9 = 0.

To solve for x in the second equation, add 9 to both sides of 4x - 9 = 0, and then divide both sides of the equation by 4. Thus,

$$4x - 9 + 9 = 0 + 9$$
$$4x = 9$$
$$\frac{4x}{4} = \frac{9}{4}$$

$$x = \frac{9}{4}.$$

Therefore, the two roots to the given equation are x = 0 and x = $\frac{9}{4}$. To check these two roots, do the following: Substitute x = 0 in the equation, $4x^2 = 9x$. Then:

$$4(0)^2 = 9(0)$$
$$4(0) = 0$$
$$0 = 0.$$

Now, substitute x = $\frac{9}{4}$ in the equation, $4x^2 = 9x$. Then:

$$4\left(\frac{9}{4}\right)^2 = 9\left(\frac{9}{4}\right)$$

$$4\left(\frac{81}{16}\right) = \frac{81}{4}$$

$$\frac{81}{4} = \frac{81}{4} .$$

Therefore, the two values x = 0, x = $\frac{9}{4}$ are valid.

● **PROBLEM 434**

Solve the equation $6x^2 = 2 - x$.

<u>Solution:</u> Write the equation in standard quadratic form by adding x - 2 to both sides of the equation. Then we have $6x^2 + x - 2 = 0$. In factored form this becomes $(3x + 2)(2x - 1) = 0$. The values of x that make this product = 0 satisfy

$$
\begin{array}{lll}
3x + 2 = 0 & \text{or} & 2x - 1 = 0 \\
3x = -2 & \text{or} & 2x = 1 \\
x = -\tfrac{2}{3} & \text{or} & x = \tfrac{1}{2}
\end{array}
$$

Check:

for x = $-\frac{2}{3}$ for x = $\frac{1}{2}$

$$6\left(-\tfrac{2}{3}\right)^2 \overset{?}{=} 2 - \left(-\tfrac{2}{3}\right) \qquad 6\left(\tfrac{1}{2}\right)^2 \overset{?}{=} 2 - \left(\tfrac{1}{2}\right)$$

$$6\left(\tfrac{4}{9}\right) \overset{?}{=} \tfrac{18}{9} - \left(-\tfrac{6}{9}\right) \qquad 6\left(\tfrac{1}{4}\right) \overset{?}{=} \tfrac{8}{4} - \left(\tfrac{2}{4}\right)$$

$$\frac{24}{9} = \frac{24}{9} \qquad\qquad\qquad \frac{6}{4} = \frac{6}{4}$$

Therefore the solution set is $\{-\tfrac{2}{3}, \tfrac{1}{2}\}$.

● **PROBLEM 435**

Solve the equation $2x^2 = x + 6$ by the factoring method.

<u>Solution:</u>

$$2x^2 = x + 6 \qquad \text{given equation}$$

$$2x^2 - x - 6 = 0 \qquad \text{adding } -x - 6 \text{ to each member}$$

$$(2x + 3)(x - 2) = 0 \qquad \text{factoring left member}$$

Whenever the product of 2 numbers ab = 0 either a = 0 or

276

b = 0. Thus either 2x + 3 = 0 or x - 2 = 0

$$2x + 3 = 0 \qquad \text{setting the first}$$
$$\text{factor equal to 0}$$
$$2x = -3$$
$$x = -\frac{3}{2} \qquad \text{solving for x}$$
$$x - 2 = 0 \qquad \text{setting second factor}$$
$$\text{equal to 0}$$
$$x = 2 \qquad \text{solving for x}$$

Consequently the solution set is $\left\{-\frac{3}{2}, 2\right\}$.

The solution set can be verified by replacing x in the given equation by each element in the set.

Check:

$$x = -\frac{3}{2} \qquad\qquad\qquad x = 2$$
$$2x^2 = x + 6 \qquad\qquad\qquad 2x^2 = x + 6$$
$$2\left(\frac{-3}{2}\right)^2 = \frac{-3}{2} + 6 \qquad\qquad 2(2)^2 = 2 + 6$$
$$2\left(\frac{9}{4}\right) = \frac{-3}{2} + \frac{12}{2} \qquad\qquad 2 \cdot 4 = 2 + 6$$
$$\frac{9}{2} = \frac{9}{2} \checkmark \qquad\qquad\qquad 8 = 8 \checkmark$$

● **PROBLEM 436**

Solve the equation $4x^2 = 4x - 1$. (1)

Solution: Subtract 4x from both sides of equation (1).

$$4x^2 - 4x = 4x - 1 - 4x$$

Therefore: $4x^2 - 4x = -1$ (2).

Add 1 to both sides of equation (2).

$$4x^2 - 4x + 1 = -1 + 1$$
$$4x^2 - 4x + 1 = 0 \quad (3)$$

Factor the left side of equation (3) as a product of two polynomials. Therefore,

$$(2x - 1)(2x - 1) = 0.$$

When ab = 0, either a = 0 or b = 0, where a and b are real numbers. Therefore, (2x - 1) = 0, (2x - 1) = 0.

$$x = \frac{1}{2} , \quad x = \frac{1}{2} \quad (\text{Both roots are equal to } \tfrac{1}{2}).$$

In order to check that $\tfrac{1}{2}$ is a solution to the given

277

equation, substitute $x = \frac{1}{2}$ in the given equation:

$$4\left(\frac{1}{2}\right)^2 = 4\left(\frac{1}{2}\right) - 1$$

$$4\left(\frac{1}{4}\right) = 2 - 1$$

$$1 = 1$$

● **PROBLEM** 437

Solve the equation $4x^2 = 100$.

Solution: To solve this equation we must find the values for x which satisfy the equation. To do this we proceed as follows: Subtract 100 from both sides of the given equation. Thus,

$$4x^2 - 100 = 100 - 100$$

$$4x^2 - 100 = 0. \tag{1}$$

Now, factor 4 from the left side of Equation (1):

$$4\left(x^2 - 25\right) = 0.$$

Next, factor $x^2 - 25$ into a product of two polynomials. To do this, notice that $x^2 - 25$ is the difference between two squares, that is, $x^2 - 5^2$. Thus, the two factors are $(x - 5)(x + 5)$. Thus, we have:

$$4(x - 5)(x + 5) = 0.$$

Dividing both sides of this equation by 4, we obtain:

$$\frac{4(x - 5)(x + 5)}{4} = \frac{0}{4}.$$

Therefore, $(x - 5)(x + 5) = 0$.

When $ab = 0$, where a and b are any numbers, either $a = 0$ or $b = 0$. Therefore, either $x - 5 = 0$, or $x + 5 = 0$.

To solve for x in the first equation add 5 to both sides of $x - 5 = 0$. Thus,

$$x - 5 + 5 = 0 + 5$$

$$x = 5.$$

To solve for x in the second equation subtract 5 from both sides of $x + 5 = 0$. Thus,

$$x + 5 - 5 = 0 - 5$$

$$x = -5.$$

Therefore, the solution of the given equation is $x = 5$, $x = -5$.

To check these two solutions, do the following:

Substituting x = 5 in the given equation, we find:

$$4(5)^2 = 100$$

$$100 = 100.$$

Substituting x = -5 in the given equation, we find:

$$4(-5)^2 = 100$$

$$100 = 100.$$

Thus, the obtained values of x are valid.

● **PROBLEM** 438

Solve the following equations by factoring.

(a) $2x^2 + 3x = 0$ (c) $z^2 - 2z - 3 = 0$

(b) $y^2 - 2y - 3 = y - 3$ (d) $2m^2 - 11m - 6 = 0$

<u>Solution:</u>   (a) $2x^2 + 3x = 0$.   Factoring out the common factor of x from the left side of the given equation,

$$x(2x + 3) = 0.$$

Whenever a product ab = 0, where a and b are any two numbers, either a = 0 or b = 0. Then, either

$$x = 0 \quad or \quad 2x + 3 = 0$$

$$2x = -3$$

$$x = \frac{-3}{2}$$

Hence, the solution set to the original equation $2x^2 + 3x = 0$ is: $\left\{ -\frac{3}{2}, \ 0 \right\}$

(b) $y^2 - 2y - 3 = y - 3$. Subtract (y - 3) from both sides of the given equation:

$$y^2 - 2y - 3 - (y - 3) = y - 3 - (y - 3)$$
$$y^2 - 2y - \cancel{3} - y + \cancel{3} = \cancel{y} - \cancel{3} - \cancel{y} + \cancel{3}$$
$$y^2 - 3y = 0.$$

Factor out a common factor of y from the left side of this equation:

$$y(y - 3) = 0.$$

Thus, y = 0 or y - 3 = 0

$$y = 3$$

Therefore, the solution set to the original equation $y^2 - 2y - 3 = y - 3$ is: {0,3}.

(c) $z^2 - 2z - 3 = 0$. Factor the original equation into a product of two polynomials:

$$z^2 - 2z - 3 = (z - 3)(z + 1) = 0$$

Hence,

$(z - 3)(z + 1) = 0$; and $z - 3 = 0$ or $z + 1 = 0$
$$z = 3 \qquad z = -1$$

Therefore, the solution set to the original equation $z^2 - 2z - 3 = 0$ is: $\{-1, 3\}$.

(d) $2m^2 - 11m - 6 = 0$. Factor the original equation into a product of two polynomials:

$$2m^2 - 11m - 6 = (2m + 1)(m - 6) = 0$$

Thus, $\qquad 2m + 1 = 0$ or $m - 6 = 0$

$$2m = -1 \qquad m = 6$$

$$m = \frac{-1}{2}$$

Therefore, the solution set to the original equation $2m^2 - 11m - 6 = 0$ is: $\left\{ -\frac{1}{2}, 6 \right\}$.

● **PROBLEM 439**

Solve the equation $a^2x + 4c^2x - 10c = 5a - 4acx$ for x.

Solution: Add $4acx$ and $10c$ to both sides.
$$a^2x + 4acx + 4c^2x = 5a + 10c,$$

Factor out x from the left-hand side and 5 from the right-hand side.
$$\left(a^2 + 4ac + 4c^2 \right)x = 5(a + 2c),$$

Factor the left-hand side which is a trinomial perfect square. Take the square roots of the first and last terms, and join them by the sign of the middle term.
$$(a + 2c)^2 x = 5(a + 2c),$$

Solve for x.

$$x = \frac{5(a + 2c)}{(a + 2c)^2} = \frac{5}{a + 2c} .$$

The solution just obtained is also the zero of the function
$$a^2x + 4c^2x - 10c - (5a - 4acx) = 0 .$$

● **PROBLEM 440**

Solve $\dfrac{x}{x - 2} + \dfrac{x - 1}{2} = x + 1$.

Solution: First eliminate the fractions to facilitate solution. This is done by multiplying both sides of the equation by the Least Common Denominator, LCD. The LCD is obtained by multiplying the denominators of every fraction: LCD = $2(x - 2)$; and multiplying each side by this, the equation becomes:

$$2(x - 2)\left[\frac{x}{x - 2} + \frac{x - 1}{2}\right] = (x + 1)2(x - 2)$$

$$2x + (x - 1)(x - 2) = 2(x + 1)(x - 2)$$

$$2x + x^2 - 3x + 2 = 2x^2 - 2x - 4$$

$$x^2 - x - 6 = 0$$

This can be solved by factoring and setting each factor equal to zero.

$$(x - 3)(x + 2) = 0$$

$$x - 3 = 0 \qquad\qquad x + 2 = 0$$

$$x = 3 \qquad\qquad x = -2$$

Since both of these solutions are admissible values of x, they both should satisfy the original equation.

Check for x = 3:

$$\frac{3}{1} + \frac{2}{2} = 3 + 1$$

$$3 + 1 = 3 + 1$$

Check for x = -2:

$$\frac{-2}{-4} + \frac{-3}{2} = -2 + 1$$

$$\frac{1}{2} + \frac{-3}{2} = -1$$

$$-1 = -1$$

● **PROBLEM** 441

Solve the equation $\frac{4}{x + 1} + \frac{3}{x} = 2$.

Solution: Our two denominators are x and x + 1. They have no common factors, thus our least common multiple, LCM, is x(x+1). We multiply both members by x(x+1) to obtain

$$x(x+1)\left[\frac{4}{x + 1} + \frac{3}{x}\right] = 2[x(x+1)]$$

$$\frac{x(x+1)4}{x + 1} + \frac{x(x+1)3}{x} = 2x(x+1)$$

$$4x + 3(x + 1) = 2x(x + 1)$$

and then

$$4x + 3x + 3 = 2x^2 + 2x$$

$$7x + 3 = 2x^2 + 2x$$

We add $-(7x + 3)$ to both sides

$$0 = 2x^2 + 2x - (7x + 3)$$
$$2x^2 + 2x - 7x - 3 = 0$$

We thus have to solve the quadratic equation

$$2x^2 - 5x - 3 = 0$$

Factoring, we have

$$(2x + 1)(x - 3) = 0 \ ;$$

and whenever a product of two numbers $ab = 0$ either $a = 0$ or $b = 0$. Therefore either

$$2x + 1 = 0 \quad \text{or} \quad x - 3 = 0$$

and
$$x = -\tfrac{1}{2} \quad \text{or} \quad x = 3$$

Thus the only possible roots are $(-\tfrac{1}{2})$ and 3. We must check if these values are indeed roots. Replace x by $(-\tfrac{1}{2})$ in the original equation,

$$\frac{4}{x + 1} + \frac{3}{x} = 2$$

$$\frac{4}{(-\tfrac{1}{2}) + 1} + \frac{3}{(-\tfrac{1}{2})} = 2$$

$$\frac{4}{(\tfrac{1}{2})} + \frac{3}{(-\tfrac{1}{2})} = 2$$

Since division by a fraction is equivalent to multiplication by its reciprocal,

$$4 \cdot 2 + 3 \cdot (-2) = 2$$
$$8 + (-6) = 2$$
$$2 = 2$$

Now replace x by 3 in the original equation,

$$\frac{4}{x + 1} + \frac{3}{x} = 2$$

$$\frac{4}{3 + 1} + \frac{3}{3} = 2$$

$$\frac{4}{4} + \frac{3}{3} = 2$$

$$1 + 1 = 2$$

$$2 = 2$$

Thus $-\tfrac{1}{2}$ and 3 are both roots and our solution set is $\{-\tfrac{1}{2}, 3\}$.

● **PROBLEM 442**

Solve

$$\frac{x + 1}{x^2 - 5x + 6} + \frac{x + 2}{x^2 - 7x + 12} = \frac{6}{x^2 - 6x + 8} \ .$$

<u>Solution</u>: Factor the denominator of each fraction to obtain

$$\frac{x + 1}{(x - 2)(x - 3)} + \frac{x + 2}{(x - 3)(x - 4)} = \frac{6}{(x - 2)(x - 4)} \ .$$

Obtain the Least Common Denominator, LCD, by multiplying the denominators of each fraction together and using the highest power of each factor only once, that is,

$$(x - 2)(x - 3)(x - 4)(x - 3)(x - 2)(x - 4)$$

$$\text{LCD} = (x - 2)(x - 3)(x - 4).$$

Multiply both sides of the equation by the LCD to remove all fractions and obtain:

$$(x - 2)(x - 3)(x - 4)\left[\frac{x + 1}{(x - 2)(x - 3)} + \frac{x + 2}{(x - 3)(x - 4)}\right]$$

$$= \left[\frac{6}{(x - 2)(x - 4)}\right](x - 2)(x - 3)(x - 4)$$

$$(x + 1)(x - 4) + (x + 2)(x - 2) = 6(x - 3)$$

$$(x + 1)(x - 4) + (x + 2)(x - 2) = 6x - 18$$

$$\left(x^2 - 3x - 4\right) + \left(x^2 - 4\right) = 6x - 18$$

$$2x^2 - 3x - 8 = 6x - 18$$

$$2x^2 - 9x + 10 = 0$$

$$(2x - 5)(x - 2) = 0$$

$2x - 5 = 0$	$x - 2 = 0$
$2x = 5$	$x = 2$
$x = \dfrac{5}{2}$	

Substituting $x = 2$ into the original equation shows that $x = 2$ is an extraneous root, since it is not an admissible value for the original equation. (It makes two of the denominators $(x - 2)(x - 3)$ and $(x - 2)(x - 4)$, equal to zero.)

$x = \dfrac{5}{2}$ is an admissible value of x for the original equation and is a solution if it will satisfy the original equation.

Check: $x = \dfrac{5}{2}$

$$\frac{\frac{7}{2}}{\left(\frac{1}{2}\right)\left(-\frac{1}{2}\right)} + \frac{\frac{9}{2}}{\left(-\frac{1}{2}\right)\left(-\frac{3}{2}\right)} = \frac{6}{\left(\frac{1}{2}\right)\left(-\frac{3}{2}\right)}$$

$$\frac{14}{-1} + \frac{18}{3} = \frac{24}{-3}$$

$$-14 + 6 = -8$$
$$-8 = -8$$

EQUATIONS WITH RADICALS

● **PROBLEM 443**

Solve the equation $\sqrt{2x^2 - 9} = x$.

Solution: Squaring both sides, we have
$$2x^2 - 9 = x^2$$
$$x^2 = 9$$

$$x = 3 \quad \text{or} \quad x = -3$$

Both 3 and -3 will satisfy the equation $2x^2 - 9 = x^2$ since $2(3)^2 - 9 = 9 = (3)^2$ and $2(-3)^2 - 9 = 9 = (-3)^2$. However, -3 does not satisfy the original equation since $\sqrt{2(-3)^2 - 9} = \sqrt{9} = 3 \neq -3$. An extraneous root was introduced by squaring. Thus the solution set is $\{3\}$.

● **PROBLEM 444**

Solve $3\sqrt{x} + 4 = x$.

Solution: Adding (-4) to both sides of the given equation,
$$3\sqrt{x} = x - 4.$$

Squaring both sides
$$(3\sqrt{x})^2 = (x - 4)^2$$
$$3^2(\sqrt{x})^2 = (x - 4)(x - 4)$$

Since $(\sqrt{a})^2 = \sqrt{a} \cdot \sqrt{a} = \sqrt{a \cdot a} = \sqrt{a^2} = a$, $(\sqrt{x})^2 = x$, and we obtain:
$$9x = (x - 4)(x - 4)$$
$$9x = x^2 - 8x + 16$$

Adding (-9x) to both sides,
$$x^2 - 17x + 16 = 0$$

Factoring, $\quad (x - 1)(x - 16) = 0$

Whenever the product of two numbers $ab = 0$, either $a = 0$ or $b = 0$. Thus
$$x - 1 = 0 \quad \text{or} \quad x - 16 = 0$$
$$x = 1 \quad \text{or} \quad x = 16$$

Hence, the possible roots are 1 and 16.

Check, replacing x by 1:
$$3\sqrt{1} + 4 = 1,$$
$$3(1) + 4 = 7 = 1, \text{ which is false.}$$

Hence 1 is an extraneous root.

Check, replacing x by 16:

$3\sqrt{16} + 4 = 16,$

$3(4) + 4 = 12 + 4 = 16,$ which is true.

Therefore, the only root of the given equation is 16.

● PROBLEM   445

Solve the equation $2y = \sqrt{2y + 5} + 1.$

Solution: $2y = \sqrt{2y + 5} + 1$

Isolate the radical term by adding -1 to both sides:

$2y - 1 = \sqrt{2y + 5}$

Eliminate the radical by squaring both sides,

$4y^2 - 4y + 1 = 2y + 5$

Put in standard quadratic form,

$4y^2 - 6y - 4 = 0$

Dividing both sides by 2, this reduces to

$2y^2 - 3y - 2 = 0$

Solve by factoring;

$(2y + 1)(y - 2) = 0$

and set each factor = 0 to find all values of y which make this product = 0.

$$(2y + 1) = 0 \quad | \quad (y - 2) = 0$$
$$y = -\tfrac{1}{2} \quad | \quad y = 2$$

Check: for $y = -\tfrac{1}{2}$

$2(-\tfrac{1}{2}) \overset{?}{=} \sqrt{2(-\tfrac{1}{2}) + 5} + 1$

$-1 \overset{?}{=} \sqrt{4} + 1$

$-1 \neq 3$

for $y = 2$

$2(2) \overset{?}{=} \sqrt{2(2) + 5} + 1$

$4 \overset{?}{=} \sqrt{9} + 1$

$4 = 4$

Therefore $y = -\tfrac{1}{2}$ is an extraneous root, and the solution set is {2}.

● PROBLEM   446

Solve the equation $\sqrt{x^2 - 3x} = 2x - 6.$

Solution: Remove the radical by squaring both sides of the equation, and obtain:

$$\sqrt{x^2 - 3x})^2 = (2x - 6)^2 \qquad \text{or}$$

$$x^2 - 3x = 4x^2 - 24x + 36$$

Writing in standard form, move every term to one side of the equation.

$3x^2 - 21x + 36 = 0$

Dividing all terms by 3, and factoring,

285

$$\frac{3x^2}{3} - \frac{21x}{3} + \frac{36}{3} = \frac{0}{3}$$

$$x^2 - 7x + 12 = 0$$

$$(x - 3)(x - 4) = 0$$

The roots are: x = 3, x = 4.

Check: Substituting x = 3 in the original equation

$$\sqrt{9} - 9 = 6 - 6$$

$$0 = 0$$

Substituting x = 4 in the original equation

$$\sqrt{4} = 8 - 6$$

$$2 = 2.$$

Observe that both x = 3 and x = 4 satisfy the original equation, and there are no extraneous roots.

● **PROBLEM** 447

Find the solution set of $\sqrt{x + 2} + 4 = x$.

Solution: Subtract 4 from both sides of the given equation.

$$\sqrt{x + 2} + \cancel{4} - \cancel{4} = x - 4$$

$$\sqrt{x + 2} = x - 4$$

Square both sides of this equation.

$$(\sqrt{x + 2})^2 = (x - 4)^2$$

Since

$$(\sqrt{x + 2})^2 = \sqrt{x + 2}\ \sqrt{x + 2} = \sqrt{(x + 2)(x + 2)} =$$

$$\sqrt{(x + 2)^2} = x + 2$$

$$x + 2 = (x - 4)^2$$

$$x + 2 = x^2 - 8x + 16$$

Subtract (x + 2) from both sides of this equation

$$x + 2 - (x + 2) = x^2 - 8x + 16 - (x + 2)$$

$$\cancel{x} + \cancel{2} - \cancel{x} - \cancel{2} = x^2 - 8x + 16 - x - 2$$

$$0 = x^2 - 9x + 14$$

Factor the right side of this equation into a product of two polynomials.

$$0 = (x - 7)(x - 2)$$

Whenever a product ab = 0, where a and b are any two numbers, either a = 0 or b = 0. Hence, either

x - 7 = 0	or	x - 2 = 0
x = 7	or	x = 2

To check whether the solutions, 2 and 7, are indeed solutions, x will be replaced by both values in the original equation. Check: Replacing x by 2 in the original equation,

$$\sqrt{x + 2} + 4 = x$$

$$\sqrt{2 + 2} + 4 = 2$$

$$\sqrt{4} + 4 = 2$$

$$2 + 4 = 2$$

$$6 = 2,$$

Which is false. Therefore, 2 is not a solution. Replacing x by 7 in the original equation,

$$\sqrt{x + 2} + 4 = x$$

$$\sqrt{7 + 2} + 4 = 7$$

$$\sqrt{9} + 4 = 7$$

$$3 + 4 = 7$$

$$7 = 7$$

Which is true. Therefore, 7 is a solution. As a result, the solution set to the original equation is { 7 }.

● **PROBLEM 448**

Find the solution of the equation
$$\sqrt{3 - 2x} = 3 - \sqrt{2x + 2} .$$

Solution: Assume that there is a number x such that $\sqrt{3 - 2x} = 3 - \sqrt{2x + 2}$. Squaring both sides, we have

$$(\sqrt{3 - 2x})^2 = (3 - \sqrt{2x + 2})^2$$

$$(\sqrt{3 - 2x})^2 = 9 - 6\sqrt{2x + 2} + (\sqrt{2x + 2})^2$$

Since $\left(\sqrt{a}\right)^2 = \left(a^{\frac{1}{2}}\right)^2 = a^{2/2} = a^1 = a$

$$\left(\sqrt{3 - 2x}\right)^2 = 3 - 2x \quad \text{and} \quad \left(\sqrt{2x + 2}\right)^2 = 2x + 2$$

Thus we obtain

$$3 - 2x = 9 - 6\sqrt{2x + 2} + 2x + 2$$

Adding $6\sqrt{2x + 2}$ to both sides,

$$(3 - 2x) + 6\sqrt{2x + 2} = 9 + 2x + 2$$

$$(3 - 2x) + 6\sqrt{2x + 2} = 11 + 2x$$

Adding $-(3 - 2x)$ to both sides,

$$6\sqrt{2x + 2} = 11 + 2x - (3 - 2x)$$

$$6\sqrt{2x + 2} = 11 + 2x - 3 + 2x$$

$$6\sqrt{2x + 2} = 4x + 8$$

Dividing both sides by 2,

$$3\sqrt{2x + 2} = 2x + 4$$

Squaring both sides of this new equation, we have

$$\left(3\sqrt{2x + 2}\right)^2 = (2x + 4)^2$$

$$3^2 \left(\sqrt{2x + 2}\right)^2 = (2x + 4)(2x + 4)$$

$$9\left(\sqrt{2x + 2}\right)^2 = 4x^2 + 16x + 16$$

Recall $\left(\sqrt{2x + 2}\right)^2 = 2x + 2$, thus

$$9(2x + 2) = 4x^2 + 16x + 16$$

$$18x + 18 = 4x^2 + 16x + 16$$

Dividing both sides by 2,

$$9x + 9 = 2x^2 + 8x + 8$$

Adding $-(9x + 9)$ to both sides,

$$0 = 2x^2 + 8x + 8 - (9x + 9)$$

$$2x^2 + 8x + 8 - 9x - 9 = 0$$

$$2x^2 - x - 1 = 0$$

Factoring, $(2x + 1)(x - 1) = 0$

Whenever a product of two numbers $ab = 0$ either $a = 0$ or $b = 0$. Thus, either $2x + 1 = 0$ or $x - 1 = 0$

$$x = -\tfrac{1}{2} \text{ or } x = 1$$

so that the only possible roots are $-\tfrac{1}{2}$ and 1. We must check if these values are indeed roots. Replace x by $(-\tfrac{1}{2})$ in our original equation

$$\sqrt{3 - 2x} = 3 - \sqrt{2x + 2}$$

$$\sqrt{3 - 2(-\tfrac{1}{2})} = 3 - \sqrt{2(-\tfrac{1}{2}) + 2}$$

$$\sqrt{3 + \tfrac{2}{2}} = 3 - \sqrt{-\tfrac{2}{2} + 2}$$

$$\sqrt{3 + 1} = 3 - \sqrt{-1 + 2}$$

$$\sqrt{4} = 3 - \sqrt{1}$$

$$2 = 3 - 1$$

$$2 = 2$$

Thus $(-\tfrac{1}{2})$ is a root.

Now replace x by 1 in our original equation

$$\sqrt{3 - 2x} = 3 - \sqrt{2x + 2}$$

$$\sqrt{3 - 2(1)} = 3 - \sqrt{2(1)+ 2}$$

$$\sqrt{3 - 2} = 3 - \sqrt{2 + 2}$$

$$\sqrt{1} = 3 - \sqrt{4}$$

$$1 = 3 - 2$$

$$1 = 1$$

Therefore 1 is also a root, and the solution set is $\{-\frac{1}{2}, 1\}$.

● **PROBLEM** 449

Solve the equation $\sqrt{x + 7} + x = 13$.

<u>Solution:</u> Subtract x from both sides of the equation which gives $\sqrt{x + 7} = 13 - x$. Then square both sides, obtaining

$$x + 7 = (13 - x)^2 = 169 - 26x + x^2,$$

Since we have just shown $169 - 26x + x^2 = x + 7$, we may subtract x + 7 from both members to obtain:

$$169 - 7 - 26x - x + x^2 = (x + 7) - (x + 7)$$

Thus, $x^2 - 27x + 162 = 0$

Factor to obtain, $(x - 9)(x - 18) = 0$.

When we have a product, ab = 0, either a = 0 or b = 0; thus with $(x -9)(x - 18) = 0$, either x - 9 = 0   or   x - 18 = 0.

Thus, x = 9 or x = 18.

Checking the value x = 9 in the original equation, we find

$$\sqrt{9 + 7} + 9 = 13$$

$$\sqrt{16} + 9 = 13$$

$$4 + 9 = 13$$

$$13 = 13$$

and x = 9 is seen to be a root. However, if we try to check x = 18, we find

$$\sqrt{18 + 7} + 18 \neq 13$$

Since $\sqrt{25} + 18 \neq 13$

$$5 + 18 \neq 13$$

$$23 \neq 13;$$

so that x = 18 is not a root of the original equation. Hence, there is only one solution of the problem: x = 9.

Solve the equation $\sqrt{11 - x} - \sqrt{x + 6} = 3$.

Solution: The process of eliminating the radical is simpler if the equation is rewritten with one radical on each side of the equal sign before squaring. Thus,

$$\sqrt{11 - x} = 3 + \sqrt{x + 6}$$

Squaring both sides:

$$(\sqrt{11 - x})^2 = (3 + \sqrt{x + 6})^2$$

$$11 - x = 9 + 6\sqrt{x + 6} + x + 6 \qquad \text{or}$$

$$6\sqrt{x + 6} = -4 - 2x$$

Dividing both sides by 2 and squaring:

$$3\sqrt{x + 6} = -2 - x$$

$$(3\sqrt{x + 6})^2 = (-2 - x)^2$$

$$9(x + 6) = x^2 + 4x + 4$$

$$9x + 54 = x^2 + 4x + 4$$

Writing in standard form, collect all the terms on one side of the equal sign:

$$x^2 - 5x - 50 = 0.$$

In factored form $\quad (x - 10)(x + 5) = 0$

$$x - 10 = 0, \qquad\qquad x + 5 = 0.$$

Therefore, the roots are $x = 10$ and $x = -5$.

Check: Substituting $x = -5$ in the original equation

$$\sqrt{16} - \sqrt{1} \overset{?}{=} 3$$

$$4 - 1 = 3$$

$$3 = 3$$

Substituting $x = 10$ in the original equation

$$\sqrt{1} - \sqrt{16} \overset{?}{=} 3$$

$$1 - 4 \neq 3$$

$$-3 \neq 3$$

Observe that $x = -5$ satisfies the original equation, but $x = 10$ does not and is therefore an extraneous root.

Solve the equation

$$\sqrt{x^2 + 24x + 3} = 2x + 3.$$

Solution: Squaring both members,

$$\left(\sqrt{x^2 + 24x + 3}\right)^2 = (2x + 3)^2$$

Since $\left(\sqrt{a}\right)^2 = \sqrt{a}\sqrt{a} = \sqrt{a \cdot a} = \sqrt{a^2} = a$, $\left(\sqrt{x^2 + 24x + 3}\right)^2 = x^2 + 24x + 3$.

Thus, $x^2 + 24x + 3 = (2x + 3)(2x + 3)$

$$x^2 + 24x + 3 = 4x^2 + 12x + 9$$

Adding $-(x^2 + 24x + 3)$ to both sides,

$$3x^2 - 12x + 6 = 0$$

Dividing both members by 3,

$$x^2 - 4x + 2 = 0$$

Adding 2 to both members,

$$x^2 - 4x + 4 = 2$$

$$(x - 2)^2 = 2$$

Taking the square root of both sides,

$$\sqrt{(x - 2)^2} = \pm \sqrt{2}$$

$$x - 2 = \pm \sqrt{2}$$

Adding 2 to both sides, $x = 2 \pm \sqrt{2}$.

To check that the roots of this equation are indeed $2 + \sqrt{2}$ and $2 - \sqrt{2}$ we replace x by these values in our original equation. Replacing x by $2 + \sqrt{2}$,

$$\sqrt{x^2 + 24x + 3} = 2x + 3$$

$$\sqrt{(2 + \sqrt{2})^2 + 24(2 + \sqrt{2}) + 3} = 2(2 + \sqrt{2}) + 3$$

$$\sqrt{4 + 4\sqrt{2} + 2 + 48 + 24\sqrt{2} + 3} = 4 + 2\sqrt{2} + 3$$

$$\sqrt{28\sqrt{2} + 57} = 7 + 2\sqrt{2}$$

$$28\sqrt{2} + 57 = (7 + 2\sqrt{2})^2$$

$$28\sqrt{2} + 57 = 49 + 28\sqrt{2} + 4(2)$$

$$28\sqrt{2} + 57 = 28\sqrt{2} + 49 + 8$$

$$28\sqrt{2} + 57 = 28\sqrt{2} + 57$$

Replacing x by $2 - \sqrt{2}$,

$$\sqrt{x^2 + 24x + 3} = 2x + 3$$

$$\sqrt{(2 - \sqrt{2})^2 + 24(2 - \sqrt{2}) + 3} = 2(2 - \sqrt{2}) + 3$$

$$\sqrt{4 - 4\sqrt{2} + 2 + 48 - 24\sqrt{2} + 3} = 4 - 2\sqrt{2} + 3$$

$$\sqrt{-28\sqrt{2} + 57} = 7 - 2\sqrt{2}$$

$$-28\sqrt{2} + 57 = \left(7 - 2\sqrt{2}\right)^2$$

$$-28\sqrt{2} + 57 = 49 - 28\sqrt{2} + 4(2)$$

$$-28\sqrt{2} + 57 = -28\sqrt{2} + 49 + 8$$

$$-28\sqrt{2} + 57 = -28\sqrt{2} + 57$$

Thus our solution set is $\{2 + \sqrt{2}, \ 2 - \sqrt{2}\}$.

Find the solution set of the equation $\sqrt{x + 7} = 2x - 1$.

Solution: Assume that there is a number x such that $\sqrt{x + 7} = 2x - 1$. Squaring both sides, we have

$$(\sqrt{x + 7})^2 = (2x - 1)^2$$

Note $(\sqrt{a})^2 = \left(a^{\frac{1}{2}}\right)^2 = a^{2/2} = a^1 = a$ thus $(\sqrt{x + 7})^2 = x + 7$.

Replacing $(\sqrt{x + 7})^2$ by $x + 7$ we obtain

$$x + 7 = (2x - 1)(2x - 1)$$

$$x + 7 = 4x^2 - 4x + 1$$

Adding $-(x + 7)$ to both members,

$$0 = 4x^2 - 4x + 1 - (x + 7)$$

$$4x^2 - 4x + 1 - x - 7 = 0$$

$$4x^2 - 5x - 6 = 0$$

Thus factoring

$$4x^2 - 5x - 6 = (4x + 3)(x - 2) = 0$$

Whenever a product of two numbers $ab = 0$ either $a = 0$ or $b = 0$, thus either $4x + 3 = 0$ or $x - 2 = 0$ and

$$x = \frac{-3}{4} \quad \text{or} \quad x = 2$$

Note that at this point we have not proved that either $x = 2$ or $x = -\frac{3}{4}$ is a solution of our equation, but simply that if there is any solution it must be either 2 or $-\frac{3}{4}$. Thus we must check our values by substituting them into our original equation. Replacing x by 2 in $\sqrt{x + 7} = 2x - 1$ we obtain

$$\sqrt{2 + 7} = 2(2) - 1$$

$$\sqrt{9} = 4 - 1$$

$$3 = 3.$$

So 2 is indeed a solution of our equation. On the other hand, $-\frac{3}{4}$ is not a solution since

$$\sqrt{-\tfrac{3}{4} + 7} = 2(-\tfrac{3}{4}) - 1$$

$$\sqrt{6\tfrac{1}{4}} = \frac{-6}{4} - 1$$

$$\sqrt{\frac{25}{4}} = \frac{-6}{4} - \frac{4}{4}$$

$$\frac{\sqrt{25}}{\sqrt{4}} = \frac{-10}{4}$$

$$\frac{5}{2} \neq \frac{-10}{4}$$

Thus the solution set is $\{2\}$.

A number such as $-\frac{3}{4}$ obtained in this way is sometimes called an extraneous root — a term we prefer not to use since it implies that we do have a root of some kind or another.

Note also that if the equation had been given in the form

$$\sqrt{x + 7} + 1 = 2x$$

and we had squared both sides, we would have obtained

$$x + 7 + 2\sqrt{x + 7} + 1 = 4x^2$$

and would not have eliminated the radical. For this reason we always "isolate" a radical on one side of the equation before squaring.

● **PROBLEM** 453

Solve the equation

$$\sqrt{x^2 - 3x + 27} = 2x + 3 .$$

<u>Solution:</u> Squaring both members,

$$\left(\sqrt{x^2 - 3x + 27}\right)^2 = (2x + 3)^2 \qquad (1)$$

Since $\left(\sqrt{a}\right)^2 = \sqrt{a}\,\sqrt{a} = \sqrt{a \cdot a} = \sqrt{a^2} = a,$

$$\left(\sqrt{x^2 - 3x + 27}\right)^2 = x^2 - 3x + 27.$$

Thus equation (1) becomes:

$$x^2 - 3x + 27 = (2x + 3)(2x + 3)$$

$$x^2 - 3x + 27 = 4x^2 + 12x + 9$$

Adding $-\left(x^2 - 3x + 27\right)$ to both members,

$$x^2 - 3x + 27 - \left(x^2 - 3x + 27\right) = 4x^2 + 12x + 9 - \left(x^2 - 3x + 27\right)$$

$$0 = 4x^2 + 12x + 9 - x^2 + 3x - 27$$

$$4x^2 - x^2 + 12x + 3x + 9 - 27 = 0$$

$$3x^2 + 15x - 18 = 0$$

Dividing both members by 3, $x^2 + 5x - 6 = 0$

$$(x + 6)(x - 1) = 0;$$

Whenever the product of two numbers $ab = 0$, either $a = 0$, or $b = 0$. Thus

$$x + 6 = 0 \quad \text{or} \quad x - 1 = 0$$

and

$$x = -6 \quad \text{or} \quad x = 1$$

Before we can conclude that the roots to this equation are -6 and 1 we must perform the following check: Replacing x by -6 in the original equation,

$$\sqrt{x^2 - 3x + 27} = 2x + 3$$

$$\sqrt{(-6)^2 - 3(-6) + 27} \overset{?}{=} 2(-6) + 3$$

$$\sqrt{36 + 18 + 27} \overset{?}{=} -12 + 3$$

$$\sqrt{81} \neq -9$$

Since substitution of (-6) for x results in a statement which isn't true, $\sqrt{81} = +9$ not -9 (unless the negative square root is indicated), (-6) is not part of our solution. (-6) is an extraneous root. Replacing x by 1 in the original equation,

$$\sqrt{x^2 - 3x + 27} = 2x + 3$$

$$\sqrt{1^2 - 3(1) + 27} = 2(1) + 3$$

$$\sqrt{1 - 3 + 27} = 2 + 3$$

$$\sqrt{25} = 5$$

$$5 = 5$$

Thus the solution set is $\{1\}$.

Solve the equation $\sqrt{5x - 11} - \sqrt{x - 3} = 4$.

<u>Solution:</u>

Add $\sqrt{x - 3}$ to both sides, $\sqrt{5x - 11} = \sqrt{x - 3} + 4$.

Square both members to eliminate one of the radicals:

$$(\sqrt{5x - 11})^2 = (\sqrt{x - 3} + 4)^2$$

$$(\sqrt{5x - 11})^2 = (\sqrt{x - 3} + 4)(\sqrt{x - 3} + 4)$$

$$(\sqrt{5x - 11})^2 = (\sqrt{x - 3})^2 + 4\sqrt{x - 3} + 4\sqrt{x - 3} + 16.$$

Since $(\sqrt{a})^2 = \sqrt{a} \cdot \sqrt{a} = \sqrt{a^2} = a$,

$$(\sqrt{5x - 11})^2 = 5x - 11$$

and $(\sqrt{x - 3})^2 = x - 3$.

Therefore, $5x - 11 = x - 3 + 8\sqrt{x - 3} + 16$.

Combine terms, $\qquad\qquad 5x - 11 = x + 8\sqrt{x - 3} + 13$

Add (-13) to both sides, $5x - 11 - 13 = x + 8\sqrt{x - 3}$

Add $(-x)$ to both sides, $5x - 24 - x = 8\sqrt{x - 3}$

$$4x - 24 = 8\sqrt{x - 3}$$

Factoring 4 from each member,

$$4(x - 6) = 4(2\sqrt{x - 3}).$$

Divide both sides by 4, $\qquad x - 6 = 2\sqrt{x - 3}$.

Square both sides, $\qquad (x - 6)^2 = (2\sqrt{x - 3})^2$

Since $(ab)^2 = a^2 b^2$,

$$(x - 6)^2 = (2)^2 (\sqrt{x - 3})^2$$

$$(x - 6)(x - 6) = 4(\sqrt{x - 3})^2$$

Recall: $(\sqrt{a})^2 = \sqrt{a}\sqrt{a} = \sqrt{a^2} = a$;

Thus, $(\sqrt{x - 3})^2 = x - 3$.

Replacing $(\sqrt{x - 3})^2$ by $x - 3$ we obtain,

$$x^2 - 12x + 36 = 4(x - 3).$$

Distribute, $\qquad\qquad x^2 - 12x + 36 = 4x - 12.$

Add $(-4x)$ to both sides,

$$x^2 - 12x + 36 - 4x = -12$$

Combine terms, $\qquad\qquad x^2 - 16x + 36 = -12$

Add 12 to both sides,

$$x^2 - 16x + 36 + 12 = 0$$

Combine terms, $\qquad x^2 - 16x + 48 = 0$

Factor, $\qquad (x - 12)(x - 4) = 0$

Whenever a product $ab = 0$, either $a = 0$, or $b = 0$; thus either $(x - 12) = 0$, or $(x - 4) = 0$, therefore $\qquad x = 12 \qquad$ or $\qquad x = 4$.

Now we must check to verify that 12 and 4 are indeed roots of the given equation.

Check: To check if 12 is a root, we replace x by 12 in the original equation,

$$\sqrt{5x - 11} - \sqrt{x - 3} = 4$$

$$\sqrt{5(12) - 11} - \sqrt{12 - 3} = 4$$

$$\sqrt{60 - 11} - \sqrt{9} = 4$$

$$\sqrt{49} - \sqrt{9} = 4$$

$$7 - 3 = 4$$

$$4 = 4$$

Thus 12 is a root of the equation.

Now to check if 4 is a root, we replace x by 4 in the original equation,

$$\sqrt{5x - 11} - \sqrt{x - 3} = 4$$

$$\sqrt{5(4) - 11} - \sqrt{4 - 3} = 4$$

$$\sqrt{20 - 11} - \sqrt{1} = 4$$

$$\sqrt{9} - \sqrt{1} = 4$$

$$3 - 1 = 2 \neq 4.$$

Since substitution of x by 4 does not result in a valid equation, 4 is not a root, and our solution set is {12}.

● PROBLEM 455

Solve and check: $\sqrt{x + 10} + \sqrt[4]{x + 10} = 2$.

Solution: Let $y = \sqrt[4]{x + 10}$ then $y^2 = \sqrt{x + 10}$.

Substituting, the original equation may be written

$$y^2 + y - 2 = 0.$$

Factor $(y + 2)(y - 1) = 0.$

Set each factor = 0 to find all values of x which can make the product = 0

$$y + 2 = 0 \quad | \quad y - 1 = 0$$
$$y = -2 \quad | \quad y = 1$$

for y = -2   for y = 1

$$y = \sqrt[3]{x + 10} = -2 \qquad y = \sqrt[3]{x + 10} = 1$$

$$x + 10 = 16 \qquad\qquad x + 10 = 1$$

$$x = 6 \qquad\qquad\qquad x = -9$$

Check: for x = 6: $\sqrt{6 + 10} + \sqrt[3]{6 + 10} \overset{?}{=} 2$

$$\sqrt{16} + \sqrt[3]{16} = 2$$

$$4 + 2 \neq 2$$

This root does not check.

for x = -9: $\sqrt{-9 + 10} + \sqrt[3]{-9 + 10} = 2$

$$1 + \sqrt[3]{1} = 2$$

$$1 + 1 = 2$$

$$2 = 2.$$

This root checks.

● **PROBLEM** 456

Solve the equation: $x^2 - x + 2\sqrt{x^2 - x - 5} = 8.$

Solution: We notice that the first two terms of the left member, $x^2 - x$, appear under the radical also. Hence we add -5 to both members to get

$$x^2 - x - 5 + 2\sqrt{x^2 - x - 5} = 3.$$

$$x^2 - x - 5 + 2\sqrt{x^2 - x - 5} - 3 = 0.$$

Then setting

$$y = \sqrt{x^2 - x - 5}$$

for brevity, we have

$$y^2 + 2y - 3 = 0.$$

Factoring:

$$(y + 3)(y - 1) = 0.$$

Setting each factor equal to zero we obtain:

$$y = -3, \quad y = 1.$$

The positively signed radical denoted by y cannot have the negative real value, -3. But y = 1 is permissible; we obtain:

$$\sqrt{x^2 - x - 5} = 1.$$

Squaring both sides:

$$x^2 - x - 5 = 1.$$

Subtracting 1 from both sides:

$$x^2 - x - 6 = 0.$$

Factoring and setting each factor equal to zero:

$$(x - 3)(x + 2) = 0$$

$$x - 3 = 0 \quad x + 2 = 0$$

$$x = 3 \qquad x = -2$$

Thus there are two solutions of the given equation.

Check: To verify this result we replace x by 3 in our original equation,

$$x^2 - x + 2\sqrt{x^2 - x - 5} = 8$$

$$3^2 - 3 + 2\sqrt{3^2 - 3 - 5} = 8$$

$$9 - 3 + 2\sqrt{9 - 3 - 5} = 8$$

$$6 + 2\sqrt{1} = 8$$

$$6 + 2(1) = 8$$

$$6 + 2 = 8$$

$$8 = 8.$$

Now we replace x by -2 in our original equation

$$x^2 - x + 2\sqrt{x^2 - x - 5} = 8$$

$$(-2)^2 - (-2) + 2\sqrt{(-2)^2 - (-2) - 5} = 8$$

$$4 + 2 + 2\sqrt{4 + 2 - 5} = 8$$

$$6 + 2\sqrt{1} = 8$$

$$6 + 2 = 8$$

$$8 = 8.$$

Solve $2\sqrt{\dfrac{x}{a}} + 3\sqrt{\dfrac{a}{x}} = \dfrac{b}{a} + \dfrac{6a}{b}$.

Solution: Let $\sqrt{\dfrac{x}{a}} = y$; then $\sqrt{\dfrac{a}{x}} = \dfrac{1}{y}$;

Hence, $2y + \dfrac{3}{y} = \dfrac{b}{a} + \dfrac{6a}{b}$;

$$yab\left(2y + \dfrac{3}{y}\right) = \left(\dfrac{b}{a} + \dfrac{6a}{b}\right)yab$$

$$2y^2ab + 3ab = b^2y + 6a^2y$$

$$2aby^2 - 6a^2y - b^2y + 3ab = 0,$$

$$(2ay - b)(by - 3a) = 0; \qquad\qquad by - 3a = 0$$

$$2ay - b = 0$$

$$2ay = b \qquad\qquad\qquad\qquad by = 3a$$

$$y = \dfrac{b}{2a} , \quad\text{or}\qquad \dfrac{3a}{b} ;$$

Substitute these two values of y:

$$\sqrt{\dfrac{x}{a}} = y$$

$$\sqrt{\dfrac{x}{a}} = \dfrac{b}{2a} \qquad\qquad\qquad \sqrt{\dfrac{x}{a}} = \dfrac{3a}{b}$$

square both sides.   square both sides

$$\dfrac{x}{a} = \dfrac{b^2}{4a^2} \qquad\qquad\qquad \dfrac{x}{a} = \dfrac{9a^2}{b^2}$$

multiply both sides by a.   multiply both sides by a.

$$x = \dfrac{b^2 a}{4a^2} = \dfrac{b^2}{4a} \qquad\qquad x = \dfrac{9a^2 \cdot a}{b^2} = \dfrac{9a^3}{b^2}$$

The solution is:

$$x = \left\{\dfrac{b^2}{4a} , \dfrac{9a^3}{b^2}\right\}$$

Find the real solutions of the equation

$$\sqrt{3x + 1} + \sqrt{x - 4} = 3.$$

Solution: Adding $(-\sqrt{x-4})$ to both sides,

$$\sqrt{3x + 1} = 3 - \sqrt{x - 4} .$$

Squaring both members,

$$\left(\sqrt{3x + 1}\right)^2 = \left(3 - \sqrt{x - 4}\right)^2.$$

Since $\left(\sqrt{a}\right)^2 = \sqrt{a} \cdot \sqrt{a} = \sqrt{a \cdot a} = \sqrt{a^2} = a,$

$$\left(\sqrt{3x + 1}\right)^2 = 3x + 1 .$$

Thus,

$$3x + 1 = (3 - \sqrt{x - 4})(3 - \sqrt{x - 4})$$
$$3x + 1 = 9 - 6\sqrt{x - 4} + x - 4.$$

Adding $-(3x + 1)$ to both sides we obtain,

$$3x + 1 - (3x + 1) = 9 - 6\sqrt{x - 4} + x - 4 - (3x + 1)$$
$$0 = 9 - 6\sqrt{x - 4} + x - 4 - 3x - 1$$
$$9 - 4 - 1 + x - 3x - 6\sqrt{x - 4} = 0$$
$$4 - 2x - 6\sqrt{x - 4} = 0$$
$$4 - 2x = 6\sqrt{x - 4}$$

At this point it may be observed that x must be $\geqq 4$ if the right member is to be real, and that $x \leqq 2$ if the left member is to be positive.

Squaring both members,
$$(4 - 2x)^2 = (6\sqrt{x - 4})^2$$
$$(4 - 2x)(4 - 2x) = (6)^2(\sqrt{x - 4})^2$$
$$16 - 16x + 4x^2 = 36(x - 4)$$
$$16 - 16x + 4x^2 = 36x - 144$$

Subtract $(36x - 144)$ from both sides:

$$16 - 16x + 4x^2 - (36x - 144) = 36x - 144 - (36x - 144)$$
$$16 - 16x + 4x^2 - 36x + 144 = 0$$
$$4x^2 - 52x + 160 = 0$$

Dividing by 4, $\qquad\qquad x^2 - 13x + 40 = 0$

Factoring, $\qquad\qquad (x - 5)(x - 8) = 0$

Whenever the product of two numbers $ab = 0$, either $a = 0$ or $b = 0$; thus either

$$x - 5 = 0 \quad \text{or} \quad x - 8 = 0;$$

hence
$$x = 5 \qquad \text{or} \qquad x = 8 .$$

These are the solutions of this last quadratic equation, but they may not be solutions of the original equation. Therefore, we must check these roots.

Check: Replace x by 5 in the original equation.

$$\sqrt{3x + 1} + \sqrt{x - 4} = 3$$
$$\sqrt{3(5) + 1} + \sqrt{5 - 4} = 3$$
$$\sqrt{15 + 1} + \sqrt{1} = 3$$
$$\sqrt{16} + 1 = 3$$
$$4 + 1 = 3$$
$$5 = 3$$

Since substitution of x by 5 results in a statement which isn't true $(5 = 3)$, $x = 5$ is not a root of the original equation.

Now replace x by 8 in the original equation.

$$\sqrt{3x + 1} + \sqrt{x - 4} = 3$$
$$\sqrt{3(8) + 1} + \sqrt{8 - 4} = 3$$
$$\sqrt{24 + 1} + \sqrt{4} = 3$$
$$\sqrt{25} + 2 = 3$$
$$5 + 2 = 3$$
$$7 = 3$$

Substitution of x by 8 also results in a statement which isn't true (7 = 3), therefore x = 8 is not a root of the original equation. Thus, there are no real solutions to the given equation.

Solve $2(x + 2)^{\frac{1}{2}} = (x + 1)^{\frac{1}{2}} - 2$.

Solution: Squaring gives

$$4(x + 2) = x + 1 - 4(x + 1)^{\frac{1}{2}} + 4$$

$$4x + 8 = x + 5 - 4(x + 1)^{\frac{1}{2}}.$$

Transposing $-4(x + 1)^{\frac{1}{2}} = 3x + 3.$

Squaring again $16(x + 1) = 9x^2 + 18x + 9.$

$$16x + 16 = 9x^2 + 18x + 9.$$

Transposing again $9x^2 + 2x - 7 = 0.$

Factoring $(9x - 7)(x + 1) = 0.$

Set each factor = 0 to find all values of x for which the product = 0

$$9x - 7 = 0 \qquad x + 1 = 0$$

$$x = \frac{7}{9} \qquad x = -1.$$

Check: for $x = \frac{7}{9}$

$$2\left(\frac{7}{9} + 2\right)^{\frac{1}{2}} \overset{?}{=} \left(\frac{7}{9} + 1\right)^{\frac{1}{2}} - 2$$

$$2\left(\frac{7}{9} + \frac{18}{9}\right)^{\frac{1}{2}} \overset{?}{=} \left(\frac{7}{9} + \frac{9}{9}\right)^{\frac{1}{2}} - 2$$

$$2\sqrt{\frac{25}{9}} \overset{?}{=} \sqrt{\frac{16}{9}} - 2$$

$$2\left(\frac{5}{3}\right) \overset{?}{=} \frac{4}{3} - 2$$

$$\frac{10}{3} \neq -\frac{2}{3}$$

for $x = -1$

$$2(-1 + 2)^{\frac{1}{2}} \overset{?}{=} (-1 + 1)^{\frac{1}{2}} - 2$$

$$2\sqrt{1} \overset{?}{=} \sqrt{0} - 2$$

$$2 \neq -2.$$

Neither of these values is a root of the given equation.
The above example illustrates that:

1. Two expressions involving radicals may not be equal for any value of the unknown.
2. Extraneous roots may be introduced by squaring.
3. Results must always be checked. There is no other way to determine whether or not a result is a root of the given equation.

● PROBLEM 460

Solve $(5x - 4)^{\frac{1}{2}} = (2x + 1)^{\frac{1}{2}} + 1$.

Solution: Squaring gives $5x - 4 = 2x + 1 + 2(2x + 1)^{\frac{1}{2}} + 1$.

Transposing $3x - 6 = 2(2x + 1)^{\frac{1}{2}}$

Squaring again $9x^2 - 36x + 36 = 4(2x + 1)$

$$9x^2 - 36x + 36 = 8x + 4.$$

Transposing again $9x^2 - 44x + 32 = 0$.

Factoring $(9x - 8)(x - 4) = 0$.

Set each factor $= 0$ to find all values of x for which the product $= 0$.

$$9x - 8 = 0 \qquad x - 4 = 0$$

$$x = \frac{8}{9} \qquad x = 4$$

Check: for $x = \frac{8}{9}$

$$\left[5\left(\frac{8}{9}\right) - 4\right]^{\frac{1}{2}} \overset{?}{=} \left[2\left(\frac{8}{9}\right) + 1\right]^{\frac{1}{2}} + 1$$

$$\left[\frac{40}{9} - \frac{36}{9}\right]^{\frac{1}{2}} \overset{?}{=} \left[\frac{16}{9} + \frac{9}{9}\right]^{\frac{1}{2}} + 1$$

$$\sqrt{\frac{4}{9}} \overset{?}{=} \sqrt{\frac{25}{9}} + 1$$

$$\frac{2}{3} \overset{?}{=} \frac{5}{3} + 1$$

$$\frac{2}{3} \neq \frac{8}{3}.$$

for $x = 4$

$$[5(4) - 4]^{\frac{1}{2}} \overset{?}{=} [2(4) + 1]^{\frac{1}{2}} + 1$$

$$\sqrt{16} = \sqrt{9} + 1$$

$$4 \overset{?}{=} 3 + 1$$

$$4 = 4.$$

The value 4 satisfies the given equation, but $\frac{8}{9}$ does not.

● **PROBLEM** 461

Solve the equation $2x^{2/5} + 5x^{1/5} - 3 = 0$.

Solution: This equation may be solved as a quadratic equation if we let $P(x) = y = x^{1/5}$. Then $x^{2/5} = y^2$ and, by substituting these expressions in the equation to be solved, we have $2y^2 + 5y - 3 = 0$. We can solve this equation for y by factoring:

$$2y^2 + 5y - 3 = 0$$
$$(2y - 1)(y + 3) = 0$$
$$(2y - 1) = 0 \quad \text{or} \quad (y + 3) = 0$$

therefore, $\qquad\qquad y = \tfrac{1}{2} \quad \text{or} \quad\qquad y = -3$

Now recall that $y = x^{1/5}$. Hence $x^{1/5} = \tfrac{1}{2}$ or $x^{1/5} = -3$.

Hence, by raising both sides of each of these equations to the fifth power, we have $x = 1/32$ or $x = -243$.

Therefore the solution set is $\{1/32, -243\}$.

● **PROBLEM** 462

If $\dfrac{x}{a} = \dfrac{y}{b} = \dfrac{z}{c}$, prove that

$$\dfrac{x^2 + a^2}{x+a} + \dfrac{y^2 + b^2}{y+b} + \dfrac{z^2 + c^2}{z+c} = \dfrac{(x+y+z)^2 + (a+b+c)^2}{x+y+z+a+b+c}.$$

Solution: Let $\dfrac{x}{a} = \dfrac{y}{b} = \dfrac{z}{c} = k$, so that $x = ak$, $y = bk$, $z = ck$; then, substituting for x we obtain:

$$\frac{x^2 + a^2}{x + a} = \frac{(ak)^2 + a^2}{ak + a} = \frac{a^2 k^2 + a^2}{ak + a} = \frac{a^2(k^2 + 1)}{a(k + 1)}$$

$$= \frac{a(k^2 + 1)}{k + 1}.$$

Similarly, by substituting for y and z we obtain:

$$\frac{y^2 + b^2}{y + b} = \frac{b(k^2 + 1)}{k + 1}, \quad \frac{z^2 + c^2}{z + c} = \frac{c(k^2 + 1)}{k + 1}.$$

Therefore, $\dfrac{x^2 + a^2}{x + a} + \dfrac{y^2 + b^2}{y + b} + \dfrac{z^2 + c^2}{z + c} =$

$$\frac{a(k^2 + 1)}{k + 1} + \frac{b(k^2 + 1)}{k + 1} + \frac{c(k^2 + 1)}{k + 1} =$$

$$\frac{a(k^2 + 1) + b(k^2 + 1) + c(k^2 + 1)}{k + 1}. \text{ Now,}$$

factoring $(k^2 + 1)$ from the numerator we obtain:

$$\frac{(k^2 + 1)(a + b + c)}{k + 1} \; .$$

Now, performing the multiplication in the numerator we have:

$$\frac{ak^2 + a + bk^2 + b + ck^2 + c}{k + 1} =$$

$$\frac{k^2(a + b + c) + (a + b + c)}{k + 1} \; .$$

Multiplying both numerator and denominator by $\frac{a + b + c}{a + b + c}$, which equals 1 and does not change the value of the fraction, we then obtain:

$$\frac{k^2(a + b + c)^2 + (a + b + c)^2}{k(a + b + c) + a + b + c} \; .$$

Since the first term in the numerator can be re-written as

$$k^2\left(a^2 + ab + ac + ab + b^2 + bc + ac + bc + c^2\right) =$$

$$k^2a^2 + k^2ab + k^2ac + k^2ab + k^2b^2 + k^2bc + k^2ac$$

$$+ k^2bc + k^2c^2 \; =$$

$(ka + kb + kc)^2$, we can rewrite the fraction as

$$\frac{(ka + kb + kc)^2 + (a + b + c)^2}{(ka + kb + kc) + a + b + c} \; ,$$

and substituting for ka, kb, kc, we obtain:

$$\frac{(x + y + z)^2 + (a + b + c)^2}{x + y + z + a + b + c} \; .$$

SOLVING BY COMPLETING THE SQUARE

● **PROBLEM** 463

Complete the square in $x^2 + x - 1$.

Solution: We proceed adding the square of half the coefficient of x and, also subtracting it. That is, we write

$$x^2 + x - 1 = x^2 + x - 1 + \tfrac{1}{4} - \tfrac{1}{4}$$

$$= x^2 + x - 1 + \left(\tfrac{1}{2}\right)^2 - \left(\tfrac{1}{2}\right)^2$$

Associating, $= \left[x^2 + x + \left(\tfrac{1}{2}\right)^2\right] - 1 - \left(\tfrac{1}{2}\right)^2$

$$= [x + \tfrac{1}{2}]^2 - 1 - \tfrac{1}{4}$$

$$= \left(x + \tfrac{1}{2}\right)^2 - \tfrac{4}{4} - \tfrac{1}{4}$$

$$= \left(x + \tfrac{1}{2}\right)^2 - \tfrac{5}{4}$$

Solve $x^2 - 6x + 8 = 0$.

Solution: This problem may be solved by the method of completing the square: Arrange the equation with the constant term in the right member

$$x^2 - 6x = -8.$$

Take $\frac{1}{2}$ of the coefficient of x, square this, and add the result to both members. Thus, $\frac{1}{2}$ of -6 is -3, and $(-3)^2 = 9$. Add 9 to both members:

$$x^2 - 6x + 9 = -8 + 9 = 1.$$

This procedure makes the left member a perfect square. Factor,

$$(x - 3)^2 = 1.$$

Extract the square root of both members,

$$x - 3 = \pm 1.$$

When x - 3 = +1, then x = 4 and when x - 3 = -1, then x = 2.

Check: for x = 4: for x = 2:

$$4^2 - 6(4) + 8 = 0 \qquad 2^2 - 6(2) + 8 = 0$$

$$16 - 24 + 8 = 0 \qquad 4 - 12 + 8 = 0$$

$$0 = 0 \qquad\qquad 0 = 0.$$

Sol: x = {4,2}.

Solve $2x^2 + 8x + 4 = 0$ by completing the square.

Solution: $2x^2 + 8x + 4 = 0$

Divide both members by 2, the coefficient of x^2.

$$x^2 + 4x + 2 = 0$$

Subtract the constant term, 2, from both members.

$$x^2 + 4x = -2$$

Add to each member the square of one-half the coefficient of the term in x.

$$x^2 + 4x + 4 = -2 + 4$$

Factor

$$(x + 2)^2 = 2$$

Set the square root of the left member (a perfect square) equal to $\pm$ the square root of the right member and solve for x.

$$x + 2 = \sqrt{2} \quad \text{or} \quad x + 2 = -\sqrt{2}$$

The roots are $\sqrt{2} - 2$ and $-\sqrt{2} - 2$. Check each solution.

$$2(\sqrt{2} - 2)^2 + 8(\sqrt{2} - 2) + 4 = 2(2 - 4\sqrt{2} + 4) + 8\sqrt{2} - 16 + 4$$
$$= 4 - 8\sqrt{2} + 8 + 8\sqrt{2} - 16 + 4$$
$$= 0$$

$$2(-\sqrt{2} - 2)^2 + 8(-\sqrt{2} - 2) + 4 = 2(2 + 4\sqrt{2} + 4) - 8\sqrt{2} - 16 + 4$$
$$= 4 + 8\sqrt{2} + 8 - 8\sqrt{2} - 16 + 4$$
$$= 0$$

● **PROBLEM 466**

Solve the equation $3x^2 + 6x - 7 = 0$.

Solution: This quadratic equation cannot be solved by factoring, but may be solved by the method of completing the square. Adding 7 to both sides, we have $3x^2 + 6x = 7$. Multiplying both sides now by $\frac{1}{3}$, we have $x^2 + 2x = 7/3$. We are now in a position to complete the square. The computation can be arranged in the following manner: Add the square of $\frac{1}{2}$ the coefficient of x to both sides, i.e., 1. Then rewrite as the equality of two squares.

$$x^2 + 2x + 1 = \frac{7}{3} + 1$$
$$(x + 1)^2 = \left(\sqrt{\frac{7}{3} + 1}\right)^2$$
$$(x + 1)^2 = \left(\sqrt{\frac{10}{3}}\right)^2$$

Adding $-\left(\sqrt{\frac{10}{3}}\right)^2$ to both sides,

$$(x + 1)^2 - \left(\sqrt{\frac{10}{3}}\right)^2 = 0$$

Factoring:

$$\left[(x + 1) + \sqrt{\frac{10}{3}}\right]\left[(x + 1) - \sqrt{\frac{10}{3}}\right] = 0$$

Hence

$$x + 1 + \sqrt{\frac{10}{3}} = 0 \quad \text{or} \quad x + 1 - \sqrt{\frac{10}{3}} = 0$$
$$x = -1 - \sqrt{\frac{10}{3}} \quad \text{or} \quad x = -1 + \sqrt{\frac{10}{3}}$$

Therefore the solution set is

$$\left\{-1 + \sqrt{\frac{10}{3}}, -1 - \sqrt{\frac{10}{3}}\right\}.$$

● **PROBLEM 467**

Solve by completing the square: $-2x^2 + 3x + 5 = 0$.

Solution: $-2x^2 + 3x + 5 = 0$. Divide by -2, the coefficient of x^2,

$$x^2 - \frac{3}{2}x - \frac{5}{2} = 0$$

Add $+\frac{5}{2}$ to both sides of the equation

$$x^2 - \frac{3}{2}x - \frac{5}{2} + \frac{5}{2} = 0 + \frac{5}{2}$$

$$x^2 + \frac{3}{2}x = \frac{5}{2}$$

Take $\frac{1}{2}$ the coefficient of x and square it to make $x^2 - \frac{3}{2}x$ a perfect square trinomial.

$$x^2 - \frac{3}{2}x + \left[\frac{1}{2}\left(-\frac{3}{2}\right)\right]^2 = \frac{5}{2} + \left[\frac{1}{2}\left(-\frac{3}{2}\right)\right]^2 \quad \text{or} \quad x^2 - \frac{3}{2}x + \frac{9}{16} = \frac{5}{2} + \frac{9}{16}$$

$$x^2 - \frac{3}{2}x + \frac{9}{16} = \frac{40}{16} + \frac{9}{16} = \frac{49}{16} .$$

To factor the trinomial perfect square, $x^2 - \frac{3}{2}x + \frac{9}{16}$, take the square roots of the first and last terms, and join these square roots by the sign of the middle term. Therefore,

$$x^2 - \frac{3}{2}x + \frac{9}{16} = \left(x - \frac{3}{4}\right)^2$$

Then,

$$\left(x - \frac{3}{4}\right)^2 = \frac{49}{16} .$$

Taking square roots we obtain:

$$x - \frac{3}{4} = \pm \frac{7}{4}$$

Solving for x:

$$x = +\frac{7}{4} + \frac{3}{4} = \frac{10}{4} = \frac{5}{2}$$

$$x = -\frac{7}{4} + \frac{3}{4} = -\frac{4}{4} = -1$$

The two solutions are $x = \frac{5}{2}$ and $x = -1$.

● **PROBLEM** 468

Solve $3x^2 + 5x - 2 = 0$.

Solution: This problem may be solved by the method of completing the square: First, divide both members by 3, the coefficient of the x^2 term,

$$x^2 + \frac{5}{3}x - \frac{2}{3} = 0.$$

Arrange the equation with the constant term in the right member. Thus,

$$x^2 + \frac{5}{3}x = \frac{2}{3}.$$

Take $\frac{1}{2}$ the coefficient of x, square this, and add the result to both members. Thus, $\frac{1}{2}$ of $\frac{5}{3}$ is $\frac{5}{6}$ and $\left(\frac{5}{6}\right)^2 = \frac{25}{36}$.

Add $\frac{25}{36}$ to both members,

$$x^2 + \frac{5}{3}x + \frac{25}{36} = \frac{2}{3} + \frac{25}{36} = \frac{49}{36}.$$

This makes the left member a perfect square. Factor:

$$\left[x + \frac{5}{6}\right]^2 = \frac{49}{36}.$$

Extract the square root of both members

$$x + \frac{5}{6} = \pm \frac{7}{6}.$$

When $x + \frac{5}{6} = \frac{7}{6}$, then $x = \frac{7}{6} - \frac{5}{6} = \frac{2}{6} = \frac{1}{3}.$

When $x + \frac{5}{6} = -\frac{7}{6}$, then $x = -\frac{7}{6} - \frac{5}{6} = -\frac{12}{6} = -2.$

Check: for $x = \frac{1}{3}$:

$$3\left(\frac{1}{3}\right)^2 + 5\left(\frac{1}{3}\right) - 2 = 0$$

$$3\left(\frac{1}{9}\right) + 5\left(\frac{1}{3}\right) - 2 = 0$$

$$\frac{3}{9} + \frac{15}{9} - \frac{18}{9} = 0$$

$$0 = 0.$$

For $x = -2$:

$$3(-2)^2 + 5(-2) - 2 = 0$$

$$3(4) + 5(-2) - 2 = 0$$

$$12 - 10 - 2 = 0$$

$$0 = 0.$$

Sol: $x = \left\{\frac{1}{3}, -2\right\}.$

● **PROBLEM 469**

Complete the square in both x and y in x^2+2x+y^2-3y.

Solution: To complete the square in x, take half the coefficient of x and square it. Add and subtract this value from the given expression. Therefore:

$\left[\frac{1}{2}(2)\right]^2 = [1]^2 = 1$, and $x^2+2x+y^2-3y = x^2+2x+y^2-3y+1-1.$

Commuting, $x^2+2x+y^2-3y = x^2+2x+1+y^2-3y-1 = (x+1)^2+y^2-3y-1.$ (1)

Now, take half the coefficient of y and square it. Add and subtract this value from equation (1).

$\left[\frac{1}{2}(-3)\right]^2 = \left[-\frac{3}{2}\right]^2 = \frac{9}{4}$, and

$x^2 + 2x + y^2 - 3y = (x+1)^2 + y^2 - 3y - 1 + \frac{9}{4} - \frac{9}{4}.$ Commuting,

$$x^2 + 2x + y^2 - 3y = (x+1)^2 + y^2 - 3y + \tfrac{9}{4} - 1 - \tfrac{9}{4}$$

$$= (x+1)^2 + \left(y - \tfrac{3}{2}\right)^2 - 1 - \tfrac{9}{4} = (x+1)^2 + \left(y - \tfrac{3}{2}\right)^2 - \tfrac{4}{4} - \tfrac{9}{4}. \quad \text{Hence, } x^2 + 2x + y^2 - 3y$$

$$= (x+1)^2 + \left(y - \tfrac{3}{2}\right)^2 - \tfrac{13}{4}.$$

● **PROBLEM 470**

Solve the following equations by completing the square.

(a) $x^2 + 2x - 1 = 0$ (c) $3t^2 - 2t + 1 = 0$

(b) $x^2 - 8x + 20 = 0$

<u>Solution</u>: To complete the square of any equation, take half the coefficient of the variable term (i.e., the term in which the variable is raised to the first power) and square it. Then, add and subtract the resulting value from both sides of the original equation.

(a) $x^2 + 2x - 1 = 0$. In this case, the variable term is 2x and the coefficient of this term is 2. Then, completing the square:

$\left[\tfrac{1}{2}(2)\right]^2 = (1)^2 = 1$. Now, the original equation becomes:

$$x^2 + 2x - 1 + 1 - 1 = 0 + 1 - 1$$
$$\left(x^2 + 2x + 1\right) - 1 - 1 = 0$$
$$(x + 1)^2 - 2 = 0$$

Adding 2 to both sides:

$$(x + 1)^2 - \cancel{2} + \cancel{2} = 0 + 2$$
$$(x + 1)^2 = 2$$

Taking the square root of both sides:

$$\sqrt{(x + 1)^2} = \pm\sqrt{2}$$
$$x + 1 = \pm\sqrt{2}$$

Subtracting 1 from both sides:

$$x + \cancel{1} - \cancel{1} = \pm\sqrt{2} - 1$$
$$x = \pm\sqrt{2} - 1$$

Therefore, the solution set to the original equation, $x^2 + 2x - 1 = 0$, is: $\{\sqrt{2} - 1, -\sqrt{2} - 1\}$.

(b) $x^2 - 8x + 20 = 0$. In this case, the variable term is -8x and the coefficient of this term is -8. Then, completing the square:

$\left[\tfrac{1}{2}(-8)\right]^2 = (-4)^2 = 16$. Now, the original equation becomes:

$$x^2 - 8x + 20 + 16 - 16 = 0 + 16 - 16.$$
$$\left(x^2 - 8x + 16\right) + 20 - 16 = 0$$
$$(x - 4)^2 + 4 = 0$$

Subtracting 4 from both sides:

$$(x - 4)^2 + \cancel{4} - \cancel{4} = 0 - 4$$
$$(x - 4)^2 = -4$$

Taking the square root of both sides:

$$\sqrt{(x - 4)^2} = \pm\sqrt{-4}$$
$$x - 4 = \pm\sqrt{-4} = \pm\sqrt{(-1)(4)} = \pm\sqrt{-1}\sqrt{4}$$
$$= \pm\, i(2)$$

or
$$x - 4 = \pm\, 2i$$

Adding 4 to both sides:

$$x - \cancel{4} + \cancel{4} = \pm\, 2i + 4$$
$$x = \pm\, 2i + 4$$

Therefore, the solution set to the original equation, $x^2 - 8x + 20 = 0$, is: $\{2i + 4,\ -2i + 4\}$.

(c) $3t^2 - 2t + 1 = 0$. In this case, the variable term is $-2t$ and the coefficient is -2. Then, completing the square:

$$\left[\tfrac{1}{2}(-2)\right]^2 = (-1)^2 = 1. \quad \text{Now, the original equation becomes:}$$

$$3t^2 - 2t + 1 + 1 - 1 = 0 + 1 - 1$$
$$\left(3t^2 - 2t + 1\right) + 1 - 1 = 0$$
$$3t^2 - 2t + 1 = 0 \qquad\qquad (1)$$

The roots of equation (1) can be found by using the quadratic formula, $x = \dfrac{-b \pm \sqrt{b^2 - 4ac}}{2a}$, since equation (1) is a quadratic equation with $a = 3$, $b = -2$, and $c = 1$. therefore,

$$t = \frac{-(-2) \pm \sqrt{(-2)^2 - 4(3)(1)}}{2(3)}$$

$$= \frac{2 \pm \sqrt{4 - 12}}{6}$$

$$= \frac{2 \pm \sqrt{-8}}{6}$$

$$= \frac{2 \pm \sqrt{(-1)(8)}}{6}$$

$$= \frac{2 \pm \sqrt{(-1)(4)(2)}}{6}$$

$$= \frac{2 \pm \sqrt{-1}\sqrt{4}\sqrt{2}}{6}$$

$$= \frac{2 \pm (i)(2)(\sqrt{2})}{6}$$

$$= \frac{2 \pm 2\sqrt{2}i}{6}$$

$$t = \frac{1}{3} \pm \frac{\sqrt{2}i}{3}$$

Therefore, the solution set to the original equation, $3t^2 - 2t + 1 = 0$, is: $\left\{ \frac{1}{3} + \frac{\sqrt{2}}{3}i, \ \frac{1}{3} - \frac{\sqrt{2}}{3}i \right\}$.

CHAPTER 17

SOLUTIONS BY QUADRATIC FORMULA

> **Basic Attacks and Strategies for Solving Problems in this Chapter. See pages 311 to 358 for step-by-step solutions to problems.**

The quadratic formula is a very useful tool in mathematics because it allows one to solve all types of quadratic equations. The procedure for using the quadratic formula, given by

$$x = \frac{-b \pm \sqrt{b^2 - 4ac}}{2a},$$

is as follows:

(1) Eliminate any fraction from the equation.

(2) Write the equation in standard form,

$$ax^2 + bx + c = 0.$$

(3) Substitute the coefficients a, b, and c of the quadratic equation in standard form into the formula.

(4) Calculate the results to obtain the solution set.

There is an interrelationship of the roots of a quadratic equation which allows one to have a rapid method for verifying the roots of the equation. In particular, the sum of the roots of any quadratic equation is $-b/a$ and the product of the roots is c/a. For example, consider the equation

$$x^2 - 2x - 8 = 0.$$

The sum of the roots is

$$-b/a = -(-2)/1 = 2,$$

and the product of the roots is

$$c/a = -8/1 = -8.$$

The roots of the equation are 4 and − 2. Thus, the sum and product of the roots are consistent with the above results.

The expression $b^2 - 4ac$ in the quadratic formula is called the discriminant. It provides a procedure for determining the character of the roots of the equation. In particular, if the discriminant is negative, then the roots are two complex numbers; if the discriminant is a positive number that is also a perfect square, then the two roots are rational numbers; if the discriminant is a positive number that is *not* a perfect square, then the two roots are irrational numbers; and if the discriminant is zero, then there is only one rational solution.

Step-by-Step Solutions to Problems in this Chapter, "Solutions by Quadratic Formula"

COEFFICIENTS WITH INTEGERS, FRACTIONS, RADICALS, AND VARIABLES

● **PROBLEM** 471

Solve for x: $4x^2 - 7 = 0$.

Solution: This quadratic equation can be solved for x using the quadratic formula, which applies to equations in the form $ax^2 + bx + c = 0$ (in our equation b = 0). There is, however, an easier method that we can use:

Adding 7 to both members, $\qquad 4x^2 = 7$

dividing both sides by 4, $\qquad x^2 = \dfrac{7}{4}$

taking the square root of both sides, $\quad x = \pm\sqrt{\dfrac{7}{4}} = \pm\dfrac{\sqrt{7}}{2}$.

The double sign $\pm$ (read "plus or minus") indicates that the two roots of the equation are

$$+\frac{\sqrt{7}}{2} \quad \text{and} \quad -\frac{\sqrt{7}}{2} .$$

● **PROBLEM** 472

Obtain the quadratic equation in standard form that is equivalent to $4x - 3 = 5x^2$.

Solution: The standard form of a quadratic equation is $ax^2 + bx + c = 0$. Starting with our given equation $4x - 3 = 5x^2$, we add $(-5x^2)$ to both members,

$$(4x - 3) + (-5x^2) = 5x^2 + (-5x^2)$$

$$(4x - 3) + (-5x^2) = 0$$

commuting we obtain $\qquad -5x^2 + 4x - 3 = 0$

This is the required equation with a = -5, b = 4, and c = -3.

● **PROBLEM** 473

Find the roots of the equation $x^2 + 12x - 85 = 0$.

Solution: The roots of this equation may be found using the quadratic formula

$$x = \frac{-B \pm \sqrt{B^2 - 4AC}}{2A}$$

In this equation A = 1, B = 12, and C = -85. Hence, by the quadratic formula,

$$x = \frac{-12 + \sqrt{144 + 340}}{2} \qquad \text{or} \qquad x = \frac{-12 - \sqrt{144 + 340}}{2}$$

$$x = \frac{-12 + 22}{2} \qquad \text{or} \qquad x = \frac{-12 - 22}{2}$$

Therefore x = 5 or x = -17. This is equivalent to the statement that the solution set is $\{-17, 5\}$.

● **PROBLEM** 474

Use the quadratic formula to solve for x in the equation

$x^2 - 5x + 6 = 0.$

Solution: The quadratic formula, $x = \dfrac{-b \pm \sqrt{b^2 - 4ac}}{2a}$, is used to solve equations in the form $ax^2 + bx + c = 0$. Here $a = 1$, $b = -5$, and $c = 6$. Hence

$$x = \frac{-(-5) \pm \sqrt{(-5)^2 - 4 \cdot 1 \cdot 6}}{2 \cdot 1} = \frac{5 \pm \sqrt{25 - 24}}{2}$$

$$= \frac{5 \pm \sqrt{1}}{2}$$

$$= \frac{5 \pm 1}{2}$$

$$= \frac{5 + 1}{2} \quad \text{or} \quad \frac{5 - 1}{2}$$

$$= \frac{6}{2} \quad \text{or} \quad \frac{4}{2}$$

$$= 3 \quad \text{or} \quad 2$$

Thus the roots of the equation $x^2 - 5x + 6 = 0$ are $x = 3$ and $x = 2$.

● **PROBLEM** 475

Solve the equation $x^2 + 5x + 6 = 0$ by the quadratic formula.

Solution: We use the quadratic formula, which states

$$x = \frac{-b \pm \sqrt{b^2 - 4ac}}{2a} \quad , \text{ for cases}$$

where $ax^2 + bx + c = 0$. For this equation, $a = 1$, $b = 5$, $c = 6$. Therefore the solutions are

$$x = \frac{-5 \pm \sqrt{25 - 4(1)(6)}}{2 \cdot 1}$$

$$= \frac{-5 \pm \sqrt{25 - 24}}{2} = \frac{-5 \pm \sqrt{1}}{2}$$

or $\quad x_1 = \dfrac{-5 + 1}{2} = -2, \quad x_2 = \dfrac{-5 - 1}{2} = -3.$

● **PROBLEM** 476

Solve $6x^2 - 7x - 20 = 0$.

Solution: $6x^2 - 7x - 20 = 0$ is not factorable. Therefore, find the roots of the quadratic equation $ax^2 + bx + c$ using:

$$x = \frac{-b \pm \sqrt{b^2 - 4ac}}{2a},$$

where $a = 6$, $b = -7$, $c = -20$.

$$x = \frac{7 \pm \sqrt{49 - 4(6)(-20)}}{12}$$

$$x = \frac{7 \pm \sqrt{529}}{12} = \frac{7 \pm 23}{12}.$$

Therefore,

$$x_1 = \frac{7 + 23}{12} = \frac{30}{12} = \frac{5}{2}$$

$$x_2 = \frac{7 - 23}{12} = -\frac{16}{12} = -\frac{4}{3}.$$

● **PROBLEM 477**

Solve the equation $2x^2 - 5x + 3 = 0$.

Solution:

(1) $2x^2 - 5x + 3 = 0$

Equation (1) is a quadratic equation of the form $ax^2 + bx + c = 0$ in which $a = 2$, $b = -5$, and $c = 3$. Therefore, the quadratic formula $x = \dfrac{-b \pm \sqrt{b^2 - 4ac}}{2a}$ may be used to find the solutions of equation (1). Substituting the values for a, b, and c in the quadratic formula:

$$x = \frac{-(-5) \pm \sqrt{(-5)^2 - 4(2)(3)}}{2(2)}$$

$$x = \frac{5 \pm \sqrt{1}}{4}$$

$$x = \frac{5 + 1}{4} = \frac{3}{2}, \quad \text{and} \quad x = \frac{5 - 1}{4} = 1$$

Check: Substituting $x = \frac{3}{2}$ in the given equation,

$$2\left(\frac{3}{2}\right)^2 - 5\left(\frac{3}{2}\right) + 3 = 0$$

$$0 = 0$$

Substituting $x = 1$ in the given equation,

$$2(1)^2 - 5(1) + 3 = 0$$

$$0 = 0$$

● **PROBLEM 478**

Solve $x^2 - 7x + 10 = 0$.

313

<u>Solution:</u> $x^2 - 7x + 10 = 0$ is a quadratic equation of the form $ax^2 + bx + c = 0$ with a = 1, b = -7, c = 10. The roots of the equation may be found using the quadratic formula:

$$x = \frac{-b \pm \sqrt{b^2 - 4ac}}{2a} .$$

Substituting values for a,b, and c

$$x = \frac{-(-7) \pm \sqrt{(-7)^2 + 4(1)(10)}}{2(1)} .$$

$$x = \frac{7 \pm \sqrt{49 - 4(1)(10)}}{2}$$

$$x = \frac{7 \pm \sqrt{9}}{2} = \frac{7 \pm 3}{2}$$

$$x = \frac{7 + 3}{2} = 5; \quad x = \frac{7 - 3}{2} = 2.$$

Check: for x = 5, $(5)^2 - 7(5) + 10 = 0$

$$25 - \quad 35 + 10 = 0$$

$$0 = 0$$

for x = 2, $(2)^2 - 7(2) + 10 = 0$

$$4 - \quad 14 + 10 = 0$$

$$0 = 0.$$

More simply, the problem could have been solved by factoring:

$$x^2 - 7x + 10 = 0$$

$$(x - 5)(x-2) = 0.$$

Set each factor equal to zero to find all values of x which make the product = 0.

$$x - 5 = 0 \quad \bigg| \quad x - 2 = 0$$

$$x = 5 \quad \bigg| \quad x = 2$$

● **PROBLEM**   479

Solve the equation $3x^2 - 5x + 2 = 0$ by means of the quadratic formula.

<u>Solution:</u> The quadratic formula, $x = \frac{-b \pm \sqrt{b^2 - 4ac}}{2a}$, applies to equations of the form $ax^2 + bx + c = 0$. The equation $3x^2 - 5x + 2 = 0$ is in this form with a = 3, b = -5, and c = 2. Substituting these values into our quadratic formula we obtain

$$x = \frac{-(-5) \pm \sqrt{(-5)^2 - 4(3)(2)}}{2(3)}$$

$$= \frac{5 \pm \sqrt{25 - 24}}{6} = \frac{5 \pm 1}{6}$$

$$= \frac{6}{6} \text{ and } \frac{4}{6}$$

$$x = 1 \text{ and } \frac{2}{3}$$

Hence the solution set is $\left\{1, \frac{2}{3}\right\}$. We can verify that the elements of $\left\{1, \frac{2}{3}\right\}$ are the roots of the given equation by means of the following check: We replace x by 1 in our original equation

$$3(1)^2 - 5(1) + 2 = 0$$

$$3 - 5 + 2 = 0$$

$$-2 + 2 = 0$$

$$0 = 0$$

Now we replace x by $\frac{2}{3}$ in the original equation

$$3\left(\frac{2}{3}\right)^2 - 5\left(\frac{2}{3}\right) + 2 = 0$$

$$3\left(\frac{4}{9}\right) - \frac{10}{3} + 2 = 0$$

$$\frac{12}{9} - \frac{10}{3} \cdot \frac{3}{3} + 2 = 0$$

$$\frac{12}{9} - \frac{30}{9} + 2 = 0$$

$$\frac{-18}{9} + 2 = 0$$

$$-2 + 2 = 0$$

$$0 = 0$$

Thus $\left\{1, \frac{2}{3}\right\}$ are indeed the roots of the given equation.

● **PROBLEM** 480

Solve for the roots of the equation $6x^2 + 5x - 2 = 0$ and for the roots of the equation $3x^2 + 4x - 4 = 0$.

Solution: To solve for the roots of an equation in the form $ax^2 + bx + c = 0$ we use the quadratic formula

$$x = \frac{-b \pm \sqrt{b^2 - 4ac}}{2a} .$$

In our first case $a = 6$, $b = 5$, and $c = -2$.

Substituting these values into the quadratic formula we obtain,

$$x = \frac{-5 \pm \sqrt{25 - 4(6)(-2)}}{12} = \frac{-5 \pm \sqrt{73}}{12} .$$

In our second case $a = 3$, $b = 4$, and $c = -4$. Substituting these values into the quadratic formula we obtain

$$x = \frac{-4 \pm \sqrt{4^2 - 4(3)(-4)}}{2(3)} = \frac{-4 \pm \sqrt{16 + 48}}{6} = \frac{-4 \pm \sqrt{64}}{6} = \frac{-4 \pm 8}{6}$$

$$= \frac{-4 + 8}{6} = \frac{4}{6} = \frac{2}{3}$$

or

$$= \frac{-4 - 8}{6} = \frac{-12}{6} = -2$$

Thus $x = -2$ or $2/3$.

Solve the quadratic equation

$$6x^2 - x - 35 = 0 .$$

Solution: Here we have a quadratic equation of the form $ax^2 + bx + c = 0$ with $a = 6$, $b = -1$, $c = -35$. Substituting in the quadratic formula,

$$x = \frac{-b \pm \sqrt{b^2 - 4ac}}{2a}$$

we find

$$x = \frac{-(-1) \pm \sqrt{(-1)^2 - 4\cdot 6(-35)}}{2 \cdot 6} = \frac{1 \pm \sqrt{1 + 840}}{12} = \frac{1 \pm 29}{12}$$

Hence the roots are

$$x = \frac{1 + 29}{12} = \frac{30}{12} = \frac{5}{2} , \quad x = \frac{1 - 29}{12} = - \frac{28}{12} = - \frac{7}{3} .$$

The quadratic equation of this example could also be solved by factoring. We find that

$$6x^2 - x - 35 \equiv (2x - 5)(3x + 7),$$

and since the equation will be satisfied if either of the two linear factors is set equal to zero, we get the two solutions found above. That is,

$$\begin{array}{ll} 2x - 5 = 0 & \quad 3x + 7 = 0 \\ 2x = 5 & \quad 3x = -7 \\ x = 5/2 & \quad x = - 7/3 \end{array}$$

To verify these results perform the following check.

Check: Replace x by $5/2$ in the original equation

$$6x^2 - x - 35 = 0$$
$$6\left(\frac{5}{2}\right)^2 - \frac{5}{2} - 35 = 0$$
$$6\left(\frac{25}{4}\right) - \frac{5}{2} - 35 = 0$$
$$\frac{150}{4} - \frac{2}{2}\left(\frac{5}{2}\right) - 35 = 0$$
$$\frac{150}{4} - \frac{10}{4} - 35 = 0$$
$$\frac{140}{4} - 35 = 0$$
$$35 - 35 = 0$$
$$0 = 0$$

Now replace x by $-\frac{7}{3}$ in the original equations

$$6x^2 - x - 35 = 0$$

$$6\left(-\frac{7}{2}\right)^2 - \left(-\frac{7}{3}\right) - 35 = 0$$

$$6\left(\frac{49}{9}\right) + \frac{7}{3} - 35 = 0$$

$$\frac{294}{9} + \frac{3 \cdot 7}{3 \cdot 3} - 35 = 0$$

$$\frac{294}{9} + \frac{21}{9} - 35 = 0$$

$$\frac{315}{9} - 35 = 0$$

$$35 - 35 = 0$$

$$0 = 0$$

● **PROBLEM** 482

Solve $x^2 + 2x - 5 = 0$.

<u>Solution</u>: $x^2 + 2x - 5 = 0$ is a nonfactorable quadratic equation of the form $ax^2 + bx + c = 0$. Therefore, to find the roots of the equation use the formula:

$$x = \frac{-b \pm \sqrt{b^2 - 4ac}}{2a}$$

with a = 1, b = 2, c = -5.

$$x = \frac{-2 \pm \sqrt{4 - 4(1)(-5)}}{2}$$

$$x = \frac{-2 \pm \sqrt{24}}{2} = \frac{-2 \pm \sqrt{4 \cdot 6}}{2} = \frac{-2 \pm \sqrt{4} \cdot \sqrt{6}}{2} .$$

This may be simplified as follows:

$$x = \frac{-2 \pm 2\sqrt{6}}{2} = \frac{2(-1 \pm \sqrt{6})}{2} = -1 \pm \sqrt{6}.$$

● **PROBLEM** 483

Use the Quadratic Formula to solve the following equation:
$x^2 - 7x - 7 = 0$.

<u>Solution</u>: The quadratic formula, $x = \frac{-b \pm \sqrt{b^2 - 4ac}}{2a}$, is used to solve equations in the form $ax^2 + bx + c = 0$. $x^2 - 7x - 7 = 0$ is in this form, with a = 1, b = -7, and c = -7. Thus,

$$x = \frac{-(-7) \pm \sqrt{(-7)^2 - 4(1)(-7)}}{2(1)}$$

$$x = \frac{7 \pm \sqrt{49 + 28}}{2}$$

$$x = \frac{7 \pm \sqrt{77}}{2}$$

Thus, the solution to the given equation is $x = \frac{7 + \sqrt{77}}{2}$, $x = \frac{7 - \sqrt{77}}{2}$.

● **PROBLEM 484**

Solve the equation $3x^2 + 5x - 7 = 0$.

Solution: In order to solve a quadratic of the form $ax^2 + bx + c = 0$, we employ the quadratic formula,

$$x = \frac{-b \pm \sqrt{b^2 - 4ac}}{2a} .$$

In our example $a = 3$, $b = 5$, $c = -7$. Substituting these values in our formula we obtain,

$$x = \frac{-5 \pm \sqrt{5^2 - 4(-7)(3)}}{2(3)}$$

$$= \frac{-5 \pm \sqrt{25 + 84}}{6}$$

$$= \frac{-5 \pm \sqrt{109}}{6}$$

Thus the two solutions to the equation $3x^2 + 5x - 7 = 0$ are $\frac{-5 + \sqrt{109}}{6}$ and $\frac{-5 - \sqrt{109}}{6}$ which can be verified by direct substitution in the original equation.

● **PROBLEM 485**

Use the quadratic formula to solve the equation

$$3x^2 + 4x - 5 = 0.$$

Solution: The quadratic formula,

$$x = \frac{-b \pm \sqrt{b^2 - 4ac}}{2a} ,$$ applies to the situation

where $ax^2 + bx + c = 0$. A comparison of the given equation $3x^2 + 4x - 5 = 0$ with the equation $ax^2 + bx + c = 0$ shows that $a = 3$, $b = 4$, and $c = -5$. Substituting these values in the quadratic formula we obtain:

$$x = \frac{-4 \pm \sqrt{16 - 4 \cdot 3(-5)}}{2 \cdot 3}$$

$$= \frac{-4 \pm \sqrt{16 + 60}}{6}$$

$$= \frac{-4 \pm \sqrt{76}}{6}$$

Since $76 = 4 \cdot 19$

$$= \frac{-4 \pm \sqrt{4 \cdot 19}}{6}$$

318

Recall $\sqrt{a \cdot b} = \sqrt{a} \cdot \sqrt{b}$

$$= \frac{-4 \pm \sqrt{4} \cdot \sqrt{19}}{6}$$

$$= \frac{-4 \pm 2\sqrt{19}}{6}$$

Factoring 2 out of the numerator and denominator

$$= \frac{\cancel{2}\,(-2 \pm \sqrt{19})}{\cancel{2}\,(3)}$$

Cancelling 2 from numerator and denominator we conclude

$$x = \frac{-2 \pm \sqrt{19}}{3}$$

● **PROBLEM 486**

Solve $t^2 - 8t + 3 = 0$ by the quadratic formula.

<u>Solution:</u> Recall the quadratic formula:

$$x = \frac{-b \pm \sqrt{b^2 - 4ac}}{2a} \, ,$$

which applies to the situation where $ax^2 + bx + c = 0$.
In our case, a = 1, b = - 8, c = 3, and

$$x = \frac{-(-8) \pm \sqrt{(-8)^2 - (4 \cdot 1 \cdot 3)}}{2 \cdot 1}$$

$$= \frac{8 \pm \sqrt{64 - 12}}{2 \cdot 1} = \frac{8 \pm \sqrt{52}}{2}$$

$$= \frac{8}{2} \pm \frac{\sqrt{52}}{2} \quad \text{by the definition of addition of fractions}$$

$$= 4 \pm \frac{\sqrt{13 \cdot 4}}{2} \quad \text{because } \frac{8}{2} = 4, \text{ and } 52 = 13 \cdot 4$$

$$= 4 \pm \frac{\sqrt{13} \cdot \sqrt{4}}{2} \quad \text{Recall } \sqrt{ab} = \sqrt{a} \cdot \sqrt{b}$$

$$= 4 \pm \frac{2\sqrt{13}}{2} \quad \text{because } \sqrt{4} = 2$$

$$= 4 \pm \sqrt{13}$$

● **PROBLEM 487**

Solve the equation $x^2 = 4x - 1$.

<u>Solution:</u> Subtract 4x from both sides of the given equation:

$$x^2 - 4x = 4x - 1 - 4x$$

$$x^2 - 4x = -1 \qquad\qquad (1)$$

Add 1 to both sides of equation (1).

$$x^2 - 4x + 1 = -1 + 1$$

$$x^2 - 4x + 1 = 0 \qquad (2)$$

An equation of the form $ax^2 + bx + c = 0$ where a, b, and c are real numbers, and a $\neq$ 0, is called a second degree or quadratic equation over the real numbers. The following formula, called the quadratic formula, may be used to find the roots or solutions to quadratic equations:

$$x = \frac{-b \pm \sqrt{b^2 - 4ac}}{2a}.$$

Therefore, equation (2) is a quadratic equation where a = 1, b = - 4, and c = 1. Substituting these values into the quadratic formula, then:

$$x = \frac{-(-4) \pm \sqrt{(-4)^2 - 4(1)(1)}}{2(1)}. \qquad (3)$$

Thus:

$$x = \frac{4 \pm \sqrt{12}}{2} = \frac{4 \pm \sqrt{4}\sqrt{3}}{2} = \frac{4 \pm 2\sqrt{3}}{2}$$

$$= \frac{\cancel{2}(2 \pm \sqrt{3})}{\cancel{2}} = 2 \pm \sqrt{3}.$$

$$x = 2 + \sqrt{3} \quad \text{and} \quad x = 2 - \sqrt{3} \qquad (4)$$

If these values of x are substituted in the original equation it will be found that they satisfy the given equation, so that $x = 2 + \sqrt{3}$ and $x = 2 - \sqrt{3}$ are roots of the equation.

● **PROBLEM** 488

Use the quadratic formula to solve the equation

$$8z(z + 1) = 1 \quad \text{for} \quad z .$$

Solution: Distributing, $\qquad 8z(z) + 8z(1) = 1$

$$8z^2 + 8z = 1$$

Adding (-1) to both sides, $\qquad 8z^2 + 8z - 1 = 0$

We use the quadratic formula,

$$x = \frac{-b + \sqrt{b^2 - 4ac}}{2a},$$

to solve equations in the form $ax^2 + bx + c = 0$. In our case a = 8, b = 8, and c = -1. Applying the quadratic formula to solve for z we obtain

$$z = \frac{-8 \pm \sqrt{8^2 - 4(8)(-1)}}{2(8)}$$

$$= \frac{-8 \pm \sqrt{64 + 32}}{16}$$

$$= \frac{-8 \pm \sqrt{96}}{16}$$

$$= \frac{-8 \pm \sqrt{16 \cdot 6}}{16}$$

$$= \frac{-8 \pm \sqrt{16}\sqrt{6}}{16}$$

$$= \frac{-8 \pm 4\sqrt{6}}{16}$$

$$= \frac{4(-2 \pm 1\sqrt{6})}{4(4)}$$

$$= \frac{-2 \pm \sqrt{6}}{4}$$

$$= \frac{-2}{4} \pm \frac{\sqrt{6}}{4}$$

$$= -\frac{1}{2} \pm \frac{\sqrt{6}}{4}$$

Hence, $z = -\frac{1}{2} + \frac{\sqrt{6}}{4}$, $-\frac{1}{2} - \frac{\sqrt{6}}{4}$.

● **PROBLEM** 489

Solve $\frac{1}{x} + \frac{1}{x + 2} = 2$.

<u>Solution:</u> In order to eliminate the fractions in this equation, we multiply both sides of the equation by the lowest common multiple (L.C.M), the expression of lowest degree into which each of the original expressions can be divided without a remainder. The L.C.M. is the product obtained by taking each factor to the highest degree. Thus in our case the L.C.M. is $(x')(x + 2)'$ and we multiply each member by $x(x + 2)$.

$$x(x + 2)\left[\frac{1}{x} + \frac{1}{x + 2}\right] = 2\left[x(x + 2)\right]$$

Distributing, $\quad x(x + 2)\left(\frac{1}{x}\right) + x(x + 2)\left(\frac{1}{x + 2}\right) = 2x \cdot (x + 2)$

Cancelling, $\quad x + 2 + x = 2x^2 + 4x$

Combining, $\quad 2x + 2 = 2x^2 + 4x$

Dividing both sides by 2, $x + 1 = x^2 + 2x$

Adding $-(x + 1)$ to both sides, $\quad 0 = x^2 + 2x - (x + 1)$

$$x^2 + 2x - x - 1 = 0$$

$$x^2 + x - 1 = 0$$

Since this is an expression in the form $ax^2 + bx + c = 0$ we may use the quadratic formula,

$$x = \frac{-b \pm \sqrt{b^2 - 4ac}}{2a}$$

to find its roots. In our case $a = 1$, $b = 1$, and $c = -1$. Hence

$$x = \frac{-1 \pm \sqrt{(1)^2 - 4(1)(-1)}}{2(1)}$$

$$= \frac{-1 \pm \sqrt{1 + 4}}{2}$$

$$= \frac{-1 \pm \sqrt{5}}{2}$$

Thus, $\qquad x = \dfrac{-1 + \sqrt{5}}{2}$ or $\dfrac{-1 - \sqrt{5}}{2}$.

Check: In order to verify these solutions, we substitute them for x in our original equation.

(a) Replace x by $\dfrac{-1 + \sqrt{5}}{2}$:

$$\frac{1}{x} + \frac{1}{x + 2} = 2$$

$$\frac{1}{\dfrac{-1 + \sqrt{5}}{2}} + \frac{1}{\dfrac{-1 + \sqrt{5}}{2} + 2} = 2$$

Since $\sqrt{5} \approx 2.24$ replace $\sqrt{5}$ by 2.24

$$\frac{1}{\dfrac{-1 + 2.24}{2}} + \frac{1}{\dfrac{-1 + 2.24}{2} + 2} \approx 2$$

$$\frac{1}{\dfrac{1.24}{2}} + \frac{1}{\dfrac{1.24}{2} + 2} \approx 2$$

$$\frac{1}{.62} + \frac{1}{.62 + 2} \approx 2$$

$$\frac{1}{.62} + \frac{1}{2.62} \approx 2$$

$$1.61 + .38 \approx 2$$

$$1.99 \approx 2$$

(b) Replace x by $\dfrac{-1 - \sqrt{5}}{2}$

$$\frac{1}{x} + \frac{1}{x + 2} = 2$$

$$\frac{1}{\dfrac{-1 - \sqrt{5}}{2}} + \frac{1}{\dfrac{-1 - \sqrt{5}}{2} + 2} = 2$$

Again replace $\sqrt{5}$ by 2.24

$$\frac{1}{\dfrac{-1 - 2.24}{2}} + \frac{1}{\dfrac{-1 - 2.24}{2} + 2} \approx 2$$

$$\frac{1}{\dfrac{-3.24}{2}} + \frac{1}{\dfrac{-3.24}{2} + 2} \approx 2$$

$$\frac{1}{-1.62} + \frac{1}{-1.62 + 2} \approx 2$$

$$\frac{1}{-1.62} + \frac{1}{.38} \approx 2$$

$$-.62 + 2.63 \approx 2$$

$$2.01 \approx 2$$

Therefore $x = \dfrac{-1 + \sqrt{5}}{2}$ are indeed solutions to our equation, and our solution set is $\left\{\dfrac{-1 + \sqrt{5}}{2}, \dfrac{-1-\sqrt{5}}{2}\right\}$.

● **PROBLEM** 490

Solve the equation $x^2 - x + 1 = 0$.

Solution: (1) $x^2 - x + 1 = 0$

Equation (1) is a quadratic equation of the form $ax^2 + bx + c = 0$ in which $a = 1$, $b = -1$, and $c = 1$.

Therefore, the quadratic formula $x = \dfrac{-b \pm \sqrt{b^2 - 4ac}}{2a}$

may be used to find solutions of equation (1). Substituting the values for a, b, and c in the quadratic formula:

(2) $x = \dfrac{-(-1) \pm \sqrt{(-1)^2 - (4)(1)(1)}}{2(1)}$

(3) $x = \dfrac{1 \pm \sqrt{-3}}{2}$

(4) $x = \dfrac{1 + \sqrt{-3}}{2}$ and $x = \dfrac{1 - \sqrt{-3}}{2}$

Substitution of each of these roots in the original Equation 1 will show that they satisfy the equation.

● **PROBLEM** 491

Solve the equation $\sqrt{x + 1} + \sqrt{2x + 3} - \sqrt{8x + 1} = 0$.

Solution: Add $\sqrt{8x + 1}$ to both sides,

$$\sqrt{x + 1} + \sqrt{2x + 3} = \sqrt{8x + 1}.$$

Square both sides of the equation,

$$(\sqrt{x + 1} + \sqrt{2x + 3})(\sqrt{x + 1} + \sqrt{2x + 3}) = (\sqrt{8x + 1})^2$$

$$(\sqrt{x + 1})^2 + 2\sqrt{x + 1}\,\sqrt{2x + 3} + (\sqrt{2x + 3})^2 = (\sqrt{8x + 1})^2$$

Since $\sqrt{a} \cdot \sqrt{b} = \sqrt{ab}$,

$$(\sqrt{x + 1})^2 + 2\sqrt{(x + 1)(2x + 3)} + (\sqrt{2x + 3})^2 = (\sqrt{8x + 1})^2.$$

Recall: $(\sqrt{a})^2 = \sqrt{a} \cdot \sqrt{a} = \sqrt{a^2} = a$.

Thus, $(\sqrt{x + 1})^2 = x + 1$

$(\sqrt{2x + 3})^2 = 2x + 3$

323

and $(\sqrt{8x + 1})^2 = 8x + 1$

Substituting these values we obtain,

$x + 1 + 2\sqrt{(x + 1)(2x + 3)} + 2x + 3 = 8x + 1$.

Combine terms, $3x + 4 + 2\sqrt{(x + 1)(2x + 3)} = 8x + 1$

Add (-4) to both sides, $3x + 2\sqrt{(x + 1)(2x + 3)} = 8x - 3$

Add (-3x) to both sides $2\sqrt{(x + 1)(2x + 3)} = 5x - 3$

Multiply the terms within the radical,

$2\sqrt{2x^2 + 5x + 3} = 5x - 3$

Square both members,

$(2\sqrt{2x^2 + 5x + 3})^2 = (5x - 3)(5x - 3)$.

Since $(ab)^2 = a^2b^2$,

$(2)^2 (\sqrt{2x^2 + 5x + 3})^2 = (5x - 3)(5x - 3)$

$4(\sqrt{2x^2 + 5x + 3})^2 = 25x^2 - 30x + 9$.

Once again recall: $(\sqrt{2x^2 + 5x + 3}) = (2x^2 + 5x + 3)$. Substituting this value, we obtain

$4(2x^2 + 5x + 3) = 25x^2 - 30x + 9$

Distribute,

$8x^2 + 20x + 12 = 25x^2 - 30x + 9$

Add $(-8x^2)$ to both sides,

$20x + 12 = 25x^2 - 30x + 9 - 8x^2$

$20x + 12 = 17x^2 - 30x + 9$

Add (- 20x) to both sides,

$12 = 17x^2 - 50x + 9$

Add (- 12) to both sides,

$17x^2 - 50x - 3 = 0$

We can find the roots of this equation using the quadratic formula $x = \dfrac{-b \pm \sqrt{b^2 - 4ac}}{2a}$, which applies to the situation $ax^2 + bx + c = 0$. In our case $a = 17$, $b = -50$, and $c = -3$. Thus

$x = \dfrac{50 \pm \sqrt{2500 + 204}}{34}$

$= \dfrac{50 \pm \sqrt{2704}}{34}$

$= \dfrac{50 \pm 52}{34} = \dfrac{50}{34} \pm \dfrac{52}{34}$

$= \dfrac{102}{34}$ and $-\dfrac{2}{34}$;

Thus, $x = 3$,

and $x = -\frac{1}{17}$.

Check: To verify that 3 and $-1/17$ are indeed roots of the given equation we replace x by these values in the original equation,

$$\sqrt{x + 1} + \sqrt{2x + 3} - \sqrt{8x + 1} = 0$$

(a) Substituting 3 for x:

$$\sqrt{3 + 1} + \sqrt{2(3) + 3} - \sqrt{8(3) + 1} = 0$$
$$\sqrt{4} + \sqrt{9} - \sqrt{25} = 0$$
$$2 + 3 - 5 = 0$$
$$0 = 0$$

Thus, 3 is a root of the equation.

(b) Substitute $-\frac{1}{17}$ for x:

$$\sqrt{-\frac{1}{17} + 1} + \sqrt{2\left[-\frac{1}{17}\right] + 3} - \sqrt{8\left[-\frac{1}{17}\right] + 1} = 0$$

$$\sqrt{-\frac{1}{17} + \frac{17}{17}} + \sqrt{-\frac{2}{17} + \frac{51}{17}} - \sqrt{-\frac{8}{17} + \frac{17}{17}} = 0$$

$$\sqrt{\frac{16}{17}} + \sqrt{\frac{49}{17}} - \sqrt{\frac{9}{17}} = 0$$

Since $\sqrt{\frac{a}{b}} = \frac{\sqrt{a}}{\sqrt{b}}$,

$$\frac{4}{\sqrt{17}} + \frac{7}{\sqrt{17}} - \frac{3}{\sqrt{17}} = 0$$

$$\frac{8}{\sqrt{17}} \neq 0$$

Since substitution of x by $-1/17$ does not result in a valid equation, $-1/17$ is not a root, and our solution set is {3}.

● PROBLEM 492

Solve for y if $6x^2 + 9y^2 + x - 6y = 0$.

Solution: Note: The standard form of a quadratic equation is $az^2 + bz + c = 0$ where $a \neq 0$. This type of equation can be solved by using the quadratic formula:

$$z = \frac{-b \pm \sqrt{b^2 - 4ac}}{2a}$$

Thus, we first put the equation in standard form,

$$9y^2 - 6y + (6x^2 + x) = 0$$

where $a = 9$, $b = -6$, and $c = 6x^2 + x$. Therefore the solutions are

$$y = \frac{6 \pm \sqrt{36 - 36(6x^2 + x)}}{18}$$

$$= \frac{6 \pm \sqrt{36[1 - (6x^2 + x)]}}{18}$$

$$= \frac{6 \pm 6\sqrt{1 - 6x^2 - x}}{18} = \frac{1 \pm \sqrt{1 - 6x^2 - x}}{3}$$

or

$$y_1 = \frac{1 + \sqrt{1 - 6x^2 - x}}{3}, \quad y_2 = \frac{1 - \sqrt{1 - 6x^2 - x}}{3}$$

• **PROBLEM** 493

Solve for y if $2x^2 + y^2 + 2xy - 2x = 0$.

Solution: In standard form, the equation becomes

$$y^2 + (2x)y + (2x^2 - 2x) = 0 \qquad (1)$$

Equation (1) is now in the standard form of a quadratic equation, $az^2 + bz + c = 0$, where $a \neq 0$. This type of equation can be solved by using the quadratic formula:

$$z = \frac{-b \pm \sqrt{b^2 - 4ac}}{2a}$$

Then, $a = 1$, $b = 2x$, and $c = 2x^2 - 2x$. Therefore the solutions are

$$y = \frac{-2x \pm \sqrt{4x^2 - 4(2x^2 - 2x)}}{2}$$

$$= \frac{-2x \pm \sqrt{4[x^2 - (2x^2 - 2x)]}}{2}$$

$$= \frac{-2x \pm 2\sqrt{x^2 - 2x^2 + 2x}}{2} = -x \pm \sqrt{2x - x^2}$$

or

$$y_1 = -x + \sqrt{2x - x^2}, \quad y_2 = -x - \sqrt{2x - x^2}$$

• **PROBLEM** 494

Solve for x by using the quadratic formula.

(a) $3x^2 = x + 6$ (b) $5x^2 - 6x + 7 = 0$.

Solution: The quadratic formula,

$$x = \frac{-b \pm \sqrt{b^2 - 4ac}}{2a},$$

is employed to solve equations in the form $ax^2 + bx + c = 0$.

(a) $3x^2 = x + 6$

In order to transform this equation into the desired form, we add $-(x + 6)$ to both sides,

$$3x^2 - (x + 6) = (x + 6) - (x + 6)$$

$$3x^2 - x - 6 = 0$$

Thus $a = 3$, $b = -1$, and $c = -6$. Substituting these values into the quadratic formula, we obtain

$$x = \frac{-(-1) \pm \sqrt{(-1)^2 - 4(3)(-6)}}{2(3)}$$

$$x = \frac{1 \pm \sqrt{1 + 72}}{6}$$

$$x = \frac{1 \pm \sqrt{73}}{6}$$

(b) $5x^2 - 6x + 7 = 0$

This equation is already in the form $ax^2 + bx + c = 0$ with $a = 5$, $b = -6$, and $c = 7$. Therefore

$$x = \frac{-(-6) \pm \sqrt{(-6)^2 - 4(5)(7)}}{2(5)}$$

$$x = \frac{6 \pm \sqrt{36 - 140}}{10}$$

$$x = \frac{6 \pm \sqrt{-104}}{10}$$

$$x = \frac{6 \pm \sqrt{(-1)(4)(26)}}{10}$$

$$x = \frac{6 \pm \sqrt{(-1)} \cdot \sqrt{4} \cdot \sqrt{26}}{10}$$

$$x = \frac{6 \pm i(2)\sqrt{26}}{10}$$

Factoring out a 2 from each term:

$$x = \frac{2(3 \pm i\sqrt{26})}{2(5)}$$

$$x = \frac{3 \pm i\sqrt{26}}{5}$$

● **PROBLEM 495**

Solve the equation

$$4\sqrt{\frac{3-x}{3+x}} - \sqrt{\frac{3+x}{3-x}} = \sqrt{2} .$$

Solution: Although this equation is not in quadratic form, the fact that the two radicals in the left member are reciprocal to each other suggests that the equation may be reduced to a tractable form. For brevity, let

$$y = \sqrt{\frac{3-x}{3+x}}$$

Then the equation becomes

$$4y - \frac{1}{y} = \sqrt{2},$$

Multiplying by y and transferring terms to one side, we obtain:

$$4y^2 - 1 = \sqrt{2} \; y$$

$$4y^2 - \sqrt{2}y - 1 = 0, \text{ which is}$$

a quadratic in y. The solutions of this equation are found from the
quadratic formula,

$$y = \frac{-b \pm \sqrt{b^2 - 4ac}}{2a}$$

where $a = 4$, $b = -\sqrt{2}$ and $c = -1$. Solving, we find

$$y = \frac{\sqrt{2} \pm \sqrt{2 + 16}}{8}$$

$$= \frac{\sqrt{2} \pm \sqrt{18}}{8} = \frac{\sqrt{2} \pm 3\sqrt{2}}{8}$$

$$= \frac{\sqrt{2}(1 \pm 3)}{8}$$

Thus,

$$y = \frac{\sqrt{2}}{2}, \quad \text{or} \quad y = -\frac{\sqrt{2}}{4}.$$

Now, since $y = \sqrt{\frac{3-x}{3+x}}$, denotes a square root with a prefixed positive sign
understood. Hence y may be either real and positive or complex, but
not real and negative. Consequently the value $y = -\sqrt{2}/4$ can lead only
to extraneous solutions and may therefore be discarded. Therefore, we
have:

$$\sqrt{\frac{3 - x}{3 + x}} = \frac{\sqrt{2}}{2} = \frac{1}{\sqrt{2}}$$

Squaring both sides:

$$\frac{3 - x}{3 + x} = \frac{1}{2},$$

Cross-multiplying, we have:

$$6 - 2x = 3 + x,$$

Collecting terms so that the variable   x   is on one side and the
numerical quantities on the other side:

$$3x = 3,$$

Dividing by three:

$$x = 1.$$

This is the only solution of the given equation; its correctness may
readily be checked as follows: Substituting x = 1,

$$4\sqrt{\frac{3 - 1}{3 + 1}} - \sqrt{\frac{3 + 1}{3 - 1}} = \sqrt{2}$$

$$4\frac{\sqrt{2}}{\sqrt{4}} - \frac{\sqrt{4}}{\sqrt{2}} = \sqrt{2}$$

$$4\frac{\sqrt{2}}{2} - \frac{2}{\sqrt{2}} = \sqrt{2}$$

$$2\sqrt{2} - \frac{2}{\sqrt{2}} = \sqrt{2}$$

Multiply by $\sqrt{2}$: $2\sqrt{2}\sqrt{2} - 2 = \sqrt{2}\sqrt{2}$

$$4 - 2 = 2$$

$$2 = 2$$

● **PROBLEM** 496

Solve for x: $3x^2 + 5 = 0$.

Solution: This quadratic equation can be solved for x using the quadratic formula, which applies to equations in the form $ax^2 + bx + c = 0$ (in our equation b = 0). There is, however, an easier method that we can use: adding -5 to both sides,

$$3x^2 = -5$$

dividing both sides by 3,

$$x^2 = -\frac{5}{3}$$

Taking the square root of both sides,

$$x = \pm\sqrt{-\frac{5}{3}} = \pm\sqrt{(-1)\left(\frac{5}{3}\right)} = \pm\sqrt{-1}\sqrt{\frac{5}{3}}$$

By definition $\sqrt{-1} = i$. Thus,

$$\pm\sqrt{-1}\sqrt{\frac{5}{3}} = \pm i\sqrt{\frac{5}{3}} = \pm i\sqrt{\frac{5}{3}}\cdot\frac{3}{3} = \pm i\sqrt{\frac{15}{9}} = \pm i\frac{\sqrt{15}}{\sqrt{9}} = \pm\frac{i\sqrt{15}}{3} .$$

Thus, $x = \frac{i\sqrt{15}}{3} , -\frac{i\sqrt{15}}{3} .$

497

Use the quadratic formula to solve $2x^2 - 5x + 8 = 0$.

Solution: Recall the quadratic formula:

$$x = \frac{-b \pm \sqrt{b^2 - 4ac}}{2a}$$

which applies to the situation where $ax^2 + bx + c = 0$. In our case a = 2, b = - 5, and c = 8. Substituting these values into the quadratic formula we obtain:

$$x = \frac{5 \pm \sqrt{25 - 4\cdot 2\cdot 8}}{2\cdot 2}$$

$$= \frac{5 \pm \sqrt{25 - 64}}{4} = \frac{5 \pm \sqrt{-39}}{4}$$

Note that the $\sqrt{-39}$, the square root of a negative number, is not defined for real numbers, hence we must use the imaginary number system.

Since $\sqrt{ab} = \sqrt{a}\cdot\sqrt{b}$, $\sqrt{-39} = \sqrt{(-1)\cdot(39)} = \sqrt{-1}\cdot\sqrt{39}$

By definition $i = \sqrt{-1}$, so $\sqrt{-39} = i\sqrt{39}$

Therefore $x = \frac{5 \pm i\sqrt{39}}{4}$

● **PROBLEM** 498

Use the quadratic formula to solve the following equation for x:

$$x^2 + 2x + 4 = 0$$

Solution: The quadratic formula,

$$x = \frac{-b \pm \sqrt{b^2 - 4ac}}{2a} ,$$

is used to solve equations in the form $ax^2 + bx + c = 0$. Consider the equation $x^2 + 2x + 4 = 0$. Here $a = 1$, $b = 2$, $c = 4$. Hence the roots are

$$\frac{-2 \pm \sqrt{2^2 - 4 \cdot 1 \cdot 4}}{2 \cdot 1} = \frac{-2 \pm \sqrt{4 - 16}}{2} = \frac{-2 \pm \sqrt{-12}}{2}$$

$$= \frac{-2 \pm \sqrt{(-3)4}}{2} = \frac{-2 \pm \sqrt{-3} \sqrt{4}}{2} = \frac{-2 \pm \sqrt{4} \sqrt{-3}}{2} = \frac{-2 \pm 2\sqrt{-3}}{2}$$

$$= \frac{2(-1 \pm \sqrt{-3})}{2} = \begin{cases} -1 + \sqrt{-3} = -1 + \sqrt{-1 \cdot 3} = -1 + \sqrt{-1} \sqrt{3} \\ -1 - \sqrt{-3} = -1 - \sqrt{-1 \cdot 3} = -1 - \sqrt{-1} \sqrt{3} \end{cases}$$

$$= -1 + i\sqrt{3}$$
$$= -1 - i\sqrt{3}$$

Notice that the roots of a quadratic equation may be imaginary, even though the coefficients are real.

● **PROBLEM 499**

Solve $x^2 + 2x + 5 = 0$.

Solution: $x^2 + 2x + 5 = 0$ is a nonfactorable quadratic equation of the form $ax^2 + bx + c = 0$. Therefore, to find the roots of the equation use the formula:

$$x = \frac{-b \pm \sqrt{b^2 - 4ac}}{2a} \text{ with } a = 1, b = 2, c = 5.$$

$$x = \frac{-2 \pm \sqrt{4 - 4(1)(5)}}{2}$$

$$x = \frac{-2 \pm \sqrt{-16}}{2} = \frac{-2 \pm \sqrt{-1} \cdot \sqrt{16}}{2} .$$

In this case the roots involve imaginary numbers. The result can be simplified by using $i = \sqrt{-1}$ to give

$$x = \frac{-2 \pm 4i}{2} = -1 \pm 2i.$$

● **PROBLEM 500**

Solve the equation $2x^2 + 5x + 8 = 0$.

Solution: Letting A = 2, B = 5, and C = 8, we substitute these values in the quadratic formula in the following manner:

$$x = \frac{-B + \sqrt{B^2 - 4AC}}{2A} \quad \text{or} \quad x = \frac{-B - \sqrt{B^2 - 4AC}}{2A}$$

$$x = \frac{-(5) + \sqrt{(5)^2 - 4(2)(8)}}{2(2)} \quad \text{or} \quad x = \frac{-(5) - \sqrt{(5)^2 - 4(2)(8)}}{2(2)}$$

$$x = \frac{-5 + \sqrt{-39}}{4} \quad \text{or} \quad x = \frac{-5 - \sqrt{-39}}{4}$$

Therefore the solution set is

$$\left\{ \frac{-5 + \sqrt{-39}}{4} , \frac{-5 - \sqrt{-39}}{4} \right\}$$

Since $\sqrt{-39}$ is not a real number, we recognize that the members of the solution set of the equation are mixed imaginary numbers. Furthermore, since $\sqrt{-39} = i\sqrt{39}$, the solution set may be rewritten as

$$\left\{ \frac{-5 + i\sqrt{39}}{4} , \frac{-5 - i\sqrt{39}}{4} \right\}$$

This emphasizes a need for the extension of the set of real numbers into the set of complex numbers.

● **PROBLEM** 501

Solve $x^2 + 2x + 5 = 0$, by the quadratic formula.

Solution: Recall the quadratic formula:

$$x = \frac{-b \pm \sqrt{b^2 - 4ac}}{2a} , \text{ which applies to the situation}$$

where $ax^2 + bx + c = 0$. In our case, $x^2 + 2x + 5 = 0$,

$a = 1$, $b = 2$, and $c = 5$; hence $x = \dfrac{-2 \pm \sqrt{(2)^2 - 4(1)(5)}}{2 \cdot 1}$

$$= \frac{-2 \pm \sqrt{4 - 20}}{2} = \frac{-2 \pm \sqrt{-16}}{2}$$

Note $\sqrt{ab} = \sqrt{a} \cdot \sqrt{b}$, therefore $\sqrt{-16} = \sqrt{16 \cdot (-1)}$

$$= \sqrt{16} \cdot \sqrt{(-1)}$$

$$= 4 \sqrt{(-1)}$$

By definition $i = \sqrt{(-1)}$,

$$= 4i$$

hence, $\qquad\qquad = \dfrac{-2 \pm 4i}{2}$. Therefore,

$$x = -1 \pm 2i.$$

As a check we substitute $x = -1 \pm 2i$ into $x^2 + 2x + 5$:

$$(-1 \pm 2i)^2 + 2(-1 \pm 2i) + 5$$

$$= 1 \mp 4i + 4i^2 - 2 \pm 4i + 5$$

$$= \mp 4i \pm 4i = 0 \text{ by the additive inverse}$$

property and since $i^2 = -1$ by definition, $4i^2 = 4(-1)$
$= -4$ hence,

$$= 1 - 4 - 2 + 5$$

$$= 6 - 6$$

$$= 0$$

Solve the equation $4x^2 = 8x - 7$ by means of the quadratic formula.

Solution: The quadratic formula, $x = \dfrac{-b \pm \sqrt{b^2 - 4ac}}{2a}$, applies to equations of the form $ax^2 + bx + c = 0$. If we add $(-8x + 7)$ to both sides of our given equation we obtain $4x^2 - 8x + 7 = 8x - 7 = 0$ which is an equation in the form $ax^2 + bx + c = 0$ with $a = 4$, $b = -8$, and $c = 7$. Substituting these values into the quadratic formula we obtain

$$x = \frac{-(-8) \pm \sqrt{(-8)^2 - 4(4)(7)}}{2(4)}$$

$$= \frac{8 \pm \sqrt{64 - 112}}{8}$$

$$= \frac{8 \pm \sqrt{-48}}{8}$$

Since $\sqrt{ab} = \sqrt{a} \cdot \sqrt{b}$, $\sqrt{-48} = \sqrt{-1 \cdot 48} = \sqrt{-1}\sqrt{48}$. Recall $\sqrt{-1} = i$.

Thus $\sqrt{-48} = \sqrt{-1}\sqrt{48} = i\sqrt{48}$ and

$$x = \frac{8 \pm i\sqrt{48}}{8}.$$

We can further break down this radical by noting

$$\sqrt{48} = \sqrt{16 \cdot 3} = \sqrt{16} \cdot \sqrt{3} = 4\sqrt{3}.$$

Thus,

$$x = \frac{8 \pm 4i\sqrt{3}}{8}$$

$$x = \frac{8}{8} \pm \frac{4i\sqrt{3}}{8}$$

$$x = 1 \pm \frac{i\sqrt{3}}{2}$$

Hence the solution set is $\left\{ 1 + \dfrac{i\sqrt{3}}{2} , 1 - \dfrac{i\sqrt{3}}{2} \right\}$.

We can verify that these two complex numbers are the roots of the given equation by means of the following check: We replace x by $1 + \dfrac{i\sqrt{3}}{2}$ in the original equation:

$$4\left(1 + \frac{i\sqrt{3}}{2}\right)^2 = 8\left(1 + \frac{i\sqrt{3}}{2}\right) - 7$$

$$4\left[1 + \frac{2i\sqrt{3}}{2} + \left(\frac{i\sqrt{3}}{2}\right)^2\right] = 8 + 8\frac{i\sqrt{3}}{2} - 7$$

$$4\left[1 + \frac{2i\sqrt{3}}{2} + i^2\left(\frac{\sqrt{3}}{2}\right)^2\right] = 1 + 4i\sqrt{3}$$

$$4\left[1 + i\sqrt{3} - \left(\frac{\sqrt{3}}{2} \cdot \frac{\sqrt{3}}{2}\right)\right] = 1 + 4i\sqrt{3}$$

$$4\left[1 + i\sqrt{3} - \frac{3}{4}\right] = 1 + 4i\sqrt{3}$$

$$4 + 4i\sqrt{3} - 3 = 1 + 4i\sqrt{3}$$

$$1 + 4i\sqrt{3} = 1 + 4i\sqrt{3}$$

Now we replace x by $1 - \frac{i\sqrt{3}}{2}$ in the original equation:

$$4\left(1 - \frac{i\sqrt{3}}{2}\right)^2 = 8\left(1 - \frac{i\sqrt{3}}{2}\right) - 7$$

$$4\left[1 - \frac{2i\sqrt{3}}{2} + \left(\frac{i\sqrt{3}}{2}\right)^2\right] = 8 - \frac{8i\sqrt{3}}{2} - 7$$

$$4\left[1 - i\sqrt{3} + i^2\left(\frac{\sqrt{3}}{2}\right)^2\right] = 1 - 4i\sqrt{3}$$

$$4\left[1 - i\sqrt{3} - 1\left(\frac{\sqrt{3}}{2}\right)^2\right] = 1 - 4i\sqrt{3}$$

$$4\left(1 - i\sqrt{3} - \frac{3}{4}\right) = 1 - 4i\sqrt{3}$$

$$4 - 4i\sqrt{3} - 3 = 1 - 4i\sqrt{3}$$

$$1 - 4i\sqrt{3} = 1 - 4i\sqrt{3}$$

● **PROBLEM 503**

Find the roots of the function F whose rule of correspondence is $F(x) = 2x^2 + 8x + 4$.

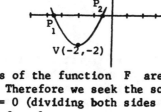

V(-2,-2)

<u>Solution:</u> The roots of the function F are those values of x which satisfy $F(x) = 0$. Therefore we seek the solution set of $2x^2 + 8x + 4 = 0$, or $x^2 + 4x + 2 = 0$ (dividing both sides of the equation by 2). Using the quadratic formula,

$$x = \frac{-b \pm \sqrt{b^2 - 4ac}}{2a}$$

with $a = 1$, $b = 4$, and $c = 2$, we have:

$$x = \frac{-4 \pm \sqrt{(4)^2 - 4(1)(2)}}{2(1)}$$

$$x = \frac{-4 \pm \sqrt{16 - 8}}{2}$$

$$x = \frac{-4 \pm \sqrt{4 \cdot 2}}{2}$$

$$x = \frac{-4 \pm 2\sqrt{2}}{2} = -2 \pm \sqrt{2}$$

Hence the roots are $x_1 = -2 + \sqrt{2}$ and $x_2 = -2 - \sqrt{2}$.

The graph of the quadratic function F is a parabola and intersects the domain axis at the points $P_1(-2-\sqrt{2}, 0)$ and $P_2(-2+\sqrt{2}, 0)$.

Solve $3x^2 - 5x + 4 = 0$ by the quadratic formula.

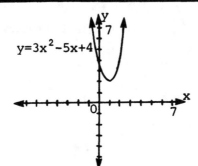

$y = 3x^2 - 5x + 4$

Solution: Recall the quadratic formula, $x = \dfrac{-b \pm \sqrt{b^2 - 4ac}}{2a}$, which applies to equations in the form $ax^2 + bx + c = 0$. In our case

$$a = 3 \qquad b = -5 \qquad c = 4 \, .$$

Substituting these values into the quadratic formula we obtain,

$$x = \frac{-(-5) \pm \sqrt{(-5)^2 - 4(3)(4)}}{2 \cdot 3}$$

$$x = \frac{5 \pm \sqrt{25 - 48}}{6} = \frac{5 \pm \sqrt{-23}}{6}$$

Since $\sqrt{-23}$ is not a real number, x is not a real number, and, consequently, the equation has no real roots. The graph of $y = 3x^2 - 5x + 4$ is shown in the accompanying figure. Notice that the graph does not cross the x-axis. This is because on the x-axis $y = 0$. Hence $y = 0 = 3x^2 - 5x + 4$, which is the equation we have just shown to have no real roots.

INTERRELATIONSHIPS OF ROOTS: SUMS; PRODUCTS

Determine the quadratic equation whose roots are

$$\frac{1}{2} \quad \text{and} \ - \left(\frac{2}{3} \right) .$$

Solution: $\quad x = \dfrac{1}{2}$ (1) $\qquad x = -\dfrac{2}{3}$ (2)

Subtract $\dfrac{1}{2}$ from both sides of equation (1).

$$x - \frac{1}{2} = \frac{1}{2} - \frac{1}{2}$$

Therefore: $x - \frac{1}{2} = 0 \qquad (3).$

Add $\frac{2}{3}$ to both sides of equation (2).

$$x + \frac{2}{3} = -\frac{2}{3} + \frac{2}{3}$$

Therefore: $x + \frac{2}{3} = 0 \qquad (4).$

Hence, from equations (3) and (4):

$$\frac{2x - 1}{2} = 0 \quad (5) \quad \text{and} \quad \frac{3x + 2}{3} = 0 \ (6).$$

Multiply both sides of equation (5) by 2 and multiply both sides of equation (6) by 3.

Therefore: $2x - 1 = 0 \qquad$ and $\qquad 3x + 2 = 0$

Hence, $(2x - 1)(3x + 2) = (0)(0) = 0$

or $\qquad\qquad\qquad 6x^2 - 3x + 4x - 2 = 0$

or $\qquad\qquad\qquad\qquad 6x^2 + x - 2 = 0.$

● **PROBLEM** 506

Show that the roots of the quadratic equation $x^2 - x - 3 = 0$ are

$$x_1 = \frac{1 + \sqrt{13}}{2} \quad \text{and} \quad x_2 = \frac{1 - \sqrt{13}}{2}$$

Solution: We use the quadratic formula derived from the quadratic equation, $ax^2 + bc + c = 0$:

$$x = \frac{-b \pm \sqrt{b^2 - 4ac}}{2a}$$

For $x^2 - x - 3 = 0$, $a = 1$, $b = -1$, and $c = -3$. Replacing these values in the quadratic formula,

$$x = \frac{-(-1) \pm \sqrt{(-1)^2 - 4(1)(-3)}}{2(1)}$$

$$x = \frac{1 \pm \sqrt{13}}{2}$$

$$x_1 = \frac{1 + \sqrt{13}}{2} \qquad\qquad x_2 = \frac{1 - \sqrt{13}}{2}$$

According to the Factor Theorem: If r is a root of the equation $f(x) = 0$, i.e., if $f(r) = 0$, then $(x - r)$ is a factor of $f(x)$. x_1 and x_2 are roots of $x^2 - x - 3 = 0$. Thus,

$$\left(x - \frac{1 + \sqrt{13}}{2}\right)\left(x - \frac{1 - \sqrt{13}}{2}\right)$$

are factors, and

$$x^2 - x - 3 = \left(x - \frac{1 - \sqrt{13}}{2}\right)\left(x - \frac{1 - \sqrt{13}}{2}\right)$$

Find the sum and product of the roots of the equation

$$3x^2 - 2x + 1 = 0.$$

Solution: The given equation is a quadratic equation in which a = 3, b = - 2, and c = 1. Using the quadratic formula $x = \dfrac{-b \pm \sqrt{b^2 - 4ac}}{2a}$ to find the roots of the

given equation:

$$x = \frac{-(-2) \pm \sqrt{(-2)^2 - 4(3)(1)}}{2(3)} = \frac{2 \pm \sqrt{-8}}{6}$$

$$= \frac{1}{3} \pm \frac{\sqrt{4}\sqrt{-2}}{6} = \frac{1}{3} \pm \frac{2\sqrt{-2}}{6}$$

$$= \frac{1 \pm \sqrt{-2}}{3}$$

The sum of the roots is:

$$\frac{1 + \sqrt{-2}}{3} + \frac{1 - \sqrt{-2}}{3} = \frac{1 + \sqrt{-2} + 1 - \sqrt{-2}}{3} = \frac{2}{3}.$$

The product of the root is:

$$\left(\frac{1 + \sqrt{-2}}{3}\right)\left(\frac{1 - \sqrt{-2}}{3}\right) = \frac{1 + \sqrt{-2} - \sqrt{-2} - (-2)}{9}$$

$$= \frac{1}{3}.$$

Find the sum and product of the roots of the equation $3x^2 + 13x - 10 = 0.$

Solution: The roots of an equation of the form $ax^2 + bx + c = 0$ can be found using the quadratic formula

$$x = \frac{-b \pm \sqrt{b^2 - 4ac}}{2a}.$$

For the given equation, a = 3, b = 13, c = -10. Therefore, the roots of the equation $3x^2 + 13x - 10 = 0$ are

$$x = \frac{-13 \pm \sqrt{169 + 120}}{6} = \frac{-13 \pm 17}{6} = -5 \text{ or } \frac{2}{3}.$$

The sum of the roots is $-5 + \frac{2}{3} = -\frac{13}{3}$. The product of the roots is $(-5)\left(\frac{2}{3}\right) = -\frac{10}{3}$.

Check. The sum of the roots of a quadratic equation is $r_1 + r_2 = \frac{-b}{a}$ and the product of the roots of a quadratic equation is $r_1 \cdot r_2 = \frac{c}{a}$. From the equation, $-\frac{b}{a} = -\frac{13}{3}$ and $\frac{c}{a} = -\frac{10}{3}$.

● **PROBLEM** 509

Without solving, find the sum and product of the roots of $8x^2 = 2x + 3$.

Solution: Given a quadratic equation in standard form, $ax^2 + bx + c = 0$, the sum of the roots is given by $\frac{-b}{a}$ and the product of the roots by $\frac{c}{a}$. Adding $-(2x + 3)$ to both sides of the given equation, we obtain $8x^2 - 2x - 3 = 0$, a quadratic equation in standard form with $a = 8$, $b = -2$, and $c = -3$. Thus:

Sum of roots $= -\frac{b}{a} = \left(\frac{-2}{8}\right) = \frac{1}{4}$.

Product of roots $= \frac{c}{a} = \frac{-3}{8}$.

● **PROBLEM** 510

Find the sum and the product of the roots in each of the following equations: $x^2 - 3x + 2 = 0$, $2x^2 + 8x - 5 = 0$, and $\sqrt{2}x^2 + 5x - \sqrt{8} = 0$

Solution: There are two relations between the roots and coefficients of a quadratic equation. When we want to find the roots of the quadratic function, $f(x) = ax^2 + bx + c$, we set $f(x) = 0$. Then $ax^2 + bx + c = 0$. By the quadratic formula, the roots are

$$r_1 = \frac{-b + \sqrt{b^2 - 4ac}}{2a} \qquad r_2 = \frac{-b - \sqrt{b^2 - 4ac}}{2a}$$

Adding $r_1 + r_2 = \dfrac{-b + \sqrt{b^2 - 4ac}}{2a} + \dfrac{-b - \sqrt{b^2 - 4ac}}{2a}$

$$= \frac{-2b}{2a} = \frac{-b}{a}$$

Multiplying $r_1 \cdot r_2 = \left(\dfrac{-b + \sqrt{b^2 - 4ac}}{2a}\right)\left(\dfrac{-b - \sqrt{b^2 - 4ac}}{2a}\right)$

$$= \frac{b^2 - (b^2 - 4ac)}{4a^2} = \frac{4ac}{4a^2} = \frac{c}{a}$$

Therefore, the sum of the roots is $-b/a$ and the product of the roots is c/a. Thus in the following tabulation $-b/a$ is the sum of the roots, and c/a is the product of the roots.

Equation	Sum of roots	Product of roots
$x^2 - 3x + 2 = 0$ Thus $a = 1$, $b = -3$ $c = 2$	$-\dfrac{b}{a} = \dfrac{-(-3)}{1} = 3$	$\dfrac{c}{a} = \dfrac{2}{1} = 2$
$2x^2 + 8x - 5 = 0$ Thus $a = 2$, $b = 8$ $c = -5$	$-\dfrac{b}{a} = \dfrac{-8}{2} = -4$	$\dfrac{c}{a} = \dfrac{-5}{2}$
$\sqrt{2}x^2 + 5x - \sqrt{8} = 0$ Thus $a = \sqrt{2}$, $b = 5$ $c = -58$	$-\dfrac{b}{a} = \dfrac{-5}{\sqrt{2}}$	$\dfrac{c}{a} = \dfrac{-\sqrt{8}}{\sqrt{2}} = -\sqrt{4} = -2$

These two relations provide a rapid method for verifying the roots of a quadratic equation. For the first equation, $x^2 - 3x + 2 = 0$, we can solve for the roots by factoring.

$$x^2 - 3x + 2 = 0$$
$$(x - 2)(x - 1) = 0$$
$$x_1 = 2 \qquad x_2 = 1$$

The sum was found by the formula to be 3.

$$x_1 + x_2 = 2 + 1 = 3$$

The product was found to be 2.

$$x_1 \cdot x_2 = 2 \cdot 1 = 2$$

Similarly for the last two equations.

$$2x^2 + 8x - 5 = 0; \quad -b/a = -4; \quad c/a = -5/2$$

$$x_1 = \frac{-8 + \sqrt{8^2 - 4(2)(-5)}}{2(2)} \qquad x_2 = \frac{-8 - \sqrt{8^2 - 4(2)(-5)}}{2(2)}$$

$$= \frac{-8 + \sqrt{104}}{4} \qquad = \frac{-8 - \sqrt{104}}{4}$$

$$x_1 + x_2 = \frac{-8 + \sqrt{104}}{4} + \frac{-8 - \sqrt{104}}{4} = \frac{-16}{4} = -4 = -b/a$$

$$x_1 \cdot x_2 = \left(\frac{-8 + \sqrt{104}}{4}\right)\left(\frac{-8 - \sqrt{104}}{4}\right) = \frac{64 - 104}{16} = \frac{-40}{16} = \frac{-5}{2} = c/a$$

$$\sqrt{2}\, x^2 + 5x - \sqrt{8} = 0$$

$$x_1 = \frac{-5 + \sqrt{25 - 4(-\sqrt{8})\sqrt{2}}}{2\sqrt{2}} \qquad x_2 = \frac{-5 - \sqrt{25 - 4(-\sqrt{8})\sqrt{2}}}{2\sqrt{2}}$$

$$= \frac{-5 + \sqrt{25 - 4(-4)}}{2\sqrt{2}} \qquad = \frac{-5 - \sqrt{25 - 4(-4)}}{2\sqrt{2}}$$

$$= \frac{-5 + \sqrt{41}}{2\sqrt{2}} \qquad = \frac{-5 - \sqrt{41}}{2\sqrt{2}}$$

$$x_1 + x_2 = \left(\frac{-5 + \sqrt{41}}{2\sqrt{2}}\right) + \left(\frac{-5 - \sqrt{41}}{2\sqrt{2}}\right) = \frac{-10}{2\sqrt{2}} = \frac{-5}{\sqrt{2}} = \frac{-b}{a}$$

$$x_1 \cdot x_2 = \left(\frac{-5 + \sqrt{41}}{2\sqrt{2}}\right)\left(\frac{-5 - \sqrt{41}}{2\sqrt{2}}\right) = \frac{25 - 41}{2 \cdot 2 \cdot 2} = \frac{-16}{8} = -2 =$$

$$= \frac{c}{a}$$

● **PROBLEM** 511

Determine the quadratic equation whose roots are $x = 2 + \sqrt{3}$ and $x = 2 - \sqrt{3}$.

<u>Solution</u>: We can determine the quadratic equation from the sum and the product of the roots. A quadratic equation whose roots are x_1 and x_2 may be written in the form

$$x^2 - \left(x_1 + x_2\right)x + x_1 \cdot x_2 = 0$$

where the sum of the roots is $x_1 + x_2 = -\frac{b}{a}$ and the product of the roots is $x_1 \cdot x_2 = \frac{c}{a}$. Here,

$$x_1 = 2 + \sqrt{3} \text{ and } x_2 = 2 - \sqrt{3}.$$

Then, $x_1 + x_2 = 2 + \sqrt{3} + 2 - \sqrt{3} = 4$ and

and $\quad x_1 \cdot x_2 = (2 + \sqrt{3})(2 - \sqrt{3}) = 4 - 3 = 1.$

Hence, the equation is:

$$x^2 - 4x + 1 = 0.$$

● **PROBLEM** 512

Find the equation whose roots are $3 + \sqrt{2}$ and $3 - \sqrt{2}$.

<u>Solution</u>: The roots of a quadratic equation $ax^2 + bx + c$ can be characterized by the following:

the sum of the roots

$$r_1 + r_2 = \frac{-b}{a} \text{ and}$$

the product of the roots

$$r_1 \cdot r_2 = \frac{c}{a}.$$

The sum of the roots is $(3 + \sqrt{2}) + (3 - \sqrt{2}) = 6.$ Hence,

$$-\frac{b}{a} = 6 \text{ or } \frac{b}{a} = -6.$$

The product of the roots is

$$(3 + \sqrt{2})(3 - \sqrt{2}) = 3^2 - (\sqrt{2})^2 = 9 - 2 = 7.$$

This is the constant term of the required equation. We obtain this from the quadratic function

$ax^2 + bx + c = 0$. Divide by a.

$$x^2 + \frac{b}{a}x + \frac{c}{a} = 0.$$

Then $\frac{b}{a}$ is the coefficient of x and $\frac{c}{a}$ is the constant term. Thus, here $\frac{b}{a} = -6$ and $\frac{c}{a} = 7$. Hence, the equation is

$$x^2 - 6x + 7 = 0.$$

Check: $x^2 - 6x + 7 = 0$ with a = 1, b = -6, c = 7.

$$X = \frac{-b \pm \sqrt{b^2 - 4ac}}{2a} = \frac{-(-6) \pm \sqrt{(-6)^2 - 4(1)(7)}}{2(1)} = \frac{6 \pm \sqrt{3b - 28}}{2}$$

$$= \frac{6 \pm \sqrt{8}}{2} = \frac{6 \pm 2\sqrt{2}}{2} = 3 + \sqrt{2} \text{ and } 3 - \sqrt{2}.$$

● **PROBLEM** 513

Find a quadratic equation whose roots are $3 + 2\sqrt{3}$ and $3 - 2\sqrt{3}$.

Solution: A quadratic equation is an equation of the form $ax^2 + bx + c = 0$, where a, b, and c are constants and a ≠ 0. If both sides of this quadratic equation are divided by a, then:

$$\frac{ax^2 + bx + c}{a} = \frac{0}{a}$$

$$x^2 + \frac{b}{a}x + \frac{c}{a} = 0 \qquad (1)$$

Note that this last result is valid since a≠0. If r_1 and r_2 are the roots of a quadratic equation, then the sum of these roots, S, is,

$$S = r_1 + r_2 = \frac{-b}{a} \text{ and the product of these roots, P,}$$

is: $P = r_1 \cdot r_2 = c/a.$

Note that the coefficient of the x-term in equation (1) is $\frac{b}{a}$. In relation to the sum of the roots, S, this coefficient

$$= \frac{b}{a} = -\left(\frac{b}{a}\right) = -(S) = -S. \text{ Hence, equation (1) can be rewritten as,}$$

$$x^2 + (-S)x + \frac{c}{a} = 0$$

or

$$x^2 - Sx + \frac{c}{a} = 0 \qquad (2)$$

340

Also, note that the constant term on the left side of equation (1), or $\frac{c}{a}$, is also the product, P, of the roots. Hence, equation (2) can be rewritten as:

$$x^2 - Sx + P = 0 \qquad\qquad (3)$$

The sum of the roots is:

$$S = r_1 + r_2, \text{ and here } r_1 \text{ and } r_2 \text{ are } 3 + 2\sqrt{3}, 3 - 2\sqrt{3}$$

Thus,

$$S = (3 + 2\sqrt{3}) + (3 - 2\sqrt{3})$$

$$= 3 + 2\sqrt{3} + 3 - 2\sqrt{3}$$

$$= 3 + 3$$

$$= 6$$

The product of the roots is:

$$P = r_1 \cdot r_2$$

$$= (3 + 2\sqrt{3})(3 - 2\sqrt{3})$$

$$= 9 + 6\sqrt{3} - 6\sqrt{3} - 4(3)$$

$$= 9 - 12$$

$$= -3$$

Then, replacing S and P by 6 and -3 respectively in equation (3):

$$x^2 - Sx + P = x^2 - 6x + (-3) = 0$$

or

$$x^2 - 6x - 3 = 0,$$

which is in the form $ax^2 + bx + c = 0$ of a quadratic equation.

● **PROBLEM 514**

Form the equation whose roots are $2 + \sqrt{3}$ and $2 - \sqrt{3}$.

<u>Solution:</u> The roots are $2 + \sqrt{3}$ and $2 - \sqrt{3}$. Hence, $x = 2 + \sqrt{3}$ and $x = 2 - \sqrt{3}$. Subtract $(2 + \sqrt{3})$ from the first equation:

$$x - (2 + \sqrt{3}) = (2 + \sqrt{3}) - (2 + \sqrt{3}) = 0,$$

or

$$x - (2 + \sqrt{3}) = 0.$$

Subtract $(2 - \sqrt{3})$ from the second equation:

$$x - (2 - \sqrt{3}) = (2 - \sqrt{3}) - (2 - \sqrt{3}) = 0,$$

or

$$x - (2 - \sqrt{3}) = 0.$$

Therefore,
$$[x - (2 + \sqrt{3})][x - (2 - \sqrt{3})] = (0)(0) = 0 ,$$
or
$$[x - (2 + \sqrt{3})][x - (2 - \sqrt{3})] = 0 . \tag{1}$$

Equation (1) is in the form $(x - a)(x - b) = 0$ where a corresponds to $(2 + \sqrt{3})$ and b corresponds to $(2 - \sqrt{3})$. Also:

$$(x - a)(x - b) = x^2 - ax - bx + ab$$
$$= x^2 - (a + b)x + ab . \tag{2}$$

Notice that a and b are the roots; that is, $2 + \sqrt{3}$ and $2 - \sqrt{3}$. The sum of the roots is:

$$a + b = (2 + \sqrt{3}) + (2 - \sqrt{3}) = 2 + \sqrt{3} + 2 - \sqrt{3}$$
$$= 4 .$$

The product of the roots is:

$$a \cdot b = (2 + \sqrt{3})(2 - \sqrt{3}) = 4 + 2\sqrt{3} - 2\sqrt{3} - 3$$
$$= 4 - 3$$
$$= 1 .$$

Hence, using the form of equation (2):

$$[x - (2 + \sqrt{3})][x - (2 - \sqrt{3})] = x^2 - (4)x + 1 = 0$$
or
$$x^2 - 4x + 1 = 0 ,$$
which is the equation whose roots are
$$2 + \sqrt{3} \text{ and } 2 - \sqrt{3} .$$

● **PROBLEM   515**

Find the equation whose roots are $\dfrac{\alpha}{\beta}$, $\dfrac{\beta}{\alpha}$.

Solution: The roots of the equation are $x = \dfrac{\alpha}{\beta}$ and $x = \dfrac{\beta}{\alpha}$. Subtract $\dfrac{\alpha}{\beta}$ from both sides of the first equation:

$$x - \frac{\alpha}{\beta} = \frac{\alpha}{\beta} - \frac{\alpha}{\beta} = 0 ,$$
or
$$x - \frac{\alpha}{\beta} = 0 .$$

Subtract $\dfrac{\beta}{\alpha}$ from both sides of the second equation:

$$x - \frac{\beta}{\alpha} = \frac{\beta}{\alpha} - \frac{\beta}{\alpha} = 0 ,$$
or
$$x - \frac{\beta}{\alpha} = 0 .$$

Therefore:
$$\left(x - \frac{\alpha}{\beta}\right)\left(x - \frac{\beta}{\alpha}\right) = (0)(0) = 0 ,$$
or
$$\left(x - \frac{\alpha}{\beta}\right)\left(x - \frac{\beta}{\alpha}\right) = 0 . \tag{1}$$

Equation (1) is of the form: $(x - c)(x - d) = 0$, or
$$x^2 - cx - dx + cd = 0 , \text{ or}$$
$$x^2 - (c + d)x + cd = 0 . \tag{2}$$

Note that c corresponds to the root $\dfrac{\alpha}{\beta}$ and d corresponds to the root $\dfrac{\beta}{\alpha}$. The sum of the roots is:

$$c + d = \frac{\alpha}{\beta} + \frac{\beta}{\alpha} = \frac{\alpha(\alpha)}{\alpha(\beta)} + \frac{\beta(\beta)}{\beta(\alpha)} = \frac{\alpha^2}{\alpha\beta} + \frac{\beta^2}{\alpha\beta}$$
$$= \frac{\alpha^2 + \beta^2}{\alpha\beta}$$

The product of the roots is:
$$c \cdot d = \frac{\alpha}{\beta} \cdot \frac{\beta}{\alpha} = \frac{\alpha\beta}{\beta\alpha} = \frac{\alpha\beta}{\alpha\beta} = 1 \ .$$

Using the form of equation (2):
$$\left(x - \frac{\alpha}{\beta}\right)\left(x - \frac{\beta}{\alpha}\right) = x^2 - \left(\frac{\alpha^2 + \beta^2}{\alpha\beta}\right)x + 1 = 0 \ .$$

Hence,
$$x^2 - \left(\frac{\alpha^2 + \beta^2}{\alpha\beta}\right)x + 1 = 0 \ , \tag{3}$$

Multiply both sides of equation (3) by $\alpha\beta$.
$$\alpha\beta\left[x^2 - \left(\frac{\alpha^2 + \beta^2}{\alpha\beta}\right)x + 1\right] = \alpha\beta(0)$$

Distributing,
$$\alpha\beta x^2 - (\alpha^2 + \beta^2)x + \alpha\beta = 0 \ ,$$

which is the equation whose roots are $\dfrac{\alpha}{\beta}$, $\dfrac{\beta}{\alpha}$.

● **PROBLEM   516**

Find the equation whose roots are the negatives of the roots of $x^2 + 7x - 2 = 0$.

Solution: The roots of a quadratic equation $ax^2 + bx + c = 0$ are given by the quadratic formula
$$x = \frac{-b \pm \sqrt{b^2 - 4ac}}{2a} \ .$$

Therefore:
$$r_1 = \frac{-b + \sqrt{b^2 - 4ac}}{2a}, \quad r_2 = \frac{-b - \sqrt{b^2 - 4ac}}{2a}$$

By adding r_1 and r_2:
$$r_1 + r_2 = \frac{-b + \sqrt{b^2 - 4ac}}{2a} + \frac{-b - \sqrt{b^2 - 4ac}}{2a} = \frac{-2b}{2a} = \frac{-b}{a}.$$

We see that the sum of the roots is:
$$r_1 + r_2 = \frac{-b}{a}.$$

Then multiply:
$$r_1 \cdot r_2 = \left(\frac{-b + \sqrt{b^2 - 4ac}}{2a}\right)\left(\frac{-b - \sqrt{b^2 - 4ac}}{2a}\right)$$
$$= \frac{(-b + \sqrt{b^2 - 4ac})(-b - \sqrt{b^2 - 4ac})}{4a^2}$$
$$= \frac{b^2 + b\sqrt{b^2 - 4ac} - b\sqrt{b^2 - 4ac} - (b^2 - 4ac)}{4a^2}$$
$$= \frac{4ac}{4a^2} = \frac{c}{a}.$$

For the given equation $a = 1$, $b = 7$, $c = -2$. If the roots of the given equation are r_1 and r_2, we seek an equation whose roots are $-r_1$ and $-r_2$. From the given equation, we have $r_1 + r_2 = -\frac{b}{a} = -\frac{7}{1} = -7$. Thus

$$-r_1 + (-r_2) = -(r_1 + r_2) = 7$$

and the coefficient of the first-degree term in the required equation is -7. The product of the roots of the given equation is $\frac{c}{a} = \frac{-2}{1} = -2$. Since $r_1 \cdot r_2 = (-r_1)(-r_2)$, the product of the roots of the required equation is also -2. Hence, the constant term of the required equation is -2. The required equation can be written

$$x^2 - 7x - 2 = 0.$$

● **PROBLEM 517**

Find the value of k if one root is twice the other.

$$x^2 - kx + 18 = 0.$$

Solution: If x_1 and x_2 are the roots of the quadratic equation $ax^2 + bx + c$, then $x_1 + x_2 = -b/a$ and $x_1 \cdot x_2 = c/a$. For $x^2 - kx + 18 = 0$, $a = 1$, $b = -k$, and $c = 18$. Since for this quadratic, one root is twice the other, let the roots be r and $2r$. Their sum is $r + 2r = 3r$. The sum of the roots is $-b/a = k$. Hence $k = 3r$. The product of the roots $r \cdot 2r = 2r^2$ is equal to $c/a = 18$.

Thus, $\qquad\qquad 2r^2 = 18; \ r^2 = 9; \ r = \pm 3$

Therefore, $\qquad\qquad k = \pm 3 \cdot 3 = \pm 9$

Check: The roots of $x^2 - 9x + 18 = 0$ are 3, and $2 \cdot 3 = 6$. The roots of $x^2 + 9x + 18 = 0$ are -3, and $2(-3) = -6$.

● **PROBLEM 518**

Find the values of the constant k in the equation

$$2x^2 - kx + 3k = 0$$

if the difference of the roots is $\frac{5}{2}$.

Solution: The given equation is a quadratic equation since it is in the form $ax^2 + bx + c = 0$. If both sides of the given equation are divided by 2, then:

$$\frac{2x^2 - kx + 3k}{2} = \frac{0}{2}$$

$$x^2 - \frac{k}{2}x + \frac{3}{2}k = 0 \qquad\qquad (1)$$

Equation (1) is in the form $x^2 - Sx + P = 0$, where S = sum of the roots = $\frac{k}{2}$ and P = product of the roots = $\frac{3}{2}k$.

Let the roots be r_1 and r_2. Since the difference of the roots is $\frac{5}{2}$ and the sum of the roots is $\frac{k}{2}$, the following equations result:

$$r_1 - r_2 = \frac{5}{2} \quad (2) \qquad \text{and} \qquad r_1 + r_2 = \frac{k}{2} \qquad (3)$$

Solving for r_1 by adding equations (2) and (3):

$$r_1 - r_2 = \frac{5}{2}$$

$$\underline{r_1 + r_2 = \frac{k}{2}}$$

$$2r_1 = \frac{5}{2} + \frac{k}{2}$$

$$2r_1 = \frac{5 + k}{2}.$$

Multiplying both sides by $\frac{1}{2}$:

$$\frac{1}{2}(2r_1) = \frac{1}{2}\left(\frac{5 + k}{2}\right)$$

$$r_1 = \frac{5 + k}{4}$$

$$r_1 = \frac{k + 5}{4} \qquad (4)$$

Solving for r_2 by subtracting equation (3) from equation (2):

$$r_1 - r_2 = \frac{5}{2}$$

$$\underline{\left(- r_1 + r_2 = \frac{k}{2}\right)}$$

$$-2r_2 = \frac{5}{2} - \frac{k}{2}$$

$$-2r_2 = \frac{5 - k}{2} \qquad (5)$$

Multiply both sides of equation (5) by $-\frac{1}{2}$:

$$\left(-\frac{1}{2}\right)(-2r_2) = \left(-\frac{1}{2}\right)\left(\frac{5 - k}{2}\right)$$

$$r_2 = \frac{(-1)(5 - k)}{(2)(2)}$$

$$= \frac{-5 + k}{4}$$

$$r_2 = \frac{k - 5}{4} \qquad\qquad (6)$$

Multiplying equations (4) and (6):

$$r_1 r_2 = \left(\frac{k + 5}{4}\right)\left(\frac{k - 5}{4}\right)$$

$$= \frac{k^2 + 5k - 5k - 25}{16}$$

$$r_1 r_2 = \frac{k^2 - 25}{16}.$$

However, it was found earlier that the product, P, of the roots was $\frac{3}{2}k$; that is, $r_1 r_2 = \frac{3}{2}k$. Then, setting these two expressions for $r_1 r_2$ equal:

$$\frac{k^2 - 25}{16} = \frac{3}{2}k$$

$$\frac{k^2 - 25}{16} = \frac{3k}{2}.$$

Subtract $\frac{3k}{2}$ from both sides of this equation:

$$\frac{k^2 - 25}{16} - \frac{3k}{2} = \frac{3k}{2} - \frac{3k}{2}$$

$$\frac{k^2 - 25}{16} - \frac{3k}{2} = 0.$$

Obtaining a common denominator of 16 for the two fractions on the left side of the equation:

$$\frac{k^2 - 25}{16} - \frac{8(3k)}{8(2)} = 0$$

$$\frac{k^2 - 25}{16} - \frac{24k}{16} = 0$$

$$\frac{k^2 - 25 - 24k}{16} = 0.$$

$$\frac{k^2 - 24k - 25}{16} = 0$$

Multiply both sides of this equation by 16:

$$16\left(\frac{k^2 - 24k - 25}{16}\right) = 16(0)$$

$$k^2 - 24k - 25 = 0.$$

Factor the left side of this equation into a product of two

polynomials:

$$(k - 25)(k + 1) = 0 \tag{7}$$

Whenever a product ab = 0, where a and b are any two numbers, either a = 0 or b = 0. Then, equation (7) becomes:

$$k - 25 = 0 \quad \text{or} \quad k + 1 = 0$$

$$k = 25 \quad \text{or} \quad k = -1.$$

Hence, the two solutions are $k = -1$ and $k = 25$.

● **PROBLEM** 519

Find the value of k if, in the equation $2x^2 - kx^2 + 4x + 5k = 0$, one root is the reciprocal of the other.

Solution: By using the distributive property in relation to the x^2 terms, the given equation becomes:

$$2x^2 - kx^2 + 4x + 5k = (2-k)x^2 + 4x + 5k = 0 \quad \text{or}$$

$$(2-k)x^2 + 4x + 5k = 0.$$

Divide both sides of this equation by (2-k):

$$\frac{(2-k)x^2 + 4x + 5k}{(2-k)} = \frac{0}{(2-k)}.$$

$$x^2 + \frac{4}{2-k}x + \frac{5k}{2-k} = 0. \tag{1}$$

Equation (1) is in the form $x^2 - Sx + P = 0$, where S is the sum of the roots of this equation, and P is the product of the roots of this equation. If r_1 and r_2 are the roots of equation (1), then

$$P = r_1r_2 = \frac{5k}{2-k}$$

$$r_1r_2 = \frac{5k}{2-k}. \tag{2}$$

It was also given in this problem that one root is the reciprocal of the other. Then,

$$r_1 = \frac{1}{r_2}.$$

Multiply both sides of this equation by r_2:

$$r_2(r_1) = r_2\left(\frac{1}{r_2}\right)$$

$$r_2r_1 = 1 \quad \text{or}$$

347

$$r_1 r_2 = 1 \qquad\qquad (3)$$

Since equations (2) and (3) are two expressions for $r_1 r_2$, the right sides of these equations can be set equal to each other:

$$\frac{5k}{2-k} = 1.$$

Multiply both sides of this equation by $(2-k)$:

$$(2-k)\left[\frac{5k}{2-k}\right] = (2-k)(1)$$

$$5k = 2-k.$$

Add k to both sides of this equation:

$$5k + k = 2 - k + k.$$

$$6k = 2.$$

Divide both sides of this equation by 6:

$$\frac{6k}{6} = \frac{2}{6}$$

$$k = \frac{2}{6}$$

$$\text{or } k = \frac{1}{3}.$$

Therefore, the value of k in the equation $2x^2 - kx^2 + 4x + 5k = 0$ is $\frac{1}{3}$.

● **PROBLEM 520**

If α and β are the roots of $x^2 - px + q = 0$, find the value of (1) $\alpha^2 + \beta^2$, (2) $\alpha^3 + \beta^3$.

Solution: The roots of the given equation are α and β. Hence, $x = \alpha$ and $x = \beta$. Subtract α from both sides of the first equation:

$$x - \alpha = \alpha - \alpha = 0,$$

or

$$x - \alpha = 0.$$

Subtract β from both sides of the second equation:

$$x - \beta = \beta - \beta = 0,$$

or

$$x - \beta = 0.$$

Hence,

$$(x - \alpha)(x - \beta) = (0)(0) = 0,$$

or

$$(x - \alpha)(x - \beta) = 0.$$

Also,

$$(x - \alpha)(x - \beta) = x^2 - \alpha x - \beta x + \alpha\beta = 0$$

or

$$x^2 - (\alpha + \beta)x + \alpha\beta = 0 \qquad\qquad (1)$$

Comparing the given equation with equation (1):

$$\alpha + \beta = p \quad \text{(eq.2)}, \quad \text{and} \quad \alpha\beta = q \quad \text{(eq. 3)}$$

348

Therefore, squaring both sides of equation (2):

$$(\alpha + \beta)^2 = p^2$$

$$(\alpha + \beta)(\alpha + \beta) = p^2$$

$$\alpha^2 + \alpha\beta + \alpha\beta + \beta^2 = p^2$$

$$\alpha^2 + 2\alpha\beta + \beta^2 = p^2 \qquad (4)$$

Subtract $2\alpha\beta$ from both sides of equation (4):

$$\alpha^2 + 2\alpha\beta + \beta^2 - 2\alpha\beta = p^2 - 2\alpha\beta$$

$$\alpha^2 + \beta^2 = p^2 - 2\alpha\beta$$

$$\alpha^2 + \beta^2 = p^2 - 2q ,$$

since $\alpha\beta = q$.

To obtain an expression for $\alpha^3 + \beta^3$, cube both sides of equation (2).

$$(\alpha + \beta)^3 = p^3$$

$$(\alpha + \beta)(\alpha + \beta)^2 = p^3$$

$$(\alpha + \beta)(\alpha^2 + 2\alpha\beta + \beta^2) = p^3$$

Distributing the left side of this equation:

$$(\alpha^3 + 2\alpha^2\beta + \alpha\beta^2) + (\alpha^2\beta + 2\alpha\beta^2 + \beta^3) = p^3$$

Combining terms and simplifying the left side of this equation:

$$\alpha^3 + 2\alpha^2\beta + \alpha\beta^2 + \alpha^2\beta + 2\alpha\beta^2 + \beta^3 = p^3$$

$$\alpha^3 + 3\alpha^2\beta + 3\alpha\beta^2 + \beta^3 = p^3$$

$$(\alpha^3 + \beta^3) + (3\alpha^2\beta + 3\alpha\beta^2) = p^3$$

Factor out $3\alpha\beta$ from the second term on the left side of this equation:

$$(\alpha^3 + \beta^3) + 3\alpha\beta(\alpha + \beta) = p^3$$

$$(\alpha^3 + \beta^3) + 3q(p) = p^3 ,$$

since $\alpha\beta = q$ and $\alpha + \beta = p$. Hence,

$$(\alpha^3 + \beta^3) + 3pq = p^3$$

Subtract $3pq$ from both sides of this equation.

$$(\alpha^3 + \beta^3) + 3pq - 3pq = p^3 - 3pq$$

$$(\alpha^3 + \beta^3) = p^3 - 3pq$$

Factor out p from the right side of this equation.

$$\alpha^3 + \beta^3 = p(p^2 - 3q)$$

Therefore, $\alpha^2 + \beta^2 = p^2 - 2q$, and $\alpha^3 + \beta^3 = p(p^2 - 3q)$.

● **PROBLEM 521**

Solve $(1 - a^2)(x + a) - 2a(1 - x^2) = 0.$

<u>Solution:</u> The given equation may be rewritten by finding the product of the first two terms on the left side and distributing the remaining terms on the left side:

$$(1 - a^2)(x + a) - 2a(1 - x^2) = 0$$

$$(x - a^2x + a - a^3) - (2a - 2ax^2) = 0$$

$$x - a^2x + a - a^3 - 2a + 2ax^2 = 0$$

$$2ax^2 + x - a^2x - a^3 - 2a + a = 0$$

$$2ax^2 + (1 - a^2)x - a - a^3 = 0$$

$$2ax^2 + (1 - a^2)x - a(1 + a^2) = 0 \qquad (1)$$

Equation (1) is in the form $cz^2 + dz + e = 0$, which is a quadratic equation. Therefore, equation (1) is a quadratic equation, and $2a$ corresponds to c, $(1 - a^2)$ corresponds to d and $-a(1 + a^2)$ corresponds to e. One of the roots is clearly a because, when $x = a$, the given equation is satisfied; that is,

$$(1-a^2)(x+a) - 2a(1-x^2) = (1-a^2)(a+a) - 2a(1-a^2) = (1-a^2)(2a) - 2a(1-a^2)$$

Using the commutative property:

$$(1-a^2)(a+a) - 2a(1-a^2) = (1-a^2)(2a) - 2a(1-a^2)$$
$$= 2a(1-a^2) - 2a(1-a^2)$$
$$= 0 .$$

The product of the roots $= x_1 \cdot x_2$

$$= \frac{e}{c}$$

$$= \frac{-a(1+a^2)}{2a}$$

$$= -\frac{(1+a^2)}{2}$$

Since one of the roots is a, let this root be x_1. Hence,

$$x_1 \cdot x_2 = a \cdot x_2 = -\frac{(1+a^2)}{2}$$

or

$$ax_2 = -\frac{(1+a^2)}{2}$$

or x_2, the second root,

$$= -\frac{(1+a^2)}{2}\left(\frac{1}{a}\right)$$

$$= -\frac{(1+a^2)}{2a}$$

Thus, the roots are a, and $\dfrac{-(1+a^2)}{2a}$.

● **PROBLEM** 522

Find the condition that the roots of $ax^2 + bx + c = 0$ may be (1) both positive, (2) opposite in sign, but the greater of them negative.

Solution: The given equation, $ax^2 + bx + c = 0$, is a quadratic equation. Let α and β be the roots of this equation. The product of the roots of the quadratic equation is:

$$\alpha\beta = \frac{c}{a} .$$

The sum of the roots of the quadratic equation is:

$$\alpha + \beta = \frac{-b}{a} .$$

(1) If the roots are both positive, $\alpha\beta$ is positive, and therefore c and a have like signs.
Also, since $\alpha + \beta$ is positive, $\frac{b}{a}$ is negative; therefore b and a have unlike signs.
Hence, the required condition is that the signs of a and c should be like, and b and a have unlike signs.

(2) If the roots are of opposite signs, $\alpha\beta$ is negative, and therefore c and a have unlike signs.
Also, since $\alpha + \beta$ has the sign of the greater root, it is negative, and therefore $\frac{b}{a}$ is positive; therefore b and a have like signs.
Hence, the required condition is that the signs of a and b should be like, and c and a have unlike signs.

DETERMINING THE CHARACTER OF ROOTS

● **PROBLEM** 523

Find the discriminant of $3x^2 - 7x + 5 = 0$. Then solve.

<u>Solution</u>: Recall that the discriminant is $b^2 - 4ac$, which appears under the radical in the quadratic formula

$$x = \frac{-b \pm \sqrt{b^2 - 4ac}}{2a} \quad ,$$

applying to equations in the form $ax^2 + bx + c = 0$. In our case, $a = 3$, $b = -7$, $c = 5$ and the discriminant is

$$b^2 - 4ac = (-7)^2 - (4 \cdot 3 \cdot 5) = 49 - 60 = -11.$$

A negative discriminant means there is a negative under the radical, which results in imaginary roots. Hence, $3x^2 - 7x + 5 = 0$ has two imaginary roots. To find these roots, we use the quadratic formula:

$$x = \frac{-b \pm \sqrt{b^2 - 4ac}}{2a} = \frac{7 \pm \sqrt{-11}}{6} \quad .$$

Since $\sqrt{ab} = \sqrt{a}\,\sqrt{b}$, $\sqrt{-11} = \sqrt{-1(11)} = \sqrt{(-1)}\sqrt{11}$. By definition $i = \sqrt{(-1)}$, thus $\sqrt{-11} = i\sqrt{11}$ and the roots are

$$\frac{7 \pm i\sqrt{11}}{6} \quad .$$

● **PROBLEM** 524

Compute the value of the discriminant and then determine the nature of the roots of each of the following four equations:

$$4x^2 - 12x + 9 = 0,$$

$$3x^2 - 7x - 6 = 0$$

$$5x^2 + 2x - 9 = 0$$

and $x^2 + 3x + 5 = 0.$

<u>Solution:</u> The discriminant, the term of the quadratic formula which appears under the radical, is $b^2 - 4ac$. It can be used to determine the nature of the roots of equations in the form $ax^2 + bx + c = 0$. Assuming a,b,c are real numbers, then,

(1) if $b^2 - 4ac > 0$, the roots are real and unequal

(2) if $b^2 - 4ac = 0$, the roots are real and equal

(3) if $b^2 - 4ac < 0$, the roots are imaginary

Assuming a,b,c are real and rational numbers then,

(1) if $b^2 - 4ac$ is a perfect square $\neq 0$, the roots are real, rational and unequal,

(2) if $b^2 - 4ac = 0$, the roots are real, rational, and equal,

(3) if $b^2 - 4ac > 0$, but not a perfect square, the roots are real, irrational and unequal,

(4) if $b^2 - 4ac < 0$, the roots are imaginary.

(a) $4x^2 - 12x + 9 = 0$

Here a,b,c are rational numbers,

   a = 4, b = -12 and c = 9.

Therefore,

$$b^2 - 4ac = (-12)^2 - 4(4)(9) = 144 - 144 = 0$$

Since the discriminant is 0, the roots are rational and equal.

(b) $3x^2 - 7x - 6 = 0$

Here a,b,c are rational numbers,

   a = 3, b = -7, and c = -6.

Therefore,

$$b^2 - 4ac = (-7)^2 - 4(3)(-6) = 49 + 72 = 121 = 11^2$$

Since the discriminant is a perfect square, the roots are rational and unequal.

(c) $5x^2 + 2x - 9 = 0$

Here a,b,c are rational numbers,

   a = 5, b = 2 and c = -9

Therefore,

$$b^2 - 4ac = 2^2 - 4(5)(-9) = 4 + 180 = 184$$

Since the discriminant is greater than zero, but not a perfect square, the roots are irrational and unequal.

(d) $x^2 + 3x + 5 = 0$

Here a,b,c are rational numbers,

352

$$a = 1, \; b = 3, \; \text{and} \; c = 5$$

Therefore,

$$b^2 - 4ac = 3^2 - 4(1)(5) = 9 - 20 = -11$$

Since the discriminant is negative the roots are imaginary.

● PROBLEM 525

Discuss the nature of the roots of

(a) $3x^2 - 7x + 3 = 0$ (b) $5x^2 + 3x + 1 = 0$

Solution: Equations (a) and (b) are of the form,

$$ax^2 + bx + c = 0.$$

In (a): $a = 3$, $b = -7$, $c = 3$

In (b): $a = 5$, $b = 3$, $c = 1$.

First we find the value of the discriminant, $b^2 - 4ac$, in each case.

If $b^2 - 4ac > 0$ the roots are real and unequal.

If $b^2 - 4ac = 0$ the roots are real and equal.

If $b^2 - 4ac < 0$ the roots are imaginary.

(a) $b^2 - 4ac = (-7)^2 - (4 \cdot 3 \cdot 3)$

$$= 49 - 36$$

$$= 13.$$

Thus, the roots are real and unequal.

(b) $b^2 - 4ac = (3)^2 - (4 \cdot 5 \cdot 1)$

$$= 9 - 20$$

$$= -11.$$

Thus, there are no real roots, i.e., the roots are imaginary.

● PROBLEM 526

Determine the character of the roots of the equation $2x^2 - x + 5 = 0$.

Solution: The given equation is a quadratic equation where $a = 2$, $b = -1$, and $c = 5$. The discriminant of a quadratic equation is: $b^2 - 4ac$. Therefore, the discriminant of the equation is $1 - 40 = -39$.

By the quadratic formula $x = \dfrac{-b \pm \sqrt{b^2 - 4ac}}{2a}$,

the roots of the given equation are:

$$x = \frac{-(-1) \pm \sqrt{-39}}{2(2)} = \frac{1 \pm \sqrt{-39}}{4}$$

$$= \frac{1}{4} \pm \frac{\sqrt{-39}}{4} = \frac{1}{4} \pm \frac{\sqrt{39}}{4}\sqrt{-1}$$

Therefore, $x = \dfrac{1}{4} \pm \dfrac{\sqrt{39}}{4}i$. Considering that the

discriminant is less than zero, and the roots obtained for the given equation, the roots are conjugate complex numbers.

● **PROBLEM 527**

Determine the character of the roots of the equation $4x^2 - 12x + 9 = 0$.

Solution: The given equation is a quadratic equation where $a = 4$, $b = -12$, and $c = 9$. The discriminant of this equation,

$$b^2 - 4ac, \text{ is } 144 - 144 = 0.$$

By the quadratic formula, $x = \dfrac{-b \pm \sqrt{b^2 - 4ac}}{2a}$,

the only root of the given equation is:

$$x = \frac{-(-12) \pm \sqrt{0}}{2(4)} = \frac{12}{8} = \frac{3}{2}.$$

Since the discriminant is equal to zero and by the root obtained for the given equation, there is only one real rational root.

● **PROBLEM 528**

Determine the character of the roots of the equation $x^2 - 5x + 6 = 0$.

Solution: The given equation is a quadratic equation where $a = 1$, $b = -5$, and $c = 6$. The discriminant of this equation, $b^2 - 4ac$, is $25 - 24 = 1$.

By the quadratic formula, $x = \dfrac{-b \pm \sqrt{b^2 - 4ac}}{2a}$, the roots of the given equation are:

$$x = \frac{-(-5) \pm \sqrt{1}}{2(1)} = \frac{5 \pm 1}{2}.$$

Therefore, $x = \dfrac{5 + 1}{2} = 3$ and $x = \dfrac{5 - 1}{2} = 2.$

Hence, the roots of the given equation are real, unequal, and rational. [Note: The roots are rational since $x = 3 = \dfrac{3}{1}$ and $x = 2 = \dfrac{2}{1}$].

● **PROBLEM** 529

Without solving the equation

$$2x^2 - 3x + 5 = 0,$$

determine the nature of its roots.

Solution: To determine the nature of the roots of a quadratic equation we look at the discriminant $b^2 - 4ac$ (this is the term that appears under the radical in the quadratic formula $x = \dfrac{-b \pm \sqrt{b^2 - 4ac}}{2a}$), which is used for equations in the form of $ax^2 + bx + c = 0$. In our example $a = 2$, $b = -3$, $c = 5$. Thus, the discriminant, $b^2 - 4ac = 9 - 4(2)(5) = -31 < 0$.

Our discriminant is negative. This means that we have a negative number appearing under the radical in our quadratic formula, which indicates that the roots are imaginary.

● **PROBLEM** 530

Compute the value of the discriminant and determine the nature of the roots in each of the following three equations:
$4x^2 - 4\sqrt{5}x + 5 = 0$, $\sqrt{3}x^2 - 6x + \sqrt{12} = 0$, and $\sqrt{2}x^2 + 3x + \sqrt{5} = 0$.

Solution: The discriminant, the term of the quadratic formula which appears under the radical, is $b^2 - 4ac$. It can be used to determine the nature of the roots of equations in the form $ax^2 + bx + c = 0$.
Assuming a, b, c are real numbers, then

    (1) if $b^2 - 4ac > 0$, the roots are real and unequal
    (2) if $b^2 - 4ac = 0$, the roots are real and equal
    (3) if $b^2 - 4ac < 0$, the roots are imaginary.

Assuming a, b, c are real and rational numbers, then

(1) if $b^2 - 4ac$ is a perfect square $\neq 0$, the roots are real, rational and unequal.

(2) if $b^2 - 4ac = 0$, the roots are real, rational and equal

(3) if $b^2 - 4ac > 0$, but not a perfect square, the roots are real, irrational and unequal,

(4) if $b^2 - 4ac < 0$ the roots are imaginary.

(a) $4x^2 - 4\sqrt{5}x + 5 = 0$. Here a, b, c are real, but not all rational, with a = 4, b = $-4\sqrt{5}$, and c = 5. Therefore

$$b^2 - 4ac = (-4\sqrt{5})^2 - 4(4)(5) = \left(4^2\right)\left(\sqrt{5}\right)^2 - 80 = (16 \cdot 5) - 80$$

$$= 80 - 80 = 0.$$

Since the discriminant equals zero, the roots are real and equal.

(b) $\sqrt{3}x^2 - 6x + \sqrt{12} = 0$. Here a, b, c are real, but not all rational, with a = $\sqrt{3}$, b = -6, and c = $\sqrt{12}$. Therefore

$$b^2 - 4ac = (-6)^2 - 4\sqrt{3}\sqrt{12} = 36 - 4\sqrt{3 \cdot 12} = 36 - 4\sqrt{36}$$

$$= 36 - 4(6) = 36 - 24 = 12.$$

Since the discriminant is greater than zero, the roots are real and unequal.

(c) $\sqrt{2}x^2 + 3x + \sqrt{5} = 0$. Here a, b, c are real, but not all rational, with a = $\sqrt{2}$, b = 3, and c = $\sqrt{5}$. Therefore

$$b^2 - 4ac = (3)^2 - 4(\sqrt{2})(\sqrt{5}) = 9 - 4\sqrt{2 \cdot 5} = 9 - 4\sqrt{10}$$

$$\approx 9 - 4(3 \ 2) = 9 - 12.8 < 0.$$

Since the discriminant is less than zero, the roots are imaginary.

● **PROBLEM** 531

Can the expression $16x^2 - 76x + 21$ be factored into rational factors?

<u>Solution:</u> To determine if this quadratic polynomial has rational factors we look at its discriminant, $b^2 - 4ac$ (this is the term that appears under the radical in the quadratic formula: $x = \dfrac{-b \pm \sqrt{b^2 - 4ac}}{2a}$, used for equations in the form of $ax^2 + bx + c = 0$).

In our example, a = 16, b = - 76, c = 21 and our discriminant $b^2 - 4ac = (-76)^2 - 4 \cdot 16 \cdot 21 = $

$$5776 - 1344 = 4432.$$

Now, recall what the discriminant tells us about the nature of the roots:

If the discriminant (b^2-4ac) is positive or zero, roots are real

If the discriminant (b^2-4ac) is negative
        roots are complex
If the discriminant (b^2-4ac) is a perfect square
        roots are rational
If the discriminant (b^2-4ac) is zero
        roots are equal and rational.

Hence, roots are rational only if the discriminant is zero or a perfect square.

Looking at the column of perfect squares in a table of square roots, we note that 4,432 is not a perfect square, hence the expression $16x^2 - 76x + 21$ cannot be factored into rational factors.

● **PROBLEM** 532

Show that the graph of $y = -x^2 + x - 1$ has no real zeros.

Solution: The real zeros of $y = -x^2 + x - 1$ are those real values of x for which the curve crosses the x-axis, i.e., satisfying:

$$0 = -x^2 + x - 1. \qquad (1)$$

The discriminant, $b^2 - 4ac$, of Equation (1) is equal to $(-1)^2 - 4(-1)(-1) = 1 - 4 = -3$. Since the discriminant is negative, Equation (1) has no real roots, and the given function has no real zeros.

● **PROBLEM** 533

Show that the equation $2x^2 - 6x + 7 = 0$ cannot be satisfied by any real values of x.

Solution: We want to show that the values of x, that is the roots, are not real. The discriminant, $b^2 - 4ac$, of an equation of the form $ax^2 + bx + c = 0$, is useful in describing the nature of the roots. If the discriminant is negative, the roots are not real. They are complex in the form $a \pm bi$. Here

$$a = 2, \ b = -6, \ c = 7; \text{ so that}$$

$$b^2 - 4ac = (-6)^2 - 4(2)(7) = -20.$$

Therefore the roots are imaginary.
To show that this is true, we apply the quadratic formula.

$$x = \frac{-b \pm \sqrt{b^2 - 4ac}}{2a}$$

$$x = \frac{-(-6) \pm \sqrt{(-6)^2 - 4(2)(7)}}{2(2)} = \frac{6 \pm \sqrt{36 - 56}}{4}$$

$$x = \frac{6 \pm \sqrt{-20}}{4} = \frac{6 \pm \sqrt{4}\sqrt{5}\sqrt{-1}}{4} = \frac{6 \pm 2i\sqrt{5}}{4}$$

$$x = \frac{3 \pm i\sqrt{5}}{2} ,$$

$$x = \left\{ \frac{3 + i\sqrt{5}}{2}, \ \frac{3 - i\sqrt{5}}{2} \right\}$$

● **PROBLEM 534**

If the equation $x^2 + 2(k+2)x + 9k = 0$ has equal roots, find k.

<u>Solution:</u> The given equation is a quadratic equation of the form $ax^2 + bx + c = 0$. In the given equation, $a = 1$, $b = 2(k+2)$, and $c = 9k$. A quadratic equation has equal roots if the discriminant, $b^2 - 4ac$, is zero.

$$b^2 - 4ac = [2(k+2)]^2 - 4(1)(9k) = 0$$
$$4(k+2)^2 - 36k = 0$$
$$4(k+2)(k+2) - 36k = 0$$
$$4(k^2 + 4k + 4) - 36k = 0$$

Distributing, $\qquad 4k^2 + 16k + 16 - 36k = 0$
$$4k^2 - 20k + 16 = 0 .$$

Divide both sides of this equation by 4:

$$\frac{4k^2 - 20k + 16}{4} = \frac{0}{4}$$

or

$$k^2 - 5k + 4 = 0 .$$

Factoring the left side of this equation into a product of two polynomials:

$$(k-4)(k-1) = 0.$$

When the product $ab = 0$, where a and b are any two numbers, either $a = 0$ or $b = 0$. Hence, in the case of this problem, either

$$k - 4 = 0 \quad \text{or} \quad k - 1 = 0.$$

Therefore, $\qquad\qquad k = 4 \quad \text{or} \quad k = 1.$

CHAPTER 18

SOLVING QUADRATIC INEQUALITIES

> **Basic Attacks and Strategies for Solving Problems in this Chapter. See pages 359 to 373 for step-by-step solutions to problems.**

A quadratic inequality is a mathematical statement that can be put in one of the forms

$$ax^2 + bx + c > 0,$$

$$ax^2 + bx + c < 0,$$

$$ax^2 + bx + c \geq 0, \text{ or}$$

$$ax^2 + bx + c \leq 0,$$

where $a \neq 0$. One general method for solving quadratic inequalities is by factoring, if indeed

$$ax^2 + bx + c$$

is factorable. The procedure is as follows:

(1) Arrange the inequality such that only 0 is on one side.

(2) Factor $ax^2 + bx + c$.

(3) Use the factors from Step 2 and the inequalities that satisfy the cases below and find a solution set for each.

Case I: The product of the factors is positive.

Case II: The product of the factors is negative.

(4) Obtain a complete solution set of the original inequality by combining all of the solutions found in the two cases into one set. This is done by taking the union of the solution sets of the two cases.

Graphing each of the various solution sets on a number line is useful in finding the complete solution set of the original quadratic inequality.

In the event that

$$ax^2 + bx + c$$

in a quadratic inequality is not factorable, then completing the square using one of the quadratic formulas is an appropriate solution procedure. Once the roots for the corresponding equation

$$ax^2 + bx + c = 0$$

have been determined, then in order to find the complete solution set we must determine the region into which the roots divide the x-axis and satisfy the original inequality.

Step-by-Step Solutions to
Problems in this Chapter,
"Solving Quadratic Inequalities"

● **PROBLEM** 535

Solve the inequality $(2x - 1)(x + 2) < 0$.

<u>Solution:</u> Since the two factors must be of opposite sign for their product to be negative, we have the two tentative possibilities:

$$2x - 1 < 0, \quad x + 2 > 0,$$

or $\quad 2x - 1 > 0, \quad x + 2 < 0.$

Solving the first pair of inequalities:

$$2x - 1 < 0 \qquad \text{and} \qquad x + 2 > 0$$

add 1 to both sides: | subtract 2 from both sides:

$$2x < 1 \qquad \qquad |$$

divide both sides by 2:|

$$x < \tfrac{1}{2} \qquad \qquad \text{and} \qquad x > -2$$

Thus, the first pair implies that $x < \tfrac{1}{2}$ and $x > -2$,
or $\qquad -2 < x < \tfrac{1}{2}$; the graph is as follows:

Solving the second pair of inequalities:

$$2x - 1 > 0 \qquad \qquad \text{and} \quad x + 2 < 0$$

Adding 1 to both sides: |

$$2x > 1 \qquad \qquad \qquad | \quad \text{Subtracting 2 from}$$
$$\qquad \qquad \qquad \qquad \qquad \text{both sides:}$$

Dividing both sides by 2:|

$$x > \tfrac{1}{2} \qquad \qquad \text{and} \quad x < -2$$

Thus, the second pair implies that $x > \tfrac{1}{2}$ and
$x < -2$; the graph is as follows:

Since there is no x such that $x > \frac{1}{2}$ and $x < -2$ we reject this solution.

The complete solution is thus the solution to the first pair of inequalities, $\{x : -2 < x < \frac{1}{2}\}$.

● **PROBLEM 536**

Solve the quadratic inequality $3x^2 - 13x - 10 > 0$.

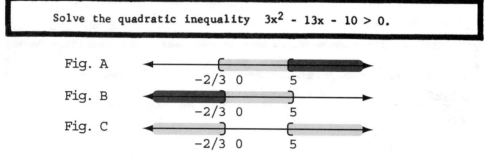

Fig. A   -2/3   0   5

Fig. B   -2/3   0   5

Fig. C   -2/3   0   5

Solution: This statement may be written in factored form as

$$(3x + 2)(x - 5) > 0$$

Hence we know that either both factors are positive or both are negative, since the product of a positive factor and a negative factor would be less than zero. This leads us to the following cases:

Case (1) $3x + 2 > 0$ and $x - 5 > 0$

                $3x > -2$ and $x > 5$

                $x > -\frac{2}{3}$ and $x > 5$

On the number line this part of the solution set may be represented as the intersection of the intervals $(-\frac{2}{3}, \infty)$ and $(5, \infty)$, as pictured in Fig. A. Hence $(-\frac{2}{3}, \infty) \cap (5, \infty) = (5, \infty)$, which is represented by the double-shaded region in diagram A.

Case (2) $3x + 2 < 0$ and $x - 5 < 0$

                $x < -\frac{2}{3}$ and $x < 5$

From Fig. B it can be seen that this appears as

$$(-\infty, -\tfrac{2}{3}) \cap (-\infty, 5) = (-\infty, -\tfrac{2}{3})$$

Finally, since Case 1 and Case 2 represent the disjunction of two propositions, we conclude that the solution set of our inequality is the union of the sets identified in these two cases. That is,

$$(5, \infty) \cup (-\infty, -\tfrac{2}{3})$$

is the solution set of the inequality, as pictured in Fig. C.

● **PROBLEM 537**

Obtain the solution set of $2x^2 > x + 6$.

Solution:

                         $2x^2 > x + 6$ given

               $2x^2 - x - 6 > 0$ adding $-x$ -6 to each member

          $(2x + 3)(x - 2) > 0$ factoring the left member

360

Now the product of the two factors on the left is greater than zero, or positive, if both factors are positive or if both are negative. Hence we seek the simultaneous solution set of the two inequalities

$$2x + 3 > 0$$

and

$$x - 2 > 0$$

and also the simultaneous set of

$$2x + 3 < 0$$

and

$$x - 2 < 0$$

The first two inequalities form Case I.

$$2x + 3 > 0 \qquad\qquad x - 2 > 0$$
$$2x > -3$$
$$x > \frac{-3}{2} \qquad \text{and} \qquad x > 2$$

Therefore, in order for $x > -3/2$ and $x > 2$, x must be greater than two. That is, $x > -3/2 \cap x > 2 = x > 2$.

Case I $\quad x > -\frac{3}{2}$

Case II concerns the second pair of inequalities.

$$2x + 3 < 0 \qquad \text{and} \qquad x - 2 < 0$$
$$2x < -3$$
$$x < -\frac{3}{2} \qquad\qquad x < 2$$

The total solution for Case II is $x < -\frac{3}{2} \cap x < 2 = x < -\frac{3}{2}$

Case II $\quad x < -\frac{3}{2}$

The solution set for $2x^2 > x + 6$ is the union of Case I and Case II; that is; $\{x \mid x > 2\} \cup \left\{x \mid x < -\frac{3}{2}\right\}$

● **PROBLEM 538**

Find the set $S = \left\{x \mid x^2 + 2x - 8 < 0\right\}$.

<u>Solution:</u> To get the required set, we find the solution set of

$$x^2 + 2x - 8 < 0$$

by the following procedure:

$(x + 4)(x - 2) < 0$ factoring left member of given
                        inequality

In order for a product to be negative (less than zero), one factor must be positive and the other must be negative. Thus there are two cases to be considered here:

Case I $x + 4 > 0$

                                $x - 2 < 0$

                                or

Case II $x + 4 < 0$

                                $x - 2 > 0$

The solution to Case I is $x > -4$ and $x < 2$. We can see this from diagram (A). Note that the solution includes all those numbers between -4 and 2, but not the endpoints themselves.

(A)

$-4 < x < 2$

For the second case, the solutions are $x < -4$ and $x > 2$. See the accompanying number line ((B)).

(B)

$x < -4$ $x > 2$

However, we can see from diagram (B) that x cannot at the same time be less than -4 and greater than 2. Hence, there is no solution for this case. That is $x < -4 \cap x > 2 = \emptyset$

Therefore, the solution set is $S = \{x \mid -4 < x < 2\}$.

● **PROBLEM** 539

Find the solution set of $(x + 1)/(x - 2) > 0$.

(A)

$x > 2$

$x > -1$

(B)

$x < 2$

$x < -1$

(C)

Solution: The fraction $(x + 1)/(x - 2)$ is positive, i.e., greater than zero, if the numerator and denominator are both positive or both negative. Hence, we seek the set of numbers that satisfies the two inequalities

$$x + 1 > 0 \qquad\qquad (1)$$
$$x - 2 > 0 \qquad\qquad (2)$$

simultaneously, and also those which satisfy

$$x + 1 < 0 \qquad\qquad (3)$$
$$x - 2 < 0 \qquad\qquad (4)$$

simultaneously.

Thus, we have two cases to consider in order to find the solution set of $(x + 1)/(x - 2) > 0$.

Case I

$x + 1 > 0 \qquad$ and $\qquad x - 2 > 0$
$x > -1 \qquad\quad$ and $\qquad x > 2$

We can see the solution from diagram (A).
Hence, for Case I the solution set is the solution

of $x > -1 \cap x > 2$, $\{x > 2\}$.

Case II

$x + 1 < 0 \qquad$ and $\qquad x - 2 < 0$

$x < -1 \qquad\quad$ and $\qquad x < 2$

Thus the solution is the solution to $x < -1 \cap x < 2$ or

$\{x \mid x < -1\}$. See (B).
Note that none of these solutions include the endpoints.

The solution set of $(x + 1)/(x - 2) > 0$ is thus the union of Case I and Case II; that is $\{x \mid x > 2\} \cup \{x \mid < -1\}$ (see diagram (C)).

● **PROBLEM 540**

Solve $x^2 - 5x + 4 \leq 0$.

Solution: We factor $x^2 - 5x + 4$ and have
$$(x - 1)(x - 4) \leq 0$$

Since the product is negative, one of the factors is positive and the other is negative. Therefore, either

$$x - 1 \geq 0 \quad \text{and} \quad x - 4 \leq 0 \qquad\qquad \text{Case 1}$$

or $\qquad$ $x - 1 \leq 0$ and $x - 4 \geq 0$ $\qquad$ Case 2

Solving for x in each inequality, we obtain

$\qquad$ $x \geq 1$ $\qquad$ and $x \leq 4$ $\qquad$ Case 1

or $\qquad$ $x \leq 1$ $\qquad$ and $x \geq 4$ $\qquad$ Case 2

The solution to the equation is the set of all x, such that $x \geq 1$ and $x \leq 4$, or

$\qquad$ $x \leq 1$ $\qquad$ and $x \geq 4$.

● **PROBLEM** 541

Find the solution set of $2x^2 - 3x - 5 > 0$.

<u>Solution:</u> Factoring, we have

$$(2x - 5)(x + 1) > 0.$$

If the product of two factors is positive, both factors must be positive or both factors must be negative. We must consider two cases.

$\qquad$ Case 1 $\qquad$ Case 2

$(2x - 5 > 0$ and $x + 1 > 0)$ or $(2x - 5 < 0$ and $x + 1 < 0)$

$\left(x > \dfrac{5}{2} \text{ and } x > -1\right)$ or $\left(x < \dfrac{5}{2} \text{ and } x < -1\right)$

$\{x: \; x > \dfrac{5}{2}\}$ $\qquad$ $\cup$ $\qquad$ $\{x: \; x < -1\}$

Since $\dfrac{5}{2} > -1$, then $x > \dfrac{5}{2}$ implies $x > -1$. Therefore the solution set of Case 1 is $\left\{x: \; x > \dfrac{5}{2}\right\}$. Similarly, $-1 < \dfrac{5}{2}$ so that $x < -1$ implies $x < \dfrac{5}{2}$. The solution set of Case 2 is $\{x: x < -1\}$. The solution set is the union of the solution sets of Case 1 and Case 2.

● **PROBLEM** 542

Solve the inequality $x^2 - x - 2 \leq 0$.

<u>Solution:</u> Factoring the left side of the given inequality,

$$(x - 2)(x + 1) \leq 0.$$

If the product of two numbers is negative, one of the numbers is positive and the other is negative. Hence, there are two cases:

Case 1: $x - 2 \geq 0, \; x + 1 \leq 0$

Solving these two inequalities,

$\qquad$ $x \geq 2, \quad x \leq -1$

Graph these new inequalities on number line (A).

(A)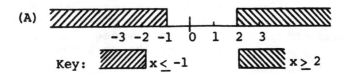

Key: ▨ x ≤ -1   ▧ x ≥ 2

Note that there is no value of x which satisfies both inequalities at the same time since these two inequalities do not intersect anywhere on the number line (A).

Thus x ≤ -1 ∩ x ≥ 2 = ∅

Case 2: x - 2 ≤ 0, x + 1 ≥ 0.

Solving these two inequalities,

x ≤ 2, x ≥ -1

Graph these inequalities on number line (B).

(B)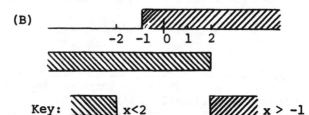

Key: ▧ x≤2   ▨ x ≥ -1

The interval of x which satisfies both inequalities at the same time is -1 ≤ x ≤ 2. Note that the two inequalities intersect in this interval on number line (B), that is

x ≥ -1 ∩ x ≤ 2 = -1 ≤ x ≤ 2.

Hence, the solution to the inequality $x^2 - x - 2 ≤ 0$ is the set:

$$\{x| \; -1 ≤ x ≤ 2\}$$

● PROBLEM 543

Find the solution set of $x^2 - 5x + 4 ≤ 0$.

Solution: First we find the solution for the equality.

$$x^2 - 5x + 4 = 0$$

$$(x - 4)(x - 1) = 0$$

$$x = 4 \text{ or } x = 1$$

For the product (x - 4)(x - 1) to be less than zero, either of the two expressions must be less than zero, but not both simultaneously. (x - 4)(x - 1) will equal zero only when

x = 4 or x = 1 so that the endpoints 1 and 4 must be included in the solution set. (x - 4)(x - 1) is less than
zero when x is greater than one but less than four. The
solution set is the union of the set whose elements are
the included endpoints and the set whose elements are values
of x between 1 and 4. Therefore the solution is the closed
interval $\{x:\ 1 \leq x \leq 4\}$.

Solve the inequality $x^2 > 4$.

<u>Solution:</u> The given relation may be replaced by an equivalent one,
by subtracting 4 from both sides of the inequality. Thus,

$$x^2 - 4 > 0 \qquad\qquad (1)$$

Now $x^2 - 4$ may be factored into two linear expressions, and we
have

$$(x + 2)(x - 2) > 0 \qquad\qquad (2)$$

This expression is not true, when either of the two factors is zero;
that is, when

$$(x + 2) = 0 \quad or \quad (x - 2) = 0,$$

or when

$$x = -2 \qquad or \qquad x = 2.$$

Hence we have the product of two factors, one of which vanishes for
x = -2 and the other for x = 2. These are therefore the critical
values: as x increases from values less than -2 to values greater
than -2, the expression x + 2 changes sign, from negative to positive; likewise, x - 2 changes from negative to positive as x passes through the critical value 2.

We now make use of the fact that the product of two quantities of like
sign (both positive or both negative) is positive, whereas the product
of two quantities of unlike sign is negative. Hence x must be such
that either both factors in (2) are positive or both are negative.
This yields the desired ranges

$$x > 2 \quad or \quad x < -2 \qquad\qquad (3)$$

It will often be found helpful to plot the critical values on a line
representation of the real numbers, and then to consider in turn those
values of x less than the left most critical point, those between
each adjacent pair of critical points, and finally those greater than
the rightmost critical point. It is possible also to plot the function
y = f(x) and thus determine the values of x for which f(x) > 0.
Thus, if we plot $y = x^2 - 4$, we find that the graph lies above the
X-axis when either of the inequalities (3) is obeyed.

Solve the inequality $2x^2 + 3x + 2 < 0$.

<u>Solution:</u> Divide the left side of the given inequality by 2:

$$\frac{2x^2 + 3x + 2}{2} < \frac{0}{2}$$

$$x^2 + \frac{3}{2}x + 1 < 0 \qquad\qquad (1)$$

To factor the left side of inequality (1), complete the square in x. This is done by taking half the coefficient of the x term and squaring this value. The result is then added to and subtracted from both sides of the inequality. Then,

$$\left[\frac{1}{2}\left(\frac{3}{2}\right)\right]^2 = \left[\frac{3}{4}\right]^2 = \frac{3}{4} \cdot \frac{3}{4} = \frac{9}{16}$$

$$x^2 + \frac{3}{2}x + 1 + \frac{9}{16} - \frac{9}{16} < 0 + \frac{9}{16} - \frac{9}{16}$$

or $\quad \left(x^2 + \frac{3}{2}x + \frac{9}{16}\right) + 1 - \frac{9}{16} < 0 + \frac{9}{16} - \frac{9}{16}$

$$\left(x + \frac{3}{4}\right)^2 + 1 - \frac{9}{16} < 0 + \frac{9}{16} - \frac{9}{16}$$

$$\left(x + \frac{3}{4}\right)^2 + \frac{7}{16} < 0 + \frac{9}{16} - \frac{9}{16}.$$

Subtract $\frac{7}{16}$ from both sides of this inequality:

$$\left(x + \frac{3}{4}\right)^2 + \frac{7}{16} - \frac{7}{16} < 0 + \frac{9}{16} - \frac{9}{16} - \frac{7}{16}$$

$$\left(x + \frac{3}{4}\right)^2 < -\frac{7}{16} \qquad (2)$$

(Note that the constant term in the squared polynomial, namely $\frac{3}{4}$, is just half the coefficient of the x-term in inequality (1)).

In reference to inequality (2), the square of any real number is always greater than or equal to zero; that is, the square of any real number cannot be negative. Therefore, there is no solution to inequality (2) and hence there is no solution to the given inequality.

● **PROBLEM** 546

Solve the inequalities
$$x^2 - 6x + 4 > 0 \quad \text{and} \quad x^2 - 6x + 4 < 0 .$$

Solution: The function which we are considering here is $x^2 - 6x + 4$; that is, $f(x) = x^2 - 6x + 4$. We must find where this function is positive or greater than zero and where it is negative or less than zero. We set $f(x) = 0$ and find the roots of this equation, $x^2 - 6x + 4 = 0$. Apply the quadratic formula. In this case $a = 1$, $b = -6$, and $c = 4$.

$$x = \frac{-(-6) \pm \sqrt{(-6)^2 - 4(1)(4)}}{2(1)} = \frac{6 \pm \sqrt{36 - 16}}{2} = \frac{6 \pm \sqrt{20}}{2}$$

$$= \frac{6 \pm \sqrt{4 \cdot 5}}{2} = \frac{6 \pm 2\sqrt{5}}{2} = 3 \pm \sqrt{5}$$

Thus, the roots are:

$$x_1 = 3 + \sqrt{5} \approx 3 + 2.2 = 5.2$$
$$x_2 = 3 - \sqrt{5} \approx 3 - 2.2 = 0.8$$

Mark the roots on the x-axis and consider the regions into which the roots divide the x-axis (see Figure). They are $x < 0.8$, $0.8 < x < 5.2$, $x > 5.2$. For each of these regions choose a value of x and see if $f(x) < 0$ or if $f(x) > 0$ holds. For the first region, we select $x = 0$. Substitute this value into $f(x)$:

$$0^2 - 6(0) + 4 = 0 - 0 + 4 = 4 > 0$$

Therefore

$$f(x) > 0 \quad \text{for} \quad x < 0.8$$

or more precisely

$$x < 3 - \sqrt{5} .$$

For the second region, $0.8 < x < 5.2$, we choose $x = 3$.

$$(3)^2 - 6(3) + 4 = 9 - 18 + 4 = -9 + 4 = -5 < 0$$

Therefore

$$f(x) < 0 \quad \text{for} \quad 0.8 < x < 5.2 \quad \text{or more exactly}$$

$$3 - \sqrt{5} < x < 3 + \sqrt{5} .$$

For the third region $x > 5.2$, we try $x = 6$.

$$6^2 - 6(6) + 4 = 36 - 36 + 4 = 0 + 4 = 4 > 0$$

Therefore

$$f(x) > 0 \quad \text{when} \quad x > 5.2;$$

that is,

$$f(x) > 0 \quad \text{when} \quad x > 3 + \sqrt{5} .$$

In conclusion, recalling that $f(x) = x^2 - 6x + 4$, we have found that $x^2 - 6x + 4 > 0$ for $x < 3 - \sqrt{5}$ and for $x > 3 + \sqrt{5}$. Furthermore, $x^2 - 6x + 4 < 0$ for $3 - \sqrt{5} < x < 3 + \sqrt{5}$. Note that the function is zero at the points $3 - \sqrt{5}$ and $3 + \sqrt{5}$.

● **PROBLEM 547**

Solve the inequality $x^2 + 9x - 7 \leq 0$.

Fig. A

$$\frac{-9-\sqrt{109}}{2} \qquad -9 \qquad \frac{-9+\sqrt{109}}{2} \quad 0$$

Fig. B

$$\frac{-9-\sqrt{109}}{2} \qquad -9 \qquad \frac{-9+\sqrt{109}}{2}$$

Solution: We solve the given inequality by the method of completing the square. To complete the square, take one half of the coefficient of x and square it. Add this quantity to both sides of the inequality. Here it is $\frac{1}{2}(9) = 9/2$ and $(9/2)^2 = 81/4$

$$x^2 + 9x + \frac{81}{4} \leq 7 + \frac{81}{4} \tag{1}$$

Write this quadratic expression as a binomial squared.

$$\left(x + \frac{9}{2}\right)^2 \leq \frac{109}{4} \tag{2}$$

Subtracting $\frac{109}{4}$ from both sides and expressing it as $\left(\sqrt{\frac{109}{4}}\right)^2$ on the left side of equation (2)

$$\left(x + \frac{9}{2}\right)^2 - \left(\sqrt{\frac{109}{4}}\right)^2 \leq 0$$

The expression $a^2 - b^2$ can be factored into $(a - b)(a + b)$. Similarly, for this example:

$$\left[\left(x + \frac{9}{2}\right) - \frac{\sqrt{109}}{2}\right]\left[\left(x + \frac{9}{2}\right) + \frac{\sqrt{109}}{2}\right] \leq 0$$

Hence we know that, since the product here is nonpositive, either

Case (1) $\left(x + \frac{9}{2}\right) - \frac{\sqrt{109}}{2} \geq 0$ and $\left(x + \frac{9}{2}\right) + \frac{\sqrt{109}}{2} \leq 0$

or

Case (2) $\left(x + \frac{9}{2}\right) - \frac{\sqrt{109}}{2} \leq 0$ and $\left(x + \frac{9}{2}\right) + \frac{\sqrt{109}}{2} \geq 0$

Case (1) $x + \frac{9}{2} - \frac{\sqrt{109}}{2} \geq 0$ and $x + \frac{9}{2} + \frac{\sqrt{109}}{2} \leq 0$

$$x \geq \frac{-9 + \sqrt{109}}{2} \quad \text{and} \quad x \leq \frac{-9 - \sqrt{109}}{2}$$

But this conjunction is logically false since no number can be larger than or equal to

$$\frac{-9 + \sqrt{109}}{2}$$

and at the same time be less than or equal to the smaller number

$$\frac{-9 - \sqrt{109}}{2}$$

(See Figure A). Thus, x cannot be a value in both sets at the same time. Therefore this case leads to the null or empty set.

Case (2) $x + \frac{9}{2} - \frac{\sqrt{109}}{2} \leq 0$ and $x + \frac{9}{2} + \frac{\sqrt{109}}{2} \geq 0$

$$x \leq \frac{-9 + \sqrt{109}}{2} \quad \text{and} \quad x \geq \frac{-9 - \sqrt{109}}{2}$$

Diagrammatically, the solution set is given in Figure B. Thus, the solution set for this inequality is

$$\left\{x \mid \frac{-9 - \sqrt{109}}{2} \leq x \leq \frac{-9 + \sqrt{109}}{2}\right\} = \left[\frac{-9 - \sqrt{109}}{2}, \frac{-9 + \sqrt{109}}{2}\right]$$

● **PROBLEM   548**

Solve the inequality

$$x - 6 > \frac{18 - 15x}{x^2 + 2x - 3} \quad .$$

Solution: We first subtract $\frac{18 - 15x}{x^2 + 2x - 3}$ from both sides of the inequality, obtaining

$$x - 6 - \frac{18 - 15x}{x^2 + 2x - 3} > 0.$$

In order to combine terms, we convert $x - 6$ into a fraction with $x^2 + 2x - 3$ as its denominator. Thus

$$\frac{(x^2 + 2x - 3)}{(x^2 + 2x - 3)} \cdot (x - 6) - \frac{18 - 15x}{x^2 + 2x - 3} > 0$$

Note that since $\frac{x^2 + 2x - 3}{x^2 + 2x - 3} = 1$, multiplication of

(x - 6) by this fraction does not alter the value of the inequality.

$$\frac{(x^2 + 2x - 3)(x - 6)}{x^2 + 2x - 3} - \frac{18 - 15x}{x^2 + 2x - 3} > 0$$

$$\frac{x^3 + 2x^2 - 3x - 6x^2 - 12x + 18}{x^2 + 2x - 3} - \frac{18 - 15x}{x^2 + 2x - 3} > 0$$

$$\frac{x^3 - 4x^2 - 15x + 18}{x^2 + 2x - 3} - \frac{18 - 15x}{x^2 + 2x - 3} > 0$$

$$\frac{x^3 - 4x^2 - 15x + 18 - 18 + 15x}{x^2 + 2x - 3} > 0$$

$$\frac{x^3 - 4x^2}{x^2 + 2x - 3} > 0$$

Now we factor numerator and denominator. Thus

$$\frac{x^2(x - 4)}{(x - 1)(x + 3)} > 0.$$

We now want all values of x which make $\frac{x^2(x - 4)}{(x - 1)(x + 3)}$

greater than zero. If (x - 1) = 0 or (x + 3) = 0 this fraction is undefined, thus we must place the restrictions

$$x - 1 \neq 0 \quad \text{and} \quad x + 3 \neq 0$$
or
$$x \neq 1 \quad \text{and} \quad x \neq -3.$$

Next we must eliminate all values of x which make

$\frac{x^2(x - 4)}{(x - 1)(x + 3)}$ equal to zero (for we only want it to

be greater than zero). The numerator will be zero if $x^2 = 0$ or x - 4 = 0, thus $x \neq 0$ and $x \neq 4$. We now have critical values x = - 3, x = 0, x = 1, x = 4.

We must test values of x in all ranges bordering on these critical values: (a) x < - 3, (b) - 3 < x < 0, (c) 0 < x < 1, (d) 1 < x < 4, (e) x > 4, to find the ranges in which the inequality holds:

(a) To test if the inequality holds for x < - 3, choose any value of x < - 3, we will use - 4, and replace x by this value in the given inequality:

$$\frac{x^2(x - 4)}{(x - 1)(x + 3)} > 0$$

$$\frac{(- 4)^2 (- 4 - 4)}{(- 4 - 1)(- 4 + 3)} > 0$$

$$\frac{16 (- 8)}{(- 5)(- 1)} > 0$$

$$\frac{-128}{5} > 0$$

Since a negative number is not greater than zero, the range $x < -3$ is not part of the solution.

(b) To test if the inequality holds for $-3 < x < 0$, choose a value of x between 0 and -3, we will use -1, and replace x by this value in the inequality:

$$\frac{x^2(x - 4)}{(x - 1)(x + 3)} > 0$$

$$\frac{(-1)^2(-1 - 4)}{(-1 - 1)(-1 + 3)} > 0$$

$$\frac{1(-5)}{(-2)(2)} > 0$$

$$\frac{-5}{-4} > 0$$

$$\frac{5}{4} > 0$$

Since 5/4 is indeed greater than zero, the range $-3 < x < 0$ is part of the solution.

(c) Testing if $0 < x < 1$ is part of the solution, we choose a value of x between 0 and 1, we will use $\frac{1}{2}$, and replace x by this value in the inequality:

$$\frac{x^2(x - 4)}{(x - 1)(x + 3)} > 0$$

$$\frac{(\frac{1}{2})^2(\frac{1}{2} - 4)}{(\frac{1}{2} - 1)(\frac{1}{2} + 3)} > 0$$

$$\frac{(1/4)(-3\frac{1}{2})}{(-\frac{1}{2})(3\frac{1}{2})} > 0$$

$$\frac{(1/4)(-7/2)}{(-\frac{1}{2})(7/2)} > 0$$

$$\frac{-7/8}{-7/4} > 0$$

$$-\frac{7}{8} \cdot -\frac{4}{7} > 0$$

$$\frac{1}{2} > 0$$

Since $\frac{1}{2}$ is indeed greater than zero, the range $0 < x < 1$ is part of the solution.

(d) Testing if $1 < x < 4$ is part of the solution, we choose a value of x between 1 and 4. We will use 2, and replace x by this value in the inequality:

$$\frac{x^2(x - 4)}{(x - 1)(x + 3)} > 0$$

$$\frac{(2)^2 (2 - 4)}{(2 - 1)(2 + 3)} > 0$$

$$\frac{4(-2)}{(1)(5)} > 0$$

$$\frac{-8}{5} > 0$$

Since a negative number is not greater than zero, the range $1 < x < 4$ is not part of the solution.

(e) Testing if $x > 4$ is part of the solution, we choose any value of x greater than 5, we will use 5, and replace x by this value in the inequality:

$$\frac{x^2 (x - 4)}{(x - 1)(x + 3)} > 0$$

$$\frac{(5)^2 (5 - 4)}{(5 - 1)(5 + 3)} > 0$$

$$\frac{25 (1)}{(4)(8)} > 0$$

$$\frac{25}{32} > 0$$

Since 25/32 is indeed greater than zero, the range $x > 4$ is part of the solution.

Thus, the permissible ranges for which the inequality $\quad x - 6 > \dfrac{18 - 15x}{x^2 + 2x - 3} \quad$ holds are

$-3 < x < 0, \qquad 0 < x < 1, \qquad x > 4.$

● **PROBLEM**   549

If   x   is a real quantity, prove that the expression $\dfrac{x^2 + 2x - 11}{2(x-3)}$ can have all numerical values except such as lie between   2   and   6.

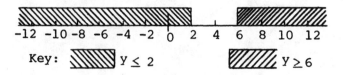

Key: ▧▧▧ $y \le 2$ ▨▨▨ $y \ge 6$

Solution: Let the given expression be represented by   y, so that

$$\frac{x^2 + 2x - 11}{2(x - 3)} = y \; ;$$

then cross-multiplying and transposing, we have

$$x^2 + 2x - 11 = 2y(x - 3)$$

$$x^2+2x-11 = 2xy - 6y$$
$$x^2+2x-11-2xy+6y = 0$$
$$x^2+2x-2xy+6y-11 = 0$$
$$x^2+2x(1-y)+6y-11 = 0,$$

or

$$x^2+2(1-y)x + (6y-11) = 0 \qquad (1)$$

Equation (1) is in the form $az^2 + bz + c = 0$, which is a quadratic equation. Hence, equation (1) is a quadratic equation, with $a = 1$, $b = 2(1-y)$, and $c = 6y-11$. In order that x may have real values, the discriminant, b^2-4ac, must be positive; that is, in order that x may have real values, $[2(1-y)]^2 - 4(1)(6y-11)$ must be positive; or $4(1-y)^2 - (24y-44)$ must be positive, i.e., $4(1-y)^2 - (24y-44) \geq 0$. Dividing by 4 and simplifying:

$$\frac{4(1-y)^2 - (24y-44)}{4} \geq \frac{0}{4}$$

$$\frac{4(1-y)^2 - 24y+44}{4} \geq 0$$

$$(1-y)^2 - 6y + 11 \geq 0$$

$$\left(1 - 2y + y^2\right)- 6y + 11 \geq 0$$

$$y^2 - 8y + 12 \geq 0 .$$

Factoring the left side of the inequality into a product of two polynomials:

$$(y - 6)(y - 2) \geq 0 .$$

Hence, the factors of this product must both be positive or both negative, since the entire product is positive.

Case 1: Both factors positive:

$$y - 6 \geq 0 \quad \text{and} \quad y - 2 \geq 0$$
$$y \geq 6 \quad \text{and} \quad y \geq 2$$

The two inequalities, $y \geq 6$ and $y \geq 2$, mean $y \geq 6 \cap y \geq 2$, and thus yield the single inequality $y \geq 6$, since this single inequality satisfies the two inequalities.

Case 2: Both factors negative:

$$y - 6 \leq 0 \quad \text{and} \quad y - 2 \leq 0$$
$$y \leq 6 \quad \text{and} \quad y \leq 2$$

The two inequalities, $y \leq 6$ and $y \leq 2$, mean $y \leq 6 \cap y \leq 2$, and thus yield the single inequality $y \leq 2$, since this single inequality satisfies the two inequalities. Therefore, y may have real values only when $y \geq 6$ (Case 1) and $y \leq 2$ (Case 2). The real values of y are indicated on the accompanying number line. Therefore, y cannot lie between 2 and 6, but y may have any other value.

CHAPTER 19

GRAPHING QUADRATIC EQUATIONS/ CONICS AND INEQUALITIES

> **Basic Attacks and Strategies for Solving Problems in this Chapter. See pages 374 to 403 for step-by-step solutions to problems.**

Parabolas, circles, ellipses, and hyperbolas are conic sections. Procedures for using formulas and obtaining graphical representation of each of these conics is given below.

PARABOLA

Equations of the standard form

$$y = ax^2 + bx + c,$$

where $a \neq 0$, have parabolas for graphs. The common method of graphing the solution set of any given quadratic equation is to first put the equation in standard form and then find a number of ordered pairs in a table that satisfy the equation and graph them.

The points associated with the intercepts and the vertex of a parabola are very useful in the procedure for graphing

$$y = ax^2 + bx + c.$$

The y-intercept is given by $y = c$ when $x = 0$ is substituted in the equation. The point associated with the intercept is $(0, c)$. The x-intercepts (if they exist) are found as a result of substituting $y = 0$ in the original equation and solving for x using the quadratic formula. The points associated with these intercepts are of the form $(x, 0)$ and $(-x, 0)$.

The vertex is the highest or lowest point on a parabola and it always occurs on the graph when the x-coordinate is $x = -b/2a$, which we use to solve for y in the equation for the parabola.

Using the points associated with the intercepts and the vertex one can sketch the graph of

$$y = ax^2 + bx + c,$$

the parabola. Note that if the coefficient of the x^2 term is positive, then the parabola is concave upward. On the other hand, if it is negative, then the parabola is concave downward.

Another way to locate the vertex of a parabola is by completing the square of the expression

$$ax^2 + bx,$$

where a must be 1 or -1 in the equation

$$y = ax^2 + bx + c.$$

For example, one can complete the square of

$$y = -x^2 - 6x + 1$$

as follows:

$$y = -(x^2 + 6x + 9) + 1 + 9 = -(x + 3)^2 + 10.$$

Hence, the x-component of the vertex is -3. The y-component is found by substituting $x = -3$ in the original equation and solve to obtain $y = 10$. The vertex is $(-3, 10)$. The parabola will be concave downward. If the equation has the form

$$x = ay^2 + b,$$

then the parabola opens to the right if a is positive and to the left if a is negative. If

$$f(x) = -f(x)$$

then the axis of symmetry is the x-axis which is a horizontal line through the vertex. Notice that the graph is not a function.

CIRCLE

The equation of a circle with center (a, b) and radius r is given by

$$(x - a)^2 + (y - b)^2 = r^2.$$

To graph the equation of a circle, the first step is to write the given equation in the above form such that the center and radius are easily determined. The next step is to plot the center and find and plot points that are r distance in horizontal and vertical positions from the center. Finally, connect the plotted points with a smooth curve to form a circle.

ELLIPSE

The ellipse is given by the equation

$$x^2/a^2 + y^2/b^2 = 1,$$

where $a \neq 0$, $b \neq 0$, in standard form. The ellipse will cross the x-axis at $(a, 0)$ and $(-a, 0)$ and cross the y-axis at $(0, b)$ and $(0, -b)$. If $a = b$ then the ellipse will be a circle. A reasonably accurate sketch of the graph of an ellipse can be obtained by plotting the above four points and connecting them with a smooth curve.

HYPERBOLA

The graph of the equation

$$x^2/a^2 - y^2/b^2 = 1$$

is the standard form of a hyperbola centered at the origin with a horizontal transverse axis. The graph will have x-intercepts at $-a$ and a. The equation of the form

$$y^2/a^2 - x^2/b^2 = 1$$

with a vertical transverse axis has a graph that is a hyperbola centered at the origin with y-intercepts at $-a$ and a. To draw a graph, we can plot the points associated with the intercepts and construct a rectangle whose sides pass through these points. An important aid in sketching either of the preceding equations is to draw the asymptotes of the graph. The asymptotes can be found by drawing a line through curves of the rectangle whose sides pass through $(-a, 0)$, $(a, 0)$, $(0, -b)$, and $(0, b)$ on the axes. These points are the vertices of hyperbola.

The graphical method for the solution set of the equation

$$ax^2 + bx + c > 0 \quad \text{or} \quad ax^2 + bx + c < 0$$

entails plotting the graph of

$$ax^2 + bx + c = 0$$

and determining what set of points satisfy the original inequality. Once this is determined, the graph is completed by shading the area that represents the solution set.

Step-by-Step Solutions to Problems in this Chapter, "Graphing Quadratic Equations/ Conics and Inequalities"

PARABOLAS

● **PROBLEM** 550

Draw the graphs of $f(x) = x^2$, $g(x) = 3x^2$, and also $h(x) = \frac{1}{2}x^2$ on one set of coordinate axes.

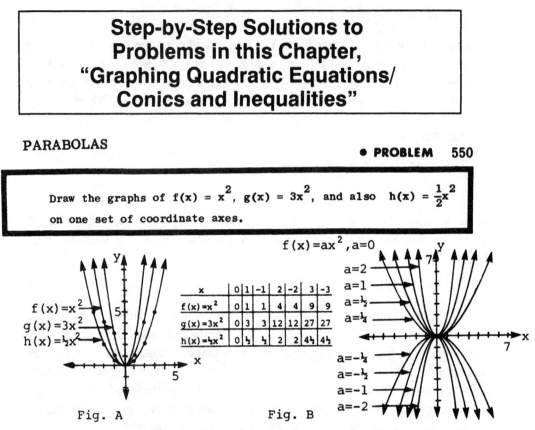

$$f(x) = ax^2, a = 0$$

x	0	1	-1	2	-2	3	-3
$f(x)=x^2$	0	1	1	4	4	9	9
$g(x)=3x^2$	0	3	3	12	12	27	27
$h(x)=\frac{1}{2}x^2$	0	$\frac{1}{2}$	$\frac{1}{2}$	2	2	$4\frac{1}{2}$	$4\frac{1}{2}$

Fig. A   Fig. B

Solution: We construct a composite table showing the values of each function corresponding to selected values for x.

In the example, we graphed three instances of the function $f(x) = ax^2$, a > 0. For different values of a, how do the graphs compare? (Fig. A). Assigning a given value to a has very little effect upon the main characteristics of the graph. The coefficient a serves as a "stretching factor" relative to the y-axis. As a increases, the two branches of the curve approach the y-axis. The curve becomes "thinner". As a decreases, the curve becomes "flatter" and approaches the x-axis.

The graph of $f(x)^2 = ax$, a ≠ 0, is called a parabola. (Fig. B). The point (0,0) is the vertex, or turning point, of the curve; the y-axis is the axis of symmetry. The value of a determines the shape of the curve. For a > 0 the parabola opens upward and for a < 0 the parabola opens downward.

● **PROBLEM** 551

Graph the function $3x^2 + 5x - 7$.

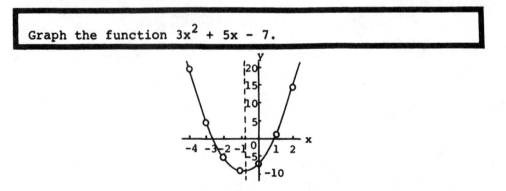

<u>Solution:</u> Let $y = 3x^2 + 5x - 7$. Substitute values for x and then find the corresponding values of y. This is done in the following table.

x	$y = 3x^2 + 5x - 7$
-4	21
-3	5
-2	-5
-1	-9
0	-7
1	1
2	15

These points are plotted and joined by a smooth curve in the figure.

● **PROBLEM** 552

Graph $y = 2x^2 - 5$.

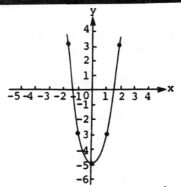

<u>Solution:</u> Graphs of the form $y = kx^2 + c$ are parabolas that stretch upward from a minimum point on the y-axis. From the table

x	- 2	- 1	0	1	2
y	3	- 3	- 5	- 3	3

we obtain the graph shown in the accompanying figure.

● **PROBLEM** 553

Sketch the graph of the function $\left\{ (x, x^2) \right\}$.

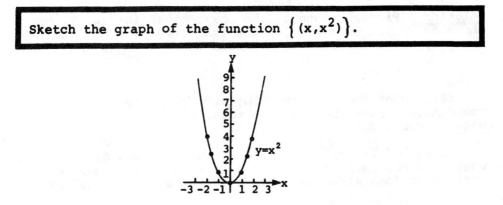

<u>Solution:</u> Substitute values of x and obtain the correspon-
ding values of $f(x) = x^2$. This was done in the following
table:

x	$f(x) = x^2$	$f(x)$
-3	$f(-3) = (-3)^2$ $= 9$	9
-2	$f(-2) = (-2)^2$ $= 4$	4
-1	$f(-1) = (-1)^2$ $= 1$	1
0	$f(0) = 0^2$ $= 0$	0
1	$f(1) = 1^2$ $= 1$	1
2	$f(2) = 2^2$ $= 4$	4
3	$f(3) = 3^2$ $= 9$	9

We plotted these points and then joined them to obtain the
part of the graph shown in the figure.

● **PROBLEM** 554

Find the graph of $y = x^2 - 3x + 2$.

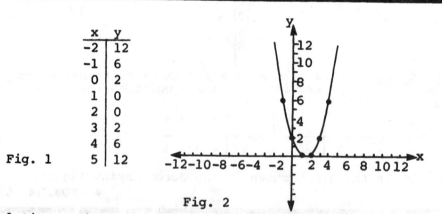

x	y
-2	12
-1	6
0	2
1	0
2	0
3	2
4	6
5	12

Fig. 1

Fig. 2

<u>Solution:</u> First, construct the chart given in Fig. 1, by
choosing values of x and finding the corresponding value
for y by substituting into the equation,

$$y = x^2 - 3x + 2.$$

Next, plot the corresponding ordered pairs on the
coordinate system. Having done this, connect the points
with a smooth curve as in Fig. 2.

If we had recognized the function as a parabola
whose general shape is that of the given curve, then we
could have done almost as well with three or four points.

Draw a graph of the set of ordered pairs which satisfy the function $f(x) = x^2 - 7$.

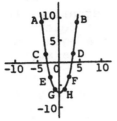

Solution: The following table lists a sufficient sequence of ordered pairs to determine the general nature of the curve:

x	0	1	2	3	4	-1	-2	-3	-4
y = f(x)	-7	-6	-3	2	9	-6	-3	2	9

Plotting these points, we obtain the curve illustrated in the figure. Note also that the function is given by a quadratic equation with the coefficient of the x^2-term positive. This implies that the graph will be a parabola opening upward. Since for this function, $f(x) = f(-x)$, the graph will be symmetric with respect to the y-axis. The point with x-coordinate 0 will be the vertex and minimum point of the parabola. The graph should confirm what the table suggests; the minimum point is (0,7). The range of the function is then limited to real numbers equal to or greater than -7. The domain of the parabola is the set of real numbers. The points A(-4,9) and B(4,9) in the figure are said to be symmetric with respect to the y-axis. Similarly, C and D, E and F, and G and H are symmetric with respect to the y-axis.

Construct the graph of the function defined by $y = x^2 - 6x + 10$.

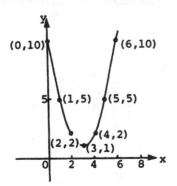

<u>Solution:</u> We are given the function $y = x^2 - 6x + 10$. The most general form of the quadratic function is $y = ax^2 + bx + c$ where a, b, and c are constants. If a is positive, the curve opens upward and it is U-shaped. If a is negative, the curve opens downward and it is inverted U-shaped.

Since $a = 1 > 0$ in the given equation, the graph is a parabola that opens upward. To determine the pairs of values (x,y) which satisfy this equation, we express the quadratic function in terms of the square of a linear function of x.

$$y = x^2 - 6x + 10 = x^2 - 6x + 9 + 1$$

$$= (x - 3)^2 + 1$$

y is least when $x - 3 = 0$ This is true because the square of any number, be it positive or negative, is a positive number. Therefore y would always be greater than or equal to one. Thus the minimum value of y is one when $x - 3 = 0$ or $x = 3$.

In order to plot the curve, we select values for x and calculate the corresponding y values. (See the table.)

x	$x^2 - 6x + 10 =$	y
0	$(0)^2 - 6(0) + 10$	10
1	$(1)^2 - 6(1) + 10$	5
2	$(2)^2 - 6(2) + 10$	2
3	$(3)^2 - 6(3) + 10$	1
4	$(4)^2 - 6(4) + 10$	2
5	$(5)^2 - 6(5) + 10$	5
6	$(6)^2 - 6(6) + 10$	10

The points and graphs determined by the table are shown in the accompanying figure.

● **PROBLEM** 557

Graph $x^2 + y - 2x = 0$.

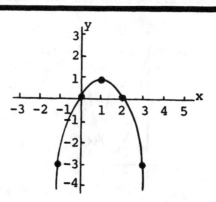

378

<u>Solution:</u> First we solve for y. Subtract x^2 from both sides of the given equation.

$$x^2 + y - 2x - x^2 = 0 - x^2$$
$$y - 2x = -x^2$$

Add 2x to both sides of this equation.

$$y - 2x + 2x = -x^2 + 2x$$
$$y = -x^2 + 2x$$

Then we construct the table by substituting values of x into this derived equation to find corresponding values of y,

x	-2	-1	0	1	2	3	4
y	-8	-3	0	1	0	-3	-8

from which we obtain the Figure.

● **PROBLEM**   558

Construct the graph of $\{(x,y) | y = \tfrac{1}{2}x^2 - 2x - 3\}$.

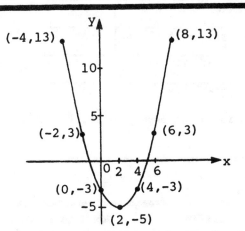

<u>Solution:</u> In general, the graph of the quadratic function $y = ax^2 + bx + c$ is a parabola. Some of its properties are:
   1. If a > 0, the graph opens upward.
   2. If a < 0, the graph opens downward.

We note that $a = \tfrac{1}{2}$ in the equation $y = \tfrac{1}{2}x^2 - 2x - 3$. Hence the graph is a parabola opening upward since a > 0. In order to determine the vertex (x_v, y_v) of the parabola, we complete the square by the following procedure.

Factor out $\tfrac{1}{2}$ from $y = \tfrac{1}{2}x^2 - 2x - 3$. Then $\tfrac{1}{2}x^2 - 2x - 3 = \tfrac{1}{2}(x^2 - 4x - 6)$ Complete the square of $x^2 - 4x$. This is done by taking $\tfrac{1}{2}$ of the coefficient of x and squaring it. Thus, the constant term is

$$[\tfrac{1}{2}(-4)]^2 = (-2)^2 = 4.$$

To keep the same equation we must retain -6. Thus, we express it as 4 - 10. Then, $\tfrac{1}{2}(x^2 - 4x - 6) = \tfrac{1}{2}(x^2 - 4x + 4 - 10)$.

Factor $(x^2 - 4x + 4)$ into $(x-2)(x-2) = (x-2)^2$.

Then, $\tfrac{1}{2}(x^2 - 4x + 4 - 10) = \tfrac{1}{2}\left[(x-2)^2 - 10\right]$.

Now, $y = \tfrac{1}{2}\left[(x-2)^2 - 10\right]$. Since the parabola opens upward, the vertex is the value where y is minimum. $(x-2)^2$ will always be positive since

379

the square of a positive or of a negative number is always positive. Therefore, the minimum value of y will occur when $(x-2) = 0$. Consequently y is least when $x-2 = 0$ or when $x = 2$. To find y, substitute $x = 2$ into $y = \frac{1}{2}\left[(x-2)^2 - 10\right]$.

$$y = \frac{1}{2}\left[(2 - 2)^2 - 10\right]$$
$$= \frac{1}{2}\left[0^2 - 10\right]$$
$$= \frac{1}{2}[-10]$$
$$= -5 .$$

Hence, the vertex is $(2,-5)$.

We now assign numbers to x and calculate the corresponding y-values to obtain points on the parabola. This is done in the following table:

x	$\frac{1}{2}x^2 - 2x - 3$	y
-4	$\frac{1}{2}(-4)^2 - 2(-4) - 3$	13
-2	$\frac{1}{2}(-2)^2 - 2(-2) - 3$	3
0	$\frac{1}{2}(0)^2 - 2(0) - 3$	-3
2	$\frac{1}{2}(2)^2 - 2(2) - 3$	-5
4	$\frac{1}{2}(4)^2 - 2(4) - 3$	-3
6	$\frac{1}{2}(6)^2 - 2(6) - 3$	3
8	$\frac{1}{2}(8)^2 - 2(8) - 3$	13

The graph is shown in the accompanying figure.

● **PROBLEM** 559

Show that the quadratic equation $y = 2x^2 - 20x + 25$ is the equation of a parabola.

Solution: If we can write $y = 2x^2 - 20x + 25$ in the standard form $y = a(x - h)^2 + k$, we will show that its graph is a parabola. The axis of symmetry will be the line $x = h$. The vertex will be at (h,k). Notice that the standard form has a term with a perfect square in it. We use a method known as completing the square to rewrite

$$y = 2x^2 - 20x + 25$$

(a) Subtract the constant term, 25, from both members.

$$y - 25 = 2x^2 - 20x$$

(b) Factor 2 from the right member.

$$y - 25 = 2\left(x^2 - 10x\right)$$

(c) We need to add 25 within the parentheses since 25 is the square of one-half the coefficient of the term in x. If we do, then we must add 50 to the left member to maintain equality, since we have in fact added two times the quantity added in the parentheses $(2 \times 25 = 50)$ to the right member.

$$y + 25 = 2\left(x^2 - 10x + 25\right)$$

(d) The expression within the parentheses is now a perfect square.

$$y + 25 = 2(x - 5)^2$$

(e) Next we subtract 25 from each member.

$$y = 2(x - 5)^2 - 25$$

This is now in standard form $y = a(x - h)^2 + k$. Here $a = 2$, $h = 5$, and $k = -25$. Thus, the graph of $y = 2x^2 - 20x + 25$ is a parabola. Its axis of symmetry is the line $x = 5$; its vertex is at $(5,-25)$.

● **PROBLEM** 560

Sketch the graph of the quadratic function $y = x^2 - 5x + 4$.

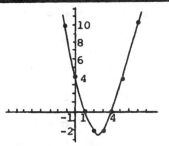

Solution: Since the coefficient of the x^2-term of the quadratic function is positive, its graph will be a parabola opening upward. Therefore the minimum point occurs at the vertex. The x-coordinate of the vertex is given by $-\dfrac{\text{coefficient of } x}{2 \cdot \text{coefficient of } x^2} = -\dfrac{(-5)}{2(1)} = \dfrac{5}{2}$. The y-coordinate is then $y = \left(\dfrac{5}{2}\right)^2 - 5\left(\dfrac{5}{2}\right) + 4 = -\dfrac{9}{4}$. The minimum point is $(5/2 , -9/4)$. The axis of symmetry is the vertical line through the point $(5/2, -9/4)$. The y-intercept, found by setting $x = 0$, is $y = 0^2 - 5(0) + 4 = 4$, i.e., $(0,4)$. The point symmetric to $(0,4)$ is a point with the same y-coordinate, 4, but 5/2 units to the right of the axis of symmetry, line $x = \dfrac{5}{2}$. The point is therefore $(5,4)$. The x-intercepts satisfy

$$0 = x^2 - 5x + 4$$

$$0 = (x - 4)(x - 1)$$

$$x - 4 = 0 \quad \big| \quad x - 1 = 0$$

$$x = 4 \quad \big| \quad x = 1$$

The x-intercepts are therefore $(1,0)$ and $(4,0)$. These points are also symmetric with respect to the axis of symmetry. Each point has y-coordinate 0 and is either $\dfrac{3}{2}$ units to the left or right of the axis line $x = \dfrac{5}{2}$.

● **PROBLEM** 561

The graph of the quadratic function $y = ax^2 + bx + c$ is a parabola. Find the equation of a parabola passing through the points $(-1, 11)$, $(1, 3)$, and $(2, 5)$, by determining the values of a, b, and c from the given data.

Solution: Each of the three points given lies on the parabola and therefore each one must satisfy the quadratic function for a parabola,

$$ax^2 + bx + c = y.$$

For each point we substitute the coordinates of x and y into the quadratic function,

For $(-1, 11)$ $a(-1)^2 + b(-1) + c = 11$

$$a \quad\quad - b \quad\quad + c = 11 \qquad\qquad (1)$$

For $(1, 3)$ $a(1)^2 + b(1) + c = 3$

$$a \quad\quad + b \quad\quad + c = 3 \qquad\qquad (2)$$

For $(2, 5)$ $a(2)^2 + b(2) + c = 5$

$$4a \quad\quad + 2b \quad\quad + c = 5 \qquad\qquad (3)$$

We have obtained a system of three equations with three unknowns.

$$a - b + c = 11 \qquad\qquad (1)$$

$$a + b + c = 3 \qquad\qquad (2)$$

$$4a + 2b + c = 5 \qquad\qquad (3)$$

We can eliminate b by by adding (1) and (2). We obtain a new equation in a and c.

$$a - b + c = 11 \qquad\qquad (1)$$

$$\underline{a + b + c = 3} \qquad\qquad (2)$$

$$2a \quad\quad + 2c = 14 \qquad\qquad (4)$$

We have one equation in two unknowns. We need another equation in a and c before we can solve for a or c.

Eliminate b from two other equations. Let us choose (2) and (3).

$$a + b + c = 3 \qquad\qquad (2)$$

$$4a + 2b + c = 5 \qquad\qquad (3)$$

Multiply (2) by -2 in order to eliminate the variable b. Then add (5) and (3)

$$-2a - 2b - 2c = -6 \qquad\qquad (5)$$

$$\underline{4a + 2b + \quad c = \quad 5} \qquad\qquad (3)$$

$$2a \quad\quad - c = -1 \qquad\qquad (6)$$

Now we have two equations (4) and (6) in two unknowns, a and c.

$$2a + 2c = 14 \qquad\qquad (4)$$

$$2a - c = -1 \qquad\qquad (6)$$

Subtract equation (6) from (4) to eliminate a.

$$2a + 2c = 14 \qquad (4)$$
$$- \quad \underline{2a - c = -1} \qquad (6)$$
$$3c = 15 \qquad (7)$$

Divide (7) by 3

$$c = 5$$

Substitute c into either (4) or (6) to find the value of a. Choose (4)

$$2a + 2c = 14 \qquad (4)$$
$$2a + 2(5) = 14$$
$$2a + 10 = 14$$
$$2a = 4 \qquad (8)$$

Divide (8) by 2

$$a = 2$$

To find b substitute a and c into any of the three original equations (1), (2), or (3). Let us choose (2).

$$a = 2; \ c = 5$$
$$a + b + c = 3 \qquad (2)$$
$$2 + b + 5 = 3$$
$$b + 7 = 3$$
$$b = -4$$

Therefore the solution of the original system is
$$a = 2$$
$$b = -4$$
$$c = 5$$

The equation of a parabola is
$$y = ax^2 + bx + c$$
For this particular parabola the equation is
$$y = 2x^2 - 4x + 5,$$

by substituting the values of a, b, and c.

Each of the three given points satisfy this equation.

Check

For $(-1, 11)$

$$11 = 2(-1)^2 - 4(-1) + 5$$
$$11 = 2 + 4 + 5$$
$$11 = 11$$

For (1, 3)

$$3 = 2(1)^2 - 4(1) + 5$$

$$3 = 2 - 4 + 5$$

$$3 = 3$$

For (2, 5)

$$5 = 2(2)^2 - 4(2) + 5$$

$$5 = 8 - 8 + 5$$

$$5 = 5$$

● **PROBLEM   562**

What is the minimum value of the expression $2x^2 - 20x + 17$?

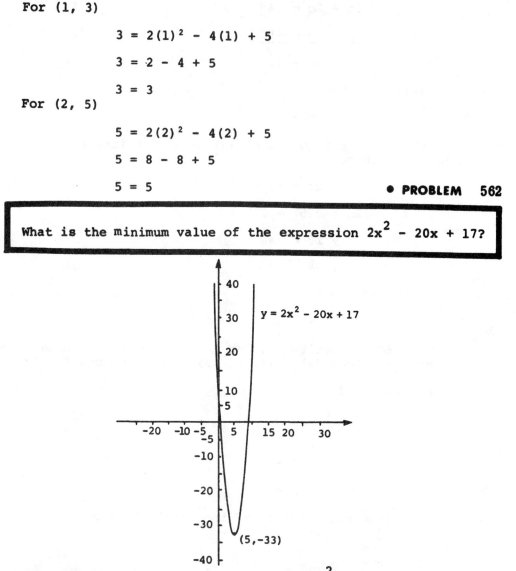

$$y = 2x^2 - 20x + 17$$

(5,-33)

Solution: Consider the function $y = 2x^2 - 20x + 17$. This function is defined by a second degree equation. The coefficient of its x^2 term is positive. Hence the curve is a parabola opening upward. Thus, the minimum point of this curve occurs at the vertex. The x-coordinate is equal to

$$- \frac{\text{coefficient of x term}}{2 \cdot \text{coefficient of } x^2 \text{ term}} = -\frac{b}{2a} = -\frac{(-20)}{2(2)} = \frac{20}{4} = 5.$$

For $x = 5$, $y = 2(5)^2 - 20(5) + 17 = -33$. Therefore the minimum value of the expression $2x^2 - 20x + 17$ for any value of x is -33. This minimum value is assumed only when $x = 5$.

● **PROBLEM   563**

Find the coordinates of the maximum point of the curve

384

$y = -3x^2 - 12x + 5$, and locate the axis of symmetry.

<u>Solution</u>: The curve is defined by a second degree equation.
The coefficient of the x^2 term is negative. Hence, the
graph of this curve is a parabola opening downward. The
maximum point of the curve occurs at the vertex and has the
x-coordinate:

$$- \frac{\text{coefficient of x term}}{2\left(\text{coefficient of x}^2 \text{ term}\right)} = - \frac{b}{2a} = - \frac{-12}{2(-3)} = \frac{12}{-6} = -2.$$

For $x = -2$, $y = -3(-2)^2 - 12(-2) + 5 = 17$. Hence the
coordinates of the vertex are $(-2,17)$. The curve is symmetric with respect to the vertical line through its vertex.
The axis of symmetry of this curve is the vertical line
through the point $(-2,17)$, i.e., the line $x = -2$.

● **PROBLEM**   564

Find the equation of the tangent to the parabola
$y = x^2 - 6x + 9$, if the slope of the tangent equals 2.

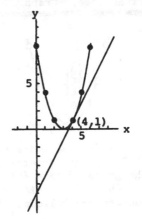

<u>Solution</u>: The equation of a straight line is $y = mx + k$
where m is the slope and k is the y-intercept. The
equation $y = 2x + k$ (1)

represents a family of parallel lines with slope 2, some
of which intersect the parabola in two points, others
which have no point of intersection with the parabola,
and just one which intersects the parabola in only one
point. The problem is to find the value of k so that the
graph of Equation 1 intersects the parabola in just one
point. If we solve the system

$$y = 2x + k \qquad\qquad (1)$$

$$y = x^2 - 6x + 9 \qquad\qquad (2)$$

by substitution, we get for the first step

$$2x + k = x^2 - 6x + 9 \qquad\text{or}$$

$$x^2 - 8x + 9 - k = 0 \qquad\qquad (3)$$

This is a quadratic equation of the form $ax^2 + bx + c = 0$. The discriminant determines the nature of the roots when $ax^2 + bx + c = 0$. The condition that Equation 3 has but one solution is that the discriminant, $b^2 - 4ac$ equals 0. Therefore, if $a = 1$, $b = -8$, $c = 9 - k$, then $b^2 - 4ac = 64 - 4(9 - k) = 0$ or $k = -7$.

Substituting this value of k in equation 1, we have $y = 2x - 7$ which is the equation of the tangent to the given parabola when the slope of the tangent is equal to 2. The figure is the graph of the parabola and the tangent. The student may verify that the point of contact is (4, 1). This is shown by substituting (4, 1) into

$$y = 2x - 7 = x^2 - 6x + 9$$

$$1 = 2(4) - 7 = 4^2 - 6(4) + 9$$

$$1 = \qquad\qquad 1 = 1$$

● **PROBLEM 565**

The surface S of a cube is given by the formula $S = 6x^2$, where x represents the length of an edge. Graph S as a function of x.

Solution: In the formula $S = 6x^2$, negative values of x may be used as well as positive ones. The points determined by these negative values of x belong to the graph of $S = 6x^2$, although they have no meaning in relation to the cube. The table of values from which the graph in the figure was constructed follows:

x	$S = 6x^2$	S
-2	$S = 6(-2)^2$ $= 6(4)$ $= 24$	24
$-\frac{3}{2}$	$S = 6\left(-\frac{3}{2}\right)^2$ $= 6\left(\frac{9}{4}\right)$ $= \frac{27}{2}$	$\frac{27}{2}$
-1	$S = 6(-1)^2$ $= 6(1)$ $= 6$	6

x	$S = 6x^2$	S
0	$S = 6(0)^2$ $= 6(0)$ $= 0$	0
1	$S = 6(1)^2$ $= 6(1)$ $= 6$	6
$\frac{3}{2}$	$S = 6\left(\frac{3}{2}\right)^2$ $= 6\left(\frac{9}{4}\right)$ $= \frac{27}{2}$	$\frac{27}{2}$
2	$S = 6(2)^2$ $= 6(4)$ $= 24$	24

Note that the axis of ordinates is labeled S. Moreover, units of different size are used on the two axes; this may be done if convenient, but should be avoided where possible.

● PROBLEM 566

Discuss the graph of the equation $y^2 = 12x$.

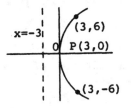

Solution: The equation written as $x = \frac{1}{12}y^2$ is a quadratic equation with the coefficient of the y^2 term positive. Therefore the graph is a parabola opening to the right. Since $f(x) = -f(x)$ the parabola is symmetric with respect to the x-axis. Point (0,0) satisfies the equation and lies on the axis of symmetry. Hence the vertex of the parabola is at (0,0), (see figure). The focus of the parabola lies on the axis of symmetry, $y = 0$, at the point (p,0) where $4p$ = coefficient of x in the original equation: $4p = 12$, $p = 3$. Therefore the focus is at (3,0). The directrix is the vertical line $x = -p = -3$. When $x = 3$, $y = 12x = 12(3) = +6$. Therefore the points (3,6) and (3,-6) are points on the graph. The graph of this parabola is not the graph of a function since for any given value of x there is more than one corresponding value of y.

● PROBLEM 567

Discuss the rational integral equation
$$x^2 - 2xy + y^2 + 2x - 3 = 0,$$
and plot its graph.

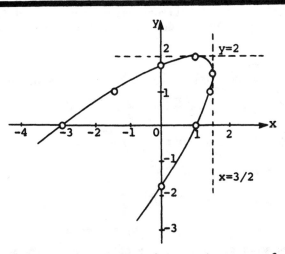

Solution: We solve this equation for y in terms of x.

387

$$x^2 - 2xy + y^2 + 2x - 3 = 0 \qquad (1)$$

Subtract x^2 from both sides of equation (1):

$$\cancel{x^2} - 2xy + y^2 + 2x - 3 - \cancel{x^2} = 0 - x^2$$

$$-2xy + y^2 + 2x - 3 = -x^2 \qquad (2)$$

Subtract $(2x-3)$ from both sides of equation (2):

$$-2xy + y^2 + 2x - 3 - (2x - 3) = -x^2 - (2x - 3)$$

$$-2xy + y^2 + \cancel{2x} - \cancel{3} - \cancel{2x} + \cancel{3} = -x^2 - 2x + 3$$

$$y^2 - 2xy = -x^2 - 2x + 3 \qquad (3)$$

To complete the square on the left side of equation (3), take half the coefficient of y and square it.

$$[\tfrac{1}{2}(-2x)]^2 = [-x]^2 = x^2 .$$

Add this value to both sides of equation (3):

$$y^2 - 2xy + x^2 = -\cancel{x^2} - 2x + 3 + \cancel{x^2}$$

$$y^2 - 2xy + x^2 = -2x + 3$$

$$y^2 - 2xy + x^2 = 3 - 2x$$

or

$$(y - x)^2 = 3 - 2x .$$

Take the square root of each side:

$$\sqrt{(y-x)^2} = \pm\sqrt{3 - 2x}$$

$$y - x = \pm\sqrt{3 - 2x}$$

Add x to both sides:

$$y - \cancel{x} + \cancel{x} = x \pm\sqrt{3 - 2x}$$

$$y = x \pm\sqrt{3 - 2x} \qquad (4)$$

This explicit relation shows immediately that: (a) for every real value of x less than $3/2$, there are two distinct real values of y; (b) for $x = 3/2$, there is only one y-value; namely, $y = 3/2$. (c) for $x > 3/2$, y is complex. Hence, we know that the graph of equation (1) will not extend to the right of the line $x = 3/2$. In addition, relation (4) enables us to compute the two values of y corresponding to each permissible value of x; thus, when $x = 0$, $y = \pm\sqrt{3}$, and when $x = 1$, we get $y = 2$ and $y = 0$. Substitute values for x and find the corresponding value of y. This is done in the following table:

x	$x \pm \sqrt{3-2x} = y$
-3	$-3 \pm \sqrt{3-2(-3)} = -3 \pm \sqrt{3+6} = -3 \pm \sqrt{9} = -3 \pm 3 = 0, -6$
$-\frac{3}{2}$	$-\frac{3}{2} \pm \sqrt{3-2\left(-\frac{3}{2}\right)} = -\frac{3}{2} \pm \sqrt{3+3} = -\frac{3}{2} \pm \sqrt{6} = -\frac{3}{2} \pm 2.45 = 0.95, -3.95$
0	$0 \pm \sqrt{3-2(0)} = 0 \pm \sqrt{3-0} = 0 \pm \sqrt{3} = \pm \sqrt{3} = \pm 1.73$
1	$1 \pm \sqrt{3-2(1)} = 1 \pm \sqrt{3-2} = 1 \pm \sqrt{1} = 1 \pm 1 = 2, 0$
$\frac{3}{2}$	$\frac{3}{2} \pm \sqrt{3-2\left(\frac{3}{2}\right)} = \frac{3}{2} \pm \sqrt{3-3} = \frac{3}{2} \pm \sqrt{0} = \frac{3}{2} \pm 0 = \frac{3}{2}$

These points may be plotted and joined by a smooth curve.

The graph obtained with the help of the above discussion is shown in the figure. By the methods of analytic geometry, it may be shown that

388

this curve is a parabola, and additional characteristics of the curve may be determined.

CIRCLES, ELLIPSES AND HYPERBOLAS

● **PROBLEM** 568

Discuss the graph of the equation $x^2 + y^2 = 25$.

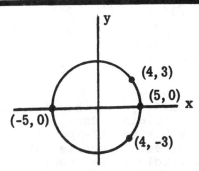

Solution: This is an equation of the form $x^2 + y^2 = r^2$, and therefore its graph is a circle with radius $r = 5$ and center at the origin (see figure). Note that the graph does not represent a function since, except for $x = -5$ or $x = 5$, each permissible value of x is associated with two values of y. For example, for $x = 4$, we have the ordered pairs (4,3) and (4,-3). The domain of this function is $\{x \mid -5 < x < 5\}$. The range of this function is $\{y \mid -5 \leq y \leq 5\}$.

● **PROBLEM** 569

Sketch the graph of the equation $x^2 + y^2 = 4$.

Solution: Substitute values for x and then find the corresponding values of y. This is done in the following table:

x	$y = \pm \sqrt{4 - x^2}$
-1	$\pm \sqrt{3} = \pm 1.73$
0	± 2
1	$\pm \sqrt{3} = \pm 1.73$
$\sqrt{2} = 1.41$	$\pm \sqrt{2} = \pm 1.41$

These points have been plotted and a smooth curve has been drawn through them in the figure. We could plot more points and then "fill in" the rest of the graph, but instead we will use some of our knowledge of the coordinate plane. Our given equation is equivalent to the equation $\sqrt{x^2 + y^2} = 2$. In the distance formula $d = \sqrt{(x_2 - x_1)^2 + (y_2 - y_1)^2}$, where d is the distance be-

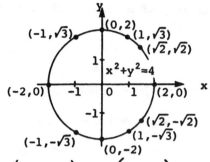

tween the points $\left(x_1, y_1\right)$ and $\left(x_2, y_2\right)$, let the origin or $(0,0) = \left(x_1, y_1\right)$ and let $(x,y) = \left(x_2, y_2\right)$. Therefore,

$$d = \sqrt{(x - 0)^2 + (y - 0)^2}$$
$$d = \sqrt{x^2 + y^2}.$$

Hence, the number $\sqrt{x^2 + y^2}$ is the distance between the point (x,y) and the origin. Thus, in words, our equation says, "The distance between the point (x,y) and the origin is 2." Clearly, the set of points that are two units from the origin is the circle whose center is the origin and whose radius is 2. This circle is therefore the graph of our given equation, and we have drawn it in the figure.

Most of the graphs of equations that we deal with in mathematics are "one-dimensional" figures, such as the circle we just drew. In these cases, we say that the graph is a curve. Not all simple looking equations in x and y, however, have graphs that are simple curves.

● PROBLEM 570

Graph the equation $2x^2 + 2y^2 - 13 = 0$.

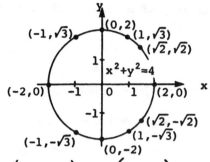

Solution: In order to verify that this is the equation of a circle, we put the equation in the standard form: add 13 to both sides of the given equation.

$$2x^2 + 2y^2 - 13 + 13 = 0 + 13$$
$$2x^2 + 2y^2 = 13$$

Divide both sides of this equation by 2.

$$\frac{2x^2 + 2y^2}{2} = \frac{13}{2}$$

or

$$x^2 + y^2 = \frac{13}{2}, \text{ which is the}$$

standard form for the equation of a circle with its center at the origin (0,0) and

$$\text{radius} = r = \sqrt{\frac{13}{2}} = \frac{\sqrt{13}}{\sqrt{2}} = \frac{\sqrt{13}}{\sqrt{2}} \frac{\sqrt{2}}{\sqrt{2}} = \frac{\sqrt{26}}{2} \ .$$

Therefore, the radius of the circle is approximately $\frac{5.1}{2}$ or 2.55. The graph is represented in the Figure.

● **PROBLEM** 571

Find the equation for the circle whose center is at the origin and whose radius is 3.

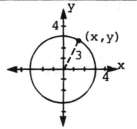

Solution: A circle is the set of all points in a plane at a given distance from a fixed point. The fixed point is called the center of the circle and the measure of the given distance is called the radius of the circle. Thus to find the equation for the circle whose center is at the origin and whose radius is 3, we wish to find the equation of all points at a distance of 3 from the origin, (0,0).

The distance formula for the distance between two points $\left(x_1, y_1 \right)$ and $\left(x_2, y_2 \right)$ is

$$d = \sqrt{\left(x_2 - x_1 \right)^2 + \left(y_2 - y_1 \right)^2} \ ,$$

In our case $d = 3, \left(x_1, y_1 \right) = (0,0)$. Thus

$$3 = \sqrt{(x - 0)^2 + (y - 0)^2}$$

$$3 = \sqrt{x^2 + y^2}$$

Squaring both sides, $x^2 + y^2 = 9$.

Hence the equation for the circle whose center is at the origin, with radius 3, is $x^2 + y^2 = 9$.

● **PROBLEM** 572

Discuss the graph of $\frac{x^2}{25} + \frac{y^2}{9} = 1.$

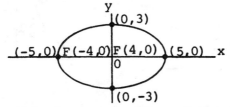

Solution: Since this is an equation of the form $\frac{x^2}{a^2} + \frac{y^2}{b^2} = 1$, with a = 5 and b = 3, it represents an ellipse. The simplest way to sketch the curve is to find its intercepts.

If we set x = 0, then

$$y = \sqrt{\left[1 - \frac{x^2}{25}\right]9} = \sqrt{\left[1 - \frac{0^2}{25}\right]9} = \pm\ 3$$

so that the y-intercepts are at (0,3) and (0,-3). Similarly, the x-intercepts are found for y = 0:

$$x = \sqrt{\left[1 - \frac{y^2}{9}\right]25}$$

$$= \sqrt{\left[1 - \frac{0^2}{9}\right]25}$$

$$= \pm 5$$

to be at (5,0) and (-5,0) (see figure). To locate the foci
we note that

$$c^2 = a^2 - b^2 = 5^2 - 3^2$$

$$c^2 = 25 - 9 = 16$$

$$c = \pm\ 4.$$

The foci lie on the major axis of the ellipse. In this
case it is the x-axis since a = 5 is greater than b = 3.
Therefore, the foci are (+c,0), that is, at (-4,0) and (4,0).
Therefore, the foci are at (-4,0) and (4,0). The sum of the
distances from any point on the curve to the foci is
2a = 2(5) = 10.   ● PROBLEM   573

Sketch the graph of the function $\{(x, 1/x^2)\}$.

Solution: Substitute different values of x and find the corresponding
value of y or $1/x^2$.

x	$y = 1/x^2$
-2	$\frac{1}{(-2)^2} = \frac{1}{4}$
-1	$\frac{1}{(-1)^2} = \frac{1}{1} = 1$
$-\frac{1}{2}$	$\frac{1}{(-\frac{1}{2})^2} = \frac{1}{\frac{1}{4}} = 4$
$\frac{1}{2}$	$\frac{1}{(\frac{1}{2})^2} = \frac{1}{\frac{1}{4}} = 4$
1	$\frac{1}{1^2} = \frac{1}{1} = 1$
2	$\frac{1}{2^2} = \frac{1}{4}$

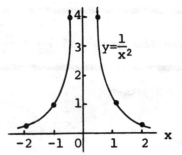

The graph is shown in the figure.

Draw the graph of $xy = 6$.

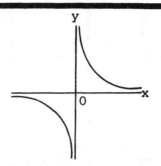

Solution: Since the product is positive the values of x and y must have the same sign, that is, when x is positive y must also be positive and when x is negative then y is also negative. Moreover, neither x nor y can be zero(or their product would be zero not 6), so that the graph never touches the coordinate axes. Solve for y and we obtain $y = 6/x$. Substituting values of x into this equation we construct the following chart:

x:	-6	-3	-2	-1	1	2	3	6
y:	-1	-2	-3	-6	6	3	2	1

The graph is obtained by plotting the above points and then joining them with a smooth curve, remembering that the curve can never cross a co-ordinate axis. The graph of the equation, $xy = k$, is a hyperbola for all nonzero real values of k. If k is negative, then x and y must have opposite signs, and the graph is in the second and fourth quadrants as opposed to the first and third.

Graph the equation $xy = -4$.

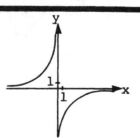

Solution: The graph of the equation, $xy = c$, is a hyperbola for all non-zero real values of c. In this case, $c = -4$ and thus the particular equation is $xy = -4$. Since the product is negative, the values of x and y must have different signs. If x is positive, then y is negative and so part of the graph lies in quadrant IV. On the other hand, if x is negative, then y is positive and the other part of the hyperbola is located in quadrant II. If we solve for y, then $y = -4/x$ and $x \neq 0$. Then the graph never touches the y-axis. Thus the line $x = 0$ is an asymptote of the graph. On the other hand, solving for x, we have $x = -4/y$ and $y \neq 0$. Thus the graph never crosses the x-axis and the line $y = 0$ is an asymptote. We now pre-pare a table of values by selecting values for x and finding the cor-responding values for y. See table and graph.

x	$\dfrac{-4}{x}$ =	y
-3	$\dfrac{-4}{-3}$	$\dfrac{4}{3} = 1\dfrac{1}{3}$
-2	$\dfrac{-4}{-2}$	2
-1	$\dfrac{-4}{-1}$	4
1	$\dfrac{-4}{1}$	-4
2	$\dfrac{-4}{2}$	-2
3	$\dfrac{-4}{3}$	$\dfrac{-4}{3} = -1\dfrac{1}{3}$

● **PROBLEM 576**

Sketch the graph of the equation y = 2/x.

<u>Solution</u>: Substitute values for x and then find the cor-
responding values for y. This is done in the following
table.

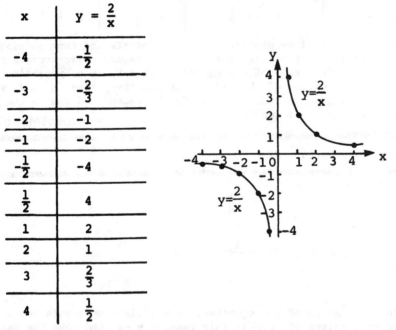

x	$y = \dfrac{2}{x}$
-4	$-\dfrac{1}{2}$
-3	$-\dfrac{2}{3}$
-2	-1
-1	-2
$-\dfrac{1}{2}$	-4
$\dfrac{1}{2}$	4
1	2
2	1
3	$\dfrac{2}{3}$
4	$\dfrac{1}{2}$

The graph is shown in the figure. This graph is an example of an equilateral hyperbola. Notice in the graph that,
when x takes on larger and larger positive values, y gets
closer and closer to 0. When x takes on larger and larger
negative values, y also gets closer and closer to 0. Also,
when x gets closer and closer to 0, y either takes on
larger and larger positive values or larger and larger
negative values. Note also that x cannot be 0, since $y = \dfrac{2}{x}$
$= \dfrac{2}{0}$ is not defined.

Discuss the graph of $\frac{x^2}{9} - \frac{y^2}{9} = 1$.

Solution: $\frac{x^2}{9} - \frac{y^2}{9} = 1$ is an equation of the form

$\frac{x^2}{a^2} - \frac{y^2}{b^2} = 1$ with a = 3 and b = 3. Therefore the graph is

a hyperbola. The x-intercepts are found by setting y = 0:

$$\frac{x^2}{9} - \frac{0^2}{9} = 1$$

$$x^2 = 9$$

$$x = \pm 3.$$

Thus, the x-intercepts are at (-3,0) and (3,0). There are no y-intercepts since for x = 0 there are no real values of y satisfying the equation, i.e., no real value of y satisfies

$$\frac{0^2}{9} - \frac{y^2}{9} = 1$$

$$y^2 = -9, \quad y = \sqrt{-9}.$$

Solving the original equation for y:

$$y = \sqrt{\left(1 - \frac{x^2}{9}\right)(-9)} \quad \text{or} \quad y = \sqrt{x^2 - 9}$$

shows that there will be no permissible values of x in the interval -3 < x < 3. Such values of x do not yield real values for y. For x = 5 and x = -5 use the equation for y to obtain the ordered pairs (5,4), (5,-4), (-5,4), and (-5,-4) as indicated in the figure. The foci of the hyperbola are located at ($\pm$c,0), where

$$c^2 = a^2 + b^2$$

$$c^2 = 3^2 + 3^2 = 9 + 9 = 18$$

$$c = \pm\sqrt{18} = \pm 3\sqrt{2}.$$

Therefore, the foci are at $(-3\sqrt{2}, 0)$ and $(3\sqrt{2}, 0)$.

Discuss the graph of the function $y = \dfrac{12}{x^2}$.

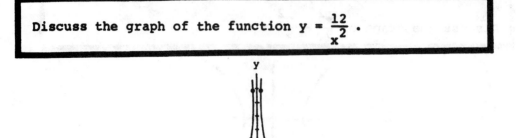

Solution: Intercepts: Since division by zero is not defined, x cannot = 0, hence there is no y-intercept. Similarly, there is no x-intercept since y cannot = 0, because
no value of x allowed in the given equation yields a value
of y = 0. Symmetry: The curve is symmetric with respect
to the y-axis since the x-term appears squared in the given
function and hence $f(x) = f(-x)$. Domain: There is no
limitation on x, except that $x \neq 0$. Range: Since

$y = \dfrac{12}{x^2}$ is a positive number divided by a positive number,

y must be positive. Therefore, the curve exists only in
the first and second quadrants. Plotting: We note that,
in the first quadrant, as x increases y decreases. Several
points to illustrate this are listed in the following table.

x	0.5	1	2	3	...	10
y	48	12	3	1.3	...	0.12

After plotting these points, and tracing the curve in the
first quadrant, the second branch is drawn in quadrant II,
using the principle of symmetry. The curve is illustrated
in the figure.

Graph the equation $x^2 - 2y^2 = 4$.

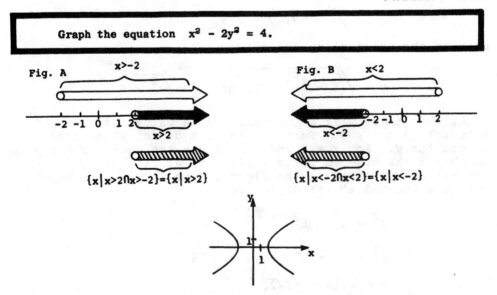

<u>Solution:</u> First, we shall determine the intercepts. Set $x = 0$, and find $-2y^2 = 4$, $y^2 = -2$, $y = \pm i\sqrt{2}$. This means that the curve does not meet the y-axis. Set $y = 0$, and find $x = \pm 2$. Thus, the graph only crosses the x-axis. To avoid fractions and negative radicands, we solve <u>for x</u> in terms of y rather than for y in terms of x: $x = \pm\sqrt{2y^2 + 4}$. Assigning values to y and computing corresponding values of x, we get the accompanying table.

y	$\pm\sqrt{2y^2 + 4}$	x
-3	$\pm\sqrt{2(-3)^2 + 4}$	$\pm\sqrt{22} = \pm 4.7$
-2	$\pm\sqrt{2(-2)^2 + 4}$	$\pm\sqrt{12} = \pm 3.5$
-1	$\pm\sqrt{2(-1)^2 + 4}$	$\pm\sqrt{6} = \pm 2.4$
0	$\pm\sqrt{2(0)^2 + 4}$	$\pm\sqrt{4} = \pm 2$
1	$\pm\sqrt{2(1)^2 + 4}$	$\pm\sqrt{6} = \pm 2.4$
2	$\pm\sqrt{2(2)^2 + 4}$	$\pm\sqrt{12} = \pm 3.5$
3	$\pm\sqrt{2(3)^2 + 4}$	$\pm\sqrt{22} = \pm 4.7$

Now plot the points. Note for each value of y, there are two values of x. For example, if $y = -3$, $x = + 4.7$ and $x = -4.7$. Thus the two points are $(4.7, -3)$ and $(-4.7, -3)$. The curve of $x^2 - 2y^2 = 4$ is called a hyperbola and consists of two parts, or branches. See Figure. If we solve the equation for y, we obtain:

$$x^2 - 2y^2 = 4$$
$$- 2y^2 = 4 - x^2$$
$$y^2 = \frac{4 - x^2}{-2}$$
$$= \frac{-4 + x^2}{2} = \frac{x^2 - 4}{2}$$
$$y = \pm\sqrt{\frac{x^2 - 4}{2}}$$

Now to obtain real values for y, the radicand cannot be negative. Therefore, $x^2 - 4 \geq 0$, and factoring $(x - 2)(x + 2) \geq 0$. There are two statements $(x - 2)(x + 2) = 0$ and $(x - 2)(x + 2) > 0$ to be considered. The roots are $x = 2$ and $x = -2$ when the factors are set equal to zero.
If we consider the inequality $(x - 2)(x + 2) > 0$, the two factors must be both positive or both negative.

Case I: both are positive

$$x - 2 > 0 \quad \text{and} \quad x + 2 > 0$$
$$x > 2 \quad \text{and} \quad x > -2$$

Therefore, $x > 2$ (see number line A)

Case II: both are negative

$$x - 2 < 0 \quad \text{and} \quad x + 2 < 0$$
$$x < 2 \quad \text{and} \quad x < -2$$

Therefore, $x < -2$ (see number line B)

$$\{x \mid x < -2 \cap x < 2\} = \{x \mid x < -2\}$$

To summarize $x = -2$ and $x = 2$, and $x < -2$ or $x > 2$. More simply, the domain of x is $\{x \mid x \leq -2$ or $x \geq 2\}$. From

$$y = \pm \sqrt{\frac{x^2 - 4}{2}},$$

the range is the set of all real numbers.

● **PROBLEM** 580

Find the inverse function of f if $y = f(x) = 2\sqrt{9 - x^2}$ and f has domain $\{x \mid -3 \leq x \leq 0\}$ and range $\{y \mid 0 \leq y \leq 6\}$.

Solution: The equation $y = f(x)$ determines the number y in the range of f, $\{y \mid 0 \leq y \leq 6\}$, from a given number x of the domain, $\{x \mid -3 \leq x \leq 0\}$. Now we want to know if this equation also determines x when y is given. If a function f is given by a simple formula $y = f(x)$, we can often obtain f^{-1} by solving this for x so that $x = f^{-1}(y)$. Therefore,

$$y = 2\sqrt{9 - x^2} = 2\left(9 - x^2\right)^{\frac{1}{2}}$$

Squaring both sides,

$$y^2 = 2^2\left[\left(9 - x^2\right)^{\frac{1}{2}}\right]^2 = 4\left(9 - x^2\right)$$

Distributing,

$$y^2 = 36 - 4x^2$$

$$y^2 - 36 = -4x^2$$

$$x^2 = \frac{y^2 - 36}{-4} = \frac{36 - y^2}{4}$$

$$x = \frac{\pm\sqrt{36 - y^2}}{2}$$

In order to obtain a function of y, we must have for each value of y one and only one value of x. Thus, we choose only one sign. Now the domain of the inverse function is the range of f, and the range of the inverse function is the domain of f. Therefore, the domain of $f^{-1}(y)$ is to be $0 \leq y \leq 6$ and the range is $-3 \leq x \leq 0$. Now since x can assume negative values then we must consider the negative values of $\pm \sqrt{36 - y^2}$. Hence, the inverse function is

$$x = \frac{-\sqrt{36 - y^2}}{2}.$$

Plot each function by selecting values in the domain and finding the corresponding values in the range.

For $y = 2\sqrt{9 - x^2}$

Domain $\{x \mid -3 \leq x \leq 0\}$

x	$2\sqrt{9 - x^2}$	y
-3	$2\sqrt{9 - (-3)^2}$	0
-2	$2\sqrt{9 - (-2)^2}$	$2\sqrt{5}$
-1	$2\sqrt{9 - (-1)^2}$	$4\sqrt{2}$
0	$2\sqrt{9 - (0)^2}$	6

For $x = -\sqrt{36 - y^2}/2$

Domain $\{y \mid 0 \leq y \leq 6\}$

y	$-\sqrt{36 - y^2}/2$	x
0	$-\sqrt{36 - 0^2}/2$	-3
1	$-\sqrt{36 - 1^2}/2$	$-\sqrt{35}/2$
2	$-\sqrt{36 - 2^2}/2$	$-2\sqrt{2}$
3	$-\sqrt{36 - 3^2}/2$	$-3\sqrt{3}/2$
4	$-\sqrt{36 - 4^2}/2$	$-\sqrt{5}$
5	$-\sqrt{36 - 5^2}/2$	$-\sqrt{11}/2$
6	$-\sqrt{36 - 6^2}/2$	0

See graph

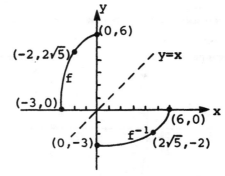

The graphs of f and f^{-1} show that if (a,b) is a point of the graph of f, then (b,a) is a point of the graph of f^{-1}. Furthermore, this is true by the definition of an inverse function which states that, given the function f such that no two of its ordered pairs have the same second element, the inverse function f^{-1} is the set of ordered pairs obtained from f by interchanging in each ordered pair the first and second elements. The graphs of f and f^{-1} are symmetric with respect to the line y = x; that is, each is the image of the other in the mirror y = x.

INEQUALITIES

● PROBLEM 581

Find the solution set of $x^2 - 6x + 10 > 0$ by the graphical method.

Solution: First we graph the function $y = x^2 - 6x + 10$. Assign values to x and then calculate y-values.

x	$x^2 - 6x + 10$	y
- 3	$(-3)^2 - 6(-3) + 10$	37
- 2	$(-2)^2 - 6(-2) + 10$	26
- 1	$(-1)^2 - 6(-1) + 10$	17
0	$(0)^2 - 6(0) + 10$	10
1	$(1)^2 - 6(1) + 10$	5
2	$(2)^2 - 6(2) + 10$	2
3	$(3)^2 - 6(3) + 10$	1
4	$(4)^2 - 6(4) + 10$	2

See graph. The curve is the graph of $y = x^2 - 6x + 10$. Since the graph is entirely above the X axis, the

solution set of $x^2 - 6x + 10 > 0$ is the set of all real numbers.

Find the solution set of $-x^2 - 4x - 5 > 0$.

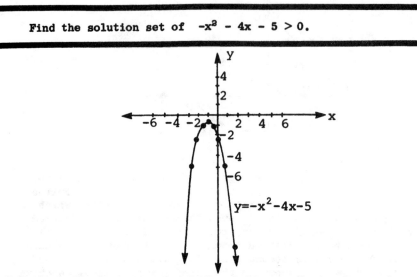

$$y = -x^2 - 4x - 5$$

Solution: To find the graphical solution, select values of x and find the corresponding y values. See the table. Note that $y = -x^2 - 4x - 5 = -[x^2 + 4x + 5]$.

x	$-[x^2 + 4x + 5] =$	y
-3	$-\left[(-3)^2 + 4(-3) + 5\right] = -(9 - 12 + 5) =$	-2
-2	$-\left[(-2)^2 + 4(-2) + 5\right] = -(4 - 8 + 5) =$	-1
-1	$-\left[(-1)^2 + 4(-1) + 5\right] = -(1 - 4 + 5) =$	-2
0	$-\left[(0)^2 + 4(0) + 5\right] = -(0 + 0 + 5) =$	-5
1	$-\left[(1)^2 + 4(1) + 5\right] = -(1 + 4 + 5) =$	-10
2	$-\left[(2)^2 + 4(2) + 5\right] = -(4 + 8 + 5) =$	-17
3	$-\left[(3)^2 + 4(3) + 5\right] = -(9 + 12 + 5) =$	-26

The graph of $y = -x^2 - 4x - 5$, as shown in the figure, lies entirely below the x-axis. Consequently

$$\{x \mid -x^2 - 4x - 5 > 0\} = \emptyset, \text{ the empty set.}$$

Solve the inequality $-2x^2 - 15x + 27 > 0$.

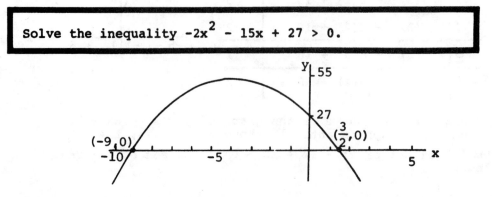

<u>Solution:</u> The graph of the corresponding quadratic function $y = -2x^2 - 15x + 27$ is a parabola which is concave down since the coefficient of the x^2-term is negative. Its zeros are the solution set of the quadratic equation $-2x^2 - 15x + 27 = 0$.

$(2x - 3)(-x - 9) = 0$ factor

$2x - 3 = 0 \ \Big| \ -x - 9 = 0$ set each factor = 0 to find all x's which make the product 0.

$\qquad x = \dfrac{3}{2} \ \Big| \qquad x = -9$

Therefore, the solution set of the given inequality is the set of all x's such that the graph of $-2x^2 - 15x + 27$ lies above the x-axis, i.e., $\{x \mid y = -2x^2 - 15 + 27 > 0\}$. This set is $\{x \mid -9 < x < \dfrac{3}{2}\}$. (see Figure).

● **PROBLEM 584**

Solve: $x^2 < 2x + 2$.

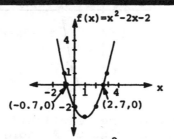

$f(x) = x^2 - 2x - 2$

$(-0.7,0)$ $(2.7,0)$

<u>Solution:</u> To solve the inequality $x^2 < 2x + 2$, transfer all the terms to one side. Subtracting $(2x+2)$ from both sides,

$$x^2 - 2x - 2 < 0 .$$

We shall solve this inequality graphically. The function that we are dealing with is $f(x) = x^2 - 2x - 2$. We must find the region where it is less than zero. Thus, we graph the left side as a function of x. Choose values of x and calculate the corresponding $f(x)$ values, as shown in the following table:

x	$x^2 - 2x - 2$	$f(x)$
-3	$(-3)^2 - 2(-3) - 2 = 9 + 6 - 2$	13
-2	$(-2)^2 - 2(-2) - 2 = 8 - 2$	6
-1	$(-1)^2 - 2(-1) - 2 = 1 + 2 - 2$	1
0	$(0)^2 - 2(0) - 2 = 0 - 0 - 2$	-2
1	$(1)^2 - 2(1) - 2 = 1 - 2 - 2$	-3
2	$(2)^2 - 2(2) - 2 = 4 - 4 - 2$	-2
3	$(3)^2 - 2(3) - 2 = 9 - 6 - 2$	1

401

The graph of $f(x) = x^2 - 2x - 2$ is shown in the accompanying figure. $f(x) < 0$ when the curve lies below the x-axis, that is, for values of x between -0.7 and 2.7. (see figure) Hence, $-0.7 < x < 2.7$.

Since these results were read from the curve, they are only approximations. If the student should read 2.6 or 2.8 instead of 2.7, his result would be acceptable.

In case more accuracy is desired, we can solve the corresponding equation $x^2 - 2x - 2 = 0$ and find $x = 1 \pm \sqrt{3}$. Thus the curve crosses the x-axis when $x = 1 - \sqrt{3}$ and $x = 1 + \sqrt{3}$, and the inequality is true for $1 - \sqrt{3} < x < 1 + \sqrt{3}$. Note $x = 1 + \sqrt{3} \approx 2.7$ and $x = 1 - \sqrt{3} \approx -0.7$.

● **PROBLEM 585**

Construct a graphical representation of the inequality $x^2 - 2x - 8 \leq 0$ and identify the solution set.

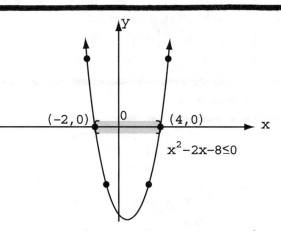

Solution: The graph of the relation $\{(x, y) | y = x^2 - 2x - 8\}$ is sketched in the figure. A table of values can be constructed.

x	$x^2 - 2x - 8$	y
-3	$(-3)^2 - 2(-3) - 8$	7
-2	$(-2)^2 - 2(-2) - 8$	0
-1	$(-1)^2 - 2(-1) - 8$	-5
0	$(0)^2 - 2(0) - 8$	-8
1	$(1)^2 - 2(1) - 8$	-9
2	$(2)^2 - 2(2) - 8$	-8
3	$(3)^2 - 2(3) - 8$	-5
4	$(4)^2 - 2(4) - 8$	0
5	$(5)^2 - 2(5) - 8$	7

We have to find the values of x for which $y \leq 0$ where $y = x^2 - 2x - 8$. First we consider the case $y = 0$. Factor y into $(x - 4)(x + 2)$. Set $y = 0$ and find the roots of this equation.

$$x - 4 = 0 \qquad\qquad x + 2 = 0$$

$$x_1 = 4 \qquad\qquad x_2 = -2$$

Now, we must find where $x^2 - 2x - 8 < 0$.

We mark the roots on the x-axis and consider the regions into which the roots divide the x-axis. They are $x < -2$, $-2 < x < 4$, and $x > 4$. For each region choose an x value and investigate the algebraic signs of the factors of $f(x)$ and also their product sign, $f(x)$. See the following table.

Regions	$x < -2$	$-2 < x < 4$	$x > 4$
Factors of $f(x)$	$(x-4)(x+2)$	$(x-4)(x+2)$	$(x-4)(x+2)$
x-value	-3	0	5
Signs of factors	$(-)\quad(-)$	$(-)\quad(+)$	$(+)\quad(+)$
$\therefore$	$y > 0$	$y < 0$	$y > 0$

Thus, $y < 0$ for $-2 < x < 4$.

Furthermore, $y \le 0$ when $-2 \le x \le 4$. See Graph.

The darkened portion is the part which represents the inequality. The solution set is the interval $[-2, 4]$.

CHAPTER 20

SYSTEMS OF QUADRATIC EQUATIONS

> **Basic Attacks and Strategies for Solving Problems in this Chapter. See pages 404 to 463 for step-by-step solutions to problems.**

A system of two equations in two variables, in which at least one equation is quadratic, is considered to be a quadratic system of equations. The solution set for such a system consists of all ordered pairs that satisfy both equations. In most cases, the solution procedure used for quadratic systems is the substitution method; however, the addition method can also be used if the like variables are raised to the same power in both equation. In a case where the resulting equation is quadratic, then the usual procedure for solving a quadratic equation is used. The final step is to check the original system of equations to ensure that the obtained solution set is really a true solution set. These methods are the same ones that are well-known in the solution of systems of linear equations in two variables.

As a precursor to the above solution procedure, it is helpful to graph each equation in the system on the same $x - y$ coordinate system in order to anticipate the number, approximate position, and value of the solution(s). Sometimes it is absolutely necessary to construct a graphical solution of the system. This is especially true if the system contains quadratic inequalities. Such a system requires that the solution set, if it exists, is a shaded region in the plane.

If both of the equations in a system of two equations in two variables are of the second degree in both variables, then the use of linear combinations of the equations often provides a simpler means of solution than does substitution. For example, solve the following system of equations:

$$3x^2 - 7y^2 + 15 = 0 \tag{1}$$

$$3x^2 - 4y^2 - 12 = 0 \tag{2}$$

By forming the linear combination of -1 times equation (1) and 1 times equation (2) we obtain

$$3y^2 - 27 = 0$$

$$y^2 = 9$$

$$y = \pm 3$$

Substituting 3 for y in equation (1) or (2), we have

$$x^2 = 16 \quad \text{or} \quad x = \pm 4.$$

Hence, (4, 3) and (– 4, 3) are solutions of the system. Now substitute – 3 for y to obtain

$$x^2 = 16 \quad \text{or} \quad x = \pm 4.$$

Hence, (4, – 3) and (– 4, – 3) are also solutions.

The solutions of some systems require the application of both linear combinations and substitution. Also, it may be necessary, after substitution does not yield a simplified equation, to replace an expression in the equation with a new variable u and solve the system with it included. Once the solution has been found, the replacement variable can be removed by substituting back into the original expression and simplifying to obtain the solution for the original variable.

Step-by-Step Solutions to Problems in this Chapter, "Systems of Quadratic Equations"

QUADRATIC/LINEAR COMBINATIONS

● **PROBLEM** 586

Solve the system

$$xy = 24, \qquad\qquad (1)$$

$$y - 2x + 2 = 0. \qquad\qquad (2)$$

<u>Solution:</u> This system is most easily solved by the method of substitution. Solve (2) for y in terms of x:

$$y = 2x - 2.$$

Substitute 2x - 2 for y in (1):

$$x(2x - 2) = 24,$$

$$2x^2 - 2x = 24,$$

$$2x^2 - 2x - 24 = 0,$$

$$x^2 - x - 12 = 0,$$

or factoring $(x+3)(x-4) = 0,$

Set each factor = 0 to find all values of x for which the product = 0.

$$
\begin{array}{c|c}
x + 3 = 0 & x - 4 = 0 \\
x = -3 & x = 4.
\end{array}
$$

In Equation (1): for x = -3, (-3)y = 24 or y = -8; for x = 4, (4)y = 24 or y = 6.

In Equation (2): for x = -3, y - 2(-3) + 2 = 0 or y = -8; for x = 4, y - 2(4) + 2 = 0 or y = 6.

● **PROBLEM** 587

Solve the system
$$2x^2 - 3xy - 4y^2 + x + y - 1 = 0,$$
$$2x - y = 3.$$

<u>Solution:</u> A system of equations consisting of one linear and one

quadratic is solved by expressing one of the unknowns in the linear equation in terms of the other, and substituting the result in the quadratic equation. From the second equation, $y = 2x - 3$. Replacing y by this linear function of x in the first equation, we find

$$2x^2 - 3x(2x - 3) - 4(2x - 3)^2 + x + 2x - 3 - 1 = 0.$$

$$2x^2 - 3x(2x - 3) - 4\left(4x^2 - 12x + 9\right) + x + 2x - 3 - 1 = 0$$

Distribute, $2x^2 - 6x^2 + 9x - 16x^2 + 48x - 36 + x + 2x - 3 - 1 = 0$

Combine terms, $\qquad - 20x^2 + 60x - 40 = 0$

Divide both sides by –20,

$$\frac{-20x^2}{-20} + \frac{60x}{-20} - \frac{40}{-20} = \frac{0}{-20}$$

$$x^2 - 3x + 2 = 0$$

Factoring, $\qquad (x - 2)(x - 1) = 0$

Setting each factor equal to zero, we obtain:

$$x - 2 = 0 \qquad\qquad x - 1 = 0$$
$$x = 2 \qquad\qquad\qquad x = 1$$

To find the corresponding y-values, substitute the x-values in $y = 2x - 3$:

when $x = 1$, $\qquad\qquad$ when $x = 2$,

$$y = 2(1) - 3 \qquad\qquad y = 2(2) - 3$$
$$= 2 - 3 \qquad\qquad\qquad = 4 - 3$$
$$y = -1 \qquad\qquad\qquad y = 1$$

Therefore the two solutions of the system are

$$(1,-1), \qquad\qquad\qquad (2,1),$$

and the solution set is $\left\{(1,-1),(2,1)\right\}$.

● **PROBLEM** 588

Solve the following system:

$$x^2 + y^2 = 25 \qquad\qquad (1)$$

$$2x + y = 10 \qquad\qquad (2)$$

<u>Solution:</u> We solve equation (2) for y by adding - 2x to both sides:

$$y = 10 - 2x \qquad\qquad (3)$$

Replacing y by 10 - 2x in equation (1):

$$x^2 + (10 - 2x)^2 = 25$$

$$x^2 + 100 - 40x + 4x^2 = 25$$

$$5x^2 - 40x + 100 = 25$$

$5x^2 - 40x + 75 = 0.$ Factoring out 5,

$$5\left(x^2 - 8x + 15\right) = 0$$

Dividing both sides by 5,

$$x^2 - 8x + 15 = 0$$

Factoring,

$$(x - 5)(x - 3) = 0$$

Whenever the product of two numbers ab = 0, either a = 0 or b = 0; therefore

$$x - 5 = 0 \quad \text{or} \quad x - 3 = 0$$

$$x = 5 \quad \text{or} \quad x = 3$$

Replacing x by 5 in equation (3):

$$y = 10 - 2(5)$$

$$y = 10 - 10 = 0$$

Replacing x by 3 in equation (3):

$$y = 10 - 2(3)$$

$$y = 10 - 6$$

$$y = 4$$

Thus our solutions are (5,0) and (3,4). Check: Replacing (x,y) by (5,0) in each equation:

$$x^2 + y^2 = 25 \qquad\qquad (1)$$

$$5^2 + 0^2 = 25$$

$$25 + 0 = 25$$

$$25 = 25$$

$$2x + y = 10 \qquad\qquad (2)$$

$$2(5) + 0 = 10$$

$$10 + 0 = 10$$

$$10 = 10$$

Replacing (x,y) by (3,4) in each equation:

$$x^2 + y^2 = 25 \qquad\qquad (1)$$

$$3^2 + 4^2 = 25$$

$$9 + 16 = 25$$

$$25 = 25$$

$$2x + y = 10 \qquad\qquad (2)$$

$$2(3) + 4 = 10$$

$$6 + 4 = 10$$

$$10 = 10$$

Therefore the solution set is $\{(5,0),\ (3,4)\}$.

● PROBLEM 589

Obtain the simultaneous solution set of

$$x^2 + 2y^2 = 54 \qquad\qquad (1)$$

$$2x - y = -9. \qquad\qquad (2)$$

Solution: Equation (2) is readily solvable for y in terms of x, and so we proceed as follows:

$$-y = -9 - 2x$$

$$y = 2x + 9. \qquad\qquad (3)$$

$x^2 + 2(2x+9)^2 = 54$ replacing y by (2x+9) in equation (1)

$x^2 + 2\left(4x^2+36x+81\right) = 54$ squaring (2x+9)

$x^2 + 8x^2 + 72x + 162 = 54$ by the distributive law

$9x^2 + 72x + 108 = 0$ adding -54 to each member and combining terms

$x^2 + 8x + 12 = 0$ dividing by 9

$(x+6)(x+2) = 0$ factoring.

Whenever a product of two numbers ab = 0, either a = 0 or b = 0; thus either

$$x + 6 = 0 \quad \text{or} \quad x + 2 = 0 \quad \text{and}$$

$$x = -6 \quad \text{or} \quad x = -2.$$

407

To find y, we proceed as follows:

$y = 2(-6) + 9 = -12 + 9 = -3$ replacing x by -6 in equation (3)

$y = 2(-2) + 9 = -4 + 9 = 5$ replacing x by -2 in equation (3)

Therefore the simultaneous solution set is $\{(-6,-3), (-2,5)\}$.

Check. Replacing x and y by (-6) and (-3) in equations (1) and (2),

$$x^2 + 2y^2 = 54 \tag{1}$$

$$(-6)^2 + 2(-3)^2 = 54$$

$$36 + 2(9) = 54$$

$$36 + 18 = 54$$

$$54 = 54$$

$$2x - y = -9 \tag{2}$$

$$2(-6) - (-3) = -9$$

$$-12 + 3 = -9$$

$$-9 = -9.$$

Now replacing x and y by (-2) and (5) in equations (1) and (2),

$$x^2 + 2y^2 = 54 \tag{1}$$

$$(-2)^2 + 2(5)^2 = 54$$

$$4 + 2(25) = 54$$

$$4 + 50 = 54$$

$$54 = 54$$

$$2x - y = -9 \tag{2}$$

$$2(-2) - (5) = -9$$

$$-4 - 5 = -9$$

$$-9 = -9.$$

● PROBLEM 590

Solve the system
$$2x + y = 1$$
$$3x^2 - xy - y^2 = -2.$$

Solution: Solving the first equation for y in terms of x:

$$2x + y = 1$$

Subtracting $2x$ from both sides:

$$\cancel{2x} + y - \cancel{2x} = 1 - 2x$$

$$y = 1 - 2x$$

When we substitute this value for y in the second equation we obtain:

$$3x^2 - x(1 - 2x) - \left(1 - 2x\right)^2 = -2$$

$$3x^2 - \left(x - 2x^2\right) - \left(1 - 4x + 4x^2\right) = -2$$

$$3x^2 - x + 2x^2 - 1 + 4x - 4x^2 = -2$$

$$3x^2 + 2x^2 - 4x^2 - x + 4x - 1 = -2$$

$$x^2 + 3x - 1 = -2$$

Adding 2 to both sides:

$$x^2 + 3x - 1 + 2 = -2 + 2$$

$$x^2 + 3x + 1 = 0 \tag{1}$$

Equation (1) is a quadratic equation since it is in the form $ax^2 + bx + c = 0$. This equation can be solved by using the quadratic formula,

$$x = \frac{-b \pm \sqrt{b^2 - 4ac}}{2a} \,.$$

In this case, $a = 1$, $b = 3$, and $c = 1$. Then,

$$x = \frac{-3 \pm \sqrt{(3)^2 - 4(1)(1)}}{2(1)}$$

$$= \frac{-3 \pm \sqrt{9 - 4}}{2}$$

$$= \frac{-3 \pm \sqrt{5}}{2}$$

$$x = \frac{-3}{2} \pm \frac{1}{2}\sqrt{5}$$

To obtain the corresponding y-values we use the equation $y = 1 - 2x$. When $x = -3/2 + 1/2\sqrt{5}$,

$$y = 1 - 2\left(-\frac{3}{2} + \frac{1}{2}\sqrt{5}\right)$$

$$= 1 - 2\left(-\frac{3}{2}\right) - 2\left(\frac{1}{2}\sqrt{5}\right)$$

$$= 1 + 3 - \sqrt{5}$$

$$y = 4 - \sqrt{5}$$

When $x = -\frac{3}{2} - \frac{1}{2}\sqrt{5}$,

$$y = 1 - 2\left(-\frac{3}{2} - \frac{1}{2}\sqrt{5}\right)$$

$$= 1 - 2\left(-\frac{3}{2}\right) - 2\left(-\frac{1}{2}\sqrt{5}\right)$$

$$= 1 + 3 + \sqrt{5}$$

$$y = 4 + \sqrt{5}$$

Therefore, the solution set to the original system of equations is:

$$\left\{ \left(-\frac{3}{2} + \frac{1}{2}\sqrt{5}, \ 4 - \sqrt{5} \right), \left(-\frac{3}{2} - \frac{1}{2}\sqrt{5}, \ 4 + \sqrt{5} \right) \right\}.$$

● **PROBLEM 591**

Find the solution set for the system:

$$3x - 5y = 13 \qquad (1)$$
$$y^2 = 4x \qquad (2)$$

Solution: Use the method of substitution to solve the system. From equation (2) we obtain:

$$x = \frac{y^2}{4}$$

Then upon substitution for x, equation (1) becomes:

$$3\left(\frac{y^2}{4}\right) - 5y = 13$$

Multiplying both sides of the equation by 4:

$$3y^2 - 20y = 52$$

The equation in standard quadratic form is:

$$3y^2 - 20y - 52 = 0$$
$$(3y - 26)(y + 2) = 0$$
$$3y - 26 = 0 \quad \text{or} \quad y + 2 = 0$$
$$y = \frac{26}{3} \quad \text{or} \quad y = -2$$

Returning to $x = \frac{y^2}{4}$, we see that

$y = \frac{26}{3}$ implies that $x = \frac{(26/3)^2}{4} = \frac{169}{9}$ and

$y = 2$ implies that $x = \frac{(-2)^2}{4} = 1$

Check to see if equation (1) is satisfied.

for $x = \frac{169}{9}$, $y = \frac{26}{3}$:

$$3\left(\frac{169}{9}\right) - 5\left(\frac{26}{3}\right) = \frac{169}{3} - \frac{130}{3} = \frac{39}{3} = 13$$

for $x = 1$, $y = -2$:

$$3(1) - 5(-2) = 3 + 10 = 13$$

Hence the solution set is

$$\left\{ \left(\frac{169}{9}, \ \frac{26}{3} \right), \ (1, -2) \right\} .$$

● **PROBLEM 592**

Solve the following linear-quadratic system:

$$\begin{cases} 2x + 3y = 1, & (1) \\ x^2 - 5xy - 8y^2 + 6y = 0. & (2) \end{cases}$$

Solution: To solve a system of equations consisting of one linear and one quadratic, express one of the unknowns in the linear equation in terms of the other variable. Substitute the result in the quadratic equation. We solve for x in terms of y.

410

$$2x + 3y = 1 \quad (1) \quad \text{(adding } -3y \text{ to both sides)}$$
$$2x \quad = 1 - 3y \quad \text{(dividing both sides by 2)}$$
$$x \quad = \frac{1 - 3y}{2} \quad\quad\quad (3)$$

Substitute in the quadratic equation:

$$\left(\frac{1 - 3y}{2}\right)^2 - 5\left(\frac{1 - 3y}{2}\right)y - 8y^2 + 6y = 0.$$

$$\left(\frac{1 - 3y}{2}\right)\left(\frac{1 - 3y}{2}\right) - 5\left(\frac{1 - 3y}{2}\right)y - 8y^2 + 6y = 0$$

$$\frac{1 - 6y + 9y^2}{4} - \frac{5y - 15y^2}{2} - 8y^2 + 6y = 0.$$

Multiplying the equation by 4.

$$4\left[\frac{1 - 6y + 9y^2}{4} - \frac{5y - 15y^2}{2} - 8y^2 + 6y\right] = [0]\,4$$

$$1 - 6y + 9y^2 - \overset{2}{\cancel{4}}\left(\frac{5y - 15y^2}{\underset{1}{\cancel{2}}}\right) - 4 \cdot 8y^2 + 4 \cdot 6y = 0$$

$$1 - 6y + 9y^2 - 10y + 30y^2 - 32y^2 + 24y = 0$$

$$9y^2 + 30y^2 - 32y^2 - 6y - 10y + 24y + 1 = 0$$

$$7y^2 + 8y + 1 = 0$$

Factoring: $(7y + 1)(y + 1) = 0$.
Set each factor equal to zero. Solve for y.

$$7y + 1 = 0 \quad\quad\quad\quad\quad y + 1 = 0$$
$$7y = -1$$
$$y = -\frac{1}{7} \quad\quad\quad\quad\quad\quad y = -1$$

Substitute in the linear equation (3). For $y = -\frac{1}{7}$;

$$x = \frac{1 - 3y}{2} = \frac{1 - 3\left(-\frac{1}{7}\right)}{2} = \frac{1 + \frac{3}{7}}{2} = \frac{\frac{10}{7}}{2} = \frac{\overset{5}{\cancel{10}}}{7} \cdot \frac{1}{\underset{1}{\cancel{2}}} = \frac{5}{7}$$

For $y = -1$;

$$x = \frac{1 - 3y}{2} = \frac{1 - 3(-1)}{2} = \frac{1 + 3}{2} = \frac{4}{2} = 2 .$$

The solutions are (2,-1) and (5/7, -1/7). Each solution should be checked by substitution in both of the given equations.

Check: For (2,-1) | For $\left(\frac{5}{7}, -\frac{1}{7}\right)$

$$2x + 3y = 1$$
$$2(2) + 3(-1) = 1$$
$$4 - 3 = 1$$
$$1 = 1$$

$$2x + 3y = 1$$
$$2(5/7) + 3(-1/7) = 1$$
$$10/7 - 3/7 = 1$$
$$7/7 = 1$$
$$1 = 1$$

$$x^2 - 5xy - 8y^2 + 6y = 0$$
$$(2)^2 - 5(2)(-1) - 8(-1)^2 + 6(-1) = 0$$
$$4 + 10 - 8 - 6 = 0$$
$$6 - 6 = 0$$
$$0 = 0$$

$$x^2 - 5xy - 8y^2 + 6y = 0$$
$$\left(\frac{5}{7}\right)^2 - 5\left(\frac{5}{7}\right)\left(-\frac{1}{7}\right) - 8\left(\frac{1}{7}\right)^2$$
$$+ 6\left(-\frac{1}{7}\right) = 0$$
$$\frac{25}{49} + \frac{25}{49} - 8\left(\frac{1}{49}\right) - \frac{6}{7} = 0$$

$$\frac{25}{49} + \frac{25}{49} - \frac{8}{49} - \frac{7(6)}{7(7)} = 0$$

$$\frac{25}{49} + \frac{25}{49} - \frac{8}{49} - \frac{42}{49} = 0$$

$$\frac{25 + 25 - 8 - 42}{49} = 0$$

$$\frac{50 - 8 - 42}{49} = 0$$

$$\frac{42 - 42}{49} = 0$$

$$\frac{0}{49} = 0$$

$$0 = 0$$

In choosing the variable to be eliminated, it is advisable to avoid fractions if possible. For example, the linear equation $6x - y = 7$ should be solved for y in terms of x: $y = 6x - 7$, rather than for x in terms of y:

$$x = \frac{y + 7}{6}.$$

● **PROBLEM** 593

Solve for x and y: $\begin{cases} x^2 + 4y^2 - 8y + 2x - 3 = 0, & (1) \\ 3y - 2x = 12. & (2) \end{cases}$

<u>Solution:</u> From the second equation, $3y - 2x = 12$, add $2x$ to both sides,

$$3y = 12 + 2x$$

divide both sides by 3 to obtain,

$$y = 4 + \tfrac{2}{3}x.$$

Substitution of this in the first equation gives

$$x^2 + 4(4 + \tfrac{2}{3}x)^2 - 8(4 + \tfrac{2}{3}x) + 2x - 3 = 0,$$

$$x^2 + 4(4 + \tfrac{2}{3}x)(4 + \tfrac{2}{3}x) - 32 - \tfrac{16}{3}x + 2x - 3 = 0$$

$$x^2 + 4\left(16 + 2\left(\tfrac{8}{3}\right)x + \tfrac{4}{9}x^2\right) - 32 - \tfrac{16}{3}x + 2x - 3 = 0$$

$$x^2 + 64 + \tfrac{4 \cdot 16}{3}x + \tfrac{16}{9}x^2 - 32 - \tfrac{16}{3}x + 2x - 3 = 0$$

Group terms raised to the same exponent of x,

$$x^2 + \tfrac{16}{9}x^2 + \tfrac{64}{3}x - \tfrac{16}{3}x + 2x + 64 - 32 - 3 = 0$$

$$\tfrac{9}{9}x^2 + \tfrac{16}{9}x^2 + \tfrac{48}{3}x + 2x + 29 = 0$$

$$\tfrac{25}{9}x^2 + 16x + 2x + 29 = 0$$

$$\tfrac{25}{9}x^2 + 18x + 29 = 0$$

Multiply both members by 9,

$$9\left(\tfrac{25}{9}x^2 + 18x + 29\right) = 9 \cdot 0$$

$$25x^2 + 162x + 261 = 0$$

Factor
$$(25x + 87)(x + 3) = 0.$$

When the product of two numbers $ab = 0$ either $a = 0$ or $b = 0$, thus with $(25x + 87)(x + 3) = 0$ either

$$25x + 87 = 0 \quad \text{or} \quad x + 3 = 0$$
$$25x = -87$$
$$x = \frac{-87}{25} \quad \text{or} \quad x = -3$$

To obtain the corresponding y values to each x value, replace the x values in equation (2) and solve for y,

$$\text{for } x = \frac{-87}{25}: \quad 3y - 2\left(\frac{-87}{25}\right) = 12$$
$$3y + \frac{174}{25} = 12$$

multiply both members by 25,

$$25\left(3y + \frac{174}{25}\right) = 25(12)$$
$$75y + 174 = 300$$
$$75y = 126$$
$$y = \frac{126}{75} = \frac{42}{25}$$

for $x = -3$:
$$3y - 2(-3) = 12$$
$$3y + 6 = 12$$
$$3y = 6$$
$$y = 2$$

Hence, our two pairs of solutions are $(-3,2)$ and $\left(\frac{-87}{25}, \frac{42}{25}\right)$.

Check: A) Replace x and y in (1) and (2) by $(-3,2)$,

(1) $x^2 + 4y^2 - 8y + 2x - 3 = 0$
$$(-3)^2 + 4\left(2^2\right) - 8(2) + 2(-3) - 3 = 0$$
$$9 + 16 - 16 - 6 - 3 = 0$$
$$3 - 3 = 0$$
$$0 = 0$$

(2) $\quad 3y - 2x = 12$
$$3(2) - 2(-3) = 12$$
$$6 + 6 = 12$$
$$12 = 12$$

B) Now replace x and y in (1) and (2) by $\left(\frac{-87}{25}, \frac{42}{25}\right)$,

(1) $\quad x^2 + 4y^2 - 8y + 2x - 3 = 0$
$$\left(\frac{-87}{25}\right)^2 + 4\left(\frac{42}{25}\right)^2 - 8\left(\frac{42}{25}\right) + 2\left(\frac{-87}{25}\right) - 3 = 0$$
$$12.1104 + 4(2.8224) - 8(1.68) + 2(-3.48) - 3 = 0$$
$$12.1104 + 11.2896 - 13.4400 - 6.96 - 3 = 0$$
$$23.4 - 6.96 - 16.44 = 0$$
$$23.4 - 23.4 = 0$$
$$0 = 0$$

413

(2) $3y - 2x = 12$

$$3\left(\frac{42}{25}\right) - 2\left(\frac{-87}{25}\right) = 12$$

$$\frac{126}{25} + \frac{174}{25} = 12$$

$$\frac{300}{25} = 12$$

$$12 = 12$$

● **PROBLEM** 594

Solve the system

$$y = -x^2 + 7x - 5 \qquad\qquad (1)$$

$$y - 2x = 2 \qquad\qquad (2)$$

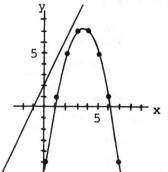

Solution: Solving Equation (2) for y yields an expression
for y in terms of x. Substituting this expression in
Equation (1),

$$2x + 2 = -x^2 + 7x - 5 \qquad\qquad (3)$$

We have a single equation, in terms of a single variable,
to be solved. Writing Equation (3) in standard quadratic
form,

$$x^2 - 5x + 7 = 0 \qquad\qquad (4)$$

Since the equation is not factorable the roots are not
found in this manner. Evaluating the discriminant will
indicate whether Equation (4) has real roots. The discriminant, $b^2 - 4ac$, of Equation (4) equals

$(-5)^2 - 4(1)(7) = 25 - 28 = -3$. Since the discriminant
is negative, Equation (4) has no real roots, and therefore
the system has no real solution. In terms of the graph,
the figure shows that the parabola and the straight line
have no point in common.

● **PROBLEM** 595

Solve the system

$$y = 3x^2 - 2x + 5 \qquad\qquad (1)$$

$$y = 4x + 2. \qquad\qquad (2)$$

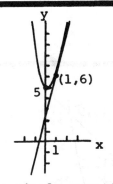

Solution: To obtain a single equation with one unknown variable, x, substitute the value of y from Equation (2) in Equation (1),

$$4x + 2 = 3x^2 - 2x + 5. \qquad\qquad (3)$$

Writing Equation (3) in standard quadratic form,

$$3x^2 - 6x + 3 = 0. \qquad\qquad (4)$$

We may simplify Equation (4) by dividing both members by 3, which is a factor common to each term:

$$x^2 - 2x + 1 = 0. \qquad\qquad (5)$$

To find the roots, factor and set each factor = 0. This may be done since a product = 0 implies one or all of the factors must = 0.

$$(x - 1)(x - 1) = 0 \qquad\qquad (6)$$

$x - 1 = 0$	$x - 1 = 0$
$x = 1$	$x = 1$

Equation (5) has two equal roots, each equal to 1. For x = 1, from Equation (2) we have y = 4(1) + 2 = 6. Therefore the system has but one common solution:

$$x = 1, \ y = 6.$$

The figure indicates that our solution is probably correct. We may also check to see if our values satisfy Equation (1) as well:

Substituting in: $y = 3x^2 - 2x + 5$

$$6 \overset{?}{=} 3(1)^2 - 2(1) + 5$$

$$6 \overset{?}{=} 3 - 2 + 5$$

$$6 = 6.$$

Solve the system of equations

$$y = x^2 - 6x + 9 \qquad\qquad (1)$$

$$y = x + 3. \qquad\qquad (2)$$

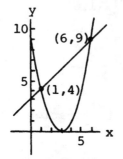

Solution: A single equation in terms of the one variable x may be obtained by the method of substitution. Substitute the value of y, x + 3, from Equation (2) for y in Equation (1).

$$x + 3 = x^2 - 6x + 9. \qquad\qquad (3)$$

Writing Equation (3) in the standard form of a quadratic equation,

$$x^2 - 7x + 6 = 0. \qquad\qquad (4)$$

Use the usual method of solving quadratic equations. Factor the equation. Then find the values of x for which each factor may = 0.

$$(x - 1)(x - 6) = 0 \qquad\qquad (5)$$

$$x - 1 = 0 \quad | \quad x - 6 = 0$$

$$x = 1 \quad | \qquad x = 6$$

The roots of Equation (4) are x = 1 and x = 6. Since y = x + 3, the solution of the given system is

$$x = 1, \; y = 4 \text{ and } x = 6, \; y = 9.$$

In the figure, the graph of the system, indicates that our solution is probably correct. We may prove that the solution is correct by substituting each solution in both equations of the given system, as usual. The values of y were obtained by satisfying equation (2). Now for Equation (1):

check for x = 1, y = 4 | check for x = 6, y = 9

$$y = x^2 - 6x + 9$$ | $$y = x^2 - 6x + 9$$

$$4 \overset{?}{=} (1)^2 - 6(1) + 9 \qquad\qquad 9 \overset{?}{=} (6)^2 - 6(6) + 9$$

$$4 \overset{?}{=} 1 - 6 + 9 \qquad\qquad 9 \overset{?}{=} 36 - 36 + 9$$

$$4 = 4 \qquad\qquad 9 = 9$$

● **PROBLEM** 597

Solve the system

$$x^2 + y^2 = 10 \qquad\qquad (1)$$

$$x + 2y = 1 \qquad\qquad (2)$$

Solution: We solve the linear equation for x in terms of y by adding -2y to both sides to obtain x = 1 - 2y. We substitute the result, 1 - 2y, for x in the quadratic equation to obtain

$$(1 - 2y)^2 + y^2 = 10$$

Then we have

$$1 - 4y + 4y^2 + y^2 = 10$$

We add (-10) to both sides and combine terms:

$$5y^2 - 4y - 9 = 0$$

We factor, $\qquad\qquad (5y - 9)(y + 1) = 0$

Whenever the product of two numbers ab = 0 either a = 0 or b = 0. Thus (5y - 9)(y + 1) = 0 implies either 5y - 9 = 0 or y + 1 = 0
$$5y = 9 \qquad \text{or} \qquad y = -1$$

Thus, $y = \frac{9}{5}$ or y = -1 .

Substituting these values in turn in the linear equation, we find the corresponding values for x: x + 2y = 1, for $y = \frac{9}{5}$

$$x + 2\left(\frac{9}{5}\right) = 1, \ x = 1 - \frac{18}{5}, \ x = \frac{-13}{5}$$

and for y = -1

$$x + 2(-1) = 1, \ x - 2 = 1, \ x = 3.$$

The solutions of the system are therefore

$$\left(x = -\frac{13}{5}, \ y = \frac{9}{5} \right) \quad \text{and} \quad (x = 3, \ y = -1).$$

To consider the corresponding graphs of this system, we notice that $x^2 + y^2 = 10$ represents a circle with radius $\sqrt{10}$ and x + 2y = 1 is the line passing through the points (1,0) and (0,½).

417

The two points where the circle and line intersect are the solution to our problem, $\left(\frac{-13}{5}, \frac{9}{5}\right)$ and $(3,-1)$, the points where $x^2 + y^2 = 10$ and $x + 2y = 1$ simultaneously.

Solve the system

$$x^2 + y^2 = 25, \tag{1}$$

$$x - y = 1. \tag{2}$$

Solution: Solve (2) for y (the problem can be done similarly for x instead): The method of substitution is most easily employed in this example to solve the system.

$$y = x - 1. \tag{3}$$

Substitute x - 1 for y in (1):

$$x^2 + (x - 1)^2 = 25. \tag{4}$$

$$x^2 + x^2 - 2x + 1 = 25.$$

From (4) $\qquad 2x^2 - 2x - 24 = 0,$

or $\qquad x^2 - x - 12 = 0. \tag{5}$

Solve (5) by factoring: $\quad (x - 4)(x + 3) = 0$

$$x - 4 = 0 \qquad x + 3 = 0$$

$$x = 4 \text{ or } -3.$$

Substituting 4 for x in (2), we obtain

$$4 - y = 1 \text{ or } y = 3.$$

Substituting -3 for x in (2), we obtain

$$-3 - y = 1 \text{ or } y = -4.$$

This gives $\quad \begin{aligned} x &= 4 \\ y &= 3 \end{aligned} \quad$ and $\quad \begin{aligned} x &= -3 \\ y &= -4 \end{aligned} \quad$ for the solutions.

Check:

    for x = 4, y = 3

in Eq. (1): $(4)^2 + (3)^2 = 25$

                $16 + \quad 9 = 25$

                     $25 = 25$

in Eq. (2): $(4) - (3) = 1$

                    $1 = 1$

for x = -3, y = -4

in Eq. (1): $(-3)^2 + (-4)^2 = 25$

                 $9 + \quad 16 = 25$

                     $25 = 25$

in Eq. (2): $(-3) - (-4) = 1$

                $-3 + \quad 4 = 1$

                    $1 = 1.$

Graphical meaning of the two solutions. We may plot the
graph for each of the equations (1) and (2). The graph of

$$x - y = 1$$

is the straight line shown in the figure, and the graph of

$$x^2 + y^2 = 25$$

is the circle there shown. To draw the graph of (1), the
student may give various values to x and calculate the corresponding values for y from $y = \pm \sqrt{25 - x^2}$.

    Any point on the straight line (2) has coordinates
that satisfy Equation (2). Any point on the circle (1) has
coordinates that satisfy Equation (1). The points (4,3)
and (-3,-4) lie on both graphs and satisfy both Equations
(1) and (2). That is to say, each point of intersection of
the graph of (1) with the graph of (2) gives a pair of
numbers that is a solution of the system.

● **PROBLEM** 599

Solve the system

$$x^2 + y^2 = 25, \tag{1}$$

$$x + y = 10, \tag{2}$$

and draw the graph to explain the fact that the solutions
are not real.

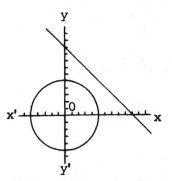

Solution: Use the method of substitution to obtain a single equation in terms of either one of the variables x or y. Solve Equation (2) for y (we could have chosen to solve for x). We get

$$y = 10 - x. \tag{3}$$

Substitute this expression for y in Equation (1)

$$x^2 + (10 - x)^2 = 25$$

$$x^2 + 100 - 20x + x^2 = 25 \qquad \text{Expand the equation}$$

$$2x^2 - 20x + 100 = 25 \qquad \text{Combine like terms}$$

$$2x^2 - 20x + 75 = 0 \qquad \text{Put in standard quadratic form.}$$

This is a nonfactorable quadratic equation of the form $ax^2 + bx + c$ with a = 2, b = -20, c = 75. To find the roots of this equation use the formula

$$x = \frac{-b \pm \sqrt{b^2 - 4ac}}{2a}$$

substituting our values of a, b, and c we get

$$x = \frac{-(-20) \pm \sqrt{(-20)^2 - 4(2)(75)}}{2(2)}$$

$$x = \frac{20 \pm \sqrt{400 - 600}}{4} = \frac{20 \pm \sqrt{-200}}{4}$$

$$x = \frac{20 \pm \sqrt{(100)(2)(-1)}}{4} = \frac{20 \pm \sqrt{100} \cdot \sqrt{2} \cdot \sqrt{-1}}{4}$$

square roots of a product.

Recall $i = \sqrt{-1}$ then $x = \dfrac{20 \pm 10i\sqrt{2}}{4}$ reduces to $x = 5 \pm \dfrac{5i}{2}\sqrt{2}$.

Using Equation (3):

420

for $x = 5 + \frac{5i}{2}\sqrt{2}$, $y = 10 - x = 10 - \left(5 + \frac{5i}{2}\sqrt{2}\right)$

$$= 10 - 5 - \frac{5i}{2}\sqrt{2} = 5 - \frac{5i}{2}\sqrt{2}$$

for $x = 5 - \frac{5i}{2}\sqrt{2}$, $y = 10 - x = 10 - \left(5 - \frac{5i}{2}\sqrt{2}\right)$

$$= 10 - 5 + \frac{5i}{2}\sqrt{2} = 5 + \frac{5i}{2}\sqrt{2}$$

Check:

for $x = 5 + \frac{5i}{2}\sqrt{2}$, $y = 5 - \frac{5i}{2}\sqrt{2}$

Eq. (1): $x^2 + y^2 = 25$

$$\left(5 + \frac{5i}{2}\sqrt{2}\right)^2 + \left(5 - \frac{5i}{2}\sqrt{2}\right)^2 \overset{?}{=} 25$$

remember $i^2 = -1$

$$\left(25 + \frac{50i}{2}\sqrt{2} - \frac{50}{4}\right) + \left(25 - \frac{50i}{2}\sqrt{2} - \frac{50}{4}\right) \overset{?}{=} 25$$

$$50 \neq 25$$

These roots do not check. The roots are extraneous.

for $x = 5 - \frac{5i}{2}\sqrt{2}$, $y = 5 + \frac{5i}{2}\sqrt{2}$

Eq. (1): $x^2 + y^2 = 25$

$$\left(5 - \frac{5i}{2}\sqrt{2}\right)^2 + \left(5 + \frac{5i}{2}\sqrt{2}\right)^2 \overset{?}{=} 25$$

remember $i^2 = -1$

$$\left(25 - \frac{50i}{2}\sqrt{2} + \frac{50}{4}\right) + \left(25 + \frac{50i}{2}\sqrt{2} - \frac{50}{4}\right) \overset{?}{=} 25$$

$$50 \neq 25.$$

These roots do not check.

● PROBLEM 600

Solve for x and y:
$$\begin{cases} 9x^2 - 16y^2 = 144 & \text{(1)} \\ x - 2y = 4. & \text{(2)} \end{cases}$$

<u>Algebraic solution:</u> We solve equation (2) for x,

$$x = 4 + 2y \qquad\qquad (3)$$

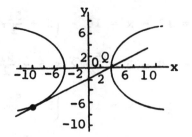

and substitute this expression for x in equation (1).
This gives

$$9(4 + 2y)^2 - 16y^2 = 144,$$

$$9\left(16 + 16y + 4y^2\right) - 16y^2 = 144$$

Dividing both sides by 4, $9\left(4 + 4y + y^2\right) - 4y^2 = 36.$

Distributing, $36 + 36y + 9y^2 - 4y^2 = 36$

Combining terms, $36 + 36y + 5y^2 = 36$

Subtracting 36 from both sides, $5y^2 + 36y = 0$

Factoring, $y(5y + 36) = 0$

Whenever the product of two numbers $ab = 0$, either $a = 0$
or $b = 0$; thus either

$y = 0$ or $5y + 36 = 0$

$$5y = -36$$

$$y = \frac{-36}{5}$$

Thus, $y = 0,$ $-\dfrac{36}{5}$.

Placing these values in linear equation (3):

when $y = 0$,

$$x = 4 + 2(0) = 4 + 0 = 4$$

when $y = -\dfrac{36}{5}$,

$$x = 4 + 2\left(-\frac{36}{5}\right) = 4 - \frac{72}{5} = \frac{20}{5} - \frac{72}{5} = -\frac{52}{5}$$

Thus the two solutions of the equations are seen to be
(4, 0) and (- 52/5, - 36/5), which are then the actual
coordinates of the points of intersection of the line
and the hyperbola to be discussed.

Geometric solution: Construct the graph of each equation
and note where the two graphs intersect. The graph of the
first equation cuts the x-axis at $x = \pm 4$, and y is
imaginary for any value of x between - 4 and 4. The graph
consists of the two curved branches in the diagram, and is

a hyperbola

The graph of the second equation is a straight line through the points (4, 0) and (0, - 2). This line intersects the hyperbola at the points P and Q, whose coordinates are approximately (4, 0) and (- 10, - 7).

Use the graphical method to find the simultaneous solution set of

$$x^2 + x - 2 > 0$$

and $\qquad \frac{3}{4} x + \frac{3}{2} < 0$

Fig. A

The slashed lines represent the simultaneous solution set.

Fig. B

<u>Solution:</u> We construct the graphs of

$$y = x^2 + x - 2 \quad \text{and} \quad y = \frac{3}{4} x + \frac{3}{2}$$

Set up tables for both equations to calculate y values.

For $y = x^2 + x - 2$

x	$x^2 + x - 2$ =	y
- 4	$(- 4)^2 + (- 4) - 2$	10
- 3	$(- 3)^2 + (- 3) - 2$	4
- 2	$(- 2)^2 + (- 2) - 2$	0
- 1	$(- 1)^2 + (- 1) - 2$	-2
0	$(0)^2 + (0) - 2$	-2
1	$(1)^2 + (1) - 2$	0
2	$(2)^2 + 2 - 2$	4
3	$(3)^2 + 3 - 2$	10

For $y = \frac{3}{4} x + \frac{3}{2}$

x	$\frac{3}{4} x + \frac{3}{2}$ =	y
-2	$\frac{3}{4}(- 2) + \frac{3}{2}$	0
0	$\frac{3}{4}(0) + \frac{3}{2}$	$\frac{3}{2}$
2	$\frac{3}{4}(2) + \frac{3}{2}$	3

423

See Figure A for graphs.

Now, find the region where the inequality $x^2 + x - 2 > 0$ holds. The function $f(x) = x^2 + x - 2$ can be factored into $(x + 2)(x - 1)$. Set $f(x) = 0$ and find the roots of this equation. Here $x = -2$ and $x = 1$. Mark the roots on the x-axis and consider the regions into which the roots divide the x-axis. They are $x < -2$, $-2 < x < 1$, and $x > 1$. See Figure B.

For each of these regions, choose a value of x and investigate the algebraic signs of the factors of the function $f(x)$. Then look at the sign of their product, $f(x)$. The table summarizes the process.

	$f(x) = (x + 2)(x - 1)$		
Regions	$x < -2$	$-2 < x < 1$	$x > 1$
x-value	$x = -3$	$x = 0$	$x = 2$
Factors of $f(x)$	$(x+2)(x-1)$	$(x+2)(x-1)$	$(x+2)(x-1)$
Signs of factors	$(-)$ $(-)$	$(+)$ $(-)$	$(+)$ $(+)$
$\therefore$	$f(x) > 0$	$f(x) < 0$	$f(x) > 0$

For our problem we now know that the graph of $x^2 + x - 2$ is greater than zero, that is, above the x-axis, for $x < -2$ or $x > 1$.

Call the second function $g(x)$. Thus, $g(x) = \frac{3}{4}x + \frac{3}{2}$. We are interested in finding the values of x for which the function $g(x)$ is negative or when it is below the x-axis. Therefore,

$$\frac{3}{4}x + \frac{3}{2} < 0 \Leftrightarrow \frac{3}{4}x < -\frac{3}{2} \Leftrightarrow 3x < -6 \Leftrightarrow x < -2$$

Hence, the solution set is $\{x \mid x < -2\}$. That is, the graph of $y = \frac{3}{4}x + \frac{3}{2}$ lies below the x-axis when $x < -2$.

Simultaneously we solve for x and we obtain $\{x \mid x < -2\}$. See Figure A.

QUADRATIC/QUADRATIC(CONIC) COMBINATIONS

● **PROBLEM** 602

Determine whether or not, the given number pair (x,y) is a solution of the accompanying system of equations.

(a) $(3,-2)$; $2x - y = 8$
      $x^2 - y^3 = 17$

(b) $(4,-1)$; $\sqrt{x} - y = -1$
      $xy^2 = 4$

Solution: A given number pair (x,y) is a solution of a system of equations if substitution of (x,y) into each equation of the system results in a valid equation.

424

(a)   Replacing $(3,-2)$ in $2x - y = 8$, with $x = 3$, $y = -2$:

$$2(3) - (-2) = 8$$

$$6 + 2 = 8$$

$$8 = 8 \text{ , which is true.}$$

Replacing $(3,-2)$ in $x^2 - y^3 = 17$:

$$3^2 - (-2)^3 = 17$$

$$9 - (-8) = 17$$

$$9 + 8 = 17$$
$$17 = 17 \text{ , which is true.}$$

Therefore $(3,-2)$ is a solution of this system of equations.

(b) Replacing $(4,-1)$ in $\sqrt{x} - y = -1$, with $x = 4$, $y = -1$:

$$\sqrt{4} - (-1) = -1$$

$$2 + 1 = -1$$

$$3 = -1 \text{ , which is not true;}$$

thus, even though replacing $(4,-1)$ in $xy^2 = 4$ yields $4 = 4$ which is true, $(4,-1)$ is <u>not</u> a solution of this system for it fails to satisfy the first equation.

● **PROBLEM**   603

Obtain the solution set of the following system of equations by substitution:

$$xy = 2 \tag{1}$$
$$15x^2 + 4y^2 = 64 \tag{2}$$

<u>Solution:</u>   Equation (1) can be solved for y in terms of x by dividing both sides by x,

$$y = \frac{2}{x} \tag{3}$$

We now replace y by 2/x in equation (2) and complete the process of solving as follows:

$$15x^2 + 4\left(\frac{2}{x}\right)^2 = 64$$

$$15x^2 + 4\left(\frac{4}{x^2}\right) = 64 \text{ ,   squaring 2/x}$$

$$15x^4 + 16 = 64x^2 \text{ ,   multiplying each member by } x^2$$

$$15x^4 - 64x^2 + 16 = 0, \text{ adding } -64x^2 \text{ to each member and arranging terms}$$

$$\left(x^2-4\right)\left(15x^2-4\right) = 0, \text{ factoring}$$

Whenever the product of two numbers $ab = 0$, either $a = 0$ or $b = 0$. Thus, either

$$x^2 - 4 = 0 \quad \text{or} \quad 15x^2 - 4 = 0$$

$$x^2 = 4 \qquad\qquad 15x^2 = 4$$

$$x = \pm\sqrt{4} \qquad\qquad x^2 = \frac{4}{15}$$

$$x = \pm 2 \qquad\qquad x = \pm\sqrt{\frac{4}{15}}$$

425

$$x \quad = \pm \frac{2}{\sqrt{15}} = \pm \frac{2}{\sqrt{15}} \cdot \frac{\sqrt{15}}{\sqrt{15}}$$

$$= \pm \frac{2\sqrt{15}}{15}$$

To obtain y, we replace x by all four of its values in eq. (3)

$$y = \frac{2}{\pm 2}, \quad \text{replacing } x = \pm 2 \text{ in (3)}$$

$$= \pm 1$$

$$y = \frac{2}{\pm 2\sqrt{15}/15}, \quad \text{replacing } x \text{ by } \pm 2\sqrt{15}/15 \text{ in (3)}$$

$$= \pm \frac{2(15)}{2\sqrt{15}} = \pm \frac{15}{\sqrt{15}} \cdot \frac{\sqrt{15}}{\sqrt{15}} = \pm \frac{15\sqrt{15}}{15} = \pm \sqrt{15}$$

Therefore the simultaneous solution set is

$$\left\{ (2, 1), (-2, -1), (2\sqrt{15}/15, \sqrt{15}), (-2\sqrt{15}/15, -\sqrt{15}) \right\}$$

The solution set can be checked by the usual method.

● **PROBLEM 604**

Obtain the simultaneous solution set of

$$4x^2 - 2xy - y^2 = -5 \tag{1}$$

$$y + 1 = -x^2 - x \tag{2}$$

Solution: Solving equation (2) for y,

$$y = -x^2 - x - 1 \tag{3}$$

Replacing y by $\left(-x^2 - x - 1 \right)$ in equation (1),

$$4x^2 - 2x\left(-x^2 - x - 1\right) - \left(-x^2 - x - 1\right)^2 = -5$$

$$4x^2 + 2x^3 + 2x^2 + 2x - \left(-x^2 - x - 1\right)\left(-x^2 - x - 1\right) = -5$$

$$4x^2 + 2x^3 + 2x^2 + 2x - \left(x^4 + x^3 + x^2 + x^3 + x^2 + x + x^2 + x + 1\right) = -5$$

$$2x^3 + 6x^2 + 2x - \left(x^4 + 2x^3 + 3x^2 + 2x + 1\right) = -5$$

$$2x^3 + 6x^2 + 2x - x^4 - 2x^3 - 3x^2 - 2x - 1 = -5$$

$$-x^4 + 3x^2 - 1 = -5$$

$$-x^4 + 3x^2 + 4 = 0 \tag{4}$$

Equation (4) is in quadratic form. This can be seen more easily by the following: divide both sides by -1; thus $x^4 - 3x^2 - 4 = 0$, and this can be rewritten as $\left(x^2\right)^2 - 3x^2 - 4 = 0$, which is in quadratic form. Now, factoring, we have:

$$\left(x^2 - 4\right)\left(x^2 + 1\right) = 0$$

Whenever the product of two numbers $ab = 0$, either $a = 0$ or $b = 0$. Thus, either

$$x^2 - 4 = 0 \qquad \text{or} \qquad x^2 + 1 = 0$$
$$x^2 = 4 \qquad\qquad\qquad x^2 = -1$$
$$x = \pm\sqrt{4} \qquad\qquad\qquad x = \pm\sqrt{-1}$$
$$x = \pm 2 \qquad\qquad\qquad x = \pm i$$

Now we obtain the value of y corresponding to each of the four values of x: replacing x by 2 in equation (3),

$$y = - x^2 - x - 1$$
$$= -(2)^2 - 2 - 1$$
$$= - 4 - 2 - 1$$
$$= - 7$$

Replacing x by $- 2$ in eq. (3),

$$y = -(-2)^2 - (-2) - 1$$
$$= -(4) + 2 - 1$$
$$= - 3$$

Replacing x by i in eq. (3),

$$y = -(i)^2 - i - 1$$
$$= -(-1) - i - 1$$
$$= - i$$

Replacing x by $-i$ in eq. (3),

$$y = -(-i)^2 - (-i) - 1$$
$$= -(-1) + i - 1$$
$$= i$$

Therefore our solution set is

$$\left\{ (2,-7), \ (-2,-3), \ (i,-i), \ (-i,i) \right\}$$

which can be verified by checking.

● **PROBLEM** 605

Solve the system:

$$x^2 + y^2 = 1 \qquad\qquad (1)$$

$$x^2 - 1 = y. \qquad\qquad (2)$$

Solution: Although the value of y in equation (2) can be substituted in equation (1), the result will be a fourth-degree equation. Hence, we substitute the value of x^2 from equation (2).

$$x^2 - 1 = y$$

$$x^2 = y + 1.$$

Replacing x^2 by $y + 1$ in equation (1),

$$y + 1 + y^2 = 1.$$

Add (-1) to both sides, $y^2 + y = 0.$
Factor out y, $\qquad\qquad y(y + 1) = 0.$

Whenever a product of two numbers ab = 0, either a = 0 or b = 0; hence

$$y = 0, \quad y + 1 = 0 \qquad \text{or}$$

$$y = 0, \qquad\quad y = -1.$$

Substituting y = 0 in equation (1),

$$x^2 + 0 = 1$$

$$x^2 = 1$$

$$x = \pm 1.$$

Two solutions are (1,0) and (-1,0).
Substituting y = -1 in equation (1),

$$x^2 + (-1)^2 = 1$$

$$x^2 + 1 = 1$$

$$x^2 = 0$$

$$x = 0.$$

The solution (0,-1) is considered a double root in the sense that if y = -1, $x = \sqrt{0}$ or $x = -\sqrt{0}$. The solution set for the system is $\{(1,0), (-1,0), (0,-1)\}$.

Check: To verify that (1,0) is a root, replace x by 1 and y by 0 in each equation.

$$x^2 + y^2 = 1 \tag{1}$$

$$(1)^2 + 0^2 = 1$$

$$1 = 1$$

$$x^2 - 1 = y \tag{2}$$

428

$$(1)^2 - 1 = 0$$

$$1 - 1 = 0$$

$$0 = 0.$$

To verify that (-1,0) is a root, replace x by -1 and y by 0 in each equation

$$x^2 + y^2 = 1 \qquad (1)$$

$$(1)^2 + 0 = 1$$

$$1 + 0 = 1$$

$$1 = 1$$

$$x^2 - 1 = y \qquad (2)$$

$$(-1)^2 - 1 = 0$$

$$1 - 1 = 0$$

$$0 = 0.$$

To verify that (0,-1) is a root, replace x by 0 and y by -1 in each equation,

$$x^2 + y^2 = 1 \qquad (1)$$

$$0 + (-1)^2 = 1$$

$$0 + 1 = 1$$

$$1 = 1$$

$$x^2 - 1 = y \qquad (2)$$

$$(0)^2 - 1 = -1$$

$$0 - 1 = -1$$

$$-1 = -1.$$

● **PROBLEM 606**

Obtain the solution set of

$$x^2 + 4xy - 7x = 12 \qquad (1)$$

$$3x^2 - 4xy + 4x = 15 \qquad (2)$$

Solution: Since the sum of the xy terms in the left members of equations (1) and (2) is zero, we proceed as follows: Adding equations (1) and (2),

$$4x^2 - 3x = 27$$

$$4x^2 - 3x - 27 = 0$$

Equations in the form $ax^2 + bx + c = 0$ can be solved using the quadratic formula, $x = \dfrac{-b \pm \sqrt{b^2 - 4ac}}{2a}$. In our case, $a = 4$, $b = -3$, and $c = -27$, thus

$$x = \frac{-(-3) \pm \sqrt{(-3)^2 - 4(4)(-27)}}{2(4)}$$

$$x = \frac{3 \pm \sqrt{9 + 432}}{8} = \frac{3 \pm \sqrt{441}}{8} = \frac{3 \pm 21}{8}$$

$$x = \frac{3 + 21}{8} = \frac{24}{8} = 3 \quad \text{or} \quad x = \frac{3 - 21}{8} = \frac{-18}{8} = -\frac{9}{4}$$

Thus $x = 3, -\dfrac{9}{4}$.

We find the second numbers in the solution pairs as follows: Replacing x by 3 in equation (1),

$$x^2 + 4xy - 7x = 12$$

$$3^2 + 4(3)y - 7(3) = 12$$

$$9 + 12y - 21 = 12$$

$$12y - 12 = 12$$

$$12y = 24$$

$$y = 2$$

Hence one simultaneous solution pair is (3,2). Then: Replacing x by -9/4 in equation (1),

$$x^2 + 4xy - 7x = 12$$

$$\left(\frac{-9}{4}\right)^2 + 4\left(\frac{-9}{4}\right)y - 7\left(\frac{-9}{4}\right) = 12$$

$$\frac{81}{16} - 9y + \frac{63}{4} = 12$$

Multiplying both sides by 16,

$$81 - 144y + 252 = 192$$

$$-144y + 333 = 192$$

$$-144y = -141$$

$$y = \frac{-141}{-144} = \frac{47}{48}$$

Therefore a second simultaneous solution pair is $\left(-\dfrac{9}{4}, \dfrac{47}{48}\right)$, and the simultaneous solution set is

$$\left\{(3,2), \left(-\frac{9}{4}, \frac{47}{48}\right)\right\}$$

Check: Using (3,2), we have:

From equation (1):

$$(3)^2 + 4(3)(2) - 7(3) = 9 + 24 - 21 = 33 - 21 = 12$$

430

From equation (2):
$$3(3)^2 - 4(3)(2) + 4(3) = 27 - 24 + 12 = \quad 3 + 12 \ = \ 15$$

Using $\left(-\frac{9}{4}, \frac{47}{48}\right)$, we obtain:

From equation (1):
$$\left(-\frac{9}{4}\right)^2 + 4\left(\frac{-9}{4}\right)\left(\frac{47}{48}\right) - 7\left(\frac{-9}{4}\right) = \frac{81}{16} - \frac{141}{16} + \frac{63}{4} = \frac{81-141+252}{16}$$
$$= \frac{192}{16} = 12$$

From equation (2):
$$3\left(\frac{-9}{4}\right)^2 - 4\left(\frac{-9}{4}\right)\left(\frac{47}{48}\right) + 4\left(\frac{-9}{4}\right) = \frac{243}{16} + \frac{141}{16} - 9 = \frac{243+141-144}{16}$$
$$= \frac{240}{16} = 15$$

Thus, the two pairs of solutions obtained are valid.

● PROBLEM 607

Find the simultaneous solution set of the equations

$$3x^2 - 2y^2 - 6x = -23 \tag{1}$$

$$x^2 + y^2 - 4x = 13 \tag{2}$$

<u>Solution:</u> Since each of the given equations contains one term in y^2 and no other term involving y, we eliminate y^2 and then complete the process of solving as follows:

$3x^2 - 2y^2 - 6x = -23$ (3)	Eq. (1) recopied
$\underline{2x^2 + 2y^2 - 8x = \quad 26}$ (4)	multiplying Eq. (2) by 2
$5x^2 \qquad -14x = \qquad 3$	adding Equations (3) and (4)
$5x^2 - 14x - 3 = \quad 0$	adding -3 to each member

We can solve equations in the form $ax^2 + bx + c = 0$ by the quadratic formula, $x = \dfrac{-b \pm\sqrt{b^2 - 4ac}}{2a}$. In our case, $a = 5$, $b = -14$, and $c = -3$, thus,

$$x = \frac{-(-14) \pm \sqrt{(-14)^2 - 4(5)(-3)}}{2(5)}$$

$$x = \frac{14 \pm \sqrt{196 + 60}}{10}$$

$$= \frac{14 \pm \sqrt{256}}{10}$$

$$= \frac{14 \pm 16}{10}$$

$$= \frac{14 + 16}{10} \qquad \text{or} \qquad \frac{14 - 16}{10}$$

431

$$= \frac{30}{10} \qquad \text{or} \qquad \frac{-2}{10}$$

Thus, x = 3 or $-\frac{1}{5}$.

Now, solve for y by substituting the two values for x in eq. (2):

$$(3)^2 + y^2 - 4(3) = 13 \qquad \text{replacing x by 3 in eq.(2)}$$

$$9 + y^2 - 12 = 13$$

$$-3 + y^2 = 13$$

$$y^2 = 16 \qquad \text{adding 3 to each number}$$

$$y = \pm 4$$

Hence, when x = 3, y = ± 4. We continue by replacing x by $-\frac{1}{5}$.

$$\left(-\frac{1}{5}\right)^2 + y^2 - 4\left(-\frac{1}{5}\right) = 13 \qquad \text{replacing x by } -\frac{1}{5} \text{ in eq.(2)}$$

$$\frac{1}{25} + y^2 + \frac{4}{5} = 13$$

$$1 + 25y^2 + 20 = 325 \qquad \text{multiplying each member by 25}$$

$$25y^2 = 304 \qquad \text{adding } - 21 \text{ to each member}$$

$$y^2 = \frac{304}{25}$$

$$y = \pm \sqrt{\frac{304}{25}}$$

$$y = \pm \sqrt{\frac{(16)(19)}{25}} = \pm \frac{\sqrt{16}\ \sqrt{19}}{\sqrt{25}} = \pm \frac{4\ \sqrt{19}}{5}$$

Consequently, if x = $-\frac{1}{5}$, y = ± 4 $\sqrt{19}/5$. Therefore the simultaneous solution set is

$$\left\{ (3,4), (3, -4), \left(-\frac{1}{5}, 4\sqrt{19}/5\right), \left(-\frac{1}{5}, -4\sqrt{19}/5\right) \right\}.$$

Each solution pair can be checked by replacing x and y in the given equations by the appropriate number from the solution pair.

● **PROBLEM** 608

Solve the system
$$2x^2 - 3xy + 4y^2 = 3 \qquad (1)$$
$$x^2 + xy - 8y^2 = -6 \qquad (2)$$

<u>Solution:</u> Multiply both sides of the first equation by 2.
$$2\left(2x^2 - 3xy + 4y^2\right) = 2(3)$$

$$4x^2 - 6xy + 8y^2 = 6 \qquad (3)$$

Add equation (3) to equation (2):

$$x^2 + xy - 8y^2 = -6$$
$$\underline{4x^2 - 6xy + 8y^2 = 6}$$
$$5x^2 - 5xy = 0 \qquad (4)$$

Factoring out the common factor, 5x, from the left side of equation (4):

$$5x(x - y) = 0$$

Whenever a product $ab = 0$, where a, b are any two numbers, either $a = 0$ or $b = 0$. Hence, either

$$5x = 0 \quad \text{or} \quad x - y = 0$$
$$x = 0/5 \qquad\qquad x = y$$
$$x = 0$$

Substituting $x = 0$ in equation (1):

$$2(0)^2 - 3(0)y + 4y^2 = 3$$
$$0 - 0 + 4y^2 = 3$$
$$4y^2 = 3$$
$$y^2 = 3/4$$
$$y = \pm\sqrt{3/4}$$
$$= \pm\sqrt{3}/\sqrt{4}$$
$$= \pm\frac{\sqrt{3}}{2}$$

Hence, two solutions are: $\left(0, +\frac{\sqrt{3}}{2}\right)$, $\left(0, -\frac{\sqrt{3}}{2}\right)$.

Substituting x for y ($x = y$) in equation (1):

$$2x^2 - 3x(x) + 4(x)^2 = 3$$
$$2x^2 - 3x^2 + 4x^2 = 3$$
$$-x^2 + 4x^2 = 3$$
$$3x^2 = 3$$
$$x^2 = 3/3$$
$$x^2 = 1$$
$$x = \pm\sqrt{1} = \pm 1 .$$

Therefore, when $x = 1$, $y = x = 1$. Also, when $x = -1$, $y = x = -1$. Hence, two other solutions are: $(1,1)$ and $(-1,-1)$. Thus the four solutions of the system are

$$\left(0, \frac{\sqrt{3}}{2}\right), \left(0, -\frac{\sqrt{3}}{2}\right), (1,1) \text{ and } (-1,-1).$$

● **PROBLEM** 609

Obtain the simultaneous solution set of the equations

$$2x^2 + 3y^2 = 21 \qquad (1)$$
$$3x^2 - 4y^2 = 23 \qquad (2)$$

Solution: We arbitrarily select y as the unknown to be eliminated and proceed: Multiplying equation (1) by 4,

$$8x^2 + 12y^2 = 84 \qquad (3)$$

Multiplying equation (2) by 3,

$$9x^2 - 12y^2 = 69 \qquad (4)$$

Adding equations (3) and (4),

$$17x^2 = 153$$

Dividing both sides by 17,

$$x^2 = 9$$

Taking the square root of both members,

$$x = \pm 3$$

We now replace x by 3 and -3 in equation (1) and solve for y. We get

$$2(3)^2 + 3y^2 = 21 \quad \text{or} \quad 2(-3)^2 + 3y^2 = 21$$
$$2(9) + 3y^2 = 21 \qquad\qquad 2(9) + 3y^2 = 21$$

Thus in either case we obtain

$$18 + 3y^2 = 21$$

Adding -18 to each member,

$$3y^2 = 3$$
$$y^2 = 1$$
$$y = \pm 1$$

Consequently, the solution set is $\{(3,1),(3,-1),(-3,1),(-3,-1)\}$.

Check: To verify these solutions we replace x and y by each pair in equations (1) and (2). Checking (3,1) in equation (1):

$$2x^2 + 3y^2 = 21$$
$$2(3)^2 + 3(1)^2 = 21$$
$$2(9) + 3(1)^2 = 21$$
$$2(9) + 3(1) = 21$$
$$18 + 3 = 21$$
$$21 = 21$$

in equation (2):

$$3x^2 - 4y^2 = 23$$
$$3(3)^2 - 4(1)^2 = 23$$
$$3(9) - 4(1) = 23$$
$$27 - 4 = 23$$
$$23 = 23$$

Checking (3,-1) in equation (1):

$$2x^2 + 3y^2 = 21$$
$$2(3)^2 + 3(-1)^2 = 21$$
$$2(9) + 3(1) = 21$$
$$18 + 3 = 21$$
$$21 = 21$$

in equation (2):

$$3x^2 - 4y^2 = 23$$

$$3(3)^2 - 4(-1)^2 = 23$$
$$3(9) - 4(1) = 23$$
$$27 - 4 = 23$$
$$23 = 23$$

Checking $(-3,1)$ in equation (1):
$$2x^2 + 3y^2 = 21$$
$$2(-3)^2 + 3(1)^2 = 21$$
$$2(9) + 3(1) = 21$$
$$18 + 3 = 21$$
$$21 = 21$$

in equation (2):
$$3x^2 - 4y^2 = 23$$
$$3(-3)^2 - 4(1)^2 = 23$$
$$3(9) - 4(1) = 23$$
$$27 - 4 = 23$$
$$23 = 23$$

Checking $(-3,-1)$ in equation (1):
$$2x^2 + 3y^2 = 21$$
$$2(-3)^2 + 3(-1)^2 = 21$$
$$2(9) + 3(1) = 21$$
$$18 + 3 = 21$$
$$21 = 21$$

in equation (2):
$$3x^2 - 4y^2 = 23$$
$$3(-3)^2 - 4(-1)^2 = 23$$
$$3(9) - 4(1) = 23$$
$$27 - 4 = 23$$
$$23 = 23$$

● **PROBLEM** 610

Obtain the simultaneous solution set of

$$3x^2 + 3y^2 + x - 2y = 20 \tag{1}$$

$$2x^2 + 2y^2 + 5x + 3y = 9 \tag{2}$$

Solution: Multiplying equation (1) by 2,

$$6x^2 + 6y^2 + 2x - 4y = 40 \tag{3}$$

Multiplying equation (2) by 3,

$$6x^2 + 6y^2 + 15x + 9y = 27 \tag{4}$$

Subtracting equation (4) from (3),

$$- 13x - 13y = 13 \tag{5}$$

Now we solve Equation (5) simultaneously with Equation (2) and complete the process of solving as indicated below.

Dividing equation (5) by 13, $y = -x - 1$ (6)

$2x^2 + 2(-x - 1)^2 + 5x + 3(-x - 1) = 9$, replacing

                 y by $-x - 1$ in Eq.(2)

$2x^2 + 2(x^2 + 2x + 1) + 5x - 3x - 3 = 9$

$2x^2 + 2x^2 + 4x + 2 + 5x - 3x - 3 - 9 = 0$,   distributing

              and adding -9 to each member

    $4x^2 + 6x - 10 = 0$,   combining similar terms

    $2x^2 + 3x - 5 = 0$,   dividing by 2

    $(x - 1)(2x + 5) = 0$,   factoring

Whenever the product of two numbers $ab = 0$, either $a = 0$ or $b = 0$; thus either

    $x - 1 = 0$ or $2x + 5 = 0$,

    and $x = 1$ or $x = -\dfrac{5}{2}$

We find the corresponding values of y as follows:

    $y = -1 \cdot\cdot 1 = -2$, repacing x by 1 in Eq.(6)

    $y = \dfrac{5}{2} - 1 = \dfrac{3}{2}$,   replacing x by $-\dfrac{5}{2}$ in Eq. (6).

Consequently the solution set is $\{(1, -2), (-5/2, 3/2)\}$.

Check: Replacing x and y by $(1, -2)$ in equations (1) and (2);

    $3x^2 + 3y^2 + x - 2y = 20$ (1)

    $3(1)^2 + 3(-2)^2 + 1 - 2(-2) \stackrel{\leq}{=} 20$

        $3(1) + 3(4) + 1 + 4 = 20$

           $3 + 12 + 5 = 20$

              $20 = 20$

    $2x^2 + 2y^2 + 5x + 3y = 9$ (2)

    $2(1)^2 + 2(-2)^2 + 5(1) + 3(-2) = 9$

        $2(1) + 2(4) + 5 - 6 = 9$

           $2 + 8 + 5 - 6 = 9$

             $15 - 6 = 9$

Replacing x and y by (- 5/2, 3/2) in equations (1) and (2):

$$3x^2 + 3y^2 + x - 2y = 20 \tag{1}$$

$$3\left(-\frac{5}{2}\right)^2 + 3\left(\frac{3}{2}\right)^2 + \left(-\frac{5}{2}\right) - 2\left(\frac{3}{2}\right) = 20$$

$$3\left(\frac{25}{4}\right) + 3\left(\frac{9}{4}\right) + \frac{-5}{2} - \frac{6}{2} = 20$$

$$\frac{75}{4} + \frac{27}{4} - \frac{10}{4} - \frac{12}{4} = 20$$

$$\frac{80}{4} = 20$$

$$20 = 20$$

$$2x^2 + 2y^2 + 5x + 3y = 9 \tag{2}$$

$$2\left(-\frac{5}{2}\right)^2 + 2\left(\frac{3}{2}\right)^2 + 5\left(-\frac{5}{2}\right) + 3\left(\frac{3}{2}\right) = 9$$

$$2\left(\frac{25}{4}\right) + 2\left(\frac{9}{4}\right) - \frac{25}{2} + \frac{9}{2} = 9$$

$$\frac{50}{4} + \frac{18}{4} - \frac{50}{4} + \frac{18}{4} = 9$$

$$\frac{36}{4} = 9$$

$$9 = 9$$

● **PROBLEM 611**

Solve

$$x^2 + y^2 = 5 \tag{1}$$

$$x^2 - xy + y^2 = 7 \tag{2}$$

Solution: Subtracting the second equation from the first, we have

$$x^2 + y^2 = 5$$

$$\frac{-(x^2 - xy + y^2 = 7)}{xy = -2} \tag{3}$$

Thus let us consider the system

$$x^2 + y^2 = 5$$

$$xy = -2.$$

Solving the second equation for y, we obtain

$$y = -\frac{2}{x}$$

Substituting this result in equation (1), we have,

$$x^2 + \left(-\frac{2}{x}\right)^2 = 5$$

$$x^2 + \frac{(-2)^2}{(x)^2} = 5$$

$$x^2 + \frac{4}{x^2} = 5.$$

Then we multiply both sides by x^2 to obtain

$$x^2\left(x^2 + \frac{4}{x^2}\right) = 5(x^2)$$

$$x^2 \cdot x^2 + x^2\left(\frac{4}{x^2}\right) = 5x^2$$

$$x^4 + 4 = 5x^2$$

or $\qquad x^4 - 5x^2 + 4 = 0.$

Hence $(x^2 - 1)(x^2 - 4) = 0$. Whenever the product of two numbers $ab = 0$, either $a = 0$ or $b = 0$; therefore $x^2 = 1$ or $x^2 = 4$. Thus $x = \sqrt{1} = \pm 1$, or $x = \sqrt{4} = \pm 2$. Substituting these values in turn in the equation $xy = -2$ we obtain the solutions

For $x = 1$, $\quad 1(y) = -2$, $\quad y = -2$

For $x = -1$, $-1(y) = -2$, $y = 2$

For $x = 2$, $\qquad 2y = -2$, $y = -1$

For $x = -2$, $\quad -2y = -2$, $y = 1$.

Therefore the solution to this system of equations is

$$(1,-2), \; (-1,2), \; (2,-1), \; (-2,1).$$

Check: To verify that these four pairs are indeed solutions we replace x and y by each pair in equations (1) and (2). Thus checking (1,-2) in (1),

$$x^2 + y^2 = 5$$

$$1^2 + (-2)^2 = 5$$

$$1 + 4 = 5$$

$$5 = 5$$

and in (2), $\quad x^2 - xy + y^2 = 7$

438

$$(1)^2 - 1(-2) + (-2)^2 = 7$$

$$1 + 2 + 4 = 7$$

$$7 = 7.$$

Checking $(-1,2)$ in (1),

$$x^2 + y^2 = 5$$

$$(-1)^2 + (2)^2 = 5$$

$$1 + 4 = 5$$

$$5 = 5$$

and in (2), $\qquad x^2 - xy + y^2 = 7$

$$(-1)^2 - (-1)(2) + (2)^2 = 7$$

$$1 + 2 + 4 = 7$$

$$7 = 7.$$

Checking $(2,-1)$ in (1),

$$x^2 + y^2 = 5$$

$$(2)^2 + (-1)^2 = 5$$

$$4 + 1 = 5$$

$$5 = 5$$

and in (2),

$$x^2 - xy + y^2 = 7$$

$$(2)^2 - (2)(-1) + (-1)^2 = 7$$

$$4 + 2 + 1 = 7$$

$$7 = 7.$$

Checking $(-2,1)$ in (1),

$$x^2 + y^2 = 5$$

$$(-2)^2 + (1)^2 = 5$$

$$4 + 1 = 5$$

$$5 = 5$$

and in (2),

$$x^2 - xy + y^2 = 7$$

$$(-2)^2 - (-2)(1) + (1)^2 = 7$$

$$4 + 2 + 1 = 7$$

$$7 = 7.$$

Thus our 4 pairs are all valid solutions and the solution set is { (1,-2), (-1,2), (2,-1)(-2,1) }.

● **PROBLEM 612**

Solve the following system of equations completely:

$$\begin{cases} x^2 - 5xy + 6y^2 = 0, \\ xy - y^2 = 2. \end{cases}$$

Solution: Upon factoring the left member of the first equation, we obtain $(x - 3y)(x - 2y) = 0$. Whenever the product $ab = 0$ where a and b are any two numbers, either $a = 0$ or $b = 0$. Hence, either $x - 3y = 0$ or $x - 2y = 0$.

Each of these is taken with the second given equation, and we obtain the following systems of equations:

System 1: $\begin{cases} xy - y^2 = 2, \\ x - 3y = 0, \end{cases}$ and System 2: $\begin{cases} xy - y^2 = 2, \\ x - 2y = 0. \end{cases}$

Any solution of either system will be a solution of the given system. For system 1:

$$\begin{cases} xy - y^2 = 2 & \text{(a)} \\ x - 3y = 0 & \text{(b)}. \end{cases}$$

Multiply both sides of equation (b) by y:

$$y(x - 3y) = y(0)$$

$$yx - 3y^2 = 0 \text{ or, by the commutative}$$

property of multiplication,

$$xy - 3y^2 = 0 \qquad \text{(c)}$$

Subtract equation (c) from equation (a):

$$\begin{array}{r} xy - y^2 = 2 \\ -\left(xy - 3y^2\right) = -0 \\ \hline 2y^2 = 2 \end{array}$$

Divide both sides by 2:

$$\frac{2y^2}{2} = \frac{2}{2}$$

$$\text{or } y^2 = 1$$

Take the square root of both sides

$$y = \pm \sqrt{1}$$

$$y = \pm 1$$

Substitute $y = -1$ in equation (a):

440

$$x(-1) - (-1)^2 = 2$$
$$-x - 1 = 2$$

Add x to both sides:

$$-x - 1 + x = 2 + x$$
$$-1 = 2 + x$$

Subtract 2 from both sides:

$$-1 - 2 = 2 + x - 2$$
$$-3 = x.$$

Hence, one solution is (-3, -1).

Substitute y = 1 in equation (a):

$$x(1) - (1)^2 = 2$$
$$x - 1 = 2$$

Add 1 to both sides:

$$x - 1 + 1 = 2 + 1$$
$$x = 3$$

Hence, another solution is (3,1).

For System 2:

$$\begin{cases} xy - y^2 = 2 & \text{(d)} \\ x - 2y = 0 & \text{(e)} \end{cases}$$

Multiply both sides of equation (e) by y:

$$y(x - 2y) = y(0)$$
$$yx - 2y^2 = 0 \text{ or, by the commutative property}$$

of multiplication,

$$xy - 2y^2 = 0 \text{ (f)}$$

Subtract equation (f) from equation (d):

$$xy - y^2 = 2$$
$$-\left(xy - 2y^2\right) = -0$$
$$\overline{ y^2 = 2}$$

Take the square root of both sides:

$$y = \pm\sqrt{2}.$$

Substitute $y = -\sqrt{2}$ in equation (d):

$$x(-\sqrt{2}) - (-\sqrt{2})^2 = 2$$

$$x(-\sqrt{2}) - (2) = 2 \text{ or, by the commutative property}$$

of multiplication,

$$-\sqrt{2}\,x - 2 = 2$$

Add 2 to both sides:

$$-\sqrt{2}\,x - 2 + 2 = 2 + 2$$

$$-\sqrt{2}\,x = 4$$

Divide both sides by $(-\sqrt{2})$:

$$\frac{-\sqrt{2}\ x}{-\sqrt{2}} = \frac{4}{-\sqrt{2}}$$

$$x = \frac{4}{-\sqrt{2}}$$

Multiply the numerator and denominator of the fraction on the right side by $\sqrt{2}$ in order to remove the radical sign in the denominator. (Note: This process is called "rationalizing the denominator")

$$x = \frac{4}{-\sqrt{2}} = \frac{\sqrt{2}(4)}{\sqrt{2}(-\sqrt{2})} = \frac{4\sqrt{2}}{-2} = -2\sqrt{2}$$

Hence, another solution is $(-2\sqrt{2}, -\sqrt{2})$.

Substitute $y = +\sqrt{2} = \sqrt{2}$ in equation (d):

$$x(\sqrt{2}) - (\sqrt{2})^2 = 2$$

$x(\sqrt{2}) - 2 = 2$ or, by the commutative property of multiplication,

$$\sqrt{2}x - 2 = 2.$$

Add 2 to both sides.

$$\sqrt{2}x - 2 + 2 = 2 + 2$$

$$\sqrt{2}x = 4$$

Divide both sides by $\sqrt{2}$:

$$\frac{\sqrt{2}x}{\sqrt{2}} = \frac{4}{\sqrt{2}}$$

$$x = \frac{4}{\sqrt{2}}$$

Rationalize the denominator on the right side:

$$x = \frac{4}{\sqrt{2}} = \frac{\sqrt{2}(4)}{\sqrt{2}(\sqrt{2})} = \frac{4\sqrt{2}}{2} = 2\sqrt{2}.$$

Therefore, another solution is $(2\sqrt{2}, \sqrt{2})$.

Hence, the four solutions are

$$(3, 1), \quad (-3, -1), \quad (2\sqrt{2}, \sqrt{2}), \quad (-2\sqrt{2}, -\sqrt{2}),$$

where the first two are the solutions of the first system, and the second two are the solutions of the second system.

● **PROBLEM** 613

Solve the system of equations,

$$\begin{cases} x^2 + y^2 = 25, \\ xy = 12. \end{cases}$$

Solution: If the second equation of the system is multiplied by 2 and subtracted from the first we obtain $x^2 - 2xy + y^2 = 1$, which can be written as $(x - y)^2 = 1$. This last equation is equivalent to the two equations

$x - y = 1$ and $x - y = -1$ (since squaring both of these gives us back the original). Thus the solutions of the given system may be found by solving the two systems

$$(1) \begin{cases} xy = 12 \\ x - y = 1, \end{cases} \quad \text{and} \quad (2) \begin{cases} xy = 12 \\ x - y = -1. \end{cases}$$

For system (1):

$$xy = 12 \qquad \text{(a)}$$
$$x - y = 1 \qquad \text{(b)}$$

Solving equation (b) for x and then substituting this expression for x in equation (a), we obtain:

$$x = 1 + y \qquad \text{(b')}$$
$$(1 + y)y = 12 \qquad \text{(a')}$$
$$y + y^2 = 12$$
$$y^2 + y - 12 = 0$$

Factoring,
$$(y + 4)(y - 3) = 0$$

Finding all values of y which make this product zero, we set each factor equal to zero:

$$(y + 4) = 0 \qquad\qquad (y - 3) = 0$$
$$y = -4 \qquad \text{or} \qquad y = 3$$

Then from (b'), $x = 1 + y = 1 + (-4) = -3$ or

$$x = 1 + y = 1 + 3 = 4$$

The results of the original system thus far are $(-3, -4)$ and $(4, 3)$. For system (2):

$$xy = 12 \qquad \text{(A)}$$
$$x - y = -1 \qquad \text{(B)}$$

Then, as for system (1):

$$x = y - 1 \qquad \text{(B')}$$
$$(y - 1)y = 12 \qquad \text{(A')}$$
$$y^2 - y = 12$$
$$y^2 - y - 12 = 0$$

Factoring, $(y - 4)(y + 3) = 0$

Setting each factor $= 0$,

$$(y - 4) = 0 \qquad \text{or} \qquad (y + 3) = 0$$
$$y = 4 \qquad \text{or} \qquad y = -3$$

Then from (B'), $x = y - 1 = 4 - 1 = 3$
$$x = y - 1 = -3 - 1 = -4$$

The results for system (2) are $(3, 4)$ and $(-4, -3)$.

Thus the results for systems (1) and (2) and for the original system are

$$(4, 3); \qquad (-3, -4); \qquad (-4, -3); \qquad (3, 4).$$

Solve the system
$$3x^2 + 4y^2 = 8 \qquad\qquad (1)$$
$$x^2 - y^2 = 5. \qquad\qquad (2)$$

<u>Solution:</u> Substituting u for x^2 and v for y^2 leads to the system of linear equations

$$3u + 4v = 8 \qquad\qquad (3)$$
$$u - v = 5 \qquad\qquad (4)$$

Multiplying both sides of equation (4) by 3 we obtain:

$$3(u - v) = 3(5)$$
$$3u - 3v = 15 \qquad\qquad (5)$$

Subtracting equation (5) from equation (3):

$$3u + 4v = 8$$
$$-(3u - 3v = 15)$$
$$\overline{\qquad 7v = -7}$$

Dividing both sides by 7:

$$\frac{7v}{7} = \frac{-7}{7}$$
$$v = -1$$

Since $v = y^2$,

$$v = -1 = y^2$$
$$\pm\sqrt{-1} = y$$

Since $i^2 = -1$ or $i = \sqrt{-1}$,

$$\pm i = y .$$

Substituting the value $y = i$ into equation (2):

$$x^2 - (i)^2 = 5$$
$$x^2 - i^2 = 5$$
$$x^2 - (-1) = 5$$
$$x^2 + 1 = 5$$

Subtracting 1 from both sides:

$$x^2 + 1 - 1 = 5 - 1$$
$$x^2 = 4$$
$$x = \pm\sqrt{4} = \pm 2$$

Hence, two solutions of the original system of equations are:
$$(2,i), \ (-2,i)$$

Substituting the value $y = -i$ into equation (2):

$$x^2 - (-i)^2 = 5$$
$$x^2 - (i^2) = 5$$
$$x^2 - (-1) = 5$$
$$x^2 + 1 = 5$$

Subtracting 1 from both sides:

$$x^2 + 1 - 1 = 5 - 1$$

$$x^2 = 4$$
$$x = \pm \sqrt{4} = \pm 2$$

Hence, two other solutions of the original system of equations are:

$$(2,-1), \quad (-2,-1).$$

Therefore, the solution set of the original system of equations is:

$$\{(2,1), (-2,1), (2,-1), (-2,-1)\}.$$

Other systems that involve quadratic equations may be solved by replacing the given system with an equivalent system that is easier to solve.

• **PROBLEM 615**

Solve $\quad x^2 + 3xy = 28,$

$$x^2 + y^2 = 20.$$

Solution: Let $y = mx$ and substitute in both equations. From the first equation, we have

$$x^2 + 3mx^2 = 28.$$

Solving this equation for x^2: use the distributive law in relation to the x^2 terms,

$$x^2 + 3mx^2 = 28$$

$$(1 + 3m)x^2 = 28.$$

Divide both sides of this equation by $(1 + 3m)$:

$$\frac{(1 + 3m)x^2}{(1 + 3m)} = \frac{28}{(1 + 3m)}$$

$$x^2 = \frac{28}{1 + 3m} \qquad (1)$$

From the second equation, we have

$$x^2 + m^2x^2 = 20.$$

Solving this equation for x^2: use the distributive law in relation to the x^2 terms,

$$x^2 + m^2x^2 = 20$$

$$(1 + m^2)x^2 = 20.$$

Divide both sides of this equation by $(1 + m^2)$:

$$\frac{(1 + m^2)x^2}{(1 + m^2)} = \frac{20}{(1 + m^2)}$$

445

$$x^2 = \frac{20}{1 + m^2}.$$ (2)

Equating the values obtained for x^2 in equations (1) and (2):

$$\frac{28}{(1 + 3m)} = \frac{20}{1 + m^2}$$

Multiply both sides of this equation by $(1 + 3m)$:

$$(1 + 3m)\left(\frac{28}{1 + 3m}\right) = (1 + 3m)\left(\frac{20}{1 + m^2}\right)$$

$$28 = \frac{(1 + 3m)(20)}{(1 + m^2)}$$

Multiply both sides of this equation by $\left(1 + m^2\right)$:

$$\left(1 + m^2\right)(28) = \left(1 + m^2\right)\left(\frac{1 + 3m}{1 + m^2}\right)(20)$$

$$\left(1 + m^2\right)(28) = (1 + 3m)(20)$$

$$28 + 28m^2 = 20 + 60m.$$

Subtract $(20 + 60m)$ from both sides of this equation:

$$28 + 28m^2 - (20 + 60m) = 20 + 60m - (20 + 60m)$$

$$28 + 28m^2 - 20 - 60m = 20 + 60m - 20 - 60m$$

$$28 + 28m^2 - 20 - 60m = 0$$

or $$28m^2 - 60m + 8 = 0.$$

Divide both sides of this equation by 4:

$$\frac{28m^2 - 60m + 8}{4} = \frac{0}{4}$$

$$7m^2 - 15m + 2 = 0.$$

Factor the left side of this equation into a product of two polynomials:

$$(7m - 1)(m - 2) = 0.$$ (3)

Whenever a product of two numbers $ab = 0$, where a and b are any two numbers, either $a = 0$ or $b = 0$. Equation (3) can be rewritten as:

$$7m - 1 = 0 \quad \text{or} \quad m - 2 = 0$$

$$7m = 1 \quad \text{or} \quad m = 2$$

$$m = \frac{1}{7}$$

Substituting these values of m in equation (1):

$$\text{for } m = \tfrac{1}{7}, \quad x^2 = \frac{28}{1 + 3\left(\tfrac{1}{7}\right)}$$

$$= \frac{28}{1 + \tfrac{3}{7}}$$

$$= \frac{28}{\tfrac{7}{7} + \tfrac{3}{7}}$$

$$= \frac{28}{\tfrac{10}{7}} \qquad (4)$$

Since division by a fraction is equivalent to multiplication of the numerator by the reciprocal of the denominator, equation (4) becomes:

$$x^2 = \overset{14}{\cancel{(28)}} \, \frac{7}{\underset{5}{\cancel{10}}}$$

$$x = \frac{98}{5}$$

Taking the square root of both sides ot this equation:

$$\sqrt{x^2} = \pm\sqrt{\frac{98}{5}}$$

$$x = \frac{\pm\sqrt{98}}{\sqrt{5}}$$

$$= \pm\frac{\sqrt{49\sqrt{2}}}{\sqrt{5}}$$

$$x = \pm\frac{7\sqrt{2}}{\sqrt{5}}$$

Rationalizing the denominator by multiplying the numerator and denominator by $\sqrt{5}$:

$$x = \pm\frac{\sqrt{5}\,(7\sqrt{2})}{\sqrt{5}\,(\sqrt{5})}$$

$$x = \pm\frac{7\sqrt{10}}{5}$$

$$x = \pm\frac{7}{5}\sqrt{10}.$$

To calculate the y-values that correspond to these x-values, use the equation $y = mx$ (with $m = 1/7$):

When $x = \frac{7}{5}\sqrt{10}$, $y = mx = \left(\tfrac{1}{7}\right)\left(\tfrac{7}{5}\sqrt{10}\right) = \tfrac{1}{5}\sqrt{10}$,

When $x = -\frac{7}{5}\sqrt{10}$, $y = mx = \left(\frac{1}{7}\right)\left(-\frac{7}{5}\sqrt{10}\right) = -\frac{1}{5}\sqrt{10}$.

Then, two solutions are: $\left(\frac{7}{5}\sqrt{10}, \frac{1}{5}\sqrt{10}\right)$ and $\left(-\frac{7}{5}\sqrt{10}, -\frac{1}{5}\sqrt{10}\right)$.

For m = 2 (using equation (1)):

$$x^2 = \frac{28}{1 + 3(2)} = \frac{28}{1 + 6} = \frac{28}{7} = 4$$

or $\quad x^2 = 4.$

Taking the square root of both sides of this equation:

$$\sqrt{x^2} = \pm\sqrt{4}$$
$$x = \pm 2.$$

To calculate the y-values that correspond to these x-values, use the equation y = mx (with m = 2):

When x = 2, y = mx = (2)(2) = 4,

When x = -2, y = mx = (2)(-2) = -4.

Then, two other solutions are: (2,4) and (-2,-4). Therefore, the solutions to the original pair of equations are:

$\left(\frac{7}{5}\sqrt{10}, \frac{1}{5}\sqrt{10}\right)$, $\left(-\frac{7}{5}\sqrt{10}, -\frac{1}{5}\sqrt{10}\right)$, (2,4), and (-2,-4).

● **PROBLEM 616**

Solve graphically

$$\begin{cases} x^2 + y^2 = 13, & (1) \\ y = x^2 - 1. & (2) \end{cases}$$

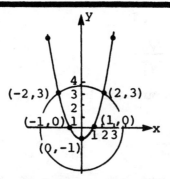

Solution: First, we must find the x and y intercepts. Set y = 0 to find the x-intercept or where the curve crosses the x-axis. Set x = 0 to find the y-intercept or where the curve crosses the y-axis. In Eq. (1), set x = 0, and find $y = \pm\sqrt{13} = \pm 3.6$. Then set y = 0, and find $x = \pm\sqrt{13}$. To get additional points, we solve for y.

$$x^2 + y^2 = 13$$
$$y^2 = 13 - x^2$$

$$y = \pm \sqrt{13 - x^2}$$

Then, set up a table. Choose various x values and calculate the corresponding y values. See Graph.

x	$\pm \sqrt{13 - x^2}$ =	y
-3.6	$+ \sqrt{13 - (-3.6)^2}$	≈ 0
-3	$\pm \sqrt{13 - (-3)^2}$	± 2
-2	$+ \sqrt{13 - (-2)^2}$	$+ 3$
-1	$+ \sqrt{13 - (-1)^2}$	± 3.5
0	$+ \sqrt{13 - (0)^2}$	$+ 3.6$
1	$\pm \sqrt{13 - (1)^2}$	± 3.5
2	$+ \sqrt{13 - (2)^2}$	± 3
3	$\pm \sqrt{13 - (3)^2}$	± 2
3.6	$+ \sqrt{13 - (3.6)^2}$	≈ 0

To find the domain of the relation, $+ \sqrt{13 - x^2}$, we know that the expression, $13 - x^2$, under the square root sign must be positive in order for the expression to be real, not imaginary.

$$(13 - x^2) \geq 0$$

subtract 13 from both sides, $-x^2 \geq -13$

multiply by -1 and reverse the inequality sign,

$$x^2 \leq 13$$

Take the square root of both sides.

$$|x| \leq \sqrt{13}$$

Another way to express $|b| \leq a$ is $-a \leq b \leq +a$. Thus,

$$-\sqrt{13} \leq x \leq + \sqrt{13}$$

Thus, for the relation $y = \pm \sqrt{13-x^2}$, the domain is $\{x | -\sqrt{13} \leq x \leq \sqrt{13}\}$. The curve is a circle. The general equation of a circle is $(x-h)^2 + (y-k)^2 = r^2$, where (h,k) is the center and r is the radius. In this case (0,0) or the origin is the center and r^2 is 13. Therefore, the radius = $+ \sqrt{13}$.

In Eq. (2), y is a quadratic function of x; hence the graph is a parabola. Set up a similar table for the quadratic function, $y = x^2 - 1$.

x	$x^2 - 1$ =	y
-3	$(-3)^2 - 1$	8
-2	$(-2)^2 - 1$	3
-1	$(-1)^2 - 1$	0
0	$(0)^2 - 1$	-1
1	$(1)^2 - 1$	0
2	$(2)^2 - 1$	3
3	$(3)^2 - 1$	8

From the graphs we read the real solutions (2,3) and (-2,3). These are points of intersection for both curves.

To find the solutions algebraically substitute equation (2) into (1).

$$x^2 + y^2 = 13 \qquad (1)$$

$$y = x^2 - 1 \qquad (2)$$

449

$$x^2 + (x^2 - 1)^2 = 13$$
$$x^2 + x^4 - 2x^2 + 1 = 13$$
$$x^4 - x^2 = 12$$
$$x^4 - x^2 - 12 = 0$$

Substitute z for x^2, i.e., $z = x^2$ to obtain a quadratic equation in z.

$$(x^2)^2 - (x^2) - 12 = 0$$
$$z^2 - z - 12 = 0$$
$$(z - 4)(z + 3) = 0$$
$$z - 4 = 0 \qquad z + 3 = 0$$
$$z = 4 \qquad z = -3$$

Therefore

$$x^2 = 4 \qquad x^2 = -3$$
$$x = \pm 2 \qquad x = \pm \sqrt{-3} = \pm \sqrt{3}i$$

Find the corresponding y-values by substituting into $y = x^2 - 1$.

$x = 2$	$x = -2$	$x = i\sqrt{3}$	$x = -i\sqrt{3}$
$y = (2)^2 - 1$	$y = (-2)^2 - 1$	$y = (i\sqrt{3})^2 - 1$	$y = (-i\sqrt{3})^2 - 1$
$= 3$	$= 3$	$y = i^2(3) - 1$	$= 3(i)^2 - 1$
		$= (-1)(3) - 1$	$= 3(-1) - 1$
		$= -4$	$= -4$

The algebraic solution gives $(2,3)$, $(-2,3)$, $(i\sqrt{3}, -4)$, and $(-i\sqrt{3}, -4)$. Notice that the imaginary solutions do not appear on the graph.

● **PROBLEM 617**

Obtain the simultaneous solution set of the equations
$$y = x^2 - 4 \qquad (1)$$
$$3x^2 + 8y^2 = 75 \qquad (2)$$
by the graphical method.

Solution: Equation (1) is in the form of the function

$ax^2 + bx + c$. Its graph is a parabola. When a is positive, the curve opens upward. If it is negative, the curve opens downward. In this case $a = 1$, which is positive. Hence the graph is a parabola opening upward. We construct the parabola by means of the following table of corresponding values and show the graph in the accompanying figure.

x	$x^2 - 4 =$	y
-3	$(-3)^2 - 4 = 9 - 4 =$	5
-2	$(-2)^2 - 4 = 4 - 4 =$	0
-1	$(-1)^2 - 4 = 1 - 4 =$	-3
0	$0^2 - 4 = 0 - 4 =$	-4
1	$(1)^2 - 4 = 1 - 4 =$	-3
2	$(2)^2 - 4 = 4 - 4 =$	0
3	$(3)^2 - 4 = 9 - 4 =$	5

Equation (2) is of the type $ax^2 + by^2 = c$, with a = 3, b = 8, and c = 75. The graph is therefore an ellipse. To find the x-intercepts, set y = 0.

$3x^2 + 8(0)^2 = 75$

Solve for x.

$3x^2 = 75$

$x^2 = 25$

$x = \pm 5$

We have (-5,0) and (+5,0) as the x- intercepts
To find the y-intercepts, set x = 0

$$3(0)^2 + 8y^2 = 75$$
$$8y^2 = 75$$
$$y^2 = \frac{75}{8} = 9\frac{3}{8} = 9.375$$
$$y \approx \pm 3.1$$

We obtain (0,3.1) and (0,-3.1) for the y-intercepts.
We construct the graph and obtain the ellipse.

We now solve for the points of intersection, indicated on the graph.

$$y = x^2 - 4 \qquad\qquad (1)$$
$$3x^2 + 8y^2 = 75 \qquad\qquad (2)$$

Substitute the value of y in (1) into (2)

$3x^2 + 8\left(x^2 - 4\right)^2 = 75$

$3x^2 + 8\left(x^2 - 8x^2 + 16\right) = 75$

$3x^2 + 8x^4 - 64x^2 + 128 = 75$

$8x^4 - 61x^2 + 53 = 0$

Let $z = x^2$

$8z^2 - 61z + 53 = 0$

Apply the quadratic formula $z = \dfrac{-b \pm \sqrt{b^2 - 4ac}}{2a}$ with a = 8, b = -61 and c = 53.

$$z = \frac{-(-61) \pm \sqrt{(-61)^2 - 4(8)53}}{2(8)} = \frac{61 \pm \sqrt{3721 - 1696}}{16}$$

$$z = \frac{61 \pm 45}{16}$$

$$z = 6.625, \ 1$$

$$z = x^2 = 6.625, \ 1$$

$$x = \pm 2.57 \approx \pm 2.6, \ x = \pm 1$$

To solve for the y-values, substitute the x values.

$$y = x^2 - 4$$

For x = ± 1

$$y = 1^2 - 4$$

$$y = -3$$

For x = ± 2.6

$$y = (2.57)^2 - 4$$

$$y = 6.6 - 4 = 2.6$$

Therefore the points of intersection are
{(1,-3), (-1,-3), (-2.6,2.6), (+2.6,+2.6)}.

● **PROBLEM   618**

Solve   xy = 3

$$x^2 + y^2 = 10$$

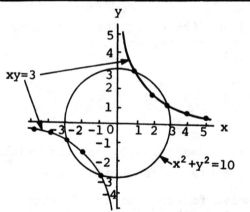

Solution: We solve the first equation for y to obtain
y = 3/x and substitute in the second equation to obtain

$$x^2 + \left(\frac{3}{x}\right)^2 = 10$$

Squaring: $x^2 + \dfrac{9}{x^2} = 10$

Multiplying both sides by x^2: $x^2\left(x^2 + \dfrac{9}{x^2}\right) = x^2 (10)$

Distributing: $x^4 + 9 = 10x^2$

Subtracting $10x^2$ from both sides: $x^4 - 10x^2 + 9 = 0$

Factoring: $(x^2 - 1)(x^2 - 9) = 0$

Thus, $x^2 - 1 = 0$ or $x^2 - 9 = 0$

$$x^2 = 1 \qquad\qquad x^2 = 9$$

$$x = \pm 1 \qquad\qquad x = \pm 3$$

Therefore, $x = 1, -1, 3,$ or -3.

Since $y = 3/x$, substituting these values in turn in this equation, we obtain the corresponding values for y:

$x = 1$	$x = -1$	$x = 3$	$x = -3$
$y = 3$	$y = -3$	$y = 1$	$y = -1$

To consider this system graphically, we notice that the second equation is the equation of a circle with radius $\sqrt{10}$, whereas the graph of the first equation is a hyperbola obtained from the following table. Also, the points $(1, 3)$ $(-1, -3)$, $(3, 1)$, $(-3, -1)$ belong to both the circle and the hyperbola, and the two graphs intersect at these points.

x	-5	-4	-3	-2	-1	1	2	3	4	5
y	$-\frac{3}{5}$	$-\frac{3}{4}$	-1	$-\frac{3}{2}$	-3	3	$\frac{3}{2}$	1	$\frac{3}{4}$	$\frac{3}{5}$

Plotting these points, and the circle with radius $\sqrt{10}$ (approximately equal to 3.16) we have the accompanying diagram.

● **PROBLEM 619**

Solve for x and y: $\begin{cases} x^2 + y^2 = 25, \\ x^2 - y^2 = 7. \end{cases}$

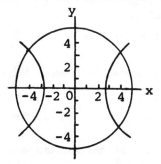

<u>Solution:</u> Add the following two equations to eliminate y^2.

$$(1) \quad x^2 + y^2 = 25$$
$$(2) \quad \underline{x^2 - y^2 = 7}$$
$$2x^2 \qquad = 32$$

453

Divide by 2

$$x^2 = 16$$

Take the square root of both sides.

$$x = \pm 4$$

Substitute x^2 into $x^2 + y^2 = 25$ to solve for y^2.

$$16 + y^2 = 25$$
$$y^2 = 25 - 16$$
$$y^2 = 9$$
$$y = \pm 3$$

$$x = 4, \quad y = -3; \qquad x = -4, \quad y = -3,$$
$$x = 4, \quad y = 3; \qquad x = -4, \quad y = 3;$$

as can be verified by substitution in the given equations.

Graphical solution. If the graph of the first equation is constructed by finding pairs of values (x,y) which satisfy the equation and plotting the corresponding points, the circle shown is obtained. In a similar manner, the hyperbola shown in the figure is obtained as the graph of the second equation. The circle and hyperbola are seen to intersect in the four points $(4,3), (-4,3), (-4,-3), (4,-3)$.

There is another way to graph these equations. The standard form of the equation of the circle whose center is at the point $c(h,k)$ and whose radius is the constant r is

$$(x - h)^2 + (y - k)^2 = r^2.$$

In this case the center is $(0,0)$ and the radius is 5. Therefore, move out 5 units from the center in all directions. The circle will then intersect the axes at $(5,0), (-5,0), (0,5)$, and $(0,-5)$. Write the hyperbola in the general form,

$$\frac{x^2}{a^2} - \frac{y^2}{b^2} = 1.$$

The hyperbola intersects one of its lines of symmetry, the x-axis, in the points $(-a,0)$ and $(a,0)$. Rewriting $x^2 - y^2 = 7$, we obtain

$$\frac{x^2}{7} - \frac{y^2}{7} = 1. \quad a^2 = 7 \quad \text{and} \quad a = \pm \sqrt{7}.$$

Therefore the points of intersection on the x-axis are $(-\sqrt{7},0)$ and $(+\sqrt{7},0)$. This is equivalent to

$$(-2.65,0) \quad \text{and} \quad (+2.65,0).$$

● PROBLEM 620

Solve $9x^2 + 16y^2 = 288,$ (1)

$\qquad x^2 + y^2 = 25.$ (2)

Solution: Solving for y^2 first (a similar method solving for x^2 first is equally good): Multiply Equation (2) by 9 and then subtract Equation (2) from Equation (1)

$$9x^2 + 16y^2 = 288$$
$$\underline{-9x^2 + 9y^2 = 225}$$
$$7y^2 = 63$$

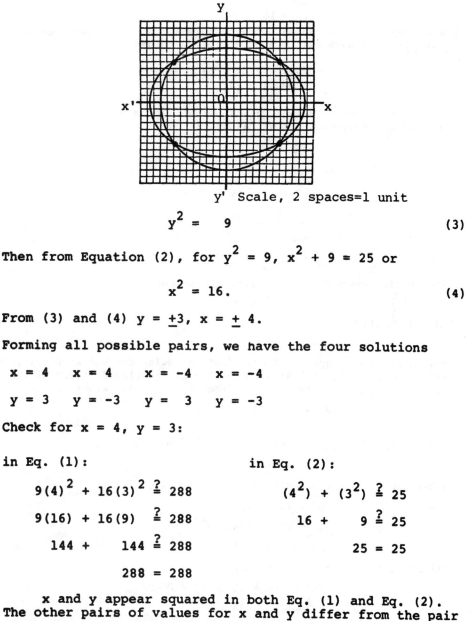

y' Scale, 2 spaces=1 unit

$$y^2 = 9 \tag{3}$$

Then from Equation (2), for $y^2 = 9$, $x^2 + 9 = 25$ or

$$x^2 = 16. \tag{4}$$

From (3) and (4) $y = \pm 3$, $x = \pm 4$.

Forming all possible pairs, we have the four solutions

$x = 4$ $x = 4$ $x = -4$ $x = -4$

$y = 3$ $y = -3$ $y = 3$ $y = -3$

Check for $x = 4$, $y = 3$:

in Eq. (1):

$9(4)^2 + 16(3)^2 \stackrel{?}{=} 288$

$9(16) + 16(9) \stackrel{?}{=} 288$

$144 + 144 \stackrel{?}{=} 288$

$288 = 288$

in Eq. (2):

$(4^2) + (3^2) \stackrel{?}{=} 25$

$16 + 9 \stackrel{?}{=} 25$

$25 = 25$

x and y appear squared in both Eq. (1) and Eq. (2). The other pairs of values for x and y differ from the pair checked only in sign. Therefore the other pairs also satisfy Equation (1) and Equation (2).

The Equation (1) has for its locus an oval-shaped figure called an ellipse. (Fig.) The equation (2) has a circle for its locus. The four points of intersection represent graphically the four solutions.

MULTIVARIABLE COMBINATIONS

● **PROBLEM** 621

Solve $x + y + z = 13$ (1),

$x^2 + y^2 + z^2 = 65$ (2),

$$xy = 10 \qquad\qquad (3).$$

__Solution:__ We note that the product of two binomials is

$$(x+y)^2 = (x+y)(x+y) = x^2 + xy + xy + y^2$$
$$= x^2 + 2xy + y^2 .$$

Therefore, if we were just to add (2) and (3) we would be missing an xy term. Thus, multiplying (3) by 2,

$$2xy = 20$$

Adding this to (2),

$$x^2 + y^2 + z^2 = 65$$
$$+ \underline{\qquad\quad 2xy = 20 \qquad}$$
$$x^2 + 2xy + y^2 + z^2 = 85$$

$$(x+y)^2 + z^2 = 85$$

Put u for x+y; then this equation becomes

$$u^2 + z^2 = 85 .$$

Also from (1),

$$u + z = 13;$$

Now to solve this system of one linear equation and one quadratic, we express one of the variables in the linear equation in terms of the other and substitute this result into the quadratic equation.

$$u^2 + z^2 = 85 \qquad\qquad (4)$$
$$u + z = 13 \qquad\qquad (5)$$

Solving (5) for z by subtracting u from both sides:

$$z = 13 - u \qquad\qquad (6)$$

Substituting this expressios for z into (4),

$$u^2 + (13 - u)^2 = 85$$
$$u^2 + 169 - 26u + u^2 = 85$$
$$2u^2 - 26u + 169 = 85.$$

Subtracting 85 from both sides,

$$2u^2 - 26u + 169 - 85 = 0$$
$$2u^2 - 26u + 84 = 0$$

Dividing by 2, $\qquad u^2 - 13u + 42 = 0$

Factor, $\qquad\qquad (u - 6)(u - 7) = 0$

Whenever the product of two factors is 0, then either one or the other must equal zero. Then, u - 6 = 0 or u - 7 = 0.

Solve for u,

$$u = 6 \text{ or } u = 7 .$$

Substituting these values of u into equation (6) to find the corresponding z values

	u = 6		u = 7
(6)	z = 13 - u	(6)	z = 13 - u
	z = 13 - 6		z = 13 - 7
	z = 7		z = 6

whence we obtain u = 7 or 6; z = 6 or 7. Given that z = 7 or z = 6, substituting this into equation (1):

For z = 7		For z = 6
x + y + 7 = 13	or	x + y + 6 = 13
therefore x + y = 6	or	x + y = 7

Thus we have

$$x + y = 7, \atop xy = 10 \Big\} \quad \text{and} \quad x + y = 6, \atop xy = 10 \Big\}$$

For z = 6   For z = 7

$$x + y = 7 \atop xy = 10 \Big\} \qquad\qquad x + y = 6 \atop xy = 10 \Big\}$$

Solve for y or x. We choose to solve for x.
$$xy = 10$$
$$x = \frac{10}{y} \qquad\qquad (7)$$

Substituting this expression into x + y = 7 or x + y = 6:

$$\left(\frac{10}{y}\right) + y = 7 \qquad\qquad \left(\frac{10}{y}\right) + y = 6$$

$$10 + y^2 = 7y \qquad\qquad\qquad 10 + y^2 = 6y$$

$y^2 - 7y + 10 = 0$ $\qquad\quad$ $y^2 - 6y + 10 = 0$. Solving by
$(y - 5)(y - 2) = 0$ $\qquad\quad$ the quadratic formula,
$y - 5 = 0,\ y - 2 = 0$
$\qquad\quad y = 5$ or $y = 2$

$$y = \frac{-(-6) \pm \sqrt{36 - 4(1)(10)}}{2}$$

For z = 6 and y = 5 substituting
into (7)
$$x = \frac{10}{5} = 2.$$

$$y = \frac{6 \pm \sqrt{-4}}{2}$$

$$= \frac{6 \mp 2\sqrt{-1}}{2}$$

For z = 6 and y = 2 substituting
into (7)
$$x = \frac{10}{2} = 5.$$

$$= 3 \pm \sqrt{-1}$$

For z = 7 and y = 3 + $\sqrt{-1}$
Substituting into (7)
$$x = \frac{10}{3 + \sqrt{-1}}$$

$$x = \frac{10}{(3 + \sqrt{-1})} \frac{(3 - \sqrt{-1})}{(3 - \sqrt{-1})}$$

$$= \frac{10}{9 - (-1)} (3 - \sqrt{-1})$$

$$= \frac{10}{10} (3 - \sqrt{-1})$$

$$= 3 - \sqrt{-1}$$

For z = 7 and y = 3 - $\sqrt{-1}$:
Substitute into (7)
$$x = \frac{10}{3 - \sqrt{-1}} \frac{(3 + \sqrt{-1})}{(3 + \sqrt{-1})}$$

$$= \frac{10(3 + \sqrt{-1})}{9 - (-1)} = \frac{10}{10} (3 + \sqrt{-1})$$

$$= 3 + \sqrt{-1}$$

Hence, the solutions are:

$x = 2$	$x = 5$	$x = 3 - \sqrt{-1}$	$x = 3 + \sqrt{-1}$
$x = 5$;	$y = 2$;	$y = 3 + \sqrt{-1}$;	$y = 3 - \sqrt{-1}$
$z = 6$	$z = 6$	$z = 7$	$z = 7$

● **PROBLEM** 622

Eliminate ℓ, m from the equations

$$\ell x + my = a, \quad mx - \ell y = b, \quad \ell^2 + m^2 = 1.$$

Solution: By squaring the first two equations and adding, we obtain:

$$(\ell x + my)^2 = a^2 \quad \text{and} \quad (mx - \ell y)^2 = b^2$$

$$(\ell x + my)(\ell x + my) = a^2; \quad (mx - \ell y)(mx - \ell y) = b^2$$

$$\ell^2 x^2 + my\ell x + my\ell x + m^2 y^2 = a^2 \quad \text{and}$$

$$m^2 x^2 - \ell ymx - \ell ymx + \ell^2 y^2 = b^2.$$

Then,

$$\ell^2 x^2 + 2\, my\ell x + m^2 y^2 + (m^2 x^2 - 2\ell ymx + \ell^2 y) = a^2 + b^2$$

$$\ell^2 x^2 + 2\, my\ell x + m^2 y^2 + m^2 x^2 - 2\ell ymx + \ell^2 y = a^2 + b^2.$$

We observe that $my\ell x$ and ℓymx are the same; thus

$$\ell^2 x^2 + m^2 x^2 + m^2 y^2 + \ell^2 y^2 = a^2 + b^2.$$

Factoring, we obtain:

$$(\ell^2 + m^2)(x^2 + y^2) = a^2 + b^2;$$

From the third given equation we know that $\ell^2 + m^2 = 1$. Therefore, $x^2 + y^2 = a^2 + b^2$.

If $\ell = \cos\theta$, $m = \sin\theta$, the third equation is satisfied identically; that is,

$$\ell^2 + m^2 = 1 \text{ becomes } \cos^2\theta + \sin^2\theta = 1,$$

which is a known identity. Again, by squaring the first two equations, which now are as follows:

$$x\cos\theta + y\sin\theta = a, \quad x\sin\theta - y\cos\theta = b$$

we have:

$$(x\cos\theta + y\sin\theta)^2 = a^2$$

$$(x\sin\theta - y\cos\theta)^2 = b^2.$$

Expanding, we have $(x\cos\theta + y\sin\theta)^2 =$

$$(x\cos\theta + y\sin\theta)(x\cos\theta + y\sin\theta) \quad =$$

458

$x^2 \cos^2 \theta + x \cos \theta\, y \sin \theta + x \cos \theta\, y \sin \theta + y^2 \sin^2 \theta$

$= a^2$ and, $(x \sin \theta - y \cos \theta)^2 =$

$(x \sin \theta - y \cos \theta)(x \sin \theta - y \cos \theta)$

$= x^2 \sin^2 \theta - x \sin \theta\, y \cos \theta - x \sin \theta\, y \cos \theta + y^2 \cos^2 \theta$

$= b^2.$

Adding both equations we obtain:

$x^2 \cos^2 \theta + 2(x \cos \theta\, y \sin \theta) + y^2 \sin^2 \theta +$

$$\left[x^2 \sin^2 \theta - 2(x \sin \theta\, y \cos \theta) + y^2 \cos^2 \theta \right]$$

$= a^2 + b^2.$

Observe that $x \cos \theta\, y \sin \theta$ and $x \sin \theta\, y \cos \theta$ are identical, thus they cancel and we have:

$x^2 \cos^2 \theta + y^2 \sin^2 \theta + x^2 \sin^2 \theta + y^2 \cos^2 \theta = a^2 + b^2.$

Factoring we obtain:

$x^2 \left(\sin^2 \theta + \cos^2 \theta \right) + y^2 \left(\sin^2 \theta + \cos^2 \theta \right) = a^2 + b^2.$

Now, since $\sin^2 \theta + \cos^2 \theta = 1$, we have:

$x^2 + y^2 = a^2 + b^2.$

● **PROBLEM 623**

Eliminate x, y from the equations

$x^2 - y^2 = px - qy, \quad 4xy = qx + py, \quad x^2 + y^2 = 1.$

<u>Solution:</u> Mutliplying the first equation by x, and the second by y, will allow us to eliminate a term from both equations. Thus, we obtain:

$x^3 - xy^2 = px^2 - qxy$

$\underline{+\ \ 4xy^2 \qquad = qxy\ \ + py^2}$

$x^3 - xy^2 + 4xy^2 \quad = px^2 - qxy + qxy + py^2$

The last equation can be rewritten as:

$x^3 + 3xy^2 = px^2 + py^2$

or, $x^3 + 3xy^2 = p\left(x^2 + y^2 \right).$

By the third equation, we have $x^2 + y^2 = 1$.
Thus, $p = x^3 + 3xy^2.$

Similarly, we multiply the first equation by y, and the second by x. This again allows us to eliminate a term from both equations. Thus, we obtain:

$$x^2y - y^3 = pxy - qy^2$$
$$-\quad \underline{(4x^2y \quad = qx^2 + pxy)}$$
$$x^2y - y^3 - 4x^2y = pxy - qy^2 - (qx^2 + pxy)$$

The last equation can be rewritten as:

$$-y^3 - 3x^2y = pxy - qy^2 - qx^2 - pxy$$

$-y^3 - 3x^2y = -qy^2 - qx^2$. Multiplying each term by -1 we obtain:

$$y^3 + 3x^2y = qy^2 + qx^2$$
$$y^3 + 3x^2y = q(x^2 + y^2).$$

Now, since $x^2 + y^2 = 1$, from the third equation, we have:

$$q = 3x^2y + y^3.$$

Now, adding our values for p and q we obtain:

$p + q = (x^3 + 3xy^2) + (3x^2y + y^3)$. But, this is the same as $(x + y)^3$, since

$$(x + y)^3 = (x + y)(x + y)(x + y)$$
$$= (x^2 + 2xy + y^2)(x + y)$$
$$= x^3 + 2x^2y + xy^2 + x^2y + 2xy^2 + y^3$$
$$= x^3 + 3x^2y + 3xy^2 + y^3.$$

Thus, $p + q = (x + y)^3$. Similarly, we subtract q from p and we obtain:

$$p - q = (x^3 + 3xy^2) - (3x^2y + y^3)$$
$$= x^3 + 3xy^2 - 3x^2y - y^3. \text{ But, this is the}$$
same as $(x - y)^3$, since

$$(x - y)^3 = (x - y)(x - y)(x - y)$$
$$= (x^2 - 2xy + y^2)(x - y)$$
$$= x^3 - 2x^2y + xy^2 - x^2y + 2xy^2 - y^3$$
$$= x^3 - 3x^2y + 3xy^2 - y^3.$$

Thus, $p - q = (x - y)^3$.

Therefore, we now have, $p + q = (x + y)^3$ and $p - q = (x - y)^3$.

Taking the cube root of both sides of each equation we have:

$$(p + q)^{\frac{1}{3}} = x + y$$

$$(p - q)^{\frac{1}{3}} = x - y$$

Now, squaring both sides of each equation gives us:

$$\left[(p + q)^{\frac{1}{3}}\right]^2 = (x + y)^2$$

$$\left[(p - q)^{\frac{1}{3}}\right]^2 = (x - y)^2$$

or

$$(p + q)^{\frac{2}{3}} = (x + y)^2; \quad (p - q)^{\frac{2}{3}} = (x - y)^2$$

Adding these equations we obtain:

$$(p + q)^{\frac{2}{3}} + (p - q)^{\frac{2}{3}} = (x + y)^2 + (x - y)^2 =$$

$$(x + y)(x + y) + (x - y)(x - y) =$$

$$x^2 + 2xy + y^2 + x^2 - 2xy + y^2 =$$

$$2x^2 + 2y^2 =$$

$$2(x^2 + y^2).$$

From the third equation we know that $x^2 + y^2 = 1$, therefore,

$$(p + q)^{\frac{2}{3}} + (p \neg q)^{\frac{2}{3}} = 2.$$

● **PROBLEM 624**

Eliminate x, y, z from the equations

$$y^2 + z^2 = ayz, \quad z^2 + x^2 = bzx, \quad x^2 + y^2 = cxy.$$

Solution: We wish to isolate a, b, and c on the right side of the three given equations. Dividing, we have:

$$\frac{y^2 + z^2}{yz} = a, \quad \frac{z^2 + x^2}{zx} = b, \quad \frac{x^2 + y^2}{xy} = c.$$

These can be rewritten as

$$\frac{y}{z} + \frac{z}{y} = a, \quad \frac{z}{x} + \frac{x}{z} = b, \quad \frac{x}{y} + \frac{y}{x} = c.$$

Multiplying together these three equations we obtain,

$$\left(\frac{y}{z} + \frac{z}{y}\right)\left(\frac{z}{x} + \frac{x}{z}\right)\left(\frac{x}{y} + \frac{y}{x}\right) = abc$$

$$\left(\frac{y}{z} \cdot \frac{z}{x} + \frac{z}{y} \cdot \frac{z}{x} + \frac{y}{z} \cdot \frac{x}{z} + \frac{z}{y} \cdot \frac{x}{z}\right)\left(\frac{x}{y} + \frac{y}{x}\right) = abc$$

$$\left(\left(\frac{y}{z} \cdot \frac{z}{x} \cdot \frac{x}{y}\right) + \left(\frac{z}{y} \cdot \frac{z}{x} \cdot \frac{x}{y}\right) + \left(\frac{y}{z} \cdot \frac{x}{z} \cdot \frac{x}{y}\right) + \left(\frac{z}{y} \cdot \frac{x}{z} \cdot \frac{x}{y}\right)\right.$$

$$+ \left(\frac{y}{z} \cdot \frac{z}{x} \cdot \frac{y}{x}\right) + \left(\frac{z}{y} \cdot \frac{z}{x} \cdot \frac{y}{x}\right) + \left(\frac{y}{z} \cdot \frac{x}{z} \cdot \frac{y}{x}\right) +$$

$$\left.\left(\frac{z}{y} \cdot \frac{x}{z} \cdot \frac{y}{x}\right)\right) = abc$$

Now, simplifying each term by multiplication, and reducing, we have:

$$1 + \frac{z^2}{y^2} + \frac{x^2}{z^2} + \frac{x^2}{y^2} + \frac{y^2}{x^2} + \frac{z^2}{x^2} + \frac{y^2}{z^2} + 1 = abc$$

or

$$2 + \frac{y^2}{z^2} + \frac{z^2}{y^2} + \frac{z^2}{x^2} + \frac{x^2}{z^2} + \frac{x^2}{y^2} + \frac{y^2}{x^2} = abc.$$

Now, since $\frac{y}{z} + \frac{z}{y} = a$, then $\left(\frac{y}{z} + \frac{z}{y}\right)^2 = a^2$.

But we can rewrite $\left(\frac{y}{z} + \frac{z}{y}\right)^2$ as:

$$\left(\frac{y}{z} + \frac{z}{y}\right)^2 = \left(\frac{y}{z} + \frac{z}{y}\right)\left(\frac{y}{z} + \frac{z}{y}\right) =$$

$$\frac{y^2}{z^2} + \left(\frac{z}{y} \cdot \frac{y}{z}\right) + \left(\frac{y}{z} \cdot \frac{z}{y}\right) + \frac{z^2}{y^2} =$$

$$\frac{y^2}{z^2} + 1 + 1 + \frac{z^2}{y^2} = \frac{y^2}{z^2} + \frac{z^2}{y^2} + 2.$$

Similarly, since $\frac{z}{x} + \frac{x}{z} = b$, and $\frac{x}{y} + \frac{y}{x} = c$,

then $\left(\frac{z}{x} + \frac{x}{z}\right)^2 = b^2$ and $\left(\frac{x}{y} + \frac{y}{x}\right)^2 = c^2$. But writing these

we have:

$$\left(\frac{z}{x} + \frac{x}{z}\right)^2 = \left(\frac{z}{x} + \frac{x}{z}\right)\left(\frac{z}{x} + \frac{x}{z}\right) = \frac{z^2}{x^2} + 1 + 1 + \frac{x^2}{z^2}$$

$$= \frac{z^2}{x^2} + \frac{x^2}{z^2} + 2$$

and, $\left(\frac{x}{y} + \frac{y}{x}\right)^2 = \left(\frac{x}{y} + \frac{y}{x}\right)\left(\frac{x}{y} + \frac{y}{x}\right) = \frac{x^2}{y^2} + 1 + 1 + \frac{y^2}{x^2}$

$$= \frac{x^2}{y^2} + \frac{y^2}{x^2} + 2$$

Thus, $\left(\dfrac{y}{z} + \dfrac{z}{y}\right)^2 = a^2 = \dfrac{y^2}{z^2} + \dfrac{z^2}{y^2} + 2$

$$\left(\dfrac{z}{x} + \dfrac{x}{z}\right)^2 = b^2 = \dfrac{z^2}{x^2} + \dfrac{x^2}{z^2} + 2$$

$$\left(\dfrac{x}{y} + \dfrac{y}{x}\right)^2 = c^2 = \dfrac{x^2}{y^2} + \dfrac{y^2}{x^2} + 2.$$

From these three equations we obtain:

$$\dfrac{y^2}{z^2} + \dfrac{z^2}{y^2} = a^2 - 2$$

$$\dfrac{z^2}{x^2} + \dfrac{x^2}{z^2} = b^2 - 2$$

$$\dfrac{x^2}{y^2} + \dfrac{y^2}{x^2} = c^2 - 2.$$

We can now substitute into the equation:

$$2 + \dfrac{y^2}{z^2} + \dfrac{z^2}{y^2} + \dfrac{z^2}{x^2} + \dfrac{x^2}{z^2} + \dfrac{x^2}{y^2} + \dfrac{y^2}{x^2} = abc.$$

Doing this we obtain:

$$2 + \left(a^2 - 2\right) + \left(b^2 - 2\right) + \left(c^2 - 2\right) = abc;$$

therefore, $\quad a^2 + b^2 + c^2 - 4 = abc.$

CHAPTER 21

EQUATIONS AND INEQUALITIES OF DEGREE GREATER THAN TWO

> **Basic Attacks and Strategies for Solving Problems in this Chapter. See pages 464 to 490 for step-by-step solutions to problems.**

Solving polynomial equations and inequalities of degree greater than two can be very difficult. However, rational solutions of polynomial equations of degree greater than two with integral coefficients can be solved as follows:

DEGREE 3

(1)   If the equation is of degree 3 and has a linear factor, then the procedure is clear. Put the equation in the form

$$ax^3 + bx^2 + cx + d = 0.$$

Then, $-b/a$ = the sum of the roots,

c/a = the sum of the products of the roots taken two at a time, and

$-d/a$ = the product of the roots.

(2)   If the equation is the difference of cubes, e.g.,

$$x^3 - 1 = 0,$$

then $(x-1)(x^2 + x + 1) = 0.$

Set $x - 1 = 0$ and $x^2 + x + 1 = 0$

and solve using standard techniques.

(3)   If the equation of degree 3 has a common variable factor, then factor it out and use the zero-factor property to solve the equation using standard techniques. For example,

$$ax^3 + bx^2 + cx = 0$$

$$x(ax^2 + bx + c) = 0$$

$$x = 0 \quad \text{and} \quad ax^2 + bx + c = 0$$

A method for solving an inequality of degree 3 involves the procedure outlined above, except it is necessary to consider and solve each of the possible cases for which the inequality in factored form is satisfied. The intersection of the various case solutions gives the final solution of the original inequality of degree 3. Appropriately shaded areas of the various case solutions on a graph usually aid in representing the final solution.

DEGREE 4

The solution of fourth degree equations can be solved by the same methods applied to quadratic equations. The basic approach is to first define a replacement variable z as $z = x^2$, and then substitute it in the given equation of degree 4. This step converts the equation of degree 4 to a quadratic equation involving variable z. After the solutions have been obtained for the variable z, then substitute the values of z into the equation $z = x^2$ and solve for x. The values of x represent the solutions of the original equation of degree 4.

Inequalities of degree 4 are handled in a manner similar to equations of the same degree, except all the various cases that satisfy the inequality must be examined and the final solution represented as the intersection of the solutions of the cases.

DEGREE 3

● **PROBLEM 625**

Remove fractional coefficients from the equation

$$2x^3 - \frac{3}{2}x^2 - \frac{1}{8}x + \frac{3}{16} = 0.$$

Solution: To rewrite this equation without fractional coefficients we must find a common denominator for all the terms of the equation. Observe that a common denominator is 16. Thus,

$$2x^3 - \frac{3}{2}x^2 - \frac{1}{8}x + \frac{3}{16} = 0$$

can be rewritten as:

$$\frac{2x^3}{1} - \frac{3x^2}{2} - \frac{x}{8} + \frac{3}{16} = 0 \qquad \text{or,}$$

$$\frac{32x^3 - 24x^2 - 2x + 3}{16} = 0. \text{ Multiplying both}$$

sides of the equation by 16 we obtain:

$$32x^3 - 24x^2 - 2x + 3 = 0. \quad \text{This is the required}$$

equation without fractional coefficients.

● **PROBLEM 626**

Solve the equation $x^3 - 16x = 0$.

Solution: Multiplying both sides by $\frac{1}{x}$, we have $x^2 - 16 = 0$. Factoring, we have $(x - 4)(x + 4) = 0$. Then all values of x which make this product equal to 0 satisfy either $x - 4 = 0$ or $x + 4 = 0$. Thus $x = 4$ or $x = -4$. Both 4 and -4 satisfy the original equation since $(4)^3 - 16(4) = 0$ and $(-4)^3 - 16(-4) = 0$. However, so does the number 0, since $(0)^3 - 16(0) = 0$. From where did this root come?

There are several logical flaws in this solution. First, we cannot multiply both sides by $1/x$ if $x = 0$. But, basically, what is wrong is that $x^3 - 16x$ is not an equivalent expression to $x^2 - 16$.

We may undo this error by writing $x^3 - 16x = 0$ in factored form as $x(x + 4)(x - 4) = 0$.

Then $x = 0$ or $x + 4 = 0$ or $x - 4 = 0$

Hence $x = 0$ or $x = -4$ or $x = 4$

Therefore the solutions set is $\{-4,0,4\}$.

● **PROBLEM** 627

Find all solutions of the equation $x^3 - 3x^2 - 10x = 0$.

Solution: Factor out the common factor of x from the terms on the left side of the given equation. Therefore,

$$x^3 - 3x^2 - 10x = x(x^2 - 3x - 10) = 0.$$

Whenever $ab = 0$ where a and b are any two numbers, either $a = 0$ or $b = 0$. Hence, either $x = 0$ or $x^2 - 3x - 10 = 0$. The expression $x^2 - 3x - 10$ factors into $(x - 5)(x + 2)$. Therefore, $(x - 5)(x + 2) = 0$. Applying the above law again:

$$\text{either} \quad x - 5 = 0 \quad \text{or} \quad x + 2 = 0$$
$$x = 5 \quad \text{or} \quad x = -2.$$

Hence,
$$x^3 - 3x^2 - 10x = x(x - 5)(x + 2) = 0$$

$$\text{Either} \quad x = 0 \quad \text{or} \quad x = 5 \quad \text{or} \quad x = -2.$$

The solution set is $X = \{0, 5, -2\}$.

We have shown that, if there is a number x such that $x^3 - 3x^2 - 10x = 0$, then $x = 0$ or $x = 5$ or $x = -2$. Finally, to see that these three numbers are actually solutions, we substitute each of them in turn in the original equation to see whether or not it satisfies the equation $x^3 - 3x^2 - 10x = 0$.

Check: Replacing x by 0 in the original equation,
$$(0)^3 - 3(0)^2 - 10(0) = 0 - 0 - 0 = 0 \checkmark$$

Replacing x by 5 in the original equation,
$$(5)^3 - 3(5)^2 - 10(5) = 125 - 3(25) - 50$$
$$= 125 - 75 - 50$$
$$= 50 - 50 = 0 \checkmark$$

Replacing x by -2 in the original equation,
$$(-2)^3 - 3(-2)^2 - 10(-2) = -8 - 3(4) + 20$$
$$= -8 - 12 + 20$$
$$= -20 + 20 = 0 \checkmark$$

● **PROBLEM** 628

Find the solutions of the equation $x^3 - 1 = 0$.

Solution: The equation $x^3 - 1 = x^3 - 1^3$, which is the difference of two cubes.

We note that the difference of two cubes:
$$(x^3 - y^3) = (x - y)(x^2 + xy + y^2). \quad \text{Then,}$$

$$x^3 - 1 = x^3 - 1^3 = (x - 1)(x^2 + x + 1) = 0.$$

Hence, $(x - 1) = 0$

$$x = 1;$$

$$\left(x^2 + x + 1\right) = 0$$

Solve by the quadratic formula since this equation is of the form $ax^2 + bx + c = 0$ with $a = 1$, $b = 1$, $c = 1$.

$$x = \frac{-b \pm \sqrt{b^2 - 4ac}}{2a}$$

$$x = \frac{-1 \pm \sqrt{1 - 4(1)}}{2}$$

$$x = -\frac{1}{2} \pm \frac{\sqrt{-3}}{2}$$

$$x = -\frac{1}{2} \pm \frac{i\sqrt{3}}{2}.$$

The solutions are

$$x = \left\{ 1, \ \frac{-1 + i\sqrt{3}}{2}, \ \frac{-1 - i\sqrt{3}}{2} \right\}.$$

● **PROBLEM** 629

Form the equation whose roots are 2, −3, and $\frac{7}{5}$.

Solution: The roots of the equation are 2, −3, and $\frac{7}{5}$.

Hence, $x = 2$, $x = -3$, and $x = \frac{7}{5}$. Subtract 2 from both sides of the first equation:

$$x - 2 = 2 - 2 = 0.$$

Add 3 to both sides of the second equation:

$$x + 3 = -3 + 3 = 0.$$

Subtract $\frac{7}{5}$ from both sides of the third equation:

$$x - \frac{7}{5} = \frac{7}{5} - \frac{7}{5} = 0.$$

Hence, $(x - 2)(x + 3)\left(x - \frac{7}{5}\right) = (0)(0)(0) = 0$ or

$$(x - 2)(x + 3)\left(x - \frac{7}{5}\right) = 0.$$

Multiply both sides of this equation by 5:

$$5(x - 2)(x + 3)\left(x - \frac{7}{5}\right) = 5(0) \text{ or}$$

$$(x - 2)(x + 3)5\left(x - \frac{7}{5}\right) = 0 \text{ or}$$

$$(x - 2)(x + 3)(5x - 7) = 0$$

$$(x^2 + x - 6)(5x - 7) = 0$$

$$5x^3 - 7x^2 + 5x^2 - 7x - 30x + 42 = 0$$

$$5x^3 - 2x^2 - 37x + 42 = 0.$$

● **PROBLEM** 630

Solve the equation $24x^3 - 14x^2 - 63x + 45 = 0$, one root being double another.

Solution: A cubic equation, with the properties stated, has three roots which may be denoted by a, 2a, b.

Then we can use the following known relations between roots and coefficients of an equation to obtain equations involving the three roots. Since we can convert the given equation into one in which the coefficient of the first term is 1, we can transform the equation into one of the form, $x^3 + b_1 x^2 + b_2x + b_3 = 0$, and then, $- b_1$ = sum of the roots

b_2 = sum of products of the roots taken two at a time

$(- 1)^3 b_3$ = product of roots.

Dividing our given equation by 24 we obtain:

$$x^3 - \frac{14}{24} x^2 - \frac{63}{24} x + \frac{45}{24} = 0.$$

Thus, $- b_1 = - \left(- \frac{14}{24}\right) = \frac{7}{12}$ = sum of roots = a + 2a + b

$b_2 = - \frac{63}{24} = - \frac{21}{8}$ = sum of products of roots

taken two at a time

$$= (a)(2a) + (a)(b) + (2a)(b)$$

$(- 1)^3 b_3 = (- 1) \frac{45}{24} = - \frac{15}{8}$ = product of roots

$$= (a)(2a)(b).$$

Therefore, $3a + b = \frac{7}{12}$, $2a^2 + 3ab = - \frac{21}{8}$,

$$2a^2b = - \frac{15}{8}.$$

We can now solve the first two equations simultaneously as follows: Transpose the constant term in each equation from the right to the left side of the equal sign. Multiply each term of the first equation by 3a, and then subtract the second equation from the first.

467

We then obtain:

$$9a^2 + 3ab - \frac{21}{12} a = 0$$

$$- \left(2a^2 + 3ab + \frac{21}{8} \quad = 0\right)$$

$$7a^2 - \frac{21}{12} a - \frac{21}{8} \quad = 0.$$

This last equation can be rewritten as,

$$7a^2 - \frac{7}{4} a - \frac{21}{8} = 0 \quad \text{or,} \quad 7a^2 - \frac{14}{8} a - \frac{21}{8} = 0.$$

Now, multiplying each term by 8 we obtain:

$56a^2 - 14a - 21 = 0$, and dividing each term by 7 gives us,

$8a^2 - 2a - 3 = 0$. Factoring, we obtain:

$(4a - 3)(2a + 1) = 0.$

Therefore, $a = \frac{3}{4}$ or $-\frac{1}{2}$ and substituting these values of a into the first equation we find that

$$b = -\frac{5}{3} \quad \text{or} \quad \frac{25}{12}.$$

It will be found on trial that the values $a = -\frac{1}{2}$, $b = \frac{25}{12}$ do not satisfy the third equation $2a^2b = -\frac{15}{8}$; hence, we are restricted to the values

$$a = \frac{3}{4}, \quad b = -\frac{5}{3}.$$

Now, since the roots were denoted by a, 2a, b, our three roots are

$$\frac{3}{4}, \ 2\left(\frac{3}{4}\right), \ -\frac{5}{3} \quad \text{or} \quad \frac{3}{4}, \frac{3}{2}, \ -\frac{5}{3}.$$

● PROBLEM 631

Solve the equation $4x^3 - 24x^2 + 23x + 18 = 0$, having given that the roots are in arithmetical progression.

Solution: Since the roots are in A.P. they may be denoted by a - b, a, a + b.

Then we can use the following known relations between roots and coefficients of an equation to obtain equations involving the three roots denoted by a - b, a, a + b. Since we can convert the given equation into one in which the coefficient of the first term is 1, we can transform the equation into one of the form,

$x^3 + b_1 x^2 + b_2 x + b_3 = 0$; and then,

$- b_1$ = sum of the roots

b_2 = sum of the products of the roots taken two at a time

$(- 1)^3 b_3$ = product of the roots

Dividing each term of our given equation by 4 we obtain:

$$x^3 - \frac{24}{4} x^2 + \frac{23}{4} x + \frac{18}{4} = 0$$

Thus, $- b_1 = - \left(- \frac{24}{4}\right) = 6$ = sum of roots

$$= (a - b) + a + (a + b)$$

$b_2 = \frac{23}{4}$ = sum of products of roots taken two at

a time $= (a-b)(a)+(a-b)(a+b)+(a)(a+b)$

$(- 1)^3 b_3 = (- 1) \frac{18}{4} = - \frac{9}{2}$ = product of roots

$$= (a - b)(a)(a + b)$$

Therefore,

$$3a = 6, \quad 3a^2 - b^2 = \frac{23}{4}, \quad a\left(a^2 - b^2\right) = - \frac{9}{2} .$$

From the first equation we find a = 2, and substituting this value in the second equation we find $b = \pm \frac{5}{2}$, and since these values satisfy the third, the three equations are consistent.

Now, since the roots were denoted by a - b, a, a + b, our three roots are $\left(2 \pm \frac{5}{2}\right)$, 2, $\left(2 \pm \frac{5}{2}\right)$, or, $- \frac{1}{2}$, 2, $\frac{9}{2}$.

● **PROBLEM   632**

Graph the function $y = x^3 - 9x$.

Solution: Choosing values of x in the interval $-4 \leq x \leq 4$, we have for $y = x^3 - 9x$,

x	- 4	-3	- 2	-1	0	1	2	3	4
y	-28	0	10	8	0	-8	-10	0	28

Notice that for each ordered pair (x,y) listed in the table there exists a pair (-x,-y) which also satisfies the equation, indicating symmetry with respect to the origin. To

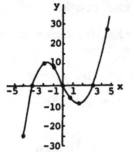

prove that this is true for all points on the curve, we substitute (-x,-y) for (x,y) in the given equation and show that the equation is unchanged. Thus

$$-y = (-x)^3 - 9(-x) = -x^3 + 9x$$

or, multiplying each member by -1,

$$y = x^3 - 9x$$

which is the original equation.

The curve is illustrated in the figure. The domain and range of the function have no restrictions in the set of real numbers. The x-intercepts are found from

$$y = 0 = x^3 - 9x$$

$$0 = x(x^2 - 9)$$

$$0 = x(x - 3)(x + 3)$$

$$x = 0 \quad | \quad x - 3 = 0 \quad | \quad x + 3 = 0$$
$$\quad | \quad x = 3 \quad | \quad x = -3.$$

The curve has three x-intercepts at x = -3, x = 0, x = 3. This agrees with the fact that a cubic equation has three roots. The curve has a single y-intercept at y = 0 since for x = 0, $y = 0^3 - 9(0) = 0$.

● **PROBLEM 633**

Locate the roots of $x^3 - 3x^2 - 6x + 9 = 0$.

Solution: If we let f(x) be a function, then a solution of the equation f(x) = 0 is called a root of the equation.

In this particular case let the function f(x) = $x^3 - 3x^2 - 6x + 9$ and set it equal to zero to find its roots. When f(x) = 0, the graph of this equation crosses the x-axis. These x-values are the roots of the function.

To locate the roots of $x^3 - 3x^2 - 6x + 9 = 0$, we consider the function $y = x^3 - 3x^2 - 6x + 9$, assign consecutive integers from - 3 to 5 to x, compute each corresponding value of y, and record the results.

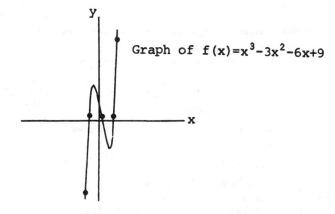

Graph of $f(x)=x^3-3x^2-6x+9$

Table of Results

x	$x^3 - 3x^2 - 6x + 9$	=	y
- 3	$(- 3)^3 - 3(- 3)^2 - 6(- 3) + 9 =$		- 27
- 2	$(- 2)^3 - 3(- 2)^2 - 6(- 2) + 9 =$		1
- 1	$(- 1)^3 - 3(- 1)^2 - 6(- 1) + 9 =$		11
0	$(0)^3 - 3(0)^2 - 6(0) + 9$	=	9
1	$(1)^3 - 3(1)^2 - 6(1) + 9$	=	1
2	$(2)^3 - 3(2)^2 - 6(2) + 9$	=	- 7
3	$(3)^3 - 3(3)^2 - 6(3) + 9$	=	- 9
4	$(4)^3 - 3(4)^2 - 6(4) + 9$	=	1
5	$(5)^3 - 3(5)^2 - 6(5) + 9$	=	29

Since $f(- 3) = - 27$ and $f(- 2) = 1$, there is an odd number of roots between $x = - 3$ and $x = - 2$. Since $f(- 3) = - 27$ which is negative and $f(- 2) = 1$ is positive, the graph must cross the x-axis at least once. The function is continuous; thus the curve must connect the two points. To do this, the curve must cross from the negative to the positive side of the x-axis. By the definition of continuity, in order for the curve to traverse the axis it must intersect the axis. Each intersection point is called a zero or a root of the function.

Note that the curve must intersect the x-axis an odd number of times if it is to pass from the negative side to the positive, for if it traversed the axis an even number of times it would end up on the side on which it started.

Similarly, there is an odd number of roots between $x = 1$ and $x = 2$, and between $x = 3$ and $x = 4$. Furthermore, since the equation is of degree 3, it has exactly three roots. Observe that the curve crosses the x-axis three times, indicating the three roots of the equation. Therefore, exactly one root lies in each of the above intervals.

Solve the inequality $(x + 2)(x - 1)(2x - 3) > 0$.

Fig. A

Fig. B

-2 -1 0 $1\frac{3}{2}$ 2 3

$\blacksquare$ x>-2

▨ x>1

▧ x>$\frac{3}{2}$

-2 -1 0 $1\frac{3}{2}$ 2

$\blacksquare$ x>-2

▨ x<1

▧ x<$\frac{3}{2}$

Solution: If we have a positive number which is a product of three factors, then all three factors are positive or two are negative and one is positive (since a negative multiplied by a negative is positive). The tentative possibilities are:

(1) x+2 > 0, and x-1 > 0, and 2x-3 > 0, ➡ x > -2, and x > 1, and x > 3/2

(2) x+2 > 0,   x-1 < 0,   2x-3 < 0, ⇒ x > -2,   x < 1,   x < 3/2

(3) x+2 < 0,   x-1 > 0,   2x-3 < 0, ⇒ x < -2,   x > 1,   x < 3/2

(4) x+2 < 0,   x-1 < 0,   2x-3 > 0, ⇒ x < -2,   x < 1,   x > 3/2

Then, if "and" means intersection (∩), we must find,

$$(x > -2) \cap (x > 1) \cap (x > 3/2).$$

(See the number line in Figure A). Thus, the inequalities in (1) yield, as the range satisfying all three linear inequalities, x > 3/2.
For (2)

$$x > -2, \quad x < 1, \text{ and } x < 3/2;$$

thus we must find,

$$(x > -2) \cap (x < 1) \cap (x < 3/2)$$

(See Figure B).
Thus, the inequalities in (2) yield, as the range satisfying all three linear inequalities, -2 < x < 1.
In Case (3), x < -2, x > 1, and x < 3/2, x > 1 is inconsistent with x < -2. Thus, there are no values on the number line common to all three inequalities.
For the last alternative, (4), x < -2, x < 1, and x > 3/2, the last inequality, x > 3/2, is inconsistent with x < -2, and x < 1. Thus, again the intersection of these three inequalities is the empty set. Hence the complete solution consists of the ranges

$$-2 < x < 1 \quad \text{and} \quad x > 3/2 .$$

Solve $x^3 + x^2 - 2x > 0$.

Solution: We factor $x^3 + x^2 - 2x$ and have $x(x^2 + x - 2) > 0$. Again, factor the left side of this inequality:

$$x^2 + x - 2 = (x + 2)(x - 1).$$

Therefore, $x(x^2 + x - 2) > 0$ becomes $x(x + 2)(x - 1) > 0$. Now we have four possibilities: (I) all factors positive; (II) the last two negative; (III) the first and third negative; and (IV) the first two negative. In tabular form this gives:

I	II	III	IV
x > 0	x > 0	x < 0	x < 0
and	and	and	and
x + 2 > 0 or	x + 2 < 0 or	x + 2 > 0 or	x + 2 < 0

and	and	and	and
$x - 1 > 0$	$x - 1 < 0$	$x - 1 < 0$	$x - 1 > 0$

Solving for x in each inequality we obtain:

I	II	III	IV
$x > 0$	$x > 0$	$x < 0$	$x < 0$
and	and	and	and
$x > -2$ or	$x < -2$ or	$x > -2$ or	$x < -2$
and	and	and	and
$x > 1$	$x < 1$	$x < 1$	$x > 1$

However, (I) simplifies to $x > 1$, for if $x > 1$, it certainly follows that $x > 0$ and $x > -2$ also. On the other hand, since we cannot have $x > 0$ and $x < -2$ it follows that (II) is impossible. Similarly, (IV) is also impossible since we cannot have $x < 0$ and $x > 1$. Finally in (III), we note that $x < 0$ certainly implies $x < 1$ so that we are left with $x < 0$ and $x > -2$. This can be written as $-2 < x < 0$. Hence,

$$x^3 + x^2 - 2x > 0 \text{ if and only if}$$

$$x > 1 \quad \text{or} \quad -2 < x < 0$$

That is, the solution set of $x^3 + x^2 - 2x > 0$ is

$$X = \{x \,|\, x > 1\} \cup \{x \,|\, -2 < x < 0\}$$

$$= \{x \,|\, x > 1 \text{ or } -2 < x < 0\}$$

● **PROBLEM** 636

Determine the real values of x for which $\sqrt{x^3 - 3x^2 + 2x}$ is real.

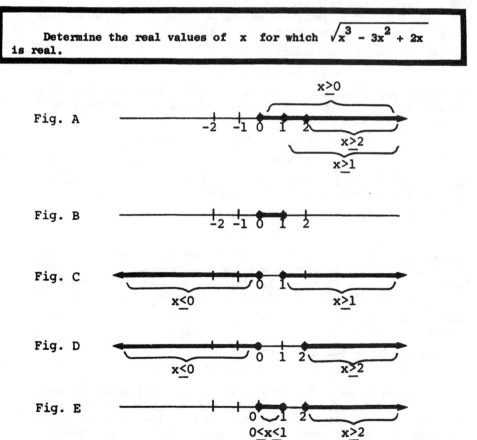

Solution: The given expression will be real for those real values of x yielding a radicand which is greater than or equal to zero. Thus, we have to solve the inequality

$$x^3 - 3x^2 + 2x \geqq 0 .$$

There is a common factor, namely x, of every term in the left side of this inequality. Hence, x is factored out from the left side:

$$x(x^2 - 3x + 2) \geqq 0$$

Factoring the expression in parenthesis into a product of two polynomials:

$$x(x - 2)(x - 1) \geqq 0 .$$

Only if all three factors are positive or two are negative and one positive will the product be positive. Thus, the following four cases result:

   Case 1: $x \geqq 0$, $x - 2 \geqq 0$, $x - 1 \geqq 0$.

   Case 2: $x \geqq 0$, $x - 2 \leqq 0$, $x - 1 \leqq 0$.

   Case 3: $x \leqq 0$, $x - 2 \leqq 0$, $x - 1 \geqq 0$.

   Case 4: $x \leqq 0$, $x - 2 \geqq 0$, $x - 1 \leqq 0$.

For Case 1: Solve the inequalities,

$$x \geqq 0, \ x \geqq 2, \ x \geqq 1.$$

These three inequalities are satisfied by the single inequality $x \geqq 2$, that is, the intersection of these three inequalities is the set $\{x \,|\, x \geqq 2\}$. This intersection can be noted in Diagram A. Hence, the solution set for Case 1 is $\{x \,|\, x \geqq 2\}$.

For Case 2: Solve the inequalities,

$$x \geqq 0, \ x \leqq 2, \ x \leqq 1.$$

The two inequalities $x \leqq 2$ and $x \leqq 1$ are satisfied by the single inequality $x \leqq 1$. Putting this inequality, $x \leqq 1$ and the remaining inequality $x \geqq 0$ together:

$$x \geqq 0 \ \text{or} \ 0 \leqq x, \ x \leqq 1; \ \text{that is,}$$

$$0 \leqq x \leqq 1.$$

The set $\{x \,|\, 0 \leqq x \leqq 1\}$ is indicated in Diagram B. Hence, the solution set for Case 2 is $\{x \,|\, 0 \leqq x \leqq 1\}$.

For Case 3: Solving the inequalities,

$$x \leqq 0, \ x \leqq 2, \ x \geqq 1 .$$

The two inequalities $x \leqq 0$ and $x \leqq 2$ are satisfied by the single inequality $x \leqq 0$. However, combining this inequality, $x \leqq 0$, with the remaining inequality $x \geqq 1$, there is no value of x which is less than or equal to zero, and, at the same time, greater than or

equal to 1. This is illustrated in Diagram C. Hence, there is no solution set for Case 3.

For Case 4: Solve the inequalities,

$$x \leqq 0, \ x \geqq 2, \ x \leqq 1.$$

The two inequalities $x \leqq 0$ and $x \leqq 1$ are satisfied by the single inequality $x \leqq 0$. However, combining this inequality, $x \leqq 0$, with the remaining inequality $x \geqq 2$, there is no value of x which is less than or equal to zero, and, at the same time, greater than or equal to 2. This is illustrated in Diagram D. Hence, there is no solution set for Case 4.

Therefore, the solution set consists of the two sets: $\{x|x \geqq 2\}$ and $\{x|0 \leqq x \leqq 1\}$; that is, the solution set is:

$$\{x|x \geqq 2\} \cup \{x|0 \leqq x \leqq 1\}$$

The solution set is pictured in Diagram E.

● **PROBLEM** 637

Solve $4x^3 + 3x^2y + y^3 = 8$,
$2x^3 - 2x^2y + xy^2 = 1$.

<u>Solution:</u> Put $y = mx$, and substitute in both equations. Thus,

$$4x^3 + 3x^2(mx) + (mx)^3 = 8 ,$$
$$2x^3 - 2x^2(mx) + x(mx)^2 = 1;$$

and factoring:

$$x^3(4 + 3m + m^3) = 8 \qquad\qquad (1)$$
$$x^3(2 - 2m + m^2) = 1 \qquad\qquad (2)$$

Divide equation (1) by equation (2):

$$\frac{\cancel{x^3}(4 + 3m + m^3)}{\cancel{x^3}(2 - 2m + m^2)} = \frac{4 + 3m + m^3}{2 - 2m + m^2} = \frac{8}{1} = 8 \qquad (2a)$$

Hence, $\dfrac{4 + 3m + m^3}{2 - 2m + m^2} = 8$. $\qquad\qquad (3)$

Multiply both sides of equation (3) by $(2 - 2m + m^2)$. Then:

$$(2 - \cancel{2m + m^2})\left(\frac{4 + 3m + m^3}{2 - \cancel{2m + m^2}}\right) = (2 - 2m + m^2)8$$
$$4 + 3m + m^3 = (2 - 2m + m^2)8 \qquad\qquad (4)$$

Distributing on the right side of equation (4):

$$4 + 3m + m^3 = 16 - 16m + 8m^2 \qquad\qquad (5)$$

Subtract $(16 - 16m + 8m^2)$ from both sides of equation (5). Then,

$$4 + 3m + m^3 - (16 - 16m + 8m^2) = 16 - 16m + 8m^2 - (16 - 16m + 8m^2)$$
$$4 + 3m + m^3 - 16 + 16m - 8m^2 = 0 .$$

Combining like terms:

$$m^3 - 8m^2 + 19m - 12 = 0 \qquad\qquad (6)$$

The possible rational roots of equation (6) are all numbers $\pm p/q$ in which the values of p are the positive divisors 1,2,3,4,6,12 of the constant term -12 and the values of q are the positive divisors of the leading coefficient 1. Hence, the only value for q is 1. Thus, the possible rational roots of equation (6) are:

$$\pm 1/1, \pm 2/1, \pm 3/1, \pm 4/1, \pm 6/1,$$

and

$$\pm 12/1, \text{ or } \pm 1, \pm 2, \pm 3, \pm 4, \pm 6, \text{ and } \pm 12.$$

Substitution of the values 1,3, and 4 for m makes equation (6) a true statement. Hence, the roots of m are 1,3, and 4. Then, $m = 1$, $m = 3$, and $m = 4$. Therefore, $m - 1 = 0$, $m - 3 = 0$ and $m - 4 = 0$. Then,

$$(m - 1)(m - 3)(m - 4) = (0)(0)(0) = 0 \qquad (7)$$

Also, $\qquad (m - 1)(m - 3)(m - 4) = \Big[(m - 1)(m - 3)\Big](m - 4)$

475

$$= [m^2 - 4m + 3](m - 4)$$

distributing,
$$= \left(m^3 - 4m^2 + 3m\right) - \left(4m^2 - 16m + 12\right)$$
$$= m^3 - 4m^2 + 3m - 4m^2 + 16m - 12$$
$$(m - 1)(m - 3)(m - 4) = m^3 - 8m^2 + 19m - 12 \qquad (8)$$

From equations (6) and (7):
$$m^3 - 8m^2 + 19m - 12 = (m - 1)(m - 3)(m - 4) = 0$$

(i) Take m = 1, and substitute in either (1) or (2). From (2),
$$x^3\left(2 - 2(1) + (1)^2\right) = 1$$
$$x^3(2 - 2 + 1) = 1$$
$$x^3(0 + 1) = 1$$
$$x^3(1) = 1$$
$$x^3 = 1$$

Take the cube root of each side:
$$\sqrt[3]{x^3} = \sqrt[3]{1}$$
$$x = \sqrt[3]{1} = 1 .$$

Also, y = mx = 1(1) = 1.

(ii) Take m = 3, and substitute in (2):

thus
$$5x^3 = 1. \text{ Then,}$$
$$x^3 = 1/5 .$$

Take the cube root of each side:
$$\sqrt[3]{x^3} = \sqrt[3]{1/5}$$
$$x = \sqrt[3]{1/5}$$

and $y = mx = 3x = 3\sqrt[3]{1/5} .$

(iii) Take m = 4; we obtain from (2):
$$10x^3 = 1.$$

Then, $x^3 = \dfrac{1}{10}$. Take the cube root of each side:
$$\sqrt[3]{x^3} = \sqrt[3]{1/10} ,$$

or
$$x = \sqrt[3]{1/10} .$$

and
$$y = mx = 4x = 4\sqrt[3]{1/10} .$$

Hence the complete solution is
$$x = 1, \sqrt[3]{1/5} , \sqrt[3]{1/10}.$$
$$y = 1,3 \sqrt[3]{1/5} , 4\sqrt[3]{1/10} .$$

● PROBLEM 638

Show that $x^3 > y^3$ if $x > y$.

Solution: If we subtract y^3 from both sides of the
inequality $x^3 > y^3$ we obtain:

$$x^3 - y^3 > 0$$

Recall the formula for the difference of two cubes:

$$a^3 - b^3 = (a - b)(a^2 + ab + b^2)$$

Thus, $x^3 - y^3 = (x - y)(x^2 + xy + y^2)$, by substitution.

If $x^3 - y^3 > 0$, then

$$(x - y)(x^2 + xy + y^2) > 0 \qquad (1)$$

Since we are given $x > y$, subtracting y from both sides of this inequality we obtain $x - y > 0$.

Now that we know $(x - y) > 0$ we can divide both sides of inequality (1) by $(x - y)$ without reversing the inequality:

$$\frac{(x - y)(x^2 + xy + y^2)}{(x - y)} > \frac{0}{(x - y)}$$

Cancelling $(x - y)$ in numerator and denominator we arrive at

$$x^2 + xy + y^2 > 0. \qquad (2)$$

Note that

$$x^2 + xy + y^2 = \left(x + \frac{1}{2} y\right)^2 + \frac{3}{4} y^2;$$

thus, we have written inequality (2) as the sum of two squares. The square of any number is positive, and if we add two positives we obtain a positive; hence:

$$x^2 + xy + y^2 = \left(x + \frac{1}{2} y\right)^2 + \frac{3}{4} y^2 > 0$$

Because inequality (2) is always valid, and the steps are all reversible, the given inequality, $x^3 > y^3$, has been proven.

● **PROBLEM 639**

Which of the points $(1,0)$, $(-1,0)$, $(4,4)$, and $(9,17)$ belong to the graph of the equation $y = x^{3/2} - x$?

<u>Solution:</u> The point (a,b) belongs to the graph of the equation $y = x^{3/2} - x$ if substitution of x by a in the equation results in b as its y value.

To check $(1,0)$ we replace x by 1:

$$y = x^{3/2} - x$$

$$y = (1)^{3/2} - 1 = \left(1^{1/2}\right)^3 - 1 = 1^3 - 1 = 1 - 1 = 0$$

Thus the point $(1,0)$ is on the graph $y = x^{3/2} - x$.

To check $(-1,0)$ we replace x by -1:

$$y = x^{3/2} - x$$

$$y = (-1)^{3/2} - (-1) = \left(-1^{1/2}\right)^3 + 1 = (i)^3 + 1 = \cdot i^2 (i) + 1$$

$$= -1(i) + 1 = -i + 1 \neq 0 .$$

Thus the point (-1,0) is not on the graph $y = x^{3/2} - x$.

To check (4,4) we replace x by 4:

$$y = x^{3/2} - x$$

$$y = (4)^{3/2} - 4 = \left(4^{1/2}\right)^3 - 4 = 2^3 - 4 = 8 - 4 = 4$$

Thus the point (4,4) is on the graph $y = x^{3/2} - x$.

To check (9,17) we replace x by 9:

$$y = x^{3/2} - x$$

$$y = (9)^{3/2} - 9 = \left(9^{1/2}\right)^3 - 9 = 3^3 - 9 = 27 - 9 = 18 \neq 17$$

Thus the point (9,17) is not on the graph $y = x^{3/2} - x$.

Therefore, the points (1,0) and (4,4) belong to the graph.

DEGREE 4

● **PROBLEM** 640

Solve the equation $x^4 - 5x^2 - 36 = 0$.

<u>Solution:</u> This is a fourth degree equation, but it can be solved by the same methods applied to quadratic equations.

To solve $x^4 - 5x^2 - 36 = 0$, we let $z = x^2$, substitute in the given equation, and get

$$z^2 - 5z - 36 = 0$$

This is now a quadratic equation in the variable z. We solve this equation by factoring.

$$z^2 - 5z - 36 = 0$$

$$(z - 9)(z + 4) = 0 \qquad \text{Factoring}$$

$$z - 9 = 0, \; z + 4 = 0 \qquad \text{Setting both factors equal to zero}$$

$$z = 9, \qquad z = -4 \qquad \text{Solving for z.}$$

Hence the solution set of the equation in z is {- 4,9}. Now we replace z in $z = x^2$ by - 4 and then by 9 and get

$$x^2 = -4$$

Taking the square root of each member,

$$x = \pm \sqrt{-4} = \pm \sqrt{4(-1)} = \pm \sqrt{4} \sqrt{-1}$$

$$x = \pm 2i$$

Also $x^2 = 9$

$$x = \pm 3$$

Consequently the solution set of the original equation is $\{2i, -2i\} \cup \{3, -3\} = \{2i, -2i, 3, -3\}$.

● **PROBLEM 641**

Solve the equation
$$2x^4 + 7x^2 - 4 = 0 .$$

<u>Solution:</u> Substitute $y = x^2$ in the given equation to obtain
$$2x^4 + 7x^2 - 4 = 2(x^2)^2 + 7(x^2) - 4$$
$$= 2y^2 + 7y - 4 = 0$$

Factoring this quadratic equation in y; we obtain
$$(2y - 1)(y + 4) = 0$$
$$2y - 1 = 0 \qquad y + 4 = 0$$
$$2y = 1$$
$$y = \tfrac{1}{2} \text{ or } y = -4$$

Then $\qquad\qquad x^2 = \tfrac{1}{2} \text{ or } x^2 = -4$

$$x = \pm \sqrt{\tfrac{1}{2}} \qquad x = \pm \sqrt{-4}$$

The solution is: $x = \pm \sqrt{\tfrac{1}{2}} \dfrac{\sqrt{2}}{2} = \pm \dfrac{\sqrt{2}}{2}$, $x = \pm 2i$

Thus the given equation of fourth degree has four solutions.

● **PROBLEM 642**

Solve for x: $3x^4 - 4x^2 - 7 = 0$.

<u>Solution:</u> Let $v = x^2$, then $3v^2 - 4v - 7 = 0$, which is quadratic in v, so that the original equation is quadratic in x^2. By factoring, we have $(3v - 7)(v + 1) = 0$, and the values of v are
$$v = \frac{7}{3} , -1 .$$

Since $v = x^2$ and the values obtained for v are $\frac{7}{3}$ and -1, $x^2 = \frac{7}{3}$ and $x^2 = -1$. Hence,

$$x = \pm \sqrt{\frac{7}{3}} \qquad \text{and} \qquad x^2 = -1$$
$$= \pm \frac{\sqrt{7}}{\sqrt{3}} \qquad\qquad x = \pm \sqrt{-1} = \pm i$$
$$= \pm \frac{\sqrt{7}\sqrt{3}}{\sqrt{3}\sqrt{3}}$$
$$= \pm \frac{\sqrt{21}}{3}$$

Therefore, the four roots are $x = +\dfrac{\sqrt{21}}{3}$, $-\dfrac{\sqrt{21}}{3}$, $+i$, and $-i$.

Solve for x.

$$x^4 - 10x^2 + 9 = 0$$

Solution: The given equation can be written as

$$\left(x^2\right)^2 - 10x^2 + 9 = 0.$$

Set $p = x^2$ and solve for p.

$$p^2 - 10p + 9 = 0.$$

Factor the left side of this equation into a product of two polynomials. Hence,

$$(p - 9)(p - 1) = 0$$

Whenever a product ab = 0, where a and b are any two numbers, either a = 0 or b = 0. Hence, either,

$$p - 9 = 0 \qquad \text{or} \qquad p - 1 = 0$$

$$p = 9 \qquad \text{or} \qquad p = 1$$

Set each value of $p = x^2$ and solve for x.

$$p = x^2 = 9$$

$$x^2 = 9$$

Take the square root of both sides.

$$\sqrt{x^2} = \sqrt{9}$$

$$x = \pm 3$$

Also,

$$p = x^2 = 1$$

$$x^2 = 1$$

Take the square root of both sides.

$$\sqrt{x^2} = \sqrt{1}$$

$$x = \pm 1$$

The solution set is { 3, -3, 1, -1}.

Solve the equation $x^4 = 4x^2 - 4.$

Solution: We first add $4 - 4x^2$ to both sides, obtaining $x^4 - 4x + 4 = 0$. Rewriting this equation as $\left(x^2\right)^2 - 4\left(x^2\right) + 4 = 0$, we see that this equation is in quadratic form, where $P(x) = x^2$. It is not necessary to substitute a new letter for $P(x)$; instead we may simply carry the expression x^2 throughout the computation in the following manner:

$$\left(x^2\right)^2 - 4\left(x^2\right) + 4 = 0$$

Factor as
$$\left(x^2 - 2\right)\left(x^2 - 2\right) = 0$$

To find each distinct value of x which makes this product $= 0$, let

$$x^2 - 2 = 0$$

therefore,
$$x^2 = 2$$

Hence $x = \sqrt{2}$ or $x = -\sqrt{2}$.

Check: for $x = \sqrt{2}$, for $x = -\sqrt{2}$

$$\left(\sqrt{2}\right)^4 \overset{?}{=} 4\left(\sqrt{2}\right)^2 - 4 \qquad \left(-\sqrt{2}\right)^4 \overset{?}{=} 4\left(-\sqrt{2}\right)^2 - 4$$

$$4 \overset{?}{=} 4(2) - 4 \qquad\qquad 4 \overset{?}{=} 4(2) - 4$$

$$4 = 4 \qquad\qquad\qquad 4 = 4$$

The solution set is $\{\sqrt{2}, -\sqrt{2}\}$.

● **PROBLEM 645**

Solve $x^4 - 2x^2 - 3 = 0$ as a quadratic in x^2.

Solution: We can write the equation as $\left(x^2\right)^2 - 2x^2 - 3 = 0$.

Let $x^2 = z$ and we have
$$z^2 - 2z - 3 = 0,$$

which is a quadratic equation. We can solve a quadratic equation in the form $ax^2 + bx + c = 0$ using the quadratic formula,

$$x = \frac{-b \pm \sqrt{b^2 - 4ac}}{2a}.$$

In our case $a = 1$, $b = -2$, and $c = -3$. Thus,

$$z = \frac{-(-2) \pm \sqrt{(-2)^2 - 4(1)(-3)}}{2(1)}$$

$$= \frac{2 \pm \sqrt{4 + 12}}{2} = \frac{2 \pm 4}{2} = 1 \pm 2.$$

Therefore, $z = 3$ or $z = -1$. Since $z = x^2$, we have $x^2 = 3$ or $x^2 = -1$. If $x^2 = 3$, $x = \pm \sqrt{3}$. If $x^2 = -1$, $x = \pm i$. Hence, the solution set of the original equation is

$$\{\sqrt{3}, -\sqrt{3}, i, -i\}.$$

● **PROBLEM 646**

Solve the equation $\left(x^2 - 3x\right)^2 - 2\left(x^2 - 3x\right) - 8 = 0$

Solution: This equation is greater than degree two.

However, it can be solved by the methods used for quadratic equations. An equation of the type $A[P(x)]^2 + B[P(x)] + C = 0$, where A, B, and C are real numbers, $A \neq 0$ and $P(x)$ is an expression in variable x, is said to be in quadratic form.

We let $z = x^2 - 3x$, substitute in the given equation, and get

$$z^2 - 2z - 8 = 0$$

We then complete the solution as follows:

$(z - 4)(z + 2) = 0$ factoring $z^2 - 2z - 8$

Setting each factor equal to zero and solving:

$z - 4 = 0$	$z + 2 = 0$
$z = 4$	$z = -2$

Hence the solution set is $\{4, -2\}$.

Now we replace z in $z = x^2 - 3x$ with 4 and then with -2, solve each of the resulting equations, and thus obtain

$$x^2 - 3x = 4$$
$$x^2 - 3x - 4 = 0$$
$$(x - 4)(x + 1) = 0$$
$$x = 4$$
$$x = -1$$

and
$$x^2 - 3x = -2$$
$$x^2 - 3x + 2 = 0$$
$$(x - 2)(x - 1) = 0$$
$$x = 2$$
$$x = 1$$

Therefore the solution set of the given equation is $\{4, -1\} \cup \{2, 1\} = \{4, -1, 2, 1\}$.

● **PROBLEM** 647

Solve for x: $x^4 + 6x^3 + 2x^2 - 21x - 18 = 0$.

Solution: This is a polynomial equation of the fourth degree since the power of the highest term is four. We complete the square from the x^4 and x^3 terms. This is done by taking one half the coefficient of the x^3 term and squaring it. This will then be the coefficient of the x^2 term. We obtain 9. Add and subtract $9x^2$ to maintain the same equation. Then:

$$\left(x^4 + 6x^3 + 9x^2\right) - 9x^2 + 2x^2 - 21x - 18 = 0.$$

Convert $\left(x^4 + 6x^3 + 9x^2\right)$ into a binomial square: $\left(x^2 + 3x\right)^2 -$

$9x^2 + 2x^2 - 21x - 18 = 0.$

Combine like terms: $\left(x^2 + 3x\right)^2 - 7x^2 - 21x - 18 = 0.$

Factor -7 from the second and third terms; then we obtain:

$$\left(x^2 + 3x\right)^2 - 7\left(x^2 + 3x\right) - 18 = 0.$$

By substituting $v = x^2 + 3x$, we have a quadratic equation in v:

$$v^2 - 7v - 18 = 0.$$

Factor this quadratic equation in v, in terms of a product of two binomials:

$$(v - 9)(v + 2) = 0.$$

Whenever we have a product of two numbers such that $ab = 0$, then either $a = 0$ or $b = 0$. Thus, (1) $v - 9 = 0$ or (2) $v + 2 - 0$. Substitute the expression $\left(x^2 + 3x\right)$ for v in equations (1) and (2). Then:

(3) $x^2 + 3x - 9 = 0$ or   (4) $x^2 + 3x + 2 = 0$

Solve for x by the quadratic formula, $x = \dfrac{-b \pm \sqrt{b^2 - 4ac}}{2a}$,

from the equation $ax^2 + bx + c = 0$. For equation (3), $a = 1$, $b = 3$, $c = -9$. Thus,

$$x = \frac{-3 \pm \sqrt{9 - 4(1)(-9)}}{2(1)}$$

$$x = \frac{-3 \pm 3\sqrt{5}}{2}$$

For equation (4), $a = 1$, $b = 3$, $c = 2$. Thus,

$$x = \frac{-3 \pm \sqrt{9 - 4(1)(2)}}{2(1)}$$

$$x = \frac{-3 \pm 1}{2} = \frac{-3 + 1}{2} \text{ or } \frac{-3 - 1}{2}$$

Therefore, the four solutions are

$$x = -1, -2, \frac{-3 \pm 3\sqrt{5}}{2}, \text{ or } x = -1, -2, \frac{-3 + 3\sqrt{5}}{2}, \frac{-3 - 3\sqrt{5}}{2}$$

● **PROBLEM** 648

Form the equation whose roots are 0, $\pm a$, $\dfrac{c}{b}$.

<u>Solution:</u> The roots of the equation are 0, $+a$, $-a$, and $\dfrac{c}{b}$. Hence, $x = 0$, $x = a$, $x = -a$, and $x = \dfrac{c}{b}$. Subtract 0 from both sides of the first equation:

$$x - 0 = 0 - 0 = 0.$$

Subtract a from both sides of the second equation:

$$x - a = a - a = 0.$$

Add a to both sides of the third equation:

483

$$x + a = -a + a = 0.$$

Subtract $\frac{c}{b}$ from both sides of the fourth equation:

$$x - \frac{c}{b} = \frac{c}{b} - \frac{c}{b} = 0.$$

Hence, $(x - 0)(x - a)(x + a)\left(x - \frac{c}{b}\right) = (0)(0)(0)(0) = 0$ or

$(x - 0)(x - a)(x + a)\left(x - \frac{c}{b}\right) = 0.$ Therefore,

$$x(x - a)(x + a)(x - \frac{c}{b}) = 0$$

$$x\left(x^2 - a^2\right)\left(x - \frac{c}{b}\right) = 0.$$

Multiply both sides by b.

$$x\left(x^2 - a^2\right)b\left(x - \frac{c}{b}\right) = 0 \cdot b$$

$$x\left(x^2 - a^2\right)(bx - c) = 0$$

$$\left(x^3 - a^2x\right)(bx - c) = 0.$$

Thus, the equation is:

$$bx^4 - cx^3 - a^2bx^2 + a^2cx = 0.$$

● **PROBLEM 649**

Form the equation of the fourth degree with rational co-efficients, one of whose roots is $\sqrt{2} + \sqrt{-3}$.

Solution: Here we must have $\sqrt{2} + \sqrt{-3}$, $\sqrt{2} - \sqrt{-3}$ as one pair of roots, and $-\sqrt{2} + \sqrt{-3}$, $-\sqrt{2} - \sqrt{-3}$ as another pair.

Recall that every quadratic equation may be written in the form:

$$ax^2 + bx + c = 0$$

Also, recall the well-known formula that the sum of the roots of a quadratic equation $= \frac{-b}{a}$, and the product of the roots is $\frac{c}{a}$. Thus, finding the sum and product of our known roots will give us their corresponding equation.

Our first pair of roots are $\sqrt{2} + \sqrt{-3}$ and $\sqrt{2} - \sqrt{-3}$. Their sum is:

$$\sqrt{2} + \sqrt{-3} + (\sqrt{2} - \sqrt{-3}) = 2\sqrt{2}, \text{ and their product}$$

is:

$$(\sqrt{2} + \sqrt{-3})(\sqrt{2} - \sqrt{-3}) = 2 + \sqrt{2}\sqrt{-3} - \sqrt{2}\sqrt{-3}$$
$$- \sqrt{-3}\sqrt{-3}$$
$$= 2 - (-3).$$
$$= 5.$$

So our sum $= 2\sqrt{2} = \dfrac{-b}{a}$, and our product $= 5 = \dfrac{c}{a}$.

Therefore, $a = 1$, $b = -2\sqrt{2}$, $c = 5$, and the quadratic factor corresponding to this pair of roots is $x^2 - 2\sqrt{2}x + 5$.

The second pair of roots are $- \sqrt{2} + \sqrt{-3}$ and $- \sqrt{2} - \sqrt{-3}$. Their sum is:

$$- \sqrt{2} + \sqrt{-3} + (- \sqrt{2} - \sqrt{-3}) = - 2\sqrt{2},$$ and their product is:

$$(- \sqrt{2} + \sqrt{-3})(- \sqrt{2} - \sqrt{-3}) = 2 - \sqrt{2}\sqrt{-3} + \sqrt{2}\sqrt{-3}$$
$$- \sqrt{-3}\sqrt{-3}$$
$$= 2 - (-3)$$
$$= 2 + 3$$
$$= 5.$$

So our sum $= - 2\sqrt{2} = \dfrac{-b}{a}$, and our product $= 5 = \dfrac{c}{a}$. Therefore, $a = 1$, $b = 2\sqrt{2}$, $c = 5$, and the quadratic factor corresponding to the second pair of roots is $x^2 + 2\sqrt{2}x + 5$.

Thus the required equation is

$$(x^2 + 2\sqrt{2}x + 5)(x^2 - 2\sqrt{2}x + 5) = 0,$$

or $$x^4 + 2x^2 + 25 = 0.$$

● **PROBLEM** 650

Solve the equation $6x^4 - 13x^3 - 35x^2 - x + 3 = 0$, having given that one root is $2 - \sqrt{3}$.

<u>Solution:</u> Since $2 - \sqrt{3}$ is a root, $2 + \sqrt{3}$ is also a root. Recall that every quadratic equation may be written in the form:

$$ax^2 + bx + c = 0.$$

Also, recall the well-known formula that the sum of the roots of a quadratic equation $= - \dfrac{b}{a}$, and the product of the roots is $\dfrac{c}{a}$. Thus, finding the sum and product of our known roots will give us their corresponding equation.

tion. Our roots are $2 - \sqrt{3}$ and $2 + \sqrt{3}$. Their sum is

$2 - \sqrt{3} + (2 + \sqrt{3}) = 4$ and their product is

$(2 - \sqrt{3})(2 + \sqrt{3}) = 4 - 2\sqrt{3} + 2\sqrt{3} - 3 = 1$. So our

sum $= 4 = -\dfrac{b}{a}$, and our product $= 1 = \dfrac{c}{a}$. Therefore,

$a = 1$, $b = -4$, $c = 1$, and the equation is $x^2 - 4x + 1$.

Now, dividing this new equation into the given equation we obtain:

$$
\begin{array}{r}
6x^2 + 11x + 3 \\
x^2 - 4x + 1 \overline{\smash{\big)}\ 6x^4 - 13x^3 - 35x^2 - x + 3} \\
\underline{6x^4 - 24x^3 + 6x^2} \\
11x^3 - 41x^2 - x + 3 \\
\underline{11x^3 - 44x^2 + 11x} \\
3x^2 - 12x + 3 \\
\underline{3x^2 - 12x + 3} \\
0
\end{array}
$$

Thus, $6x^4 - 13x^3 - 35x^2 - x + 3 = \left(x^2 - 4x + 1\right)$

$\left(6x^2 + 11x + 3\right)$;

Since $\qquad 6x^4 - 13x^3 - 35x^2 - x + 3 = 0$,

then $\quad \left(x^2 - 4x + 1\right)\left(6x^2 + 11x + 3\right) = 0$,

and this means that $6x^2 + 11x + 3 = 0$.

Factoring this equation we obtain $(3x + 1)(2x + 3)$
$= 0$. Thus, $3x + 1 = 0$ or $x = -\dfrac{1}{3}$; and $2x + 3 = 0$, or

$x = -\dfrac{3}{2}$. Therefore, the roots of the given equation are

$-\dfrac{1}{3}$, $-\dfrac{3}{2}$, $2 + \sqrt{3}$, $2 - \sqrt{3}$.

● **PROBLEM** 651

Find all rational roots of the equation
$x^4 - 4x^3 + x^2 - 5x + 4 = 0$.

Solution: This is a fourth degree equation. We can solve
it by synthetic division. Guess at a root by trying to
find an x-value which will make the equation equal to zero.
$x = 4$ works.

Now write the coefficients of the equation in
descending powers of x. Note that if a term is missing,
its coefficient is zero. In the corner box, the root 4
is placed. Bring the first coefficient down and multiply
it by the root. Place the result below the next coefficient and add. Multiply the result by the root and continue as before.

$$
\begin{array}{rrrrr|r}
1 & -4 & +1 & -5 & +4 & \underline{4} \\
 & +4 & 0 & +4 & -4 & \\
\hline
1 & 0 & 1 & -1 & 0 &
\end{array}
$$

The last result is zero which indicates (x - 4) is a factor and x = 4 is a root. The other results are the coefficients of the third degree expression when (x - 4) is factored.

$$(x - 4)(x^3 + 0x^2 + x - 1) = 0$$
$$(x - 4)(x^3 + x - 1) = 0$$

To find the roots of the third degree equation, call it g(x), we must set it equal to zero.

$$g(x) = x^3 + x - 1 = 0$$

Try to find where the curve of the equation crosses the x-axis which is when y = 0.

x	-2	-1	0	1	2
y	-11	-3	-1	1	9

It crosses the x-axis between x = 0 and x = 1. It is an irrational root.

Since the given equation is a fourth degree equation, it has 4 roots. All of the real roots, namely the rational number 4, and an irrational number between 0 and 1, have been found. Therefore, the two remaining roots are not real; that is, they are complex numbers.

● **PROBLEM** 652

Approximate the real roots of the equation
$$x^4 + 2x^3 - 5x^2 - 4x + 6 = 0.$$

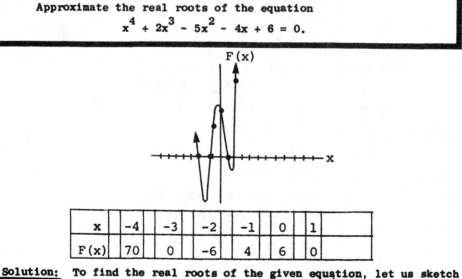

x	-4	-3	-2	-1	0	1
F(x)	70	0	-6	4	6	0

Solution: To find the real roots of the given equation, let us sketch the graph of the related polynomial function defined by

$$F(x) = x^4 + 2x^3 - 5x^2 - 4x + 6. \quad \text{(See Fig.)}$$

When f(x) = 0, x is a root of the equation. From the table, two real roots are x = -3 and x = 1. Reducing F(x) by dividing by (x + 3)(x - 1) and solving, we find that $-\sqrt{2}$ and $+\sqrt{2}$ are also roots.

Solve $x^4 + y^4 = 82$ (1),

$x - y = 2$ (2).

Solution: Let $x = u + v$ and $y = u - v$. From equation (2)

$$x - y = 2$$
$$u + v - (u - v) = 2$$
$$u + v - u + v = 2$$
$$2v = 2$$
$$v = 1$$

Substituting in equation (1),

$$x^4 + y^4 = 82$$
$$(u + v)^4 + (u - v)^4 = 82$$

Replacing v by 1,

$$(u + 1)^4 + (u - 1)^4 = 82 \qquad (3)$$

Now, simplify the left side of equation (3),

$$(u + 1)^2 = (u+1)(u+1) = u^2 + 2u + 1$$
$$(u+1)^3 = (u+1)(u+1)^2 ,$$

or

$$(u+1)^3 = (u+1)\left(u^2+2u+1\right)$$

Distributing on the right side of this equation:

$$(u+1)^3 = \left(u^3+2u^2+u\right) + \left(u^2+2u+1\right)$$
$$= u^3 + 3u^2 + 3u + 1 .$$
$$(u+1)^4 = (u+1)(u+1)^3$$

or

$$(u+1)^4 = (u+1)\left(u^3+3u^2+3u+1\right) .$$

Distributing on the right side of this equation:

$$(u+1)^4 = \left(u^4+3u^3+3u^2+u\right) + \left(u^3+3u^2+3u+1\right)$$
$$(u+1)^4 = u^4 + 4u^3 + 6u^2 + 4u + 1 \qquad (4)$$

Also, $(u-1)^2 = (u-1)(u-1) = u^2 - 2u + 1$

$$(u-1)^3 = (u-1)(u-1)^2$$

or

$$(u-1)^3 = (u-1)\left(u^2-2u+1\right) .$$

Distributing on the right side of this equation:

$$(u-1)^3 = \left(u^3-2u^2+u\right) - \left(u^2-2u+1\right)$$
$$= u^3 - 3u^2 + 3u - 1$$
$$(u-1)^4 = (u-1)(u-1)^3$$

or

$$(u-1)^4 = (u-1)\left(u^3-3u^2+3u-1\right)$$

Distributing on the right side of this equation:

$$(u-1)^4 = \left(u^4-3u^3+3u^2-u\right) - \left(u^3-3u^2+3u-1\right)$$
$$(u-1)^4 = u^4 - 4u^3 + 6u^2 - 4u + 1 \qquad (5)$$

Using equations (4) and (5) to simplify equation (3):

$$(u+1)^4 + (u-1)^4 = \left(u^4+4u^3+6u^2+4u+1\right) + \left(u^4-4u^3+6u^2-4u+1\right)$$

$$= u^4 + 4u^3 + 6u^2 + 4u + 1 + u^4 - 4u^3 + 6u^2 - 4u + 1$$
$$= 2u^4 + 12u^2 + 2 = 82.$$

Thus, $2u^4 + 12u^2 + 2 = 82$.

Subtract 82 from both sides of this equation:
$$2u^4 + 12u^2 + 2 - 82 = 82 - 82$$
$$2u^4 + 12u^2 - 80 = 0$$

Divide both sides of this equation by 2:
$$\frac{2u^4 + 12u^2 - 80}{2} = \frac{0}{2},$$

or
$$u^4 + 6u^2 - 40 = 0$$

Factoring the left side of this equation into a product of two polynomials:
$$\left(u^2 + 10\right)\left(u^2 - 4\right) = 0.$$

Whenever a product $ab = 0$, where a and b are any two numbers, either $a = 0$ or $b = 0$. Hence, either
$$u^2 + 10 = 0 \quad \text{or} \quad u^2 - 4 = 0$$
$$u^2 = -10 \quad \text{or} \quad u^2 = 4$$
$$u = \pm \sqrt{-10} \quad \text{or} \quad u = \pm \sqrt{4} = \pm 2$$

Using the equations relating x and y to u and v in order to solve for x and y:

when $u = \sqrt{-10}$, $x = u + v = \sqrt{-10} + 1 = 1 + \sqrt{-10}$ and
$$y = u - v = \sqrt{-10} - 1 = -1 + \sqrt{-10} .$$

when $u = -\sqrt{-10}$, $x = u + v = -\sqrt{-10} + 1 = 1 - \sqrt{-10}$ and
$$y = u - v = -\sqrt{-10} - 1 = -1 - \sqrt{-10} .$$

when $u = 2$, $x = u + v = 2 + 1 = 3$ and
$$y = u - v = 2 - 1 = 1 .$$

when $u = -2$, $x = u + v = -2 + 1 = -1$ and
$$y = u - v = -2 - 1 = -3 .$$

Thus, the pairs of solutions are:
$$x = 1 + \sqrt{-10} , \quad y = -1 + \sqrt{-10}$$
$$x = 1 - \sqrt{-10} , \quad y = -1 - \sqrt{-10}$$
$$x = 3 , y = 1$$
$$x = -1 , y = -3 .$$

● **PROBLEM** 654

If $\frac{a}{b} = \frac{c}{d} = \frac{e}{f}$, show that

$$\frac{a^3b + 2c^2e - 3ae^2f}{b^4 + 2d^2f - 3bf^3} = \frac{ace}{bdf} .$$

<u>Solution:</u> Let $\frac{a}{b} = \frac{c}{d} = \frac{e}{f} = k;$

then $a = bk$, $c = dk$, $e = fk$.

489

Now, substituting bk for a, dk for c, and fk for e in the given fraction we obtain:

$$\frac{a^3b + 2c^2e - 3ae^2f}{b^4 + 2d^2f - 3bf^3}$$

$$= \frac{(bk)^3b + 2(dk)^2fk - 3(bk)(fk)^2f}{b^4 + 2d^2f - 3bf^3}$$

$$= \frac{b^3k^3b + 2d^2k^2fk - 3bkf^2k^2f}{b^4 + 2d^2f - 3bf^3}$$

$$= \frac{b^4k^3 + 2d^2fk^3 - 3bf^3k^3}{b^4 + 2d^2f - 3bf^3} \; .$$

Factoring k^3 from each term of the numerator, then dividing numerator and denominator by their common factor, we obtain:

$$\frac{k^3(b^4 + 2d^2f - 3bf^3)}{b^4 + 2d^2f - 3bf^3} = k^3$$

But, since $k = \frac{a}{b}$, $k = \frac{c}{d}$, and $k = \frac{e}{f}$, then

$$k \cdot k \cdot k = \frac{a}{b} \cdot \frac{c}{d} \cdot \frac{e}{f} \quad \text{or,}$$

$$k^3 = \frac{a}{b} \times \frac{c}{d} \times \frac{e}{f}$$

$$= \frac{ace}{bdf} \; .$$

CHAPTER 22

PROGRESSIONS AND SEQUENCES

> **Basic Attacks and Strategies for Solving Problems in this Chapter. See pages 491 to 529 for step-by-step solutions to problems.**

Arithmetic and geometric sequences are two types of sequences or progressions that are of special interest mathematically because they fit the real world as mathematical models in many situations. The procedure for finding the terms of an arithmetic sequence, given the first term, involves adding a constant called the common difference, to the preceding term. For example,

> 3, 10, 17, 24, 31, ...

is an arithmetic sequence because after the first term 3, the other given terms are found by adding the constant 7 to each preceding term. Any additional terms in the sequence is found in a similar manner.

The procedure for finding the terms of a geometric sequence, given the first term, involves multiplying the preceding number by a constant number called the common ratio. For example,

> 3, 6, 12, 24, 48

is a geometric sequence because the ratio of one term to the preceding term equals the constant 2.

Another important progression is called the harmonic progression. The sequence of numbers of such a progression are reciprocals of the sequence of numbers in an arithmetic progression. Thus, the procedure for finding the next term in a harmonic progression is to first take the reciprocal of each term to get the corresponding arithmetic progression. Then, find the common differences between the terms in the arithmetic progression and use it to find the next term by adding the common difference value to the last term in the sequence. Take the reciprocal of this value to find the next term in the harmonic progression. For example, find the next term in the harmonic progression

> $1/_{12}, 1/_{19}, 1/_{26}, 1/_{33}, ...$

The corresponding arithmetic progression is given by 12, 19, 26, 33, ... which has a common difference of 7. Thus, the next term in the arithmetic progression sequence is 40. The reciprocal of 40 is $1/_{40}$. So, the next term in the harmonic progression is $1/_{40}$.

Step-by-Step Solutions to Problems in this Chapter, "Progressions and Sequences"

ARITHMETIC

● **PROBLEM** 655

If the 6th term of an arithmetic progression is 8 and the 11th term is - 2, what is the 1st term? What is the common difference?

Solution: An arithmetic progression is a sequence of numbers where each term excluding the first is obtained from the preceding one by adding a fixed quantity to it. This constant amount is called the common difference. Let a = value of first term, and d = common difference

Term of sequence: 1st 2nd 3^{thrd} 4th ... n^{th}

Value of term: a   a+d   a+2d   a+3d ... a+(n-1)d

Use the formula for the nth term of the sequence to write equations for the given 6th and 11th terms, to determine a and d.

11th term: a + (11 - 1)d = - 2

6th term: a + (6 - 1)d = 8. Simplifying the above equations we obtain:

$$a+ 10d = - 2 \tag{1}$$
$$a+ 5d = 8 \tag{2}$$

$5d = - 10$ Subtracting (2) from (1)
$d = - 2$ Substituting in (1)
$a + 10(- 2) = - 2$
$a = 18$

The first term is 18 and the common difference is - 2.

● **PROBLEM** 656

Find a_n for the sequence 1, 4, 7, 10,...

Solution: An arithmetic progression (A.P.) is a sequence of numbers each of which, after the first, is obtained by adding to the preceding number a constant number, d, called the common difference. Thus 1, 4, 7, 10,... is an arithmetic progression because each term is obtained by adding 3 to the preceding number. The nth term, a_n, of an A.P. is:

$$a_n = a_1 + (n-1)d$$

where a_1 = first term of the progression; d = common difference; n = number of terms.

Thus, with $a_1 = 1$ and $d = 3$,

$$a_n = 1 + (n-1)3 = 1 + 3n - 3$$

$$a_n = 3n - 2$$

It is easily verified by substitution that $a_n = 3n - 2$ will suffice, i.e.

$$a_1 = 3(1) - 2 = 1$$
$$a_2 = 3(2) - 2 = 4$$
$$a_3 = 3(3) - 2 = 7$$
$$a_4 = 3(4) - 2 = 10$$
$$a_5 = 3(5) - 2 = 13 = a_4 + d = 10 + 3$$

by definition of an A.P. etc.

● **PROBLEM** 657

Find the first term of an arithmetic progression if the fifth term is 29 and d is 3.

Solution: The nth term, or last term, of an arithmetic progression (A.P.) is:

$$\ell = a_1 + (n-1)d \qquad (1)$$

where a_1 = first term of the progression

d = common difference
n = number of terms
$\ell = n^{th}$ term, or last term.

Using this formula we can find the first term of an A.P. whose fifth term is 29 and d is 3. Since $\ell = a_5 = 29$,

$d = 3$, and $n = 5$, substituting into equation (1) gives:

$$29 = a_1 + (5 - 1)3$$
$$29 = a_1 + 12$$
$$a_1 = 29 - 12 = 17.$$

Thus, the first term is 17.

● **PROBLEM** 658

If the first term of an arithmetic series is -4 and the twelfth term is 32, find the common difference.

Solution: Since the first and last terms and the number of terms are known, the formula for the nth term, or last term of the series

$$\ell = a_1 + (n-1)d$$

where a_1 = first term of the series

n = number of terms
d = common difference
$\ell = n^{th}$ term, or last term

can be solved for d.

$\ell = a_1 + (n-1)d$ with $a_1 = -4$, $n = 12$, $\ell = 32$ gives

$$32 = -4 + (12 - 1)d$$

$$36 = 11\,d$$

$$\frac{36}{11} = d$$

● PROBLEM 659

Find the twelfth term of the arithmetic sequence

$$2, 5, 8, \ldots .$$

__Solution:__ It is given that the sequence is an arithmetic sequence. The common difference d is obtained by subtracting any term from the succeeding term;

$$d = 5 - 2 = 3$$

The twelfth term, a_{12}, can be obtained by substituting $a_1 = 2$, $d = 3$, and $n = 12$ in the expression for the nth term:

$$a_1 + (n - 1)d$$

Thus,

$$a_{12} = a_1 + (n - 1)d = 2 + (12 - 1)3 = 35$$

This can be checked by completing the sequence to the twelfth term.

2, 5, 8, 2+(4-1)3, 2+(5-1)3, 2+(6-1)3, ... , 2+(11-1)3, 35

2, 5, 8, 11, 14, 17, 20, 23, 26, 29, 32, 35

● PROBLEM 660

Given that the first term of an arithmetic sequence is 56 and the seventeenth term is 32, find the tenth term and the twenty-fifth term.

__Solution:__ The formula for the nth term, or last term, of an arithmetic sequence is:

$$\ell = s_1 + (n-1)d$$

where s_1 = first term of the sequence

d = common difference

n = number of terms

ℓ = nth term, or last term.

Since we are given the first and seventeenth term of the sequence we can use this information, with n = 17, to find d before we proceed to find the tenth and twenty-fifth terms. Thus we find d by

$$s_{17} = s_1 + (17 - 1)d,$$

$$32 - 56 = 16d$$

$$16d = -24$$

$$d = -\frac{3}{2}$$

Now we can find the tenth term, s_{10}, as

$$s_{10} = 56 + (10 - 1)\left(-\frac{3}{2}\right) = 56 - \frac{27}{2} = \frac{85}{2}$$

and the twenty-fifth term, s_{25}, as

$$s_{25} = 56 + (25 - 1)\left(-\frac{3}{2}\right) = 56 - 36 = 20$$

● **PROBLEM** 661

The 54th and 4th terms of an arithmetic progression are - 61 and 64; find the 23rd term.

Solution: An arithmetic progression is a sequence of numbers, each of which, after the first, is obtained by adding a constant to the preceding term. This constant is called the 'common difference'. Let a be the first term, and d the common difference; then the sequence looks as follows:

a, a + d, (a + d) + d, (a + d + d) + d, ...   or,

a, a + d, a + 2d, a + 3d, ...

Notice that, 1st. term = a = a + (1 - 1)d

2nd. term = a + 1d = a + (2 - 1)d

3rd. term = a + 2d = a + (3 - 1)d

4th. term = a + 3d = a + (4 - 1)d

Thus, we obtain the formula for any term of an A.P. (arithmetic progression). Let us call this general term n. Then the nth. term = a + (n - 1)d. Therefore,

- 61 = the 54th term = a + 53d;

and   64 = the 4th term = a + 3d.

We now solve these two equations for d by subtracting the second from the first. Thus,

a + 53d = - 61

a +   3d = 64

a - a + 53d - 3d = - 61 - 64   or

$$50d = - 125, \text{ or } d = \frac{-5}{2}$$

Thus, substituting for d in equation two we have:

$$a + 3\left(-\frac{5}{2}\right) = 64$$

$$a - \frac{15}{2} = 64$$

$$a = \frac{128}{2} + \frac{15}{2} = \frac{143}{2} .$$

Thus, the 23rd term = a + 22d, and substituting for a and d we have:

$$\frac{143}{2} + 22\left(-\frac{5}{2}\right) = \frac{143}{2} - \frac{110}{2} = \frac{33}{2} .$$

● PROBLEM 662

If the first term of an arithmetic progression is 7, and the common difference is -2, find the fifteenth term and the sum of the first fifteen terms.

Solution: An arithmetic progression is a sequence of numbers each of which is obtained from the preceding one by adding a constant quantity to it, the common difference, d. If we designate the first term by a and the common difference by d, then the terms can be expressed as follows:

terms of
series   1   2   3   .....   n

value of
term   a   a + d   a + 2d ℓ = a + (n-1)d

In this example a = 7, and d = -2. To find the fifteenth term, we have n = 15. The nth term is a + (n-1)d. For n = 15, a + (n-1)d = 7 + (15 - 1)(-2) = 7 - 28 = -21. To find the sum of the first fifteen terms apply the following formula:

$$S_n = \frac{n}{2}(a + \ell)$$

$$S_{15} = \frac{15}{2}[7 + (-21)] = \frac{15}{2}(-14) = -105.$$

● PROBLEM 663

Find the sum of the first sixteen terms of the arithmetic series whose first term is $\frac{1}{4}$ and common difference is $\frac{1}{2}$.

Solution: The sum of the first n terms of an arithmetic series is

$$S_n = \frac{n}{2}\left[2a_1 + (n - 1)d\right]$$

where a_1 = first term of the series

    d = common difference
    n = number of terms
    S_n = sum of first n terms

Hence, the sum of the first sixteen terms of the arithmetic series with $a_1 = \frac{1}{4}$, d = $\frac{1}{2}$, and n = 16, is

$$S_{16} = \frac{16}{2}\left[2\left(\frac{1}{4}\right) + (16 - 1)\frac{1}{2}\right]$$

$$= 8\left(\frac{1}{2} + \frac{15}{2}\right)$$

$$= 8\left(\frac{16}{2}\right)$$

$$= 8(8)$$

$$= 64$$

● **PROBLEM 664**

Find the sum of the first 20 terms of the arithmetic progression -9, -3, 3, ...

Solution: An arithmetic progression is a sequence in which
each term after the first is formed by adding a fixed
amount, called the common difference, to the preceding term.
The common difference of -9, -3, 3, ... is 6 since $-9 + 6 = -3$,
$-3 + 6 = 3$, etc. If a is the first term, d is the common
difference, and n is the number of terms of the arithmetic
progression, then the last term (or n^{th} term) ℓ is given by

$$\ell = a + (n-1)d \qquad (1)$$

and the sum S_n of the n terms of this progression is given
by

$$S_n = \frac{n}{2}[a + \ell] \qquad (2)$$

In this example,

$a = -9$, $d = 6$, $n = 20$. By equation (1):

$\ell = -9 + (19)(6)$

$\quad = -9 + 114$

$\quad = 105.$

By equation (2): $S_{20} = \frac{20}{2}\left(-9 + 105\right)$

$$= 10(96)$$

$$= 960.$$

Thus, the sum of the first 20 terms is 960.

● **PROBLEM 665**

Find the sum of the arithmetic series

$$5 + 9 + 13 + \ldots + 401$$

Solution: The common difference is $d = 9 - 5 = 4$, and the
nth term, or last term, is $\ell = a + (n-1)d$, where
     a = first term of the progression
     d = common difference
     n = number of terms
     ℓ = nth term, or last term.

Hence, $401 = 5 + (n-1)4$. Solving for the number of terms
n, we have n = 100. The required sum is

$$S = 5 + 9 + 13 + \ldots + 393 + 397 + 401$$

Written in reverse order, this sum is

$$S = 401 + 397 + 393 + \ldots + 13 + 9 + 5$$

Adding the two expressions for S, we have

$$2S = (5 + 401) + (9 + 397) + (13 + 393) + \ldots$$
$$+ (393 + 13) + (397 + 9) + (401 + 5)$$

Each term in parentheses is equal to the sum of the first and last terms; 5 + 401 = 406. There is a parenthetic term corresponding to each term of the original series; that is, there are 100 terms. Hence,

$$2S = 100(5 + 401) = 40,600 \text{ and } S = \frac{40,600}{2} = 20,300$$

In general, the sum of the first n terms of an arithmetic series is:

$$S = \frac{n}{2}(a + \ell) = \frac{n}{2}[2a + (n-1)d]$$

For this problem,

$$S = \frac{100}{2}(5 + 401) = \frac{100}{2}[2(5) + (100-1)4] = 20,300$$

● **PROBLEM** 666

Find the sum of the first 100 positive integers.

Solution: The first 100 positive integers is an arithmetic progression (A.P.), because each number after the first is obtained by adding 1, called the common difference, to the preceding number. For an A.P., the sum of the first n terms is

$$S_n = \frac{n}{2}(a + \ell)$$

where a = first number of the progression
      n = number of terms
      ℓ = n^{th} term, or last term
      S_n = sum of first n terms.

Concerning the first 100 positive integers: there are 100 terms; hence n = 100. The first term is 1; hence, a = 1. The last term is 100; hence, ℓ = 100.

$$S_{100} = \frac{100(1 + 100)}{2} = 5050$$

● **PROBLEM** 667

Find the sum of the first 25 even positive integers.

Solution: The even integers form an arithmetic progression which is a sequence of numbers each of which is obtained from the preceding one by adding a constant quantity to it. This constant quantity is called the common difference, d. The first term of an arithmetic progression is a and the nth term is $\ell = a + (n-1)d$. In this case:

$$a = 2, n = 25, d = 2.$$

$$\ell = 2 + (25 - 1)2$$

$$= 50$$

To find the sum of the n terms of an arithmetic progression, we apply the formula

$$S_n = \frac{n}{2}(a + \ell).$$

$$S_{25} = \frac{25}{2}(2 + 50)$$

$$= 25(26)$$

$$= 650.$$

● **PROBLEM** 668

Find the sum of the first p terms of the sequence whose nth term is 3n - 1.

<u>Solution:</u> By putting n = 1, and n = p, respectively, in 3n - 1, we obtain

1st term = 3(1) - 1 = 2

last term = pth term = 3p - 1.

We can now apply the formula for the sum of the terms of an arithmetic progression, which states:

$$S_n = \frac{n}{2}(a + \ell), \text{ where}$$

n = the number of terms

a = the first term

ℓ = the last term

S_n = sum of first n terms.

By subtitution,

$$S_p = \text{sum of first p terms} = \frac{p}{2}(2 + 3p - 1) =$$

$$\frac{p}{2}(3p + 1).$$

● **PROBLEM** 669

How many terms of the sequence - 9, - 6, - 3, ... must be taken that the sum may be 66?

<u>Solution:</u> To solve this problem we apply the formula for the sum of the first n terms of an arithmetic progression. The formula states:

$$S_n = \frac{n}{2}[2a + (n - 1)d], \text{ where}$$

S_n = sum of the first n terms

n = number of terms
a = first term

d = common difference

We are given all of the above information except n. Therefore, by substituting for S_n, a, and d, we can solve for n. We are given that S_n = 66, a = - 9, and d = 3, since - 9 + 3 = - 6, - 6 + 3 = - 3, ...

Hence, $\frac{n}{2}$ [- 18 + (n - 1)3] = 66.

Now, multiplying both sides of the equation by 2 we obtain: n [-18+ (n - 1)3] = 132; and simplifying the expression in brackets, we have: n (- 18 + 3n - 3) = 132, or n(3n - 21) = 132. Therefore, we have:

$3n^2$ - 21n = 132, and dividing each term by 3 we obtain:

n^2 - 7n - 44 = 0; factoring we have,

(n - 11) (n + 4) = 0;

therefore,   n = 11 or - 4.

We can reject the negative value because there cannot be a negative number of terms in the sequence, and therefore, 11 terms must be taken so that the sum of the terms is 66.

We can check this by taking 11 terms of the series. Doing this we have:

- 9, - 6, - 3, 0, 3, 6, 9, 12, 15, 18, 21;

the sum of which is 66.

● **PROBLEM** 670

How many terms of the sequence 26, 21, 16, ... must be taken to amount to 74?

Solution: To solve this problem we apply the formula for the sum of the first n terms of an arithmetic progression. The formula states:

$S = \frac{n}{2}$ [2a + (n - 1)d] , where

S = sum of the first n terms

n = number of terms

a = first term

d = common difference

We are given all of the above information except n. Therefore, by substituting for S, a, and d, we can solve for n. We are given that S = 74, a = 26, and d = - 5, since 26 - 5 = 21, 21 - 5 = 16, ...

Hence, 74 = $\frac{n}{2}$ [2(26) + (n - 1)(- 5)] , or

$$\frac{n}{2} [52 + (n - 1)(- 5)] = 74.$$

Simplifying, and multiplying both sides of the equation by 2, we obtain:

$$\frac{n}{2} (52 - 5n + 5) = 74$$

$$n (52 - 5n + 5) = 148$$

$$n (- 5n + 57) = 148$$

$$- 5n^2 + 57n = 148, \text{ or}$$

$$5n^2 - 57n + 148 = 0; \text{ factoring we obtain:}$$

$$(n - 4)(5n - 37) = 0;$$

therefore, $n = 4$ or $\frac{37}{5}$.

We can readily reject the value $\frac{37}{5}$, since it is not possible to have $7 \frac{2}{5}$ terms. Thus, 4 terms of the sequence must be added to amount to 74.

We can check this by adding the four terms, 26, 21, 16, 11, and observing that the sum is indeed 74.

● PROBLEM 671

Insert 4 arithmetic means between 1 and 36.

Solution: The terms between any two given terms of a progression are called the means between these two terms. Inserting 4 arithmetic means between 1 and 36 requires an arithmetic progression (A.P.) of the form 1, ___, ___, ___, ___, 36. Using the formula, $\ell = a + (n - 1)d$, for the nth term, or last term, of an A.P., where a = first term of the progression, d = common difference, n = number of terms, $\ell = n^{th}$ term or last term, we can determine d. Knowing the common difference, d, we can obtain the means by adding d to each preceding number after the first.

a = 1, ℓ = 36, and n = 6 since there will be six terms

$\ell = a + (n - 1)d$

36 = 1 + 5d

5d = 35

d = 7

The arithmetic means are: 1+7, (1+7)+7, (1+7+7)+7, (1+7+7+7)+7; that is, 8, 15, 22, and 29. The arithmetic progression is 1, 8, 15, 22, 29, 36.

● PROBLEM 672

Insert five arithmetic means between 13 and 31.

<u>Solution:</u> In an arithmetic progression, the terms between any two other terms are called the arithmetic means between the two given terms. An arithmetic progression is a sequence of numbers where each is derived from the preceeding one by adding a constant quantity to it. The constant quantity is called the common difference. The first term of the A.P. is designed by a and the common difference by d. We express the terms of the series:

Term of the Series	1	2	3	4	...	n
Value of the Series	$a_1 = a$	$a_2 = a+d$	$a_3 = a_2+d$ $= (a+d)+d$ $= a+2d$	$a_4 = a_3+d$ $= a+3d$	...	$a_n =$ $a+(n-1)d$

We are concerned with seven terms here: the first term, five arithmetic means, and the last term. In order to find the arithmetic means, we need to find the common difference, d. (We know a, which is 13.) The seventh term, $31 = a + (n-1)d = 13(7-1)d = 13 + 6d$. Thus,

$$31 = 13 + 6d$$
$$18 = 6d$$
$$d = 3$$

Consequently the five arithmetic means are

$$a_2 = 13 + 3 = 16, \ a_3 = 19, \ a_4 = 22, \ a_5 = 25, \ a_6 = 28.$$

● **PROBLEM** 673

Insert 20 arithmetic means between 4 and 67.

<u>Solution:</u> 'Arithmetic means' are all the terms that fall between any two given terms in an arithmetic progression. The two given terms are called the extremes. Thus, in this example, including the extremes, the number of terms will be 22; so that we have to find a sequence of 22 terms in A.P., of which 4 is the first and 67 the last.

Let d be the common difference; then, since the general nth term of an A.P. = $a + (n - 1)d$, and 67 is the 22nd term, we have:

$$67 = a + 21d, \qquad a = \text{first term.}$$

Since the first term is 4 we obtain:

$$4 + 21d = 67. \text{ Solving for d we find:}$$

$$21d = 63$$

$$d = 3.$$

Thus, the sequence is,

$$4, \ 4 + 3, \ (4 + 3) + 3, \ ... \quad \text{or}$$

$$4, \ 7, \ 10, \ 13, \ ..., \ 67$$

and the 20 required means are,

7, 10, 13, 16, 19, 22, 25, 28, 31, 34, 37, 40, 43, 46, 49,

52, 55, 58, 61, 64.

● **PROBLEM 674**

Determine the first four terms and 12th term of the arithmetic progression generated by F(x) = 2x + 3.

Solution: Find the terms of the progression by letting x = 1, 2, 3, ... etc.

1st term = F(1) = 2(1) + 3 = 5

2nd term = F(2) = 2(2) + 3 = 7

3rd term = F(3) = 2(3) + 3 = 9

4th term = F(4) = 2(4) + 3 = 11

12th term = F(12) = 2(12) + 3 = 27

The common difference, d, is found by subtracting one term from the one that immediately follows it.

The first term is denoted by a.

Note: For this progression a = 5 and d = 2. The coefficient of x in the linear function will always be the common difference for the arithmetic progression.

● **PROBLEM 675**

If an arithmetic progression is generated by the linear function F(x) = -3x + 14, what is the first term? What is the 15th term? What is the common difference?

Solution: 1st term = F(1) = - 3 + 14 = 11

15th term = F(15) = - 3(15) + 14 = - 31

common difference = d = - 3, the coefficient of the linear term.

The coefficient of x in a linear function will always be the common difference, d. To verify that d = - 3, find the second term and subtract the first term from it.

● **PROBLEM 676**

The sum of three numbers in arithmetic progression is 27, and the sum of their squares is 293; find them.

Solution: Let a be the middle number, d the common difference: then the three numbers are a - d, a, a + d.

Since the sum of the three numbers is 27 we have:

a - d + a + a + d = 27; or 3a = 27. Hence, a = 9, and the three numbers are 9 - d, 9, 9 + d.

Now, since the sum of the squares of the numbers is 293, we can use the following equation to solve for d:

$$(9 - d)^2 + 9^2 + (9 + d)^2 = 293.$$

Squaring, we obtain:

$$(81 - 18d + d^2) + (81) + (81 + 18d + d^2) = 293$$

$$2d^2 + 243 = 293$$

$$2d^2 = 50$$

$$d^2 = 25$$

$$d = \sqrt{25} = \pm 5.$$

Therefore, the three numbers are:

$$9 \pm 5, 9, 9 \pm 5 \quad \text{or}$$

$$4, 9, 14.$$

● PROBLEM 677

The sums of n terms of two arithmetic progressions are in the ratio of 7n + 1 : 4n + 27; find the ratio of their 11th terms.

Solution: Let the first term and common difference of the two sequences be a_1, d_1, and a_2, d_2 respectively.

The sum of the first n terms of an arithmetic progression is given by the formula:

$$S = \frac{n}{2} [2a + (n - 1)d] .$$

Since the sums of the two given progressions are in the ratio $\dfrac{7n + 1}{4n + 27}$ we have:

$$\frac{\frac{n}{2} [2a_1 + (n - 1)d_1]}{\frac{n}{2} [2a_2 + (n - 1)d_2]} = \frac{7n + 1}{4n + 27} , \text{ or}$$

$$\frac{2a_1 + (n - 1)d_1}{2a_2 + (n - 1)d_2} = \frac{7n + 1}{4n + 27} .$$

Now, using the fact that the nth term of an A.P. = $a + (n - 1)d$, we find

$$11\text{th term} = a + (11 - 1)d = a + 10d.$$

Therefore, the ratio of the 11th terms of our two progressions is:

$$\frac{a_1 + 10d_1}{a_2 + 10d_2} .$$

Now, we want to transform the ratio of the sums into the above ratio of the 11th terms. We can do this by dividing each term in the following proportion by 2:

$$\frac{2a_1 + (n-1)d_1}{2a_2 + (n-1)d_2} = \frac{7n+1}{4n+27} \; .$$

Doing this we obtain:

$$\frac{a_1 + \dfrac{(n-1)}{2} d_1}{a_2 + \dfrac{(n-1)}{2} d_2} = \frac{\dfrac{7n}{2} + \dfrac{1}{2}}{\dfrac{4n}{2} + \dfrac{27}{2}} \; .$$

Now, to obtain the ratio on the left of the equal sign in the form $\dfrac{a_1 + 10d_1}{a_2 + 10d_2}$, we must have: $\dfrac{n-1}{2} = 10$, or $n - 1 =$

20; therefore, $n = 21$. Substituting this value in our proportion we obtain:

$$\frac{a_1 + \left(\dfrac{21-1}{2}\right) d_1}{a_2 + \left(\dfrac{21-1}{2}\right) d_2} = \frac{\dfrac{7(21)}{2} + \dfrac{1}{2}}{\dfrac{4(21)}{2} + \dfrac{27}{2}} \qquad \text{or,}$$

$$\frac{a_1 + 10d_1}{a_2 + 10d_2} = \frac{\dfrac{147}{2} + \dfrac{1}{2}}{\dfrac{84}{2} + \dfrac{27}{2}} = \frac{\dfrac{148}{2}}{\dfrac{111}{2}} = \frac{148}{2} \cdot \frac{2}{111} = \frac{148}{111} \; .$$

Now, since both numerator and denominator are divisible by 37, we find that the desired ratio is $4 : 3$ or $\dfrac{4}{3}$.

● **PROBLEM** 678

If S_1, S_2, S_3, ... S_p are the sums of n terms of an arithmetic progression whose first terms are 1, 2, 3, 4, ... and whose common differences are 1, 3, 5, 7, ... respectively, find the value of

$$S_1 + S_2 + S_3 + \ldots S_p$$

Solution: We can find S_1, S_2, S_3, ..., S_p by applying the formula for the sum of the first n terms of an arithmetic progression. The formula states:

$$S = \frac{n}{2} [2a + (n-1)d] \text{ , where}$$

S = the sum

n = the number of terms

a = the first term

d = the common difference

Thus, for S_1, $a = 1$ and $d = 1$, we have:

$$S_1 = \frac{n}{2} [2(1) + (n-1)1] = \frac{n}{2} (2 + n - 1) =$$

$$\frac{n}{2}(n + 1) = \frac{n(n + 1)}{2}.$$

For S_2, $a = 2$ and $d = 3$; thus,

$$S_2 = \frac{n}{2} [2(2) + (n - 1)3] = \frac{n}{2} (4 + 3n - 3) =$$

$$\frac{n}{2} (3n + 1) = \frac{n(3n + 1)}{2}.$$

For S_3, $a = 3$ and $d = 5$; thus

$$S_3 = \frac{n}{2} [2(3) + (n - 1)5] = \frac{n}{2} (6 + 5n - 5) =$$

$$\frac{n}{2}(5n + 1) = \frac{n(5n + 1)}{2}.$$

Now, to find a and d for S_p we notice that a relation exists between the sum and the first term, and the sum and the common difference. For S_1, the first term is 1, and the difference 1, or $2(1) - 1$. For S_2, the first term is 2, and the difference 3, or $2(2) - 1$. For S_3, the first term is 3, and the difference 5, or $2(3) - 1$. Similarly, for S_p, the first term is p, and the common difference is $(2p - 1)$. Thus,

$$S_p = \frac{n}{2} [2p + (n - 1)(2p - 1)]$$

$$= \frac{n}{2} [2p + (2pn - 2p - n + 1)] = \frac{n}{2} (2pn - n + 1).$$

Factoring n from the first two terms in the parentheses we have:

$$\frac{n}{2}[(2p - 1) n + 1].$$

Therefore, the required sum,

$$S_1 + S_2 + S_3 + \ldots + S_p \quad \text{is:}$$

$$\frac{n(n + 1)}{2} + \frac{n(3n + 1)}{2} + \frac{n(5n + 1)}{2} + \ldots + \frac{n[(2p - 1)n + 1]}{2}$$

Factoring $\frac{n}{2}$ from each term, we obtain:

$$\frac{n}{2}[(n + 1) + (3n + 1) + (5n + 1) + \ldots (\{2p - 1\}n + 1)]$$

Now, since we are adding 1, p times, we can write:

$$\frac{n}{2} [(n + 3n + 5n + \ldots + \{2p - 1\}n) + p].$$

Factoring n from the terms in the parentheses, we obtain:

$$\frac{n}{2} [n(1 + 3 + 5 + \ldots + \{2p - 1\}) + p].$$

Let us now examine the terms of the above series: $1 + 3 + 5 + \ldots + (2p - 1)$. Notice that we can apply the formula for the sum of the first n terms of an arithmetic progression, which states the following:

$$S = \frac{n}{2} [2a + (n - 1)d] = \frac{n}{2}(a + \ell).$$

In our case it is more efficient to use the form $S = \frac{n}{2}(a + \ell)$, where S = the sum

n = the number of terms

a = first term

ℓ = last term

We know that n = p, a = 1, ℓ = (2p - 1). Thus,

$$S = \frac{p}{2} (1 + 2p - 1) \quad \text{or,} \quad S = \frac{p}{2} (2p) = p^2.$$

Therefore,

$$S_1 + S_2 + S_3 + \ldots + S_p = \frac{n}{2}[n(1 + 3 + 5 + \ldots$$

$$\ldots + \{2p - 1\}) + p] =$$

$$\frac{n}{2} \left[n(p^2) + p \right].$$ Factoring p from both terms in the brackets we obtain: $\frac{n}{2} [p (np + 1)] =$

$$\frac{np}{2} (np + 1).$$

Note: It is of interest to observe that in the formula for the sum of n terms of an A.P.,

$$S = \frac{n}{2} [2a + (n - 1)d] = \frac{n}{2}(a + \ell),$$

we can easily derive the first formula, $\frac{n}{2} [2a + (n - 1)d]$, from the second,

$\frac{n}{2} (a + \ell)$, as follows:

Since n = the number of terms, and ℓ = last term, then we can use the fact that: $\ell = a + (n - 1)d$. Substituting this value for ℓ we obtain:

$$\frac{n}{2}(a + \ell) = \frac{n}{2} [a + (a + \{n - 1\}d)]$$

$$= \frac{n}{2} [2a + (n - 1)d],$$

which is precisely our first formula.

506

GEOMETRIC

If the first term of a geometric progression is 9 and the common ratio is $-\frac{2}{3}$, find the first five terms.

Solution: A geometric progression (G.P.) is a sequence of numbers each of which, after the first, is obtained by multiplying the preceding number by a constant number called the common ratio, r. Thus a G.P. such as a_1, a_2, a_3, a_4, a_5, ... or a_1, $a_1 r$, $a_2 r$, $a_3 r$, $a_4 r$, ... with $a_1 = 9$ and $r = -\frac{2}{3}$ is determined as follows:

$$a_1 = 9$$
$$a_2 = 9\left(-\frac{2}{3}\right) = -6$$
$$a_3 = (-6)\left(-\frac{2}{3}\right) = 4$$
$$a_4 = 4\left(-\frac{2}{3}\right) = -\frac{8}{3}$$
$$a_5 = \left(-\frac{8}{3}\right)\left(-\frac{2}{3}\right) = \frac{16}{9}$$

Thus, the first five terms are: $9, -6, 4, -\frac{8}{3}, \frac{16}{9}$

● **PROBLEM** 680

Find the next three terms of the geometric progression 1,2,4, 8,... .

Solution: First determine the common ratio of the progression by dividing any term by the term immediately preceeding it. Once the common ratio has been determined any term can be computed by multiplying the term immediately preceeding the unknown term by the common ratio. The common ratio is 2 so the 5th term is 8 X 2 = 16. The 6th term is 16 X 2 = 32. The 7th term is 32 X 2 = 64. The first seven terms of the geometric progression are:

$$1,2,4,8,16,32,64$$

● **PROBLEM** 681

Find the next three terms of the geometric progression 27, -9, 3, -1,... .

Solution: A geometric progression takes the form:

$$a_1 + a_1 r + a_1 r^2 + \ldots + a_1 r^{n-1}$$

where r is the common ratio. Once the common ratio is determined, ant term in the progression can be computed by multiplying the term immediately preceeding the unknown term by the common ratio. The common ratio obtained by dividing any term by the term preceding it is $-1/3$. The 5th term is

$$(-1)\left(-\frac{1}{3}\right) = \frac{1}{3},$$

the 6th term is
$$\left(\frac{1}{3}\right)\left(-\frac{1}{3}\right) = -\frac{1}{9} ,$$

and the 7th term is
$$\left(-\frac{1}{9}\right)\left(-\frac{1}{3}\right) = \frac{1}{27} .$$

● PROBLEM 682

Find the 10th term of the geometric progression 3, 6, 12, 24,... .

Solution: $a_1 = 3$ and $r = \frac{6}{3} = 2$
If we let ℓ represent the 10th term

$$\ell = a_1 r^{n-1}$$
$$\ell = 3(2)^{10-1}$$
$$= 3(2)^9$$
$$= 3(512)$$
$$= 1536$$

● PROBLEM 683

The seventh term of a geometric progression is 192 and r = 2. Find the first four terms.

Solution: The formula for the nth term, or last term, of a geometric progression is:

$$\ell = s_1 r^{n-1}$$

where s_1 = first term of the progression

　　r = common ratio
　　n = number of terms
　　ℓ = nth term, or last term

Since we are given the seventh term and the common ratio of the progression we can use this information, with n = 7, to find the first term:

$$s_7 = s_1 r^{n-1}$$
$$192 = s_1(2)^{7-1} = 2^6 s_1 = 64 s_1$$
$$s_1 = \frac{192}{64} = 3$$

Then, since a geometric progression is a sequence of numbers each of which, after the first, is obtained by multiplying the preceding number by a constant number called the common ratio,

$s_1 = 3,$ $s_2 = 3 \cdot 2 = 6,$ $s_3 = 6 \cdot 2 = 12,$ and $s_4 = 12 \cdot 2 = 24$

● PROBLEM 684

If the 8th term of a geometric progression is 16 and the common ratio is -3, what is the 12th term?

Solution: A geometric progression is a sequence of numbers

508

in which each term after the first is obtained by multiplying the preceding one by a fixed number which is called the common ratio. We express the first term of a geometric progression by a and the common ratio by r. We write the terms of the sequence in this manner:

Terms of
the progression   (1)   (2)   (3)   (4) ... (n)

Value of the
term   a   ar ar^2 ar^3 ... ar^{n-1}

The formula for the nth term is ar^{n-1}. We are given the 8th term which is 16 and the common ratio is -3. Then, in this case n = 8 and $ar^{n-1} = ar^{8-1} = ar^7$. Therefore, the 8th term = 16 = ar^7

When n = 12, then the nth term is $ar^{n-1} = ar^{11}$. Therefore,
the 12th term = ar^{11}

Factor out ar^7 since we know its value.

$$ar^{11} = \left(ar^7 \right) r^4$$

Set ar^7 equal to its known value.

$$ar^{11} = 16r^4$$

Furthermore, r = -3. Then,

$$ar^{11} = 16(-3)^4 = 16 \cdot 81 = 1296$$

● PROBLEM 685

The first term of a geometric progression is 27, the nth term is 32/9, and the sum of n terms is 665/9. Find n and r.

<u>Solution:</u> A geometric progression (G.P.) is a sequence of numbers in which each number, after the first, is obtained by multiplying the preceding number by a constant number called the common ratio, r. The following two formulas for geometric progressions will be helpful in finding n, which is the number of terms, and r, which is the common ratio:

(1) the nth term or last term = $\ell = ar^{n-1}$,

(2) the sum of the first n terms = $S_n = \dfrac{a\left(r^n-1\right)}{r-1}$ where a = first

term, r = common ratio, n = number of terms, ℓ = nth term, or last term, and S_n = sum of the first n terms. In this problem it is given that ℓ = 32/9, a = 27, and S_n = 665/9. Using the formula for the nth term or last term:

$$\frac{32}{9} = 27r^{n-1} \qquad\qquad (1)$$

Using the formula for the sum of the first n terms:

$$\frac{665}{9} = \frac{27\left(r^n-1\right)}{r-1} = \frac{27r^n-27}{r-1} \qquad (2)$$

Multiply both sides of equation (1) by r:

509

$$r\left(\frac{32}{9}\right) = r\left(27\,r^{n-1}\right)$$

$$\frac{32}{9}\,r = 27r\,r^{n-1}$$

$$\frac{32}{9}\,r = 27r^{1+n-1}$$

$$\frac{32}{9}\,r = 27r^{n}$$

Substituting $\frac{32}{9}\,r$ for $27r^{n}$ in equation (2),

$$\frac{665}{9} = \frac{32r/9 - 27}{r - 1}$$

$$9(r-1)\,\frac{665}{9} = 9(r-1)\left[\frac{32r/9 - 27}{r-1}\right]$$

$$(r-1)665 = 9[32r/9 - 27]$$

Distributing on the right side:

$$(r-1)665 = 32r - 243$$

Distributing on the left side:

$$665r - 665 = 32r - 243$$

Subtract $32r$ from both sides:

$$665r - 665 - 32r = 32r - 243 - 32r$$
$$633r - 665 = -243$$

Add 665 to both sides:

$$633r - 665 + 665 = -243 + 665$$
$$633r = 422$$

Divide both sides by 633:

$$\frac{663r}{663} = \frac{442}{633}$$

$$r = \frac{422}{633} = \frac{2(211)}{3(211)} = \frac{2}{3}$$

Hence, the common ratio = $r = \frac{2}{3}$.

Substituting $\frac{2}{3}$ for r in equation (1):

$$\frac{32}{9} = 27\left(\frac{2}{3}\right)^{n-1}$$

Multiply both sides by $\frac{1}{27}$:

$$\frac{1}{27}\left(\frac{32}{9}\right) = \frac{1}{27}\left[27\left(\frac{2}{3}\right)^{n-1}\right]$$

$$\frac{1}{27}\left(\frac{32}{9}\right) = \left(\frac{2}{3}\right)^{n-1}$$

$$\frac{32}{243} = \left(\frac{2}{3}\right)^{n-1} \tag{3}$$

Express the fraction on the left side as a power of 2/3. Since $32 = 2^5$ and $243 = 3^5$, equation (3) becomes:

$$\frac{32}{243} = \frac{2^5}{3^5} = \left(\frac{2}{3}\right)^5 = \left(\frac{2}{3}\right)^{n-1}.$$

Hence, $5 = n - 1$.

Add 1 to both sides:

$$5 + 1 = n - 1 + 1$$
$$6 = n .$$

Hence, the number of terms = n = 6.

● **PROBLEM** 686

Find the sum of the first ten terms of the geometric progression: 15, 30, 60, 120, ...

Solution: A geometric progression is a sequence in which each term after the first is formed by multiplying the preceding term by a fixed number, called the common ratio.

If a is the first term, r is the common ratio, and n is the number of terms, the geometric progression (G.P.) is

$$a, ar, ar^2, ..., ar^{n-1}$$

The given G.P., 15, 30, 60, 120, ..., may be written as 15, 15(2), 15(2^2), 15(2^3)... . The sum, S_n, of the first n terms of the geometric progression is given by

$$S_n = \frac{a(1 - r^n)}{1 - r} \text{ , where } a = \text{first term}$$
$$r = \text{common ratio}$$
$$n = \text{number of terms.}$$

Here a = 15, r = 2, and n = 10.

$$S_{10} = \frac{15(1 - 2^{10})}{1 - 2}$$
$$= \frac{15(1 - 1024)}{-1}$$
$$= 15(1023)$$
$$= 15,345$$

● **PROBLEM** 687

Find the sum of the first four terms of the geometric series $2 + \left(-\frac{1}{3}\right) + \frac{1}{18} + ... $.

Solution: The ratio of any number of a geometric series to the number preceding it is constant. In this example,

the common ratio $r = \frac{\left(-\frac{1}{3}\right)}{2} = -\frac{1}{6}$. Since the sum of the

first n terms of a geometric series is:

$$S_n = \frac{a_1(r^n - 1)}{r - 1}, \; r \neq 1$$

where a_1 = first term; r = common ratio; n = number of terms; S_n = sum of first n terms; then with $a_1 = 2$, $r = -\frac{1}{6}$, n = 4, the sum S_4 is

$$S_4 = \frac{2\left[\left(-\frac{1}{6}\right)^4 - 1\right]}{-\frac{1}{6} - 1}$$

$$S_4 = \frac{2\left(\frac{1}{1296} - 1\right)}{-\frac{7}{6}}$$

$$S_4 = \frac{185}{108}$$

Check: $2 + 2\left(-\frac{1}{6}\right) + 2\left(-\frac{1}{6}\right)^2 + 2\left(-\frac{1}{6}\right)^3$

$= 2 + \left(-\frac{1}{3}\right) + \left(\frac{1}{18}\right) - \left(\frac{1}{108}\right)$

$= 2 - \frac{1}{3} + \frac{1}{18} - \frac{1}{108} = \frac{185}{108}.$

● **PROBLEM** 688

Find the sum of the first eight terms of the geometric progression:
$$4, \frac{-4}{3}, \frac{4}{9}, \frac{-4}{27}.$$

Solution: $a_1 = 4$, $r = \frac{-4/3}{4} = \frac{-1}{3}$, and $n = 8$

$S_n = \frac{a_1(1 - r^n)}{1 - r}$. Therefore, $S_8 = \frac{4\left[1 - \left(-\frac{1}{3}\right)^8\right]}{1 - \left(-\frac{1}{3}\right)}$

$$= \frac{4\left[1 - \left(\frac{-1^8}{3^8}\right)\right]}{1 + \frac{1}{3}}$$

$$= \frac{4\left[1 - \frac{1}{3^8}\right]}{\frac{4}{3}}$$

$$= \frac{4\left[1 - \frac{1}{6561}\right]}{\frac{4}{3}}$$

$$= \frac{4\left[\frac{6561}{6561} - \frac{1}{6561}\right]}{\frac{4}{3}}$$

$$= \frac{4\left(\frac{6560}{6561}\right)}{\frac{4}{3}}$$

$$= 4\left(\frac{3}{4}\right)\left(\frac{6560}{6561}\right)$$

$$= 3\left(\frac{6560}{6561}\right) = \frac{6560}{2187}$$

$$\frac{}{2187}$$

512

Sum the sequence $\frac{2}{3}$, -1, $\frac{3}{2}$, ... to 7 terms.

Solution: To solve this problem we use the formula for finding the sum of the first n terms of a geometric progression, which states:

$$S = \frac{a\left(r^n - 1\right)}{r - 1}, \qquad r \neq 1, \text{ where}$$

S = the sum

a = the first term

r = the common ratio

n = the number of terms

Now, we are looking for S, when $a = \frac{2}{3}$, $n = 7$, and $r = -\frac{3}{2}$, since $\frac{2}{3}\left(-\frac{3}{2}\right) = -1$, $(-1)\left(-\frac{3}{2}\right) = \frac{3}{2}$, ...

Thus, $\qquad S = \dfrac{\frac{2}{3}\left[\left(-\frac{3}{2}\right)^7 - 1\right]}{-\frac{3}{2} - 1}$. Simplifying, we obtain:

$$S = \frac{\frac{2}{3}\left(-\frac{2187}{128} - \frac{128}{128}\right)}{-\frac{3}{2} - \frac{2}{2}} = \frac{\frac{2}{3}\left(-\frac{2315}{128}\right)}{-\frac{5}{2}} = \frac{-\frac{4630}{384}}{-\frac{5}{2}}$$

$$= \frac{-4630}{384} \cdot -\frac{2}{5} =$$

$$\frac{9260}{1920} = \frac{463}{96} .$$

Thus, the sum of the first seven terms of the above sequence is $\frac{463}{96}$.

Find the sum of the first six terms of a geometric progression whose first term is $\frac{1}{3}$ and whose second term is -1.

Solution: A geometric progression (G.P.) is a sequence of numbers each of which, after the first, is obtained by multiplying the preceding number by a constant number called the common ratio. Thus the sequence may be represented as a_1, $a_1 r$, $a_1 r^2$, $a_1 r^3$, ..., where a_1 = first term and r = common ratio.

For the G.P. whose first term is $\frac{1}{3}$ and whose second term is -1, we can find r as follows: since $a_1 r$ is the second term, and $a_1 = \frac{1}{3}$,

$$-1 = a_1 r$$

$$-1 = \frac{1}{3} r$$

$$r = -3.$$

Then, since the sum of the first n terms of a G.P. is

$$S_n = \frac{a_1 (r^n - 1)}{r - 1}$$

where n = number of terms, we can find the sum of the first six terms of the G.P.:

$$S_6 = \frac{1}{3} \frac{(-3)^6 - 1}{-3 - 1} = \frac{1}{3} \cdot \frac{729 - 1}{-4} = -\frac{182}{3}$$

● PROBLEM 691

Find the seventh term of a geometric series whose third term is $\frac{1}{8}$ and common ratio is 2. Find the sum of the first seven terms.

Solution: Formulas for geometric series include formulas for the nth term, or last term:

$$\ell = a_1 r^{n-1}$$

and the sum of the first n terms:

$$S_n = \frac{a_1 (r^n - 1)}{r - 1} , \quad r \neq 1$$

where a_1 = first term; r = common ratio; n = number of terms; $\ell = n^{th}$ term, or last term; S_n = sum of first n terms.

Since the formulas for ℓ and S_n require the value of the first term, we will use the given information to determine a_1. For our series of the form a_1, $a_1 r$, $a_1 r^2$, $a_1 r^3$, ... the third term is $\frac{1}{8}$ and the ratio is 2. The first term, a_1, can be found using the fact that the third term is $a_1 r^2$.

$$a_1 r^2 = \frac{1}{8}$$

$$a_1 (2)^2 = \frac{1}{8}$$

$$4a_1 = \frac{1}{8}$$

$$a_1 = \frac{1}{32}$$

514

The seventh term is $a_1 r^6 = \left(\frac{1}{32}\right)\left(2^6\right) = 2$.

The sum $S_7 = \dfrac{a_1 \left(r^n - 1\right)}{r - 1} = \dfrac{\frac{1}{32}\left(2^7 - 1\right)}{2 - 1} = \dfrac{127}{32}$.

● **PROBLEM 692**

Determine the fifth term and the sum of the first ten terms of the geometric progression

$$2, \ -\frac{3}{2}, \ \frac{9}{8}, \ -\ldots$$

Solution: The ratio of any term in a geometric progression (G.P.) to its preceeding term is a constant, r, called the common ratio. Therefore,

$$r = \frac{-3/2}{2} = -\frac{3}{4}.$$

The formula for the nth term, or last term, of a G.P. is:

$$\ell = a_1 r^{n-1}$$

and the formula for the sum of the first n terms of a G.P. is:

$$S_n = \frac{a_1 \left(r^n - 1\right)}{r - 1} = \frac{a_1\left(1 - r^n\right)}{1 - r}, \quad r \neq 1$$

where a_1 = first term; r = common ratio; n = number of terms; ℓ = nth term, or last term; S_n = sum of first n terms.

The terms have the common ratio $r = -\frac{3}{4}$, and $a_1 = 2$. Hence the fifth term is given by

$$\ell = a_5 = 2\left(-\frac{3}{4}\right)^4 = \frac{2 \cdot 3^4}{4^4} = \frac{81}{128}.$$

For the second part of the problem, n = 10, $a_1 = 2$, $r = -\frac{3}{4}$, and we have

$$S_{10} = \frac{a_1 - a_1 (r)^n}{1 - r} = \frac{2 - 2\left(-\frac{3}{4}\right)^{10}}{1 - \left(-\frac{3}{4}\right)} = \frac{2 - 2\left(-\frac{3}{4}\right)^{10}}{\frac{7}{4}} =$$

$$= \frac{2\left[1 - \left(-\frac{3}{4}\right)^{10}\right]}{\frac{7}{4}} = \frac{4 \cdot 2}{7}\left(1 - \frac{3^{10}}{4^{10}}\right) = \frac{4 \cdot 2}{7}\left(\frac{4^{10} - 3^{10}}{4^{10}}\right) =$$

$$= \frac{4^2}{7 \cdot 2}\left(\frac{4^{10} - 3^{10}}{4^{10}}\right) = \frac{4^{10} - 3^{10}}{7 \cdot 2 \cdot 4^8} = \frac{989,527}{917,504}.$$

● **PROBLEM 693**

The fourth term of a geometric progression is ½ and the

515

sixth term is 1/8. Find the first term and the common ratio.

Solution: A geometric progression, a_1, a_2, a_3, ..., a_n, has terms with a common ratio r, so that the sequence can be expressed by

$$a_1, \ a_1r, \ a_1r^2, \ a_1r^3, \ ... \ a_1r^{n-1}$$

Observe that $a_2 = a_1r$, $a_3 = a_1r^2$, $a_4 = a_1r^3$, ..., $a_n = a_1r^{n-1}$. We are given that the fourth term is ½ and the sixth term is 1/8 so that

$$a_1r^3 = \tfrac{1}{2}$$

$$a_1r^5 = 1/8$$

Dividing the second equation by the first,

$$\frac{a_1r^5}{a_1r^3} = \frac{1/8}{\tfrac{1}{2}} \ .$$

Now, $a_1/a_1 = 1$, and $r^5/r^3 = r^{5-3}$; also, since division by a fraction is equivalent to multiplication by its reciprocal,

$$\frac{1/8}{\tfrac{1}{2}} = 1/8 \cdot 2/1$$

Therefore $r^2 = \tfrac{1}{4}$, and taking the square of both sides, $r = \pm \tfrac{1}{2}$.

Thus, there are two possible common ratios, ½ and $-\tfrac{1}{2}$; and therefore there are two possible series that satisfy the given conditions:

For $r = \tfrac{1}{2}$, the first term, a_1, is given by:

$$a_1 = \frac{a_1r^3}{r^3} = \frac{a_4}{r^3} \ .$$

Since a_4 = the fourth term = ½, and $r = \tfrac{1}{2}$, the first term

$$= \frac{\tfrac{1}{2}}{(\tfrac{1}{2})^3} = 4.$$

For $r = -\tfrac{1}{2}$,

$$a_1 = \frac{a_4}{r^3} = \frac{\tfrac{1}{2}}{(-\tfrac{1}{2})^3} = -4.$$

● **PROBLEM 694**

Insert 4 geometric means between 160 and 5.

Solution: 'Geometric means' are the terms between any two given terms in a geometric progression. Thus, for this problem we have to find 6 terms in G.P. of which 160 is

516

the first, and 5 the sixth.

We can apply the formula:

nth term = ar^{n-1}, where a = 1st term

r = common ratio

Thus, for the sixth term we have:

sixth term = 5 = $160r^{6-1}$

$$5 = 160r^5$$

Solving for r we obtain

$$r^5 = \frac{5}{160} = \frac{1}{32}$$

$$\sqrt[5]{r} = \sqrt[5]{\frac{1}{32}}$$

$$r = \frac{1}{2}$$

Now, since $r = \frac{1}{2}$ is the common ratio, we obtain each successive term of the progression by multiplication by $\frac{1}{2}$. Thus, we have:

160, 80, 40, 20, 10, 5, and the

four required means are 80, 40, 20, 10.

● **PROBLEM** 695

Insert three geometric means between 16 and 81.

Solution: In a geometric progression, G.P., the terms between any two other terms are called the geometric means. A G.P. is a sequence of numbers in which each term after the first is obtained by multiplying the preceeding term by a fixed number called the common ratio. If we designate the first term by a and the common ratio by r, then the terms of the series can be written as follows:

a, ar, ar^2, ar^3,...,ar^{n-1}. Note that the nth term is designated by $\ell = ar^{n-1}$.

In this example a = 16 and ℓ = 81. Furthermore, n = 5, since we have three geometric means between 16 and 81. A geometric mean is just a term between any two other terms. Hence,

$$81 = ar^{n-1}$$
$$81 = 16r^{5-1}$$
$$81 = 16r^4$$
$$\frac{81}{16} = r^4$$
$$\frac{9}{4} = r^2$$
$$\pm\frac{3}{2} = r$$

Symbolically, the geometric means are ar^1, ar^2, and ar^3. If $r = +\frac{3}{2}$

then the means are

$$16\left(\frac{3}{2}\right)^1, \ 16\left(\frac{3}{2}\right)^2, \ 16\left(\frac{3}{2}\right)^3$$

or 24, 36, 54. If $r = -\frac{3}{2}$, then the means are -24, 36, -54.

● **PROBLEM 696**

Find the first four terms of the geometric progression generated by the exponential function $f(x) = 12(3/2)^x$ if the domain of the function is the set of nonnegative integers $(0,1,2,3,\ldots)$.

Solution: $f(0) = 12\left(\frac{3}{2}\right)^0 = 12(1) = 12$

$f(1) = 12\left(\frac{3}{2}\right)^1 = 18$

$f(2) = 12\left(\frac{3}{2}\right)^2 = 12\left(\frac{9}{4}\right) = 27$

$f(3) = 12\left(\frac{3}{2}\right)^3 = 12\left(\frac{27}{8}\right) = \frac{81}{2}$

The first four terms are 12, 18, 27, and $\frac{81}{2}$.

● **PROBLEM 697**

Find three numbers in geometric progression whose sum is 19, and whose product is 216.

Solution: The three numbers of the G.P. may be denoted by $\frac{a}{r}$, a, ar; then $\frac{a}{r} \times a \times ar = 216$. Carrying out the multiplication we obtain:

$$a^3 = 216$$

$$\sqrt[3]{a^3} = \sqrt[3]{216}$$

$$a = 6$$

Thus, substituting for a, we find that the numbers are $\frac{6}{r}$, 6, 6r.

Now, since the sum of the three numbers is 19, we have:

$\frac{6}{r} + 6 + 6r = 19$, and we wish to solve for r. To do this we take r as a common denominator. Thus,

$$\frac{6 + 6r + 6r^2}{r} = 19$$

$$6 + 6r + 6r^2 = 19r$$

$$6 + 6r + 6r^2 - 19r = 0$$

$$6r^2 - 13r + 6 = 0$$

518

Factoring, we have

$(3r - 2)(2r - 3) = 0$; hence

$r = \dfrac{3}{2}$ or $\dfrac{2}{3}$.

Thus the numbers are 4, 6, 9.

● **PROBLEM** 698

Express $.4\overline{23}$ as a rational fraction.

<u>Solution:</u> We know that,

$.4\overline{23} = .423232323...$

This is a repeating decimal in which 23 is the repeated portion of the decimal. This is indicated by the bar above the given decimal, that is, $.4\overline{23}$.

The decimal $.4232323...$ can be rewritten as $.4 + .023 + .00023 + .0000023 + ...$ We can easily see this by adding each term in column form. Thus, we have:

```
        .4
        .023
        .00023
    +   .0000023
        .4232323
```
; and this sum is the desired result.

Now, $.4 + .023 + .00023 + .0000023 + ...$ can be rewritten as:

$$\frac{4}{10} + \frac{23}{1000} + \frac{23}{100000} + \frac{23}{10000000} + \cdots$$

$$= \frac{4}{10} + \frac{23}{10^3} + \frac{23}{10^5} + \frac{23}{10^7} + \cdots$$

Factoring $\dfrac{23}{10^3}$ from all terms except the first, we have:

$$.4\overline{23} = \frac{4}{10} + \frac{23}{10^3}\left(1 + \frac{1}{10^2} + \frac{1}{10^4} + \cdots\cdots\right).$$

Notice that the series $1 + \dfrac{1}{10^2} + \dfrac{1}{10^4} + \cdots$ has terms which are in a geometric progression where a = first term = 1, and r = common ratio = $\dfrac{1}{10^2}$.

Since r is less than 1 and the series is an infinite one, we can state that:

$$S = \text{sum} = \frac{a}{1 - r}$$

$$= \frac{1}{1 - \frac{1}{10^2}}$$

Thus, $.4\overline{23} = \frac{4}{10} + \frac{23}{10^3} \cdot \frac{1}{1 - \frac{1}{10^2}}$.

Now, we can simplify $\frac{1}{1 - \frac{1}{10^2}}$ as follows:

$$\frac{1}{1 - \frac{1}{10^2}} = \frac{1}{\left(1 - \frac{1}{100}\right)} = \frac{1}{\left(\frac{100}{100} - \frac{1}{100}\right)} = \frac{1}{\frac{99}{100}} = \frac{100}{99} .$$

Thus, substituting $\frac{100}{99}$ for $\frac{1}{1 - \frac{1}{10^2}}$, we have:

$$4\overline{23} = \frac{4}{10} + \frac{23}{10^3} \cdot \frac{100}{99} = \frac{4}{10} + \left(\frac{23}{1000} \cdot \frac{100}{99}\right)$$

$$= \frac{4}{10} + \frac{23}{990} = \frac{396}{990} + \frac{23}{990}$$

$$= \frac{419}{990} .$$

● **PROBLEM** 699

A person has 2 parents, 4 grandparents, 8 great-grand-parents, and so on. Find the number of his ancestors during the eight generations preceding his own, provided there are no duplications.

Solution: This is a geometric progression, which is a sequence of numbers in which each term after the first is obtained by multiplying the preceding term by a fixed number. The first term is the 2 parents. It is multiplied by a fixed number 2 since each parent has two parents of his or her own. To find the number of ancestors of the eight generations preceding his own, we must add up the terms starting from the parental generation up until the eighth ancestral generation. We apply the formula for the sum of a geometric progression: the first term of it is designated by a and the fixed amount called the common ratio is r. Let S_n represent the sum of n terms of the progression.

$$S_n = \frac{a - ar^n}{1 - r}$$

In this problem, a = 2, r = 2, and n = 8. By sub-stituting these values, we get

$$S_8 = \frac{2 - 2(2^8)}{1 - 2} = \frac{2 - 2(256)}{-1} = 510$$

If $|x| < 1$, sum the series

$$1 + 2x + 3x^2 + 4x^3 + \ldots\ldots \text{ to infinity.}$$

<u>Solution:</u> The terms of the given series are nearly in the form of a geometric progression, with a common ratio of x. Since $|x| < 1$, and the required sum is to infinity, obtaining the given series in the form of a geometric progression will enable us to apply the formula for the sum of an infinite G.P. The formula states: $S = \dfrac{a}{1 - r}$,

where   S = the sum
          a = the first term
          r = common ratio

Now, to transform the given series into a G.P. we can multiply it by $(1 - x)$. Thus,

$$(1 - x)\,(1 + 2x + 3x^2 + 4x^3 + \ldots)$$

$$= (1 - x) + (2x - 2x^2) + (3x^2 - 3x^3) + (4x^3 - 4x^4) + \ldots$$

$$= 1 + (- x + 2x) + (- 2x^2 + 3x^2) + (- 3x^3 + 4x^3) - 4x^4 + \ldots$$

$$= 1 + x + x^2 + x^3 + \ldots$$

The last series, $1 + x + x^2 + x^3 + \ldots$ is our desired G.P. with $a = 1$, $r = x$. Thus,

$$S = \frac{1}{1 - x}$$

But, $\dfrac{1}{1 - x}$ represents the sum of: $(1 - x)\,(1 + 2x + 3x^2 + \ldots)$, and we want the sum of: $1 + 2x + 3x^2 + \ldots$ Therefore, dividing $\dfrac{1}{1 - x}$ by $(1 - x)$ gives us the required sum. Hence, the sum of the given series =

$$\frac{\frac{1}{1 - x}}{(1 - x)} = \frac{1}{1 - x} \cdot \frac{1}{1 - x} = \frac{1}{(1 - x)^2}.$$

Find the sum of the geometric series

$$30 + 10 + 3\tfrac{1}{3} + \ldots + 30\left(\tfrac{1}{3}\right)^{n-1} + \ldots$$

<u>Solution:</u> Rewriting the geometric series as

$$30 + 30\left(\tfrac{1}{3}\right) + 30\left(\tfrac{1}{3}\right)^2 + \ldots + 30\left(\tfrac{1}{3}\right)^{n-1} + \ldots,$$

it can be seen that the first term is $a_1 = 30$; the ratio is $r = \tfrac{1}{3}$. Hence, since the sum to infinity (S_∞) of any geometric   progression in which the common ratio r is numerically less than 1 is given by

$$S_\infty = \frac{a_1}{1-r} \text{ , where } |r| < 1,$$

then $S_\infty = \dfrac{30}{1 - \frac{1}{3}} = 45$

Note that the sum of the first n terms of this series differs from 45 by $\frac{1}{2}$ of the nth term. For example, when n = 2, $S_2 = 40$, $a_2 = 10$, and $\frac{1}{2}a_2 = 5$. Thus,

$$40 = 45 - 5$$

When n = 3, $S_3 = 43\frac{1}{3}$, $a_3 = 3\frac{1}{3}$, and $\frac{1}{2}a_3 = 1\frac{2}{3}$. Thus $43\frac{1}{3} =$ = $45 - 1\frac{2}{3}$, etc.

● **PROBLEM** 702

The sum of an infinite number of terms in geometric progression is 15, and the sum of their squares is 45; find the sequence. Assume that the common ratio of the G.P. is less than 1.

<u>Solution:</u> The sum of any infinite geometric progression in which the common ratio is less than   is:

$$S_1 = \frac{a}{1 - r}$$

Now, squaring the terms of the sequence,

$$a, \ ar, \ ar^2, \ ar^3, \ \ldots$$

we have:

$$a^2, \ (ar)^2, \ (ar^2)^2, \ (ar^3)^2, \ \ldots \qquad =$$

$$a^2, \ a^2r^2, \ a^2r^4, \ a^2r^6, \ \ldots$$

This is a new infinite geometric progression with a^2 as the first term, and r^2 as the common ratio, and $r^2 < 1$. Therefore,

$$S_2 = \frac{a^2}{1 - r^2} \quad .$$

We are given that the sum of the terms, or S_1, is 15, and the sum of the squares of the terms, or S_2, is 45. Thus,

$$\frac{a}{1 - r} = 15, \quad \text{and} \quad \frac{a^2}{1 - r^2} = 45$$

We must now solve for a and for r. This can be done in the following manner: Multiply both sides of the equation $\frac{a}{1 - r} = 15$ by $(1 - r)$. Thus, we obtain: $a = 15(1 - r)$.

Now, multiply both sides of the equation $\frac{a^2}{1 - r^2} = 45$ by $(1 - r^2)$. Thus, we obtain:

$$a^2 = 45(1 - r^2) \qquad (1)$$

Squaring the equation $a = 15(1 - r)$ will give us a value for a^2 in terms of r. Substituting this value in equation (1) will give us an equation in r alone. Then, we can solve for r. Thus,

$a^2 = (15 - 15r)^2$, and expanding we have:

$a^2 = (15 - 15r)(15 - 15r)$

$\quad = 225 - 225r - 225r + 225r^2$

$\quad = 225r^2 - 450r + 225$

Now, by substitution:

$225r^2 - 450r + 225 = 45(1 - r^2)$

$225r^2 - 450r + 225 = 45 - 45r^2$

Subtracing 45 from both sides of the equation, and adding $45r^2$ to both sides we obtain:

$225r^2 + 45r^2 - 450r + 225 - 45 = 45 - 45 - 45r^2 + 45r^2$

$270r^2 - 450r + 180 = 0$

Dividing each term by 90, we have:

$\frac{270r^2}{90} - \frac{450r}{90} + \frac{180}{90} = \frac{0}{90}$

$3r^2 - 5r + 2 = 0$

Factoring gives us:

$(3r - 2)(r - 1) = 0$, and this means that either $3r - 2 = 0$, or $r - 1 = 0$.

To solve the first equation for r we first add 2 to both sides of $3r - 2 = 0$, and then divide by 3. Thus,

$3r - 2 + 2 = 0 + 2$

$3r = 2$

$\frac{3r}{3} = \frac{2}{3}$

$r = \frac{2}{3}$

To solve the second equation we add 1 to both sides

of r - 1 = 0. Thus,

$$r - 1 + 1 = 0 + 1$$

$$r = 1$$

Therefore, we have two values for r, $r = \frac{2}{3}$, $r = 1$.

But notice that the second value, $r = 1$, can be rejected. This is so because if we substitute this into either of our original equations, $\frac{a}{1 - r} = 15$ or $\frac{a^2}{1 - r^2} = 45$, we

obtain $\frac{a}{0}$, which is an undefined expression, and also

because the formula for the sum, $S = \frac{a}{1 - r}$, holds only

for progessions where the common ratio is less than 1.

Thus, we have $r = \frac{2}{3}$, and to solve for a we sub-

stitute this value into

$$a = 15(1 - r)$$

Thus, $a = 15\left(1 - \frac{2}{3}\right)$

$$= 15 \left(\frac{3}{3} - \frac{2}{3}\right)$$

$$= 15 \left(\frac{1}{3}\right) = \frac{15}{3} = 5$$

Therefore, we have a = first term of the sequence = 5, and r = common ratio = $\frac{2}{3}$. Thus the terms of the sequence

are: $5, 5\left(\frac{2}{3}\right), \left(5 \cdot \frac{2}{3}\right)\frac{2}{3}, \ldots$ and the sequence is:

$5, \frac{10}{3}, \frac{20}{9}, \ldots \ldots$

● **PROBLEM** 703

Convert the repeating decimal .477477 ... to a fraction.

Solution: This is a geometric progression. We can compute the rational equivalent by first determining the common ratio and then using the formula for the nth partial sum of a geometric series. The common ratio is computed by dividing any term by the term immediately preceeding it. Therefore r, the common ratio, is:

$$\frac{.000477}{.477} = .001$$

Allow a_1 to be .477. Then the sum of the geometric progression, S_n, is: $a_1 + a_1 r + a_1 r^2 + \ldots + a_1 r^{n-1}$. Now we can compute S_n using the formula.

$$S_n = \frac{a_1 - a_1 r^n}{1 - r} = \frac{.477 - .477(.001)^n}{1 - r}$$

by taking the limit of S_n as $n \to \infty$. We then compute the rational expression to which this geometric progression converges.

524

$$\lim_{n \to \infty} S_n = \lim_{n \to \infty} \frac{.477 - .477(.001)^n}{1 - (.001)} = \frac{.477}{1 - (.001)}$$

because $(.001)^n = \left(\frac{1}{1000}\right)^n$ goes to zero as n goes to ∞. Then

$$\frac{.477}{1 - (.001)} = \frac{477}{999} = \frac{53}{111}$$

$\frac{53}{111}$ is the fractional equivalent of the repeating decimal $.477477 \ldots$

● **PROBLEM** 704

Find the sum of the infinite geometric progression: $2, 1, \frac{1}{2}, \frac{1}{4}, \ldots$

Solution: First compute the common ratio by dividing any term by the term immediately preceeding it. The common ratio, r, is thus $1/2$ and since we are dealing with an infinite geometric progression and $|r| < 1$ we can find the sum using the formula:

$$S_n = \frac{a_1 - a_1 r^n}{1 - r} .$$

We must take the limit as n goes to infinity to determine the value to which this progression converges.

The first term $a = 2$, $r = \frac{1}{2}$

$$\lim_{n \to \infty} S_n = \lim_{n \to \infty} \frac{2 - 2(1/2)^n}{1 - 1/2}$$

$(1/2)^n$ goes to zero as $n \to \infty$.

$$\lim_{n \to \infty} S_n = \frac{2}{1 - \frac{1}{2}}$$

$$= \frac{2}{\frac{1}{2}} = 4$$

This result can also be attained by considering the Figure.

From the figure we can see the values of the sums are: $S_1 = 2$, $S_2 = 3$, $S_3 = 3\frac{1}{2}$, and $S_4 = 3\frac{3}{4}$. To find the position on the number scale for each of the successive sums, we move from the present position to the right one-half of the distance from the present position to the position of 4. It will not take many such steps after S_6 until we are so close to 4 that we will not be able to divide the line segment remaining in half.

Repeating decimals can be converted to fractional form by considering the repeating decimal as the sum of terms of an infinite geometric progression.

HARMONIC

● **PROBLEM** 705

Find the 9th term of the harmonic progression $3, 2, \frac{3}{2}, \ldots$.

Solution: The terms of a harmonic progression that lie between two given terms are called the harmonic means between these terms. If a single harmonic mean is inserted between two numbers, it is called the harmonic mean of the numbers.

A harmonic progression (H.P.) is a sequence of numbers whose reciprocals are in arithmetic progression, (A.P.). The terms of the A.P. are $\frac{1}{3}, \frac{1}{2}, \frac{2}{3}, \ldots$. We find the ninth term of the A.P. and take its reciprocal to find the corresponding term in the H.P. The formula for the nth term, a_n, of an A.P. is $a_1 + (n-1)d$ where a_1 is the first term and d is the common difference, the constant quantity added to each term to form the progression. Hence, if $a_n = a_1 + (n-1)d$, $a_1 = 1/3$ and $n = 9$, to find d subtract the first term, $1/3$, from the second term, $1/2$.

$$d = \frac{1}{2} - \frac{1}{3} = \frac{3}{6} - \frac{2}{6} = \frac{1}{6} .$$

Thus,

$$a_9 = \frac{1}{3} + (9-1)\frac{1}{6} = \frac{1}{3} + 8\left(\frac{1}{6}\right) = \frac{1}{3} + \frac{4}{3} = \frac{5}{3} .$$

Therefore, the ninth term of the harmonic progression is $\frac{3}{5}$.

● **PROBLEM** 706

Insert three harmonic means between $\frac{1}{10}$ and $\frac{1}{42}$.

Solution: The harmonic means are those terms of a harmonic progression that lie between two given terms. These two terms are $1/10$ and $1/42$. A harmonic progression (H.P.) is a sequence of numbers whose reciprocals are in arithmetric progression. Therefore, we consider the corresponding arithmetic progression, A.P. We shall first insert three arithmetic means between 10 and 42. The formula for the nth term is $a_n = a_1 + (n-1)d$ where a_1 is the first term of the A.P. and d is the common difference, the fixed amount added to each term to form the progression. The nth term is 42 where $n = 5$ since there are 3 terms to be inserted between 10 and 42. $a_1 = 10$. Thus, $a_5 = 10 + (5-1)d = 42$. Solve for d.

$$10 + 4d = 42$$
$$4d = 42 - 10$$
$$4d = 32$$
$$d = 8$$

Now we can find the terms of the A.P. by adding 8 to the first term and continuing in the same manner with each succeeding term. Thus, we obtain:

10, 10 + 8 = 18, 18 + 8 = 26, 26 + 8 = 34, 34 + 8 = 42 or

10, 18, 26, 34, 42.

The arithmetic progression is 10, 18, 26, 34, 42 .

The harmonic progression is $\frac{1}{10}$, $\frac{1}{18}$, $\frac{1}{26}$, $\frac{1}{34}$, $\frac{1}{42}$.

The harmonic means are $\frac{1}{18}$, $\frac{1}{26}$, $\frac{1}{34}$.

● **PROBLEM** 707

Insert 40 harmonic means between 7 and $\frac{1}{6}$.

<u>Solution:</u> Recall that 'means' are the terms between any two given terms of a progression. Thus, we wish to find a harmonic progression with 7 as the first term, and $\frac{1}{6}$ as the 42nd term. But a harmonic progression is a sequence of numbers whose reciprocals form an arithmetic progression (A.P.). Thus, 6 is the 42nd term of an A.P. whose first term is $\frac{1}{7}$; let d be the common difference; then, since the nth term, or last term, of an A.P. =

$$a + (n - 1)d, \text{ where } \quad \begin{aligned} a &= \text{first term} \\ n &= \text{last term} \\ d &= \text{common difference,} \end{aligned}$$

we have:

$$6 = \frac{1}{7} + (42 - 1)d \qquad \text{or,}$$

$6 = \frac{1}{7} + 41d.$ To solve for d subtract $\frac{1}{7}$ from both sides of the equation, and then multiply both sides by $\frac{1}{41}$. Thus,

$$6 - \frac{1}{7} = 41d$$

$$\frac{42}{7} - \frac{1}{7} = 41d$$

$$\frac{41}{7} = 41d$$

$$\frac{41}{7} \cdot \frac{1}{41} = \frac{1}{7} = d$$

Thus, the arithmetic progression is:

$$\frac{1}{7}, \ \frac{1}{7} + \frac{1}{7} \ , \ \left(\frac{1}{7} + \frac{1}{7}\right) + \frac{1}{7} \ , \ \ldots \qquad =$$

$$\frac{1}{7} \ , \ \frac{2}{7} \ , \ \frac{3}{7} \ \ldots\ldots, \ \frac{41}{7} \ , \ 6$$

Therefore, the harmonic progression is:

$$7, \ \frac{7}{2}, \ \frac{7}{3}, \ \cdots, \ \frac{7}{41}, \ \frac{1}{6}$$

and the 40 harmonic means between 7 and $\frac{1}{6}$ are:

$\frac{7}{2}, \ \frac{7}{3} \ , \ \cdots \ \frac{7}{41}$.

If a^2, b^2, c^2 are in arithmetic progression, show that b + c, c + a, a + b are in harmonic progression.

Solution: We are given that a^2, b^2, c^2 are in arithmetic progression. By this we mean that each new term is obtained by adding a constant to the preceding term.

By adding (ab + ac + bc) to each term, we see that,

a^2 + (ab + ac + bc), b^2 + (ab + ac + bc),

c^2 + (ab + ac + bc)

are also in arithmetic progression. These three terms can be rewritten as

a^2 + ab + ac + bc, b^2 + bc + ab + ac,

c^2 + ac + bc + ab

Notice that:

a^2 + ab + ac + bc = (a + b)(a + c)

b^2 + bc + ab + ac = (b + c)(b + a)

c^2 + ac + bc + ab = (c + a)(c + b)

Therefore, the three terms can be rewritten as:

(a + b)(a + c), (b + c)(b + a), (c + a)(c + b),

which are also in arithmetic progression.

Now, dividing each term by (a + b)(b + c)(c + a), we obtain:

$$\frac{1}{b + c}, \frac{1}{c + a}, \frac{1}{a + b}$$, which are in

arithmetic progression.

Recall that a sequence of numbers whose reciprocals form an arithmetic progression, is called a harmonic progression. Thus, since $\frac{1}{b + c}, \frac{1}{c + a}, \frac{1}{a + b}$ is an

arithmetic progression, b + c, c + a, a + b are in harmonic progression.

Find the first six terms of the sequence determined by the function g(x), where x = 1, 2, 3, 4, 5, 6.

$$g(x) = \frac{x^2}{x!}$$, x a positive integer

Solution:

$$g(1) = \frac{1}{1!} = 1$$

$$g(2) = \frac{4}{2!} = \frac{4}{2} = 2$$

$$g(3) = \frac{9}{3!} = \frac{9}{6} = \frac{3}{2}$$

$$g(4) = \frac{16}{4!} = \frac{16}{24} = \frac{2}{3}$$

$$g(5) = \frac{25}{5!} = \frac{25}{120} = \frac{5}{24}$$

$$g(6) = \frac{36}{6!} = \frac{36}{720} = \frac{1}{20}$$

If we consider simplifying the function $g(x) = \frac{x^2}{x!}$ before we evaluate, we would write

$$g(x) = \frac{x^2}{(1)(2)\ldots(x-1)x}$$

Upon dividing the numerator and denominator by x, we would obtain

$$g(x) = \frac{x}{1(2)\ldots(x-1)} = \frac{x}{(x-1)!} .$$

If we use this form of $g(x)$ and find $g(2)$, $g(3)$, $g(4)$, $g(5)$, $g(6)$, we obtain the same results as we did before with a little less effort. However, when we try to evaluate $g(1)$, we encounter a denominator of $0!$. Since we have already determined $g(1)$ to be 1 by using the original form of $g(x)$, $\frac{1}{0!}$ should be equal to one.

The symbol $0!$ arises rather frequently and in all cases we find our results are consistent if we define it to be one. That is, $0! = 1$.

With this extension to the definition of factorial, we have:

$$0! = 1$$

$$1! = 1$$

$$2! = (1)(2)$$

$$x! = (1)(2)\ldots(x) \quad x \text{ an integer}$$

Applying: $g(x) = \frac{x}{(x-1)!}$

$$g(1) = \frac{1}{0!} = 1$$

$$g(2) = \frac{2}{1!} = \frac{2}{1} = 2$$

$$g(3) = \frac{3}{2!} = \frac{3}{2}$$

$$g(4) = \frac{4}{3!} = \frac{2}{3}$$

$$g(5) = \frac{5}{4!} = \frac{5}{24}$$

$$g(6) = \frac{6}{5!} = \frac{\cancel{6}}{5 \cdot 4 \cdot \cancel{3} \cdot \cancel{2} \cdot 1} = \frac{1}{20} .$$

529

CHAPTER 23

MATHEMATICAL INDUCTION

> **Basic Attacks and Strategies for Solving Problems in this Chapter. See pages 530 to 538 for step-by-step solutions to problems.**

The procedure for doing mathematical induction, a method of proving that a statement, formula, or theorem is true for every integer

$$N \geq N_1,$$

where N is a given integer, is as follows:

(1) Show that the statement is true for the smallest integer, N, for which it is claimed to be true.

(2) Next, assume that the statement is true for some integer, call it k, which is greater than or equal to N_1.

(3) Finally, show (using the assumption made in Step 2) that the statement is true for the next integer, $k + 1$.

For example, prove that $2^N > N$ for all positive integer values of N.

PROOF

(1) If $N = 1$, then

$$2^N > N$$

becomes $2^1 > 1$, which is a true statement.

(2) Assume that for

$$N = k, 2^k > k$$

is true.

(3) We must prove the statement, if $2^k > k$, then

$$2^{k+1} > k + 1$$

for all positive integer values of k. This means we should be able to start with $2^k > k$ and from that deduce that

$$2^{k+1} > k + 1.$$

To show this we first note that

$$2^k > k \implies 2(2^k) > 2k \implies 2^{k+1} > 2k.$$

We know that $k \geq 1$ because we are working with positive integers. So,

$$k + k \geq k \implies 2k \geq k + 1.$$

Since

$$2^{k+1} > 2k \quad \text{and} \quad 2k \geq k + 1,$$

by the transitivity property, we can conclude it is true that

$$2^{k+1} > k + 1.$$

● **PROBLEM** 710

Prove by mathematical induction

$$1^2 + 2^2 + 3^2 + \ldots + n^2 = \frac{1}{6}n(n+1)(2n+1).$$

<u>Solution:</u> Mathematical induction is a method of proof. The steps are:
(1) The verification of the proposed formula or theorem for the smallest value of n. It is desirable, but not necessary, to verify it for several values of n.
(2) The proof that if the proposed formula or theorem is true for n = k, some positive integer, it is true also for n = k+1. That is, if the proposition is true for any particular value of n, it must be true for the next larger value of n.
(3) A conclusion that the proposed formula holds true for al¹ values of n.

<u>Proof:</u> Step 1. Verify:

For n = 1: $1^2 = \frac{1}{6}(1)(1+1)[2(1)+1] = \frac{1}{6}(1)(2)(3) = \frac{1}{6}(6) = 1$

$$1 = 1 \checkmark$$

For n = 2: $1^2 + 2^2 = \frac{1}{6}(2)(2+1)[2(2)+1] = \frac{1}{6}(2)(3)(5) = \frac{1}{6}(6)(5)$

$$1 + 4 = (1)(5)$$

$$5 = 5 \checkmark$$

For n = 3: $1^2 + 2^2 + 3^2 = \frac{1}{6}(3)(3+1)[2(3)+1]$

$$1 + 4 + 9 = \frac{1}{6}(3)(4)(7) = \frac{1}{6}(12)(7) = 14$$

$$14 = 14 \checkmark$$

Step 2. Let k represent any particular value of n. For n = k, the formula becomes

$$1^2 + 2^2 + 3^2 + \ldots + k^2 = \frac{1}{6}k(k+1)(2k+1). \qquad \text{(A)}$$

For n = k+1, the formula is

$$1^2 + 2^2 + 3^2 + \ldots + k^2 + (k+1)^2 = \frac{1}{6}(k+1)[(k+1) + 1][2(k+1) + 1]$$

$$= \frac{1}{6}(k+1)(k+2)(2k+3). \qquad \text{(B)}$$

We must show that if the formula is true for n = k, then it must be true for n = k+1. In other words, we must show that (B) follows from (A). The left side of (A) can be converted into the left side of (B) by merely adding $(k+1)^2$. All that remains to be demonstrated is that when $(k+1)^2$ is added to the right side of (A), the result is the right side of (B).

$$1^2 + 2^2 + \ldots + k^2 + (k+1)^2 = \frac{1}{6}k(k+1)(2k+1) + (k+1)^2$$

Factor out (k+1):

$$1^2 + 2^2 + 3^2 + \ldots + k^2 + (k+1)^2 = (k+1)\left[\frac{1}{6}k(2k+1) + (k+1)\right]$$

$$= (k+1)\left[\frac{k(2k+1)}{6} + \frac{(k+1)6}{6}\right]$$

$$= (k+1)\frac{2k^2 + k + 6k + 6}{6}$$

$$= \frac{(k+1)(2k^2 + 7k + 6)}{6}$$

$$= \frac{1}{6}(k+1)(k+2)(2k+3),$$

since

$$2k^2 + 7k + 6 = (k+2)(2k+3).$$

Thus, we have shown that if we add $(k+1)^2$ to both sides of the equation for $n = k$, then we obtain the equation **or** formula for $n = k+1$. We have thus established that if (A) is true, then (B) must be true; that is, if the formula is true for $n = k$, then it must be true for $n = k+1$. In other words, we have proved that if the proposition is true for a certain positive integer k, then it is also true for the next greater integer $k+1$.

Step 3. The proposition is true for $n = 1,2,3$ (Step 1). Since it is true for $n = 3$, it is true for $n = 4$ (Step 2, where $k = 3$ and $k+1 = 4$). Since it is true for $n = 4$, it is true for $n = 5$, and so on, for all positive integers n.

● **PROBLEM** 711

Prove:
$$1 \cdot 2 + 2 \cdot 3 + 3 \cdot 4 + \ldots + n(n + 1) = \frac{n(n + 1)(n + 2)}{3} \, .$$

<u>Solution</u> <u>by</u> <u>mathematical</u> <u>induction</u>: The steps for a proof by mathematical induction are:

    I)   check validity of formula for $n = 1$

    II)   assume the formulation is true for $n = p$

   III)   prove it is true for $n = p + 1$.

I. For $n = 1$ the formula gives

$$1(1 + 1) = 1(2) = 2 = \frac{1(1+1)(1+2)}{3} = \frac{1 \cdot 2 \cdot 3}{3} = 2$$

which is correct and completes Step I.

II) Assume the formula is true for $n = p$:

$$1 \cdot 2 + 2 \cdot 3 + 3 \cdot 4 + \ldots + p(p + 1) = \frac{p(p + 1)(p + 2)}{3}.$$

Prove the formula is true for $n = p + 1$, that is, prove

$$1 \cdot 2 + 2 \cdot 3 + \ldots + p(p + 1) + (p + 1)(p + 2) = \frac{(p + 1)(p + 2)(p + 3)}{3} ,$$

$(p + 1)(p + 2)$ is added to both members of the first equation in this step; this gives

$$[1 \cdot 2 + 2 \cdot 3 + \ldots + p(p+1)] + (p+1)(p+2) =$$

$$= \left[\frac{p(p+1)(p+2)}{3}\right] + (p+1)(p+2)$$

Factoring out $(p+1)(p+2)$,

$$= (p+1)(p+2)\left(\frac{p}{3} + 1\right)$$

$$= (p+1)(p+2)\left(\frac{p}{3} + 1\right)\left(\frac{3}{3}\right)$$

$$= \frac{(p+1)(p+2)(p+3)}{3} ,$$

$$= \frac{(p+1)[(p+1) + 1][(p+2) + 1]}{3}$$

which is of the same form as the result we assumed to be true for p terms, p+1 taking the place of p. Since the statement is true for n = 1 and n = p + 1 assuming it was true for n = p, then the statement is true for all n.

● **PROBLEM 712**

Prove by mathematical induction that, for all positive integral values of n,
$$1 + 2 + 3 + \ldots + n = \frac{n(n+1)}{2} .$$

<u>Solution:</u> Step 1. The formula is true for n = 1, since $1 = \frac{1(1+1)}{2} = 1$.

Step 2. Assume that the formula is true for n = k. Then, adding (k+1) to both sides,
$$1 + 2 + 3 + \ldots + k + (k+1) = \frac{k(k+1)}{2} + (k+1) = \frac{(k+1)(k+2)}{2}$$
which is the value of $\frac{n(n+1)}{2}$ when (k+1) is substituted for n.

Hence if the formula is true for n = k, we have proved it to be true for n = k + 1. But the formula holds for n = 1; hence it holds for n = 1 + 1 = 2. Then, since it holds for n = 2, it holds for n = 2 + 1 = 3, and so on. Thus the formula is true for all positive integral values of n.

● **PROBLEM 713**

Prove by mathematical induction that, for all positive integral values of n,
$$\frac{1}{1 \cdot 3} + \frac{1}{3 \cdot 5} + \frac{1}{5 \cdot 7} + \ldots + \frac{1}{(2n-1)(2n+1)} = \frac{n}{2n+1} .$$

<u>Solution:</u> Step 1. The formula is true for n = 1, since
$$\frac{1}{(2-1)(2+1)} = \frac{1}{2+1} = \frac{1}{3} .$$

Step 2. Assume that the formula is true for n = k. Then
$$\frac{1}{1 \cdot 3} + \frac{1}{3 \cdot 5} + \frac{1}{5 \cdot 7} + \ldots + \frac{1}{(2k-1)(2k+1)} = \frac{k}{2k+1} .$$
Add the (k+1)th term, which is $\frac{1}{(2k+1)(2k+3)}$, to both sides of the above equation. Then
$$\frac{1}{1 \cdot 3} + \frac{1}{3 \cdot 5} + \frac{1}{5 \cdot 7} + \ldots + \frac{1}{(2k-1)(2k+1)} + \frac{1}{(2k+1)(2k+3)} = \frac{k}{2k+1} + \frac{1}{(2k+1)(2k+3)} .$$
The right hand side of this equation $= \frac{k(2k+3)+1}{(2k+1)(2k+3)} = \frac{k+1}{2k+3}$, which is the value of $\frac{n}{2n+1}$ when n is replaced by (k+1).

Hence if the formula is true for n = k, it is true for n = k + 1.

But the formula holds for $n = 1$; hence it holds for $n = 1 + 1 = 2$. Then, since it holds for $n = 2$, it holds for $n = 2 + 1 = 3$, and so on. Thus the formula is true for all positive integral values of n.

● **PROBLEM 714**

Using mathematical induction, prove that
$$x^{2n} - y^{2n} \text{ is divisible by } x + y.$$

Solution:

(1) The theorem is true for $n = 1$, since $x^2 - y^2 = (x-y)(x+y)$ is divisible by $x + y$.

(2) Let us assume the theorem true for $n = k$, a positive integer; that is, let us assume

(A) $x^{2k} - y^{2k}$ is divisible by $x + y$.

We wish to show that, when (A) is true,

(B) $x^{2k+2} - y^{2k+2}$ is divisible by $x + y$.

Now $x^{2k+2} - y^{2k+2} = \left(x^{2k+2} - x^2 y^{2k}\right) + \left(x^2 y^{2k} - y^{2k+2}\right)$

$$= x^2\left(x^{2k} - y^{2k}\right) + y^{2k}\left(x^2 - y^2\right).$$

In the first term $\left(x^{2k} - y^{2k}\right)$ is divisible by $(x+y)$ by assumption, and in the second term $\left(x^2 - y^2\right)$ is divisible by $(x+y)$ by Step (1); hence, if the theorem is true for $n = k$, a positive integer, it is true for the next one $n = k + 1$.

(3) Since the theorem is true for $n = k = 1$, it is true for $n = k + 1 = 2$; being true for $n = k = 2$, it is true for $n = k + 1 = 3$; and so on, for every positive integral value of n.

● **PROBLEM 715**

Prove by mathematical induction that

$$1 + 7 + 13 + \ldots + (6n - 5) = n(3n - 2).$$

Solution: (1) The proposed formula is true for $n = 1$, since $1 = 1(3 - 2)$.

(2) Assume the formula to be true for $n = k$, a positive integer; that is, assume

(A) $1 + 7 + 13 + \ldots + (6k - 5) = k(3k - 2).$

Under this assumption we wish to show that

(B) $1 + 7 + 13 + \ldots + (6k - 5) + (6k + 1) = (k+1)(3k+1).$

When $(6k+1)$ is added to both members of (A), we have on the right

$$k(3k-2) + (6k+1) = 3k^2 + 4k + 1 = (k+1)(3k+1);$$

hence, if the formula is true for $n = k$ it is true for $n = k + 1$.

(3) Since the formula is true for $n = k = 1$ (Step 1), it is true for $n = k + 1 = 2$; being true for $n = k = 2$ it is true for $n = k + 1 = 3$; and so on, for every positive integral value of n.

● **PROBLEM** 716

Prove by mathematical induction that
$$1 + 5 + 5^2 + \ldots + 5^{n-1} = \tfrac{1}{4}\left(5^n - 1\right).$$

Solution: (1) The proposed formula is true for $n = 1$, since $1 = \tfrac{1}{4}(5-1)$.

(2) Assume the formula to be true for $n = k$, a positive integer; that is assume

(A) $\qquad 1 + 5 + 5^2 + \ldots + 5^{k-1} = \tfrac{1}{4}\left(5^k - 1\right).$

Under this assumption we wish to show that

(B) $\qquad 1 + 5 + 5^2 + \ldots + 5^{k-1} + 5^k = \tfrac{1}{4}\left(5^{k+1} - 1\right).$

When 5^k is added to both members of (A), we have on the right
$$\tfrac{1}{4}\left(5^k - 1\right) + 5^k = \tfrac{5}{4}\left(5^k\right) - \tfrac{1}{4} = \tfrac{1}{4}\left(5 \cdot 5^k - 1\right) = \tfrac{1}{4}\left(5^{k+1} - 1\right);$$

hence, if the formula is true for $n = k$ it is true for $n = k + 1$.

(3) Since the formula is true for $n = k = 1$ (Step 1), it is true for $n = k + 1 = 2$; being true for $n = k = 2$ it is true for $n = k + 1 = 3$; and so on, for every positive integral value of n.

● **PROBLEM** 717

Prove by mathematical induction that
$$\frac{5}{1 \cdot 2 \cdot 3} + \frac{6}{2 \cdot 3 \cdot 4} + \frac{7}{3 \cdot 4 \cdot 5} + \cdots + \frac{n+4}{n(n+1)(n+2)} = \frac{n(3n+7)}{2(n+1)(n+2)} .$$

Solution:

(1) The formula is true for $n = 1$, since $\dfrac{5}{1 \cdot 2 \cdot 3} = \dfrac{1(3+7)}{2 \cdot 2 \cdot 3} = \dfrac{5}{6} .$

(2) Assume the formula to be true for $n = k$, a positive integer; that is, assume

(A) $\qquad \dfrac{5}{1 \cdot 2 \cdot 3} + \dfrac{6}{2 \cdot 3 \cdot 4} + \cdots + \dfrac{k+4}{k(k+1)(k+2)} = \dfrac{k(3k+7)}{2(k+1)(k+2)} .$

Under this assumption we wish to show that

(B) $\qquad \dfrac{5}{1 \cdot 2 \cdot 3} + \dfrac{6}{2 \cdot 3 \cdot 4} + \cdots + \dfrac{k+4}{k(k+1)(k+2)} + \dfrac{k+5}{(k+1)(k+2)(k+3)}$

$$= \dfrac{(k+1)(3k+10)}{2(k+2)(k+3)} .$$

When $\dfrac{k+5}{(k+1)(k+2)(k+3)}$ is added to both members of (A), we have on the right

$$\frac{k(3k+7)}{2(k+1)(k+2)} + \frac{k+5}{(k+1)(k+2)(k+3)} = \frac{1}{(k+1)(k+2)}\left[\frac{k(3k+7)}{2} + \frac{k+5}{k+3}\right]$$

$$= \frac{1}{(k+1)(k+2)} \frac{k(3k+7)(k+3)+2(k+5)}{2(k+3)} = \frac{1}{(k+1)(k+2)} \frac{3k^3+16k^2+23k+10}{2(k+3)}$$

$$= \frac{1}{(k+1)(k+2)} \frac{(k+1)^2(3k+10)}{2(k+3)} = \frac{(k+1)(3k+10)}{2(k+2)(k+3)} ;$$

hence, if the formula is true for $n = k$ it is true for $n = k + 1$.

(3) Since the formula is true for $n = k = 1$ (Step 1), it is true for $n = k + 1 = 2$; being true for $n = k = 2$, it is true for $n = k + 1 = 3$; and so on, for all positive integral values of n.

● **PROBLEM** 718

Let x be any real number. Show that $|\sin nx| \leq n|\sin x|$ for every positive integer n.

<u>Solution:</u> Let $\{P_n\}$ be the sequence in which P_n is the statement "$|\sin nx| \leq n|\sin x|$." Recall the following theorem: If $\{P_n\}$ is a sequence of statements that possesses the properties

(i) P_1 is true and

(ii) for each index k such that P_k is true, the statement P_{k+1} is also true, then P_n is a true statement for every positive integer n. P_1 is the statement "$|\sin x| \leq |\sin x|$," which is surely true. Now we must show that the truth of the statement P_k implies that statement P_{k+1} also is true. For any k, we have

$$|\sin(k+1)x| = |\sin(kx+x)|$$
$$= |\sin kx \cos x + \cos kx \sin x|$$
$$\leq |\sin kx \cos x| + |\cos kx \sin x|$$
$$\leq |\sin kx| + |\sin x|.$$

Hence, if P_k is true $\left(\text{that is, if } |\sin kx| \leq k|\sin x|\right)$, we see that
$$|\sin(k+1)x| \leq |\sin kx| + |\sin x| \leq k|\sin x| + |\sin x|$$
$$= (k+1)|\sin x| .$$

Thus, the statement P_{k+1} ,
$$|\sin(k+1)x| \leq (k+1)|\sin x|,$$

is true whenever P_k is true. The two conditions of the above theorem are satisfied, so we conclude that statement P_n is true for any positive integer n.

● **PROBLEM** 719

Prove by mathematical induction that the number of straight lines determined by $n > 1$ points, no 3 on the same straight line, is $\frac{1}{2}n(n-1)$.

<u>Solution:</u>

(1) The theorem is true when $n = 2$, since $\frac{1}{2} \cdot 2(2-1) = 1$ and two

points determine one line.

(2) Let us assume that k points, no 3 on the same straight line, determine $\frac{1}{2}k(k-1)$ lines.

When an additional point is added (not on any of the lines already determined) and is joined to each of the original k points, k new lines are determined. Thus, altogether we have $\frac{1}{2}k(k-1) + k = \frac{1}{2}k(k-1+2) = \frac{1}{2}k(k+1)$ lines and this agrees with the theorem when $n = k + 1$.

Hence, if the theorem is true for $n = k$, a positive integer greater than 1, it is true for the next one $n = k + 1$.

(3) Since the theorem is true for $n = k = 2$ (Step (1)), it is true for $n = k + 1 = 3$; being true for $n = k = 3$, it is true for $n = k+1 = 4$; and so on, for every possible integral value > 1 of n.

● **PROBLEM** 720

Prove by mathematical induction that the sum of n terms of an arithmetic progression a, a + d, a + 2d,... is $\frac{n}{2}\left[2a + (n-1)d\right]$, that is

$$a + (a+d) + (a+2d) + \ldots + \left[a+(n-1)d\right] = \frac{n}{2}\left[2a + (n-1)d\right].$$

<u>Solution:</u> Step 1. The formula holds for $n = 1$, since
$$a = \frac{1}{2}\left[2a + (1-1)d\right] = a.$$

Step 2. Assume that the formula holds for $n = k$. Then
$$a + (a+d) + (a+2d) + \ldots + \left[a + (k-1)d\right] = \frac{k}{2}\left[2a + (k-1)d\right].$$

Add the (k+1)th term, which is (a+kd), to both sides of the latter equation. Then
$$a + (a+d) + (a+2d) + \ldots + \left[a + (k-1)d\right] + (a+kd) = \frac{k}{2}\left[2a + (k-1)d\right] + (a+kd).$$
The right hand side of this equation $= ka + \frac{k^2 d}{2} - \frac{kd}{2} + a + kd =$
$$= \frac{k^2 d+kd+2ka+2a}{2}$$
$$= \frac{kd(k+1)+2a(k+1)}{2}$$
$$= \frac{k+1}{2}(2a+kd)$$

which is the value of $\frac{n}{2}\left[2a + (n-1)d\right]$ when n is replaced by (k+1).

Hence if the formula is true for $n = k$, we have proved it to be true for $n = k + 1$. But the formula holds for $n = 1$; hence it holds for $n = 1 + 1 = 2$. Then, since it holds for $n = 2$, it holds for $n = 2 + 1 = 3$, and so on. Thus the formula is true for all positive integral values of n.

> Prove that the sum of the cubes of the first n natural numbers is equal to
> $$\left\{ \frac{n(n+1)}{2} \right\}^2 .$$

Solution: We note by trial that the statement is true when $n = 1$, or 2, or 3 (when $n = 1$,

$$1^3 = 1 \quad \text{and} \quad \left[\frac{1(1+1)}{2} \right]^2 = \left[\frac{1(2)}{2} \right]^2 = \left(\frac{2}{2} \right)^2 = 1^2 = 1,$$

when $n = 2$,

$$1^3 + 2^3 = 1 + 8 = 9 \quad \text{and} \quad \left[\frac{2(2+1)}{2} \right]^2 = \left[\frac{2(3)}{2} \right]^2$$
$$= 3^2 = 9$$

etc.) Assume that it is true when n terms are taken; that is, suppose

$$1^3 + 2^3 + 3^3 + \ldots \text{ to n terms} = \left\{ \frac{n(n+1)}{2} \right\}^2 .$$

Add the $(n+1)^{th}$ term, that is, $(n+1)^3$ to each side; then

$$1^3 + 2^3 + 3^3 + \ldots \text{ to n+1 terms} = \left\{ \frac{n(n+1)}{2} \right\}^2 + (n+1)^3$$

$$= \frac{n^2(n+1)^2}{2^2} + (n+1)(n+1)^2$$

$$= (n+1)^2 \left(\frac{n^2}{4} \right) + (n+1)^2 (n+1)$$

$$= (n+1)^2 \left(\frac{n^2}{4} + n + 1 \right)$$

$$= (n+1)^2 \left(\frac{n^2}{4} + \frac{4n}{4} + \frac{4}{4} \right)$$

$$= (n+1)^2 \left(\frac{n^2 + 4n + 4}{4} \right)$$

$$= \frac{(n+1)^2 (n^2 + 4n + 4)}{4}$$

$$= \frac{(n+1)^2 (n+2)^2}{2^2}$$

$$= \left\{ \frac{(n+1)(n+2)}{2} \right\}^2 ;$$

$$= \left\{ \frac{(n+1)[(n+1) + 1]}{2} \right\}^2$$

which is of the same form as the result we assumed to be true for n terms, $n + 1$ taking the place of n; in other words, if the result is true when we take a certain number of terms, whatever that number may be, it is true when we increase that number by one; but we see that it is true when 3 terms are taken; therefore it is true when 4 terms are taken; it is therefore true when 5 terms are taken; and so on. Thus the result is true universally.

● **PROBLEM** 722

> Using mathematical induction, prove the binomial formula

$$(a+x)^n = a^n + na^{n-1}x + \frac{n(n-1)}{2!}a^{n-2}x^2 + \ldots + \frac{n(n-1)\ldots(n-r+2)}{(r-1)!}a^{n-r+1}x^{r-1}$$

$$+ \ldots + x^n$$

for positive integral values of n.

Solution: Step 1. The formula is true for n = 1.

Step 2. Assume the formula is true for n = k. Then

$$(a+x)^k = a^k + ka^{k-1}x + \frac{k(k-1)}{2!}a^{k-2}x^2 + \ldots + \frac{k(k-1)\ldots(k-r+2)}{(r-1)!}a^{k-r+1}x^{r-1}$$

$$+ \ldots + x^k .$$

Multiply both sides by a+x. The multiplication on the right may be written

$$a^{k+1} + ka^k x + \frac{k(k-1)}{2!}a^{k-1}x^2 + \ldots + \frac{k(k-1)\ldots(k-r+2)}{(r-1)!}a^{k-r+2}x^{r-1}$$

$$+ \ldots + ax^k$$

$$+ a^k x + ka^{k-1}x^2 + \ldots + \frac{k(k-1)\ldots(k-r+3)}{(r-2)!}a^{k-r+2}x^{r-1} + \ldots + x^{k+1} .$$

Since

$$\frac{k(k-1)\ldots(k-r+2)}{(r-1)!}a^{k-r+2}x^{r-1} + \frac{k(k-1)\ldots(k-r+3)}{(r-2)!}a^{k-r+2}x^{r-1}$$

$$= \frac{k(k-1)\ldots(k-r+3)}{(r-2)!}a^{k-r+2}x^{r-1}\left\{\frac{k-r+2}{r-1} + 1\right\} = \frac{(k+1)k(k-1)\ldots(k-r+3)}{(r-1)!}a^{k-r+2}x^{r-1},$$

the produc may be written

$$(a+x)^{k+1} = a^{k+1} + (k+1)a^k x + \ldots + \frac{(k+1)k(k-1)\ldots(k-r+3)}{(r-1)!}a^{k-r+2}x^{r-1}$$

$$+ \ldots + x^{k+1}$$

which is the binomial formula with n replaced by k+1.

Hence if the formula is true for n = k, it is true for n = k + 1. But the formula holds for n = 1; hence it holds for n = 1 + 1 = 2, and so on. Thus the formula is true for all positive integral values of n.

CHAPTER 24

FACTORIAL NOTATION

> **Basic Attacks and Strategies for Solving Problems in this Chapter. See pages 539 to 541 for step-by-step solutions to problems.**

If n is any positive integer, then the symbol $n!$, read "n factorial," represents a product of the integers from 1 up to and including n itself. Thus, in general,

$$n! = n(n - 1)\,(n - 2)\,(n - 3)\, ...\, (2)\, 1.$$

For example,

$$5! = 5(5 - 1)\,(5 - 2)\,(5 - 3)\,(5 - 4)$$

$$5! = 5 * 4 * 3 * 2 * 1$$

$$5! = 120.$$

Note that the $n!$ expansion may stop at any point. For example,

$$5! = 5 * 4 * 3!$$

$$5! = 20 * 3!$$

Thus, in general, if $r < n$, we can write

$$n! = n(n - 1) ... r!,$$

where, in the above example $n = 5$ and $r = 3$.

Find the value of $\frac{7!}{5!}$.

Solution: Apply the definition of factorial. If n is any positive integer, the symbol n! is the product of the integers from 1 up to and including n.

$$n! = 1 \cdot 2 \cdot 3 \ldots n.$$

$$\frac{7!}{5!} = \frac{(1)(2)(3)(4)(5)(6)(7)}{(1)(2)(3)(4)(5)}$$

$$= 6(7)$$

$$= 42$$

Simplify the expression $\frac{5! - 4!}{6!}$.

Solution: Since $5! = 5 \cdot 4!$ and $6! = 6 \cdot 5 \cdot 4!$, the expression $\frac{5! - 4!}{6!}$ becomes:

$$\frac{5! - 4!}{6!} = \frac{(5 \cdot 4!) - 4!}{6 \cdot 5 \cdot 4!}$$

$$= \frac{(4! \cdot 5) - 4!}{4! \cdot 6 \cdot 5} \qquad (1)$$

By the distributive property, which states that $ab + ac = a(b+c)$, equation (1) becomes:

$$\frac{5! - 4!}{6!} = \frac{4!(5-1)}{4! \cdot 6 \cdot 5}$$

$$= \frac{5 - 1}{6 \cdot 5} = \frac{4}{6 \cdot 5}$$

$$= \frac{4}{30} .$$

Thus,

$$\frac{5! - 4!}{6!} = \frac{2}{15} .$$

Simplify the following numbers.

(a) $\frac{8!}{11!}$ (b) $\frac{5! - 8!}{4! - 7!}$

Solution:

(a) Note $n! = n \cdot (n-1) \cdot (n-2) \cdot (n-3) \cdot \ldots \cdot 1$;

also $n! = n \cdot (n-1)!$ or $n \cdot (n-1) \cdot (n-2)!$, etc.

Thus, $\dfrac{8!}{11!} = \dfrac{\cancel{8!}}{11 \cdot 10 \cdot 9 \cdot \cancel{8!}} = \dfrac{1}{11 \cdot 10 \cdot 9} = \dfrac{1}{990}$.

(b) Similarly,

$$\frac{5!-8!}{4!-7!} = \frac{5 \cdot 4! - 8 \cdot 7 \cdot 6 \cdot 5 \cdot 4!}{4! - 7 \cdot 6 \cdot 5 \cdot 4!}$$

Factoring 4!, $= \dfrac{\cancel{4!}[5 - (8 \cdot 7 \cdot 6 \cdot 5)]}{\cancel{4!}[1 - (7 \cdot 6 \cdot 5)]}$

$$= \frac{5 - 1680}{1 - 210}$$

$$= \frac{- 1675}{- 209}$$

$$= \frac{1675}{209} \; .$$

● **PROBLEM 726**

Find the value of $\dfrac{5!6!}{4!7!}$.

Solution: Apply the definition of factorial: If n is any positive integer, the symbol n! is the product of the integers from 1 up to and including n.

Also if r and n are both positive integers and r is less than n, then $n! = n \cdot (n - 1) \ldots (r + 2)(r + 1) \, r!$. Use these two ideas to expand each factorial.

$$n! = 1 \cdot 2 \cdot 3 \ldots n$$

$5! = (4!)(5)$ and $7! = (6!)(7)$ Substituting the values of 5! and 7! we have

$\dfrac{5!6!}{4!7!} = \dfrac{(4!)(5)(6!)}{(4!)(6!)(7)}$. Dividing the common factors

4! and 6!

$$= \frac{5}{7} \; .$$

● **PROBLEM 727**

If n and r are positive integers, and $r < n$, show that

$$n! = r!(r + 1)(r + 2) \cdot \ldots \cdot n.$$

Solution: By definition of factorial,

$$r! = 1 \cdot 2 \cdot \ldots \cdot r \; .$$

Then,

$$r!(r+1)(r+2) \cdot \ldots \cdot n = (1 \cdot 2 \cdot \ldots \cdot r)(r+1)(r+2) \cdot \ldots \cdot n$$

$$r!(r+1)(r+2) \cdot \ldots \cdot n = 1 \cdot 2 \cdot \ldots \cdot n \qquad (1)$$

Again, by definition of factorial,

$$n! = 1 \cdot 2 \cdot \ldots \cdot n \; .$$

Hence, equation (1) becomes:

$$r!(r+1)(r+2) \cdot \ldots \cdot n = n!$$

or

$$n! = r!(r+1)(r+2) \cdot \ldots \cdot n \; .$$

CHAPTER 25

BINOMIAL THEOREM/EXPANSION

> **Basic Attacks and Strategies for Solving Problems in this Chapter. See pages 542 to 554 for step-by-step solutions to problems.**

It is possible to expand the binomial expression of

$(x + y)^n,$

where n is any positive integer, without showing all of the intermediate steps of multiplying and combining similar terms. To do this, first observe some patterns in the following examples:

$(x + y)^1 = x + y$

$(x + y)^2 = x^2 + 2xy + y^2$

$(x + y)^3 = x^3 + 3x^2y + 3xy^2 + y^3$

$(x + y)^4 = x^4 + 4x^3y + 6x^2y^2 + 4xy^3 + y^4$

The patterns which are evident are:

(1)   In each case the expansion of $(x + y)^n$ has $n + 1$ terms.

(2)   In the expansions of $(x + y)^n$, the first term is x^n and in each term thereafter the power of x decreases by 1 until finally the last term has no factors of x at all. The powers of y behave just the opposite; that is, in the first term there is no power of y and in each term thereafter the power of y increases by 1 until finally the last term is y^n.

(3)   The coefficients form a pattern known as Pascal's Triangle as follows:

```
              1
           1     1
        1     2     1
     1     3     3     1
  1     4     6     4     1
```

Note that in each row the first and last entry is 1, and each of the interior entries is the sum of the two closest numbers in the row above it. Beginning with the second row,

1   1,

one sees the coefficients of the expansion $(x + y)^n$.

Another method of obtaining the coefficients in the expansion $(x + y)^n$ is to use the formula

$$\binom{n}{r} = \frac{n!}{r!(n-r)!}.$$

For instance, expand $(x + y)^4$. Using the above formula for the coefficients and the pattern for the power the expansion is given by:

$$(x+y)^4 = \binom{4}{0} x^4 + \binom{4}{1} x^3 y + \binom{4}{2} x^2 y^2 + \binom{4}{3} xy^3 + \binom{4}{4} y^4,$$

where the coefficients are calculated using the aforementioned formula.

This method of generating an expansion of $(x + y)^n$ is often called the binomial theorem which is stated as follows: For any binomial $(x + y)$ and any positive integer n,

$$(x+y)^n = \binom{n}{0} x^n + \binom{n}{1} x^{n-1} y + \binom{n}{2} x^{n-2} y^2 + \ldots + \binom{n}{n} y^n.$$

Once the binomial theorem is known, expansion of all types of binomial expressions, even if n is a negative integer or a fraction, can be achieved.

Step-by-Step Solutions to Problems in this Chapter, "Binomial Theorem/Expansion"

Find the binomial expansion of $(2x - 5)^4$.

<u>Solution:</u> The generalized form of the binomial expansion is

$$(a+b)^n = a^n + {}_nC_1 a^{n-1}b + {}_nC_2 a^{n-2}b^2 + {}_nC_3 a^{n-3}b^3 + \ldots + {}_nC_{n-1} a^1 b^{n-1}$$
$$+ {}_nC_n a^0 b^n$$

Here we take $a = 2x$, $b = -5$, and $n = 4$.

$$(2x-5)^4 = (2x)^4 + {}_4C_1 (2x)^3(-5)^1 + {}_4C_2 (2x)^2(-5)^2 + {}_4C_3 (2x)^1(-5)^3$$
$$+ {}_4C_4 (2x)^0(-5)^4$$

$$= 16x^4 + {}_4C_1\, 8x^3 \cdot (-5) + {}_4C_2\, 4x^2(25) + {}_4C_3\, 2x(-125) + {}_4C_4 1(-5)^4$$

Note that ${}_nC_r = \dfrac{n!}{r!(n-r)!}$. Therefore

$$(2x-5)^4 = 16x^4 + \frac{4!}{1!3!}\, 8x^3(-5) + \frac{4!}{2!2!}\left(4x^2\right)(25) + \frac{4!}{3!1!}(2x)(-125)$$

$$+ \frac{4!}{4!0!}(1)(-5)^4$$

$$= 16x^4 + \frac{4 \cdot 3!}{3!}\, 8x^3(-5) + \frac{4 \cdot 3 \cdot 2!}{2 \cdot 1 \cdot 2!}\left(100x^2\right) + \frac{4 \cdot 3!}{3! 1!}(-250x) + 625$$

$$= 16x^4 - 160x^3 + 600x^2 - 1,000x + 625.$$

Expand $(2z - 3y)^4$.

<u>Solution:</u> Use the binomial theorem.

$$(u+v)^n = u^n + nu^{n-1}v + \frac{n(n-1)}{2}v^{n-2}v^2 + \frac{n(n-1)(n-2)}{2 \cdot 3}u^{n-3}v^3 + \ldots + v^n$$

where $u = 2z$
$\quad\quad\quad v = -3y$
$\quad\quad\quad n = 4$

Therefore,

$$(2z - 3y)^4 = (2z)^4 + 4(2z)^3(-3y) + 6(2z)^2(-3y)^2 + 4(2z)(-3y)^3 + (-3y)^4$$
$$= 16z^4 + 4\left(8z^3\right)(-3y) + 6\left(4z^2\right)\left(9y^2\right) + 4(2z)\left(-27y^3\right) + 81y^4$$
$$= 16z^4 - 96z^3y + 216z^2y^2 - 216zy^3 + 81y^4$$

Find the expansion of $(a - 2x)^7$.

Solution: Use the binomial formula:

$$(u + v)^n = u^n + nu^{n-1}v + \frac{n(n-1)}{2} u^{n-2}v^2$$

$$+ \frac{n(n-1)(n-2)}{2 \cdot 3} u^{n-3}v^3 + \ldots + v^n$$

and substitute a for u and (-2x) for v and 7 for n to obtain:

$$(a-2x)^7 = [a+(-2x)]^7$$

$$= a^7 + 7a^6(-2x) + \frac{7 \cdot \overset{3}{\cancel{6}}}{\underset{1}{\cancel{2}}} a^5(-2x)^2 + \frac{7 \cdot \cancel{6} \cdot 5}{\cancel{2} \cdot \cancel{3}} a^4(-2x)^3$$

$$+ \frac{7 \cdot \cancel{6} \cdot 5 \cdot \cancel{4}}{\cancel{2} \cdot \cancel{3} \cdot \cancel{4}} a^3(-2x)^4 + \frac{7 \cdot \cancel{6} \cdot \cancel{5} \cdot \cancel{4} \cdot \overset{3}{\cancel{3}}}{\cancel{2} \cdot \cancel{3} \cdot \cancel{4} \cdot \cancel{5}} a^2(-2x)^5$$

$$+ \frac{7 \cdot \cancel{6} \cdot \cancel{5} \cdot \cancel{4} \cdot \cancel{3} \cdot \cancel{2}}{\cancel{2} \cdot \cancel{3} \cdot \cancel{4} \cdot \cancel{5} \cdot \cancel{6}} a^1(-2x)^6 + \frac{\cancel{7} \cdot \cancel{6} \cdot \cancel{5} \cdot \cancel{4} \cdot \cancel{3} \cdot \cancel{2} \cdot 1}{\cancel{2} \cdot \cancel{3} \cdot \cancel{4} \cdot \cancel{5} \cdot \cancel{6} \cdot \cancel{7}} a^0(-2x)^7$$

$$(a - 2x)^7 = a^7 - 14a^6x + 84a^5x^2 - 280a^4x^3 + 560a^3x^4$$

$$- 672a^2x^5 + 448a x^6 - 128x^7.$$

● **PROBLEM 731**

Find the expansion of $(x + y)^6$.

Solution: Use the Binomial Theorem which states that

$$(a+b)^n = \frac{1}{0!} a^n + \frac{n}{1!} a^{n-1}b + \frac{n(n-1)}{2!} a^{n-2}b^2 + \ldots + nab^{n-1} + b^n.$$

Replacing a by x and b by y:

$$(x+y)^6 = \frac{1}{0!} x^6 + \frac{6}{1!} x^5y + \frac{6 \cdot 5}{2!} x^4y^2 + \frac{6 \cdot 5 \cdot 4}{3!} x^3y^3 + \frac{6 \cdot 5 \cdot 4 \cdot 3}{4!} x^2y^4$$

$$+ \frac{6 \cdot 5 \cdot 4 \cdot 3 \cdot 2}{5!} x^1y^5 + \frac{6 \cdot 5 \cdot 4 \cdot 3 \cdot 2 \cdot 1}{6!} x^0y^6$$

$$= \frac{1}{1} x^6 + \frac{6}{1} x^5y + \frac{\cancel{6} \cdot 5}{\cancel{2} \cdot 1} x^4y^2 + \frac{\cancel{6} \cdot 5 \cdot 4}{\cancel{3} \cdot \cancel{2} \cdot 1} x^3y^3 + \frac{\cancel{6} \cdot 5 \cdot \cancel{4} \cdot \cancel{3}}{\cancel{4} \cdot \cancel{3} \cdot \cancel{2} \cdot 1} x^2y^4$$

$$+ \frac{6 \cdot \cancel{5} \cdot \cancel{4} \cdot \cancel{3} \cdot \cancel{2}}{\cancel{5} \cdot \cancel{4} \cdot \cancel{3} \cdot \cancel{2} \cdot 1} xy^5 + \frac{\cancel{6} \cdot \cancel{5} \cdot \cancel{4} \cdot \cancel{3} \cdot \cancel{2} \cdot \cancel{1}}{\cancel{6} \cdot \cancel{5} \cdot \cancel{4} \cdot \cancel{3} \cdot \cancel{2} \cdot \cancel{1}} y^6$$

$$(x+y)^6 = x^6 + 6x^5y + 15x^4y^2 + 20x^3y^3 + 15x^2y^4 + 6xy^5 + y^6.$$

● **PROBLEM 732**

Expand $(x + 2y)^5$.

Solution: Apply the binomial theorem. If n is a positive integer, then

$$(a + b)^n = \binom{n}{0}a^nb^0 + \binom{n}{1}a^{n-1}b + \binom{n}{2}a^{n-2}b^2 + \ldots + \binom{n}{r}a^{n-r}b^r$$

543

$$+ \ldots + \begin{bmatrix} n \\ n \end{bmatrix} b^n.$$

Note that $\begin{bmatrix} n \\ r \end{bmatrix} = \frac{n!}{r!\,(n-r)!}$ and that $0! = 1$. Then, we obtain:

$$(x + 2y)^5 = \begin{bmatrix} 5 \\ 0 \end{bmatrix} x^5 (2y)^0 + \begin{bmatrix} 5 \\ 1 \end{bmatrix} x^4 (2y)^1 + \begin{bmatrix} 5 \\ 2 \end{bmatrix} x^3 (2y)^2$$

$$+ \begin{bmatrix} 5 \\ 3 \end{bmatrix} x^2 (2y)^3 + \begin{bmatrix} 5 \\ 4 \end{bmatrix} x^1 (2y)^4 + \begin{bmatrix} 5 \\ 5 \end{bmatrix} x^0 (2y)^5$$

$$= \frac{5!}{0!5!} x^5 + \frac{5!}{1!4!} x^4 2y + \frac{5!}{2!3!} x^3 \left(4y^2\right)$$

$$+ \frac{5!}{3!2!} x^2 \left(8y^3\right) + \frac{5!}{4!1!} x \left(16y^4\right) + \frac{5!}{5!0!} 1 \left(32y^5\right)$$

$$= x^5 + \frac{5 \cdot \cancel{4}!}{\cancel{4}!} x^4 2y + \frac{5 \cdot \overset{2}{\cancel{4}} \cdot \cancel{3}!}{\cancel{2} \cdot 1 \cdot \cancel{3}!} x^3 \left(4y^2\right)$$

$$+ \frac{5 \cdot \cancel{4} \cdot \cancel{3}!}{\cancel{3}! \underset{1}{\cancel{2}} \cdot 1} x^2 \left(8y^3\right) + \frac{5 \cdot \cancel{4}!}{\cancel{4}!1!} x \left(16y^4\right) + \frac{\cancel{5}!}{\cancel{5}!0!} \left(32y^5\right)$$

$$= x^5 + 10x^4 y + 40x^3 y^2 + 80x^2 y^3 + 80xy^4 + 32y^5.$$

● **PROBLEM** 733

Expand $\left(3p^2 - 2q^{\frac{1}{2}}\right)^4$ by means of the binomial theorem.

<u>Solution:</u> We apply the binomial theorem: If n is a positive integer, then

$$(a + b)^n = \binom{n}{0} a^n b^0 + \binom{n}{1} a^{n-1} b + \binom{n}{2} a^{n-2} b^2 + \ldots + \binom{n}{r} a^{n-r} b^r + \ldots + \binom{n}{n} a^0 b^n$$

where $\binom{n}{r} = \frac{n!}{r!\,(n-r)!}$.

We identify $3p^2$ with a, $-2q^{\frac{1}{2}}$ with b, and n with 4. The binomial theorem then gives us

$$\left(3p^2 - 2q^{\frac{1}{2}}\right)^4 = \binom{4}{0}\left(3p^2\right)^4\left(-2q^{\frac{1}{2}}\right)^0 + \binom{4}{1}\left(3p^2\right)^3\left(-2q^{\frac{1}{2}}\right) + \binom{4}{2}\left(3p^2\right)^2\left(-2q^{\frac{1}{2}}\right)^2$$

$$+ \binom{4}{3}\left(3p^2\right)^1\left(-2q^{\frac{1}{2}}\right)^3 + \binom{4}{4}\left(3p^2\right)^0\left(-2q^{\frac{1}{2}}\right)^4$$

$$= \frac{4!}{0!(4)!} 3^4 p^8 + \frac{4!}{1!3!} 3^3 p^6 \left(-2q^{\frac{1}{2}}\right) + \frac{4!}{2!2!} 3^2 p^4 (-2)^2 q$$

$$+ \frac{4!}{3!1!} 3p^2 (-2)^3 q^{3/2} + \frac{4!}{4!0!} (-2)^4 q^2$$

$$= 81p^8 + 4 \cdot \frac{3!}{3!} 27p^6 \left(-2q^{\frac{1}{2}}\right) + \frac{4 \cdot 3 \cdot \overset{2}{\cancel{2}!}}{\cancel{2!} \cancel{2}} 9p^4 4q$$

$$+ \frac{4 \cdot \cancel{3}!}{\cancel{3}! \cancel{1}!} 3p^2 (-8) q^{3/2} + 16q^2$$

544

$$= 81p^8 - 216p^6q^{\frac{1}{2}} + 216p^4q - 96p^2q^{3/2} + 16q^2 .$$

Give the expansion of $\left[r^2 - \frac{1}{s}\right]^5$.

Solution: Write the given expression as the sum of two terms raised to the 5th power:

$$\left[r^2 - \frac{1}{s}\right]^5 = \left[r^2 + \left(-\frac{1}{s}\right)\right]^5 \qquad (1)$$

The Binomial Theorem can be used to expand the expression on the right side of equation (1). The Binomial Theorem is stated as:

$$(a+b)^n = a^n + na^{n-1}b + \frac{n(n-1)}{1 \cdot 2}a^{n-2}b^2 + \frac{n(n-1)(n-2)}{1 \cdot 2 \cdot 3}a^{n-3}b^3$$

$$+ \ldots + nab^{n-1} + b^n, \text{ where a and b are any two}$$

numbers.

Let $a = r^2$, $b = -\frac{1}{s}$, and $n = 5$. Then, using the Binomial Theorem:

$$\left[r^2 - \frac{1}{s}\right]^5 = \left[r^2 + \left(-\frac{1}{s}\right)\right]^5$$

$$= \left(r^2\right)^5 + 5\left(r^2\right)^{5-1}\left(-\frac{1}{s}\right) + \frac{5(5-1)}{1 \cdot 2}\left(r^2\right)^{5-2}\left(-\frac{1}{s}\right)^2$$

$$+ \frac{5(5-1)(5-2)}{1 \cdot 2 \cdot 3}\left(r^2\right)^{5-3}\left(-\frac{1}{s}\right)^3$$

$$+ \frac{5(5-1)(5-2)(5-3)}{1 \cdot 2 \cdot 3 \cdot 4}\left(r^2\right)^{5-4}\left(-\frac{1}{s}\right)^4$$

$$+ \frac{5(5-1)(5-2)(5-3)(5-4)}{1 \cdot 2 \cdot 3 \cdot 4 \cdot 5}\left(r^2\right)^{5-5}\left(-\frac{1}{s}\right)^5$$

$$= r^{10} - \frac{5(r^2)^4}{s} + \frac{5(\cancel{4})}{1 \cdot \cancel{2}}\left(r^2\right)^3\left(\frac{1}{s^2}\right) - \frac{5(\cancel{4})(\cancel{3})}{1 \cdot \cancel{2} \cdot \cancel{3}}\left(r^2\right)^2\left(\frac{1}{s^3}\right)$$

$$+ \frac{5(\cancel{4})(\cancel{3})(\cancel{2})}{1 \cdot \cancel{2} \cdot \cancel{3} \cdot \cancel{4}}\left(r^2\right)^1\left(\frac{1}{s^4}\right) - \frac{\cancel{5}(\cancel{4})(\cancel{3})(\cancel{2})(\cancel{1})}{\cancel{1} \cdot \cancel{2} \cdot \cancel{3} \cdot \cancel{4} \cdot \cancel{5}}\left(r^2\right)^0\left(\frac{1}{s^5}\right)$$

$$= r^{10} - \frac{5r^8}{s} + \frac{10r^6}{s^2} - \frac{10r^4}{s^3} + \frac{5r^2}{s^4} - (1)(1)\left(\frac{1}{s^5}\right)$$

$$\left[r^2 - \frac{1}{s}\right]^5 = r^{10} - \frac{5r^8}{s} + \frac{10r^6}{s^2} - \frac{10r^4}{s^3} + \frac{5r^2}{s^4} - \frac{1}{s^5}$$

545

Find the first five terms of the expansion of $(1 + x)^{-2}$.

Solution: The Binomial Theorem states that:

$$(a+b)^n = \frac{1}{0!} a^n + \frac{n}{1!} a^{n-1}b + \frac{n(n-1)}{2!} a^{n-2}b^2 + \ldots + nab^{n-1} + b^n .$$

This theorem can be used to find the first five terms of the expansion of $(1 + x)^{-2}$. Replacing a by 1 and b by x, the expression $(1 + x)^{-2}$ becomes:

$$(1+x)^{-2} = \frac{1}{0!} 1^{-2} + \frac{-2}{1!} 1^{-3}x + \frac{(-2)(-3)}{2!} 1^{-4}x^2 + \frac{(-2)(-3)(-4)}{3!} 1^{-5}x^3$$

$$+ \frac{(-2)(-3)(-4)(-5)}{4!}1^{-6}x^4 + \ldots + (-2)1x^{-3} + x^{-2} .$$

Writing only the first five terms of this expansion:

$$(1+x)^{-2} = \frac{1}{0!}1^{-2} + \frac{-2}{1!}1^{-3}x + \frac{(-2)(-3)}{2!}1^{-4}x^2 + \frac{(-2)(-3)(-4)}{3!}1^{-5}x^3$$

$$+ \frac{(-2)(-3)(-4)(-5)}{4!}1^{-6}x^4 + \ldots$$

$$= \frac{1}{1}\left(\frac{1}{1^2}\right) - 2x\left(\frac{1}{1^3}\right) + \frac{6x^2}{2 \cdot 1}\left(\frac{1}{1^4}\right) + \frac{(-24)x^3}{3 \cdot 2 \cdot 1}\left(\frac{1}{1^5}\right) + \frac{120x^4 \overset{30}{}}{4 \cdot 3 \cdot 2 \cdot 1 \underset{1}{}}\left(\frac{1}{1^6}\right) + \ldots$$

$$(1+x)^{-2} = 1 - 2x + 3x^2 - 4x^3 + 5x^4 + \ldots \qquad (1)$$

Hence, the right side of equation (1) represents the first five terms of the expansion of $(1 + x)^{-2}$.

Find the first four terms of the expansion of $(2 - 1)^{\frac{1}{2}}$.

Solution: The Binomial Theorem states that:

$$(a+b)^n = \frac{1}{0!}a^n + \frac{n}{1!}a^{n-1}b + \frac{n(n-1)}{2!}a^{n-2}b^2 + \ldots + nab^{n-1} + b^n .$$

This theorem can be used to expand the expression $(2 - 1)^{1/2}$. Replacing a by 2 and b by -1, the expression $(2 - 1)^{1/2}$ becomes:

$$(2-1)^{1/2} = [2+(-1)]^{1/2} = \frac{1}{0!}2^{1/2} + \frac{1/2}{1!}2^{\frac{1}{2}-1}(-1) + \frac{\frac{1}{2}(\frac{1}{2}-1)}{2!}(2)^{\frac{1}{2}-2}(-1)^2$$

$$+ \frac{\frac{1}{2}(\frac{1}{2}-1)(\frac{1}{2}-2)}{3!}(2)^{\frac{1}{2}-3}(-1)^3 + \ldots + \frac{1}{2}(2)(-1)^{\frac{1}{2}-1} + (-1)^{\frac{1}{2}}$$

$$= \frac{1}{1}\sqrt{2} + \frac{1}{2}(2)^{-\frac{1}{2}}(-1) + \frac{\frac{1}{2}(-\frac{1}{2})}{2 \cdot 1}(2)^{-\frac{3}{2}} \qquad (1)$$

$$+ \frac{\frac{1}{2}(-\frac{1}{2})}{3 \cdot 2 \cdot 1}(2)^{-\frac{5}{2}}(-1) + \ldots + 1(-1)^{-\frac{1}{2}} + \sqrt{-1} .$$

$$= \sqrt{2} - \frac{1}{2}\left(\frac{1}{2^{\frac{1}{2}}}\right) - \frac{1}{8}\left(\frac{1}{2^{\frac{3}{2}}}\right) - \frac{3}{48}\left(\frac{1}{2^{\frac{5}{2}}}\right) + \ldots + 1\left(\frac{1}{\sqrt{-1}}\right) + \sqrt{-1} .$$

The last result is true because of the law of exponents which states that

$$(N)^{-a/b} = \frac{1}{(N)^{a/b}}$$

where a and b are any positive integers. Writing only the first four terms of the expansion:

$$(2-1)^{\frac{1}{2}} = \sqrt{2} - \tfrac{1}{2}\left(\frac{1}{2^{\frac{1}{2}}}\right) - \tfrac{1}{8}\left(\frac{1}{2^{\frac{3}{2}}}\right) - \tfrac{3}{48}\left(\frac{1}{2^{\frac{5}{2}}}\right) + \cdots$$

$$= \sqrt{2} - \tfrac{1}{2}\left(\frac{1}{\sqrt{2}}\right) - \tfrac{1}{8}\left(\frac{1}{\sqrt{2}^3}\right) - \tfrac{3}{48}\left(\frac{1}{\sqrt{2}^5}\right) + \cdots$$

$$= \sqrt{2} - \frac{1}{2\sqrt{2}} - \tfrac{1}{8}\left(\frac{1}{\sqrt{2}^2\sqrt{2}}\right) - \tfrac{3}{48}\left(\frac{1}{\sqrt{2}^2\sqrt{2}^2\sqrt{2}}\right) + \cdots$$

$$= \sqrt{2} - \frac{1}{2\sqrt{2}} - \tfrac{1}{8}\left(\frac{1}{2\sqrt{2}}\right) - \tfrac{3}{48}\left(\frac{1}{(2)(2)\sqrt{2}}\right) + \cdots$$

$$= \sqrt{2} - \frac{1}{2\sqrt{2}} - \frac{1}{16\sqrt{2}} - \tfrac{3}{48}\left(\frac{1}{4\sqrt{2}}\right) + \cdots$$

$$= \sqrt{2} - \frac{1}{2\sqrt{2}} - \frac{1}{16\sqrt{2}} - \frac{3}{192}\,\frac{1}{\sqrt{2}} + \cdots$$

$$= \sqrt{2} - \frac{1}{2\sqrt{2}} - \frac{1}{16\sqrt{2}} - \frac{1}{64}\,\frac{1}{\sqrt{2}} + \cdots$$

$$= \sqrt{2} - \frac{1}{2\sqrt{2}} - \frac{1}{16\sqrt{2}} - \frac{1}{64\sqrt{2}} + \cdots$$

Since $\sqrt{2} = 1.414$,

$$(2-1)^{\frac{1}{2}} = 1.414 - \frac{1}{2(1.414)} - \frac{1}{16(1.414)} - \frac{1}{64(1.414)}$$

$$= 1.414 - \frac{1}{2.828} - \frac{1}{22.624} - \frac{1}{90.496}$$

$$(2-1)^{\frac{1}{2}} = 1.414 - 0.354 - 0.044 - 0.011 \qquad (1)$$

Hence, the right side of equation (1) represents the first four terms of the expansion of $(2-1)^{\frac{1}{2}}$.

Note that the sum of these four terms is 1.005.

● **PROBLEM** 737

Find the fifth term of $\left(a + 2 + 2x^3\right)^{17}$.

Solution: Use the Binomial Theorem which states that

$$(c+b)^n = \frac{1}{0!}\,c^n + \frac{n}{1!}\,c^{n-1}b + \frac{n(n-1)}{2!}\,c^{n-2}b^2 + \cdots + ncb^{n-1} + b^n.$$

Replacing c by a and b by $2x^3$:

$$\left(a+2x^3\right)^{17} = \frac{1}{0!}\,a^{17} + \frac{17}{1!}\,a^{16}\left(2x^3\right) + \frac{17\cdot16}{2!}\,a^{15}\left(2x^3\right)^2$$

$$+ \frac{17\cdot16\cdot15}{3!}\,a^{14}\left(2x^3\right)^3 + \frac{17\cdot16\cdot15\cdot14}{4!}\,a^{13}\left(2x^3\right)^4$$

$$+ \cdots$$

The fifth term of this expansion is:

$$\frac{17\cdot16\cdot15\cdot14}{4!}\,a^{13}\left(2x^3\right)^4 = \frac{17\cdot\overset{4}{\cancel{16}}\cdot\overset{5}{\cancel{15}}\cdot\overset{7}{\cancel{14}}}{\underset{1}{\cancel{4}}\cdot\underset{1}{\cancel{3}}\cdot\underset{1}{\cancel{2}}\cdot1}\,a^{13}\left(2^4\right)\left(x^3\right)^4$$

$$= \frac{17\cdot4\cdot5\cdot7}{1}\,a^{13}\,16x^{12}$$

$$= 38,080\,a^{13}\,x^{12}$$

547

Find the 5th term of the expansion of $(3y - 4w)^8$.

Solution: Use the binomial formula:

$$(u+v)^n = u^n + nu^{n-1}v + \frac{n(n-1)}{2}u^{n-2}v^2 + \frac{n(n-1)(n-2)}{2\cdot 3}u^{n-2}v^3 + \ldots + v^n .$$

Applying the formula for the rth term where $r \leq n+1$; the rth term is:

$\frac{n(n-1)(n-2)\ldots(n-r+2)}{1\cdot 2\cdot 3 \ldots (r-1)}u^{n-r+1}v^{r-1}$. Let $u = 3y$, $v = -4w$, $n = 8$, $r = 5$.

Then $n - r + 1 = 8 - 5 + 1 = 4$

$\qquad r - 1 = 5 - 1 = 4$

Therefore $u^{n-r+1}v^{r-1} = u^4 v^4$

$$r - 1 = 4$$
$$n - r + 2 = 8 - 5 + 2 = 5$$

So: $\qquad n(n-1)(n-2)\ldots(5) = 8\cdot 7\cdot 6\cdot 5$

Thus the coefficient is:

$$\frac{8\cdot 7\cdot 6\cdot 5}{1\cdot 2\cdot 3\cdot 4}$$

$$u^4 = (3y)^4$$
$$v^4 = (-4w)^4$$

Therefore: $\qquad (3y - 4w)^8 = \frac{8\cdot 7\cdot 6\cdot 5}{4\cdot 3\cdot 2\cdot 1}(3y)^4(-4w)^4$

$$= \frac{\cancel{8}^2(7)(\cancel{6})(5)}{\cancel{4}(\cancel{3})(\cancel{2})}\left(81y^4\right)\left(256w^4\right)$$

$$= 70\left(81y^4\right)\left(256w^4\right)$$

$$= 1,451,520y^4w^4$$

Find the term involving y^5 in the expansion of $\left(2x^2 + y\right)^{10}$.

Solution: The formula for the binomial expansion is:

$$(a + b)^n = a^n + na^{n-1}b + \frac{n(n-1)}{1\cdot 2}a^{n-2}b^2 + \frac{n(n-1)(n-2)}{1\cdot 2\cdot 3}a^{n-3}b^3 + \ldots + nab^{n-1} + b^n$$

The rth term of the expansion of $(a + b)^n$ is

$$\text{rth term} = \frac{n(n-1)(n-2)\ldots(n-r+2)}{(r-1)!}a^{n-r+1}b^{r-1}$$

In this example,

$$b^{r-1} = y^5$$
$$r-1 = 5$$
$$r = 6 \quad \text{and} \quad n = 10$$

Thus,

$$6\text{th term} = \frac{10\cdot 9\cdot 8\cdot 7\cdot 6}{5!}\left(2x^2\right)^5 y^5$$

$$= \frac{\cancel{10}\cdot \cancel{9}^3\cdot \cancel{8}^{2\,\cancel{1}}\cdot 7\cdot 6}{\cancel{5}\cdot \cancel{4}\cdot \cancel{3}\cdot \cancel{2}\cdot 1}32x^{10}y^5$$

$$= 8064\ x^{10}y^5$$

Find the constant term in the expansion of $\left(2x^2 + \frac{1}{x}\right)^9$.

Solution: The rth term in the expansion of $(a+b)^n$ is given by:

$$\frac{n(n-1)(n-2)\ldots(n-r+2)}{1\cdot2\cdot3.\ldots(r-1)} a^{n-r+1} b^{r-1}$$

Replacing a by $2x^2$ and b by $1/x$ in this formula, the rth term in the expansion of $\left(2x^2 + 1/x\right)^9$ is given by:

$$\frac{9(8)(7)\ldots(9-r+2)}{(1)(2)(3)\ldots(r-1)}\left(2x^2\right)^{9-r+1}\left(\frac{1}{x}\right)^{r-1} = \frac{9(8)(7)\ldots(11-r)}{(1)(2)(3)\ldots(r-1)}\left(2x^2\right)^{10-r}\left(\frac{1}{x}\right)^{r-1}$$

Then the rth term in the expansion will contain the factors $\left(2x^2\right)^{10-r}$ and $\left(\frac{1}{x}\right)^{r-1}$. Hence, as far as powers of x are concerned, the rth term will involve

$$\left(x^2\right)^{10-r}\left(\frac{1}{x}\right)^{r-1} \quad \text{or} \quad \left(x^2\right)^{10-r}\frac{(1)^{r-1}}{x^{r-1}} \quad \text{or} \quad \frac{x^{20-2r}}{x^{r-1}}$$

$$\text{or} \quad x^{20-2r-(r-1)}$$

$$\text{or} \quad x^{20-2r-r+1}$$

$$\text{or} \quad x^{21-3r}$$

The desired constant term is free of x; that is, the constant term has a factor of x^0 since $kx^0 = k(1) = k$, where k is the constant. Hence,

$$21 - 3r = 0$$
$$21 = 3r$$
$$\frac{21}{3} = r$$
$$7 = r$$

The rth term or seventh term can be found by using the Binomial Theorem which states that:

$$(a+b)^n = \frac{1}{0!}a^n + \frac{n}{1!}a^{n-1}b + \frac{n(n-1)}{2!}a^{n-2}b^2 + \ldots + nab^{n-1} + b^n .$$

Replacing a by $2x^2$ and b by $1/x$, $\left(2x^2 + 1/x\right)^9$ can be expanded as:

$$\left(2x^2 + \frac{1}{x}\right)^9 = \frac{1}{0!}\left(2x^2\right)^9 + \frac{9}{1!}\left(2x^2\right)^8\left(\frac{1}{x}\right) + \frac{9(8)}{2!}\left(2x^2\right)^7\left(\frac{1}{x}\right)^2$$

$$+ \frac{9(8)(7)}{3!}\left(2x^2\right)^6\left(\frac{1}{x}\right)^3 + \frac{9(8)(7)(6)}{4!}\left(2x^2\right)^5\left(\frac{1}{x}\right)^4$$

$$+ \frac{9(8)(7)(6)(5)}{5!}\left(2x^2\right)^4\left(\frac{1}{x}\right)^5 + \frac{9(8)(7)(6)(5)(4)}{6!}\left(2x^2\right)^3\left(\frac{1}{x}\right)^6$$

$$+ \frac{9(8)(7)(6)(5)(4)(3)}{7!}\left(2x^2\right)^2\left(\frac{1}{x}\right)^7$$

$$+ \frac{9(8)(7)(6)(5)(4)(3)(2)}{8!}\left(2x^2\right)\left(\frac{1}{x}\right)^8$$

$$+ \frac{9(8)(7)(6)(5)(4)(3)(2)(1)}{9!}\left(2x^2\right)^0\left(\frac{1}{x}\right)^9$$

Hence, the rth term or 7th term of this expansion is:

$$\frac{9(8)(7)(6)(5)(4)}{6!}\left(2x^2\right)^3\left(\frac{1}{x}\right)^6$$

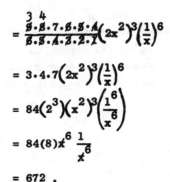

$$= \frac{\overset{3}{\cancel{9}} \cdot \cancel{8} \cdot 7 \cdot \cancel{6} \cdot \cancel{5} \cdot \overset{4}{\cancel{4}}}{\cancel{6} \cdot \cancel{5} \cdot \cancel{4} \cdot \cancel{3} \cdot \cancel{2} \cdot \cancel{1}} \left(2x^2\right)^3 \left(\frac{1}{x}\right)^6$$

$$= 3 \cdot 4 \cdot 7 \left(2x^2\right)^3 \left(\frac{1}{x}\right)^6$$

$$= 84\left(2^3\right)\left(x^2\right)^3 \left(\frac{1^6}{x^6}\right)$$

$$= 84(8)x^6 \frac{1}{x^6}$$

$$= 672 .$$

Hence, 672 is the desired constant term in the expansion of $\left(2x^2 + \frac{1}{x}\right)^9$.

● **PROBLEM** 741

Find the coefficient of $a_1^2 a_2 a_3$ in the expansion of $\left(a_1 + a_2 + a_3\right)^4$.

Solution: The binomial theorem states that, if n is a positive integer, then

$$(a + b)^n = a^n + na^{n-1}b + \frac{n(n-1)}{1 \cdot 2} a^{n-2}b^2 + \frac{n(n-1)(n-2)}{1 \cdot 2 \cdot 3} a^{n-3}b^3$$
$$+ \ldots + nab^{n-1} + b^n.$$

Use the binomial theorem, but for convenience, associate the terms $\left(a_2 + a_3\right)$, then expand the expression.

$$\left[a_1 + \left(a_2 + a_3\right)\right]^4 = a_1^4 + 4a_1^3\left(a_2 + a_3\right) + \frac{4 \cdot 3}{1 \cdot 2} a_1^2\left(a_2 + a_3\right)^2$$

$$+ \frac{4 \cdot 3 \cdot 2}{1 \cdot 2 \cdot 3} a_1\left(a_2 + a_3\right)^3 + \left(a_2 + a_3\right)^4.$$

Notice that the only term involving $a_1^2 a_2 a_3$ is the third term with coefficient $\frac{4 \cdot 3}{2}$ and, further, that $\left(a_2 + a_3\right)^2$ must be expanded also.

$$\left(a_2 + a_3\right)^2 = a_2^2 + 2a_2 a_3 + a_3^2 .$$

Therefore, the third term becomes:

$$\frac{4 \cdot 3}{1 \cdot 2} a_1^2\left(a_2 + a_3\right)^2 = \frac{4 \cdot 3}{1 \cdot 2} a_1^2\left(a_2^2 + 2a_2 a_3 + a_3^2\right)$$

$$= 6a_1^2 a_2^2 + 12a_1^2 a_2 a_3 + 6a_1^2 a_3^2$$

Hence, the coefficient of $a_1^2 a_2 a_3$ is 12.

● **PROBLEM** 742

Find the term involving $x^3 y z^2$ in the expansion $(x + 2y - 3z)^6$.

Solution: Use the binomial formula: $a^n + na^{n-1}b + \frac{n(n-1)}{1 \cdot 2}$ x
$a^{n-2}b^2 + \frac{n(n-1)(n-2)}{1 \cdot 2 \cdot 3} a^{n-3}b^3 + \ldots + nab^{n-1} + b^n = (a+b)^n$

and associate 2y - 3z and substitute it for b in the formula, x for a, and 6 for n:

$$[x + (2y - 3z)]^6 = x^6 + 6x^5(2y - 3z)^1 + \frac{6 \cdot 5}{1 \cdot 2} x^4(2y - 3z)^2$$

$$+ \frac{6 \cdot 5 \cdot 4}{1 \cdot 2 \cdot 3} x^3(2y - 3z)^3 + \frac{6 \cdot 5 \cdot 4 \cdot 3}{1 \cdot 2 \cdot 3 \cdot 4} x^2(2y - 3z)^4$$

$$+ \frac{6 \cdot 5 \cdot 4 \cdot 3 \cdot 2}{1 \cdot 2 \cdot 3 \cdot 4 \cdot 5} x(2y - 3z)^5 + (2y - 3z)^6.$$

The term that involves x^3yz^2 is: $\frac{6 \cdot 5 \cdot 4}{1 \cdot 2 \cdot 3} x^3(2y - 3z)^3$ in which $(2y - 3z)^3$ must be expanded.

$$\left(2y - 3z\right)^3 = \left(2y - 3z\right)\left(2y - 3z\right)^2 = \left(2y - 3z\right)\left(4y^2 - 12yz + 9z^2\right)$$

$$= 8y^3 - 36zy^2 + 54yz^2 - 27z^3$$

When $(2y - 3z)^3$ is multiplied by $\frac{6 \cdot 5 \cdot 4}{1 \cdot 2 \cdot 3} x^3$, the final term is

$$\frac{6 \cdot 5 \cdot 4}{1 \cdot 2 \cdot 3} x^3\left[8y^3 - 36zy^2 + 54yz^2 - 27z^3\right].$$

Distributing, notice that the term of interest involves

$$\frac{6 \cdot 5 \cdot 4}{1 \cdot 2 \cdot 3} x^3\left(54yz^2\right) = 1080x^3yz^2.$$

● PROBLEM 743

Find the value of $\left(a + \sqrt{a^2 - 1}\right)^7 + \left(a - \sqrt{a^2 - 1}\right)^7$.

Solution: Use the binomial theorem to expand the first term:

$$(a + b)^n = a^n + na^{n-1}b + \frac{n(n-1)}{1 \cdot 2} a^{n-2}b^2 + \frac{n(n-1)(n-2)}{1 \cdot 2 \cdot 3} a^{n-3}b^3$$

$$+ \ldots + b^n.$$

Now $\left(a + \sqrt{a^2 - 1}\right)^7 = a^7 + 7a^6\left(\sqrt{a^2 - 1}\right)^1 + \frac{7 \cdot 6}{1 \cdot 2}(a)^5\left(\sqrt{a^2 - 1}\right)^2$

$$+ \frac{7 \, 6 \, 5}{1 \, 2 \, 3} a^4\left(\sqrt{a^2 - 1}\right)^3 + \frac{7 \cdot 6 \cdot 5 \cdot 4}{1 \cdot 2 \cdot 3 \cdot 4} a^3\left(\sqrt{a^2 - 1}\right)^4$$

$$+ \frac{7 \cdot 6 \cdot 5 \cdot 4 \cdot 3}{1 \cdot 2 \cdot 3 \cdot 4 \cdot 5} a^2\left(\sqrt{a^2 - 1}\right)^5 + 7a\left(\sqrt{a^2 - 1}\right)^6 + \left(\sqrt{a^2 - 1}\right)^7$$

Expand $\left(a - \sqrt{a^2 - 1}\right)^7$ in the same fashion.

$$\left(a - \sqrt{a^2 - 1}\right)^7 = a^7 + 7a^6\left(-\sqrt{a^2 - 1}\right)^1 + \frac{7 \cdot 6}{1 \cdot 2}(a)^5\left(-\sqrt{a^2 - 1}\right)^2$$

$$+ \frac{7 \cdot 6 \cdot 5}{1 \cdot 2 \cdot 3} a^4\left(-\sqrt{a^2 - 1}\right)^3 + \frac{7 \cdot 6 \cdot 5 \cdot 4}{1 \cdot 2 \cdot 3 \cdot 4} a^3\left(-\sqrt{a^2 - 1}\right)^4$$

$$+ \frac{7 \cdot 6 \cdot 5 \cdot 4 \cdot 3}{1 \cdot 2 \cdot 3 \cdot 4 \cdot 5} a^2\left(-\sqrt{a^2 - 1}\right)^5 + 7a\left(-\sqrt{a^2 - 1}\right)^6 + \left(-\sqrt{a^2 - 1}\right)^7.$$

In this expansion, the odd powers of $-\sqrt{a^2 - 1}$ will cancel the corresponding terms in the previous expansion.

$$(a + \sqrt{a^2-1})^7 + (a - \sqrt{a^2-1})^7 = a^7 + 7a^6(\sqrt{a^2-1})^1$$

$$+ \frac{7 \cdot \cancel{6}^{3}}{1 \cdot \cancel{2}} (a)^5(\sqrt{a^2-1})^2 + \frac{7 \cdot 6 \cdot 5}{1 \cdot 2 \cdot 3} a^4(\sqrt{a^2-1})^3$$

$$+ \frac{7 \cdot \cancel{6} \cdot 5 \cdot \cancel{4}}{1 \cdot \cancel{2} \cdot \cancel{3} \cdot \cancel{4}} a^3(\sqrt{a^2-1})^4 + \frac{7 \cdot 6 \cdot 5 \cdot 4 \cdot 3}{1 \cdot 2 \cdot 3 \cdot 4 \cdot 5} a^2(\sqrt{a^2-1})^5$$

$$+ 7a(\sqrt{a^2-1})^6 + (\sqrt{a^2-1})^7 + a^7 - 7a^6(\sqrt{a^2-1})$$

$$+ \frac{7 \cdot \cancel{6}^{3}}{1 \cdot \cancel{2}} (a)^5(\sqrt{a^2-1})^2 - \frac{7 \cdot 6 \cdot 5}{1 \cdot 2 \cdot 3} a^4(\sqrt{a^2-1})^3 + \frac{7 \cdot \cancel{6} \cdot 5 \cdot \cancel{4}^{2}}{1 \cdot \cancel{2} \cdot 3 \cdot \cancel{4}} a^3(\sqrt{a^2-1})^4$$

$$- \frac{7 \cdot 6 \cdot 5 \cdot 4 \cdot 3}{1 \cdot 2 \cdot 3 \cdot 4 \cdot 5} a^2(\sqrt{a^2-1})^5 + 7a(\sqrt{a^2-1})^6 - (\sqrt{a^2-1})^7$$

$$= a^7 + 21a^5(a^2-1) + 35a^3(a^2-1)^2 + 7a(a^2-1)^3 + a^7 + 21a^5(a^2-1)$$

$$+ 35a^3(a^2-1)^2 + 7a(a^2-1)^3$$

$$= 2a^7 + 42a^5(a^2-1) + 70a^3(a^2-1)^2 + 14a(a^2-1)^3$$

$$= 2[a^7 + 21a^5(a^2-1) + 35a^3(a^2-1)^2 + 7a(a^2-1)^3]$$

$$= 2[a^7 + 21a^7 - 21a^5 + 35a^3(a^4-2a^2+1) + 7a(a^2-1)^2(a^2-1)]$$

$$= 2[22a^7 - 21a^5 + 35a^7 - 70a^5 + 35a^3 + 7a(a^4-2a^2+1)(a^2-1)]$$

$$= 2[57a^7 - 91a^5 + 35a^3 + 7a(a^6-2a^4+a^2-a^4+2a^2-1)]$$

$$= 2a[57a^6 - 91a^4 + 35a^2 + 7(a^6-2a^4+a^2-a^4+2a^2-1)]$$

$$= 2a[57a^6 - 91a^4 + 35a^2 + 7a^6 - 14a^4 + 7a^2 - 7a^4 + 14a^2 - 7]$$

$$= 2a[64a^6 - 112a^4 + 56a^2 - 7]$$

● PROBLEM   744

Compute the approximate value of $(1.01)^5$

<u>Solution:</u> The Binomial Theorem can be used to find the approximate value of $(1.01)^5$. The Binomial Theorem is stated as:

$$(a + b)^n = a^n + na^{n-1}b + \frac{n(n-1)}{1 \cdot 2} a^{n-2}b^2 + \frac{n(n-1)(n-2)}{1 \cdot 2 \cdot 3} \times$$

$$a^{n-3}b^3 + \cdots + nab^{n-1} + b^n,$$

where a and b are any two numbers. In order to use this

theorem, express $(1.01)^5$ as the sum of two numbers raised to the fifth power. Then,

$$(1.01)^5 = (1+ .01)^5$$

Now, let a = 1, b = .01, and n = 5 in the Binomial Theorem. Calculating the first four terms of this theorem with these substitutions:

$$(1.01)^5 = (1 + .01)^5$$

$$= (1)^5 + 5(1)^{5-1}(.01) + \frac{5(5-1)}{1 \cdot 2} \, x$$

$$(1)^{5-2}(.01)^2 + \frac{5(5-1)(5-2)}{1 \cdot 2 \cdot 3} \, x$$

$$(1)^{5-3}(.01)^3 + \cdots$$

$$= 1 + 5(1)^4(.01) + \frac{5(\cancel{4})^2}{\cancel{2}_1}(1)^3(.01)^2$$

$$+ \frac{5(4)(\cancel{3})^1}{\cancel{6}_2}(1)^2(.01)^3 + \cdots$$

$$= 1 + 5(1)(.01) + 10(1)(.0001) + 10(1)$$

$$(.000001) + \cdots$$

$$= 1 + 0.05 + 0.001 + 0.00001$$

$$= 1.05101$$

● **PROBLEM** 745

Use the binomial formula with n = 1/3 to find an approximation to $\sqrt[3]{28}$.

Solution: To apply the binomial formula, try to express $\sqrt[3]{28}$ as the sum of two numbers raised to a power. Note that the formula is simplified if one of the numbers is one.

$$\sqrt[3]{28} = (28)^{(1/3)} = (27+1)^{1/3} .$$

We can write the expansion of $(x+y)^{1/3}$ to four terms and later substitute for x and y. We write out the binomial expansion to four terms when n = 1/3.

$$(x+y)^{1/3} = x^{1/3} + \frac{1}{3}x^{(1/3-1)}y + \frac{\left(\frac{1}{3}\right)\left(\frac{1}{3}-1\right)}{1 \cdot 2}x^{(1/3-2)}y^2$$

$$+ \frac{\left(\frac{1}{3}\right)\left(\frac{1}{3}-1\right)\left(\frac{1}{3}-2\right)}{1 \cdot 2 \cdot 3}x^{(1/3-3)}y^3 + \cdots$$

$$= x^{1/3} + \frac{1}{3}x^{-2/3}y + \frac{\left(\frac{1}{3}\right)\left(\frac{-2}{3}\right)}{2}x^{-5/3}y^2$$

$$+ \frac{\left(\frac{1}{3}\right)\left(\frac{-2}{3}\right)\left(\frac{-5}{3}\right)}{1\cdot 2\cdot 3} \; x^{-8/3}\, y^3 + \ldots$$

$$= x^{1/3} + \frac{1}{3}\, x^{-2/3}\, y + \left(\frac{-2}{9}\right)\left(\frac{1}{2}\right) x^{-5/3}\, y^2$$

$$+ \frac{\overset{5}{\cancel{10}}\cdot 1}{3\cdot 3\cdot 3\cdot 1\cdot \underset{1}{\cancel{2}}\cdot 3} \; x^{-8/3}\, y^3 + \ldots$$

$$(x+y)^{1/3} \qquad = x^{1/3} + \frac{1}{3}\, x^{-2/3}\, y - \frac{1}{9}\, x^{-5/3}\, y^2$$

$$+ \frac{5}{81}\, x^{-8/3}\, y^3 + \ldots \qquad\qquad\qquad (1)$$

In this case, n is fractional. We obtain an infinite series and we can expand it for the first few terms if $|x| < |y|$. x = 27 and y = 1, and $|y| < |x|$, i.e., $|1| < |27|$. Therefore, using equation (1) with x = 27, y = 1 and n = 1/3 (writing only the first four terms):

$$\sqrt[3]{28} = (28)^{1/3}$$

$$= (27 + 1)^{1/3}$$

$$= 27^{1/3} + 1/3\left(27^{-2/3}\right)(1) - \frac{1}{9}\left(27^{-5/3}\right)\left(1^2\right)$$

$$+ \frac{5}{81}\left(27^{-8/3}\right)\left(1^3\right)$$

$$= 3 + \frac{1}{3}\left(\frac{1}{9}\right) - \frac{1}{9}\left(\frac{1}{243}\right) + \frac{5}{81}\left(\frac{1}{6,561}\right)$$

$$= 3 + 0.037037 - 0.000457 + 0.000009$$

$$= 3.036589$$

CHAPTER 26

LOGARITHMS AND EXPONENTIALS

Basic Attacks and Strategies for Solving
Problems in this Chapter. See pages 555 to
614 for step-by-step solutions to problems.

Logarithms are exponents and the properties of logarithms are actually the properties of exponents. For any positive number b other than 1

$$y = f(x) = b^x$$

is called the base b exponential function. On the other hand, for any positive number b other than 1, the logarithm to the base b of x, denoted by $\log_b x$, is defined to be the power of b that equals x. In other words,

$\log_b x = y$ if and only if $b^y = x$.

The base b exponential function and the base b logarithmic function are inverses of each other. The base b exponential expression (in exponential form) can be written in base b logarithmic form. For instance, in general,

$$b^y = x$$

is equivalent to

$\log_b x = y$,

whereas b is the base for both the exponent and log. Also, the exponent y in b^y is the value of the log, and x, the value of b^y, is the number of which the log is taken. In a similar manner, a logarithm statement can be written in exponential form. For example,

$\log_2 8 = 3$

written in exponential form is $2^3 = 8$.

To find the value of the expression $\log_b x$, first set

$\log_b x = y$,

which is equivalent to the exponential form $b^y = x$. By trial and error or by using the calculator, find the value of y that satisfies $b^y = x$.

To find the value of $\log_b x^y$, recall that

$$\log_b x^y = y(\log_b x).$$

When finding logs with particular values and bases (including base 10 and e), we can use tables, calculators, and the various properties of logarithms.

To solve a logarithmic equation, the following general steps are suggested:

(1) Write the equivalent exponential equation for the given logarithmic equation.

(2) Evaluate the power that results.

(3) Solve the resulting equation for the variable. For instance,

$$\log_3(2x - 3) = 4 \implies 2x - 3 = 3^4 \text{ or}$$

$$2x - 3 = 81 \implies x = 42.$$

In order to solve an exponential equation in which the bases on both sides of the equation can be written as powers of the same base, b, follow these steps:

(1) Write the base on both sides as powers of b.

(2) Simplify the exponents on each side, if possible.

(3) Set the exponents on each side equal.

(4) Solve the resulting equation for the variable.

For example,

$$9^2 = 27^x \implies (3^2)^2 = (3^3)^x \implies 3^4 = 3^{3x}.$$

$$\implies 4 = 3x \implies x = {}^4/_3.$$

Step-by-Step Solutions to Problems in this Chapter, "Logarithms and Exponentials"

EXPRESSIONS

● PROBLEM 746

Write $5^3 = 125$ in logarithmic form.

Solution: The statement $b^y = x$ is equivalent to the statement $\log_b x = y$, where b is the base and y is the exponent. The latter form is the logarithmic form. Thus $5^3 = 125$ in logarithmic form is $\log_5 125 = 3$, where the base is 5 and the logarithm of 125 is 3.

● PROBLEM 747

Find $\log_{10} 100$.

Solution: The following solution presents 2 methods for solving the given problem.

Method I. The statement $\log_{10} x = y$ is equivalent to $10^y = x$, hence $\log_{10} 100 = x$ is equivalent to $10^x = 100$. Since $10^2 = 100$, $\log_{10} 100 = 2$.

Method II. Note that $100 = 10 \times 10$; thus $\log_{10} 100 = \log_{10} (10 \times 10)$. Recall: $\log_x (a \times b) = \log_x a + \log_x b$, therefore

$$\log_{10} (10 \times 10) = \log_{10} 10 + \log_{10} 10$$

$$= \quad 1 \quad + \quad 1$$

$$= 2.$$

● PROBLEM 748

Find the logarithm of 3^2.

Solution: Recall that $\log_b x^y = y \log_b x$; thus

$$\log_{10} 3^2 = 2 \log_{10} 3$$

Referring to a table of common logarithms we find:

$$\log_{10} 3 = .4771; \text{ hence,}$$

$$= 2(.4771)$$

$$= .9542 .$$

Thus, $\log_{10} 3^2 = .9542$.

We could have obtained the same result by noting that $3^2 = 9$. Using a log table to find $\log_{10} 9$, we observe:

$\log_{10} 9 = .9542$, as above.

● **PROBLEM 749**

Write $\frac{1}{2} = \log_9 3$ in exponential form.

Solution: The statement $\log_b x = y$ is equivalent to the statement $b^y = x$, where b is the base and y the exponent. The latter form is the exponential form. Thus, $\frac{1}{2} = \log_9 3$ in exponential form is $9^{\frac{1}{2}} = 3$, where the base is 9 and the exponent is $\frac{1}{2}$.

● **PROBLEM 750**

If $\log_3 N = 2$, find N.

Solution: The equation $\log_x a = y$ is equivalent to the equation $x^y = a$. Thus $\log_3 N = 2$ is equivalent to the equation $3^2 = N$. $3^2 = 9$, hence $N = 9$.

● **PROBLEM 751**

Find the value of x if $\log_4 64 = x$.

Solution:
$$\log_b u = v$$

is equivalent to,

$$b^v = u,$$

thus the exponential equivalent of
$$\log_4 64 = x \qquad \text{is } 4^x = 64.$$

Since,

$$4^3 = 4 \cdot 4 \cdot 4 = 64$$

$$\log_4 64 = 3.$$

That is,

$$x = 3.$$

● **PROBLEM 752**

Find $\log_3 729$.

Solution: Since we are working with log base 3, we check whether 729 has factors of 3.

$$729 = 3 \cdot 243 = 3 \cdot (3 \cdot 81) = 3 \cdot [3 \cdot (3 \cdot 27)] = 3[3 \cdot 3(3 \cdot 9)]$$
$$= [3 \cdot 3 \cdot 3 \cdot 3 \cdot (3 \cdot 3)]$$
$$= 3^6$$

Thus $\log_3 729 = \log_3 3^6 = 6$ (because $\log_b b^a = a$ is equivalent to the statement $b^a = b^a$).

Check: $\log_3 729 = 6$ is equivalent to $3^6 = 729$ which we have just seen to be true.

● **PROBLEM** 753

Find the value of N if $\log_8 N = \frac{2}{3}$.

Solution: The inverse of the logarithmic function, $y = \log_a N$, is the exponential function $N = a^y$. Then for

$$\log_8 N = \frac{2}{3},$$

$$N = 8^{2/3} = \sqrt[3]{(8)^2} = (\sqrt[3]{8})^2 = (2)^2 = 4$$

$$N = 4.$$

● **PROBLEM** 754

Express the logarithm of 7 to the base 3 in terms of common logarithms.

Solution: By definition, if $\log_b a = n$, then $b^n = a$. Therefore, if $\log_3 7 = x$, then $3^x = 7$. Take the logarithm of both sides:

$$\log 3^x = \log 7.$$

By the law of the logarithm of a power of a positive number which states that $\log a^n = n \log a$, $\log 3^x = x \log 3$. Hence, $x \log 3 = \log 7$. Divide both sides of this equation by $\log 3$:

$$\frac{x \log 3}{\log 3} = \frac{\log 7}{\log 3}.$$

Therefore, $x = \frac{\log 7}{\log 3} = \log_3 7$ is the logarithm of 7 to the base 3 expressed in terms of common logarithms.

● **PROBLEM** 755

What is the value of b in the relation $\log_b \frac{1}{25} = -2$?

Solution: Since the statement ,

$$\log_b x = y$$

is equivalent to,

557

$$b^y = x, \quad \log_b \frac{1}{25} = -2$$

is equivalent to

$$b^{-2} = \frac{1}{25}, \quad x^{-y} = \frac{1}{x^y}$$

thus,

$$b^{-2} = \frac{1}{b^2}. \text{ Therefore } b^{-2} = \frac{1}{25} \text{ is equivalent to } b^2 = \frac{1}{25}.$$

Cross multiply to obtain the equivalent equation ,

$$b^2 = 25.$$

Take the square root of both sides. Thus,

$$b = \pm 5$$

● **PROBLEM** 756

Express the logarithm of $\dfrac{\sqrt{a^3}}{c^5 b^2}$ in terms of log a, log b and log c.

Solution: We apply the following properties of logarithms:

$$\log_b (P \cdot Q) = \log_b P + \log_b Q$$

$$\log_b (P/Q) = \log_b P - \log_b Q$$

$$\log_b (P^n) = n \log_b P$$

$$\log_b \left(\sqrt[n]{P} \right) = \frac{1}{n} \log_b P$$

Therefore,

$$\log \frac{\sqrt{a^3}}{c^5 b^2} = \log \frac{a^{3/2}}{c^5 b^2}$$

$$= \log a^{3/2} - \log \left(c^5 b^2 \right)$$

$$= 3/2 \log a - \left(\log c^5 + \log b^2 \right)$$

$$= 3/2 \log a - \log c^5 - \log b^2$$

$$= 3/2 \log a - 5 \log c - 2 \log b .$$

● **PROBLEM** 757

If $\log_{10} 3 = .4771$ and $\log_{10} 4 = .6021$, find $\log_{10} 12$.

Solution: Since $12 = 3 \times 4$,

$$\log_{10} 12 = \log_{10} (3)(4).$$

Since $\log_b (xy) = \log_b x + \log_b y$

$$\log_{10} (3 \times 4) = \log_{10} 3 + \log_{10} 4$$

$$= .4771 + .6021$$

$$= 1.0792.$$

Thus $\log_{10} 12 = 1.0792$.

● **PROBLEM** 758

Given $\log_{10}2 = 0.3010$, find $\log_{10}32$.

<u>Solution:</u> Note that,
$$32 = 2 \cdot 2 \cdot 2 \cdot 2 \cdot 2 = 2^5.$$
Thus,

$$\log _{10}32 = \log_{10}2^5.$$

Recall the logarithmic property,

$$\log_b x^y = y \log_b x.$$
Hence,

$$\log_{10}32 = \log_{10}2^5 = 5 \log_{10}2$$

$$= 5 (0.3010)$$

$$= 1.5050$$

● **PROBLEM** 759

If $\log_b 2 = .69$ and $\log_b 3 = 1.10$, find $\log_b 6$ and $\log_b 8$.

<u>Solution:</u> $6 = 2 \times 3$, therefore $\log_b 6 = \log_b (2 \times 3)$. Since
$$\log_a(y \times z) = \log_a y + \log_a z$$
$$\log_b(2 \times 3) = \log_b 2 + \log_b 3$$
Thus
$$\log_b 6 = \log_b 2 + \log_b 3 = .69 + 1.10 = 1.79.$$
$8 = 2^3$, therefore $\log_b 8 = \log_b 2^3$. Since
$$\log_a y^z = z \log_a y$$
$$\log_b 2^3 = 3 \log_b 2$$
thus
$$\log_b 8 = 3 \log_b 2 = 3(.69) = 2.07.$$

● **PROBLEM** 760

Given that $\log_{10}2 = 0.3010$ and $\log_{10}3 = 0.4771$, find $\log_{10}\sqrt{6}$.

<u>Solution:</u> $\sqrt{6} = 6^{\frac{1}{2}}$, thus $\log_{10}\sqrt{6} = \log_{10}6^{\frac{1}{2}}$. Since $\log_b x^y = y \log_b x$,
$\log_{10}6^{\frac{1}{2}} = \frac{1}{2} \log_{10}6$. Therefore $\log_{10}\sqrt{6} = \frac{1}{2} \log_{10}6$. $6 = 3 \cdot 2$, hence
$\frac{1}{2} \log_{10}6 = \frac{1}{2} \log_{10}(3 \cdot 2)$. Recall $\log_{10}(a \cdot b) = \log_{10}a + \log_{10}b$. Thus
$\frac{1}{2} \log_{10}(3 \cdot 2) = \frac{1}{2}\left(\log_{10}3 + \log_{10}2\right)$. Replace our values for $\log_{10}3$
and $\log_{10}2$,

$$= \tfrac{1}{2}(0.4771 + 0.3010)$$
$$= \tfrac{1}{2}(0.7781)$$
$$\approx 0.3890$$

Therefore $\log_{10}\sqrt{6} = 0.3890$.

● **PROBLEM** 761

If $\log_4 7 = n$, find $\log_4 \frac{1}{7}$.

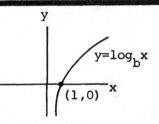

Solution: Apply the following property of logarithms
$\log_b \frac{x}{y} = \log_b x - \log_b y$. Then,

$$\log_4 \frac{1}{7} = \log_4 1 - \log_4 7.$$

We note that if $x = 1$, $\log_b x = 0$. See the sketch.
Thus $\log_4 1 = 0$, and if $\log_4 7 = n$, then $\log_4 \frac{1}{7} = 0 - n$
$= -n$.

● **PROBLEM** 762

Find $\log_{10}(10^2 \cdot 10^{-3} \cdot 10^5)$.

Solution: Recall $\log_x(a \cdot b \cdot c) = \log_x a + \log_x b + \log_x c$. Thus

$$\log_{10}(10^2 \cdot 10^{-3} \cdot 10^5) = \log_{10}10^2 + \log_{10}10^{-3} + \log_{10}10^5$$

Recall $\log_b b^x = x$, since $b^x = b^x$; therefore, $\log_{10}10^2 + \log_{10}10^{-3} +$

$\log_{10}10^5 = 2 + (-3) + 5 = 4$. Thus $\log_{10}(10^2 \cdot 10^{-3} \cdot 10^5) = 4$.

Another method of finding $\log_{10}(10^2 \cdot 10^{-3} \cdot 10^5)$ is to note

$10^2 \cdot 10^{-3} \cdot 10^5 = 10^{2+(-3)+5} = 10^4$ (because $a^x \cdot a^y \cdot a^z = a^{x+y+z}$).
Thus $\log_{10}(10^2 \cdot 10^{-3} \cdot 10^5) = \log_{10}10^4 = 4$.

● **PROBLEM** 763

Find the values of the following logarithims:

   a) $\log_{10}10$ b) $\log_{10}100$ c) $\log_{10}1$

   d) $\log_{10}0.1$ e) $\log_{10}0.01$

Solution: The logarithmic expression $N = \log_b x$ is equivalent to $b^N = x$. Hence,

560

a) Let $N_1 = \log_{10}10$. Then the logarithmic expression $N_1 = \log_{10}10$ is equivalent to $10^{N_1} = 10$. Since $10^1 = 10$, $N_1 = 1$. Therefore, $N_1 = 1 = \log_{10}10$.

b) Let $N_2 = \log_{10}100$. Then the logarithmic expression $N_2 = \log_{10}100$ is equivalent to $10^{N_2} = 100$. Since $10^2 = 100$, $N_2 = 2$. Therefore, $N_2 = 2 = \log_{10}100$.

c) Let $N_3 = \log_{10}1$. Then the logarithmic expression $N_3 = \log_{10}1$ is equivalent to $10^{N_3} = 1$. Since $10^0 = 1$, $N_3 = 0$. Therefore, $N_3 = 0 = \log_{10}1$.

d) Let $N_4 = \log_{10}0.1 = \log_{10}\frac{1}{10}$. Then the logarithmic expression $N_4 = \log_1 0.1 = \log_{10}\frac{1}{10}$ is equivalent to $10^{N_4} = \frac{1}{10}$. Since $10^{-1} = \frac{1}{10^1} = \frac{1}{10}$, $N_4 = -1$. Therefore, $N_4 = -1 = \log_{10}0.1$.

e) Let $N_5 = \log_{10}0.01 = \log_{10}\frac{1}{100}$. Then the logarithmic expression $N_5 = \log_{10}0.01 = \log_{10}\frac{1}{100}$ is equivalent to $10^{N_5} = \frac{1}{100}$. Since $10^{-2} = \frac{1}{10^2} = \frac{1}{100}$, $N_5 = -2$. Therefore, $N_5 = -2 = \log_{10}0.01$.

● **PROBLEM 764**

Find $\log_9\left(\frac{1}{27}\right)$.

Solution: $\frac{1}{27} = 27^{-1}$, thus $\log_9\left(\frac{1}{27}\right) = \log_9 27^{-1}$. Recall $\log_x a^b = b \log_x a$ so:
$$\log_9 27^{-1} = (-1)\log_9 27 = -\log_9 27 .$$

The statement $\log_x a = y$ is equivalent to $x^y = a$ thus $\log_9 27 = y$ is equivalent to $9^y = 27$. Observe that $9 = 3^2$ and $27 = 3^3$ therefore $\left(3^2\right)^y = 3^3$. Since $\left(a^x\right)^y = a^{xy}$, $3^{2y} = 3^3$. If $a^x = a^y$, $x = y$, thus
$$2y = 3$$
and
$$y = \frac{3}{2}$$

thus $\log_9 27 = \frac{3}{2}$ and $-\log_9 27 = -\frac{3}{2}$. Hence $\log_9\left(\frac{1}{27}\right) = -\frac{3}{2}$

● **PROBLEM 765**

Find the base b for which $\log_b 16 = \log_6 36$.

Solution: The statement $y = \log_b a$ is equivalent to the statement $b^y = a$. Thus, $x = \log_6 36$ is equivalent to

$6^x = 36$. $6^2 = 36$, thus $x = 2$ and $\log_6 36 = 2$. Replacing

$\log_6 36$ by 2 we obtain $\log_b 16 = 2$, or equivalently

$b^2 = 16$. Thus $b = \sqrt{16} = 4$.

• **PROBLEM** 766

Evaluate $\log_{10} \sqrt[3]{7}$.

<u>Solution:</u> Since $\sqrt[3]{x} = x^{1/a}$, $\sqrt[3]{7} = 7^{\frac{1}{3}}$, and

$$\log_{10} \sqrt[3]{7} = \log_{10} 7^{\frac{1}{3}}$$

Recall the property of logarithms: $\log_b x^a = a \log_b x$.

Thus, $\log_{10} 7^{\frac{1}{3}} = \frac{1}{3} \log_{10} 7$.

From the table of common logarithms we find that
$\log_{10} 7 = .8451$, thus

$$\frac{1}{3} \log_{10} 7 = \frac{1}{3} (.8451) = .2817$$

Therefore, $\log_{10} \sqrt[3]{7} = .2817$.

• **PROBLEM** 767

Find $\log \left(4 \frac{2}{7} \right)$.

<u>Solution:</u> $4 \frac{2}{7} = \frac{30}{7}$, thus $\log \left(4 \frac{2}{7} \right) = \log \frac{30}{7}$.

Since $\log_b \frac{x}{y} = \log_b x - \log_b y$,

$$\log \frac{30}{7} = \log 30 - \log 7.$$

Reducing 30 to prime factors,

$$\log \frac{30}{7} = \log (2 \times 3 \times 5) - \log 7$$

Recalling $\log_b xyz = \log_b x + \log_b y + \log_b z$,

$$\log \frac{30}{7} = \log 2 + \log 3 + \log 5 - \log 7.$$

From a log table we find

$$\log 2 = .3010$$
$$\log 3 = .4771$$
$$\log 5 = .6990$$
$$\log 7 = .8451$$

Hence, $\log \left(4 \frac{2}{7} \right) = .3010 + .4771 + .6990 - .8451 = .6320.$

562

Find the logarithm of 258, using a trig table.

__Solution:__ Since our log tables only give values of logarithms between 1.00 and 9.99 we must express 258 in terms of some number between 1 and 9.99 multiplied by a power of ten. Hence

$$258 = 2.58 \times 100 = 2.58 \times 10^2$$

and $\quad \log 258 = \log\left(2.58 \cdot 10^2\right)$

since $\quad \log_a BC = \log_a B + \log_a C$

$$\log_{10}\left(2.58 \cdot 10^2\right) = \log_{10} 2.58 + \log_{10} 10^2 \ .$$

Since $\log_{10} x = a$ means by definition $10^a = x$,

we note $\log_{10} 10^2 = 2$ because $10^2 = 10^2$;

hence $\quad \log_{10}\left(2.58 \cdot 10^2\right) = \log_{10} 2.58 + 2$. From our trig. table we read $\log 2.58 = .4116$. Hence $\log 258 = .4116 + 2 = 2.4116$.

Evaluate $\dfrac{\log_{10} 12}{\log_{10} 5}$.

__Solution:__ First calculate $\log_{10} 12$.

$$\log_{10} 12 = \log_{10}(1.2 \times 10)$$

By the law of logarithms which states that $\log_b(x \cdot y) = \log_b x + \log_b y$,

$$\log_{10} 12 = \log_{10}(1.2 \times 10) = \log_{10} 1.2 + \log_{10} 10$$

$$= 0.0792 + 1$$

$$= 1.0792$$

The $\log_{10} 1.2$ was obtained from a table of common logarithms, base 10. Also, $\log_{10} 5 = 0.6990$. This value was also obtained from a table of common logarithms, base 10.

$$\frac{\log_{10} 12}{\log_{10} 5} = \frac{1.0792}{0.6990} = 1.544 .$$

Find the logarithm of 30,700.

__Solution:__ First express 30,700 in scientific notation. $30,700 = 3.07 \times 10^4$. 4 is the characteristic. To find the mantissa, see a table of common logarithms of numbers The mantissa is 4871. Thus $\log 30,700 = 4 + .4871 = 4.4871$.

Find log 0.0364 .

Solution: $0.0364 = 3.64 \times 10^{-2}$. Therefore, the characteristic, the power of 10, is -2. From a table of logarithms, the mantissa for 3.64 is 0.5611. Therefore, log 0.0364 = -2 + 0.5611 = -1.4389.

Find N if log N = 0.7917 - 3.

Solution: Using a table of logarithms, the mantissa .7917 is found to correspond to the number 6.19. Therefore the antilogarithm is 6.19. Then, since the characteristic is -3,

$$N = 6.19 \times 10^{-3} = 0.00619.$$

What is the value of log 0.0148?

Solution: $0.0148 = 1.48 \times 10^{-2}$. The characteristic is the exponent of 10. Hence, the characteristic is -2. The mantissa for 148 can be found in a table of logarithms. The mantissa is 0.1703. Therefore, log 0.0148 = -2 + 0.1703 = -2.0000 + 0.1703 = -1.8297. Notice that the number 0.0148 is less than 1. Therefore, the value of log 0.0148 must be negative, as it was found to be.

Determine x when log x = 3.1818.

Solution: $\text{Log}_{10} x = 3.1818$ is equivalent to $10^{3.1818} = x$. Since $a^{x+y} = a^x \cdot a^y$, $x = 10^{3.1818} = 10^3 \cdot 10^{.1818}$

$= 1,000 \cdot 10^{.1818}$.

We look in the body of the log table for the mantissa 0.1818 and find it in row 15 and column 2, so that the digits of x are 1.52. Thus

$$x = 1,000 \times 1.52$$

$$= 1520.$$

Find $\text{Antilog}_{10}\left(0.8762 - 2\right)$.

Solution: Let $N = \text{Antilog}_{10}\left(0.8762 - 2\right)$. The following relationship between log and antilog exists: $\log_{10} x = a$ is the equivalent of $x = \text{antilog}_{10} a$. Therefore,

$$\log_{10} N = 0.8762 - 2.$$

The characteristic is -2. The mantissa is 0.8762. The number that corresponds to this mantissa is 7.52. This number is found from a table of common logarithms, base 10. Therefore,

$$N = 7.52 \times 10^{-2}$$

$$= 7.52 \times \left(\frac{1}{10^2}\right)$$

$$= 7.52 \times \left(\frac{1}{100}\right)$$

$$= 7.52(.01)$$

$$N = 0.0752 .$$

Therefore, $N = \text{Antilog}_{10} \ 0.8762 - 2 = 0.0752$.

● **PROBLEM** 776

Find the antilogarithm of 1.4349 to three significant figures.

Solution: The mantissa 4349 does not appear in a table of four place logarithms. However, it falls between the two listed mantissas:

4346, whose antilogarithm is 2.72

and   4362, whose antilogarithm is 2.73

Since the given mantissa is closer in value to the mantissa 4346, the three figure sequence of digits in the antilogarithm is 2.72. Using the characteristic 1, we have $2.72 \times 10' = 27.2$. Note that the characteristic is positive, (1), and is one less than the number of digits to the left of the decimal point. Thus there are two digits to the left of the decimal point, (27).

● **PROBLEM** 777

Find $\sqrt[5]{.2}$.

Solution: It is easier to perform this computation using logarithms.

Let $N = (.2)^{1/5}$. Then

$$\log N = \log (.2)^{1/5}$$

The characteristic of the common logarithm of any positive number less than 1 is negative and is one more than the number of zeros between the decimal point and the first digit.

To find the mantissa, see a table of common logarithms of numbers.

$$\text{Thus for log } (.2) = \underbrace{.3010}_{\text{mantissa}} \overbrace{- 1}^{\text{characteristic}}$$

Then: $\log N = \dfrac{1}{5} \log .2 = \dfrac{.3010 - 1}{5}$.

If we proceed with the arithmetic at this point, we shall

find that $\log N = \dfrac{- .6990}{5} = - .1398$. But this last number

must be written in standard form before we can solve for
N. It is easier to replace the number (.3010 - 1) with
its equivalent expression (4.3010 - 5) before we divide
by 5. Then

$$\log N = \frac{4.3010 - 5}{5} = .8602 - 1.$$

See a table to find the mantissa, 8602 and look for the
corresponding number. We find 7248. Adjust the decimal
point by its characteristic, - 1. Hence,

$$\sqrt[5]{.2} = .7248.$$

● **PROBLEM** 778

Find $\sqrt[5]{20}$.

Solution: Let $N = \sqrt[5]{20} = 20^{1/5}$. Then, taking the logarithm of both
sides:
$$\log N = \log 20^{1/5};$$
and since $\log a^x = x \log a$, $= 1/5 \log 20$. Using a table of logs of
numbers, we find that the mantissa for 20 is .3010; and the characteristic is 1, since for a number greater than 1 (in this case 20) the
characteristic is positive and one less than the number of digits before
the decimal. Thus, we have:
$$\log N = \frac{1.3010}{5} = .2602.$$

To find N look up .2602 in a Table of Mantissas of Common Logarithms.
We find the closest number is 182. Since we have 0.2602, which has a
characteristic of zero, then there is one digit to the left of the
decimal point. Thus, we adjust the decimal point and N = 1.82. Thus,
$$N = \sqrt[5]{20} = 1.82.$$

● **PROBLEM** 779

Evaluate

$$\frac{542.3\sqrt{0.1383}}{32.72} \quad \text{using logarithms.}$$

Solution: Let x denote the above expression. Then,
$$x = \frac{542.3\sqrt{0.1383}}{32.72} .$$

Take the logarithms of both sides to obtain:
$$\log x = \log \frac{542.3(0.1383)^{1/2}}{32.72}$$

Apply the following properties of logarithms:
$$\log\left(\frac{a}{b}\right) = \log a - \log b$$

566

$$\log(a \cdot b) = \log a + \log b$$

$$\log a^n = n \log a$$

Then,

$$\log x = \log 542.3(0.1383)^{1/2} - \log 32.72$$

$$= \log 542.3 + \log(0.1383)^{1/2} - \log 32.72$$

$$= \log 542.3 + \frac{1}{2}\log(0.1383) - \log 32.72$$

To find the characteristic and mantissa of each number, express each number in powers of 10. The exponent is the corresponding characteristic

$$542.3 = 5.423 \times 10^2$$
$$0.1383 = 1.383 \times 10^{-1}$$
$$32.72 = 3.272 \times 10^1$$

Note that these numbers are expressed as numbers between 1 and 10, multiplied by a power of ten. Look up the corresponding number in a table of common logarithms of numbers. This is the mantissa. Then,

$$\log 542.3 = 2.7343$$

$$\log 0.1383 = -1.1408.$$

The form 19.1408 - 20 is more convenient for computation.

$$\log 32.72 = 1.5148$$

Hence

$$\frac{1}{2}\log 0.1383 = \frac{1}{2}(19.1408 - 20) = 9.5704 - 10,$$

$$\log 542.3 = 2.7342$$

$$\frac{1}{2}\log 0.1383 + \log 542.3 = 9.5704-10 + 2.7342 = 2.3046$$

$$\log 32.72 = 1.5148$$

$$\log 542.3 + \frac{1}{2}\log(0.1383) - \log 32.72 = 2.3046 - 1.5148 = 0.7898$$

$$\log x = 0.7898$$

Look up the mantissa and we find the number is 6163. To adjust the decimal point, we note that the characteristic is 0. Since the characteristic is positive, it is one less than the number of digits to the left of the decimal point. Here, there is one digit to the left of the decimal point. Thus, the number is 6.163, and

$$x = 6.163.$$

Therefore,

$$\frac{542.3\sqrt{0.1383}}{32.72} = 6.163.$$

● **PROBLEM** 780

Evaluate $\sqrt{\dfrac{x^3 + 1}{x^3 - 1}}$ where $x = 1.47$.

<u>Solution:</u> $\log x = \log 1.47 = 0.1673$. This value can be found in a table of logarithms. By the law of the logarithm of a power of a positive number which states that

$$\log a^n = n \log a, \quad \log x^3 = 3 \log x = 3(0.1673)$$

$$= 0.5019.$$

Hence, $x^3 = $ antilog $0.5019 = 3.18$. This value is obtained from a table of logarithms by noting that the number that corresponds to the mantissa 0.5019 is approximately 3.18. Then $x^3 + 1 = 4.18$ and $x^3 - 1 = 2.18$. Let

$$N = \sqrt{\frac{x^3 + 1}{x^3 - 1}} = \sqrt{\frac{4.18}{2.18}}.$$

Take the logarithm of both sides of the above equation.

$$\log N = \log\sqrt{\frac{4.18}{2.18}} = \log\left(\frac{4.18}{2.18}\right)^{\frac{1}{2}}.$$

By the law of the logarithm of a power of a positive number which states that

$$\log a^n = n \log a, \quad \log\left(\frac{4.18}{2.18}\right)^{\frac{1}{2}} = \frac{1}{2} \log\left(\frac{4.18}{2.18}\right).$$

By the law of the logarithm of a quotient which states that $\log \frac{a}{b} = \log a - \log b$, $\log\left(\frac{4.18}{2.18}\right) = \log 4.18 - \log 2.18$. Hence,

$$\log N = \log\left(\frac{4.18}{2.18}\right)^{\frac{1}{2}} = \frac{1}{2} \log\left(\frac{4.18}{2.18}\right) = \frac{1}{2}\left(\log 4.18 - \log 2.18\right)$$

$$= \frac{1}{2}(0.6212 - 0.3385).$$

Note that the values for the two logs were found in a table of logarithms. Therefore,

$$\log N = \frac{1}{2}(0.6212 - 0.3385) = \frac{1}{2}(0.2827) = 0.1414.$$

$$N = \text{antilog } 0.1414 = 1.38.$$

Note that in a table of logarithms, the number that corresponds to the mantissa 0.1414 is approximately 1.38.

INTERPOLATIONS

● PROBLEM 781

Use linear interpolation to find log 5.723.

Solution: Since 5.723 is .3 of the way from 5.72 to 5.73, we argue that log 5.723 is approximately .3 of the way from log 5.72 to log 5.73.

This is the basic idea involved in linear interpolation. We obtain log 5.72 and log 5.73 from a table of common logarithms, and find the mantissas to be 7574 and 7582, respectively. We now use interpolation to find the mantissa for 5.723.

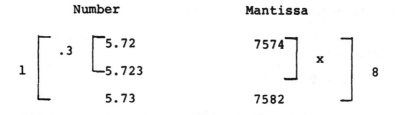

	Number		Mantissa	
1	.3	5.72 ⌐	7574 ⌐	
		5.723 ⌐		x
		5.73	7582 ⌐	8

Note: Observe that 5.73 - 5.72 = 0.1, and
5.723 - 5.72 = .003; but we can rewrite these as 1 and
.3 by shifting the decimal two places.

$$\frac{.3}{1} = \frac{x}{8}$$

$$x = 2.4 \approx 2$$

Thus, the mantissa of the given number is
7574 + 2 = 7576. Since the number is 5.723, and there is
one digit before the decimal point, we know that the
characteristic is one less than the number of digits, or
one less than one, or 0. Thus,

$$\log 5.723 = .7576.$$

● **PROBLEM** 782

Find log 0.7056.

<u>Solution:</u> First determine the characteristic by realiz-
ing that it will be one more than the number of zeros to
the right of the decimal point, with a negative sign
because the number is less than one. Thus, the characteristic is - 1. To compute the mantissa notice that the
number 7056 lies between 7050 and 7060, so that its
log will occur between the logs of those numbers. Interpolating we obtain

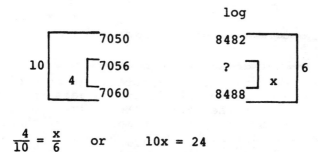

			log	
10		7050	8482	
	4	7056	?	6
		7060	8488	x

$$\frac{4}{10} = \frac{x}{6} \quad \text{or} \quad 10x = 24$$

$$x = 2.4$$

Now subtract this value from the higher mantissa

```
    8488
     2.4
   8485.6
```

The mantissa is always less than one so the decimal

point must be moved four places to the left.

Now - 1 can be written as 9 - 10 for convenience, so our final answer becomes

log 0.7056 = 9.84856-10

● PROBLEM 783

Determine log 51.83.

Solution: With the fourth digit dropped, the remaining three-digit number is 518, and the next larger one is 519. Schematically, we have

$$10 \begin{bmatrix} 3 \begin{bmatrix} 5180 & 7143 \\ 5183 & x \end{bmatrix} h \\ 5190 & 7152 \end{bmatrix} 9,$$

Number   Log

We set up the proportion $\frac{h}{9} = \frac{1}{10}$. Cross multiplying we obtain   10h = 27

$$h = \frac{27}{10} \simeq 3.$$

Thus x = 7143 + 3 = .7146.

Since   51.83 = 5.183 × 10

log 51.83 = log(5.183 × 10)

= log 5.183 + log 10

= .7146 + 1

= 1.7146.

● PROBLEM 784

Find log 513.06

Solution: First determine the characteristic according to the rule that the characteristic is one less than the number of digits to the left of the decimal point. In this case it is two. Now find the mantissa by checking the mantissa for the number 51300 which is 7101 and the mantissa for 51400 which is 7110; the mantissa for 51306 will lie between these two mantissas. Now a proportion can be set up to determine the actual mantissa for the number.

log

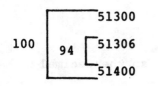

$$\frac{94}{100} = \frac{x}{9} \text{ , or cross-multiplying to obtain}$$

$$(.94)9 = x = 8.46$$

Now subtract this value from the higher mantissa to obtain:

7110.00

−   8.46

7101.54

Since the mantissa must be less than one, the decimal must be moved four places to the left, and the final answer is

log 513.06 = characteristic + mantissa

$$= 2 + 0.71015$$

$$= 2.71015$$

● **PROBLEM** 785

Using a table of logarithms, determine the logarithm of 14.57.

Solution: This logarithm can be found by interpolation.

               x   log x

                1.450   .1614

.010   .007 1.457   log 1.457   d .0030

                1.460   .1644

Set up the following proportion:

$$\frac{.007}{.010} = \frac{d}{.0030}$$

Cross multiply to obtain

$$d = (.0030)\left(\frac{.007}{.010}\right)$$

$$= (3 \times 10^{-3})\left(\frac{7 \times 10^{-3}}{1 \times 10^{-2}}\right) = \frac{21 \times 10^{-6}}{1 \times 10^{-2}} = 21 \times 10^{-4}$$

$$= (21)(0.0001)$$

$$d = 0.0021$$

Hence, log 1.457 = 0.1614 + 0.0021

$$= 0.1635$$

Note $14.57 = 1.457 \times 10^{1}$. The characteristic is the exponent of 10, thus the characteristic is 1, and log 14.57 = 1 + 0.1635 = 1.1635.

● **PROBLEM** 786

Find log 2.563.

Solution: From the table,

log 2.57 = 0.4099,

log 2.56 = 0.4082

Hence we can use the following chart to find log 2.563

Number	Log of the Number
2.5700	.4099
2.5630	?
2.5600	.4082

.010   .003   .0017   x

The difference between 2.57 and 2.56 is .0100

The difference between log 2.57 and log 2.56 is .0017.

The difference between 2.560 and 2.563 is .0030.

and the difference between log 2.560 and log 2.563 is x.

We proceed with the following ratio:

$$\frac{\text{Difference between 2.563 + 2.560}}{\text{Difference between 2.570 + 2.560}}$$

$$= \frac{\text{Difference between log 2.563 + log 2.560}}{\text{Difference between log 2.570 + log 2.560}}$$

Replacing these differences with our above data we obtain the following ratio:

$$\frac{.0030}{.0100} = \frac{x}{.0017}$$

Cross multiply: $(.0100)x = (.0030)(.0017)$

$$.0100 \ x = .0000051$$

Multiplying each side by 100 (moving each decimal point 2 digits to the right: $x = .00051$

Thus the difference between log 2.560 and log 2.563 is .00051, or rounded off to the nearest ten thousandth is .0005. Therefore

log 2.560 + .0005 = log 2.563

.4082 + .0005 = log 2.563

log 2.563 = 0.4087

● **PROBLEM 787**

Find the logarithm of 0.003124.

Solution: $0.003124 = 3.124 \times 10^{-3}$ and -3, the power of 10, is the characteristic. From a table of logarithms, .4942 and .4955 are the mantissas for 3.12 and 3.130 respectively. The mantissa for 3.124 can be found by interpolation.

$$
\begin{array}{cc}
\text{N} & \text{log N}
\end{array}
$$

$$
.01 \left[.004 \left[\begin{array}{c} 3.120 \\ 3.124 \\ \\ 3.130 \end{array} \right. \qquad \left. \begin{array}{c} .4942 \\ x \\ \\ .4955 \end{array} \right] d \right] .0013
$$

Setting up the following proportion:

$$\frac{.004}{.01} = \frac{d}{.0013}.$$ Cross multiply to obtain $(.01)d = (.004)(.0013)$

$$
d = (.0013)\left(\frac{.004}{.01}\right) = \frac{\left(1.3 \times 10^{-3}\right)\left(4 \times 10^{-3}\right)}{1 \times 10^{-2}}
$$

$$
= \frac{5.2 \times 10^{-6}}{1 \times 10^{-2}}
$$

$$
= 5.2 \times 10^{[-6-(-2)]} = 5.2 \times 10^{-4}
$$

$$
= (5.2)(0.0001)
$$

$$
d = 0.00052
$$

Therefore, $x = \log 3.124 = 0.4942 + 0.00052 = 0.49472$. Hence, the mantissa for 3.124 is 0.49472. Therefore, $\log 0.003124 = -3 + 0.49472$

$$
= -2.50528
$$

or

$$
7.49472 - 10 .
$$

● **PROBLEM** 788

Find the logarithm of 3614.0.

Solution: $3{,}614 = 3.614 \times 10^3$, hence our characteristic (exponent of 10) is 3. To determine the mantissa of 3.614, since our log tables only give us values for 3.61 and 3.62, we make the following interpolation:

$$
\begin{array}{cc}
\text{N} & \text{log N}
\end{array}
$$

$$
0.01 \left[0.004 \left[\begin{array}{c} 3.61 \\ 3.614 \\ \\ 3.62 \end{array} \right. \qquad \left. \begin{array}{c} 0.5575 \\ x \\ \\ 0.5587 \end{array} \right] d \right] .0012
$$

The following proportion is set up:

$$\frac{0.004}{0.01} = \frac{d}{0.0012},$$ cross multiplying we obtain $(.01)d = (.004)(.0012)$

or

$$
d = \frac{0.004}{0.01}(0.0012) = \frac{\left(4 \times 10^{-3}\right)\left(1.2 \times 10^{-3}\right)}{1 \times 10^{-2}} = \frac{4.8 \times 10^{-6}}{1 \times 10^{-2}}
$$

$$
= 4.8 \times 10^{-6-(-2)} = (4.8) \times 10^{-4} = (4.8)(0.0001)
$$

$$
d = 0.00048
$$

Hence, $x = \log 3.614 = 0.5575 + d$

$$
= 0.5575 + 0.00048
$$

$$
= 0.5580
$$

Therefore, the mantissa for 3614 is 0.5580. The characteristic is 3. Hence, the logarithm of 3614.0 is $3 + .5580 = 3.5580$.

If log N = 8.35721 - 10, find N.

Solution: The characteristic is - 2 (i.e. 8 - 10) and
the mantissa 35721 is located between 3560 and 3579.
These mantissas correspond to the numbers 227 and 228,
respectively. Now a proportion can be set up of the
form:

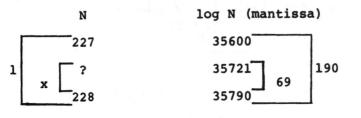

$$\frac{x}{1} = \frac{69}{190} \quad \text{or} \quad x = \frac{69}{190} = .36315 \; \tilde{\sim} \; .3632$$

Now subtract x from 228 to obtain:

    228.0000

       .3632
    _____

    227.6368

The characteristic tells us that the number must be less
than one and that there must be one zero between the
decimal point and the first nonzero number. (The characteristic is one more than the number of zeros between
the decimal point and the first nonzero digit).

Therefore the number N is: .02276368 $\tilde{\sim}$ .022764

Find N if log N = 3.6129.

Solution: As a first step, we write this number in
standard form:

$$\log N = .6129 + 3.$$

Use a table of common logarithms of numbers to find the
number whose mantissa is 6129. We must use interpolation
to do this.

Number	Mantissa

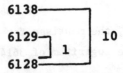

$$\frac{x}{10} = \frac{1}{10}$$

$$x = 1$$

The number is $4100 + 1 = 4101$. Since $\log N = .6129 + 3$, the characteristic of the number is $+ 3$. Recall that for a number > 1 the characteristic is positive and is one less than the number of digits before the decimal point. Thus, there are four digits before the decimal point, and the desired number is $4,101$.

● **PROBLEM** 791

Find N if $\log N = 0.8045 - 2$.

<u>Solution:</u> Note that the characteristic is -2. Therefore, the number that corresponds to the mantissa 0.8045 will be multiplied by 10^{-2}. Since the mantissa 0.8045 does not appear in a table of four place logarithms, the number that corresponds to this mantissa can be found by interpolation.

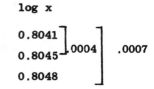

Set up the following proportion:

$$\frac{d}{.01} = \frac{.0004}{.0007}$$

Cross multiply to obtain

$$d = (.01)\left(\frac{.0004}{.0007}\right) = \left(1 \times 10^{-2}\right)\left(\frac{4 \times 10^{-4}}{7 \times 10^{-4}}\right)$$

$$= \frac{4 \times 10^{-6}}{7 \times 10^{-4}} = \frac{4}{7} \times 10^{-2}$$

$$= \left(\frac{4}{7}\right)\left(\frac{1}{100}\right)$$

$$d = \frac{1}{175} = 0.0057 \quad .$$

Thus, $y = 6.37 + d = 6.37 + .0057 = 6.3757$ and

$$N = (6.3757) \times 10^{-2}$$

$$= 6.376 \times 10^{-2} \quad \text{rounding off to 4 significant figures}$$

$$N = 0.06376$$

● **PROBLEM** 792

Find Antilog$_{10}$ 1.4850.

<u>Solution:</u> By definition, Antilog$_{10}$ $a = N$ is equivalent to $\log_{10} N = a$. Let Antilog$_{10} 1.4850 = N$. Hence, Antilog$_{10} 1.4850 = N$ is equivalent to $\log_{10} N = 1.4850$. The characteristic is 1. The mantissa is 0.4850.

Therefore, the number that corresponds to this mantissa will be multiplied by 10^1 or 10. The mantissas which appear in a table of common logarithms and are closest to the mantissa 0.4850 are 0.4843 and 0.4857. The number that corresponds to the mantissa 0.4850 will be found by interpolation.

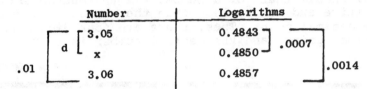

	Number	Logarithms	
	3.05	0.4843 ⌉	
d	x	0.4850 ⌋ .0007	
.01	3.06	0.4857	.0014

Set up the following proportion.

$$\frac{d}{.01} = \frac{.0007}{.0014}$$

cross-multiplying, $.0014d = (.01)(.0007)$, or $d = .01\left(\frac{.0007}{.0014}\right)$

$$= \left(1 \times 10^{-2}\right)\left(\frac{7 \times 10^{-4}}{1.4 \times 10^{-3}}\right)$$

$$= \frac{7 \times 10^{-6}}{1.4 \times 10^{-3}} = \frac{7}{1.4} \times \frac{10^{-6}}{10^{-3}} = 5 \times 10^{-6-(-3)}$$

$$= 5 \times 10^{-3} = 5 \times .001 = .005$$

Hence, $d = 0.005$

$$x = d + 3.05$$

$$= 0.005 + 3.050$$

$$= 3.055$$

Hence, $N = \text{Antilog}_{10} 1.4850 = 3.055 \times 10$

$$= 30.550$$

$$= 30.55$$

Therefore $\text{Antilog}_{10} 1.4850 = 30.55$.

● **PROBLEM** 793

Find Antilog 2.3625.

Solution: By definition, $b = \text{Antilog } a$, is equivalent to $\log b = a$. Let $N = \text{Antilog } 2.3625$. Therefore, $\log N = 2.3625$. The characteristic is 2. Hence, the number that corresponds to the mantissa 0.3625 will be multiplied by 10^2 or 100. In a table of four-place common logarithms, the mantissas 0.3617 and 0.3636 are those given that are closest to 0.3625. Therefore, since the mantissa 0.3625 does not appear in the table, the number that corresponds to this mantissa will be found through interpolation.

		Number	Log		
.01	d	2.30	0.3617 ⌉	.0008	.0019
		x	0.3625 ⌋		
		2.31	0.3636		

The following proportion is now established:

$$\frac{d}{.01} = \frac{.0008}{.0019}$$

Cross multiplying,

$$.0019d = (.01)(.0008)$$

$$d = .01\left(\frac{.0008}{.0019}\right)$$

$$= 1 \times 10^{-2}\left(\frac{8 \times 10^{-4}}{1.9 \times 10^{-3}}\right)$$

$$= \frac{8 \times 10^{-2+(-4)}}{1.9 \times 10^{-3}} = \frac{8 \times 10^{-6}}{1.9 \times 10^{-3}}$$

$$= \frac{8}{1.9} \times \frac{10^{-6}}{10^{-3}}$$

$$= 4.2 \times 10^{-6-(-3)}$$

$$= 4.2 \times 10^{-3}$$

$$d = .0042$$

Hence, x = 2.30 + 0.0042 = 2.3042 $\approx$ 2.304. Therefore, **Antilog** 2.3625 =

$$N = 2.304 \times 10^2$$

$$= 2.304 \times 100$$

$$= 230.4$$

● **PROBLEM** 794

Determine the value of x such that $10^x = 3.142$.

<u>Solution:</u> The statement $10^x = 3.142$ is equivalent by de-finition to $\log_{10} 3.142 = x$. Thus we must find log 3.142, using the following interpolation:

$$.01\left(.002\left(\frac{\begin{array}{c|c}\text{Number} & \text{Log}\\ \hline 3.140 & .4969\\ 3.142 & \\ \hline 3.150 & .4983\end{array}}{}\right) x\right).0014$$

We set up the proportion,

$$\frac{.002}{.01} = \frac{x}{.0014}$$

Cross multiply to obtain,

$$.01x = .0000028$$

$$x = .00028$$

$$x \simeq .0003$$

Thus log 3.142 = .4969 + .0003 = 0.4972
Therefore x = $\log_{10} 3.142$ = 0.4972

● **PROBLEM** 795

Find the value of $(2.154)^5$.

Solution: We will use logs, in solving the given problem.

Let $x = (2.154)^5$. Now take the log of both sides:

$\log x = \log(2.154)^5$; and now by the rule $\log a^b = b \log a$ we obtain

$$\log x = 5 \log(2.154).$$

Log(2.154) is now found by interpolation, using a table of common logs. Notice 2.154 occurs between 2.150 and 2.160 which have recorded logs.

$$.010 \left[.006 \left[\begin{array}{l} N = 2.150 \\ N = 2.154 \\ N = 2.160 \end{array} \right. \right. \qquad \begin{array}{l} \log x = .3324 \\ ? \\ \log x = .3345 \end{array} \left. \right] d \right] .0021$$

Now set up the proportion and use scientific notation.

$$\frac{.006}{.010} = \frac{d}{.0021}$$

$$\frac{6 \times 10^{-3}}{1 \times 10^{-2}} = \frac{d}{2.1 \times 10^{-3}}$$

$$\left(2.1 \times 10^{-3}\right)\left(\frac{6 \times 10^{-3}}{1 \times 10^{-2}}\right) = \left(\cancel{2.1 \times 10^{-3}}\right)\left(\frac{d}{\cancel{2.1 \times 10^{-3}}}\right)$$

$$\frac{12.6 \times 10^{-6}}{1 \times 10^{-2}} = d$$

$$12.6 \times 10^{-4} = d$$

$$(12.6)(0.0001) = d$$

$$0.00126 = d$$

or

$$d \approx 0.0013$$

Hence, $\log 2.154 = .3345 - d = .3345 - .0013$

$$\log 2.154 = .3332$$

Therefore,

$$\log x = 5 \log(2.154)$$

$$= 5(.3332)$$

$$\log x = 1.6660$$

The characteristic is 1. Therefore, the number that corresponds to the mantissa 0.6660 will be multiplied by 10^1 or 10. Using a table of common logarithms, the number that approximately corresponds to the mantissa is 4.63. Then,

$$x = (4.63)10$$

or $(2.154)^5 = x = 46.3$

Hence, $(2.154)^5 = 46.3$

● **PROBLEM** 796

Find $\sqrt[4]{36.91}$

Solution: We can use the rule for the log of a number raised to a power to find the solution.

$$\log a^b = b \log a$$

578

or $\log (36.91)^{\frac{1}{4}} = (\frac{1}{4}) \log 36.91$

log 36.91 lies between log 36.90 and log 37.00
Therefore, we set up the proportion:

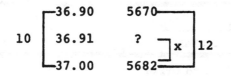

$\frac{9}{10} = \frac{x}{12}$ or cross multiply to obtain

$(.9)(12)= x,$ $x = 10.8$

Now subtract 10.8 from 5682

$$5682$$

$$- \quad 10.8$$

5671.2. Since the mantissa must be less than one, move the decimal pint four places to the left i.e. (.56712). Now that the mantissa has been determined the characteristic is determined by examining the number of decimal places to the left of the decimal point of the original number. The characteristic is one less than this number. In this case the characteristic is 1. This process is used for characteristic determination because the number is greater than one.

$(\frac{1}{4}) \log 36.91 = \frac{1}{4} (1.56712) = .39178$

Now the answer is obtained by finding the number whose log is .39178. Again consult the table of common logarithms.

$$180 \begin{bmatrix} 92 \begin{bmatrix} .39090 = \log 2.460 \\ .39178 \\ .39270 = \log 2.470 \end{bmatrix} x \end{bmatrix} 10$$

Now set up a proportion.

$\frac{92}{180} = \frac{23}{45} = \frac{x}{10}$ or cross multiply to obtain

$45x = 230$ or $x = \frac{230}{45}$,

$x = 5.1$

Now subtract 5.1 from 2470 to obtain

$$\begin{array}{r} 2470.0 \\ - \quad 5.1 \\ \hline 2464.9 \end{array}$$

Now convert to the answer by placing the decimal point between the two and four.(Note: from the above interpolation we knew the answer must be the integer 2 plus a rational part.)

$$\sqrt[4]{36.91} = 2.4649$$

Calculate the number $\dfrac{3480 \times 1265}{.00143}$.

Solution: One way to proceed is to write

$$\frac{3480 \times 1265}{.00143} = \frac{3.48 \times 10^3 \times 1.265 \times 10^3}{1.43 \times 10^{-3}} = \frac{3.48 \times 1.265 \times 10^6}{1.43 \times 10^{-3}} ,$$

and since $\dfrac{a^x}{a^y} = a^{x-y}$, we have

$$\frac{3.48 \times 1.265 \times 10^9}{1.43} .$$

Then calculate the number

$$N = \frac{3.48 \times 1.265}{1.43} ,$$

where $\dfrac{3480 \times 1265}{.00143} = N \times 10^9$. Taking logarithms of both sides of the equation:

$$\log N = \log\left(\frac{3.48 \times 1.265}{1.43}\right)$$

Use the rules of logarithms,

$$\log (ab) = \log a + \log b$$
$$\log c/d = \log c - \log d$$

then

$$\log N = \log(3.48 \times 1.265) - \log 1.43$$
$$\log N = \log 3.48 + \log 1.265 - \log 1.43$$

To find log 3.48 and log 1.43 we use a table of logarithms of numbers from the table:

$$\log 3.48 = .5416$$
$$\log 1.43 = .1553$$

To obtain log 1.265 we interpolate:

$$
.01 \left[.005 \left[\begin{array}{l} 1.26 \\ 1.265 \\ 1.27 \end{array} \right. \right.
\qquad
\left. \begin{array}{l} .1004 \\ ? \\ .1038 \end{array} \right] y \right] .0034
$$

$$\frac{.005}{.01} = \frac{y}{.0034}$$

$$\frac{.5}{1} = \frac{y}{.0034}$$

$$y = (.0034)(.5) = .00170$$

$$\log 1.265 = .1004 + .0017 = .1021$$

Therefore, substituting:

$$\log N = .5416 + .1021 - .1553 = .4884.$$

Therefore,

$$\frac{3480 \times 1265}{.00143} = 3.079 \times 10^9$$

We must now find N from $\log N$. This is done by interpolating:

$$\underline{N} \qquad\qquad\qquad \underline{\log N}$$

$$.01 \left[y \left[\begin{array}{c} 3.07 \\ ? \\ 3.08 \end{array} \right. \qquad\qquad \left. \begin{array}{c} .4871 \\ .4884 \\ .4886 \end{array} \right] .0013 \right] .0015$$

$$\frac{y}{.01} = \frac{.0013}{.0015}$$

$$\frac{y}{.01} = \frac{13}{15}$$

$$15y = .13$$

$$y = \frac{.13}{15} = .00866$$

$$N = 3.07 + .00866 = 3.07866 \approx 3.079$$

● **PROBLEM** 798

Find $\dfrac{0.3612}{456.53}$, using logarithms.

Solution: Let $x = \dfrac{0.3612}{456.53}$

We solve for x by taking the logarithm of both sides of this equation. Hence

$$\log x = \log \left[\frac{0.3612}{456.53} \right] = \log .3612 - \log 456.53,$$

from the rule log a/b = log a - log b.

We find these logarithms by interpolation.

$$\underline{x} \qquad \underline{\log x}$$

$$.001 \left[.0002 \left[\begin{array}{cc} .361 & 9.5575 - 10 \\ .3612 & ? \\ .362 & 9.5587 - 10 \end{array} \right] y \right] .0012$$

Interpolation means taking the proportions indicated above.

$$\frac{.0002}{.001} = \frac{y}{.0012}$$

We can simplify the left side by multiplying numerator and

581

denominator by 1000 (moving the decimal 3 places to the right). Thus,

$$\frac{.2}{1} = \frac{y}{.0012}$$

$y = (.2)(.0012)$, by cross multiplying. Therefore, $y = .00024$.

$$\log .3612 = (9.5575 - 10) + .00024$$

$$= 9.55774 - 10$$

We must now find log 456.53. Since 456.53 > 1, and for a number greater than 1, the characteristic is positive and is one less than the number of digits before the decimal point, the characteristic is 2. By use of the log table we find the mantissa for 456 is .6590, and for 457 is .6599. Hence, by interpolation,

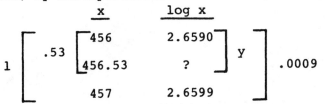

$$\frac{.53}{1} = \frac{y}{.0009}$$

$$y = (.53)(.0009)$$

$$y = .00047$$

$$\log 456.53 = 2.6590 + .00047 = 2.65947$$

Thus, $\log x = (9.55774 - 10) - (2.65947)$

$$\log x = 6.89827 - 10$$

We again use interpolation to obtain x. Observe that the characteristic is - 4. Thus, the desired number will be less than 1, and have 3 zeros following the decimal (by the rule that for a number less than 1 the characteristic is negative and is one more than the number of zeros following the decimal point). Thus, by interpolation:

$$.000001 \left[y \left[\begin{array}{cc} .000791 & 6.8982 - 10 \\ ? & 6.89827 - 10 \end{array} \right] .00007 \right] .0005$$
$$.000792 \quad 6.8987 - 10$$

$$\frac{y}{.000001} = \frac{.00007}{.0005}$$

$$\frac{y}{.000001} = \frac{.7}{5}$$

$$5y = .0000007$$

$$y = .00000014$$

Therefore, $x = .000791 + .00000014$

$$x = .00079114$$

● **PROBLEM** 799

Find $\text{Antilog}_{10} 0.5579 - 1$.

<u>Solution:</u> By definition, $\text{Antilog}_{10} a = N$ is equivalent to $\log_{10} N = a$. Let $N = \text{Antilog}_{10} 0.5579 - 1$. Therefore, $\text{Antilog}_{10} 0.5579 - 1 = N$ is equivalent to $\log_{10} N = 0.5579 - 1$. The characteristic is -1. Therefore, the number that corresponds to the mantissa 0.5579 will be multiplied by 10^{-1}. The number that corresponds to the mantissa 0.5579 must be found through interpolation since only the mantissas 0.5575 and 0.5587 appear in a table of four-place logarithms, base 10.

	Number		log	
.01	d [3.61		0.5575] .0004	.0012
	x		0.5579	
	3.62		0.5587	

Set up the following proportion.

$$\frac{d}{.01} = \frac{.0004}{.0012}$$

$$d = .01\left(\frac{.0004}{.0012}\right)$$

$$= (1 \times 10^{-2})\left(\frac{4 \times 10^{-4}}{1.2 \times 10^{-3}}\right)$$

$$= \frac{4 \times 10^{-6}}{1.2 \times 10^{-3}}$$

$$= 3.3 \times 10^{-3}$$

$$= 3.3(.001)$$

$$d = 0.0033$$

Therefore, $x = 3.6100 + d$

$$= 3.6100 + 0.0033$$

$$= 3.6133$$

Hence, $\text{Antilog}_{10} 0.5579 - 1 = N = 10^{-1}(3.6133)$

$$= 0.36133.$$

$$N \approx 0.3613.$$

● **PROBLEM** 800

Find the product 5.06×71.32 by using logs and antilogs.

<u>Solution:</u> By definition, $\text{antilog } a = N$ is equivalent to $\log N = a$. Now substitute the value for a in the antilog expression. Therefore, $\text{antilog } a = N$ becomes $\text{antilog}(\log N) = N$. Let $N = 5.06 \times 71.32$; then, $5.06 \times 71.32 = \text{antilog}[\log(5.06 \times 71.32)]$. Since $5.06 \times 71.32 = 5.06 \times 7.132 \times 10$, we write:

$$5.06 \times 71.32 = \text{antilog}[\log(5.06 \times 7.132 \times 10)] \qquad (1)$$

Evaluating the expression in the brackets:

$$\log(5.06 \times 7.132 \times 10) = \log 5.06 + \log 7.132 + \log 10.$$

This is true because of the following law of exponents:

$$\log abc = \log a + \log b + \log c.$$

Using a table of common logarithms to find the value of log 5.06 and noting that $\log 10 = 1$,

$$\log(5.06 \times 7.132 \times 10) = 0.7042 + (\log 7.132) + 1$$

$$= 1.7042 + \log 7.132 \qquad (2)$$

We now evaluate log 7.132. The numbers that appear in a table of common logarithms which are closest to the number 7.132 are 7.13 and 7.14. The mantissa that corresponds to the number 7.132 will be found by interpolation.

Now, setting up the following proportion:

$$\frac{d}{.0006} = \frac{.002}{.01}$$

Cross-multiplying, $\quad d = .0006\left(\frac{.002}{.01}\right)$

$$= \left(6 \times 10^{-4}\right)\left(\frac{2 \times 10^{-3}}{1 \times 10^{-2}}\right) = \frac{12 \times 10^{-4+(-3)}}{1 \times 10^{-2}}$$

$$= \frac{12 \times 10^{-7}}{1 \times 10^{-2}} = 12 \times 10^{-7-(-2)}$$

$$= 12 \times 10^{-5}$$

$$= \left(1.2 \times 10^{1}\right) \times 10^{-5} = 1.2 \times 10^{1+(-5)}$$

$$= 1.2 \times 10^{-4}$$

$$= 1.2 \times 0.0001$$

$$= 0.00012$$

$$\approx 0.0001$$

Hence, $\log 7.132 = x = 0.8531 + 0.0001$

$$= 0.8532.$$

Therefore, equation (2) becomes:

$$\log(5.06 \times 7.132 \times 10) = 1.7042 + 0.8532$$

$$= 2.5574$$

Equation (1) becomes:

$$5.06 \times 71.32 = \text{antilog}[2.5574] = M \qquad (3)$$

By definition, antilog[2.5574] is equivalent to log M = 2.5574. The characteristic is 2. The mantissa is 0.5574. The number that corresponds to this mantissa will be multiplied to 10^{2} or 100. The mantissas which appear in a table of logarithms and are closest to the mantissa 0.5574 are 0.5563 and 0.5575. The number that corresponds to the mantissa 0.5574 can be found by interpolation.

$$.01 \left[d \left[\begin{array}{l} 3.60 \\ x \\ 3.61 \end{array} \right. \right. \quad \left. \left. \begin{array}{l} 0.5563 \\ 0.5574 \\ 0.5575 \end{array} \right] .0011 \right] \quad .0012$$

Now setting up the following proportion:

$$\frac{d}{.01} = \frac{.0011}{.0012}$$

Cross-multiplying, $\quad d = .01\left(\frac{.0011}{.0012}\right)$

$$= \left(1 \times 10^{-2}\right)\left(\frac{1.1 \times 10^{-3}}{1.2 \times 10^{-3}}\right) = \frac{1.1 \times 10^{-2+(-3)}}{1.2 \times 10^{-3}}$$

$$= \frac{1.1 \times 10^{-5}}{1.2 \times 10^{-3}} = \frac{1.1}{1.2} \times 10^{-5-(-3)}$$

$$= 0.917 \times 10^{-2}$$

$$= 0.917 \times 0.01$$

$$= 0.00917$$

$$\approx 0.009$$

Hence, $\quad x = 3.60 + 0.009$

$$= 3.600 + 0.009$$

$$= 3.609$$

Therefore, $M = 3.609 \times 10^{2} = 360.9$. Equation (3) becomes:

$$5.06 \times 71.32 = M = 360.9$$

● **PROBLEM** 801

Calculate $\frac{50.73}{2.42}$, using logs and antilogs.

Solution: By definition, antilog $a = N$ is equivalent to $\log N = a$. Now substitute the value for a in the antilog expression. Therefore, antilog $a = N$ becomes antilog($\log N$) = N. Let

$$N = \frac{50.73}{2.42} ;$$

then

$$\frac{50.73}{2.42} = \text{antilog}\left[\log\frac{50.73}{2.42}\right].$$

Since

$$\frac{50.73}{2.42} = \frac{5.073 \times 10}{2.42} ,$$

we write:

$$\frac{50.73}{2.42} = \text{antilog}\left[\log\frac{5.073 \times 10}{2.42}\right] \tag{1}$$

Evaluating the expression in the brackets:

$$\log\frac{5.073 \times 10}{2.42} = (\log 5.073 + \log 10) - \log 2.42.$$

This statement is true by the two following laws of exponents:

$$\log ab = \log a + \log b$$

$$\log \frac{c}{d} = \log c - \log d$$

Noting that $\log 10 = 1$, and using a table of common logarithms to obtain $\log 2.42$, we obtain:

$$\log \frac{5.073 \times 10}{2.42} = (\log 5.073+1) - 0.3838,$$

or

$$\log \frac{5.073 \times 10}{2.42} = (\log 5.073)+ 1 - 0.3838 \qquad (2)$$

Now, to find $\log 5.073$, we use interpolation.

Number	Logarithm
5.07	0.7050
5.073	x
5.08	0.7059

.01   .003   d   .0009

Set up the following proportion:

$$\frac{d}{.0009} = \frac{.003}{.01}$$

Cross-multiplying, $d = .0009\left(\frac{.003}{.01}\right)$

$$= \left(9 \times 10^{-4}\right)\left(\frac{3 \times 10^{-3}}{1 \times 10^{-2}}\right) = \frac{27 \times 10^{-4+(-3)}}{1 \times 10^{-2}}$$

$$= \frac{27 \times 10^{-7}}{1 \times 10^{-2}} = \frac{27}{1} \times 10^{-7-(-2)} = 27 \times 10^{-7+2}$$

$$= 27 \times 10^{-5}$$

$$= 2.7 \times 10^{1} \times 10^{-5}$$

$$= 2.7 \times 10^{-4}$$

$$d = 0.00027 \approx 0.0003$$

Hence, $x = 0.7050 + d = 0.7050 + 0.0003$. Therefore,

$$\log 5.073 = x = 0.7053.$$

Rewriting equation (2):

$$\log \frac{5.073 \times 10}{2.42} = 0.7053 + 1 - 0.3838$$

$$= 1.7053 - 0.3838$$

$$= 1.3215$$

Therefore, equation (1) becomes:

$$\frac{50.73}{2.42} = \text{antilog}[1.3215] = M \qquad (3)$$

By definition, $\text{antilog}[1.3215]$ is equivalent to $\log M = 1.3215$. The characteristic is 1. The mantissa is 0.3215. The number that corresponds to this mantissa will be multiplied by 10^1 or 10. The mantissas which appear in a log table and are closest to the mantissa 0.3215 are 0.3201 and 0.3222. The number that corresponds to the mantissa 0.3215 will be found by interpolation.

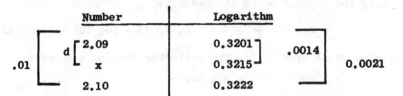

Number	Logarithm
2.09	0.3201
x	0.3215
2.10	0.3222

.01   d   .0014   0.0021

Now, set up the following proportion.

$$\frac{d}{.01} = \frac{.0014}{.0021}$$

Cross-multiplying, $d = (.01)\left(\dfrac{.0014}{.0021}\right)$

$$= \left(1 \times 10^{-2}\right)\left(\dfrac{1.4 \times 10^{-3}}{2.1 \times 10^{-3}}\right) = \dfrac{1.4 \times 10^{-2-3}}{2.1 \times 10^{-3}}$$

$$= \dfrac{1.4 \times 10^{-5}}{2.1 \times 10^{-3}} = \dfrac{1.4}{2.1} \times 10^{-5-(-3)}$$

$$= 0.67 \times 10^{-2}$$

$$= 0.67 \times 0.01$$

$$d = 0.0067 \approx 0.007$$

Hence, x = 2.09 + d

$$= 2.09 + 0.007$$

$$= 2.090 + 0.007$$

$$= 2.097$$

Therefore, M = 2.097 $\times$ 10 = 20.97. Hence, equation (3) becomes:

$$\dfrac{50.73}{2.42} = \text{antilog}[1.3215]$$

$$= M$$

$$= 20.97 \ .$$

FUNCTIONS AND EQUATIONS

● **PROBLEM** 802

Write the following equations in logarithmic form.
(a) $3^4 = 81$ (b) $10^0 = 1$ (c) $M^k = 5$ (d) $5^k = M$

<u>Solution:</u> The expression $b^y = x$ is equivalent to the logarithmic expression $\log_b x = y$. Hence,

a) $3^4 = 81$ is equivalent to the logarithmic expression
    $\log_3 81 = 4$

b) $10^0 = 1$ is equivalent to the logarithmic expression
    $\log_{10} 1 = 0$

c) $M^k = 5$ is equivalent to the logarithmic expression
    $\log_M 5 = k$

d) $5^k = M$ is equivalent to the logarithmic expression
    $\log_5 M = k$.

● **PROBLEM** 803

Show that $\log_{10} 10{,}000^{\frac{1}{2}} = \frac{1}{2} \log_{10} 10{,}000$.

<u>Solution:</u> We will evaluate both sides of this equation, and show them to be equal to the same quantity:

$$\underline{\log_{10} 10{,}000^{\frac{1}{2}}} \qquad\qquad\qquad \underline{\tfrac{1}{2} \log_{10} 10{,}000}$$

Since $x^{\frac{1}{2}} = \sqrt{x}$

$\log_{10} 10,000^{\frac{1}{2}} = \log_{10} \sqrt{10,000}$

$\hspace{3.5cm} = \log_{10} 100$

$\hspace{3.5cm} = 2$

$\frac{1}{2} \log_{10} 10,000 = \frac{1}{2} \log_{10} 10^4$

$\hspace{3.5cm} = \frac{1}{2} \cdot 4$

$\hspace{3.5cm} = 2$

Thus $\log_{10} 10,000^{\frac{1}{2}} = 2 = \frac{1}{2} \log_{10} 10,000$, and

$\log_{10} 10,000^{1/2} = \frac{1}{2} \log_{10} 10,000$.

● **PROBLEM** 804

The graph of an exponential function f contains the point (2,9). What is the base of f?

Solution: Since f is an exponential function, we know that $f(x) = b^x$, where b, the base, is a positive number that we are to determine. An exponential function f may also be written as $y = f(x) = b^x$. Since the exponential function f contains the point (2,9),

$$9 = f(2) = b^2 \quad \text{or} \quad b^2 = 9$$
$$\sqrt{b^2} = \sqrt{9}$$
$$b = 3 .$$

Note that only the positive square root was taken, since for the base b, a positive number, is desired.

● **PROBLEM** 805

If f is the logarithmic function with base 4, find f(4), $f\left(\frac{1}{4}\right)$, and f(8).

Solution: Since f is the logarithmic function with base 4, then $y = f(x) = \log_4 x$. The values $f(4)$, $f\left(\frac{1}{4}\right)$, and $f(8)$ can be found by replacing x by 4, $\frac{1}{4}$, and 8 in the logarithmic function $y = f(x) = \log_4 x$. Hence, $f(4) = \log_4$. Let $N_1 = f(4) = \log_4 4$. By definition, $\log_x a = N$ is equivalent to $x^N = a$. Therefore, $N_1 = \log_4 4$ is equivalent to $4^{N_1} = 4$. Since $4^1 = 4$, $N_1 = 1$. Then , $N_1 = 1 = f(4)$.

For the second value $f\left(\frac{1}{4}\right)$, $f\left(\frac{1}{4}\right) = \log_4 \frac{1}{4}$. Let $N_2 = f\left(\frac{1}{4}\right) = \log_4 \frac{1}{4}$. Hence, $N_2 = \log_4 \frac{1}{4}$ is equivalent to $4^{N_2} = \frac{1}{4}$. Since $4^{-1} = \frac{1}{4^1} = \frac{1}{4}$, $N_2 = -1$. Then, $N_2 = -1 = f\left(\frac{1}{4}\right)$.

For the third value $f(8)$, $f(8) = \log_4 8$. Let $N_3 = f(8) = \log_4 8$. Hence, $N_3 = \log_4 8$ is equivalent to $4^{N_3} = 8$. Since $4 = 2^2$, $N_3 = \log_4 8$ is equivalent to $\left(2^2\right)^{N_3} = 8$ or

$2^{2N_3} = 8$. Since $2^3 = 8$, $2N_3 = 3$. Dividing both sides of the equation $2N_3 = 3$ by 2:

$$\frac{2N_3}{2} = \frac{3}{2} \quad \text{or} \quad N_3 = \frac{3}{2}.$$

Then, $N_3 = \frac{3}{2} = f(8)$.

● **PROBLEM** 806

Solve the equation $\log_3\left(x^2 - 8x\right) = 2$.

Solution: The expression $\log_b a = y$ is equivalent to $b^y = a$. Hence, $\log_3\left(x^2 - 8x\right) = 2$ is equivalent to $3^2 = x^2 - 8x$. Therefore,

$$3^2 = x^2 - 8x$$
or
$$9 = x^2 - 8x .$$

Subtract 9 both sides of this equation:

$$9 - 9 = x^2 - 8x - 9$$
$$0 = x^2 - 8x - 9.$$

Factoring this equation:

$$0 = (x - 9)(x + 1).$$

Whenever the product $ab = 0$ where a and b are any two numbers, either $a = 0$ or $b = 0$. Hence, either

$$x - 9 = 0 \quad \text{or} \quad x + 1 = 0$$
$$x = 9 \quad \text{or} \quad x = -1.$$

● **PROBLEM** 807

Solve the equation $2^x = 7$ for x.

Solution: By taking logarithms of both sides of the equation, we obtain the equation

$$\log 2^x = \log 7.$$

Using the rule $\log M^x = x \log M$, we obtain:

$$x \log 2 = \log 7$$

From a table on common logarithms the $\log 2 = .3010$ and the $\log 7 = .8451$. So our equation becomes $.3010x = .8451$. Hence,

$$x = \frac{.8451}{.3010} = 2.808.$$

Remark 1. Since $2^2 = 4$ and $2^3 = 8$, it should be obvious at the start that the solution of the equation $2^x = 7$ is a number between 2 and 3.

Remark 2. Since x log 2 = log 7, it follows that x = log 7/log 2. It should be emphasized that the expression log 7/log 2 is a quotient. We do not evaluate this quotient by looking up log 2 and log 7 in the table and subtracting; we look up the two numbers and divide. We can divide with the aid of logarithms, but it still will be division.

● **PROBLEM 808**

Solve the equation $2^x = 3^{x+1}$ for x.

Solution: We take logarithms of both sides of the equation to obtain the equation:

$$\log \left(2^x\right) = \log \left(3^{x+1}\right)$$

Using the rule $\log_b M^x = x \log_b M$, we obtain:

$$x \log 2 = (x + 1) \log 3.$$

Hence, $x \log 2 = x \log 3 + \log 3$, by distributing log 3; and

$$x \log 2 - x \log 3 = \log 3,$$

or in other words $x(\log 2 - \log 3) = \log 3$

$$x = \frac{\log 3}{\log 2 - \log 3} .$$

Using a table of common logarithms we obtain:

$$x = \frac{.4771}{.3010 - .4771} = \frac{.4771}{-.1761} = -2.709.$$

● **PROBLEM 809**

Solve for x in the equation $7^{2x-1} - 5^{3x} = 0$.

Solution: Writing the equation as $7^{2x-1} = 5^{3x}$, and equating logarithms of both members, we have

$$\log 7^{2x-1} = \log 5^{3x}$$

Recall $\log x^y = y \log x$, thus,

$$(2x - 1)\log 7 = 3x \log 5$$

Looking up log 7 and log 5 in our log table and substituting,

$$(2x - 1)(0.8451) = 3x(0.6990).$$

Hence

$$1.6902x - 0.8451 = 2.097x, \quad 0.4068x = -0.8451$$

and

$$x = -2.077.$$

● **PROBLEM 810**

Solve the equation

$$2^{3x} = 4^{x+1} .$$

Solution: This is an exponential equation, an equation involving one or more unknowns in an exponent. We solve it by logarithms.

$$2^{3x} = 4^{x+1}$$

Taking logarithms of both sides:

$$\log\left(2^{3x}\right) = \log\left(4^{x+1}\right)$$

Apply $\log p^q = q \log p$: $\quad 3x \log 2 = (x+1)\log 4$

Now $\log 4 = \log 2^2 = 2 \log 2$, and consequently

$$3x \log 2 = (x+1)\log 2^2$$

$$3x \log 2 = 2(x+1)\log 2,$$

Divide by log 2: $\quad 3x = 2x + 2,$

Subtract 2x from both sides:

$$x = 2.$$

The solution can be verified by substituting the value $x = 2$ into the original equation, $2^{3x} = 4^{x+1}$. Thus,

$$2^{3\cdot 2} \overset{?}{=} 4^{2+1}$$

$$2^6 = 4^3$$

$$2^6 = \left(2^2\right)^3$$

$$2^6 = 2^6$$

Thus, the solution is $x = 2$.

● **PROBLEM** 811

Solve $5^{2x} = 7^{x+1} .$

Solution: If we take the common logarithm of each member of the given equation, we have

$$\log 5^{2x} = \log 7^{x+1}$$

Recall $\log x^y = y \log x$. Thus $\log 5^{2x} = 2x \log 5$ and $\log 7^{x+1} = (x+1) \log 7$. Making these substitutions we obtain,

$$2x \log 5 = (x + 1)\log 7$$

Distribute, $\quad 2x \log 5 = x \log 7 + \log 7$

Add (-x log 7) to both sides,

$$2x \log 5 - x \log 7 = \log 7$$

Factor x from the left member,

$$x(2 \log 5 - \log 7) = \log 7$$

Since $y \log x = \log x^y$

$$2 \log 5 = \log 5^2 = \log 25$$

Thus replacing 2 log 5 by log 25 we obtain,

$$x(\log 25 - \log 7) = \log 7$$

Divide both members by log 25 - log 7,

$$x = \frac{\log 7}{\log 25 - \log 7}$$

Looking up log 7 in a log table we observe log 7 = .8451. Since

$$25 = 2.5 \times 10^1$$

$$
\begin{aligned}
\log 25 &= \log\left(2.5 \times 10^1\right) \\
&= \log 2.5 + \log 10^1 \\
&= \log 2.5 + 1 \\
&= .3979 + 1 \\
&= 1.3979
\end{aligned}
$$

Substituting these values for our logarithms,

$$= \frac{.8451}{1.3979 - .8451}$$

$$= \frac{.8451}{.5528} = 1.529$$

● **PROBLEM** 812

Solve the "exponential" equation: $2^{0.4x} = 7$.

Solution: Taking the log of both sides of the given equation,

$$\log 2^{0.4x} = \log 7.$$

Since $\log_b y^a = a \log_b y$, $\log 2^{0.4x} = 0.4x \log 2$.

Thus $\qquad 0.4x \log 2 = \log 7$

or, $\qquad\qquad x = \dfrac{\log 7}{0.4 \log 2}$

In a log table we find $\log 7 = .8451$ and $\log 2 = .3010$. Thus

$$x = \frac{.8451}{(0.4)(.3010)} = \frac{.8451}{.1204} = 7.02 .$$

Or, if we wish, we may complete the solution by using logarithms again:

$$\log 0.8451 = \log 8.451 \times 10^{-1} = .9269 - 1$$
$$\log 0.1204 = \log 1.204 \times 10^{-1} = .0806 - 1$$

Since $\log_b \dfrac{y}{z} = \log_b y - \log_b z$,

$$\log x = \log \frac{.8451}{.1204} = \log .8451 - \log .1204$$

$$= (.9269 - 1) - (.0806 - 1) .$$

Thus $\qquad\qquad \log x = .8463$

We look up the mantissa, 0.8463, in a table of Common Logarithms and find its corresponding number to be 7020. We adjust the decimal point by noting the characteristic 0 of 0.8463 is one less than the number of digits to the left of the decimal point of the number we seek. In this case, therefore, there should be one digit to the left of the decimal point. Hence,

$$x = 7.020 .$$

Solve $8x^{\frac{3}{2n}} - 8x^{-\frac{3}{2n}} = 63$.

Solution: Multiply by $x^{\frac{3}{2n}}$ and transpose; thus

$$8x^{\frac{3}{2n}}\left(x^{\frac{3}{2n}}\right) - 8x^{-\frac{3}{2n}}\left(x^{\frac{3}{2n}}\right) = 63x^{\frac{3}{2n}}$$

$$8x^{\frac{3}{2n}+\frac{3}{2n}} - 8x^{-\frac{3}{2n}+\frac{3}{2n}} = 63x^{\frac{3}{2n}}$$

$$8x^{\frac{6}{2n}} - 8x^{0} = 63x^{\frac{3}{n}}$$

$$8x^{\frac{3}{n}} - 8\cdot 1 - 63x^{\frac{3}{2n}} = 0$$

$$8x^{\frac{3}{n}} - 63x^{\frac{3}{2n}} - 8 = 0.$$

Factor, $\left(x^{\frac{3}{2n}} - 8\right)\left(8x^{\frac{3}{2n}} + 1\right) = 0.$

Whenever a product of two numbers $ab = 0$ either $a = 0$ or $b = 0$. Hence,

$$x^{\frac{3}{2n}} - 8 = 0 \qquad \text{or} \qquad 8x^{\frac{3}{2n}} + 1 = 0$$

$$8x^{\frac{3}{2n}} = -1$$

$$x^{\frac{3}{2n}} = 8 \qquad\qquad x^{\frac{3}{2n}} = -\frac{1}{8}$$

$$x^{\frac{3}{2n}\cdot\frac{2n}{3}} = 8^{\frac{2n}{3}} \qquad\qquad x^{\frac{3}{2n}\cdot\frac{2n}{3}} = \left(\frac{-1}{8}\right)^{\frac{2n}{3}}$$

$$x = \left(2^3\right)^{\frac{2n}{3}} \qquad \text{or} \qquad x = \left(\frac{-1}{2^3}\right)^{\frac{2n}{3}}$$

$$x = \frac{(-1)^{\frac{2n}{3}}}{\left(2^3\right)^{\frac{2n}{3}}}$$

Note that $(-1)^{\frac{2n}{3}} = \sqrt[3]{(-1)^{2n}} = \sqrt[3]{1} = 1$

and $\left(2^3\right)^{\frac{2n}{3}} = \left[2^{3\left(\frac{1}{3}\right)}\right]^{2n} = 2^{2n}$. Therefore, $x = 2^{2n}$ or $x = \frac{1}{2^{2n}}$.

Express y in terms of x if

$$\log_b y = 2x + \log_b x .$$

<u>Solution:</u> Transposing $\log_b x$, we have

$$\log_b y - \log_b x = 2x,$$

A property of logarithms is that the logarithm of the quotient of two positive numbers S and T is equal to the difference of the logarithms of the numbers; that is,

$$\log_b \frac{S}{T} = \log_b S - \log_b T .$$

Therefore,

$$\log_b \frac{y}{x} = 2x .$$

Now use the definition of logarithm: The logarithm of N to the base b is $x = \log_b N$; and $b^x = N$ is an equivalent statement. Then,

$$2x = \log_b \frac{y}{x} \text{ is equivalent to}$$

$$b^{2x} = \frac{y}{x}$$

Solving for y we obtain:

$$y = x \cdot b^{2x}$$

Find x from the equation $a^x \cdot c^{-2x} = b^{3x+1}$.

<u>Solution:</u> An exponential equation is an equation involving one or more unknowns in an exponent. This can be solved by means of logarithms.

$$\log\left(a^x \cdot c^{-2x}\right) = \log b^{3x+1} .$$

Apply the following properties of logarithms. If P and Q are positive numbers, then:

(a) $\log_b (P \cdot Q) = \log_b P + \log_b Q$

(b) $\log_b\left(P^n\right) = n \log_b P$

Then, in this problem, we have:

$$\log a^x + \log c^{-2x} = \log b^{3x+1}$$

$$x \log a - 2x \log c = (3x + 1) \log b$$

$$x \log a - 2x \log c = 3x \log b + \log b$$

$$x \log a - 2x \log c - 3x \log b = \log b$$

$$x (\log a - 2 \log c - 3 \log b) = \log b$$

Hence,

$$x = \frac{\log b}{\log a - 2 \log c - 3 \log b} .$$

Solve the equation $2 \log x - \log 10x = 0$.

<u>Solution</u>: We can use a fundamental property of logarithms to simplify the left-hand side of this equation.

The logarithm of the product of two or more positive numbers is equal to the sum of the logarithms of the several numbers. If P, Q, and R are positive numbers, then log (P · Q · R) = log P + log Q + log R.

$$2 \log x - \log 10x = 2 \log x - (\log 10 + \log x)$$
$$= 2 \log x - \log 10 - \log x$$
$$= \log x - \log 10.$$

But log 10 means that base 10 raised to what power = 10, or $10^? = 10$; and $10^1 = 10$. Thus, log 10 = 1, and the equation becomes: log x - 1 = 0.

Rewriting this equation:

log x - 1 = 0

log x = 1

Since the problem is in base 10, log x = 1 can be rewritten as,

$10^1 = x$. Thus x = 10.

● **PROBLEM** 817

Solve $\log(40x - 1) - \log(x - 1) = 3$.

<u>Solution</u>: By the law of the logarithm of a quotient of two numbers which states that $\log \frac{a}{b} = \log a - \log b$,

$\log(40x - 1) - \log(x - 1) = \log \frac{40x - 1}{x - 1} = 3$. Hence,

$\log_{10} \frac{40x - 1}{x - 1} = 3$. By the definition of a logarithm, if

$\log_b N = x$, then $b^x = N$. Therefore, $\log_{10} \frac{40x - 1}{x - 1} = 3$

means $10^3 = \frac{40x - 1}{x - 1}$. Thus $1000 = \frac{40x - 1}{x - 1}$. Multiply both sides by (x - 1):

$$(x - 1)1000 = (x - 1)\left(\frac{40x - 1}{x - 1}\right)$$

$$(x - 1)1000 = 40x - 1.$$

Distributing, 1000x - 1000 = 40x -1. Subtract 40x from both sides to obtain:

$$1000x - 1000 - 40x = 40x - 1 - 40x.$$

Combining terms, 960x - 1000 = -1. Add 1000 to both sides to obtain:

$$960x - 1000 + 1000 = -1 + 1000.$$

Combining terms, $960x = 999$. Divide both sides by 960:

$$\frac{9\cancel{6}0x}{96\cancel{0}} = \frac{999}{960}$$

$$x = \frac{999}{960} = \frac{333}{320}.$$

● PROBLEM 818

Solve $\log_2(x - 1) + \log_2(x + 1) = 3$.

Solution: Applying a property of logarithms, $\log_b x + \log_b y = \log_b xy$, to

$$\log_2(x - 1) + \log_2(x + 1) = 3$$

we get $\log_2[(x - 1)(x + 1)] = 3$. $\log_b x = y$ is equivalent to $b^y = x$ by definition, thus $\log_2[(x - 1)(x + 1)] = 3$ is equivalent to

$$(x - 1)(x + 1) = 2^3 = 8$$
$$x^2 = -1 = 8$$
$$x^2 - 9 = 0$$
$$0 = x^2 - 9 = x^2 - 3^2.$$

Thus we apply the formula for the difference of two squares, $a^2 - b^2 = (a + b)(a - b)$, replacing a by x and b by 3 and obtain $0 = x^2 - 3^2 = (x + 3)(x - 3)$. Whenever the product of two numbers $ab = 0$, either $a = 0$ or $b = 0$. Thus

$$(x + 3)(x - 3) = 0 \text{ means either}$$

$$x + 3 = 0 \text{ or } x - 3 = 0$$

and $\qquad x = -3 \text{ or } \quad x = 3.$

Therefore, $\{3, -3\}$ is the possible solution set, but we must check each in the given equation. This is necessary because we have not defined the logarithm of a negative number and, consequently, must rule out any value of x which would re-quire the use of the logarithm of a negative number.

Check: Replacing x by 3 in our original equation

$$\log_2(x - 1) + \log_2(x + 1) = 3$$

$$\log_2(3 - 1) + \log_2(3 + 1) = 3$$

$$\log_2 2 + \log_2 4 = 3$$

$$1 + 2 = 3 \text{ since } 2^1 = 2 \text{ and } 2^2 = 4$$
$$3 = 3.$$

Replacing x by (-3) in our original equation

$$\log_2(x - 1) + \log_2(x + 1) = 3$$

$$\log_2(-3 - 1) + \log_2(-3 + 1) = 3$$

$$\log_2(-4) + \log_2(-2) = 3.$$

x = -3 cannot be accepted as a root because we have not defined the **logarithm** of a negative number. Thus our solution set is {3}.

● **PROBLEM** 819

Solve the equation $\log 2 + 2 \log x = \log(5x + 3)$.

__Solution__ : By the law of the logarithm of a power of a positive number which states that

$$\log a^n = n \log a, \quad 2 \log x = \log x^2.$$

Hence, $\log 2 + 2 \log x = \log 2 + \log x^2 = \log(5x + 3)$. Therefore, $\log 2 + \log x^2 = \log(5x + 3)$. By the law of the logarithm of a product of two numbers which states that $\log(a \cdot b) = \log a + \log b$, $\log 2 + \log x^2 = \log 2x^2$. Therefore, $\log 2x^2 = \log(5x + 3)$. Hence, $2x^2 = 5x + 3$. Subtract 5x from both sides to obtain:

$$2x^2 - 5x = 5x + 3 - 5x.$$

Combining terms, $2x^2 - 5x = 3$. Subtract 3 from both sides to obtain:

$$2x^2 - 5x - 3 = 3 - 3.$$

Combining terms, $2x^2 - 5x - 3 = 0$. Factoring the left side of this equation into two polynomial factors, $(2x + 1)(x - 3) = 0$. Whenever $a \cdot b = 0$ where a and b are any two real numbers, either $a = 0$ or $b = 0$. Therefore, either

$$2x + 1 = 0 \quad \text{or} \quad x - 3 = 0.$$
$$2x = -1$$
$$\text{and} \quad x = -\frac{1}{2} \quad \text{or} \quad x = 3.$$

Since the domain of the logarithmic function is the set of positive real numbers, it is important to check all proposed solutions of a logarithmic equation. In this example, the given equation is satisfied for x = 3, but $x = -\frac{1}{2}$ is not a solution since $\log\left(-\frac{1}{2}\right)$ is not defined in the relation

$$\log 2 + 2 \log\left(-\frac{1}{2}\right) = \log \frac{1}{2}.$$

● **PROBLEM** 820

Solve for x: $2 \log (3 - x) = \log 2 + \log (22 - 2x)$.

__Solution:__ We shall rewrite the equation in the form log M = log N and then state that M = N. (If two numbers have equal logarithms, the numbers must be equal.)

$$2 \log (3 - x) = \log 2 + \log (22 - 2x)$$

Since $a \log_b x = \log_b x^a$, $2 \log (3 - x) = \log (3 - x)^2$. Thus

$$\log (3 - x)^2 = \log 2 + \log(22 - 2x)$$

Since $\log_b x + \log_b y = \log_b xy$, $\log 2 + \log (22 - 2x)$ $= \log 2(22 - 2x)$. Thus

$$\log (3 - x)^2 = \log 2(22 - 2x)$$

$$(3 - x)^2 = 2(22 - 2x)$$

$$x^2 - 6x + 9 = 44 - 4x$$

Adding $- (44 - 4x)$ to both sides, $x^2 - 2x - 35 = 0$;

factoring, $(x - 7)(x + 5) = 0$.

Whenever the product of two numbers $ab = 0$, either $a = 0$ or $b = 0$; thus either

$$x - 7 = 0 \qquad \text{or} \qquad x + 5 = 0$$
$$x = 7 \qquad \text{or} \qquad x = -5$$

Check for $x = 7$: $2 \log (3 - 7) = \log 2 + \log[22 - 2(7)]$,

$2 \log (-4) = \log 2 + \log 8$. Since the logarithm of a negative number has not been defined, we discard $x = 7$.

Check for $x = -5$:

$$2 \log [3 - (-5)] = \log 2 + \log [22 - 2(-5)]$$
$$2 \log (3 + 5) = \log 2 + \log (22 + 10)$$
$$2 \log 8 = \log 2 + \log 32$$
$$\log 8^2 = \log (2.32)$$
$$\log 64 = \log 64$$

This is true, therefore $x = -5$ is the solution to the given equation.

● **PROBLEM** 821

Solve the equation
$$2 \log x - \log(30 - 2x) = 1. \qquad (1)$$

Solution: The essential point in solving an equation such as (1) is to combine all logarithmic expressions into a single logarithm with coefficient unity so that the relation takes the form $\log_a u = v$.

This form can then be immediately changed into the equivalent exponential form $u = a^v$.
Accordingly, we first replace $2 \log x$ by $\log x^2$, since by a property of logs: $a \log b = \log b^a$. Then, since the difference of two logarithms, each with the coefficient unity, is equal to the logarithm of a quotient, we have

$$\log x^2 - \log(30 - 2x) = \log \frac{x^2}{30-2x} .$$

Thus,
$$\log \frac{x^2}{30 - 2x} = 1.$$

Recall that $\log 10 = 1$ (since $10^1 = 10$). Therefore,

$$\log \frac{x^2}{30 - 2x} = \log 10.$$

Now since the logarithms of the two quantities are equal, the quantities are equal. Thus,

$$\frac{x^2}{30 - 2x} = 10.$$

Cross multiplying, we obtain:

$$x^2 = 10(30 - 2x) = 300 - 20x$$

Subtracting x^2 from both sides:

$$x^2 + 20x - 300 = 0$$

Factoring:

$$(x + 30)(x - 10) = 0$$

Setting each factor to zero:

$$x + 30 = 0, \qquad x - 10 = 0 ,$$

or

$$x = -30, \qquad x = 10$$

It is important that all solutions of a logarithmic equation be checked. Here we see that the negative number $x = -30$ is not permissible, for the first term of equation (1) has no meaning if x is negative, because $2 \log x$ becomes $2 \log(-30)$, and the log of a negative quantity is meaningless. The value $x = 10$, however, when substituted in the left member of (1) gives us $2 \log 10 - \log 10 = 2 - 1 = 1$. Therefore, $x = 10$ is a valid solution.

● **PROBLEM** 822

Solve the equation $\log_{10}(x^2 + 3x) + \log_{10} 5x = 1 + \log_{10} 2x$.

Solution: We first subtract $\log_{10} 2x$ from both sides of our equation so as to have the right-hand side free of logarithmic expressions and obtain

$$\log_{10}(x^2 + 3x) + \log_{10} 5x - \log_{10} 2x = 1$$

By the law of logarithms which states that $\log_b \frac{x}{y} = \log_b x - \log_b y$, $\log_{10} 5x - \log_{10} 2x = \log_{10} \frac{5x}{2x}$. Also, by the law of exponents which states that $\log_b (x \cdot y) = \log_b x + \log_b y$,

$$\log_{10}(x^2 + 3x) + \log_{10} 5x - \log_{10} 2x = \log_{10}(x^2 + 3x) + \log_{10} \frac{5x}{2x}$$

$$= \log_{10}(x^2 + 3x)\left(\frac{5x}{2x}\right)$$

$$= \log_{10} \frac{5x(x^2 + 3x)}{2x} = 1$$

Hence, $\log_{10} \frac{5x(x^2 + 3x)}{2x} = 1$ or $\log_{10} \frac{5(x^2 + 3x)}{2} = 1$. The expression $\log_b a = y$ is equivalent to $b^y = a$. Therefore, $\log_{10} \frac{5(x^2 + 3x)}{2} = 1$ is equivalent to $10^1 = \frac{5(x^2 + 3x)}{2}$. Hence, distributing:

$$10 = \frac{5x^2 + 15x}{2}$$

Multiply both sides of this equation by 2.

$$2(10) = 2\left(\frac{5x^2 + 15x}{2}\right)$$

$$20 = 5x^2 + 15x$$

Subtract 20 from both sides of this equation:

$$20 - 20 = 5x^2 + 15x - 20$$

$$0 = 5x^2 + 15x - 20$$

Factor out the common factor of 5 from the right side of this equation:

$$0 = 5(x^2 + 3x - 4)$$

Divide both sides of this equation by 5.

$$\frac{0}{5} = \frac{5(x^2 + 3x - 4)}{5}$$

$$0 = x^2 + 3x - 4$$

Factoring the right side of this equation:

$$0 = (x + 4)(x - 1)$$

Whenever the product ab = 0 where a and b are any two numbers, either a = 0 or b = 0. Therefore,

$$x + 4 = 0 \quad \text{or} \quad x - 1 = 0$$

$$x = -4 \quad \text{or} \quad x = 1$$

To check if these two values are indeed solutions, replace x by -4 and 1 in the original equation:
When x = -4,

$$\log_{10}\left((-4)^2 + 3(-4)\right) + \log_{10}5(-4) = 1 + \log_{10}2(-4)$$

$$\log_{10}(16 - 12) + \log_{10}(-20) = 1 + \log_{10}(-8)$$

$$\log_{10}4 + \log_{10}(-20) = 1 + \log_{10}(-8).$$

The number -4 is not a solution because the logarithm of a negative number does not exist. When x = 1,

$$\log_{10}\left((1)^2 + 3(1)\right) + \log_{10}5(1) = 1 + \log_{10}2(1)$$

$$\log_{10}(1+3) + \log_{10}5 = 1 + \log_{10}2$$

$$\log_{10}4 + \log_{10}5 = 1 + \log_{10}2$$

$$0.6021 + 0.6990 = 1.0000 + 0.3010$$

$$1.3011 \approx 1.3010.$$

Therefore, x = 1 is the only solution.

● **PROBLEM** 823

Solve the equation $x^{\log x} = 100x$.

<u>Solution:</u> By taking logarithms of both sides of the equation, we obtain the equivalent equation

$$\log\left[x^{\log x}\right] = \log 100x.$$

But, by the law of exponents which states $\log x^p = p \log x$,

$$\log\left[x^{\log x}\right] = (\log x)(\log x) = (\log x)^2.$$

Also, by another law of exponents which states

600

$$\log (x \cdot y) = \log x + \log y,$$

$$\log 100x = \log 100 + \log x.$$

Now, since log 100 can be equivalently written as $\log_{10} 100$, then $\log_{10} 100 = x$ or $10^x = 100$; and we can replace log 100 by 2 ($10^2 = 100$).

Thus, we have: $2 + \log x$.

We can therefore write our equation as

$$(\log x)^2 = 2 + \log x,$$

and so it is equivalent to the equation

$$(\log x)^2 - \log x - 2, \quad \text{and factoring:}$$

$$= (\log x - 2)(\log x + 1) = 0.$$

Now, $\qquad \{(\log x - 2)(\log x + 1) = 0\}$

$$= \{x | \log x = 2 \text{ or } \log x = -1\}$$

$$= \{\log x = 2\} \cup \{\log x = -1\}.$$

Recall that when no base is expressed it is assumed to be 10. Thus, the equation $\log x = 2$ means $10^2 = x$, or $x = 100$; and $\log x = -1$ means $10^{-1} = x$, or $x = \frac{1}{10}$. Thus,

$$\{100\} \cup \left\{\tfrac{1}{10}\right\} = \left\{100, \tfrac{1}{10}\right\},$$

and this is the set of numbers that solves the given equation.

● **PROBLEM** 824

Solve the equation $27^{x^2+1} = 243$.

Solution: We seek all numbers which satisfy the equation. If x is such a number, then

$$27^{x^2+1} = 243$$

Then, taking logarithms to the base 3 of both sides we have

$$\log_3 27^{x^2+1} = \log_3 243$$

Since $\log_b x^r = r \log_b x$, it follows that

$$(x^2 + 1)\log_3 27 = \log_3 243.$$

Note that the expression $\log_b x = y$ is equivalent to $b^y = x$. Hence, $\log_3 27 = N$ is equivalent to $3^N = 27$. Therefore, $N = 3$ and $\log_3 27 = 3$. Also, $\log_3 243 = M$ is equivalent to $3^M = 243$. Therefore, $M = 5$ and $\log_3 243 = 5$. Hence,

$$(x^2 + 1)3 = 5$$

or, by the commutative property of multiplication,

601

$$3(x^2 + 1) = 5.$$

Divide both sides of the equation by 3.

$$\frac{3(x^2 + 1)}{3} = \frac{5}{3}$$

$$x^2 + 1 = \frac{5}{3}$$

Subtract 1 from both sides of the equation.

$$x^2 + 1 - 1 = \frac{5}{3} - 1$$

$$x^2 = \frac{5}{3} - 1 = \frac{5}{3} - \frac{3}{3} = \frac{2}{3}$$

Therefore, $x = \pm \sqrt{\frac{2}{3}}$, i.e., $x = \sqrt{\frac{2}{3}}$ or $x = -\sqrt{\frac{2}{3}}$.

To check that each of these numbers satisfies the given equation, substitute each number for x in the given equation. Substituting $\sqrt{\frac{2}{3}}$ for x:

$$(27)^{(\sqrt{2/3})^2} + 1 = (27)^{(2/3)+1} = 27^{5/3} = \sqrt[3]{27^5} = \left(\sqrt[3]{27}\right)^5$$

$$= (3)^5$$

$$= 243 \checkmark$$

Substituting $-\sqrt{\frac{2}{3}}$ for x:

$$(27)^{(-\sqrt{2/3})^2 + 1} = (27)^{(2/3)+1} = 27^{5/3} = \sqrt[3]{27^5} = \left(\sqrt[3]{27}\right)^5$$

$$= (3)^5$$

$$= 243 \checkmark$$

● **PROBLEM** 825

Solve $2^{x+1} = 7^{x+2}$.

<u>Solution:</u> Take the logarithm of each side of the equation. $\log 2^{x+1} = \log 7^{x+2}$. By the law of the logarithm of a power of a positive number which states that

$\log a^n = n \log a$, $\log 2^{x+1} = (x+1)\log 2$ and $\log 7^{x+2} = (x+2)\log 7$. Therefore: $(x+1)\log 2 = (x+2)\log 7$. Distributing, $x \log 2 + \log 2 = x \log 7 + 2 \log 7$. Subtract $x \log 7$ from both sides to obtain:

$$x \log 2 + \log 2 - x \log 7 = x \log 7 + 2 \log 7 - x \log 7$$

or $x \log 2 - x \log 7 + \log 2 = 2 \log 7$. Now, subtract $\log 2$ from both sides to obtain:

$$x \log 2 - x \log 7 + \log 2 - \log 2 = 2 \log 7 - \log 2$$

or $x \log 2 - x \log 7 = 2 \log 7 - \log 2$. Factoring out the common factor x from the left side:

$$x(\log 2 - \log 7) = 2 \log 7 - \log 2.$$

Dividing both sides by $\log 2 - \log 7$:

$$\frac{x (\log 2 - \log 7)}{\log 2 - \log 7} = \frac{2 \log 7 - \log 2}{\log 2 - \log 7}$$

$$x = \frac{2 \log 7 - \log 2}{\log 2 - \log 7}.$$

By the law of the logarithm of a power of a positive number which states that $\log a^n = n \log a$, $2 \log 7 = \log 7^2 = \log 49$. Therefore,

$$x = \frac{\log 49 - \log 2}{\log 2 - \log 7}.$$

By the law of the logarithm of a quotient which states that $\log \frac{a}{b} = \log a - \log b$, $\log 49 - \log 2 = \log \frac{49}{2}$ and $\log 2 - \log 7 = \log \frac{2}{7}$. Therefore,

$$x = \frac{\log 49 - \log 2}{\log 2 - \log 7} = \frac{\log \frac{49}{2}}{\log \frac{2}{7}}.$$

Solutions may be left in logarithmic form, as above. If a decimal approximation is desired, the final expression can be evaluated by means of a table of logarithms.

● **PROBLEM** 826

Solve the equation $I = \frac{E}{R} \left(1 - e^{-Rt/L}\right)$ for t.

<u>Solution:</u> We can write the equation as

$$\frac{RI}{E} = 1 - e^{-Rt/L} \quad \text{or} \quad e^{-Rt/L} = 1 - \frac{RI}{E}.$$

Then, taking the logarithm of both sides:

$$\log e^{-Rt/L} = \log \left(1 - \frac{RI}{E}\right).$$

Apply the following fundamental law of logarithms: The logarithm of a positive number raised to a power is equal to the exponent of the power multiplied by the number. That is, if P is a positive number raised to a power n, then:

$$\log \left(P^n\right) = n \log P$$

Therefore:

$$\frac{-Rt}{L} \log e = \log \left(1 - \frac{RI}{E}\right),$$

$$\frac{-Rt}{L} = \frac{\log \left(1 - \frac{RI}{E}\right)}{\log e},$$

$$t = \frac{-L \log \left(1 - \frac{RI}{E}\right)}{R \log e}.$$

If we use logarithms to the base e, the expression for t takes a simpler form. To convert from a logarithm to the base 10 to a logarithm to the base e, employ the formula

$$\log_e x = \frac{\log_{10} x}{\log_{10} e} \ . \ \text{In our case, } x = \left(1 - \frac{RI}{E}\right).$$

Thus,

$$t = \frac{-L}{R} \left\{ \frac{\log_{10} \left(1 - \frac{RI}{E}\right)}{\log_{10} e} \right\}$$

or, in base e,

$$t = \frac{-L}{R} \ \log_e \left(1 - \frac{RI}{E}\right) \ .$$

● **PROBLEM** 827

Solve for x: $\log(x + 1) + \log x = 1.3010$.

Solution: $\log(x + 1) + \log x = 1.3010$
Recall $\log a + \log b = \log(ab)$; thus $\log(x + 1) + \log x = \log(x + 1)x$.
Hence $\log (x + 1)(x) = 1.3010$. Take the antilog of each side,

$$\text{antilog}[\log(x + 1)(x)] = \text{antilog } 1.3010.$$

Since the antilog of $\log(x + 1)(x)$ is $(x + 1)(x)$,
$$x(x + 1) = \text{antilog } 1.3010$$

Evaluate antilog 1.3010. The antilog of 1.3010 is the number whose log is 1.3010. The characteristic is 1; hence to obtain the number multiply the antilog of the mantissa by 10'. The antilog of .3010 is 2.0, therefore our number is $2.0 \times 10' = 20$.
$$x^2 + x = 20$$

Add (-20) to both sides, $x^2 + x - 20 = 0$

Factor, $(x + 5)(x - 4) = 0$

Whenever a product of numbers $ab = 0$ either $a = 0$ or $b = 0$ thus $(x + 5)(x - 4) = 0$ means either $x + 5 = 0$ or $x - 4 = 0$, that is,
$$x = -5 \ \text{ or } \ x = 4 \ .$$

The domain of the logarithmic function is the positive real numbers, thus the value -5 must be excluded, and our solution is x = 4.

Check: Replace x by 4 in our original equation,
$$\log(x + 1) + \log x = 1.3010$$
$$\log(4 + 1) + \log 4 = 1.3010$$
$$\log 5 + \log 4 = 1.3010$$
$$\log(5 \cdot 4) = 1.3010$$
$$\log 20 = 1.3010$$
$$\log(10 \cdot 2) = 1.3010$$
$$\log 10 + \log 2 = 1.3010$$
$$1 + .3010 = 1.3010$$
$$1.3010 = 1.3010$$

Find the inverse of the function

$$y = \ln\left(1 + \sqrt{1 - e^2 x^4}\right) - 2 \ln x - 1. \qquad (1)$$

Solution: Transfer all the natural logarithm functions to one side with the variables and constants on the other side. Thus,

$$y + 1 = \ln\left(1 + \sqrt{1 - e^2 x^4}\right) - 2 \ln x$$

Recall $a \log b = \log b^a$ thus $2\ln x = \ln x^2$ and

$$y + 1 = \ln\left(1 + \sqrt{1 - e^2 x^4}\right) - \ln x^2$$

Since $\ln a - \ln b = \ln \frac{a}{b}$, $\ln\left(1 + \sqrt{1 - e^2 x^4}\right) - \ln x^2 = \ln \frac{1 + \sqrt{1 - e^2 x^4}}{x^2}$

Thus,

$$y + 1 = \ln \frac{1 + \sqrt{1 - e^2 x^4}}{x^2} \quad ,$$

The inverse function of the logarithmic function is the exponential function. $\ln u = v$ and $e^v = u$. Here

$$u = \frac{1 + \sqrt{1 - e^2 x^4}}{x^2} \quad , \quad v = y + 1$$

$$e^{y+1} = \frac{1 + \sqrt{1 - e^2 x^4}}{x^2} \quad .$$

Then, multiply by x^2 and subtract 1 from both sides to obtain:

$$x^2 e^{y+1} - 1 = \sqrt{1 - e^2 x^4} \quad .$$

Rationalizing this equation, we get by squaring:

$$x^4 e^{2y+2} - 2x^2 e^{y+1} + 1 = 1 - e^2 x^4 \quad .$$

Now x must be positive in order that the term involving $\ln x$, in equation (1), have meaning. $\ln x$ exists for $x > 0$. In particular, x cannot be zero; hence, after subtracting 1 from each member of the last equation, we may further simplify by dividing by ex^2.

$$x^4 e^{2y+2} - 2x^2 e^{y+1} = -e^2 x^4 \quad \text{(subtracting 1 from each side)}$$

$$x^2 e^{2y+1} - 2e^y = -ex^2, \left(\text{dividing by } ex^2\right)$$

$$x^2 e^{2y+1} + ex^2 = 2e^y$$

$$ex^2\left(e^{2y} + 1\right) = 2e^y \quad , \quad \left(\text{factor out } ex^2\right)$$

Solve for x^2

$$x^2 = \frac{2e^y}{e\left(e^{2y} + 1\right)} = \frac{2}{e\left(e^y + e^{-y}\right)} \quad .$$

Remembering that x must be positive, we extract only the positive square root to obtain

$$x = \sqrt{\frac{2}{e\left(e^y + e^{-y}\right)}} \quad . \qquad (2)$$

This is the required inverse function.

The form of relation (2) would seem to indicate that, if $x = x_1$ is the value corresponding to $y = y_1$, then $y = -y_1$ will also yield $x = x_1$. However, equation (1) shows that y cannot be replaced by $-y$ without creating thereby a different functional relation. In fact, y as given by (1) will be non-negative for every real value of x in its permissible range, $0 < x \leqq e^{-\frac{1}{2}}$. $x > 0$ because $\ln 0$ does not exist. If $x \neq e^{-\frac{1}{2}}$, for example $x = e$, the

$$\ln\left(1 + \sqrt{1 - e^2 e^4}\right)$$

does not exist. The functional relation $y = \ln\left(1 - \sqrt{1 - e^2 x^4}\right) - 2\ln x - 1$ yields only non-positive values of y and likewise leads to the same inverse relation (2). If you select an x from the domain $0 < x \leq e^{-\frac{1}{2}}$, for example $x = e^{-\frac{1}{2}}$, and substitute it in $y = \ln\left(1 - \sqrt{1 - e^2 x^4}\right) - 2\ln x - 1$, then $y = \ln\left(1 - \sqrt{1 - e^2 e^{-4/2}}\right) - 2\ln e^{-\frac{1}{2}} - 1$

$$= \ln\left(1 - \sqrt{1 - e^0}\right) - 2(-\tfrac{1}{2}) - 1$$

$$y = \ln(1) + 1 - 1 = 0 .$$

Therefore, y is a non-positive value.

Our example thus illustrates the fact that all conclusions drawn from a deduced inverse function must be checked against the original relation.   ● **PROBLEM** 829

> Solve for x in the equation
> $$3251 = 2184(1.02)^x .$$

<u>Solution:</u> Taking the logarithm of each member of the given equation we obtain
$$\log 3251 = \log\left[2184(1.02)^x\right]$$

Since $\log(ab) = \log a + \log b$,
$$\log 3251 = \log 2184 + \log(1.02)^x$$

since $\log a^b = b \log a$,
$$\log 3251 = \log 2184 + x \log 1.02 .$$

Adding $(-\log 2184)$ to both sides,
$$x \log 1.02 = \log 3251 - \log 2184 ,$$

dividing both sides by $\log 1.02$,
$$x = \frac{\log 3251 - \log 2184}{\log 1.02}$$

Solving for our logarithms:
$$3,251 = 3.251 \times 10^3$$

Thus the characteristic is 3 and we interpolate to find $\log 3.251$:

Number	log
3.250	5119
.001 3.251	? x 13
3.260	5132

.01 (.001

we set up the following proportion:

$$\frac{.001}{.01} = \frac{x}{13}$$

Cross multiply to obtain,   .013 = .01x

$$13 = 10x$$

$$x = \frac{13}{10} = 1.3$$

Thus   log 3.251 = 5119 + 1.3 ≈ .5120, and

log 3.251 = 3 + .5120 = 3.5120

$$2,184 = 2.184 \times 10^3$$

Thus the characteristic is 3 and we interpolate to find log 2.184:

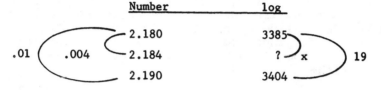

Number		log
2.180		3385
2.184		?
2.190		3404

.01 (.004 (2.180 — 2.184 — 2.190)   3385 — ? x — 3404) 19

We set up the following proportion:

$$\frac{.004}{.01} = \frac{x}{19}$$

cross multiply to obtain,

$$.076 = .01x$$

$$76 = 10x$$

$$x = 7.6$$

Thus log 2.184 = 3385 + 7.6 ≈ 3393   and

log 2.184 = 3 + .3393 = 3.3393

Last, log 1.02 can be found in a table of logarithms,

log 1.02 = 0.0086

Substitute these values into our equation for x:

$$x = \frac{3.5120 - 3.3393}{0.0086}$$

$$= \frac{0.1727}{0.0086}$$

$$= 20.08$$

● **PROBLEM   830**

Solve the inequality $(.3)^x < \frac{4}{3}$.

Solution: This is an exponential inequality which involves one or more unknowns in an exponent, and is solved by means of logarithms. If M < N, then

$\log_b$ M < $\log_b$ N, and conversely. Thus $(.3)^x < \frac{4}{3}$ if,

and only if, log $(.3)^x$ < log $\frac{4}{3}$. We can then use the

fact that:

$$\log a^b = b \log a \text{ and,}$$

$$\log \frac{a}{b} = \log a - \log b. \text{ Thus,}$$

from $\log (.3)^x < \log \frac{4}{3}$ we have:

$$x \log (.3) < \log 4 - \log 3,$$

Express $\log (.3)$. The characteristic of the common logarithm of any positive number smaller than 1 is negative and is obtained by adding one more than the number of zeros between the decimal point and the first digit. The mantissa is obtained by looking it up in a table of common logarithms. For .3, the characteristic is - 1 since it is less than one and there are no zeros between the decimal point and the first digit. Its mantissa is .4771. Thus $\log .3 = (.4771 - 1)$. The $\log 4$ and $\log 3$ can be obtained from a table of mantissas of common logarithms.

Thus, solving the inequality we obtain:

$$x (.4771 - 1) < .6021 - .4771,$$

$$- .5229x < .1250,$$

$$x > - .239.$$

Therefore, $\{x > - .239\}$ is the solution set of our given inequality.

● **PROBLEM** 831

Solve $5^{x+y} = 100$ (1)
and $2^{2x-y} = 10$ (2)
for x and y.

<u>Solution:</u> If we equate the common logarithms of the members of each of (1) and of (2), we get

$$\log 5^{x+y} = \log 100$$
$$\log 2^{2x-y} = \log 10.$$

Recalling that $\log_b a^y = y \log_b a$,

$$\log 5^{x+y} = (x + y) \log 5 \text{ and } \log 2^{2x-y} = (2x - y) \log 2.$$

We also know $\log_{10} 100 = 2$ $\left(\text{since } 10^2 = 100\right)$ and

$$\log_{10} 10 = 1 \left(\text{since } 10^1 = 10\right).$$

Substituting these values into equations (1) and (2) we obtain $(x + y) \log 5 = 2$ (3)

$$(2x - y) \log 2 = 1. \qquad (4)$$

If we solve these equations for $x + y$ and $2x - y$ we obtain

$$x + y = \frac{2}{\log 5}$$

$$2x - y = \frac{1}{\log 2}.$$

We observe in a log table that log 5 = .6990 and log 2 = .3010. Thus,
$$x + y = \frac{2}{\log 5} = \frac{2}{.6990} = 2.86 \qquad (5)$$

$$2x - y = \frac{1}{\log 2} = \frac{1}{.3010} = 3.32. \qquad (6)$$

Adding equations (5) and (6) we obtain

$$\begin{array}{rl}
x + y &= 2.86 \\
2x - y &= 3.32 \\
\hline
3x &= 6.18 \\
x &= 2.06.
\end{array}$$

Substituting 2.06 for x in (5) and solving for y gives

$$2.06 + y = 2.86$$

$$y = 2.86 - 2.06$$

$$y = .80.$$

Therefore the solution set of this system is $\{(x,y)\} = \{(2.06, .80)\}$.

● **PROBLEM** 832

Under favorable conditions a single cell of the bacterium Escherichia coli divides into two about every 20 minutes. If this same rate of division is maintained for 10 hours, how many organisms will be produced from a single cell?

Solution: The 10-hour interval may be divided into 30 periods of 20 minutes each. This is because for each hour there are three twenty minute periods. Thus in 10 hours, there are 3 X 10 = 30 periods. At the end of the first period there are 2 bacteria. These two then divide and at the end of the second period there are $2 \cdot 2 = 2^2$ bacteria. Now, each bacterium (four in total) divide in two. At the end of the third period there are $8 = 2 \cdot 2 \cdot 2 = 2^3$ bacteria. Then each of these divide in two to form $16 = 2^4$ bacteria at the end of the fourth period. A pattern has been set up where at the end of the nth period, there are 2^n bacteria. At the end of the 30th period there will be 2^{30} bacteria, which we will call N, the number we were seeking. It is easier to compute 2^{30} with logarithms.

$$N = 2^{30}$$

Taking logarithms: $\quad \log N = \log 2^{30}$

Then since $\quad \log x^y = y \log x,$

$$\log N = 30 \log 2$$

See a table of logarithms for log 2.

$$\log N = 30(0.3010) = 9.0300$$

The characteristic is 9 and its mantissa is 0.0300. Expressing this as a number, we have
$$10^{9.0300} = 10^{0.0300} \times 10^9 .$$

609

See a table of common mantissas of logarithms for the number corresponding to 0.0300. Note that the characteristic of 0.0300 which is zero, will be one less than the number of digits to the left of the decimal point of the corresponding number . Therefore, there is one digit to the left of the decimal point. Thus, the number is 1.072. Then we see that

$$N = 1.072 \times 10^9 = 1,072,000,000,$$

so a single cell is potentially capable of producing about a billion organisms in a 10-hour period.

● **PROBLEM** 833

From the given graph find as well as you can (a) $\log_e 1.5$, (b) $\log_e .5$, (c) the number x for which $\log_e x = 1.5$, and (d) the value of e.

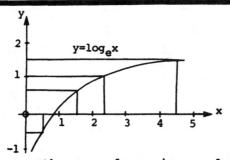

Solution: The smooth curve drawn is $y = \log_e x$. For (a) $\log_e 1.5$ and (b) $\log_e .5$, find the corresponding x-values, x = 1.5 and x = .5, and move along these vertical lines until you reach the curve, $y = \log_e x$. Then find the y-values from the corresponding projections onto the y – axis. We find that (a) $\log_e 1.5 = .4$, and (b) $\log_e .5 = -.7$. For (c), we are given the ordinate. Move along the horizontal line, y = 1.5, up to the curve of $\log_e x$, and then find its abscissa, which is 4.5. Thus, (c) x = 4.5. The number e satisfies the equation $\log_e x = 1$, and from the figure, it appears that e = 2.7 (actually, to 5 decimal places, e = 2.71828).

● **PROBLEM** 834

Construct the graph of $y = \log_2 x$.

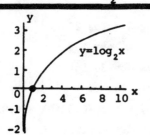

<u>Solution:</u> The equations $u = \log_b v$ and $v = b^u$ are equivalent.
Therefore, the relation $y = \log_2 x$ is equivalent to $x = 2^y$. Hence we
assume values of y and compute the corresponding values of x, getting
the table:

x:	$\frac{1}{8}$	$\frac{1}{4}$	$\frac{1}{2}$	1	2	4	8
y:	-3	-2	-1	0	1	2	3

For example, if $y = -3$, then $x = 2^y = 2^{-3} = \frac{1}{2^3} = \frac{1}{8}$.

The points corresponding to these values are plotted on the coordinate
system in the figure. The smooth curve joining these points
is the desired graph of $y = \log_2 x$. It should be noted that the graph
lies entirely to the right of the y-axis. The graph of $y = \log_b x$ for
any $b > 1$ will be similar to that in the figure. Some of the properties of this function which can be noted from the graph are:

    I. $\log_b x$ is not defined for negative values of x or zero.

    II. $\log_b 1 = 0$.

    III. If $x > 1$, then $\log_b x > 0$.

    IV. If $0 < x < 1$, then $\log_b x < 0$.   ● PROBLEM 835

Construct the graph of $y = 3^x$.

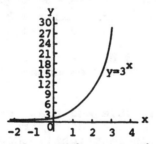

<u>Solution:</u> Assume values of x and compute the corresponding values of
y by substituting into $y = 3^x$, obtaining the following table of values:

x:	-3	-2	-1	0	1	2	3
y:	$\frac{1}{27}$	$\frac{1}{9}$	$\frac{1}{3}$	1	3	9	27

The points corresponding to these pairs of values are plotted on the
coordinate system of the figure and these points are joined by a smooth
curve, which is the desired graph of the function. Note that the values
of y are all positive. Furthermore, if $x < 0$, then y increases to
a small extent as x does. If $x > 0$, y increases at a more rapid
rate.

          ● PROBLEM 836

Graph the following functions: (A) $y = 2^x$, (B) $y = 4^x$,

(C) 4^{-x}, (D) $y = 3 \cdot 2^x$.

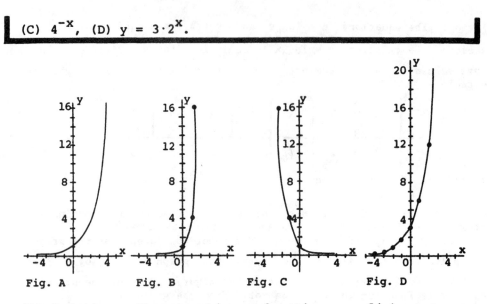

Fig. A   Fig. B   Fig. C   Fig. D

(A) Solution: When graphing a function $y = f(x)$, set up a table consisting of two columns: one for x and one for y. Choose values for x and find the corresponding value for y. In this problem if:

$$x = -4, \text{ then } y = 2^x = 2^{-4} = \frac{1}{2^4} = \frac{1}{16}.$$

Similarly, find other y values for different values of x. It is best to choose negative and positive values of x centering around and including zero to determine the nature of the graph.

x	-4	-3	-2	-1	0	1	2	3	4
y	$\frac{1}{16}$	$\frac{1}{8}$	$\frac{1}{4}$	$\frac{1}{2}$	1	2	4	8	16

Plot these points and draw a smooth curve through them. This is the graph of the exponential function $y = 2^x$.

Note that from the table and graph constructed, as you increase x by 1 each time moving from $x = -4$ to $x = 0$, the y values increase slightly. However, when you move through the positive values of x, the change in y is much greater for each unit change in x.

See Figure A.

(B) Solution: Construct a table in the same manner as problem A. The table of values for the integers -3 to 3 can be determined to be:

x	-3	-2	-1	0	1	2	3
y	$\frac{1}{64}$	$\frac{1}{16}$	$\frac{1}{4}$	1	4	16	64

Then plot these points and draw a smooth curve.

Figure B is the graph of $y = 4^x$ although it is not practical to plot the points corresponding to $x = -3$ or $x = 3$ on this coordinate system.

When this curve is compared to the graph of $y = 2^x$ (Figure A), we see that the general shape is the same.

612

Both curves pass through the point (0,1), that is, both have a y-intercept of 1. If we consider the curves to the left of the y-axes, we see that the curve of Figure B approaches the x-axis faster than the curve of Figure A. If we consider the same negative value on both curves, the y-value in Figure B is smaller than in Figure A. If $x = -3$, then $y = 2^{-3} = \frac{1}{8}$ for Figure A and $y = 4^{-3} = \frac{1}{64}$ for Figure B. Thus (the) point $\left(-3, \frac{1}{64}\right)$ is closer to the x-axis than $\left(-3, \frac{1}{8}\right)$.

(C) Solution: Obtain a table of ordered pairs as in Examples A and B. Plot the points and draw the smooth curve by connecting them.

$$f(-3) = 4^{-(-3)} = 4^3 = 64$$

$$f(-2) = 4^{-(-2)} = 4^2 = 16$$

$$f(-1) = 4^{-(-1)} = 4^1 = 4$$

$$f(0) = 4^0 = 1$$

$$f(1) = 4^{-1} = \frac{1}{4}$$

$$f(2) = 4^{-2} = \frac{1}{4^2} = \frac{1}{16}$$

$$f(3) = 4^{-3} = \frac{1}{4^3} = \frac{1}{64}$$

x	-3	-2	-1	0	1	2	3
y	64	16	4	1	$\frac{1}{4}$	$\frac{1}{16}$	$\frac{1}{64}$

Figure C is the graph of $y = 4^{-x}$.

The graph of $y = 4^x$ of the Figure B and the graph of $y = 4^{-x}$ of Figure C are mirror images of each other.

(D) Solution: In Example A we determined the values of 2^x for x an integer and $-4 \le x \le 4$. The values of y for this function then must be three times the corresponding values of y of Example A.

x	-4	-3	-2	-1	0	1	2	3	4
y	$\frac{3}{16}$	$\frac{3}{8}$	$\frac{3}{4}$	$\frac{3}{2}$	3	6	12	24	48

The graph of this function is shown in Figure D.

From these four examples we can see some of the features of the graph of $y = ab^x$, $a > 0$, and $b > 0$. The y-intercept of the function is a: If $a > 1$, the curve will approach the x-axis to the left of the y-axis and the y value increases as the x value increases. The graph will be in quadrants I and II.

The y-intercept of Examples A, B, and C is 1 since $a = 1$. It is true for Examples A and B that the curve approaches the x-axis as x becomes more negative and the y-value increases as x increases. However when a = 1 in Example C, the reverse occurs. As x decreases, y increases and the curve approaches the x-axis as x becomes more positive.

CHAPTER 27

TRIGONOMETRY

> ## Basic Attacks and Strategies for Solving Problems in this Chapter. See pages 615 to 656 for step-by-step solutions to problems.

In trigonometry it is convenient to think of an angle in terms of rotating a ray about its endpoint from some original position (vertical side) to a new position (terminal side). The size of an angle is usually described using degree or radian measure.

If an angle A is in standard position and the point (x, y) is any point on the terminal side r of A, other than the origin, then six trigonometric functions of angle A are defined as follows:

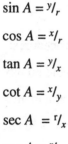

$\sin A = y/_r$

$\cos A = x/_r$

$\tan A = y/_x$

$\cot A = x/_y$

$\sec A = r/_x$

and $\csc A = r/_y$.

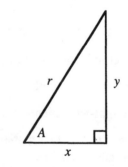

An easy way to find the trigonometric function of an angle between 0 and 90 degrees, inclusive, is to use trigonometric function tables and/or a calculator. However, it may be necessary to first change degrees-minutes-seconds measure of the angle to the decimal form or radian form before using tables or the calculator. For example, 75° 12′ 36″ in decimal form is given by

$$75° + (12/60)° + (36/3600)° = 75° + (.2)° + (.01)° = 75.21°$$

and in radian form is about 1.31 radians.

If the angle is larger than 90 degrees, then the values of the trigonometric functions of the angle are found by first determining the corresponding or related reference angle for the given angle. Then, the values of the trigonometric functions of the reference angle are equal to the trigonometric functions of the original angle. The trigonometric functions of any angle $A > 90$ degrees are equal to those of the reference angle associated with angle A, except possibly for the sign. The sign can be determined by considering the quadrant in what the terminal side of angle A lies. If the angle is larger than $360°$ or a negative angle (e.g., $- 480°$) smaller than $- 360°$, then a coterminal angle should be found before finding the reference angle. For example, to find $\tan (- 480°)$ note that the coterminal angles for $- 480°$ is $240°$ since

$$- 480° + 2(360°) = 240°.$$

The reference angle for $240°$ is $60°$ in the third quadrant and the tangent is positive in this quadrant. Thus,

$$\tan (- 480°) = + \tan 60° = \text{square root of } 3.$$

There are two general types of trigonometric equations — identities and conditional. Trigonometric identity equations, like algebraic identities, are statements which are true for all permissible replacements of the variable, this is, for any angle for which the function is defined. Examples of identity equations are

$$\sec^2 x = \tan^2 x + 1 \quad \text{and} \quad \tan (x - y) = (\tan x - \tan y) / (1 + \tan x \cdot \tan y).$$

The second type of trigonometric equations, conditional equations, are true only for specific replacements of the variable. Examples of conditional equations are

$$4 \cos^2 x = 1 \quad \text{and} \quad \sin 2x + \sin x = 0.$$

Any trigonometric identity can be proved by three different procedures:

(1) The left-hand member may be reduced to the right-hand member,

(2) The right-hand member may be reduced to the left-hand member, or

(3) Each side may be separately reduced to the same form as long as the steps are reversible.

The series of steps used in the simplification process is by no means unique. So, no matter which side you start on, there may be a choice of next steps of which some may result in a shorter proof than others. Some suggestions for handling identities are as follows:

(1) Memorize the basic trigonometric relations and be able to recognize when they should be used.

(2) It is usually better to start with the side of the identity that appears to be more complicated. If both sides of the identity appear to be equally complicated, express each function in terms of sines or cosines and simplify.

(3) Look for places to use algebraic manipulations that do not involve division by zero.

(4)   Keep in mind where you are heading and what steps might get you there.

Solving a triangle using trigonometric functions is straight forward if the triangle is a right triangle. The basic trigonometric relationships — sine, cosine, tangent and their respective reciprocals, cosecant, secant, and cotangent, are used. However, if there is an oblique triangle (that is, it either has three acute angles or one obtuse angle and two acute angles), then the solution technique involves the use of the law of sines or law of cosines. The law of sines is given by:

$$\frac{a}{\sin A} = \frac{b}{\sin B} = \frac{c}{\sin C},$$

where a, b, and c represent the sides of the triangle and A, B, and C represent the corresponding angle. In any event, we must be given three parts of the triangle, at least one of which is a side, in order to use the aforementioned laws. The law of cosines is given by:

$$a^2 = b^2 + c^2 - 2bc(\cos A);$$

$$b^2 = a^2 + c^2 - 2ac(\cos B); \text{ and}$$

$$c^2 = a^2 + b^2 - 2ab(\cos C),$$

where a, b, and c represent the sides of the triangle and A, B, and C represent the corresponding angles.

ANGLES AND TRIGONOMETRIC FUNCTIONS

● **PROBLEM** 837

If tan θ = 3.8436, find θ.

<u>Solution:</u>   Looking through a table of trigonometric functions under the vertical column marked tan, it is found that the angle θ = 75°35' corresponds to the number 3.8436.

● **PROBLEM** 838

Complete the following table:

Width of θ in radians	1	2	3	4	5	6	7	8	9
Width of θ in radians	0	$\frac{1}{6}\pi$	$\frac{1}{4}\pi$		$\frac{1}{2}\pi$	$\frac{2}{3}\pi$			π
Width of θ in degrees	0°			60°			135°	150°	

<u>Solution:</u> If an angle θ is A degrees wide and also t radians wide, then the numbers A and t are related by the equation:

$$\frac{A}{180°} = \frac{t}{\pi} \qquad (1)$$

Thus, equation (1) can be used to complete the table. For column (2):

$$\frac{A}{180°} = \frac{1/6\,\pi}{\pi}$$

$$\frac{A}{180°} = 1/6$$

Multiplying both sides by 180° ,

$$180°\left(\frac{A}{180°}\right) = 180°\,(1/6)$$

$$A = 30°$$

For column (3):

$$\frac{A}{180°} = \frac{\frac{1}{4}\,\pi}{\pi}$$

$$\frac{A}{180°} = \frac{1}{4}$$

Multiplying both sides by 180°,

$$180°\left(\frac{A}{180°}\right) = 180°\,(\tfrac{1}{4})$$

$$A = 45°$$

For column (4):

$$\frac{60°}{180°} = \frac{t}{\pi}$$

$$\frac{1}{3} = \frac{t}{\pi}$$

Multiplying both sides by π,

$$\pi(1/3) = \cancel{\pi}(t/\cancel{\pi})$$

$$1/3\ \pi = t$$

For column (5):

$$\frac{A}{180°} = \frac{\frac{1}{2}\cancel{\pi}}{\cancel{\pi}}$$

$$\frac{A}{180°} = \frac{1}{2}$$

Multiplying both sides by $180°$,

$$\cancel{180°}\left(\frac{A}{\cancel{180°}}\right) = 180°\ (\tfrac{1}{2})$$

$$A = 90°$$

For column (6):

$$\frac{A}{180°} = \frac{2/3\ \cancel{\pi}}{\cancel{\pi}}$$

$$\frac{A}{180°} = 2/3$$

Multiplying both sides by $180°$,

$$\cancel{180°}\left(\frac{A}{\cancel{180°}}\right) = 180°\ (2/3)$$

$$A = 120°$$

For column (7):

$$\frac{135°}{180°} = \frac{t}{\pi}$$

$$\frac{27}{36} = \frac{t}{\pi}$$

$$\frac{3}{4} = \frac{t}{\pi}$$

Multiplying both sides by π,

$$\pi(3/4) = \cancel{\pi}(t/\cancel{\pi})$$

$$3/4\ \pi = t$$

For column (8):

$$\frac{150°}{180°} = \frac{t}{\pi}$$

$$\frac{50}{60} = \frac{t}{\pi}$$

$$\frac{5}{6} = \frac{t}{\pi}$$

Multiplying both sides by π, $\pi(\frac{5}{6}) = \cancel{\pi}(\frac{t}{\cancel{\pi}})$.

$$\frac{5}{6}\pi = t$$

For column (9):

$$\frac{A}{180°} = \frac{\cancel{\pi}}{\cancel{\pi}}$$

$$\frac{A}{180°} = 1$$

Multiplying both sides by $180°$,

$$\cancel{180°}\left(\frac{A}{\cancel{180°}}\right) = 180°(1)$$

$$A = 180°$$

All of the computed values are now put into the table as follows:

	1	2	3	4	5	6	7	8	9
Width of θ in radians	0	$\frac{1}{6}\pi$	$\frac{1}{4}\pi$	$\frac{1}{3}\pi$	$\frac{1}{2}\pi$	$\frac{2}{3}\pi$	$\frac{3}{4}\pi$	$\frac{5}{6}\pi$	π
Width of θ in degrees	$0°$	$30°$	$45°$	$60°$	$90°$	$120°$	$135°$	$150°$	$180°$

● **PROBLEM** 839

Find $\sin 22\frac{1}{2}°$, $\cos 22\frac{1}{2}°$, $\tan 22\frac{1}{2}°$, and $\cot 22\frac{1}{2}°$.

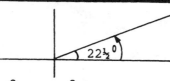

Solution: The angle $22\frac{1}{2}°$, or $22°30'$, is a first quadrant angle (see the figure). Therefore, all the required functions (sine, cosine, tangent and cotangent) are positive. Hence, from a table of trigonometric functions, $\sin 22\frac{1}{2}° = \sin 22°30' = 0.3827$, $\cos 22\frac{1}{2}° = \cos 22°30' = 0.9239$, $\tan 22\frac{1}{2}° = \tan 22°30' = 0.4142$, and $\cot 22\frac{1}{2}° = \cot 22°30' = 2.4142$.

● **PROBLEM** 840

Find $\tan 635°19'$.

Solution: The reference angle of $635°19'$ is $2(360°00')- 635°19' = 84°41'$. Therefore, the $\tan 635°19' = \tan 84°41' = 10.746$. (This value may be found from a table of trigonometric functions.) However, the angle $635°19'$ is a fourth quadrant angle and the tangent function is negative in the fourth quadrant. Hence, $\tan 635°19' = -\tan 84°41' = -10.746$.

● **PROBLEM** 841

Find $\sin 195°$, $\cos 195°$, $\tan 195°$, $\cot 195°$.

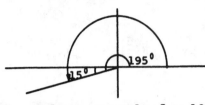

Solution: The reference angle for 195° is 15°. Also, 195° is a third quadrant angle (see Figure)

In the third quadrant, the sine and cosine functions are negative, but the tangent and the cotangent functions are positive. Therefore, sin 195° = sin 15° = -0.2588, cos 195° = cos 15° = -0.9659, tan 195° = tan 15°= 0.2679, and cot 195° = cot 15° = 3.7321. (Note that the values obtained for the trigonometric functions were found in a table of trigonometric functions.)

● **PROBLEM** 842

Find   (a) tan 30°   (b) tan 90°

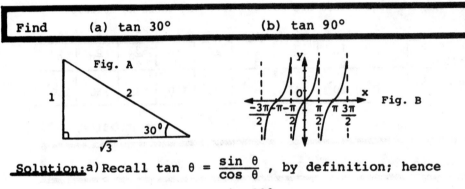

Fig. A

Fig. B

Solution: a) Recall $\tan \theta = \dfrac{\sin \theta}{\cos \theta}$, by definition; hence

$$\tan 30° = \frac{\sin 30°}{\cos 30°}$$

Looking at a 30-60 right triangle we find the values of sin 30° and cos 30°:

$$\sin = \frac{\text{opposite side}}{\text{hypotenuse}} \; ; \; \sin 30° = \tfrac{1}{2}$$

$$\cos = \frac{\text{adjacent side}}{\text{hypotenuse}} \; ; \; \cos 30° = \frac{\sqrt{3}}{2}$$

hence, $\dfrac{\sin 30°}{\cos 30°} = \dfrac{\tfrac{1}{2}}{\frac{\sqrt{3}}{2}}$

Multiplying numerator and denominator by 2:

$$= \frac{2\left(\tfrac{1}{2}\right)}{2\left(\frac{\sqrt{3}}{2}\right)} = \frac{1}{\sqrt{3}} \quad \text{therefore } \tan 30° = \frac{1}{\sqrt{3}}$$

b) To determine tan 90° we can use the same method as in part a:

$$\tan 90° = \frac{\sin 90°}{\cos 90°} \text{ , by definition of tangent.}$$

618

$$= \frac{1}{0} \ .$$

However, since the quotient $\frac{1}{0}$ is undefined, we must therefore conclude that the tangent of 90° does not exist.

We may also observe the graph of the tangent function to determine tan 90°.

$$\tan 90° = \tan \frac{90 \cdot \pi}{180} \text{ radians} = \tan \frac{\pi}{2}$$

We observe that the tangent function does not exist on the line $\frac{\pi}{2}$ (the graph of the tangent function is asymptodic to the line $x = \frac{\pi}{2}$, but never touches it).

● **PROBLEM** 843

Calculate the values of the six trigonometric functions at the point $\frac{1}{3}\pi$.

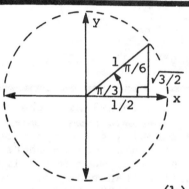

Solution: To find the trigonometric point $P\left(\frac{1}{3}\pi\right)$, proceed around the unit circle in a counterclockwise direction, since $\frac{\pi}{3}$ is a positive angle. Recall that $\sin 60°$ i.e., $\sin(\pi/3) = \sqrt{3}/2$. Now, using the Pythagorean theorem and the fact that the hypotenuse is unity because it is a unit circle we can compute the third side, which we find to be 1/2 (see figure). Therefore, the coordinates of the trigonometric point $P\left(\frac{1}{3}\pi\right)$ are $(1/2, 1/2\sqrt{3})$. Hence, we apply the following equations:

$$\cos \theta = \frac{\text{adjacent side}}{\text{hypotenuse}} \qquad \sec \theta = \frac{1}{\cos \theta} = \frac{\text{hypotenuse}}{\text{adjacent side}}$$

$$\sin \theta = \frac{\text{opposite side}}{\text{hypotenuse}} \qquad \csc \theta = \frac{1}{\sin \theta} = \frac{\text{hypotenuse}}{\text{opposite side}}$$

$$\tan \theta = \frac{\text{opposite side}}{\text{adjacent side}} \qquad \cot \theta = \frac{\cos \theta}{\sin \theta} = \frac{\text{adjacent side}}{\text{opposite side}}$$

Thus,

$$\cos \frac{1}{3}\pi = \frac{1}{2} \ , \qquad\qquad \sec \frac{1}{3}\pi = 2,$$

$$\sin \frac{1}{3}\pi = \frac{1}{2}\sqrt{3}, \qquad\qquad \csc \frac{1}{3}\pi = 2/\sqrt{3} = 2/\sqrt{3} \cdot \sqrt{3}/\sqrt{3}$$
$$= 2/3 \ \sqrt{3}$$

$$\tan \frac{1}{3}\pi = \sqrt{3}, \qquad\qquad \cot \frac{1}{3}\pi = 1/\sqrt{3} = 1/\sqrt{3} \cdot \sqrt{3}/\sqrt{3}$$
$$= 1/3 \ \sqrt{3}$$

What primary angle is coterminal with the angle of - 743°?

Solution:

$$- 743° = \alpha - n \cdot 360° \qquad (1)$$

$$- 743° = \alpha - 3 \cdot 360° = \alpha - 1080° \qquad (2)$$

Multiply both sides of equation (2) by - 1.

$$- 1(- 743°) = - 1(\alpha - 1080°)$$

$$743° = - \alpha + 1080°$$

$$743° = 1080° - \alpha \qquad (3)$$

Note that the positive integer value chosen for n results in an angle (in equation (3)) which is larger than but closest to the angle of 743°. Also,

$$0° \leq \alpha \leq 360°.$$

From equation (3),

$$\alpha = 1080° - 743° = 337°.$$

What primary angle is coterminal with the angle of 1243°?

Solution: Using the formula, $\beta = \alpha + n \cdot 360°$, obtain $1243° = \alpha + n \cdot 360°$. In this formula choose the largest positive integer for n which, when multiplied by 360, is closest to but smaller than the given angle. Also, α is an angle between 0° and 360°, that is

$$0° \leq \alpha \leq 360°.$$

Since $3 \cdot 360° = 1080°$ and $4 \cdot 360° = 1440°$, n = 3, and $1243° = \alpha + 1080°$ or $\alpha = 1243° - 1080° = 163°$. Thus, the angles 1243° and 163° are coterminal.

What primary angle is coterminal with the angle of $5\frac{1}{4}\pi$ radians?

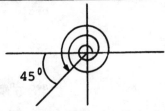

Solution: The figure illustrates the angle of $5\frac{1}{4}\pi$ radians.
Note that the angle of $5\frac{1}{4}\pi$ radians has a reference angle of 45°.
However, we seek a primary angle (an angle between 0° and 360°) which is coterminal with $5\frac{1}{4}\pi$; that is, which has the same terminal side as an angle of $5\frac{1}{4}\pi$ radians. Also, since a primary angle is a

positive angle, its initial side is the positive x-axis and the angle revolves in the counter-clockwise direction. Therefore, a primary angle with the same terminal side as an angle of $5\frac{1}{4}\pi$ radians is $(180° + 45°) = 225° = (\pi + \pi/4)$ radians $= 5/4\ \pi$ radians.

● **PROBLEM** 847

Find $\sin 202\frac{1}{2}°$ and $\tan 202\frac{1}{2}°$.

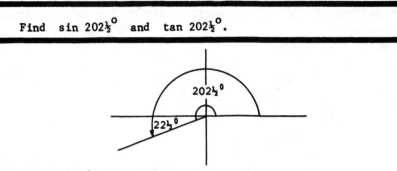

Solution: $202\frac{1}{2}°$ is a third quadrant angle. Thus its reference angle is $22\frac{1}{2}°$ ($202\frac{1}{2}° - 180° = 22\frac{1}{2}°$). See the figure.

In the third quadrant, the sine function is negative while the tangent function is positive. Therefore, using a table of trigonometric functions, $\sin 202\frac{1}{2}° = -\sin 22\frac{1}{2}° = -\sin 22°30' = -0.3827$ and $\tan 202\frac{1}{2}° = \tan 22\frac{1}{2}° = \tan 22°30' = 0.4142$.

● **PROBLEM** 848

Find the values of the trigonometric functions of an angle of 300°.

Solution: An angle of 300° is a fourth quadrant angle and its reference angle is an angle of 60°. In the fourth quadrant the sine, tangent, cotangent and cosecant functions are negative. This yields

$$\sin 300° = \sin 60° = -\frac{\sqrt{3}}{2}\ ,$$

$$\cos 300° = \cos 60° = \frac{1}{2}\ ,$$

$$\tan 300° = \tan 60° = -\sqrt{3},$$

$$\cot 300° = \cot 60° = -\frac{\sqrt{3}}{3}\ ,$$

$$\sec 300° = \sec 60° = \frac{1}{\cos 60°} = \frac{1}{\frac{1}{2}} = 2,\ \text{and}$$

$$\csc 300° = \csc 60° = \frac{1}{\sin 60°} = \frac{1}{-\sqrt{3}/2} = -\frac{2}{\sqrt{3}}$$

$$= - \frac{2\sqrt{3}}{\sqrt{3}\sqrt{3}} = - \frac{2\sqrt{3}}{3} \,.$$

• **PROBLEM 849**

Find the values of the trigonometric functions of an angle of - 510°.

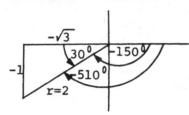

Solution: We see that angles of - 510° and - 150° are coterminal angles having the same values for the trigonometric functions. The reference angle of an angle of - 150° is an angle of 30° and - 150° is a third quadrant angle. In the third quadrant, the tangent and cotangent functions are positive while the other four functions are negative. This yields

$$\sin (- 510°) = \sin 30° = - \frac{1}{2} \,,$$

$$\cos (- 510°) = \cos 30° = - \frac{\sqrt{3}}{2} \,,$$

$$\tan (- 510°) = \tan 30° = \frac{\sqrt{3}}{3} \,,$$

$$\cot (- 510°) = \cot 30° = \frac{1}{\tan 30°} = \frac{1}{\sqrt{3}/3} = \frac{3}{\sqrt{3}}$$

$$= \frac{3\sqrt{3}}{\sqrt{3}\sqrt{3}} = \frac{3\sqrt{3}}{3} = \sqrt{3},$$

$$\sec (- 510°) = \sec 30° = \frac{1}{\cos 30°} = \frac{1}{-\sqrt{3}/2} = - \frac{2}{\sqrt{3}}$$

$$= - \frac{2\sqrt{3}}{\sqrt{3}\sqrt{3}} = - \frac{2\sqrt{3}}{3} \,,$$

$$\csc (- 510°) = \csc 30° = \frac{1}{\sin 30°} = \frac{1}{-\frac{1}{2}} = - 2.$$

• **PROBLEM 850**

Calculate the values of the six trigonometric functions at the point -3π.

Solution: To locate the trigonometric point P(-3π) we proceed 3π units around the unit circle in a clockwise direction, and we find that the coordinates of P(-3π) are (-1,0). See the accompanying figure. On the figure, proceeding from the point (1,0) to (-1,0) is π units; back to (1,0) is π more units (or 2π); and returning to (-1,0) is another π units (or 3π). Hence,

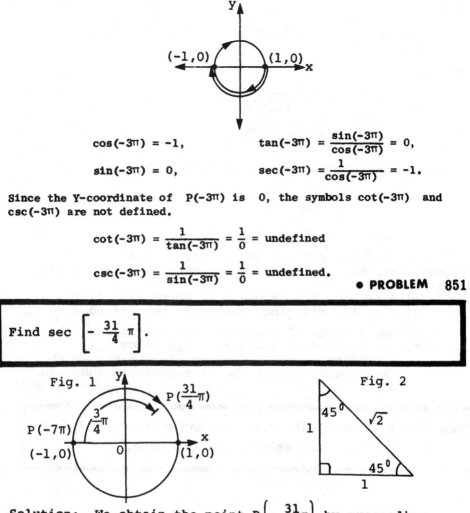

$$\cos(-3\pi) = -1, \qquad \tan(-3\pi) = \frac{\sin(-3\pi)}{\cos(-3\pi)} = 0,$$

$$\sin(-3\pi) = 0, \qquad \sec(-3\pi) = \frac{1}{\cos(-3\pi)} = -1.$$

Since the Y-coordinate of $P(-3\pi)$ is 0, the symbols $\cot(-3\pi)$ and $\csc(-3\pi)$ are not defined.

$$\cot(-3\pi) = \frac{1}{\tan(-3\pi)} = \frac{1}{0} = \text{undefined}$$

$$\csc(-3\pi) = \frac{1}{\sin(-3\pi)} = \frac{1}{0} = \text{undefined}.$$

● **PROBLEM** 851

Find $\sec\left[-\frac{31}{4}\pi\right]$.

Fig. 1

Fig. 2

Solution: We obtain the point $P\left(-\frac{31}{4}\pi\right)$ by proceeding

$\frac{31}{4}\pi = 7\pi + \frac{3}{4}\pi$ units around the unit circle from the

point $(1, 0)$ in a clockwise direction. Movement in a clockwise direction will make the value of the angle negative. Now, $7\pi = 1260°$; and starting at point $(1, 0)$, and moving in a clockwise direction, when we return to the point $(1,0)$ we have covered $360°$. Doing this three times gives us an angle of $1080°$, again leaving us at $(1, 0)$. Proceeding another $180°$ takes us to $(-1, 0)$, where we have now covered -7π, or $1260°$. Since $\frac{3}{4}\pi = 135°$, we proceed from

$(-1,0)$ to $(0, 1)$ (covering $90°$) and then another $45°$; thus we have proceeded $-\frac{31}{4}\pi$.(see Figure 1). The point

is in the first quadrant. Since each quadrant is $\frac{1}{2}\pi$,

and our desired point is $\frac{1}{4}\pi$ into the 1st quadrant, we

use the reference number $\frac{1}{4}\pi$ (or the reference angle $45°$).

Hence,

$$\sec\left(-\frac{31}{4}\,\pi\right) = \sec\frac{1}{4}\,\pi = \sqrt{2}.$$

Now, recall that $\sec x = \dfrac{1}{\cos x}$, and

$$\cos x = \frac{\text{adjacent}}{\text{hypotenuse}} \; ; \qquad \text{therefore,}$$

$$\sec x = \frac{\text{hypotenuse}}{\text{adjacent}} \; .$$

Observing the 45-45 right triangle, (see Figure 2)

$$\cos 45° = \frac{1}{\sqrt{2}} \; . \text{ Thus,}$$

$$\sec 45° = \sqrt{2}.$$

Our method for finding the values of the trigonometric functions at any number t consists of four steps:

    (i) Determine the quadrant in which P(t) lies;
    (ii) Find the reference number t_1 associated with t;

    (iii) Use a table of trigonometric functions;
    (iv) Use the proper algebraic sign (+ or -), according to which quadrant the point, which is associated with the trigonometric function, lies in.

        ● **PROBLEM 852**

Given that θ is an angle in the second quadrant and that $\tan \theta = -\dfrac{15}{8}$, find the other functions of θ.

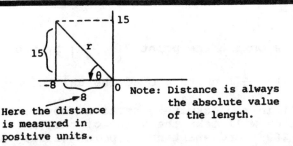

Here the distance is measured in positive units.

Note: Distance is always the absolute value of the length.

Solution: The equations which relate the trigonometric functions of an angle to the sides of a right triangle will be used to solve this problem. Therefore,

$$\tan \theta = \frac{y}{x} = -\frac{15}{8} \; .$$

Also, θ is a second quadrant angle where $y > 0$ and $x < 0$. Therefore, $y = 15$ and $x = -8$. Thus,

$$x^2 + y^2 = (-8)^2 + (15)^2 = 64 + 225 = 289 = r^2.$$

Therefore, $r = \sqrt{289} = 17$. Then

$$\sin \theta = \frac{15}{17}, \qquad\qquad \cos \theta = \frac{-8}{17} = -\frac{8}{17} \; ,$$

$$\tan \theta = -\frac{15}{8}, \qquad \cot \theta = \frac{-8}{15} = -\frac{8}{15},$$

$$\sec \theta = \frac{17}{-8} = -\frac{17}{8}, \text{ and } \csc \theta = \frac{17}{15}.$$

● **PROBLEM** 853

Given that tan θ = 2 and cos θ is negative, find the other functions of θ.

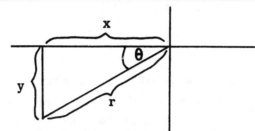

<u>Solution:</u> Since cos θ is negative, θ must be a second or third quadrant angle. In the second quadrant, the tangent function is negative. Hence, θ must be a third quadrant angle.

In the figure, the trigonometric functions have the following values:

$$\sin \theta = \frac{\text{opposite side}}{\text{hypotenuse}} = \frac{y}{r},$$

$$\cos \theta = \frac{\text{adjacent side}}{\text{hypotenuse}} = \frac{x}{r},$$

$$\tan \theta = \frac{\text{opposite side}}{\text{adjacent side}} = \frac{y}{x},$$

$$\cot \theta = \frac{1}{\tan \theta} = \frac{x}{y},$$

$$\sec \theta = \frac{1}{\cos \theta} = \frac{1}{x/r} = \frac{r}{x}, \qquad \text{and}$$

$$\csc \theta = \frac{1}{\sin \theta} = \frac{1}{y/r} = \frac{r}{y}.$$

Also, from the figure, $r^2 = x^2 + y^2$ (from the Pythagorean Theorem), or $r = \sqrt{x^2 + y^2}$. Therefore, in this problem,

$$\tan \theta = 2 = \frac{-2}{-1} = \frac{y}{x}.$$

Hence, y = - 2 and x = - 1. Also, in this problem,

$$r^2 = x^2 + y^2 = (-1)^2 + (-2)^2 = 1 + 4$$

or $r^2 = 5$ or $r = \sqrt{5}$. Therefore,

$$\sin \theta = \frac{-2}{\sqrt{5}} = -\frac{2\sqrt{5}}{5}, \qquad \cos \theta = \frac{-1}{\sqrt{5}} = -\frac{\sqrt{5}}{5},$$

$$\tan \theta = 2, \qquad \cot \theta = \frac{-1}{-2} = \frac{1}{2},$$

$$\sec \theta = \frac{\sqrt{5}}{-1} = -\sqrt{5}, \text{ and } \csc \theta = \frac{\sqrt{5}}{-2} = -\frac{\sqrt{5}}{2}.$$

If $0 < t < \frac{1}{2}\pi$, and $\tan t = \frac{1}{2}\sqrt{5}$, find $\cos t$.

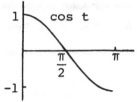

Solution: One way to find $\cos t$ is to express $\frac{1}{2}\sqrt{5}$ as a decimal, use a table of trigonometric functions to solve the equation $\tan t = \frac{1}{2}\sqrt{5}$ for t, and then again use the table to find $\cos t$. But it is much easier to use some trigonometric identities to solve this problem.

Since we are given that $\tan t = \frac{1}{2}\sqrt{5} = \frac{\sqrt{5}}{2}$, we can use the identity $\sec^2 t = 1 + \tan^2 t$ to find $\sec^2 t$. Substituting, we obtain:

$$\sec^2 t = 1 + (\tan t)^2$$

$$\sec^2 t = 1 + \left(\frac{\sqrt{5}}{2}\right)^2 \text{, or}$$

$$\sec^2 t = 1 + \frac{5}{4} = \frac{4}{4} + \frac{5}{4} = \frac{9}{4}$$

Thus, since $\cos t = \frac{1}{\sec t}$, then

$$\cos^2 t = 1/\sec^2 t = \frac{1}{\frac{9}{4}} = \frac{4}{9} \text{; therefore}$$

$$\cos t = \sqrt{\frac{4}{9}} = \pm\frac{2}{3}$$

Since $0 < t < \frac{1}{2}\pi$, we know that $\cos t > 0$ (see graph) and hence we reject the negative value, $-\frac{2}{3}$, and it follows from the last equation that $\cos t = \frac{2}{3}$.

TRIGONOMETRIC INTERPOLATIONS

Find $\cos 37°12'$.

Solution: See a table of natural trigonometric functions, which is constructed in terms of multiples of ten seconds. The cosine of 37°12' lies between 37°10' and 37°20'. Therefore, we must interpolate. The cosine decreases as the angle increases, so we form our proportion as follows, where

$$x = \text{the cosine of the angle } 37°12'$$
$$d = \text{the difference between the cos } 37°10' \text{ and cos } 37°12':$$

$$10' \begin{bmatrix} \begin{array}{c} \rule{0pt}{0pt} \\ 2' \left[\begin{array}{l} \cos 37°10' = 0.7969 \\ \cos 37°12' = \quad x \\ \cos 37°20' = 0.7951 \end{array} \right] d \end{array} \end{bmatrix} -0.0018$$

$$\frac{2}{10} = \frac{d}{-0.0018}$$

Cross multiply to obtain:

$$10d = 2(-0.0018)$$

$$d = .2(-0.0018)$$
$$= -0.00036$$
$$d \approx -0.0004$$

Thus,

$$x = 0.7969 - 0.0004$$
$$= 0.7965$$

Since the cosine is positive in the first quadrant,

$$\cos 37°12' = 0.7965$$

Remember that results obtained by interpolation are approximations. You should not use an answer that is more accurate than the original data, in this case, four significant digits.

● **PROBLEM** 856

Find the value of tan 38°46' by use of interpolation.

Solution: Since 38°46' is between 38°40' and 38°50', we assume that tan 38°36' is between tan 38°40' and tan 38°50'. In fact, since 38°46' is six-tenths of the way from 38°40' toward 38°50', we assume that tan 38°46' is six-tenths of the way from tan 38°40' = .8002 toward tan 38°50' = .8050. Using these assumptions we perform the following interpolation:

$$10' \begin{bmatrix} \begin{array}{c} 6' \left[\begin{array}{l} \tan 38°40' = .8002 \\ \tan 38°46' = \quad ? \end{array} \right] c \\ \tan 38°50 = .8050 \end{array} \end{bmatrix} .0048$$

Set up the proportion $\dfrac{c}{.0048} = \dfrac{6}{10}$

$$10c = 6(.0048)$$

$$c = \frac{6}{10}(.0048) = .0029$$

$$\tan 38°46' = .8002 + .0029 = .8031$$

Therefore, c was added because tan θ increases from

θ = 38°40' to θ = 38°50′.

● **PROBLEM** 857

Find tan 63°19.27'.

<u>Solution:</u> The value of tan 63°19.27' can be found by interpolating the values of tan 63°19' and tan 63°20'.

Degree	Value of function

$$1\begin{bmatrix}.27\begin{bmatrix}63°19' \\ 63° 19.27'\end{bmatrix} \\ 63° 20'\end{bmatrix} \quad \begin{bmatrix}\begin{bmatrix}1.9897 \\ x\end{bmatrix}d \\ 1.9912\end{bmatrix} 0.0015$$

Now, set up the following proportion:

$$\frac{d}{0.0015} = \frac{.27}{1} = .27$$

$$d = (.27)(0.0015)$$

$$d = 0.0004.$$

Therefore, x = tan 63°19.27' = tan 63°19' + d

$$= 1.9897 + 0.0004$$

$$\tan 63°19.27' = 1.9901.$$

● **PROBLEM** 858

Find θ if sin θ = .6212, and $-\frac{\pi}{2} \le \theta \le \frac{\pi}{2}$.

<u>Solution:</u> Since 0.6212 is not found in the sine table, we proceed by finding the two numbers closest to .6212, one greater and the other less than it, and interpolating.

$$10'\begin{bmatrix}c\begin{bmatrix}\sin 38°20' = .6302 \\ \sin \theta = .6212\end{bmatrix}.0010 \\ \sin 38°30' = .6225\end{bmatrix}.0023$$

We set up the proportion $\frac{c}{10'} = \frac{.0010}{.0023}$.

$$c = \frac{.0010}{.0023}(10') = \frac{10}{23}(10')$$

$$\theta = 4' \text{ to the nearest minute.}$$

Thus,   θ = 38°20' + 4'

$$= 38°24'.$$

Find θ if cos θ = -0.5731 and 0 ≤ θ ≤ 360°.

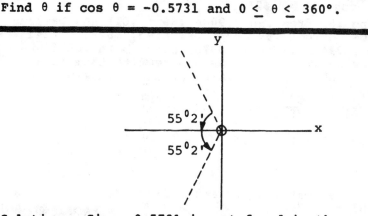

<u>Solution:</u> Since 0.5731 is not found in the cosine table, we proceed by finding the two numbers closest to .5731, one greater and the other less than it, and interpolating. Notice that the cosine function decreases as the angle increases.

$$
10' \left[\begin{array}{l} d \left[\begin{array}{l} \cos 55°0' \quad = 0.5736 \\ \cos x \qquad\; = 0.5731 \end{array} \right] 0.0005 \\ \cos 55°10' = 0.5712 \end{array} \right] 0.0024
$$

Setting up the proportion, $\dfrac{d}{10} = \dfrac{0.0005}{0.0024} = \dfrac{5}{24}$. Cross multiplying, 24d = 50

$$d = 2' \text{ to the nearest minute.}$$

Thus, x = 55°0' + 2' = 55°2'. Since we are given a negative cosine, and cosine is negative in the second and third quadrants, we know that our reference angle 55°2' appears in the second or third quadrant (see diagram). Hence

$$= 180° - 55°2' = 124°58', \text{ or}$$

$$= 180° + 55°2' = 235°2'.$$

Find a solution of the equation cos t = .6241.

<u>Solution:</u> We see that .6241 is not found in a trigonometric table; therefore, interpolation is necessary. We choose the two entries in the trigonometric table which .6241 lies between, and arrange the numbers as follows:

$$\cos .89 = .6294,$$

$$\cos t = .6241,$$

$$\cos .90 = .6216.$$

We obtain,

$$.01 \left[x \left[\begin{array}{l} \cos .89 = .6294 \\ \cos t = .6241 \\ \cos .90 = .6216 \end{array} \right] 53 \right] 78$$

by subtracting .89 from .90 (.90 - .89 = .01) and calling
the difference between .89 and t, x. Also, .6294 - 6241
= .0053, and .6294 - .6216 = .0078. Since both .0053 and
.0078 are four decimal places we can rewrite them as 53
and 78, without changing the value of the following proportion. Now,

$$\frac{x}{.01} = \frac{53}{78}$$

$$78x = 53(.01) = .53$$

$$x = \frac{.53}{78} \approx .007$$

The cost lies between .89 and .90; thus,
t = .89 + .007 = .897.

● **PROBLEM** 861

Find log sin 36°41´.

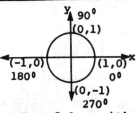

Solution: From a table of logarithms of trigonometric
functions we find log sin 36°40´ and log sin 36°50´.
Then, by the process of interpolation we find: log
sin 36°41´ = 9.77626 - 10. Since all values of the sine
function for acute angles are in the range of 0 < sin
x ≤ 1, the characteristic is negative. (Recall that for
a number less than one, the characteristic is negative.)
The range of sine can be seen by inspecting the accompanying figure. Sine is given by the y coordinate; cos is
given by the x coordinate. Observe that y value varies
from 0 to 1, as the angle varies from 0° to 90°.

● **PROBLEM** 862

Find log cos 49°13.6´.

Solution: First consult a table of logarithms of trigonometric functions.

Notice that 49°13.6´ lies between 49°10´ and 49°20´,
so that the log of 49°13.6´ will occur between the logs
of 49°10´ and 49°20´ and can be determined by interpolation.

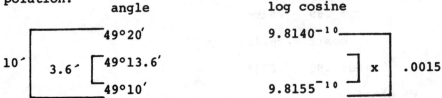

Now set up the proportion

$$\frac{3.6}{10} = \frac{x}{.0015} \quad \text{or} \quad 10x = .00540$$

$$x = .000540$$

Since cos decreases on the interval $0 \le \theta < \pi/2$, subtract x from the log cosine of $49°1\overline{0}´$.

$$
\begin{array}{r}
9.8155 \\
- \quad .000540 \\
\hline
9.814960
\end{array}
$$

Thus, log cosine $49°13.6´$ is $9.81496-10$.

TRIGONOMETRIC IDENTITIES

● **PROBLEM** 863

Change $4 + (\tan \theta - \cot \theta)^2$ to $\sec^2\theta + \csc^2\theta$.

Solution: If we square the binomial in the first expression, we have
$$4 + (\tan \theta - \cot \theta)^2 = 4 + (\tan \theta - \cot \theta)(\tan \theta - \cot \theta)$$
$$= 4 + \tan^2\theta - 2 \tan \theta \cot \theta + \cot^2\theta$$
Since $\cot \theta = \frac{1}{\tan \theta}$ the term $-2 \tan \theta \cot \theta = -2 \tan \theta\left(\frac{1}{\tan \theta}\right) = -2(1) = -2$.

Thus
$$4 + (\tan \theta - \cot \theta)^2 = 4 + \tan^2\theta - 2 + \cot^2\theta$$
$$= 2 + \tan^2\theta + \cot^2\theta$$
Since $2 = 1 + 1$,
$$= 1 + \tan^2\theta + 1 + \cot^2\theta$$
Recall $1 + \tan^2\theta = \sec^2\theta$ and $1 + \cot^2\theta = \csc^2\theta$. Replacing these values we obtain
$$4 + (\tan \theta - \cot \theta)^2 = \sec^2\theta + \csc^2\theta$$

● **PROBLEM** 864

Change $\tan \theta (\sin \theta + \cot \theta \cos \theta)$ to $\sec \theta$.

Solution: Distribute to obtain,
$$\tan \theta(\sin \theta + \cot \theta \cos \theta) = \tan \theta \sin \theta + \tan \theta \cot \theta \cos \theta$$

Recall that $\cot \theta = 1/\tan \theta$, and replace $\cot \theta$ by $1/\tan \theta$:
$$= \tan \theta \sin \theta + \tan \theta (1/\tan \theta)\cos \theta$$

$$= \tan \theta \sin \theta + \cos \theta$$

Since $\tan \theta = \sin \theta/\cos \theta$ we may replace $\tan \theta$ by $\sin \theta/\cos \theta$:
$$= \frac{\sin \theta}{\cos \theta} \sin \theta + \cos \theta$$

$$= \frac{\sin^2 \theta}{\cos \theta} + \cos \theta.$$

To combine terms, we convert $\cos \theta$ into a fraction whose denominator is $\cos \theta$, thus

$$= \frac{\sin^2 \theta}{\cos \theta} + \left(\frac{\cos \theta}{\cos \theta}\right) \cdot \cos \theta. \quad \text{(Note that}$$

$\cos \theta / \cos \theta$ equals one, so the equation is unaltered)

$$= \frac{\sin^2 \theta}{\cos \theta} + \frac{\cos^2 \theta}{\cos \theta}$$

$$= \frac{\sin^2 \theta + \cos^2 \theta}{\cos \theta}.$$

Recall the identity $\sin^2 \theta + \cos^2 \theta = 1$; hence,

$$= \frac{1}{\cos \theta}$$

$$= \sec \theta.$$

● **PROBLEM** 865

Reduce the expression $\dfrac{\tan x - \cot x}{\tan x + \cot x}$ to one involving only $\sin x$.

Solution: Since, by definition, $\tan x = \dfrac{\sin x}{\cos x}$ and

$$\cot x = \frac{1}{\tan x} = \frac{1}{\sin x/\cos x} = \frac{\cos x}{\sin x},$$

$$\frac{\tan x - \cot x}{\tan x + \cot x} = \frac{\dfrac{\sin x}{\cos x} - \dfrac{\cos x}{\sin x}}{\dfrac{\sin x}{\cos x} + \dfrac{\cos x}{\sin x}}$$

$$= \frac{\dfrac{\sin x(\sin x)}{\sin x(\cos x)} - \dfrac{\cos x(\cos x)}{\cos x(\sin x)}}{\dfrac{\sin x(\sin x)}{\sin x(\cos x)} + \dfrac{\cos x(\cos x)}{\cos x(\sin x)}}$$

$$= \frac{\dfrac{\sin^2 x - \cos^2 x}{\sin x \cos x}}{\dfrac{\sin^2 x + \cos^2 x}{\sin x \cos x}}$$

$$= \frac{\sin^2 x - \cos^2 x}{\sin x \cos x} \times \frac{\sin x \cos x}{\sin^2 x + \cos^2 x}$$

$$= \frac{\sin^2 x - \cos^2 x}{\sin^2 x + \cos^2 x}$$

Since $\sin^2 x + \cos^2 x = 1$ or $\cos^2 x = 1 - \sin^2 x$,

$$\frac{\tan x - \cot x}{\tan x + \cot x} = \frac{\sin^2 x - \cos^2 x}{\sin^2 x + \cos^2 x} = \frac{\sin^2 x - \cos^2 x}{1}$$

$$= \sin^2 x - \cos^2 x$$

$$= \sin^2 x - (1 - \sin^2 x)$$

$$= \sin^2 x - 1 + \sin^2 x$$

$$= 2 \sin^2 x - 1.$$

Find sin 105° without the use of a trig. table.

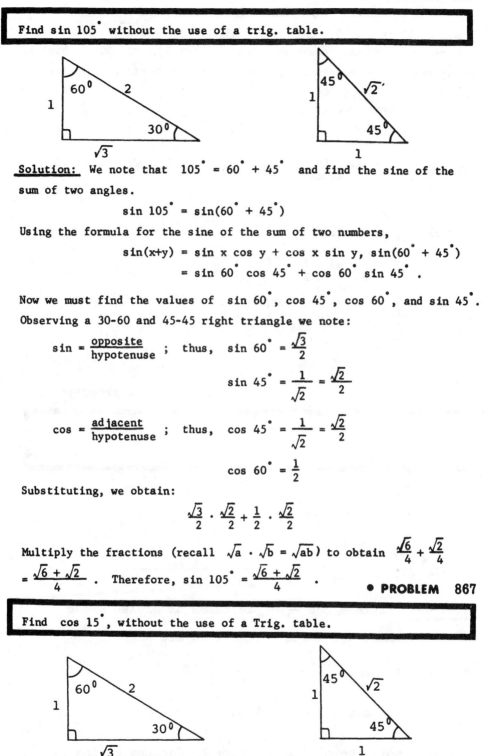

Solution: We note that 105° = 60° + 45° and find the sine of the sum of two angles.

$$\sin 105° = \sin(60° + 45°)$$

Using the formula for the sine of the sum of two numbers,

$$\sin(x+y) = \sin x \cos y + \cos x \sin y, \quad \sin(60° + 45°)$$
$$= \sin 60° \cos 45° + \cos 60° \sin 45°.$$

Now we must find the values of sin 60°, cos 45°, cos 60°, and sin 45°. Observing a 30-60 and 45-45 right triangle we note:

$$\sin = \frac{\text{opposite}}{\text{hypotenuse}} \; ; \quad \text{thus,} \quad \sin 60° = \frac{\sqrt{3}}{2}$$

$$\sin 45° = \frac{1}{\sqrt{2}} = \frac{\sqrt{2}}{2}$$

$$\cos = \frac{\text{adjacent}}{\text{hypotenuse}} \; ; \quad \text{thus,} \quad \cos 45° = \frac{1}{\sqrt{2}} = \frac{\sqrt{2}}{2}$$

$$\cos 60° = \frac{1}{2}$$

Substituting, we obtain:

$$\frac{\sqrt{3}}{2} \cdot \frac{\sqrt{2}}{2} + \frac{1}{2} \cdot \frac{\sqrt{2}}{2}$$

Multiply the fractions (recall $\sqrt{a} \cdot \sqrt{b} = \sqrt{ab}$) to obtain $\frac{\sqrt{6}}{4} + \frac{\sqrt{2}}{4}$

$= \frac{\sqrt{6} + \sqrt{2}}{4}$. Therefore, sin 105° $= \frac{\sqrt{6} + \sqrt{2}}{4}$.

Find cos 15°, without the use of a Trig. table.

Solution: We express cos 15° as the cosine of the difference of two angles whose cosine and sine we know. Since we know 15° = 45° - 30°, then cos 15° = cos(45° - 30°). Now we apply the formula for the cosine of the difference of two angles, which states:

$$\cos(u - v) = \cos u \cos v + \sin u \sin v .$$

Thus, $\cos 15° = \cos(45° - 30°) = \cos 45° \cos 30° + \sin 45° \sin 30°$.
Now, we must find the values for $\cos 45°$, $\cos 30°$, $\sin 30°$. This can
be accomplished by observing the 45-45 and 30-60 right triangles.

Since $\cos = \dfrac{\text{adjacent}}{\text{hypothenuse}}$ we find: $\cos 45° = \dfrac{1}{\sqrt{2}} = \dfrac{\sqrt{2}}{2}$

$$\cos 30° = \frac{\sqrt{3}}{2} ,$$

and since $\sin = \dfrac{\text{opposite}}{\text{hypotenuse}}$ we find: $\sin 45° = \dfrac{1}{\sqrt{2}} = \dfrac{\sqrt{2}}{2}$

$$\sin 30° = \frac{1}{2}$$

Substituting, we obtain:

$$\frac{\sqrt{2}}{2} \cdot \frac{\sqrt{3}}{2} + \frac{\sqrt{2}}{2} \cdot \frac{1}{2} .$$

Multiplying the fractions $\left(\text{recall } \sqrt{a} \cdot \sqrt{b} = \sqrt{ab}\right)$ we obtain:

$$\frac{\sqrt{6}}{4} + \frac{\sqrt{2}}{4} = \frac{\sqrt{6} + \sqrt{2}}{4} .$$

Therefore, $\cos 15° = \dfrac{\sqrt{6} + \sqrt{2}}{4} .$

● **PROBLEM** 868

Find $\cos \dfrac{1}{12} \pi$.

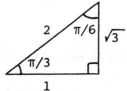
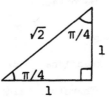

Solution: Express $\cos \dfrac{\pi}{12}$ in terms of angles whose
values of the trigonometric functions are known.

$$\cos \frac{1}{12} \pi = \cos \left(\frac{4}{12} \pi - \frac{3}{12} \pi \right)$$

$$= \cos \left(\frac{1}{3} \pi - \frac{1}{4} \pi \right)$$

Now apply the difference formula for the cosine of
two angles, α and β. $\cos(\alpha - \beta) = \cos \alpha \cos \beta + \sin \alpha \sin \beta$.
In this example, $\alpha = \dfrac{1}{3} \pi$ and $\beta = \dfrac{1}{4} \pi$.

$$\cos \left(\frac{1}{3} \pi - \frac{1}{4} \pi \right) = \cos \frac{1}{3} \pi \cos \frac{1}{4} \pi$$

$$+ \sin \frac{1}{3} \pi \sin \frac{1}{4} \pi$$

See the accompanying diagrams to find the values of these angles. We find:

$$\cos \frac{\pi}{3} = \frac{1}{2}$$

$$\cos \frac{\pi}{4} = \frac{1}{\sqrt{2}}$$

$$\sin \frac{\pi}{3} = \frac{\sqrt{3}}{2}$$

$$\sin \frac{\pi}{4} = \frac{1}{\sqrt{2}}$$

Thus, $\cos \left(\frac{1}{3} \pi - \frac{1}{4} \pi \right) = \frac{1}{2} \cdot \frac{1}{\sqrt{2}} \left(\frac{\sqrt{2}}{\sqrt{2}} \right)$

$$+ \frac{\sqrt{3}}{2} \cdot \frac{1}{\sqrt{2}} \left(\frac{\sqrt{2}}{\sqrt{2}} \right) = \frac{\sqrt{2}}{4} + \frac{\sqrt{2} \sqrt{3}}{4}$$

$$= \frac{1}{4} \sqrt{2} (1 + \sqrt{3})$$

● **PROBLEM** 869

Find the sine and cosine of 75˚ .

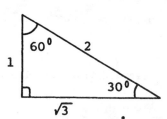
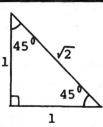

<u>Solution:</u> The angle 75˚ may be expressed as the sum or difference of two special angles, whose functions are known, in various ways: 75˚ = 45˚ + 30˚, 75˚ = 120˚ - 45˚, etc. If we choose the first of these, we use the sine of the sum of two angles,

$$\sin(\alpha + \beta) = \sin \alpha \cos \beta + \sin \beta \cos \alpha .$$

Here $\alpha = 45˚$ and $\beta = 30˚$. Then,

$$\sin 75˚ = \sin(45˚ + 30˚) = \sin 45˚ \cos 30˚ + \cos 45˚ \sin 30˚.$$

To find the sine and cosine of 45˚ and 30˚, construct 45˚, 45˚, 90˚ and 30˚, 60˚, 90˚ triangles (see Figures). Since $\sin(ϟ) = \frac{\text{side opposite}(ϟ)}{\text{hypotenuse}}$, and $\cos (ϟ) = \frac{\text{side adjacent }(ϟ)}{\text{hypotenuse}}$; then

$$\sin 45˚ = \frac{1}{\sqrt{2}} = \frac{1}{\sqrt{2}} \left(\frac{\sqrt{2}}{\sqrt{2}} \right) = \frac{\sqrt{2}}{2} , \quad \sin 30˚ = \tfrac{1}{2} ,$$

and

$$\cos 45° = \frac{1}{\sqrt{2}} = \frac{1}{\sqrt{2}}\left(\frac{\sqrt{2}}{\sqrt{2}}\right) = \frac{\sqrt{2}}{2} \ , \quad \cos 30° = \frac{\sqrt{3}}{2} \ .$$

Substituting,

$$\sin 75° = \frac{\sqrt{2}}{2} \cdot \frac{\sqrt{3}}{2} + \frac{\sqrt{2}}{2} \cdot \frac{1}{2} = \frac{\sqrt{2}\sqrt{3}}{4} + \frac{\sqrt{2}}{4} = \frac{1}{4}\sqrt{6} + \frac{1}{4}\sqrt{2}$$

$$= \frac{1}{4}\left(\sqrt{6} + \sqrt{2}\right).$$

To find $\cos 75°$, apply the formula for the cosine of the sum of two angles,

$$\cos(\alpha + \beta) = \cos \alpha \cos \beta - \sin \alpha \sin \beta \ .$$

Here $\alpha = 45°$ and $\beta = 30°$. Thus,

$$\cos 75° = \cos(45° + 30°) = \cos 45° \cos 30° - \sin 45° \sin 30°.$$

$$= \frac{\sqrt{2}}{2} \cdot \frac{\sqrt{3}}{2} - \frac{\sqrt{2}}{2} \cdot \frac{1}{2} = \frac{1}{4}\left(\sqrt{6} - \sqrt{2}\right).$$

If we use the approximate values, $\sqrt{6} = 2.4494$, $\sqrt{2} = 1.4142$, we find

$$\sin 75° = 0.9659, \quad \cos 75° = 0.2588,$$

which check with the values given in the tables.

● **PROBLEM** 870

Find an expression for $\tan(u + v)$.

Solution: By definition of the tangent function, $\tan \theta = \dfrac{\sin \theta}{\cos \theta}$.

Then,

$$\tan(u+v) = \frac{\sin(u+v)}{\cos(u+v)} \tag{1}$$

The addition formulas for the sine and cosine functions are:

$$\sin(\alpha+\beta) = \sin \alpha \cos \beta + \cos \alpha \sin \beta$$
$$\cos(\alpha+\beta) = \cos \alpha \cos \beta - \sin \alpha \sin \beta$$

Replacing α by u and β by v, and using these addition formulas, equation (1) becomes:

$$\tan(u+v) = \frac{\sin(u+v)}{\cos(u+v)} = \frac{\sin u \cos v + \cos u \sin v}{\cos u \cos v - \sin u \sin v}$$

If neither $\cos u = 0$ nor $\cos v = 0$, we can divide both the numerator and the denominator of this fraction by the product $\cos u \cos v$ to obtain a formula that involves only the tangent function:

$$\tan(u+v) = \frac{\dfrac{\sin u \cos v + \cos u \sin v}{\cos u \cos v}}{\dfrac{\cos u \cos v - \sin u \sin v}{\cos u \cos v}}$$

$$= \frac{\dfrac{\sin u \cos v}{\cos u \cos v} + \dfrac{\cos u \sin v}{\cos u \cos v}}{\dfrac{\cos u \cos v}{\cos u \cos v} - \dfrac{\sin u \sin v}{\cos u \cos v}}$$

$$= \frac{\dfrac{\sin u}{\cos u} + \dfrac{\sin v}{\cos v}}{1 - \dfrac{\sin u \sin v}{\cos u \cos v}}$$

$$= \frac{\dfrac{\sin u}{\cos u} + \dfrac{\sin v}{\cos v}}{1 - \dfrac{\sin u}{\cos u}\dfrac{\sin v}{\cos v}}$$

$$\tan(u+v) = \frac{\tan u + \tan v}{1 - \tan u \tan v} \ .$$

Find $\sin 15^\circ$, $\cos 15^\circ$, $\tan 15^\circ$, and $\cot 15^\circ$.

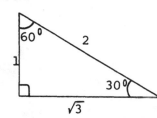 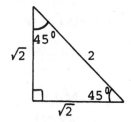

__Solution:__ To find the values of these trigonometric functions, use the following subtraction formulas for the sine, cosine, tangent, and cotangent functions.

$$\sin(\alpha - \beta) = \sin \alpha \cos \beta - \cos \alpha \sin \beta$$

$$\cos(\alpha - \beta) = \cos \alpha \cos \beta + \sin \alpha \sin \beta$$

$$\tan(\alpha - \beta) = \frac{\tan \alpha - \tan \beta}{1 + \tan \alpha \tan \beta}$$

$$\cot(\alpha - \beta) = \frac{\cot \alpha \cot \beta + 1}{\cot \beta - \cot \alpha}$$

Recall that in a $30^\circ - 60^\circ$ right triangle:

$$\sin 30^\circ = \tfrac{1}{2} \quad \text{and} \quad \cos 30^\circ = \frac{\sqrt{3}}{2} \ ,$$

$$\sin 60^\circ = \frac{\sqrt{3}}{2} \quad \text{and} \quad \cos 60^\circ = \tfrac{1}{2} \ .$$

In a $45^\circ - 45^\circ$ right triangle:

$$\sin 45^\circ = \frac{\sqrt{2}}{2} \quad \text{and} \quad \cos 45^\circ = \frac{\sqrt{2}}{2} \quad \text{(see figures)}$$

Now, substitute $\alpha = 45^\circ$ and $\beta = 30^\circ$ in the above formulas.

$$\sin 15^\circ = \sin(45^\circ - 30^\circ) = \sin 45^\circ \cos 30^\circ - \cos 45^\circ \sin 30^\circ$$

$$= \left(\frac{\sqrt{2}}{2}\right)\left(\frac{\sqrt{3}}{2}\right) - \left(\frac{\sqrt{2}}{2}\right)\left(\frac{1}{2}\right)$$

$$= \frac{\sqrt{2}\sqrt{3}}{4} - \frac{\sqrt{2}}{4} = \frac{\sqrt{2}(\sqrt{3} - 1)}{4}$$

$$\cos 15^\circ = \cos(45^\circ - 30^\circ) = \cos 45^\circ \cos 30^\circ + \sin 45^\circ \sin 30^\circ$$

$$= \left(\frac{\sqrt{2}}{2}\right)\left(\frac{\sqrt{3}}{2}\right) + \left(\frac{\sqrt{2}}{2}\right)\left(\frac{1}{2}\right)$$

$$= \frac{\sqrt{2}\sqrt{3}}{4} + \frac{\sqrt{2}}{4} = \frac{\sqrt{2}(\sqrt{3} + 1)}{4}$$

$$\tan 15^\circ = \frac{\tan 45^\circ - \tan 30^\circ}{1 + \tan 45^\circ \tan 30^\circ}$$

$$= \frac{1 - \frac{\sqrt{3}}{3}}{1 + 1\left(\frac{\sqrt{3}}{3}\right)} = \frac{1 - \frac{\sqrt{3}}{3}}{1 + \frac{\sqrt{3}}{3}}$$

Obtaining a common denominator of 3 for both the numerator and the denominator,

$$\tan 15^{\circ} = \frac{\frac{3}{3} - \frac{\sqrt{3}}{3}}{\frac{3}{3} + \frac{\sqrt{3}}{3}} = \frac{\frac{3 - \sqrt{3}}{3}}{\frac{3 + \sqrt{3}}{3}} = \frac{3 - \sqrt{3}}{\cancel{3}} \cdot \frac{\cancel{3}}{3 + \sqrt{3}} = \frac{3 - \sqrt{3}}{3 + \sqrt{3}}$$

$$\cot 15^{\circ} = \cot(45^{\circ} - 30^{\circ})$$

$$= \frac{\cot 45^{\circ} \cot 30^{\circ} + 1}{\cot 30^{\circ} - \cot 45^{\circ}}$$

$$= \frac{(1)(\sqrt{3}) + 1}{\sqrt{3} - 1}$$

$$= \frac{\sqrt{3} + 1}{\sqrt{3} - 1}$$

● **PROBLEM** 872

Find the sin of 195° using functions of 225° and 30°.

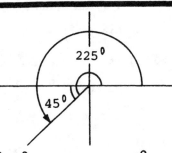

Solution: $195^{\circ} = 225^{\circ} - 30^{\circ}$. Hence, $\sin 195^{\circ} = \sin(225^{\circ} - 30^{\circ})$.
Using the subtraction formula for sines, $\sin(\alpha - \beta) = \sin \alpha \cos \beta - \cos \alpha \sin \beta$ where $\alpha = 225^{\circ}$ and $\beta = 30^{\circ}$, $\sin 195^{\circ} = \sin(225^{\circ} - 30^{\circ}) = \sin 225^{\circ} \cos 30^{\circ} - \cos 225^{\circ} \sin 30^{\circ}$.
The reference angle for 225° is 45° ($225^{\circ} - 180^{\circ} = 45^{\circ}$), and the angle 225° is a third quadrant angle. See the figure.

In the third quadrant both the sine and cosine functions are negative. From a table of trigonometric functions, $\sin 225^{\circ} = \sin 45^{\circ} = -0.7071 = -\frac{\sqrt{2}}{2}$, $\cos 225^{\circ} = \cos 45^{\circ} = -0.7071 = -\frac{\sqrt{2}}{2}$, $\cos 30^{\circ} = 0.8660 = \frac{\sqrt{3}}{2}$ and $\sin 30^{\circ} = 0.5000 = \frac{1}{2}$. Therefore,

$$\sin 195^{\circ} = \sin(225^{\circ} - 30^{\circ}) = \left(-\frac{\sqrt{2}}{2}\right)\left(\frac{\sqrt{3}}{2}\right) - \left(-\frac{\sqrt{2}}{2}\right)\left(\frac{1}{2}\right)$$

$$= \frac{-\sqrt{2}\sqrt{3}}{4} + \frac{\sqrt{2}}{4}$$

$$= \frac{-\sqrt{2}\sqrt{3} + \sqrt{2}}{4}$$

$$= \frac{\sqrt{2}(-\sqrt{3} + 1)}{4}$$

Hence, $\sin 195^\circ = \frac{\sqrt{2}(1 - \sqrt{3})}{4}$.

● **PROBLEM** 873

Derive a formula for $\sin 3\alpha$ in terms of $\sin \alpha$.

<u>Solution:</u> We may regard 3α as $\alpha + 2\alpha$, and use the addition formula for the sine of two angles.

$$\sin(a + b) = \sin a \cos a + \cos a \sin b$$

where $a = \alpha$ and $b = 2\alpha$.

$$\sin 3\alpha = \sin(\alpha + 2\alpha) = \sin \alpha \cos 2\alpha + \cos \alpha \sin 2\alpha.$$

Now replace $\sin 2\alpha$ and $\cos 2\alpha$ by the expressions

$$\cos 2a = \cos^2 a - \sin^2 a$$

and

$$\sin 2a = 2 \sin a \cos a.$$

We find that

$$\sin 3\alpha = (\sin \alpha)\left(\cos^2\alpha - \sin^2\alpha\right) + (\cos \alpha)(2 \sin \alpha \cos \alpha)$$
$$= \sin \alpha \cos^2\alpha - \sin^3\alpha + 2 \sin \alpha \cos^2\alpha$$
$$= 3 \sin \alpha \cos^2\alpha - \sin^3\alpha.$$

Finally, since we wish a result involving only $\sin \alpha$, replace $\cos^2\alpha$ by $1 - \sin^2\alpha$; then

$$\sin 3\alpha = 3(\sin \alpha)\left(1 - \sin^2\alpha\right) - \sin^3\alpha$$
$$= 3 \sin \alpha - 3 \sin^3\alpha - \sin^3\alpha$$
$$= 3 \sin \alpha - 4 \sin^3\alpha.$$

This is the desired identity.

● **PROBLEM** 874

Simplify the expression

$$2 \sin\left(\frac{3\pi}{2} - \theta\right) - 3 \cos(\pi + \theta) - \tan(-\theta) + \cot\left(\frac{\pi}{2} + \theta\right).$$

<u>Solution:</u> Apply the following laws of subtraction and addition of trigonometric functions:

$$\sin(\alpha - \beta) = \sin \alpha \cos \beta - \cos \alpha \sin \beta$$
$$\cos(\alpha + \beta) = \cos \alpha \cos \beta - \sin \alpha \sin \beta$$

Derive $\cot(\alpha + \beta)$ from the addition formula for the tangent:

$$\tan(\alpha + \beta) = \frac{\tan \alpha + \tan \beta}{1 - \tan \alpha \tan \beta}$$

Since $\cot(\alpha + \beta) = \frac{1}{\tan(\alpha + \beta)}$, we have: $\cot(\alpha + \beta)$

$$= \frac{1 - \tan \alpha \tan \beta}{\tan \alpha + \tan \beta}$$

$$= \frac{1 - \dfrac{1}{\cot \alpha \cot \beta}}{\dfrac{1}{\cot \alpha} + \dfrac{1}{\cot \beta}}$$
Multiplying all terms by $\cot \alpha \cot \beta$,

hence, $\cot(\alpha+\beta) = \dfrac{\cot \alpha \cot \beta - 1}{\cot \beta + \cot \alpha}$

Then, applying the above formulas:

$$\sin\left(\frac{3\pi}{2} - \theta\right) = \sin \frac{3\pi}{2} \cos \theta - \cos \frac{3\pi}{2} \sin \theta$$

$$= (-1)\cos \theta - (0)\sin \theta = -\cos \theta$$

$$\cos(\pi+\theta) = \cos \pi \cos \theta - \sin \pi \sin \theta = (-1)\cos \theta - (0)\sin \theta$$

$$= -\cos \theta$$

$$\cot\left(\frac{\pi}{2} + \theta\right) = \frac{\cot \frac{\pi}{2} \cot \theta - 1}{\cot \theta + \cot \frac{\pi}{2}} = \frac{(0)\cot \theta - 1}{\cot \theta + 0} = \frac{-1}{\cot \theta} = -\tan \theta$$

Recall: $\tan(-\theta) = -\tan \theta$, that is, the tangent of a negative angle is equal to minus the tangent of a positive angle. Thus, substituting:

$$2 \sin\left(\frac{3\pi}{2} - \theta\right) - 3 \cos(\pi + \theta) - \tan(-\theta) + \cot\left(\frac{\pi}{2} + \theta\right)$$

$$= 2(-\cos \theta) - 3(-\cos \theta) - (-\tan \theta) + (-\tan \theta)$$

$$= -2 \cos \theta + 3 \cos \theta + \tan \theta - \tan \theta$$

$$= \cos \theta$$

● **PROBLEM 875**

(a) Derive a formula for $\cos 3\theta$ which involves only the cosine function. (b) Also, derive a formula for $\sin 3\theta$ which involves only the sine function.

Solution: (a) $\cos 3\theta = \cos(2\theta + \theta)$

Using the addition formula for the cosine function which states that $\cos(\alpha + \beta) = \cos \alpha \cos \beta - \sin \alpha \sin \beta$ and replacing α by 2θ and β by θ:

$$\cos 3\theta = \cos(2\theta + \theta) = \cos 2\theta \cos \theta - \sin 2\theta \sin \theta . \quad (1)$$

The double-angle formulas for $\cos 2\alpha$ and $\sin 2\alpha$ yield:

$$\cos 2\alpha = 2 \cos^2\alpha - 1 \quad \text{and} \quad \sin 2\alpha = 2 \sin \alpha \cos \alpha .$$

Hence, replacing α by θ in both cases:

$$\cos 3\theta = \left(2 \cos^2\theta - 1\right)\cos \theta - (2 \sin \theta \cos \theta)\sin \theta$$

$$= 2 \cos^3 \theta - \cos \theta - 2 \sin^2 \theta \cos \theta \quad (2)$$

Since $\sin^2\theta = 1 - \cos^2\theta$, equation (2) becomes:

$$\cos 3\theta = 2 \cos^3\theta - \cos \theta - 2\left(1 - \cos^2\theta\right)\cos \theta$$

$$= 2 \cos^3\theta - \cos \theta - 2 \cos \theta + 2 \cos^3\theta$$

$$\cos 3\theta = 4 \cos^3\theta - 3 \cos \theta .$$

(b) $\sin 3\theta = \sin(2\theta + \theta)$

Using the addition formula for the sine function which states that $\sin(\alpha + \beta) = \sin \alpha \cos \beta + \cos \alpha \sin \beta$ and replacing α by 2θ and β by θ:

$$\sin 3\theta = \sin(2\theta + \theta) = \sin 2\theta \cos \theta + \cos 2\theta \sin \theta \quad (3)$$

Again, using the double-angle formulas for $\cos 2\alpha$ and $\sin 2\alpha$ and replacing α by θ in both cases:

$$\cos 2\theta = 2 \cos^2\theta - 1 \quad \text{and} \quad \sin 2\theta = 2 \sin \theta \cos \theta.$$

Hence, equation (3) becomes:

$$\sin 3\theta = (2 \sin \theta \cos \theta)\cos \theta + \left(2 \cos^2\theta - 1\right)\sin \theta$$

$$= 2 \cos^2\theta \sin\theta + 2 \cos^2\theta \sin\theta - \sin\theta \qquad (4)$$

Since $\cos^2\theta = 1 - \sin^2\theta$, equation (4) becomes:

$$\sin 3\theta = 2(1 - \sin^2\theta)\sin\theta + 2(1 - \sin^2\theta)\sin\theta - \sin\theta$$

$$= 2\sin\theta - 2\sin^3\theta + 2\sin\theta - 2\sin^3\theta - \sin\theta$$

$$\sin 3\theta = 3\sin\theta - 4\sin^3\theta .$$

SOLVING TRIANGLES

● **PROBLEM** 876

Given the right triangle with a = 3, b = 4, and c = 5, find the values of the trigonometric functions of α.

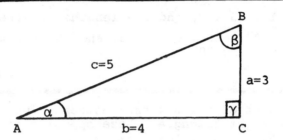

Solution: In the accompanying figure, a is the side opposite angle α, b is the side opposite angle β, and c is the side opposite angle γ. The values of the trigonometric functions of α are:

$$\cos\alpha = \frac{\text{adjacent side}}{\text{hypotenuse}}, \quad \sin\alpha = \frac{\text{opposite side}}{\text{hypotenuse}},$$

$$\tan\alpha = \frac{\text{opposite side}}{\text{adjacent side}}, \quad \cot\alpha = \frac{1}{\tan\alpha},$$

$$\sec\alpha = \frac{1}{\cos\alpha}, \quad \text{and} \quad \csc\alpha = \frac{1}{\sin\alpha}.$$

Therefore:

$$\cos\alpha = \frac{4}{5}, \qquad \sin\alpha = \frac{3}{5},$$

$$\tan\alpha = \frac{3}{4}, \qquad \cot\alpha = \frac{1}{3/4} = \frac{4}{3},$$

$$\sec\alpha = \frac{1}{4/5} = \frac{5}{4}, \quad \csc\alpha = \frac{1}{3/5} = \frac{5}{3}.$$

● **PROBLEM** 877

Solve the oblique triangle ABC for side c, and the two unknown angles, where a = 20, b = 40, α = 30 ; and α is the angle between sides b and c.

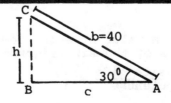

<u>Solution:</u> If we draw an altitude h from b to c, as in the accompanying diagram, we can find the length of this altitude by trigonometry.

$$\sin 30° = \frac{h}{b}$$

$$h = b \sin 30°$$

Since $\sin 30° = \frac{1}{2}$, and $b = 40$,

$$h = 40 \left(\frac{1}{2}\right) = 20 = \text{side a}$$

Thus, the triangle must have the altitude as one of its sides; therefore, we have a right triangle with angles 30°, 60°, and 90°. Sides of such right triangles are in proportion $1 : \sqrt{3} : 2$, and the lengths are therefore $20 : 20\sqrt{3} : 40$. Hence, $c = 20\sqrt{3}$, and the two unknown angles are 60° and 90°.

● **PROBLEM**   878

Find the values of the six trigonometric functions of an angle, in a right triangle, whose opposite side is 3 and hypotenuse 5.

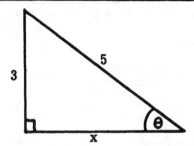

<u>Solution:</u> We are given that, in a right triangle, the side opposite the angle is 3 and the hypotenuse is 5. To determine the adjacent side, x (see figure), we recall the **Pythagorean Theorem** which states that the sum of the square of the legs of a right triangle equals the square of the hypotenuse. Thus,

$$3^2 + x^2 = 5^2$$

$$9 + x^2 = 25$$

$$x^2 = 16$$

$$x = 4$$

Now that we know the value of each side of the triangle, we can find the values of the six trigonometric functions:

$$\sin \theta = \frac{\text{opposite side}}{\text{hypotenuse}} = \frac{3}{5}$$

$$\cos \theta = \frac{\text{adjacent side}}{\text{hypotenuse}} = \frac{4}{5}$$

$$\tan \theta = \frac{\text{opposite side}}{\text{adjacent side}} = \frac{3}{4}$$

$$\csc \theta = \frac{1}{\sin \theta} = \frac{\text{hypotenuse}}{\text{opposite side}} = \frac{5}{3}$$

$$\sec \theta = \frac{1}{\cos \theta} = \frac{\text{hypotenuse}}{\text{adjacent side}} = \frac{5}{4}$$

$$\cot \theta = \frac{1}{\tan \theta} = \frac{\text{adjacent side}}{\text{opposite side}} = \frac{4}{3}$$

● **PROBLEM** 879

Given a = 8, c = 7, β = 135°, find b.

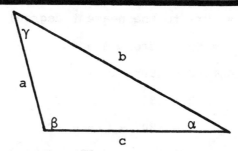

<u>Solution:</u> Use the law of cosines to find one side given two sides and an included angle.

$$b^2 = a^2 + c^2 - 2ac \cos(\text{included angle}) \text{ where the included}$$

angle is the angle between the two given sides.

$$b^2 = a^2 + c^2 - 2ac \cos \beta$$

$$b^2 = 64 + 49 - 2 \cdot 8 \cdot 7 \cos 135°$$

$$= 113 - 112 \cdot \left(\frac{\sqrt{2}}{2}\right) = 113 + 79.184$$

$$= 192.184$$

$$b = 13.863$$

● **PROBLEM** 880

Solve triangle ABC, given a = 30, b = 50, ∠C = 25°.

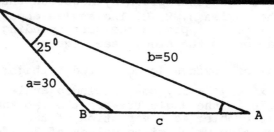

<u>Solution:</u> Two of the sides of △ABC, and their included angle are given. We wish to find the third side, c. Therefore use the law of cosines to find c.

$$c^2 = a^2 + b^2 - 2ab \cos C$$

$$c^2 = 30^2 + 50^2 - 2(30)(50) \cos 25°$$

$$c^2 = 900 + 2500 - 2(30)(50)(0.9063)$$

$$c^2 = 681.1$$

$$c = 26 \text{ (to two significant digits)}$$

Use the law of sines to find one of the remaining angles.

$$\frac{\sin A}{30} = \frac{\sin 25°}{26}$$

$$\sin A = \frac{30 \sin 25°}{26} = \frac{30(0.4226)}{26}$$

$$\sin A = 0.4876$$

$$\angle A = 29° \text{ (to the nearest degree)}$$

$\angle B$ can be found from $\angle A$ and $\angle C$.

$$\angle A + \angle B + \angle C = 180°$$

$$B = 180° - \angle A - \angle C$$

$$= 180° - (\angle A + \angle C)$$

$$= 180° - (29° + 25°)$$

$$= 180° - (54°)$$

$$\angle B = 126°$$

● **PROBLEM** 881

Solve the triangle ABC, given a = 17, b = 23, c = 32.

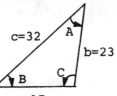

a=17

Solution: Use the generalized Pythagorean theorem to determine the angles A, B, C. The angles appear in the upper case directly opposite the corresponding lower case designation of the side, (see figure).

The Law of cosines (generalized Pythagorean) states $z^2 = x^2 + y^2 - 2xy \cos$ (included angle). The included angle is the angle between the two known sides, x and y. This problem is a variation of the usual Law of Cosines problem, which gives values of 2 sides and their included angle, and instead supplies the three sides and asks for the angles opposite those sides Solve the equation for cos (included angle) to obtain:

$$\cos \text{ (included angle)} = \frac{x^2 + y^2 - z^2}{2xy}$$

$$\cos A = \frac{b^2 + c^2 - a^2}{2bc} \qquad \cos B = \frac{a^2 + c^2 - b^2}{2ac}$$

$$\cos A = \frac{23^2 + 32^2 - 17^2}{(2)(23)(32)} \qquad \cos B = \frac{17^2 + 32^2 - 23^2}{(2)(17)(32)}$$

$$\cos A = 0.8587 \qquad\qquad \cos B = 0.7206.$$

∢ A = 31° (nearest degree)   ∢ B = 44° (nearest degree)

$$\cos C = \frac{a^2 + b^2 - c^2}{2ab}$$

$$\cos C = \frac{17^2 + 23^2 - 32^2}{(2)(17)(23)}$$

$$\cos C = -0.2634$$

∢ C = 105°   (nearest degree)

It is not necessary to calculate angle C using the law of cosines when angles A and B are known. Angle C can be found instead by the following:

∢A + ∢B + ∢C = 180°   since the sum of the angles of a triangle is 180°,

or 31° + 44° + ∢C = 180°

75° + ∢C = 180°

∢C = 180° - 75° = 105°.

● **PROBLEM** 882

Solve triangle ABC for ∢ C, a, and b, given
∢ A = 52°20', ∢ B = 28°10', c = 87.6

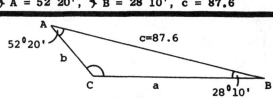

Solution: Given two angles, the third is determined, since the three angles of a triangle equal 180°. Thus,

∢A + ∢B + ∢C = 180°.

∢C = 180° - (∢A + ∢B)

∢C = 180° - 80°30' = 179°60' - 80°30' = 99°30'.

Sides a and b can be determined from the law of sines. In any triangle, the sides are proportional to the sines of the opposite angles (see figure). Then:

$$\frac{a}{\sin A} = \frac{c}{\sin C} \qquad\qquad \frac{b}{\sin B} = \frac{c}{\sin C}$$

Now, since ∢A, ∢B, and side C were given, and we previously determined ∢C, substituting into these proportions, and using a table of trigonometric functions, will give us a and b. Thus,

$$a = \frac{c \sin A}{\sin C} = \frac{c \sin 52^\circ 20'}{\sin 99^\circ 30'}$$

$$b = \frac{c \sin B}{\sin C} = \frac{c \sin 28^\circ 10'}{\sin 99^\circ 30'}$$

$$a = \frac{87.6(0.7916)}{(0.9863)} = 70.3$$

$$b = \frac{87.6(0.4720)}{(0.9863)} = 41.9$$

● **PROBLEM** 883

Given $a = 9$, $b = 6$, $c = 5$, find γ .

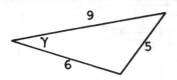

Solution: Since the lengths of the three sides of the triangle are known, the only unknown is $\cos \gamma$. If this value is known, the value of γ can be found. Since triangles can have angles no larger than 180°, if $\cos \gamma$ is positive, γ is a first quadrant angle, and if $\cos \gamma$ is negative, γ is a second quadrant angle. Use the law of cosines to determine the angle since three sides are given, i.e., $c^2 = a^2 + b^2 - 2ab \cos(\text{included angle})$. Here all is known except the included angle.

$$5^2 = 9^2 + 6^2 - 2 \cdot 9 \cdot 6 \cos \gamma$$

$$\cos \gamma = \frac{25 - 81 - 36}{-108} = \frac{92}{108} = 0.85185$$

$$\gamma = 31^\circ 35.2'$$

● **PROBLEM** 884

In right triangle ABC, if $a = 3$, and $b = 4$, find c, α, β .

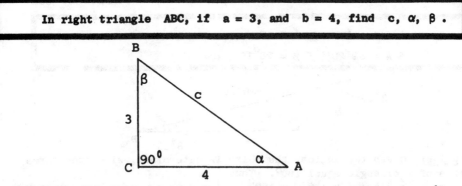

Solution: The figure can be drawn with accuracy in this case. (See the Figure).

The Pythagorean Theorem $c^2 = a^2 + b^2$ such that $c^2 = 3^2 + 4^2 = 25$ shows that $c = 5$. The ratio of the two known sides is, indeed, a value of a trigonometric function of one of the acute angles. In fact,

$$\tan \alpha = \frac{\text{side opposite } \measuredangle \alpha}{\text{side adjacent } \measuredangle \alpha} = \frac{3}{4}$$

Similarly for $\measuredangle \beta$, $\tan \beta = \frac{4}{3}$

$$\cot \alpha = \frac{1}{\tan \alpha} = \frac{1}{\dfrac{\text{side opposite } \measuredangle \alpha}{\text{side adjacent } \measuredangle \alpha}} = \frac{\text{side adjacent } \measuredangle \alpha}{\text{side opposite } \measuredangle \alpha} = \frac{4}{3}$$

Similarly for $\sphericalangle \beta$, $\cot \beta = \frac{3}{4}$. It is merely a matter of choosing the preferred one of these ratios. Let us choose $\tan \alpha = \frac{3}{4}$. Thus,

$$\tan \alpha = \frac{3}{4} = 0.75000$$
$$\alpha = 36°52.2'$$
$$\alpha + \beta + 90° = 180°$$
$$\alpha + \beta = 180° - 90° = 90°$$

Therefore, $\qquad \beta = 90° - \alpha = 90° - 36°52.2' = 53°7.8'$

● **PROBLEM** 885

In triangle ABC, if a = 675, $\alpha = 48°36'$, find b, c, β.

Solution: If we wish to work with α, the functions of α involving a and one other side are

$$\sin \alpha = \frac{\text{opposite}}{\text{hypotenuse}}$$

$\sin \alpha = a/c$

$$\tan \alpha = \frac{\text{opposite}}{\text{adjacent}}$$

$\tan \alpha = a/b$, and

$$\cot \alpha = \frac{1}{\tan \alpha} = \frac{\text{adjacent}}{\text{opposite}}$$

$\cot \alpha = b/a$. Either of these ratios can be chosen. If the first is chosen, $\sin \alpha = a/c$ or $c \sin \alpha = a$ and $c = a/\sin \alpha$. Thus,

$$c = \frac{675}{\sin 48°36'} = \frac{675}{0.75011} = 899.85$$

Using $\tan \alpha = a/b$ or $b = a/\tan \alpha$, we obtain

$$b = \frac{675}{\tan 48°36'} = 595.08$$

To find β, $\alpha + \beta + 90° = 180°$

$$\alpha + \beta = 180° - 90° = 90°$$
$$\beta = 90° - \alpha = 90° - 48°36'$$
$$\beta = 41°24'$$

● **PROBLEM** 886

Find all the sides and angles of triangle ABC, given a = 137, c = 78.0, $< C = 23°0'$.

Solution: Draw triangle ABC, filling in the given information. Thus

647

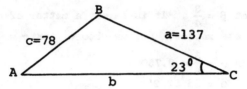

we divide the problem into 3 parts:

(1)   find angle A.
(2)   find angle B .
(3)   find side b .

In order to find angle A we may use the law of sines, $\dfrac{\sin A}{\text{side } a} = \dfrac{\sin C}{\text{side } c}$,

because we are given side a = 137, side c = 78, and sin C = sin 23° ;

thus, $\dfrac{\sin A}{137} = \dfrac{\sin 23°0'}{78}$.

Using our Trig. table we find $\sin 23° = 0.3907$. Thus

$$\frac{\sin A}{137} = \frac{0.3907}{78}$$

Multiplying both sides by 137 we obtain

$$\sin A = \frac{137(0.3907)}{78}$$

$$\sin A = 0.6862 .$$

Using our Trig. table we find that ∢ A = 43°20' . Our Trig. table
only gives values of sine between 0° and 90° . Since we are dealing
with an angle in a triangle, which can take on values greater than 90°
(recall there are 180° in a triangle) we must examine what happens
to the sine function in the second quadrant, that is between 90° and
180°

We use our trignometric identity $\sin \theta = \sin(180-\theta)$:
$$\sin 43°20' = \sin(180-43°20') = \sin 136°40'$$

Hence   ∢ A = 43°20' <u>or</u> 136°40'

Thus there are two solutions. We now proceed to the next part of our
problem, finding angle B, and side b.

∢ A + ∢ B ∢ C = 180°   (There are 180° in a
∢ A + ∢ B + 23° = 180°   triangle)
∢ A + ∢ B = 157°

If ∢ A = 43°20' then

$$43°20' + ∢ B = 157°$$
$$∢ B = 157° - 43°20' = 113°40'$$

Since we now know ∢ B, we may apply the law of sines:

648

$$\frac{\sin B}{\text{side } b} = \frac{\sin C}{\text{side } c}$$

to find side b:

$$\frac{\sin 113°40'}{b} = \frac{\sin 23°}{78}$$

Cross multiplying we obtain

$$b \sin 23° = 78 \sin 113°40'$$

$$b = \frac{78 \sin 113°40'}{\sin 23°}$$

Substituting in the values $\sin 113°40' = 0.9159$ and $\sin 23° = 0.3907$

we obtain

$$b = 78\left(\frac{0.9159}{0.3907}\right) = 183$$

Hence if we choose $< A = 43°20'$ then

$$< B = 113°40' \quad \text{and}$$

$$\text{side } b = 183$$

If, however, we choose $< A = 136°40'$, then since

$$< A + < B = 157°$$
$$136°40' + < B = 157°$$

and

$$< B = 157° - 136°40' = 20°20'$$

Applying the law of sines to find side b gives us:

$$\frac{\sin B}{\text{side } b} = \frac{\sin C}{\text{side } c}$$

$$\frac{\sin 20°20'}{b} = \frac{\sin 23°}{78}$$

Cross multiplying gives us

$$b \sin 23° = 78 \sin 20°20'$$

$$b = \frac{78 \sin 20°20'}{\sin 23°}$$

Substituting in $\sin 20°20' = 0.3475$ and $\sin 23° = 0.3907$ we obtain

$$b = \frac{78(0.3475)}{0.3907} = 69$$

Hence if we choose $< A = 136°40'$, then $< B = 20°20'$, and

side $b = 69$.

● **PROBLEM** 887

Find all the sides and angles of triangle ABC, given $a = 43$, $b = 32$, $< B = 67°$.

Solution: Draw triangle ABC, filling in the given information. Thus

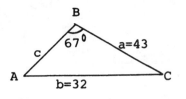

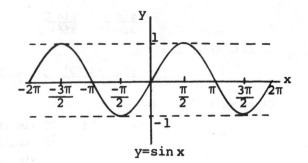

$y = \sin x$

we divide the problem into 3 parts:

    (1)   find angle A

    (2)   find angle C

    (3)   find side c

In order to find angle A we may use the law of sines, $\frac{\sin A}{\text{side } a} = \frac{\sin B}{\text{side } b}$, because we are given, side a = 43, side b = 32, and sin B = sin 67 . Thus, $\frac{\sin A}{43} = \frac{\sin 67}{32}$, multiplying both sides by 43,

$$\sin A = \frac{43 \sin 67°}{32}$$

In our trig table we find sin 67° = 0.9205. Thus,

$$\sin A = \frac{43(0.9205)}{32} = 1.2369$$

We found sin A = 1.2369; however the sine function is only defined

on the interval [-1,1] — that is, it can only take on values between

-1 and 1, which can be seen from the graph y = sin x, as shown.

Thus there is no angle A such that sin A = 1.2369, and no triangle

can contain such an angle, hence triangle ABC is non-existent, and

we cannot solve for the other sides or angles.

                                    ● **PROBLEM**   888

In the accompaning figure, given b = 16.351, c = 11.189, $\alpha = 42°19.8'$; find a, β, γ.

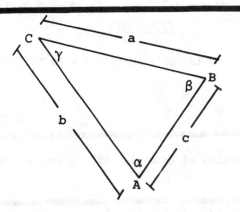

<u>Solution</u>:   The Law of Tangents states:

$$\frac{b - c}{b + c} = \frac{\tan \frac{1}{2} (\beta - \gamma)}{\tan \frac{1}{2} (\beta + \gamma)}$$

We use this relation here to find the remaining angles, β and γ. We know two sides, b and c, and we can derive the two unknown angles from the law of tangents and the fact that the sum of the angles in a triangle is 180°. From this last fact, we know the sum of the unknown angles:

$\beta + \gamma = 180° - \alpha$, or $\beta + \gamma = 180° - 42°19.8´$

$\qquad = 179°60.0´ - 42°19.8´ = 137°40.2´$

$\frac{1}{2}(\beta + \gamma) = \frac{1}{2} (137°40.2´)$.   Now dividing by 2

we have 68°, with 1° remaining. Since 1° = 60´, we have 60´ + 40.2´ = 100.2´, and divided by 2 = 50.1´. Thus,

$$\frac{1}{2} (\beta + \gamma) = 68°50.1´.$$

Now, since b - c = 16.351 - 11.189 = 5.162, b+c=27.540,

$$\frac{1}{2}(\beta + \gamma) = 68°50.1´,$$

substituting into the law of tangents and solving for tan $\frac{1}{2}$ (β - γ):

$$\tan \frac{1}{2} (\beta - \gamma) = \frac{5.162}{27.54} \tan 68°50.1´$$

Now, using a table of values of trigonometric functions we find that tan 68°50´ = 2.583. We accept this value also as tan 68°50.1´, because the two angles differ only by .1, and even interpolation will not give us a different value. Thus, we have:

$$\tan \frac{1}{2} (\beta - \gamma) = \frac{5.162}{27.54} (2.583),$$

and performing the multiplication:

$\tan \frac{1}{2} (\beta - \gamma) = .4841$

Now, we again refer to a table of trigonometric functions. We are looking for the angle whose tangent is .4841 (or the arctangent of .4841). We find that, tan 25°50´= .4841. Thus,

$\frac{1}{2} (\beta - \gamma) = $ Arctan .4841, or

$\frac{1}{2} (\beta - \gamma) = 25°50´$

We now solve the equations

$\frac{1}{2} (\beta + \gamma) = 68°50.1´$

$\frac{1}{2} (\beta - \gamma) = 25°50´$

simultaneously for β and γ by distributing $\frac{1}{2}$, and then adding:

$$\tfrac{1}{2} \beta + \tfrac{1}{2} \gamma = 68°50.1´$$

$$\tfrac{1}{2} \beta - \tfrac{1}{2} \gamma = 25°50.0´$$

$$\beta = 93°100.1´$$

Since 60´ = 1°, we have: β = 94°40.1´

We obtain the value for γ by substituting into the equation, β + γ = 137°40.2´, and obtain: γ = 43°0.1´.

To obtain the value of side a, we use the law of sines to form the proportion:

$$\frac{a}{\sin \alpha} = \frac{c}{\sin \gamma}$$

Since we know c, and can find sin γ and sin α, this law enables us to compute the value of a. Solving for a,

$$a = \sin \alpha \left(\frac{c}{\sin \gamma} \right)$$

$$a = \sin 42°19.8´ \left(\frac{11.189}{\sin 43°0.1´} \right)$$

We must now obtain values for sin 42°19.8´ and 43°0.1´. Since the former is very close to 42°40´ and the latter is very close to 43°, we do not need interpolation. (In fact, interpolating will give us the same values as those for sin 42°20´ and 43°). Using a table we find:

sin 42°20´ = .6734 ≃ sin 42°19.8´

sin 43°= .6820 ≃ sin 43°0.1´;

and substituting:

$$a = .6734 \left(\frac{11.189}{.6820} \right) = 11.0479 \simeq 11.048$$

Therefore,   a = 11.048

$$\beta = 94°40.1´$$

$$\gamma = 43°0.1´$$

● **PROBLEM** 889

Solve the given oblique triangle ABC for a, c and β, where b = 47, α = 20°43´, γ = 153°44´.

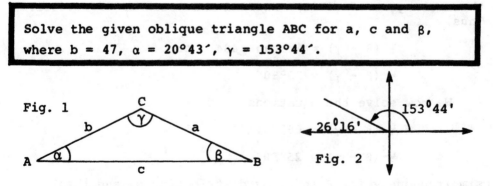

Fig. 1

Fig. 2

Solution: The given triangle can be seen in Figure 1. Since the sum of the angles in a triangle is $180°$, $\alpha + \beta + \gamma = 180°$, or $\beta = 180° - \alpha - \gamma$; and substituting:

$$\beta = 180° - 20°43' - 153°44'$$

$$= (179°60' - 20°43') - 153°44'$$

$$= 159°17' - 153°44'$$

$$= 158°77' - 153°44' = 5°33'$$

Using the law of sines, we set up the proportion,

$$\frac{a}{\sin \alpha} = \frac{b}{\sin \beta} \; .$$ Since we are given α and b, and have just found β, we employ this law to obtain a. We have:

$$\frac{a}{\sin 20°43'} = \frac{47}{\sin 5°33'}$$

$$a = \sin 20°43' \left(\frac{47}{\sin 5°33'} \right)$$

Taking the logarithm of both sides of this equation, and recalling the fact that $\log (a/b) = \log a - \log b$, and $\log (ab) = \log a + \log b$, we obtain:

$$\log a = \log \sin 20°43' + (\log 47 - \log \sin 5°33').$$

To solve for log a we use a table of logs of trig functions, a table of logs of numbers, and interpolation. First let us find log sin $20°43'$. To do this we interpolate:

$$10 \left[3 \begin{bmatrix} \begin{array}{cc} \underline{\quad x \quad} & \underline{\quad \log \sin x \quad} \\ 20°40' & 9.5477 - 10 \\ 20°43' & ? \end{array} \end{bmatrix} y \right] .0033$$

$$\begin{array}{cc} 20°50' & 9.5510 - 10 \end{array}$$

$$\frac{3}{10} = \frac{y}{.0033}$$

$$10y = .0099$$

$$y = .00099$$

$$\log \sin 20°43' = 9.5477 - 10 + .00099 = 9.54869 - 10$$

Now, we must find log 47. From a table of logarithms of numbers we find the mantissa to be 67210. To find the characteristic, recall that for a number greater than 1, the characteristic is positive and is one less than the number of digits before the decimal. Since the number 47 has two digits before the decimal, the characteristic is 1. Thus, $\log 47 = 1.67210$.

To obtain log sin $5°33'$ we again interpolate:

$$
10\left[\begin{array}{c} 3\left[\begin{array}{cc} \begin{array}{c} \underline{x} \\ 5°30´ \\ 5°33´ \\ 5°40´ \end{array} & \begin{array}{c} \underline{\log \sin x} \\ 8.9816 - 10 \\ ? \\ 8.9945 - 10 \end{array} \end{array}\right]y \end{array}\right].0129
$$

$$\frac{3}{10} = \frac{y}{.0129}$$

$$10y = .0387$$

$$y = .00387$$

log sin 5°33´ = 8.9816 - 10 + .00387 = 8.98547 - 10

Thus, substituting we obtain:

log a = 9.54869 - 10 + [1.67210 - (8.98547 - 10)],

and since 1.67210 can be rewritten as 11.67210 - 10,

log a = 9.54869 - 10 + [11.67210 - 10 - (8.98547 - 10)];

performing the subtraction in the brackets:

$$
\begin{array}{r}
11.67210 - 10 \\
- \quad (8.98547 - 10) \\
\hline
2.68663 \qquad = 12.68663 - 10
\end{array}
$$

performing the addition:
$$
\begin{array}{r}
9.54869 - 10 \\
+ \ (12.68663 - 10) \\
\hline
22.23532 - 20 = 2.23532
\end{array}
$$

Thus, log a = 2.23532

To find a we interpolate; observe that the characteristic is 2, and the mantissa is 23532. Recall that if the characteristic is positive, it is one less than the number of digits before the decimal in the number. Thus, there are 3 digits before the decimal, and interpolating:

$$
1\left[\begin{array}{c} y\left[\begin{array}{cc} \begin{array}{c} \underline{N} \\ 171. \\ a \\ 172. \end{array} & \begin{array}{c} \underline{\log N} \\ 2.2330 \\ 2.23532 \\ 2.2355 \end{array} \end{array}\right].00232 \end{array}\right].0025
$$

$$\frac{y}{1} = \frac{.00232}{.0025}$$

$$y = \frac{232}{250} = .928$$

a = 171. + .928 = 171.928 $\approx$ 171.93

To find side c we again employ the law of sines, and set up the following proportion:

$$\frac{c}{\sin \gamma} = \frac{b}{\sin \beta}, \quad \text{and substituting:}$$

$$\frac{c}{\sin 153°44´} = \frac{47}{\sin 5°33´}$$

$$c = \sin 153°44´ \left(\frac{47}{\sin 5°33´} \right) \quad . \text{ Taking the}$$

logarithm of both sides, we obtain:

$$\log c = \log \sin 153°44´ + (\log 47 - \log \sin 5°33´)$$

To solve for log c we must find log sin 153°44´, and substitute this value, and those found previously for log 47 and log sin 5°33´. To obtain sin 153°44´ we first observe the coordinate axis, (see fig. 2), and discover that sin 153°44´ = sin 26°16´. We now use interpolation:

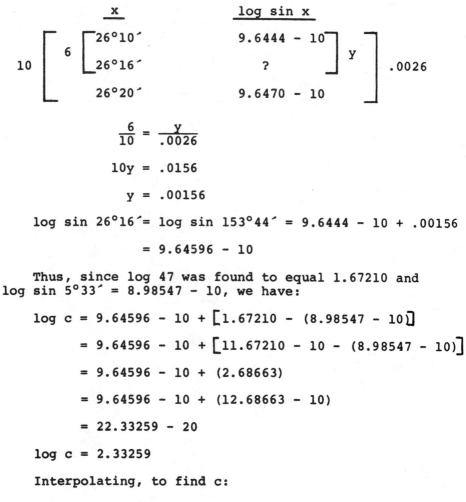

$$\frac{6}{10} = \frac{y}{.0026}$$

$$10y = .0156$$

$$y = .00156$$

$$\log \sin 26°16´ = \log \sin 153°44´ = 9.6444 - 10 + .00156$$

$$= 9.64596 - 10$$

Thus, since log 47 was found to equal 1.67210 and log sin 5°33´ = 8.98547 - 10, we have:

$$\log c = 9.64596 - 10 + \left[1.67210 - (8.98547 - 10) \right]$$

$$= 9.64596 - 10 + \left[11.67210 - 10 - (8.98547 - 10) \right]$$

$$= 9.64596 - 10 + (2.68663)$$

$$= 9.64596 - 10 + (12.68663 - 10)$$

$$= 22.33259 - 20$$

$$\log c = 2.33259$$

Interpolating, to find c:

$$y = \frac{.00019}{.0021} = \frac{19}{210} = .0904761$$

Thus, c = 215 + .090 = 215.09

Therefore,   a = 171.93

                c = 215.09

                β =   5°33´

CHAPTER 28

INVERSE TRIGONOMETRIC FUNCTIONS

> **Basic Attacks and Strategies for Solving Problems in this Chapter. See pages 657 to 670 for step-by-step solutions to problems.**

The inverses of functions defined by formulas involving trigonometric functions can be determined in the same way we determine the inverses of other functions. That is, interchange the independent and dependent variables and solve for the new dependent variable. Note that the elements of the ranges of the inverse trigonometric functions are angles in standard position.

When evaluating the inverse function of the sine, cosine, and tangent, it helps to first remember that the arc (sine/cosine/tangent), respectively, of x is the angle (or number) whose sine/cosine/tangent, respectively, is x. Secondly, to find the value of an inverse trigonometric function, apply its definition and determine which quadrant(s) the angle terminates. Then, calculate or find the y value (angle) which is the solution. The use of a sketched graph is helpful to visualize the positions of the angle.

For example, to find the value of

$$\sin^{-1}\left(\frac{1}{2}\right) \quad \text{or} \quad \arcsin\left(\frac{1}{2}\right).$$

Set $y = \sin^{-1}\left(\frac{1}{2}\right)$. Then, by definition of the inverse sine function, $\sin y = \frac{1}{2}$. Since

$$\sin \frac{\pi}{6} = \frac{1}{2} \quad \text{and} \quad \frac{\pi}{6}$$

is in the range of $y = \sin^{-1}x$, then

$$\sin\left(\frac{1}{2}\right) = \frac{\pi}{6}.$$

A calculator can be used to easily find the values of inverse trigonometric functions.

The domains of the other trigonometric functions: $\csc x$, $\sec x$, and $\cot x$, can be restricted so that each has an inverse function. A calculator can be used to find the value of the functions. If appropriate keys are not on the calculator, then we can use the sine, cosine, and tangent keys to find the values.

To find the value of composite function, first find the value of the argument function. After this value is determined, then take the trigonometric function of this value. For example, evaluate $\sin(\cos^{-1} 1/2)$. This expression means the sine of the angle between 0 and π, inclusive, whose cosine is $1/2$. The angle $\pi/3$ is the value whose cosine is $1/2$. Then,

$$\sin\left(\frac{\pi}{3}\right) = \frac{\sqrt{3}}{2}.$$

Therefore,

$$\sin(\cos^{-1} 1/2) = \frac{\sqrt{3}}{2}.$$

Step-by-Step Solutions to Problems in this Chapter, "Inverse Trigonometric Functions"

Calculate the following numbers.

(a) Arctan $\sqrt{3}$ (c) Tan^{-1} 1.871

(b) Tan^{-1}.2027

Solution: a) The expression tan y = x is equivalent to arctan x = $\tan^{-1}$ x = y. Let the expression arctan $\sqrt{3}$ = y. Hence, the expression arctan $\sqrt{3}$ = y is equivalent to tan y = $\sqrt{3}$ = 1.7321. In a table of trigonometric functions, the number y that corresponds to tan y = 1.7321 is approximately 1.05.

b) Note that the expression $\tan^{-1}$.2027 = arctan .2027 . Let the expression $\tan^{-1}$ .2027 = y. Hence, the expression $\tan^{-1}$.2027 = arctan .2027 = y is equivalent to tan y = .2027. In a table of trigonometric functions, the number y that corresponds to tan y = .2027 is .20.

c) Note that the expression $\tan^{-1}$ 1.871 = arctan 1.871. Let the expression $\tan^{-1}$1.871 = y. Hence, the expression $\tan^{-1}$1.871 = arctan 1.871 = y is equivalent to tan y = 1.871. In a table of trigonometric functions, the number y that corresponds to tan y = 1.871 is 1.08.

Find arcsin $\frac{1}{2}$.

Solution: Let y = arcsin $\frac{1}{2}$; then sin y = $\frac{1}{2}$. Since sin 30° = $\frac{1}{2}$, we have y = 30°, 150° . Because the sine has a period of 360°, any angle which we can obtain by adding any integral multiple of 360° to 30° or 150° would also satisfy this equation. We thus get

$$\text{arcsin } \tfrac{1}{2} = \begin{cases} 2n\pi + \tfrac{1}{6}\pi \\ 2n\pi + \tfrac{5}{6}\pi \end{cases} \text{ or } \begin{cases} n(360°) + 30° \\ n(360°) + 150° \end{cases},$$

where n may be any integer, positive, negative, or zero.

Find arctan (- $\sqrt{3}$).

$y = \arctan(-\sqrt{3})$; $\tan y = -\sqrt{3}$; $y = 120°, 300°$. Since the tangent has period π, we have

$$\arctan(-\sqrt{3}) = n\pi + \frac{2}{3}\pi .$$

● PROBLEM 893

Does the sine function have an inverse?

Solution: The domain of the sine function is the set of real numbers, and its range is the interval $\{-1 \le y \le 1\}$. For each number y in this interval, the equation $y = \sin x$ has infinitely many solutions. For example, for the number 0 in $\{-1 \le y \le 1\}$,

$$0 = \sin 0° = \sin 360° = \sin 720° \ldots .$$

Recall that a function has an inverse if for each number y in the range of the function there is only one number x in the domain of the function such that $y = f(x)$. Thus, the sine function does not have an inverse.

● PROBLEM 894

In $\triangle ABC$, $A = \text{arc } \cos\left(-\frac{\sqrt{3}}{2}\right)$. What is the value of A expressed in radians?

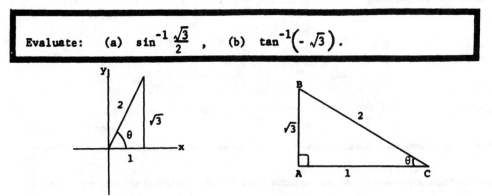

Solution: The expression "arc $\cos\left(-\frac{\sqrt{3}}{2}\right)$" means "the angle whose cosine equals $-\frac{\sqrt{3}}{2}$." Angles whose cosine equals $-\frac{\sqrt{3}}{2}$ are

$150°, 210°, -150°,$ and $-210°$.

Since the principal value of an arc cosine of an angle is the

positive angle having the smallest numerical value of the angle,

$150°$, or $\frac{5\pi}{6}$, is the principal value of angle A. .

● PROBLEM 895

Evaluate: (a) $\sin^{-1} \frac{\sqrt{3}}{2}$, (b) $\tan^{-1}\left(-\sqrt{3}\right)$.

<u>Solution:</u> (a) Recall that inverse sines are angles. Thus we are looking for the angle whose sin is $\frac{\sqrt{3}}{2}$. $\sin^{-1}\frac{\sqrt{3}}{2} = x$ means $\sin x = \frac{\sqrt{3}}{2}$ where $\sin = \frac{\text{opposite}}{\text{hypotenuse}}$.

We note that triangle ABC is a 30-60 right triangle, and angle $x = 60^\circ$. Since $\sin 60^\circ = \frac{\sqrt{3}}{2}$,

$$\sin^{-1}\frac{\sqrt{3}}{2} = 60^\circ .$$

(b) Recall that inverse tangents are angles. Thus we are looking for the angle whose tangent is $-\sqrt{3}$. $\text{Tan}^{-1}\left(-\sqrt{3}\right) = \theta$ means $\tan\theta = -\sqrt{3}$ where $\tan = \frac{\text{opposite}}{\text{adjacent}}$.

Since tangent is negative in the 4th quadrant, we draw our triangle there, and note it is a 30-60 right triangle, and angle $\theta = (-60^\circ)$. Since $\tan(-60^\circ) = \frac{-\sqrt{3}}{1}$, $\tan^{-1}\left(-\sqrt{3}\right) = -60^\circ$.

● **PROBLEM** 896

Evaluate $\arctan\frac{5}{12} - \arccos\frac{3}{5}$.

<u>Solution:</u> Let $\alpha = \arctan\frac{5}{12}$, $\beta = \arccos\frac{3}{5}$. We are to evaluate the angle $\alpha - \beta$, where $-\frac{\pi}{2} < \alpha < \frac{\pi}{2}$ and $0 \le \beta \le \pi$. We first find some function of $\alpha - \beta$, say $\sin(\alpha - \beta)$. Since $\tan\alpha = \frac{5}{12}$, using the Pythagorean theorem and the fact that $\tan\alpha = \frac{\text{opposite}}{\text{adjacent}}$, we obtain hypotenuse $= 13$; thus $\sin\alpha = \frac{5}{13}$ and $\cos\alpha = \frac{12}{13}$. Also, since $\cos\beta = \frac{3}{5}$, $\sin\beta = \frac{4}{5}$. Then, using the formula for the sin of the difference of two angles, $\sin(\alpha - \beta) = \sin\alpha\cos\beta - \cos\alpha\sin\beta$

$$= \frac{5}{13}\cdot\frac{3}{5} - \frac{12}{13}\cdot\frac{4}{5}$$

$$= \frac{15}{65} - \frac{48}{65} = -\frac{33}{65} = -.5077$$

approximately. Now, observe that α is less than $\frac{\pi}{4}$ and β is greater than $\frac{\pi}{4}$ $\left(\text{this can be seen by observing that } \sin\alpha = \frac{5}{13} \text{ and } \sin\beta = \frac{4}{5}, \text{ and also observing the sin function from } 0 \text{ to } \frac{\pi}{2}\right)$. Hence $\alpha - \beta$ is a negative angle. Reference to tables gives us

$$\alpha - \beta = -\arcsin .5077 = -.5325 \text{ approximately.}$$

● **PROBLEM** 897

If principal values are used, then what does the ratio $\frac{\text{arc sin } \frac{1}{2}}{\text{arc tan } 1}$ equal?

<u>Solution:</u> The expression "arc sin ½" means "the angle whose sine equals ½"; similarly, "arc tan 1" means "the angle whose tangent equals 1."

Since the principal value of an arc sine or an arc tangent of an angle is the smallest numerical value of the angle,

$$\text{arc sin } \tfrac{1}{2} = 30° \text{ and arc tan } 1 = 45°.$$

Hence, the ratio

$$\frac{\text{arc sin } \tfrac{1}{2}}{\text{arc tan } 1} = \frac{30°}{45°} = \frac{2}{3} .$$

● **PROBLEM 898**

Find $\sin^{-1} 0.4075$.

Solution: The inverse sine of 0.4075 is the angle, x, whose sine is 0.4075. Since 0.4075 is not found in the sine table, we proceed by finding the two numbers closest to .4075 (one greater and the other less than) which do appear in the table, and interpolating.

$$10' \left[d \left[\begin{array}{l} \sin 24°0' = 0.4067 \\ \sin x = 0.4075 \end{array} \right] 0.0008 \right. \left. \begin{array}{l} \\ \end{array} \right] 0.0027$$
$$\sin 24°10' = 0.4094$$

Setting up the proportion, $\dfrac{d}{10} = \dfrac{0.0008}{0.0027} = \dfrac{8}{27}$. Cross multiplying, $27d = 80$

$$d = 3' \text{ to the nearest minute.}$$

Therefore,

$$\sin^{-1} 0.4075 = 24°0' + 3' = 24°3'$$

Each function value is the function value of many angles. For example, a negative cosine is the cosine of an angle in the second quadrant and also of an angle in the third quadrant. But we defined the inverse relation so that it is a function. Hence, $\sin^{-1} 0.4075$ defines a unique angle, namely, 24°3'.

● **PROBLEM 899**

Find Arccot 10.365.

Solution: The problem is to find the angle θ whose cotangent function has the value 10.365 . In a table of trigonometric functions, the number 10.365 does not appear in the vertical column headed by cot. We find two numbers, 10.385 and 10.354, one of which is larger and the other smaller than 10.365.

In such a case, we must interpolate to obtain the proper angle. It must lie between 5°30' and 5°31'. The form for interpolation can be used.

Angles	Values of Function

$$1'=60'' \left[x \left[\begin{array}{l} 5°30' \\ \theta \\ 5°31' \end{array} \right. \right. \qquad \left. \left. \begin{array}{l} 10.385 \\ 10.365 \\ 10.354 \end{array} \right] -0.020 \right] -0.031$$

Note that the negative values found in the values of the cotangent function were obtained by subtracting a number from the one below it. The same was done for the angles. Illustrating this point: 5°31' - 5°30' = 1' = 60", θ-5°30' = x, 10.354 - 10.385 =-0.031, 10.365 - 10.385 = -0.020. Now, the following proportion is set up:

$$\frac{x}{60"} = \frac{-0.020}{-0.031}$$

or $x = \frac{-0.020}{-0.031}$ (60") = (.645)(60") = 38.7"

Therefore, θ = 5°30'0" + 0°0'38.7" = 5°30'38.7".

Hence, Arccot 10.365 = θ =5° 30'38.7".

● **PROBLEM 900**

Find Arccos 0.74652.

<u>Solution:</u>   The problem is to find the angle θ whose cosine function has value 0.74652.   In a table of trigonometric functions, the number 0.74652 does not appear in the column headed by cos. We find two numbers, 0.74664 and 0.74644, one of which is larger and the other smaller than 0.74652. In such a case, we must interpolate to obtain the proper angle. It must lie between 41°42' and 41°43'. The form for interpolation may be used.

Angles   Values of Function

$$1'\begin{bmatrix} x\begin{bmatrix} 41°42' \\ \theta \end{bmatrix} \\ 41°43' \end{bmatrix} \quad \begin{bmatrix} \begin{bmatrix} 0.74664 \\ 0.74652 \end{bmatrix}-0.00012 \\ 0.74644 \end{bmatrix}-0.00020$$

Note that the negative values found in the values of the cosine function were obtained by subtracting a number from the one below it. The same was done for the angles. Illustrating this point: 41°43' - 41°42' = 1', θ - 41°42' = x, 0.74652 - 0.74664 = -0.00012, and 0.74644 - 0.74664 = -0.00020.

Now the following proportion is set up:

$$\frac{x}{1'} = \frac{-0.00012}{-0.00020}$$

or $x = \frac{-0.00012}{-0.00020}$ (1') = $\frac{-0.00012}{-0.00020}$ (60")

$$= (.6)(60") = 36.0" = 36"$$

Therefore, x = 41°42' 0" + 0° 0'36" = 41°42'36".

Hence, Arccos 0.74652 = θ = 41°42'36".

● **PROBLEM 901**

Evaluate cos[arc sin (-1)].

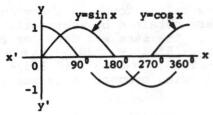

Solution: The expression "arc sin (-1)" means "the angle whose sine equals -1." Between $0°$ and $360°$, the only angle whose sine equals -1 is $270°$.

Hence, cos [arc sin(-1)] = cos $270°$.

The value of cos $270°$ = 0.

Note: In problems of this type, a sketch of y = sin x and y = cos x is very useful.

● PROBLEM 902

Evaluate $\cos\left(\sin^{-1}\frac{1}{2}\right)$.

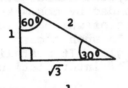

Solution: Inverse sine of $\frac{1}{2}$ is the angle whose sin is $\frac{1}{2}$. From our diagram of a 30° - 60° - 90° right triangle we observe sin 30° = $\frac{1}{2}$ so $\sin^{-1}\frac{1}{2}$ = 30°. Thus $\cos\left(\sin^{-1}\frac{1}{2}\right)$ = cos(30°). Consulting our diagram we see cos 30° = $\frac{\sqrt{3}}{2}$. Therefore $\cos\left(\sin^{-1}\frac{1}{2}\right)$ = $\frac{\sqrt{3}}{2}$.

● PROBLEM 903

Express in radical form the positive value of sin(arc cos ½).

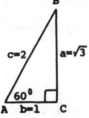

Solution: The expression "arc cos ½" means "the angle whose cosine equals ½." In the diagram, note that A is the angle whose cosine is ½.

Since $60°$ is the angle whose cosine equals ½, A = $60°$. The positive value of sin(arc cos ½) = sin $60°$ = $\frac{\sqrt{3}}{2}$.

662

Find $\sin\left(\text{Arc tan } \frac{3}{4}\right)$.

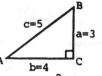

Solution: The expression "arc tan $\frac{3}{4}$" means "the angle whose tangent equals $\frac{3}{4}$." Spelled with a capital "A", "Arc tan" refers to the principal value of that angle, that is, its measure as a positive acute angle.

Draw a right triangle ABC with an acute angle A such that the leg adjacent to A is 4 and the leg opposite A is 3. Thus, $\tan A = \frac{3}{4}$ and the expression "Arc tan $\frac{3}{4}$" can be replaced by A.

Hence, $\sin\left(\text{Arc tan } \frac{3}{4}\right) = \sin A.$

Apply the Law of Pythagoras: if the legs of a right triangle are 3 and 4, its hypotenuse is 5. Hence, c = 5.

Therefore, $\sin A = \frac{3}{5}$, or $\sin\left(\text{Arc tan } \frac{3}{4}\right) = \frac{3}{5}$.

Find the value of $\sin\left(\text{Arc cos } \frac{8}{17}\right)$.

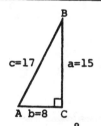

Solution: The expression "arc cos $\frac{8}{17}$ " means "the angle whose cosine equals $\frac{8}{17}$." Spelled with a capital "A", "Arc cos" refers to the principal value of that angle, that is, its measure as a positive acute angle.

Draw a right $\triangle ABC$ with an acute angle A such that the leg adjacent to A is 8 and the hypotenuse 17. Thus, $\cos A = \frac{8}{17}$, and the expression "Arc cos $\frac{8}{17}$" can be replaced by A.

Hence, $\sin\left(\text{Arc cos } \frac{8}{17}\right) = \sin A = \frac{a}{c}$.

Since c = 17 and b = 8, then $a^2 = 17^2 - 8^2$

$$= 289 - 64 = 225$$

Hence, $a = \sqrt{225} = 15$

Therefore, $\dfrac{a}{c} = \dfrac{15}{17}$ or

$$\sin\!\left(\text{Arc cos } \dfrac{8}{17}\right) = \dfrac{15}{17} \ .$$

● **PROBLEM 906**

What is the value of $\tan\!\left(\text{Arc cos } \dfrac{\sqrt{2}}{2}\right)$?

Solution: The expression "arc cos $\dfrac{\sqrt{2}}{2}$ " means "the angle whose cosine is $\dfrac{\sqrt{2}}{2}$ " . Spelled with a capital "A," "Arc cos" refers to the principle value of that angle. If the cosine is positive, the principle value of an arc cosine is its measure as a positive acute angle.

Hence, Arc cos $\dfrac{\sqrt{2}}{2} = 45°$.

Therefore, $\tan\!\left(\text{Arc cos } \dfrac{\sqrt{2}}{2}\right) = \tan 45° = 1.$

● **PROBLEM 907**

Find $\sin \text{arccos } \dfrac{4}{5}$, if $\text{arccos } \dfrac{4}{5}$ is in quadrant I.

Solution: Let $\theta = \text{arccos } \dfrac{4}{5}$; then $\cos \theta = \dfrac{4}{5}$. We can then construct the triangle of the figure. From the triangle we get $\sin \theta = \dfrac{3}{5}$; therefore $\sin \text{arccos } \dfrac{4}{5} = \dfrac{3}{5}$.

● **PROBLEM 908**

Calculate $\sin\!\left(\cos^{-1} \dfrac{3}{5}\right).$

Solution: Let $t = \cos^{-1} \dfrac{3}{5}$. This means that $\cos t = \dfrac{3}{5}$. We restrict t to $0 \le t \le \pi$ to obtain the inverse of the cos function also as a function. But since $\cos t > 0$ we know that $0 \le t < \tfrac{1}{2}\pi$. Now, $\cos^2 t = \left(\dfrac{3}{5}\right)^2 = \dfrac{9}{25}$. We are looking for $\sin t$, and by use of the identity, $\sin^2 t + \cos^2 t = 1$, we have:

$$\sin^2 t + \dfrac{9}{25} = 1$$
$$\sin^2 t = 1 - \dfrac{9}{25} = \dfrac{16}{25} \ .$$

Since $0 \le t < \tfrac{1}{2}\pi$, we know that $\sin t \ge 0$, and thus $\sin t = \dfrac{4}{5}$.

Hence,
$$\left(\cos^{-1} \dfrac{3}{5}\right) = \dfrac{4}{5} \ .$$

Find $\sin\left(\frac{1}{2} \text{ Arccos } \frac{5}{15}\right)$.

Solution: Let Arccos $\frac{5}{13} = \alpha$; then we have

$$\sin\left(\frac{1}{2}\alpha\right) = \sqrt{\frac{1 - \cos\alpha}{2}} = \sqrt{\frac{1 - \frac{5}{13}}{2}} = \sqrt{\frac{8}{26}} = \frac{2}{13}\sqrt{13}.$$

Find $\sin(\text{Arctan } x)$, where x may be any real number.

Solution: Let t = Arctan x. Then $\tan t = x$, and $-\frac{1}{2}\pi < t < \frac{1}{2}\pi$. Also t and x have the same sign (see Figure). Now, we express $\sin t$ in terms of x as follows:

since $\tan t = x = \dfrac{\sin t}{\cos t}$, and

$$\sin t = \frac{\sin t}{\cos t} \cdot \cos t,$$

we write

$$\sin t = x \cos t$$

$$\sin^2 t = x^2 \cos^2 t,$$

and by the identity

$$\sin^2 t + \cos^2 t = 1,$$

$$\sin^2 t = x^2(1 - \sin^2 t).$$

Solving for $\sin^2 t$,

$$1 = \frac{x^2(1 - \sin^2 t)}{\sin^2 t}$$

$$1 = \frac{x^2 - x^2 \sin^2 t}{\sin^2 t}$$

$$1 = \frac{x^2}{\sin^2 t} - \frac{x^2 \sin^2 t}{\sin^2 t}$$

$$1 = \frac{x^2}{\sin^2 t} - x^2$$

$$1 + x^2 = \frac{x^2}{\sin^2 t}$$

$$\sin^2 t(1 + x^2) = x^2$$

$$\sin^2 t = \frac{x^2}{1 + x^2}.$$

Hence,

$$\sin t = \frac{x}{\sqrt{1 + x^2}} \quad \text{or} \quad \frac{-x}{\sqrt{1 + x^2}} \quad .$$

Since $-\frac{1}{2}\pi < t < \frac{1}{2}\pi$, $\sin t$ and t have the same sign. From above we know that t and x have the same sign. Thus, $\sin t$ and x must have the same sign. Now

$$\frac{x}{\sqrt{1 + x^2}}$$

has the same sign as x, while

$$\frac{-x}{\sqrt{1 + x^2}}$$

has the opposite sign; thus $\sin t = \sin(\text{Arctan } x) = \frac{x}{\sqrt{1 + x^2}}$

● **PROBLEM** 911

Find the positive value of $\tan\left(\text{arc sin } \frac{\sqrt{2}}{2}\right)$.

Solution: The expression within the parentheses, "arc sin $\frac{\sqrt{2}}{2}$,"

means "the angle whose sine equals $\frac{\sqrt{2}}{2}$." Angles whose sine equals

$\frac{\sqrt{2}}{2}$ are $45°$, $135°$, $-225°$, $-315°$, and other angles that can be obtained

from these by adding or subtracting $360°$ or a multiple of $360°$.

Hence, the expression "$\tan\left(\text{arc sin } \frac{\sqrt{2}}{2}\right)$" may mean $\tan 45°$, $\tan 135°$,

or the tangent of any of the other angles. The value of these tangents

is either $+1$ or -1. However, only the positive value is required,

thus $\tan\left(\text{arc sin } \frac{\sqrt{2}}{2}\right) = 1$.

● **PROBLEM** 912

Evaluate $\tan\left[\frac{1}{2} \arcsin\left(-\frac{8}{17}\right)\right]$.

Solution: Let $\theta = \arcsin\left(-\frac{8}{17}\right)$. Then $\sin \theta = -\frac{8}{17}$. Thus, we
wish to evaluate $\tan\left(\frac{\theta}{2}\right)$. Using the half-angle formula for $\tan$ we
obtain:
$$\tan\left(\frac{\theta}{2}\right) = \frac{1 - \cos \theta}{\sin \theta} \quad .$$

Thus, we must find, in addition to $\sin \theta = -\frac{8}{17}$, the corresponding

value of $\cos \theta$. Observe that when $\sin \theta$ is negative, θ must be a

negative angle in the fourth quadrant, since we have the restriction

on the inverse sin function $-\frac{\pi}{2} \leq \theta \leq \frac{\pi}{2}$. Thus, $\cos \theta$ will be posi-

tive (since cos is positive in the fourth quadrant), and using the
identity $\sin^2\theta + \cos^2\theta = 1$ we obtain:

$$\cos\theta = \sqrt{1 - \sin^2\theta} = \sqrt{1 - \left(-\frac{8}{17}\right)^2} = \sqrt{1 - \frac{64}{289}}$$

$$= \sqrt{\frac{225}{289}} = \frac{15}{17} .$$

Therefore the desired value is

$$\tan\left(\frac{\theta}{2}\right) = \frac{1 - \frac{15}{17}}{-\frac{8}{17}} = \frac{2}{17} \cdot \left(-\frac{17}{8}\right) = -\frac{1}{4} ,$$

or

$$\tan\left[\frac{1}{2}\arcsin\left(-\frac{8}{17}\right)\right] = -\frac{1}{4} .$$

● **PROBLEM** 913

Find cos (Arccos x - Arccos 3y).

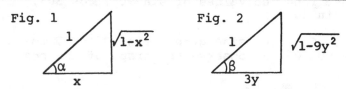

Fig. 1   Fig. 2

Solution: Let Arccos x = α and Arccos 3y = β. We can then construct
the triangles of Figs. 1 and 2.

$$\cos(\alpha - \beta) = \cos\alpha\cos\beta + \sin\alpha\sin\beta$$
$$= (x)(3y) + (\sqrt{1 - x^2})(\sqrt{1 - 9y^2})$$
$$= 3xy + \sqrt{(1 - x^2)(1 - 9y^2)} .$$

Note: sin α and sin β are both positive since α and β must be
between 0 and π; we thus take only the positive radical.

● **PROBLEM** 914

Find sin $\left(\sin^{-1} 1 + \sin^{-1} \frac{1}{2}\right)$.

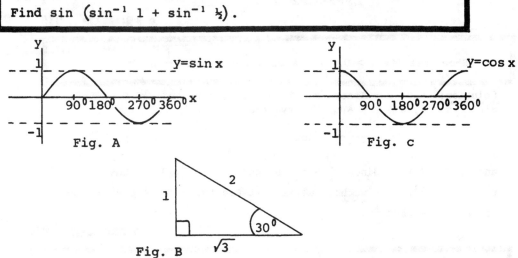

Fig. A   Fig. c

Fig. B

Recall that inverse sines are angles. Hence $\sin^{-1} 1$ is the angle whose sine is 1.

From the diagram of the sine function, we see that the angle whose sine is 1 is $90°$.

Similarly, $\sin^{-1} \frac{1}{2}$ is the angle whose sine is $\frac{1}{2}$; that is, the angle whose opposite side is 1 and whose hypotenuse is 2.

We note this is a 30-60 right triangle and the angle whose sine is $\frac{1}{2}$ is a $30°$ angle hence,

$$\sin \left(\sin^{-1} 1 + \sin^{-1} \tfrac{1}{2}\right) = \sin \left(90° + 30°\right)$$

Using the sum of the sines formula

$$\sin (x + y) = \sin x \cos y + \cos x \sin y:$$

$$= \sin 90° \cos 30° + \cos 90° \sin 30°$$

Now we must find the values of $\sin 90°$, $\cos 30°$, $\cos 90°$, and $\sin 30°$.

We observed from our sine graph, $\sin 90° = 1$. To calculate $\cos 90°$ we can observe the graph of the cos function.

Thus we see $\cos 90° = 0$.

To calculate $\cos 30°$ and $\sin 30°$ we look at a 30-60 right triangle:

$$\cos \theta = \frac{\text{adjacent side}}{\text{hypotenuse}} \text{ , hence } \cos 30° = \frac{\sqrt{3}}{2}$$

$$\sin \theta = \frac{\text{opposite side}}{\text{hypotenuse}} \text{ , hence } \sin 30° = \tfrac{1}{2}$$

$$= 1 \cdot \frac{\sqrt{3}}{2} + 0 \cdot \frac{1}{2} = \frac{\sqrt{3}}{2} .$$

● **PROBLEM   915**

Show that if $x > 0$, then $\text{Arctan } x = \text{Arccot } \frac{1}{x}$.

Solution: Let $u = \text{Arctan } x$. Then $\tan u = x$, and $-\frac{1}{2}\pi < u < \frac{1}{2}\pi$. Since $x > 0$, $0 < u < \frac{1}{2}\pi$. Observe that

$$\frac{1}{x} = \frac{1}{\tan u} = \cot u \quad ,$$

and $0 < u < \frac{1}{2}\pi$. Since $\frac{1}{x} = \cot u$ and $0 < u < \frac{1}{2}\pi$ we have $u = \text{Arccot} \left(\frac{1}{x}\right)$. But we have already stated that $u = \text{Arctan } x$. Hence $\text{Arctan } x = \text{Arccot } \frac{1}{x}$.

● **PROBLEM   916**

Show that $\text{Arcsin } x + \text{Arccos } x = \frac{1}{2}\pi$ for any number x such that $-1 \le x \le 1$.

<u>Solution:</u> Let $u = $ Arcsin x. Then $\sin u = x$. We restrict u such that $-\frac{1}{2}\pi \leq u \leq \frac{1}{2}\pi$ so that the inverse of the sin function is also a function. Now let $v = \frac{1}{2}\pi - u$; then $0 \leq v \leq \pi$, and $\cos v = \cos(\frac{1}{2}\pi - u)$. Recall the formula for the cosine of the difference of two angles, $\cos(a - b) = \cos a \cos b + \sin a \sin b$. Using this we obtain:

$$\cos(\tfrac{1}{2}\pi - u) = \cos \tfrac{1}{2}\pi \cos u + \sin \tfrac{1}{2}\pi \sin u,$$

and since $\cos \frac{1}{2}\pi = 0$ and $\sin \frac{1}{2}\pi = 1$, we have:

$$\cos(\tfrac{1}{2}\pi - u) = \sin u = x.$$

Since $0 \leq v \leq \pi$, and $\cos v = x$, we have $v = $ Arccos x. Hence,

$$\text{Arcsin } x + \text{Arccos } x = u + v = u + (\tfrac{1}{2}\pi - u) = \tfrac{1}{2}\pi .$$

● **PROBLEM** 917

Solve the equation $\arcsin x + \arccos(1 - x) = 0$.

<u>Solution:</u> Let $\alpha = \arcsin x$, $\beta = \arccos(1 - x)$. Then we must solve the equation

$$\alpha + \beta = 0,$$

where $\sin \alpha = x$ and $\cos \beta = 1 - x$. Now, by use of the identity $\sin^2 a + \cos^2 a = 1$ we obtain:

$$\cos \alpha = \underset{-}{+} \sqrt{1 - \sin^2 \alpha} = \underset{-}{+} \sqrt{1 - x^2} ,$$

and

$$\sin \beta = \underset{-}{+} \sqrt{1 - \cos^2 \beta} = \underset{-}{+} \sqrt{1 - (1-x)^2} = \underset{-}{+} \sqrt{2x - x^2} .$$

We now make use of the formula for the sin of the sum of two angles which states,

$$\sin(\alpha + \beta) = \sin \alpha \cos \beta + \cos \alpha \sin \beta ;$$

and this equals 0 since $\alpha + \beta = 0$, and $\sin 0 = 0$. Substituting the above values for $\sin \alpha$, $\cos \beta$, $\cos \alpha$, $\sin \beta$ we have:

$$\sin \alpha \cos \beta + \cos \alpha \sin \beta = 0$$
$$x(1 - x) \underset{-}{+} \sqrt{1 - x^2} \sqrt{2x - x^2} = 0$$
$$x^2(1 - x)^2 = (1 - x^2)(2x - x^2)$$
$$x^2(1 - 2x + x^2) = (1 - x^2)(2x - x^2)$$

Observe that $x = 0$ satisfies this equation. Substituting this value in the given equation we obtain:

$$\arcsin 0 + \arccos(1 - 0) = 0.$$

Now, since $\arcsin 0 = 0$ and $\arccos 1 = 0$ we have:

$$0 + 0 = 0;$$
$$0 = 0$$

thus this value of x also satisfies the original equation. Removing the factor x from $x^2(1 - 2x + x^2) = (1 - x^2)(2x - x^2)$ yields:

$$x^2(1 - 2x + x^2) = (1 - x^2)(2 - x)x$$
$$x(1 - 2x + x^2) = (1 - x^2)(2 - x)$$
$$x - 2x^2 + x^3 = 2 - x - 2x^2 + x^3$$
$$2x = 2$$
$$x = 1$$

Thus, $x = 1$ is another possible solution to the original equation. But $\arcsin 1 = \pi/2$ and $\arccos 0 = \pi/2$, thus

$$\arcsin 1 + \arccos(1 - 1) = \frac{\pi}{2} + \frac{\pi}{2} = \pi \neq 0.$$

Thus, x = 1 does not satisfy the given equation. (Notice that outside the restricted values for arccos we have arccos $0 = -\pi/2$
also, which, together with arcsin $1 = \pi/2$, makes x = 1 a solution.)

Hence x = 0 is the only solution of the given equation when only
the restricted values are permitted (recall that for the inverse sine
function,

$$-\frac{\pi}{2} \leq (\arcsin x = \alpha) \leq \frac{\pi}{2}$$

and for the inverse cos function

$$0 \leq \left(\arccos(1 - x) = \beta\right) \leq \pi\Big).$$

CHAPTER 29

TRIGONOMETRIC EQUATIONS

> **Basic Attacks and Strategies for Solving Problems in this Chapter. See pages 671 to 705 for step-by-step solutions to problems.**

The process of solving trigonometric equations is very similar to the process of solving algebraic equations. Because of the periodic nature of trigonometric functions, trigonometric equations will not typically have a finite number of solutions unless one restricts the search for solutions to some specified interval.

The simplest kind of trigonometric equation has the linear form

$$\text{trg } x = c,$$

where trg is one of the trigonometric functions, c is a given number, and x is the variable whose value is to be determined. Such equations can be solved using memorized values or the inverse trigonometric function operations on a calculator.

If the trigonometric equation is more complex, then the first step in the solution process is to set the equation equal to zero and factor (if possible) the other side of the equation. Then set each factor equal to zero and solve each resulting equation.

If the equation is not factorable and is in quadratic form, then other techniques for solving algebraic equations should be tried, such as, the quadratic formula and completing the square. If more than one trigonometric function is involved in a trigonometric equation and/or different multiples of the angle and the factoring method is not applicable, then it is necessary to use identities to obtain an equation involving only one trigonometric function. After identities have been applied, then solve the equation using the techniques mentioned above.

General guidelines for proving trigonometric identities involve the following steps:

(1) Begin work on the side of the statement which is more complicated.

(2) Look for trigonometric substitutions involving the basic identities that may help simplify things.

(3) Look for algebraic operations, such as adding fractions, the distributive property, or factoring, that may simplify the side you are working with or that will at least lead to an expression that will be easier to simplify.

(4) If you cannot think of anything else to do, change everything to sines and cosines and see if that helps.

(5) Always keep an eye on the side that you are not working with to be sure you are working toward it. There is a certain sense of direction that accompanies a successful proof.

Step-by-Step Solutions to Problems in this Chapter, "Trigonometric Equations"

FINDING SOLUTIONS TO EQUATIONS

● **PROBLEM** 918

Find the solution set for sin x = 0.

<u>Solution:</u> The solution set consists of all distinct values of the angle x which satisfy the equation. Note that x = 0 and x = π are both elements of the solution set. Similarly, any angle nπ, n = 0, ±1, ±2,..., satisfies the equality. Thus, the solution set is {x | x = nπ,n = 0, ±1, ±2,...}. This set is an infinite set.

● **PROBLEM** 919

Find the solution set of cos x = 3.

<u>Solution:</u> Since the values of the cosine function only range from - 1 to +1; that is, since -1≤ cos x ≤ 1, there is no value of x which satisfies the equality cos x = 3. Therefore, the solution set is the empty set.

● **PROBLEM** 920

Find the solution set on [0,2π) for sin x = cos x.

<u>Solution:</u> Since the equation does not lend itself to factoring, we divide both sides by cos x, obtaining sin x/cos x = 1, or tan x = 1. The solution set of this new equation is {π/4, 5π/4}. However, in dividing the original equation by cos x, it was assumed that cos x ≠ 0 because division by 0 is not permitted. However, cos x may be equal to zero. When cos x = 0, x = π/2 or x = 3π/2. Checking the four solutions X = π/4, π/2, 5π/4, 3π/2 in the original equation:

$$\sin \pi/4 = 0.7071 = \cos \pi/4 \checkmark$$

$$\sin \pi/2 = 1 \neq \cos \pi/2 \text{ since } \cos \pi/2 = 0.$$

$$\sin 5\pi/4 = -0.7071 = \cos 5\pi/4 \checkmark$$

$$\sin 3\pi/2 = -1 \neq \cos 3\pi/2 \text{ since } \cos 3\pi/2 = 0.$$

Therefore, the solution set of the original equation is:

$$\{\pi/4, \ 5\pi/4\}.$$

● **PROBLEM** 921

Find the solution set on [0,2π] of the equation

$$\sqrt{1 + \sin^2 x} = \sqrt{2} \sin x.$$

<u>Solution:</u>   Since the unknown quantity is involved in the radicand, squaring of both sides to eliminate the radical is suggested. Thus, we obtain $1 + \sin^2 x =$

$2 \sin^2 x$. Hence, $\sin^2 x - 1 = 0$,

or $\sin^2 x = 1$

$$\sqrt{\sin^2 x} = \pm\sqrt{1}$$

$$\sin x = \pm 1$$

When $\sin x = 1$ on $[0,2\pi]$, $x = \pi/2$. When $\sin x = -1$ on $[0,2\pi]$, $x = 3\pi/2$.

The complete solution set seems to be $\{\pi/2, 3\pi/2\}$. Since we squared both sides of the equation, we should try each element in the original equation. When $x =\pi/2$, we obtain $\sqrt{1+1} =\sqrt{2}\cdot 1$. When $x = 3\pi/2$, we obtain $\sqrt{1+1}=\sqrt{2}(-1)$. The second element does not satisfy the original equation, hence does not belong to the solution set. An extraneous root was introduced by squaring the equation, it would seem. Thus, the solution set is $\{\pi/2\}$.   ● PROBLEM 922

Find one number in the solution set of the equation $\tan t = \sqrt{3}$.

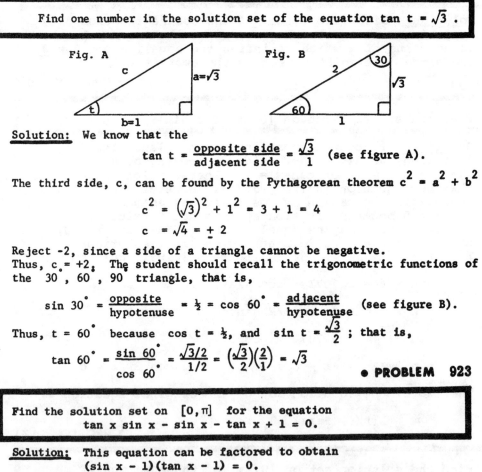

<u>Solution:</u> We know that the

$$\tan t = \frac{\text{opposite side}}{\text{adjacent side}} = \frac{\sqrt{3}}{1} \quad \text{(see figure A).}$$

The third side, c, can be found by the Pythagorean theorem $c^2 = a^2 + b^2$

$$c^2 = \left(\sqrt{3}\right)^2 + 1^2 = 3 + 1 = 4$$

$$c = \sqrt{4} = \pm 2$$

Reject -2, since a side of a triangle cannot be negative.
Thus, $c = +2$. The student should recall the trigonometric functions of the $30°$, $60°$, $90°$ triangle, that is,

$$\sin 30° = \frac{\text{opposite}}{\text{hypotenuse}} = \tfrac{1}{2} = \cos 60° = \frac{\text{adjacent}}{\text{hypotenuse}} \quad \text{(see figure B).}$$

Thus, $t = 60°$ because $\cos t = \tfrac{1}{2}$, and $\sin t = \frac{\sqrt{3}}{2}$; that is,

$$\tan 60° = \frac{\sin 60°}{\cos 60°} = \frac{\sqrt{3}/2}{1/2} = \left(\frac{\sqrt{3}}{2}\right)\left(\frac{2}{1}\right) = \sqrt{3}$$

● PROBLEM 923

Find the solution set on $[0,\pi]$ for the equation
$\tan x \sin x - \sin x - \tan x + 1 = 0$.

<u>Solution:</u> This equation can be factored to obtain
$(\sin x - 1)(\tan x - 1) = 0$.
The values of x satisfying this equation may be found by setting each factor equal to zero.

$$\sin x - 1 = 0 \; ; \quad \tan x - 1 = 0$$

or $\sin x = 1 \; ; \qquad \tan x = 1$

Keeping in mind that our solution set cannot contain values exceeding π or less than zero. We find that $x = \pi/2$, is our only acceptable solution for the first equation and $x = \pi/4$ is the only acceptable solution for the second, i.e.,

$$\sin(k \cdot \pi/2) = 1 \quad \text{where} \quad k = 1,3,5,\ldots \quad \text{but } k = 3,5,\ldots \quad \text{is}$$
unacceptable

$$\tan(k\pi/4) = \frac{\sin(k \cdot \pi/4)}{\cos(k \cdot \pi/4)} = \frac{\sqrt{2}/2}{\sqrt{2}/2}, k = 1,3,5,\ldots \quad \text{but}$$

$k = 3,5,\ldots$ is also unacceptable.

By substituting $\pi/2$ into $(\sin x - 1)(\tan x - 1) = 0$ we arrive at the undefined quantity $0 \times (1/0 - 1)$ so we must specify that $\pi/4$ is our only solution as its substitution leads to a valid identity.

● **PROBLEM**   924

Find all angles on $[0^\bullet, 360^\bullet)$ which satisfy $\sin 2x - \sqrt{2} \sin x = 0$.

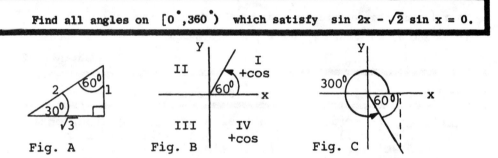

Fig. A   Fig. B   Fig. C

Solution: The equation contains mixed expressions and must be expressed as an equation involving only one multiple of the angle. It seems to be practical to express $\sin 2x$ in terms of the angle x; by using the double angle formula we can express $\sin 2x$ in terms of an angle multiplied by unity.

$$\sin(x + x) = \sin x \cos x + \sin X \cos x = 2 \sin x \cos x$$

$$\sin 2x - \sqrt{2} \sin x = 2 \sin x \cos x - \sqrt{2} \sin x = 0$$

$$\sin x(2 \cos x - \sqrt{2}) = 0$$

Setting each factor = 0,

$$\sin x = 0$$
$$x = 0, \pi$$
$$2 \cos x - \sqrt{2} = 0$$

$$\cos x = \frac{\sqrt{2}}{2}$$

$$x = \frac{\pi}{4}, \frac{7\pi}{4}$$

The solution set is $\left\{ 0, \frac{\pi}{4}, \pi, \frac{7\pi}{4} \right\}$

● **PROBLEM**   925

Solve $2 \sin^2\theta + 3 \cos\theta - 3 = 0$ for θ if $0 \le \theta < 360^\circ$.

Solution: The solution to the equation can be found by expressing the equation in terms of one trigonometric function. Here the convenient function is $\cos\theta$. Using the identity $\sin^2\theta + \cos^2\theta = 1$, we can eliminate $\sin^2\theta$ from the equation by substituting $1 - \cos^2\theta$:

$$2\left(1 - \cos^2\theta\right) + 3 \cos \theta - 3 = 0$$

distributing: $2 - 2 \cos^2\theta + 3 \cos \theta - 3 = 0$

adding: $-2\cos^2\theta + 3 \cos \theta - 1 = 0$

multiply by -1: $2 \cos^2\theta - 3 \cos \theta + 1 = 0$

factoring: $(2 \cos \theta - 1)(\cos \theta - 1) = 0$

Hence, $\cos \theta = \frac{1}{2}$ or $\cos \theta = 1$

Observe Figure A, of a 30-60 right triangle. From it we find

$\cos 60^\circ = \dfrac{\text{adjacent side}}{\text{hypotenuse}} = \dfrac{1}{2}$.

The cosine is positive in quadrants I and IV; thus, θ can be 60° or 300°. (See Figures B and C). Then,

$$\theta = 60^\circ, 300^\circ \quad \text{or} \quad \theta = 0^\circ (\text{since } \cos 0^\circ = 1)$$

Substitution verifies all three. The solution set is

$$\{0^\circ, 60^\circ, 300^\circ\}$$

By removing the restriction that $0 \le \theta < 360^\circ$, the solution set is $\{0^\circ + 360^\circ k, \; 60^\circ + 360^\circ k, \; 300^\circ k\}$, k an integer.

● **PROBLEM** 926

Find the solution set of $\sin^2 \theta + \sin \theta = 0$.

<u>Solution:</u> Factoring the left side of the equation, obtain $\sin \theta (\sin \theta + 1) = 0$. Setting each factor equal to zero and solving for $\sin \theta$, obtain $\sin \theta = 0$ and $\sin \theta = -1$. For $\sin \theta = 0$, $\theta = 0$, π, ... and all integral multiples of π.

For $\sin \theta = -1$, note that the sign is negative and that the value of the sine is one. Thus, $\theta = 3\pi/2$ and all integral multiples of 2π plus $3\pi/2$. Therefore, from the first equation, the solution set contains the elements $n\pi$, n = 0, ±1, ... From the second, the solution set contains the elements $3\pi/2 + 2n\pi$, n = 0, ±1,... The solution set of the original equation is then

$$\left\{\theta \mid \theta = n\pi, \text{ or } 3\pi/2 + 2n\pi, \; n = 0, \; \pm 1, \; \pm 2, \ldots\right\}$$

● **PROBLEM** 927

Find the solution set on $[0, 2\pi]$ for the equation $\sin x \cos x = \cos x$.

<u>Solution:</u> Dividing by cos x we obtain $\sin x = 1$. However, this operation assumes that $\cos x \ne 0$, since division by 0 is not permitted. Hence we have that cos x may also be equal to zero. Therefore, the solution set consists of all values of x which satisfy the two equations $\sin x = 1$ and $\cos x = 0$. When $\sin x = 1$ on $[0, 2\pi]$ $x = \pi/2$. When $\cos x = 0$ on $[0, 2\pi]$, $x = \pi/2$ or $x = 3\pi/2$.

Note that one of the values of x obtained from the cosine function ($x = \pi/2$) is the same as the value of x obtained from the sine function. Therefore, the complete solution set is $\{\pi/2, \; 3\pi/2\}$.

Find the solution set on $[0,2\pi)$ for the equation $\cot^2\theta + (1 - \sqrt{3})\cot\theta - \sqrt{3} = 0$.

Solution: Factoring, we obtain $(\cot\theta - \sqrt{3})(\cot\theta + 1) = 0$. Hence, $\cot\theta - \sqrt{3} = 0$ or $\cot\theta + 1 = 0$
$$\cot\theta = \sqrt{3} \text{ or } \cot\theta = -1.$$

The first equation gives the set $\{\pi/6,\ 7\pi/6\}$ while the second gives $\{3\pi/4,\ 7\pi/4\}$. Note that the solutions to the first equation $\cot\theta = \sqrt{3}$ are angles which lie in the first and third quadrants, since the cotangent function is positive in these two quadrants. Also, the solutions to the second equation $\cot\theta = -1$ are angles which lie in the second and fourth quadrants, since the cotangent function is negative in these quadrants. The complete solution set on $[0,2\pi)$ is then $\{\pi/6,\ 7\pi/6,\ 3\pi/4,\ 7\pi/4\}$.

Find the solution set on $[0,2\pi)$ of $2\tan x + \sqrt{3}\sin x$
$\sec^2 x = 0$.

Solution: Use the following facts concerning trigonometric functions to rewrite the given equation:

$$\tan x = \frac{\sin x}{\cos x}, \quad \sec x = \frac{1}{\cos x}.$$

Therefore, the given equation becomes:

$$2\tan x + \sqrt{3}\sin x \sec^2 x = \frac{2\sin x}{\cos x} + \frac{\sqrt{3}\sin x}{\cos^2 x} = 0 \qquad (1)$$

Multiply both sides of equation (1) by $\cos^2 x$. Hence,

equation (1) becomes: $\cos^2 x \left[\dfrac{2\sin x}{\cos x} + \dfrac{\sqrt{3}\sin x}{\cos^2 x} \right] = \cos^2 x \ (0)$

$$\text{or } 2\sin x \cos x + \sqrt{3}\sin x = 0$$

$$\text{or } \sin x (2\cos x + \sqrt{3}) = 0$$

Therefore : $\sin x = 0$ or $2\cos x + \sqrt{3} = 0$

$$2\cos x = -\sqrt{3}$$

$$\cos x = -\frac{\sqrt{3}}{2}$$

When $\sin x = 0$ on $[0,2\pi)$, $x = 0$ or $x = \pi$. When $\cos x = -\frac{\sqrt{3}}{2}$ on $[0,2\pi)$, $x = 5\pi/6$ or $x = 7\pi/6$. Therefore, the complete solution set is $\{0,\ 5\pi/6,\ \pi,\ 7\pi/6\}$.

Solve for θ : $\sin\theta + 2\tan\theta = 0$, $0 \le \theta \le 2\pi$.

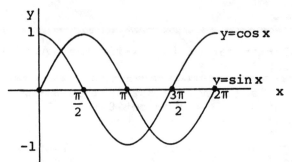

Solution: $\sin\theta + 2\tan\theta = 0$

Since $\tan\theta = \dfrac{\sin\theta}{\cos\theta}$, by substitution: $\sin\theta + 2\dfrac{\sin\theta}{\cos\theta} = 0$

Multiplying both sides by $\cos\theta$ gives us $\cos\theta\left(\sin\theta + 2\dfrac{\sin\theta}{\cos\theta}\right) = 0(\cos\theta)$

Distributing we obtain,

$$\sin\theta\cos\theta + 2\sin\theta = 0$$

Factoring out $\sin\theta$ gives us

$$\sin\theta(\cos\theta + 2) = 0$$

If $xy = 0$, either $x = 0$ or $y = 0$, hence either $\sin\theta = 0$ or

$\cos\theta + 2 = 0$. Subtracting 2 from each side of the latter gives us

$\cos\theta = -2$ Thus $\sin\theta = 0$ or $\cos\theta = -2$

On the given interval, $0 \le \theta \le 2\pi$, $\sin\theta = 0$ when $\theta = 0$, $\theta = \pi$ or $\theta = 2\pi$; and $\cos\theta = -2$ for no angles of θ ($\cos$ is only defined on the interval $|-1,1|$).

Thus the solution set is $\{0,\pi,2\pi\}$. Check these values. If we do not restrict θ as we have done, then the solutions may be expressed as $\theta = 0 + k\pi$, where k is any integer. That is, by adding any integral multiple of π to the angle θ, we obtain an angle coterminal with either zero or π. (By coterminal we mean the initial sides of the angles lie on the positive branch of the x-axis and the terminal sides coincide.)

● **PROBLEM** 931

Solve the equation

$$\sin^2\theta + 2\cos\theta - 1 = 0$$

for non-negative values of θ less than 2π.

Solution: Two trigonometric functions of the unknown θ itself appear in this equation. Accordingly, we make use of the identity connecting these functions, namely, $\sin^2\theta + \cos^2\theta = 1$, to transform it into an equation involving only one function of θ. Replace $\sin^2$ by $1 - \cos^2 \theta$.

$$\sin^2 \theta + 2 \cos \theta - 1 = 0$$

$$\cancel{1} - \cos^2 \theta + 2 \cos \theta - \cancel{1} = 0.$$

Factor out $\cos \theta$

$$\cos \theta (2 - \cos \theta) = 0.$$

Whenever a product of two numbers $ab = 0$, either $a = 0$ or $b = 0$, hence $\cos \theta = 0$ or $2 - \cos \theta = 0$. Thus $\cos \theta = 0$ or $\cos \theta = 2$.

Now there are two angles in the range $0 \leq \theta < 2\pi$ for which $\cos \theta = 0$ namely

$$\theta = \frac{\pi}{2}, \quad \theta = \frac{3\pi}{2}.$$

But, since a cosine of an angle can never exceed unity, the relation $\cos \theta = 2$ does not yield a value of θ. Hence we have just two solutions, as given above. It is easy to check these solutions.

Check: $\sin^2 \theta + 2 \cos \theta - 1 = 0$

$$(\sin \theta)^2 + 2 \cos \theta - 1 = 0$$

For $\theta = \frac{\pi}{2}$

$$(\sin \pi/2)^2 + 2 \cos \frac{\pi}{2} - 1 = 0$$

$$1 + 2 \cdot 0 - 1 = 0$$
$$0 = 0 \checkmark$$

For $\theta = \frac{3}{2}\pi$

$$(\sin 3/2\pi)^2 + 2 \cos \frac{3}{2}\pi - 1 = 0$$

$$(-1)^2 + 2 \cdot 0 - 1 = 0$$
$$0 = 0 \checkmark$$

● **PROBLEM   932**

Find the solution set of $2 \cos^2 x - 5 \cos x + 2 = 0$.

<u>Solution:</u> Factoring, we obtain $(\cos x - 2)(2 \cos x - 1) = 0$. Setting each factor equal to zero, we obtain $\cos x = 2$ and $\cos x = 1/2$. There is no value of x satisfying the first factor because the range of values for $\cos x$ is from -1 to $+1$; that is, $-1 \leq \cos x \leq 1$. Therefore, the solution set of the first factor is the empty set. For the second factor the solution set is $\{x \mid x = \pi/3 + 2n\pi$ or $5\pi/3 + 2n\pi, n = 0, \pm 1, \pm 2, \ldots\}$. Since the first solution set is the empty set, the second set is the complete solution set.

Solve the equation $\sin^2 x - 4 \sin x + 3 = 0$.

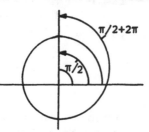

Solution: Factoring the left side of the given equation into a product of two trigonometric functions

$$(\sin x - 3)(\sin x - 1) = 0 \qquad\qquad (1)$$

Whenever a product $ab = 0$ (where a and b are any two numbers) either $a = 0$ or $b = 0$. Hence, either

$$\sin x - 3 = 0 \quad \text{or} \quad \sin x - 1 = 0$$
$$\sin x = 3 \quad \text{or} \qquad \sin x = 1 \;,$$

and so our desired solution set is the union

$$\{\sin x = 3\} \cup \{\sin x = 1\}. \qquad\qquad (2)$$

The first of these sets is the empty set because $\sin x$ takes on values only between -1 and 1; that is, $-1 \leq \sin x \leq 1$. For the second set, $\sin \pi/2 = 1$. Note that any integral multiple of 2π added to $\pi/2$ will result in the same reference angle. Hence,

$$\sin \frac{\pi}{2} = \sin\left(\frac{\pi}{2} + 2\pi\right) = \sin\left(\frac{\pi}{2} - 2\pi\right) = \sin\left(\frac{\pi}{2} + 4\pi\right) = 1 \;.$$

Therefore, the solution set for the second set is $\left\{\frac{\pi}{2} + 2\pi k, \text{ where } k \text{ is } 0, \pm 1, \pm 2, \ldots\right\}$. Since the first set has no solution set, the solution set for the second set is the solution set of the given equation. Hence, $\left\{\frac{\pi}{2} + 2 k, \text{ where } k = 0, \pm 1, \pm 2\right\}$ is the solution set of the given equation.

Determine all angles x, $0° \leq x < 360°$, such that $\sin 2x = -\frac{1}{2}$.

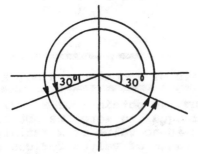

Solution: To determine all values of x such that $0° \leq x < 360°$ and $\sin 2x = -\frac{1}{2}$, we must determine all values of $2x$ such that

$$2 \cdot 0° \leq 2 \cdot x < 2 \cdot 360° \quad \text{and } \sin 2x = -\frac{1}{2} \;,$$

or all values of 2x must be determined such that

$$0° \leq 2x < 720° \quad \text{and} \quad \sin 2x = -\tfrac{1}{2} .$$

Since the sine function is negative in only the third and fourth quadrants, the angle 2x may lie in only these two quadrants. Also, since the sin 30° = -½ in the third and fourth quadrants, any angle with a reference angle of 30° will satisfy the equation sin 2x = -½ . The angles that satisfy this equation and which are in the range of 0° ≤ 2x < 720° are 180° + 30° = 210°, 360° - 30° = 330°,360° + 210° = 570°, and 720° - 30° = 690° (see diagram). Therefore, 2x = 210° or x = 105°, 2x = 330° or x = 165°, 2x = 570° or x = 285°, and 2x = 690° or x = 345°. Hence, the solutions of the equation are x = 105°, x = 165°, x = 285°, and x = 345°. These solutions are checked by substituting each of them into the equation.

$$\sin 2(105°) = \sin 210° = \sin 30° = -\tfrac{1}{2}\sqrt{}$$
$$\sin 2(165°) = \sin 330° = \sin 30° = -\tfrac{1}{2}\sqrt{}$$
$$\sin 2(285°) = \sin 570° = \sin 30° = -\tfrac{1}{2}\sqrt{}$$
$$\sin 2(345°) = \sin 690° = \sin 30° = -\tfrac{1}{2}\sqrt{} .$$

Note that the angles 210°, 333°, 570°, and 690° lie in either the third or fourth quadrant, in which the sine function is always negative.

● PROBLEM 935

Determine all values of x such that 0°≤ x < 360° and tan 2x = - 1.

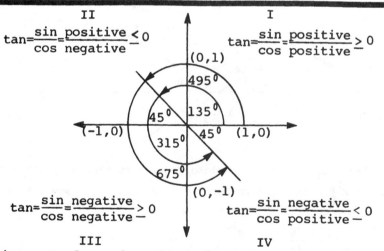

Solution: To determine all angles x such that 0° < x < 360°, we must find all angles 2x such that 2 · ⁻0° < 2 · x < 2 · 360° or 0° ≤ 2x · 720°. Now determine in what quadrants tan is negative, to conform to the equation tan θ = - 1. (See Figure). Thus, the desired angles lie in the second and fourth quadrants. For tan θ = sin θ/cos θ = 1 to be true, either sin θ and cos θ both equal 1, or sin θ = cos θ. Since sine and cosine are never equal to 1 simultaneously, we seek the angle for which sin and cos are equal. 45° satisfies this relation: $\dfrac{\sin 45°}{\cos 45°} = \dfrac{\sqrt{2}/2}{\sqrt{2}/2} = 1.$

We must pick angles of 45° in the 2nd and 4th quadrants.

They are 135° in the second quadrant, and 315° in the fourth quadrant. This is through one revolution or 360°. Through a second revolution (720°) the chosen angles are 495° in the second quadrant and 675° in the fourth (see Figure).

The angles are 135°, 315°, 495°, 675°. According to the equation each angle is twice the required angle. Therefore, the values of x which we are interested in are

$$\frac{135°}{2} = 67.5°$$

$$\frac{315°}{2} = 157.5°$$

$$\frac{495°}{2} = 247.5°$$

$$\frac{675°}{2} = 337.5°$$

● **PROBLEM** 936

Solve the equation
$$3 \tan \theta + \sec \theta + 1 = 0$$
for non-negative values of θ less than 2π, that is, $0 \leq \theta < 2\pi$.

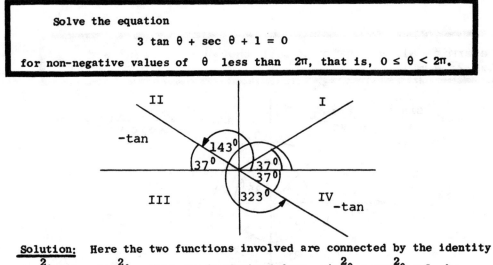

Solution: Here the two functions involved are connected by the identity $\sec^2\theta = 1 + \tan^2\theta$. This can be derived from $\sin^2\theta + \cos^2\theta = 1$, by dividing by $\cos^2\theta$.

$$\frac{\sin^2\theta}{\cos^2\theta} + \frac{\cos^2\theta}{\cos^2\theta} = \frac{1}{\cos^2\theta}$$
$$\tan^2\theta + 1 = \sec^2\theta$$

Since both functions appear to the first degree in the equation, the introduction of an irrationality is unavoidable. Therefore a redundancy may arise, and all solutions obtained must be checked.

We choose to eliminate $\sec \theta$; thus, solving for $\sec \theta$ in the given equation,

$$3 \tan \theta + \sec \theta + 1 = 0$$
$$3 \tan \theta + 1 = -\sec \theta$$

Solving for $\sec \theta$ in the identity $\sec^2\theta = \tan^2\theta + 1$, and substituting:

$$\sec \theta = \pm \sqrt{1 + \tan^2\theta}$$

$$3 \tan \theta + 1 = -\sec \theta = \pm \sqrt{1 + \tan^2 \theta}$$

Squaring both sides:

$$(3 \tan \theta + 1)^2 = \left(\pm \sqrt{1 + \tan^2 \theta} \right)^2$$

$$9 \tan^2 \theta + 6 \tan \theta + 1 = 1 + \tan^2 \theta$$

Subtract $\tan^2 \theta + 1$ from both sides to obtain:

$$9 \tan^2 \theta + 6 \tan \theta + 1 - \left(\tan^2 \theta + 1 \right) = 1 + \tan^2 \theta$$
$$- \left(\tan^2 \theta + 1 \right)$$

$$8 \tan^2 \theta + 6 \tan \theta = 0$$

Factor $2 \tan \theta$ from the left side:

$$(2 \tan \theta)(4 \tan \theta + 3) = 0. \quad \text{Thus,}$$

$$2 \tan \theta = 0, \quad \text{or} \quad 4 \tan \theta + 3 = 0$$

$$\tan \theta = 0, \quad \text{or} \quad \tan \theta = -3/4$$

Therefore, we must find the angles whose tangent is 0, and the angle whose tangent is $(-3/4) = -.75$, where the angle is greater than or equal to 0 and less than 2π (or $360°$).

The angle whose tangent is 0 is $0°$ and $180°$. Thus, for $\tan \theta = 0$ we find $\theta = 0°$, and $\theta = 180°$.

Now, we must find θ such that $\tan \theta = -.75$. Referring to a table of trigonometric functions, we find $\tan 37° \approx .75$. Thus, $37°$ is our reference angle. Since the tangent function is negative in the second and fourth quadrants, $\theta = 143°, 323°$ (see figure).

Therefore, the roots of the equation are $\theta = 0°, 180°, 143°, 323°$. But, before accepting these roots as solutions to the original equation, we substitute each root into the equation as a check for validity. Thus, when $\theta = 0°$, $3 \tan \theta + \sec \theta + 1 = 0$ becomes,

$$3 \tan 0° + \sec 0° + 1 = 0, \quad \text{and}$$

since $\tan 0° = 0$ and $\sec 0° = \dfrac{1}{\cos 0°} = 1$, we have:

$$3(0) + 1 + 1 \overset{?}{=} 0 \; ; \quad 2 \neq 0$$

Thus, $\theta = 0°$ is an extraneous root.

when $\theta = 180°$, we have

$3 \tan 180° + \sec 180° + 1 = 0$, since

$\tan 180° = 0$, and $\sec 180° = \cos 180° = -1$

we have $\qquad 0 + (-1) + 1 = 0$

Therefore, $\theta = 180°$ is a solution of the given equation.

When $\theta = 143°$, $3 \tan \theta + \sec \theta + 1 = 0$ becomes,

$$3 \tan 143° + \sec 143° + 1 = 0; \quad \text{and}$$

since $\sec 143° = -\sec 37°$ (the sign is negative because sec is negative in the second quadrant), we have:

$$3 \tan 143° - \sec 37° + 1 = 0$$

From a table of trig functions we find that $\sec 37° \approx 1.25$, and we found previously that $\tan 143° \approx -.75$; thus:

$$3(-.75) - 1.25 + 1 \overset{?}{=} 0$$
$$-2.25 - .25 \overset{?}{=} 0$$
$$-2.5 \neq 0$$

Therefore, $\theta = 143°$ is not a solution of the given equation,

When $\theta = 323°$, we have: $3 \tan 323° + \sec 323° + 1 = 0$;

and since sec is positive in Quadrant IV, $\sec 323° = \sec 37°$. We have previously found that $\tan 323° \approx - .75$. Thus,

$$3 \tan (323°) + \sec 37° + 1 \stackrel{?}{=} 0$$

$$3(- .75) + 1.25 + 1 = 0$$

$$-2.25 + 2.25 = 0$$

$$0 = 0$$

Therefore, $\theta = 323°$ is a solution of the given equation.
Thus, the solutions of the given equation are

$$\theta = 180°, 323° \text{ for } 0 \le \theta < 2\pi.$$

● PROBLEM 937

Solve the equation

$$2 \sin 2\theta + \cos 2\theta + 2 \sin \theta = 1$$

for non-negative values of θ less than 2π.

<u>Solution:</u> Here we have functions of both θ itself and 2θ.
Hence we first transform so as to get an equivalent equation
involving functions of θ only. We set $\sin 2\theta = 2 \sin \theta \cos \theta$
and $\cos 2\theta = 1 - 2 \sin^2 \theta$, the latter form being chosen so
that the constant terms in the equation will cancel.
   This leads to

$$2 \sin 2\theta + \cos 2\theta + 2 \sin \theta = 1$$

$$2(2 \sin \theta \cos \theta) + \left(1 - 2 \sin^2 \theta\right) + 2 \sin \theta = 1$$

$$4 \sin \theta \cos \theta + 1 - 2 \sin^2 \theta + 2 \sin \theta = 1$$

$$4 \sin \theta \cos \theta - 2 \sin^2 \theta + 2 \sin \theta = 0 \text{ (subtracting 1}$$
$$\text{from both sides)}$$

$$(2 \sin \theta)(2 \cos \theta - \sin \theta + 1) = 0 \quad \text{(Factoring out 2 sin } \theta)$$

$$2(\sin \theta)(2 \cos \theta - \sin \theta + 1) = 0.$$

Whenever a product of two numbers $ab = 0$ either $a = 0$ or
$b = 0$. Hence $2 \sin \theta = 0$ or $2 \cos \theta - \sin \theta + 1 = 0$. When
the factor $\sin \theta$ is set equal to zero, the last equation is
satisfied, whence we get

$$\theta = 0, \quad \theta = \pi.$$

It is easy to verify these solutions in the original equation.

Check:

   $\theta = 0$

   $2 \sin 2(0) + \cos 2(0) + 2 \sin (0) = 1$

   $2 \sin (0) + \cos (0) + 2 \sin (0) = 1$

   $2 \cdot 0 \qquad 1 \quad + \quad 2 \cdot 0 \quad = 1$

   $1 \qquad\qquad = 1$

682

$$\theta = \pi$$

$$2 \sin 2(\pi) + \cos 2(\pi) + 2 \sin \pi = 1$$

$$2 \sin 2\pi \quad + \cos 2\pi \quad + 2 \sin \pi = 1$$

$$2(0) \qquad + \quad 1 \quad + 2(0) \quad = 1$$

$$1 \qquad \qquad = 1$$

To find possible solutions of the equation $2 \cos \theta - \sin \theta + 1 = 0$, we transpose to get $2 \cos \theta + 1 = \sin \theta$.

From the Pythagorean relation $\cos^2 \theta + \sin^2 \theta = 1$, we obtain an expression for $\sin \theta$.

$$\sin^2 \theta = 1 - \cos^2 \theta$$

$$\sin \theta = \pm \sqrt{1 - \cos^2 \theta}.$$

Substitute this into the factor $2 \cos \theta - \sin \theta + 1 = 0$.

$$2 \cos \theta - \sin \theta + 1 = 2 \cos \theta \pm \sqrt{1 - \cos^2 \theta} + 1 = 0$$

Transpose $\pm \sqrt{1 - \cos^2 \theta}$,

$$2 \cos \theta + 1 = \pm \sqrt{1 - \cos^2 \theta}.$$

To solve this radical equation, we square both sides:

$$4 \cos^2 \theta + 4 \cos \theta + 1 = 1 - \cos^2 \theta.$$

Subtract 1 and add $\cos^2 \theta$ to both sides to obtain:

$$5 \cos^2 \theta + 4 \cos \theta = 0$$

$$\cos \theta (5 \cos \theta + 4) = 0.$$

Set both factors equal to zero.

$$\cos \theta = 0 \quad 5 \cos \theta + 4 = 0$$

$$5 \cos \theta = -4$$

$$\cos \theta = 0 \qquad \cos \theta = -\frac{4}{5}.$$

Note that all values of θ obtained must be checked because the process of rationalizing a radical equation may lead to extraneous roots.

Corresponding to $\cos \theta = 0$, $\theta = \frac{\pi}{2}$ and $\theta = \frac{3}{2}\pi$. Substitute these values into the given equation, $2 \cos \theta - \sin \theta + 1 = 0$.

Verify both values of θ:

For $\theta = \frac{\pi}{2}$.

$$2 \cos \theta - \sin \theta + 1 = 0$$

$$2 \cos \frac{\pi}{2} - \sin \frac{\pi}{2} + 1 = 0$$

$$2(0) - 1 + 1 = 0$$

$$0 = 0.$$

For $= \frac{3\pi}{2}$

$$2 \cos \frac{3\pi}{2} - \sin \frac{3\pi}{2} + 1 \stackrel{?}{=} 0$$

$$- (-1) + 1 \neq 0$$

$$2 \neq 0.$$

Thus $\theta = \frac{3\pi}{2}$ is an extraneous root.

If $\cos \theta = -\frac{4}{5}$, θ may be in either the second or third quadrant; from tables we then get $\theta = 143° \ 8'$ and $\theta = 216° \ 52'$ (approx.). Only the latter of these is found to satisfy the given equation by similarly substituting these values of θ into $2 \cos \theta - \sin \theta + 1 = 0$. For $\theta = 143° \ 8'$, the reference angle is $36° \ 52'$ and it is in quadrant II.

$$2 \cos 143° \ 8' - \sin 143° \ 8' + 1 = 0$$

$$-2 \cos 36° \ 52' - \sin 36° \ 52' + 1 = 0.$$

Approximate $36° \ 52'$ by $37°$.

$$-2(.7986) - (.6018) + 1 = 0$$

$$-1.5972 \quad - \quad .6018 \quad + 1 = 0$$
$$-2.199 + 1 \quad \neq 0$$
$$-1.199 \quad \neq 0.$$

For $\theta = 216° \ 52'$, the reference angle is $36° \ 52'$ and θ is in quadrant III. We shall approximate by $37°$.

$$2 \cos 216° \ 52' - \sin 216° \ 52' + 1 = 0$$

$$-2 \cos 37° \quad - (-\sin 37°) \quad + 1 = 0$$

$$-2(.7986) \quad + (.6018) \quad + 1 = 0$$

$$-1.5972 \quad + 1.6018 \quad = 0$$

$$.0046 \quad \approx 0$$

Hence there are two more valid solutions:

$$\theta = \frac{\pi}{2}, \quad \theta = 216° \ 52'.$$

● PROBLEM 938

Find the solution set of $5\tan^2\alpha - 2\tan \alpha - 1 = 0$.

Solution: Use the quadratic formula $x = \dfrac{-b \pm \sqrt{b^2 - 4ac}}{2a}$,

684

to find $\tan \alpha$. For the given equation, $a = 5$, $b = -2$, $c = -1$, and we solve for $x = \tan \alpha$.

$$\tan \alpha = \frac{-(-2) \pm \sqrt{(-2)^2 - 4(5)(-1)}}{2(5)} = \frac{2 \pm \sqrt{4 + 20}}{10} = \frac{2 \pm \sqrt{24}}{10}$$

$$= \frac{2 \pm \sqrt{4}\sqrt{6}}{10}$$

$$\tan \alpha = \frac{2 \pm 2\sqrt{6}}{10} = \frac{2 \pm 2(2.449)}{10}$$

$$\tan \alpha = \frac{2 \pm 4.8989}{10}$$

Therefore, $\tan \alpha = \dfrac{2 + 4.899}{10}$ and $\tan \alpha = \dfrac{2 - 4.899}{10}$

$$= \frac{6.899}{10} \qquad\qquad = \frac{-2.899}{10}$$

$$\tan \alpha = 0.6899 \quad (1) \qquad \tan \alpha = -0.2899 \quad (2)$$

For the $\tan \alpha = 0.6899$, α lies in the first and third quadrants since $\tan \alpha > 0$ in these quadrants. For $\tan \alpha = 0.6899$, $\alpha = 34° 36'$ (an angle in the first quadrant) and $\alpha = 214° 36'$ (an angle in the third quadrant). Note that the reference angle of $214° 36'$ is $34° 36'$, $214° 36' - 180° = 34° 36'$. Hence, the solution set given by equation (1) is
$$\{\alpha \mid \alpha = 34°36' + n\pi, \; n = 0, \pm 1, \pm 2, \dots\}.$$

For the $\tan \alpha = -0.2899$, α lies in the second and fourth quadrants since $\tan \alpha < 0$ in these quadrants. For $\tan \alpha = -0.2899$, $\alpha = 163°50'$. Note that the angle $163°50'$ is a reference angle for $16°10'$, $180° - 163°50' = 16°10'$. Hence, the solution set given by equation (2) is
$$\{\alpha \mid \alpha = 163°50' + n\pi, \; n = 0, \pm 1, \pm 2, \dots\}.$$

Therefore, the entire solution set is
$$\{\alpha \mid \alpha = 34°36' + n\pi \; \text{ or } \; \alpha = 163°50' + n\pi, \; n = 0, \pm 1, \pm 2, \dots\}.$$

● **PROBLEM 939**

Solve: $2 \cos 3x + 1 = 0$.

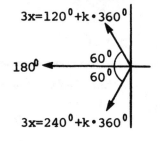

<u>Solution:</u> $2 \cos 3x + 1 = 0$

$$2 \cos 3x = -1$$

$$\cos 3x = -\frac{1}{2}.$$

The cosine is negative in quadrants II and III. The reference angle is 60° since $\cos 60° = \frac{1}{2}$. Therefore, the angle $3x = 120° + k \cdot 360°$ or $240° + k \cdot 360°$, k is an integer.

To convert to radians, set up the proportion:

$$\frac{120°}{180°} = \frac{y}{\pi}$$

$$y = \frac{2}{3}\pi = \pi - \frac{\pi}{3} = 120°$$

$$\frac{240°}{180°} = \frac{z}{\pi}$$

$$z = \frac{4}{3}\pi = \pi + \frac{\pi}{3} = 240°.$$

Substitute these radian measurements for 120° and 240°. In terms of radians:

$$3x = \pi - \frac{\pi}{3} + k \cdot 2\pi \text{ or } \pi + \frac{\pi}{3} + k \cdot 2\pi.$$

Divide by 3

Solution in radians:

$$x = \left\{\frac{2\pi}{9}, \frac{4\pi}{9}\right\} + \frac{2k\pi}{3}$$

Convert these measures back to degrees.

$$\frac{\frac{2}{9}\pi}{\pi} = \frac{m}{180°} \qquad \frac{\frac{4}{9}\pi}{\pi} = \frac{n}{180°}$$

$$m = 40° = \frac{2}{9}\pi \text{ and } n = \frac{4}{9}\pi = 80°.$$

From before, $120° = \frac{2}{3}\pi$.

Solution: $x = \{40°, 80°\} + k \cdot 120°$.

● PROBLEM 940

Determine the non-negative values of x less than 2π for which

$$2 \cos^2 x + \sin x - 2 > 0.$$

Solution: We first transform the left member by means of the identity $\sin^2 x + \cos^2 x = 1$, thus

$$\cos^2 x = 1 - \sin^2 x$$

$$2 \cos^2 x + \sin x - 2 > 0$$

$$2\left(1 - \sin^2 x\right) + \sin x - 2 > 0$$

$$2 - 2 \sin^2 x + \sin x - 2 > 0,$$

a relation involving only one trigonometric function. Multiplying by -1 (or transposing), we have

$$2 \sin^2 x - \sin x < 0,$$

or $(\sin x)(2 \sin x - 1) < 0.$

Now for a product to be negative, then one factor must be negative and the other must be positive. There are two cases to be considered.

Case I		Case II
$\sin x > 0$	or	$\sin x < 0$
$2 \sin x - 1 < 0$		$2 \sin x - 1 > 0.$

For Case I, when $\sin x > 0$, the angle is in quadrant I or II and $x > 0$. Consider the second restriction, $2 \sin x - 1 < 0$. Then $\sin x < \frac{1}{2}$. Thus, $x < \frac{\pi}{6}$. Combine these two restrictions: $0 < x < \frac{\pi}{6}$. Since $\frac{\pi}{6}$ is a reference angle in quadrant II, then $\frac{5\pi}{6} < x < \pi$ since $0 < \sin x < \frac{1}{2}$.

For Case II, when $\sin x < 0$, the angle is in quadrant III or IV. Also, if $2 \sin x - 1 > 0$, then $\sin x > \frac{1}{2}$. Thus $x > \frac{7}{6}\pi$ and $x < \frac{11}{6}\pi$, $\frac{7}{6}\pi < x < \frac{11}{6}\pi$.

● **PROBLEM**   941

Sketch three periods of the graph y = 3 cos 2x.

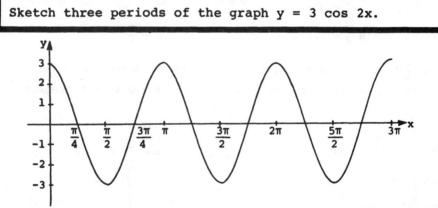

Solution: The coefficient of the function is 3, which means that the maximum and minimum values are 3 and - 3, respectively. The period of the cosine function is the coefficient of x multiplied by $\frac{\pi}{2}$ radians. Therefore, the period of the cosine function given in this problem is

$$2 \left(\frac{\pi}{2} \text{ radians} \right) = \pi \text{ radians}$$

and with this knowledge, we sketch the curve as in the Figure.

Graph $y = \csc x$, $0 \leq x \leq 2\pi$.

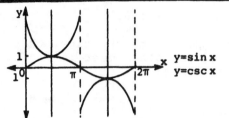

Solution: To plot points for the function cosecant of x, first find the y-values of the reciprocal function, the sine of the angle x.

x	0	$\frac{\pi}{2}$	π	$\frac{3\pi}{2}$	2π
sin x	0	1	0	-1	0
csc x	not defined	1	not defined	-1	not defined

Since the sine and cosecant are reciprocals, we state the following conclusions based on properties of real numbers.

(1) For $0 \leq x \leq \frac{\pi}{2}$, $0 \leq \sin x \leq 1$ and $\csc x \geq 1$.

In fact, as sin x increases, csc x decreases. For example:

$\sin 0^{\circ} = \sin 0 = 0$ $\csc 0 = \text{undefined}$

$\sin 30^{\circ} = \sin \frac{\pi}{6} = \frac{1}{2} = .5000$ $\csc \frac{\pi}{6} = 2$

$\sin 45^{\circ} = \sin \frac{\pi}{4} = \frac{1}{\sqrt{2}} = \frac{\sqrt{2}}{2} \approx \frac{1.414}{2} = .7070$ $\csc \frac{\pi}{4} = \sqrt{2} \approx 1.414$

$\sin 90^{\circ} = \sin \frac{\pi}{2} = 1.000$ $\csc \frac{\pi}{2} = 1$

(2) For $\frac{\pi}{2} \leq x \leq \pi$, sin x decreases from 1 to 0. Hence, csc x will increase from 1 to very large values. We can observe this from specific examples:

$\sin \frac{\pi}{2} = 1$ $\csc \frac{\pi}{2} = 1$

$\sin \frac{2}{3}\pi = \frac{\sqrt{3}}{2} \approx .87$ $\csc \frac{2}{3}\pi \approx 1.15$

$\sin \frac{3}{4}\pi \approx .707$ $\csc \frac{3}{4}\pi \approx 1.4$

$\sin \frac{5}{6}\pi = .500$ $\csc \frac{5}{6}\pi = 2$

$\sin \pi = 0$ $\csc \pi = \text{undefined}$

(3) For $\pi \leq x \leq \frac{3\pi}{2}$, sin x decreases from 0 to -1. Hence, csc x will be increasing and will increase from very large negative values to -1.

(4) For $\frac{3\pi}{2} \leq x \leq 2\pi$, $-1 \leq \sin x \leq 0$, and the graph will be increasing. Hence, csc x will decrease from -1 to very large negative values. The student can verify conclusions (3) and (4) in a similar manner to that used for (1) and (2), that is, choosing specific angles between $\frac{\pi}{2}$ and $\frac{3\pi}{2}$, and then between $\frac{3\pi}{2}$ and 2π.

The graphs of both functions are shown in the accompanying figure. Note that the range of the sine function is $-1 \le \sin x \le 1$, but the range of the cosecant function is $\csc x \ge 1$ or $\csc x \le -1$.

PROVING TRIGONOMETRIC IDENTITIES

● **PROBLEM 943**

Show that $\dfrac{\cos \theta}{\sin \theta} = \cot \theta$ is an identity.

Solution: In the diagram $\cos \theta = \dfrac{\text{adjacent side}}{\text{hypotenuse}} = \dfrac{b}{a}$, and

$$\sin \theta = \frac{\text{opposite side}}{\text{hypotenuse}} = \frac{c}{a} .$$

Thus $\dfrac{\cos \theta}{\sin \theta} = \dfrac{\frac{b}{a}}{\frac{c}{a}}$. Multiplying this fraction by $\dfrac{a}{a}$ $(= 1)$ gives us

$$\frac{\frac{b}{a}}{\frac{c}{a}} \cdot \frac{a}{a} = \frac{b}{c} .$$

We observe that b is the adjacent side of angle θ and c is the opposite side. Hence, $\dfrac{b}{c} = \dfrac{\text{adjacent}}{\text{opposite}} = \cot \theta$, by definition.

● **PROBLEM 944**

Show that $\sin(\tfrac{1}{2}\pi + t) = \cos t$ for every number t.

Solution: Using the addition formula for the sine function, which states that $\sin(\alpha+\beta) = \sin \alpha \cos \beta + \cos \alpha \sin \beta$ and replacing α by $\tfrac{1}{2}\pi$ and β by t,

$$\sin(\tfrac{1}{2}\pi + t) = \sin \tfrac{1}{2}\pi \cos t + \cos \tfrac{1}{2}\pi \sin t \qquad (1)$$

Since $\sin \tfrac{1}{2}\pi = \sin 90° = 1$, and $\cos \tfrac{1}{2}\pi = \cos 90° = 0$, equation (1) becomes:

$$\sin(\tfrac{1}{2}\pi + t) = (1) \cos t + (0) \sin t$$

$$= \cos t + 0.$$

Thus,

$$\sin(\tfrac{1}{2}\pi + t) = \cos t .$$

● **PROBLEM 945**

Show that $\sec^2\theta - \tan^2\theta = 1$ is an identity.

Solution: Given a right triangle (see the accompanying figure) we have:

$$\tan \theta = \frac{\text{opposite side}}{\text{adjacent side}} = \frac{y}{x}$$

689

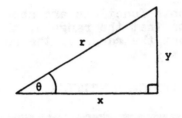

$$\sec \theta = \frac{1}{\cos \theta} = \frac{\text{hypotenuse}}{\text{adjacent side}} = \frac{r}{x}$$

Substitute these expressions into the identity:

$$\sec^2\theta - \tan^2\theta = \left(\frac{r}{x}\right)^2 - \left(\frac{y}{x}\right)^2$$

$$= \frac{r^2}{x^2} - \frac{y^2}{x^2}$$

$$= \frac{r^2 - y^2}{x^2}$$

By the Pythagorean Theorem, $r^2 = x^2 + y^2$; substitute x^2 for $r^2 - y^2$; since $x^2 + y^2 = r^2$,

$$x^2 = r^2 - y^2.$$

Thus, $\sec^2\theta - \tan^2\theta$

$$= \frac{x^2}{x^2}$$

$$= 1$$

● **PROBLEM** 946

Prove the following two identities:

(1) $\cos\left(\frac{\pi}{2} - \theta\right) = \sin \theta$ (2) $\cos \theta = \sin\left(\frac{\pi}{2} - \theta\right)$.

<u>Solution:</u> To prove identity (1), we use the cosine difference formula, which states: $\cos(u - v) = \cos u \cos v + \sin u \sin v$. Thus,

$$\cos\left(\frac{\pi}{2} - \theta\right) = \cos\frac{\pi}{2} \cos \theta + \sin\frac{\pi}{2} \sin \theta$$

Now, we must find the values for $\cos\frac{\pi}{2}$ and $\sin\frac{\pi}{2}$. Since $\frac{\pi}{2} = 90°$, and we know that $\cos 90° = 0$ and $\sin 90° = 1$, by substitution we have:

$$\cos\left(\frac{\pi}{2} - \theta\right) = 0\cdot\cos \theta + 1\cdot\sin \theta$$

$$= \sin \theta, \text{ the desired result.}$$

To prove identity (2) note that

$$\cos\left[\frac{\pi}{2} - \left(\frac{\pi}{2} - \theta\right)\right] = \cos\left(\frac{\pi}{2} - \frac{\pi}{2} + \theta\right) = \cos \theta .$$

Since $\cos \theta = \cos\left[\frac{\pi}{2} - \left(\frac{\pi}{2} - \theta\right)\right]$ we can use the difference formula, which states: $\cos(u-v) = \cos u \cos v + \sin u \sin v$. Thus, we have:

$$\cos \theta = \cos\left[\frac{\pi}{2} - \left(\frac{\pi}{2} - \theta\right)\right] = \cos\frac{\pi}{2} \cos\left(\frac{\pi}{2} - \theta\right) + \sin\frac{\pi}{2} \sin\left(\frac{\pi}{2} - \theta\right)$$

Now since $\cos\frac{\pi}{2} = 0$ and $\sin\frac{\pi}{2} = 1$, by substitution we obtain:

$$\cos \theta = \theta \cdot \cos\left(\frac{\pi}{2} - \theta\right) + 1 \cdot \sin\left(\frac{\pi}{2} - \theta\right)$$
$$= \sin\left(\frac{\pi}{2} - \theta\right)$$

which is the desired result.

● **PROBLEM** 947

If u and v are two numbers such that $u + v = \frac{1}{2}\pi$,

show that $\sin^2 u + \sin^2 v = 1$.

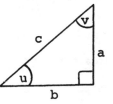

Solution: We know the following trigonometric identity,

$\sin^2 u + \cos^2 u = 1$.

Now, if $u + v = \frac{\pi}{2}$, these angles are complementary,

that is, the sum of these angles is $90° = \frac{\pi}{2}$. From the

accompanying figure, it is seen that

$$\sin v = \frac{b}{c} = \cos u$$

Substitute this relation into the identity to obtain:

$$\sin^2 u + \cos^2 u = (\sin u)^2 + (\cos u)^2$$

$$= (\sin u)^2 + (\sin v)^2$$

$$= \sin^2 u + \sin^2 v = 1.$$

● **PROBLEM** 948

Show that $\tan^2 t + 1 = \sec^2 t$.

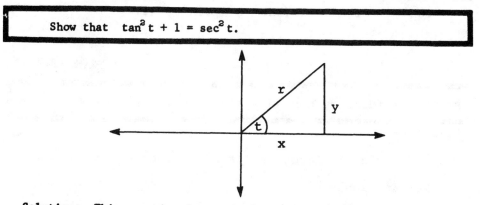

Solution: This equation is meaningless if the X-coordinate of P(t)

is 0, since, from the accompanying figure,

$$\tan t = \frac{\text{opposite side}}{\text{adjacent side}} = \frac{y}{x} = \text{undefined when} \quad x = 0$$

and

$$\sec t = \frac{1}{\cos t} = \frac{1}{\frac{\text{adjacent side}}{\text{hypotenuse}}} = \frac{1}{\frac{x}{r}}$$

$$= \frac{r}{x} = \text{undefined when} \quad x = 0$$

Furthermore, if the $\cos t$ is zero, then the $\sec t$ is undefined. We assume that it is not. In other words, $\cos t \neq 0$ for each number in which we are interested. Thus we may divide both sides of the elementary identity $\sin^2 t + \cos^2 t = 1$ by $\cos^2 t$ to obtain the equation:

$$\frac{\sin^2 t}{\cos^2 t} + 1 = \frac{1}{\cos^2 t}$$

Substitute $\tan t = \frac{\sin t}{\cos t}$ and $\sec t = \frac{1}{\cos t}$, to obtain:

$$\tan^2 t + 1 = \sec^2 t.$$

● **PROBLEM 949**

Prove that $\cos^4 B - \sin^4 B = \cos^2 B - \sin^2 B$ is an identity.

Solution: Note that $\cos^4 B = (\cos^2 B)^2$ and

$$\sin^4 B = (\sin^2 B)^2.$$

Thus, $\cos^4 B - \sin^4 B = (\cos^2 B)^2 - (\sin^2 B)^2$, which is the difference of two squares. Therefore, we may use the formula for the difference of two squares, $x^2 - y^2 = (x - y)(x + y)$, replacing x by $\cos^2 B$ and y by $\sin^2 B$, obtaining:

$$\cos^4 B - \sin^4 B = (\cos^2 B)^2 - (\sin^2 B)^2$$

$$= (\cos^2 B - \sin^2 B)(\cos^2 B + \sin^2 B).$$

Now, recall the trigonometric identity $\cos^2 B + \sin^2 B = 1$. Thus,

$$\cos^4 B - \sin^4 B = (\cos^2 B - \sin^2 B)(1)$$

$$= \cos^2 B - \sin^2 B.$$

● **PROBLEM 950**

Prove the identity $\cos^4 \beta - \sin^4 \beta = 1 - 2 \sin^2 \beta$.

Solution:

$$\cos^4 \beta - \sin^4 \beta = (\cos^2 \beta + \sin^2 \beta)(\cos^2 \beta - \sin^2 \beta)$$

For any angle θ,

$$\cos^2 \theta + \sin^2 \theta = 1 \text{ or } \cos^2 \theta = 1 - \sin^2 \theta.$$

Therefore,

$$\cos^4 \beta - \sin^4 \beta = (\cos^2 \beta + \sin^2 \beta)(\cos^2 \beta - \sin^2 \beta)$$

$$= (1)(\cos^2 \beta - \sin^2 \beta)$$

$$= \cos^2 \beta - \sin^2 \beta$$

$$= (1 - \sin^2 \beta) - \sin^2 \beta$$

$$= 1 - 2 \sin^2 \beta$$

Therefore, $\cos^4 \beta - \sin^4 \beta = 1 - 2 \sin^2 \beta$.

● **PROBLEM** 951

Prove the identity $1 + \sin 2x = (\sin x + \cos x)^2$.

<u>Solution:</u>   To prove this identity, start with the right side of the equation.

$$(\sin x + \cos x)^2 = (\sin x + \cos x)(\sin x + \cos x)$$

$$= \sin^2 x + 2 \sin x \cos x + \cos^2 x$$

$$= \sin^2 x + \cos^2 x + 2 \sin x \cos x$$

$$= 1 + 2 \sin x \cos x, \text{ since}$$

$$\left(\sin^2 x + \cos^2 x\right) = 1.$$

But $\sin 2x = 2 \sin x \cos x$, and $(\sin x + \cos x)^2 =$

$1 + 2 \sin x \cos x$. Therefore $(\sin x + \cos x)^2 = 1 + \sin 2x$.

● **PROBLEM** 952

Prove the identity $\csc 2x = \dfrac{\csc x}{2 \cos x}$.

<u>Solution:</u> Starting with the right side of the identity,

$$\frac{\csc x}{2 \cos x} = \frac{1/\sin x}{2 \cos x}, \text{ since } \csc x = \frac{1}{\sin x}. \quad \text{Hence,}$$

$$\frac{\csc x}{2 \cos x} = \frac{1}{\sin x} \cdot \frac{1}{2 \cos x} = \frac{1}{2 \sin x \cos x}.$$

Using the double-angle formula, $\sin 2\alpha = 2 \sin \alpha \cos \alpha$,

$$\frac{\csc x}{2 \cos x} = \frac{1}{2 \sin x \cos x} = \frac{1}{\sin 2x}.$$

Again, since $\csc x = \dfrac{1}{\sin x}$, $\dfrac{\csc x}{2 \cos x} = \dfrac{1}{\sin 2x} = \csc 2x$. Since we

have proved the left side equal to the right,

$$\frac{\csc x}{2 \cos x} = \csc 2x.$$

● **PROBLEM** 953

Prove the identity $\dfrac{\sin^2 \theta + \cos^2 \theta}{\cos^2 \theta} = \sec^2 \theta$.

<u>Solution:</u> The fraction $\dfrac{a+c}{b} = \dfrac{a}{b} + \dfrac{c}{b}$, by the definition of addition

693

of fractions. Thus,

$$\frac{\sin^2\theta + \cos^2\theta}{\cos^2\theta} = \frac{\sin^2\theta}{\cos^2\theta} + \frac{\cos^2\theta}{\cos^2\theta}$$

$$= \left(\frac{\sin\theta}{\cos\theta}\right)^2 + 1 ,$$

since the fraction $\frac{a^2}{b^2} = \left(\frac{a}{b}\right)^2$ by definition, and $\frac{\cos^2\theta}{\cos^2\theta} = 1$.

$\frac{\sin\theta}{\cos\theta} = \tan\theta$ by definition. Thus we obtain:

$$\tan^2\theta + 1 ;$$

and using the trignometric identity $\tan^2\theta + 1 = \sec^2\theta$, we have:

$$\frac{\sin^2\theta + \cos^2\theta}{\cos^2\theta} = \sec^2\theta ,$$

our desired result.

● **PROBLEM 954**

Prove the identity $\dfrac{1 - \sin^2\alpha}{\sin^2\alpha} = \cot^2\alpha$.

Solution: The left member is the more complicated of the two sides of the identity. Operating only on the left member, we obtain

$$\frac{1 - \sin^2\alpha}{\sin^2\alpha} = \frac{1}{\sin^2\alpha} - \frac{\sin^2\alpha}{\sin^2\alpha}$$

$$= \frac{1}{\sin^2\alpha} - 1.$$

Since $\csc\theta = \dfrac{1}{\sin\theta}$ or $\csc^2\theta = \dfrac{1}{\sin^2\theta}$ for any angle θ, then

$$\frac{1 - \sin^2\alpha}{\sin^2\alpha} = \csc^2\alpha - 1.$$

Also, since $\csc^2\theta - \cot^2\theta = 1$ or $\csc^2\theta - 1 = \cot^2\theta$ for any angle θ, then

$$\frac{1 - \sin^2\alpha}{\sin^2\alpha} = \cot^2\alpha.$$

● **PROBLEM 955**

Prove that

$$\frac{\cos^3 x - \cos x + \sin x}{\cos x} = \tan x - \sin^2 x$$

is an identity.

Solution: If we put each term of the numerator separately over the denominator, we have

$$\frac{\cos^3 x - \cos x + \sin x}{\cos x} = \frac{\cos^3 x}{\cos x} - \frac{\cos x}{\cos x} + \frac{\sin x}{\cos x} \quad.$$

But $\dfrac{\cos^3 x}{\cos^1 x} = \cos^{3-1} x = \cos^2 x$

$\dfrac{\cos x}{\cos x} = 1$

and $\dfrac{\sin x}{\cos x} = \tan x$.

Thus, replacing these values we obtain:

$$= \cos^2 x - 1 + \tan x.$$

Recall the identity $\sin^2 \theta + \cos^2 \theta = 1$.

Subtracting $\sin^2 \theta$ from both sides gives us

$$\cos^2 \theta = 1 - \sin^2 \theta,$$

and subtracting 1 from both sides we obtain

$$\cos^2 \theta - 1 = - \sin^2 \theta;$$

thus replacing $\cos^2 x - 1$ by $- \sin^2 x$ we have:

$$\frac{\cos^3 x - \cos x + \sin x}{\cos x} = - \sin^2 x + \tan x$$

$$= \tan x - \sin^2 x.$$

● **PROBLEM** 956

Show that $\tan t + \cot t = \csc t \sec t$.

<u>Solution:</u> Since $\tan t = \dfrac{\sin t}{\cos t}$ and $\cot t = \dfrac{\cos t}{\sin t}$,
by substitution we have:

$$\tan t + \cot t = \frac{\sin t}{\cos t} + \frac{\cos t}{\sin t}$$

Since multiplying by $1 \equiv \dfrac{\sin t}{\sin t}$ and $1 \equiv \dfrac{\cos t}{\cos t}$

does not alter the value of either fraction, we perform this multiplication, and obtain:

$$\frac{\sin t}{\cos t} \left(\frac{\sin t}{\sin t} \right) + \frac{\cos t}{\sin t} \left(\frac{\cos t}{\cos t} \right)$$

$$= \frac{\sin^2 t}{\sin t \cos t} + \frac{\cos^2 t}{\sin t \cos t}$$

$$= \frac{\sin^2 t + \cos^2 t}{(\sin t)(\cos t)}$$

Now, using the trigonometric identity,
$\sin^2 t + \cos^2 t = 1$, we obtain:

$$\frac{1}{(\sin t)(\cos t)} \; ;$$

and, since $\frac{1}{\sin t} = \csc t$ and $\frac{1}{\cos t} = \sec t$, we obtain
the desired result:

$$\tan t + \cot t = \csc t \sec t.$$

● PROBLEM 957

Prove that

$$\frac{1}{\sec A - \tan A} = \sec A + \tan A$$

is an identity.

Solution: If we multiply numerator and denominator of
the left member of the given equation by $\sec A + \tan A$,
we have

$$\frac{1}{\sec A - \tan A} = \left(\frac{1}{\sec A - \tan A}\right)\left(\frac{\sec A + \tan A}{\sec A + \tan A}\right)$$

(Note that $(\sec A + \tan A)/(\sec A + \tan A) = 1$, and therefore does not alter the equation)

$$= \frac{\sec A + \tan A}{(\sec A - \tan A)(\sec A + \tan A)}$$

$$= \frac{\sec A + \tan A}{\sec^2 A - \tan^2 A}$$

Recall the trigonometric identity $1 + \tan^2 \theta = \sec^2 \theta$.
Subtracting $\tan^2 \theta$ from both sides we obtain $1 = \sec^2 \theta - \tan^2 \theta$. Thus replacing $\sec^2 A - \tan^2 A$ by 1,
we obtain:

$$\frac{1}{\sec A - \tan A} = \frac{\sec A + \tan A}{1} = \sec A + \tan A.$$

● PROBLEM 958

Prove the identity $\frac{1 - \cos \theta}{\sin \theta} = \frac{\sin \theta}{1 + \cos \theta}$.

Solution: One side of this identity is as complicated as
the other so that it makes no difference which side is
used. The illustration uses both sides.

(a) $$\frac{1 - \cos \theta}{\sin \theta} = \frac{(1 - \cos \theta)}{\sin \theta}\frac{(1 + \cos \theta)}{(1 + \cos \theta)}$$

$$= \frac{1 - \cos \theta + \cos \theta - \cos^2 \theta}{\sin \theta(1 + \cos \theta)}$$

$$= \frac{1 - \cos^2 \theta}{\sin \theta (1 + \cos \theta)}$$

Since $\cos^2 \theta + \sin^2 \theta = 1$ or $\sin^2 \theta = 1 - \cos^2 \theta$, then

$$\frac{1 - \cos \theta}{\sin \theta} = \frac{1 - \cos^2 \theta}{\sin \theta (1 + \cos \theta)} = \frac{\sin^2 \theta}{\sin \theta (1 + \cos \theta)}$$

$$= \frac{\sin \theta}{1 + \cos \theta}$$

Note that this method starts with the left side of the identity to be proved. The following is another method which can be used to prove the identity.

(b)
$$\frac{\sin \theta}{1 + \cos \theta} = \frac{\sin \theta (1 - \cos \theta)}{(1 + \cos \theta)(1 - \cos \theta)}$$

$$= \frac{\sin \theta (1 - \cos \theta)}{1 + \cos \theta - \cos \theta - \cos^2 \theta}$$

$$= \frac{\sin \theta (1 - \cos \theta)}{1 - \cos^2 \theta}$$

Again, since $\sin^2 \theta = 1 - \cos^2 \theta$,

$$\frac{\sin \theta}{1 + \cos \theta} = \frac{\sin \theta (1 - \cos \theta)}{1 - \cos^2 \theta}$$

$$= \frac{\sin \theta (1 - \cos \theta)}{\sin^2 \theta}$$

$$= \frac{1 - \cos \theta}{\sin \theta}$$

Note that this second method starts with the right side of the identity to be proved.

● **PROBLEM** 959

Prove that

$$\frac{\cos A}{\csc A - 1} + \frac{\cos A}{\csc A + 1} = 2 \tan A$$

is an identity.

Solution: The left member is the more complicated; hence, we shall work with it and begin by performing the indicated addition. The lowest common denominator is $(\csc A - 1)(\csc A + 1) = \csc^2 A - 1$. Thus

$$\frac{\cos A}{\csc A - 1} + \frac{\cos A}{\csc A + 1} = \left(\frac{\csc A + 1}{\csc A + 1}\right) \frac{\cos A}{\csc A - 1}$$

$$+ \left(\frac{\csc A - 1}{\csc A - 1}\right) \frac{\cos A}{\csc A + 1}$$

$$= \frac{(\csc A + 1)\cos A}{\csc^2 A - 1} + \frac{(\csc A - 1)\cos A}{\csc^2 A - 1}$$

$$= \frac{\csc A \cos A + \cos A}{\csc^2 A - 1} + \frac{\csc A \cos A - \cos A}{\csc^2 A - 1}$$

$$= \frac{\csc A \cos A + \cos A + \csc A \cos A - \cos A}{\csc^2 A - 1}$$

$$= \frac{\csc A \cos A + \csc A \cos A}{\csc^2 A - 1}$$

Recall the trignometric identity $\csc^2 A - 1 = \cot^2 A$,

$$= \frac{2 \cos A \csc A}{\cot^2 A}$$

replace $\csc A$ by $\frac{1}{\sin A}$,

$$= \frac{(2 \cos A)/(\sin A)}{\cot^2 A}$$

replace $\frac{\cos A}{\sin A}$ by $\cot A$,

$$= \frac{2 \cot A}{\cot^2 A}$$

cancelling out $\cot A$,

$$= \frac{2}{\cot A}$$

Replace $\cot A$ by $\frac{1}{\tan A}$, $= \frac{2}{\frac{1}{\tan A}}$

Multiply numerator and denominator by $\tan A$,

$$= 2 \tan A$$

We have thus proved

$$\frac{\cos A}{\csc A - 1} + \frac{\cos A}{\csc A + 1} = 2 \tan A \quad \text{is an identity.}$$

● **PROBLEM** 960

Prove that the following equation is an identity:

$$\frac{1 - \sin x}{\cos x} = \frac{\cos x}{1 + \sin x}$$

Solution: This problem may be approached in a variety of ways. One method is based on the fact that two fractions are equal if their cross products are equal. That is,

$$\frac{1 - \sin x}{\cos x} = \frac{\cos x}{1 + \sin x} \quad \text{if } (1 - \sin x)(1 + \sin x)$$
$$= (\cos x)(\cos x).$$

$$(1 - \sin x)(1 + \sin x) = 1 - \sin^2 x$$

Now, recall the trigonometric identity $\cos^2 \theta = 1 - \sin^2 \theta$; thus $(1 - \sin x)(1 + \sin x) = \cos^2 x = (\cos x) \cdot (\cos x)$, and since the cross products are equal, we have proven the original fractions equivalent.

● **PROBLEM** 961

Prove the identity: $\sec A \csc A = \tan A + \cot A$.

Solution: One approach to the proof of identities, when many functions are involved, is to express the given functions in terms of fewer functions. In this case, suppose we express each of the given trigonometric functions in terms of sine and cosine functions. We will work in parallel columns, with each side of the given equation:

Since sec A = 1/cos A and csc A = 1/sin A,

  sec A csc A

$$= \frac{1}{\cos A} \cdot \frac{1}{\sin A}$$

$$= \frac{1}{\cos A \sin A}$$

Since tan A = sin A/cos A and cot A = 1/tan A = 1/sin A/cos A = cos A/sin A, tan A + cot A

$$= \frac{\sin A}{\cos A} + \frac{\cos A}{\sin A} \; .$$

To combine these fractions, we convert them into fractions with the least common denominator (LCD) cos A sin A; thus:

$$= \left(\frac{\sin A}{\sin A} \; \frac{\sin A}{\cos A} \right) +$$

$$\left(\frac{\cos A}{\cos A} \; \frac{\cos A}{\sin A} \right)$$

$$= \frac{\sin^2 A + \cos^2 A}{\cos A \sin A} \; .$$

Recall the trigonometric identity $\sin^2 A + \cos^2 A = 1$; replacing $\sin^2 A + \cos^2 A$ by 1 we obtain:

$$= \frac{1}{\cos A \sin A}$$

Now, since we have proved that both sides of the given equation are equal to the same expression, we are tempted to say that they are therefore equal to each other, and that we have therefore proved what we set out to prove.

We have indeed, except for one detail. We have not considered the values of A for which the given expressions and those which we substituted are meaningful.

This aspect of the proof of a trigonometric identity rarely leads to trouble, and may therefore usually be omitted. The careful student, however, will want to be prepared to investigate this question.

Thus we note that sec A and tan A are defined if and only if A is a real number of degrees not an odd multiple of 90; csc A and cot A are defined if and only if A is a real number of degrees not an even multiple of 90. Both sides of the equation are therefore defined if and only if A is a real number of degrees not an integer multiple of 90.

Each of the substitutions made in the parallel columns above is valid if A is such a number.

Therefore sec A csc A and tan A cot A are equal whenever both are defined, and the equation sec A csc A = tan A + cot A is an identity.

● **PROBLEM** 962

Show that tan (- v) = - tan v for every number v in the domain of the tangent function.

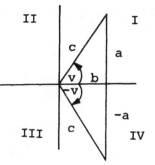

<u>Solution:</u> We have

$$\tan (- v) = \frac{\sin (- v)}{\cos (- v)}$$

Now, let us examine sin (- v) and cos (- v). We know that in drawing the negative angles we rotate clockwise on the coordinate axes.

Thus, the value of the negative angle is the same as that of the reference angle in quadrant I, but it may have a different sign. Therefore, sin v and sin (- v) have the value $\frac{a}{c}$, but for angle - v, the opposite side is - a, thus sin (- v) = $\frac{- a}{c}$ = - sin v. Cos v and cos (- v) have the value $\frac{b}{c}$, and cos v = $\frac{b}{c}$, cos (- v) = $\frac{b}{c}$; thus, cos (- v) = cos v. Therefore, $\frac{\sin (- v)}{\cos (- v)}$ can be written as:

$$\frac{- \sin v}{\cos v} = - \tan v.$$

● **PROBLEM** 963

Prove $\sin(45° + x) + \sin(45° - x) = \sqrt{2} \cos x$.

<u>Solution:</u> Use the formula for the sine of the sum and the difference of two angles.

$$\sin(A + B) = \sin A \cos B + \cos A \sin B$$

$$\sin(A - B) = \sin A \cos B - \cos A \sin B$$

Adding these two shows that cos A sin B will cancel and the result becomes

$$2 \sin A \cos B \quad \text{with} \quad A = 45° .$$

Therefore we have

$$\sin(45° + x) + \sin(45° - x) = 2 \sin 45° \cos x$$
$$= \frac{2\sqrt{2}}{2} \cos x = \sqrt{2} \cos x .$$

● **PROBLEM** 964

Prove that $\dfrac{\cos 2\theta}{\cos \theta} = \dfrac{1 - \tan^2 \theta}{\sec \theta}$.

Solution: Working on the left side of the equation, we note that cos 2θ can be rewritten as cos (θ + θ), to which we apply the formula for the cosine of the sum of two angles, cos (u + v) = cos u cos v - sin u sin v. Thus, cos 2θ = cos (θ + θ) = cos θ cos θ - sin θ sin θ = $\cos^2 \theta - \sin^2 \theta$.

Replacing cos 2θ by $\cos^2 \theta - \sin^2 \theta$ we obtain,

$$\frac{\cos 2\theta}{\cos \theta} = \frac{\cos^2 \theta - \sin^2 \theta}{\cos \theta}$$

Divide numerator and denominator by $\cos^2 \theta$,

$$= \frac{\dfrac{\cos^2 \theta - \sin^2 \theta}{\cos^2 \theta}}{\dfrac{\cos \theta}{\cos^2 \theta}}$$

$$= \frac{\dfrac{\cos^2 \theta}{\cos^2 \theta} - \dfrac{\sin^2 \theta}{\cos^2 \theta}}{\dfrac{\cos \theta}{\cos^2 \theta}}$$

$$\frac{1 - \dfrac{\sin^2 \theta}{\cos^2 \theta}}{\dfrac{\cos \theta}{\cos^2 \theta}}$$

Recall that tan θ = sin θ/cos θ; thus

$$\tan^2 \theta = \left(\frac{\sin \theta}{\cos \theta}\right)^2 = \frac{\sin^2 \theta}{\cos^2 \theta} \ .$$

Substituting $\tan^2 \theta$ for $\sin^2 \theta / \cos^2 \theta$ we obtain,

$$= \frac{1 - \tan^2 \theta}{\dfrac{\cos \theta}{\cos^2 \theta}}$$

$$= \frac{1 - \tan^2 \theta}{\dfrac{1}{\cos \theta}} \ .$$

Since 1/cos θ = sec θ, replace 1/cos θ by sec θ; thus

$$\frac{1 - \tan^2 \theta}{\sec \theta} \ .$$

Therefore, we have shown that

$$\frac{\cos 2\theta}{\cos \theta} = \frac{1 - \tan^2 \theta}{\sec \theta} \ .$$

● PROBLEM 965

Prove that $\dfrac{\cos \theta}{1 - \sin \theta} = \dfrac{1 + \sin \theta}{\cos \theta}$.

Solution: $\dfrac{\cos \theta}{1 - \sin \theta} = \dfrac{\cos \theta}{1 - \sin \theta} \cdot \dfrac{1 + \sin \theta}{1 + \sin \theta}$, since multiplication by 1 does not change the value of the fraction. Performing the multiplication we obtain:

$$\frac{\cos \theta (1 + \sin \theta)}{(1 - \sin \theta)(1 + \sin \theta)}$$

$$= \frac{\cos \theta (1 + \sin \theta)}{1 - \sin^2 \theta} .$$

Since we are dividing by $1 - \sin^2 \theta$, we must be sure $1 - \sin^2 \theta \neq 0$, that is $1 \neq \sin^2 \theta$. Thus $\sin \theta \neq 1, -1$.

We use the identity $1 - \sin^2 \theta = \cos^2 \theta$ to rewrite the fraction as follows: $\frac{\cos \theta (1 + \sin \theta)}{\cos^2 \theta} = \frac{\cos \theta (1 + \sin \theta)}{(\cos \theta)(\cos \theta)}$, since

$$\cos^2 \theta = (\cos \theta)(\cos \theta) .$$

Finally, $\frac{\cos \theta}{1 - \sin \theta} = \frac{1 + \sin \theta}{\cos \theta}$. Since we are dividing by $\cos \theta$ we must exclude all $\cos \theta = 0$. Hence $\cos \theta \neq 0$. Thus, we have proven that $\frac{\cos \theta}{1 - \sin \theta} = \frac{1 + \sin \theta}{\cos \theta}$, with the restrictions

$$\cos \theta \neq 0 , \sin \theta \neq 1, -1 .$$

● **PROBLEM** 966

Prove the identity $\sin^2 \theta + \tan^2 \theta = \sec^2 \theta - \cos^2 \theta$.

Solution: Since $\tan \theta = \frac{\sin \theta}{\cos \theta}$, and $\sec \theta = \frac{1}{\cos \theta}$, we can express this equation in terms of sine and cosine:

$$\sin^2 \theta + \left(\frac{\sin \theta}{\cos \theta}\right)^2 = \left(\frac{1}{\cos \theta}\right)^2 - \cos^2 \theta .$$

Since $\left(\frac{\sin \theta}{\cos \theta}\right)^2 = \frac{\sin^2 \theta}{\cos^2 \theta}$, and $\left(\frac{1}{\cos \theta}\right)^2 = \frac{1}{\cos^2 \theta}$ we obtain

$$\sin^2 \theta + \frac{\sin^2 \theta}{\cos^2 \theta} = \frac{1}{\cos^2 \theta} - \cos^2 \theta$$

We also know that $\cos^2 \theta = 1 - \sin^2 \theta$ (a trigonometric identity). Substituting $1 - \sin^2 \theta$ for $\cos^2 \theta$:

$$\sin^2 \theta + \frac{\sin^2 \theta}{1 - \sin^2 \theta} = \frac{1}{1 - \sin^2 \theta} - \left(1 - \sin^2 \theta\right)$$

In order to combine terms we multiply $\sin^2 \theta$ and $\left(1 - \sin^2 \theta\right)$ by $\frac{1 - \sin^2 \theta}{1 - \sin^2 \theta}$ (which equals 1, and therefore does not change the value of the terms). Thus:

$$\frac{\sin^2 \theta \left(1 - \sin^2 \theta\right)}{1 - \sin^2 \theta} + \frac{\sin^2 \theta}{1 - \sin^2 \theta} = \frac{1}{1 - \sin^2 \theta} - \frac{\left(1 - \sin^2 \theta\right)\left(1 - \sin^2 \theta\right)}{1 - \sin^2 \theta}$$

Adding the fractions we obtain:

$$\frac{\sin^2 \theta \left(1 - \sin^2 \theta\right) + \sin^2 \theta}{1 - \sin^2 \theta} = \frac{1 - \left(1 - \sin^2 \theta\right)\left(1 - \sin^2 \theta\right)}{1 - \sin^2 \theta}$$

Multiplying the terms in the numerator we obtain:

$$\frac{\sin^2\theta - \sin^4\theta + \sin^2\theta}{1-\sin^2\theta} = \frac{1-\left(1- 2\sin^2\theta + \sin^4\theta\right)}{1-\sin^2\theta}$$

$$\frac{2\sin^2\theta - \sin^4\theta}{1-\sin^2\theta} = \frac{1-\left(1-2\sin^2\theta + \sin^4\theta\right)}{1-\sin^2\theta}$$

$$\frac{2\sin^2\theta - \sin^4\theta}{1-\sin^2\theta} = \frac{1-1+2\sin^2\theta - \sin^4\theta}{1-\sin^2\theta}$$

$$\frac{2\sin^2\theta - \sin^4\theta}{1-\sin^2\theta} = \frac{2\sin^2\theta - \sin^4\theta}{1-\sin^2\theta}$$

Since $\sin^2\theta + \tan^2\theta = \dfrac{2\sin^2\theta - \sin^4\theta}{1-\sin^2\theta}$ and

$$\sec^2\theta - \cos^2\theta = \frac{2\sin^2\theta - \sin^4\theta}{1-\sin^2\theta}$$

$$\sin^2\theta + \tan^2\theta = \sec^2\theta - \cos^2\theta$$

because two expressions equal to the same expression are equal to each other.

● **PROBLEM** 967

Prove the identity $\dfrac{\sec x + 1}{\sec x - 1} = \cot^2 \dfrac{x}{2}$.

<u>Solution:</u> Starting with the left side of the identity,

$$\frac{\sec x + 1}{\sec x - 1} = \frac{\dfrac{1}{\cos x} + 1}{\dfrac{1}{\cos x} - 1} \quad \text{since } \sec x = \frac{1}{\cos x}$$

Hence,

$$\frac{\sec x + 1}{\sec x - 1} = \frac{\dfrac{1}{\cos x} + \dfrac{\cos x}{\cos x}}{\dfrac{1}{\cos x} - \dfrac{\cos x}{\cos x}} \quad \text{for } 1 = \frac{\cos x}{\cos x}$$

Combining fractions,

$$= \frac{\dfrac{1 + \cos x}{\cos x}}{\dfrac{1 - \cos x}{\cos x}}$$

Dividing by a fraction is equivalent to multiplying by its reciprocal, hence,

$$= \frac{1 + \cos x}{\cos x} \cdot \frac{\cos x}{1 - \cos x}$$

Cancelling $\cos x$ from numerator and denominator we obtain:

$$\frac{\sec x + 1}{\sec x - 1} = \frac{1 + \cos x}{1 - \cos x} = \left(\pm \sqrt{\frac{1 + \cos x}{1 - \cos x}}\right)^2$$

Looking at the right side of the identity:

from the formula for the tangent of a half angle,

$$\tan \tfrac{1}{2}\theta = \tan \frac{\theta}{2} = \pm \sqrt{\frac{1 - \cos \theta}{1 + \cos \theta}} , \quad \text{and}$$

$$\cot \tfrac{1}{2}x = \cot \frac{x}{2} = \frac{1}{\tan \dfrac{x}{2}} = \frac{1}{\pm \sqrt{\dfrac{1 - \cos x}{1 + \cos x}}}$$

$$= \frac{1}{\pm \sqrt{\dfrac{1 - \cos x}{1 + \cos x}}}$$

$$= (1)\left(\pm \frac{\sqrt{1 + \cos x}}{\sqrt{1 - \cos x}} \right) = \pm \frac{\sqrt{1 + \cos x}}{\sqrt{1 - \cos x}}$$

Hence, $\quad \cot \dfrac{x}{2} = \pm \sqrt{\dfrac{1 + \cos x}{1 - \cos x}}$

Therefore, $\dfrac{\sec x + 1}{\sec x - 1} = \left(\cot \dfrac{x}{2} \right)^2 = \cot^2 \dfrac{x}{2} .$

● **PROBLEM** 968

Prove the identity

$$\tan \left(\frac{\pi}{4} + \frac{\theta}{2} \right) = \sec \theta + \tan \theta.$$

<u>Solution:</u> Factor out $\tfrac{1}{2}$.

$$\tan \left(\frac{\pi}{4} + \frac{\theta}{2} \right) = \tan \frac{1}{2}\left(\frac{\pi}{2} + \theta \right).$$

Then, apply the half-angle formula for the tangent.

$$\tan \tfrac{1}{2}\theta_1 = \pm \sqrt{\frac{1 - \cos \theta_1}{1 + \cos \theta_1}}.$$

Rationalize the denominator to obtain

$$\tan \tfrac{1}{2}\theta_1 = \pm \sqrt{\frac{1 - \cos \theta_1}{1 + \cos \theta_1} \frac{1 - \cos \theta_1}{1 - \cos} } = \frac{1 - \cos \theta_1}{\pm\sqrt{1 - \cos^2 \theta_1}}$$

$$= \frac{1 - \cos \theta_1}{\sqrt{\sin^2 \theta_1}}$$

$$\tan \tfrac{1}{2}\theta_1 = \frac{1 - \cos \theta_1}{\sin \theta_1}.$$

Replace θ_1 by $\dfrac{\pi}{2} + \theta$.

$$\tan \tfrac{1}{2}\left(\frac{\pi}{2} + \theta \right) = \frac{1 - \cos \left(\frac{\pi}{2} + \theta \right)}{\sin \left(\frac{\pi}{2} + \theta \right)}$$

Apply the formula for the sum of the sine of two angles

and the cosine of the sum of two angles.

$$\cos(\alpha + \beta) = \cos\alpha \cos\beta - \sin\alpha \sin\beta$$

$$\sin(\alpha + \beta) = \sin\alpha \cos\beta + \cos\alpha \sin\beta$$

$$\cos\left(\frac{\pi}{2} + \theta\right) = \cos\frac{\pi}{2}\cos\theta - \sin\frac{\pi}{2}\sin\theta$$

$$= 0(\cos\theta) - \sin\theta = -\sin\theta.$$

$$\sin\left(\frac{\pi}{2} + \theta\right) = \sin\frac{\pi}{2}\cos\theta + \cos\frac{\pi}{2}\sin\theta$$

$$= 1\cos\theta + 0(\sin\theta)$$

$$= \cos\theta$$

Substitute these two results.

$$\tan\frac{1}{2}\theta_1 = \frac{1 - \cos\theta_1}{\sin_1} = \tan\frac{1}{2}\left(\frac{\pi}{2} + \theta\right) = \frac{1 - \cos\left(\frac{\pi}{2} + \theta\right)}{\sin\left(\frac{\pi}{2} + \theta\right)}$$

$$= \frac{1 - (-\sin\theta)}{\cos\theta} = \frac{1 + \sin\theta}{\cos\theta}$$

$$= \frac{1}{\cos\theta} + \frac{\sin\theta}{\cos\theta}$$

$$= \sec\theta + \tan\theta.$$

CHAPTER 30

POLAR COORDINATES

> **Basic Attacks and Strategies for Solving Problems in this Chapter. See pages 706 to 711 for step-by-step solutions to problems.**

Some problems in analytic geometry, especially involving motion about a point, are difficult to solve using the *x-y* coordinate system. In fact, such equations as

$$x^2 + y^2 - 2x = 2\sqrt{x^2 + y^2}$$

are unwieldly in the rectangular coordinate system, but are manageable in the polar coordinate system.

To set up the polar coordinate system in a plane, begin with a fixed point O (called the origin or pole) and a directed half-line (called the polar axis) with its endpoint at O drawn horizontally to the right. To each point P in the plane, the polar coordinates (r, θ) can be assigned. The angle θ has the polar axis as its initial side and the half-line OP as its terminal side. The number r indicates the distance from the pole to the point P. There is an infinite number of ordered pairs associated with (r, θ)

For instance, a point whose coordinates are (3, 45°) may be described by

(3, 405°), (3, 765°), (− 3, 225°), (− 3, − 135°), (3, − 315°)

to name the function. Polar coordinate paper is useful in plotting points.

We can superimpose an *xy*-plane on the *rθ*-plane with the positive *x*-axis coinciding with the polar axis. Then, the following polar-rectangular relationships can be deduced.

$$x = r \cos \theta \qquad\qquad y = r \sin \theta$$

$$\tan \theta = {}^y\!/_x \qquad\qquad r^2 = x^2 + y^2$$

These relationships provide the basis for changing polar equations to equations in rectangular form and vice versa. For example, using the above relationships

$$x^2 + y^2 - 2x = 2\sqrt{x^2 + y^2}$$

in rectangular form may be written in polar form as

$$r = 2 + 2 \cos \theta$$

and vice versa.

The graph of a polar equation is the set of all points whose ordered pairs are solutions of the equation. For example, the graph of the polar equation $r = 2$ is given by a circle of radius 2 from the pole 0. All the points on the circle satisfy the equation.

To graph a polar equation, begin choosing various values of θ and finding the corresponding values of r. However, before doing so, three observations will reduce the number of points we must choose.

(1) Tests for symmetry:

    (a) The polar equation exhibits symmetry with respect to the polar axis if replacing θ with $-\theta$ or replacing r with $-r$ and θ with $\pi - \theta$ produces an equivalent equation;

    (b) The polar equation exhibits symmetry with respect to the $(\pi/2)$–axis if replacing r with $-r$ and θ with $-\theta$ or replacing θ with $\pi - \theta$ produces an equivalent equation; and

    (c) The polar equation exhibits pole symmetry if replacing r with $-r$ or θ with $\pi + \theta$ produces an equivalent equation.

(2) If a trigonometric function is in the polar equation, determine if the function is a periodic function and the period. For instance, if $\cos \theta$ is in the equation, then we know that it is periodic with period 2π. So, it is only necessary to choose points between 0 and 2π.

(3) If a trigonometric function is in the polar function, determine in what interval within the period the trigonometric function with respect to θ is negative. This means that r will be negative in that interval. (For example, in the interval

$$\pi/2 < \theta < 3\pi/2,$$

$\cos \theta$ yields negative values.) So, we need only consider angles θ in the first and fourth quadrants. However, since the graph is symmetric with respect to the polar axis, it is only necessary to choose values θ between 0 and $\pi/2$.

Often changing from polar to rectangular form or vice versa can yield a simple equation to graph.

Step-by-Step Solutions to Problems in this Chapter, "Polar Coordinates"

● **PROBLEM** 969

What is the graph of ρ = 3?

Solution: This equation says that for all values of θ, ρ = 3. Thus for θ = 0˙, ρ = 3; θ = 30˙, ρ = 3; etc. The graph is a circle of radius 3 with the center at the origin.

● **PROBLEM** 970

Draw the graph of ρ = 2 cos θ.

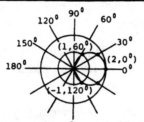

Solution: We assign values to θ and find the corresponding values of ρ , giving the following table:

θ	cos θ	ρ=2cos θ
0⁰	1	2
30⁰	.87	1.74
60⁰	.5	1
90⁰	0	0
120⁰	-.5	-1
150⁰	-.87	-1.74
180⁰	-1	-2
Values from 180⁰ to 360⁰ give the same points. (Check this.)		

We then plot the points (ρ,θ) and draw a smooth curve through them. We get the graph of the figure. The equation which defines the path of P may involve only one of the variables (ρ,θ). In that case the variable which is not mentioned may have any and all values.

● **PROBLEM** 971

Transform the equation $x^2 + y^2 - x + 3y = 3$ to a polar equation.

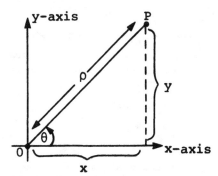

y-axis

P

ρ

y

θ

0

x

x-axis

Solution: Ordinarily, when we wish to locate a point in a plane, we draw a pair of perpendicular axes and measure specified signed distances from the axes. The points are designated by pairs in terms of (x,y). These are called rectangular coordinates.

Another way is to designate a point in terms of polar coordinates. (ρ,θ) are the polar coordinates of a point P where ρ is the radius vector of P and θ is the angle that is made with the positive x-axis and the radius vector, OP. (See diagram.)

If P is designated by the coordinates (x,y) in rectangular coordinates and by (ρ,θ) in polar coordinates, then the following relationships hold:

$$\text{Cos } \theta = \frac{\text{adjacent side}}{\text{hypotenuse}} = \frac{x}{\rho} \quad \text{or} \quad x = \rho \text{ Cos } \theta$$

$$\text{Sin } \theta = \frac{\text{opposite side}}{\text{hypotenuse}} = \frac{y}{\rho} \quad \text{or} \quad y = \rho \text{ Sin } \theta$$

Now in this example, we replace x by ρ Cos θ and y by ρ Sin θ to obtain:

$$(\rho \text{ Cos } \theta)^2 + (\rho \text{ Sin } \theta)^2 - (\rho \text{ Cos } \theta) + 3(\rho \text{ Sin } \theta) = 3$$

$$\rho^2 \text{ Cos}^2 \theta + \rho^2 \text{ Sin}^2 \theta - \rho \text{ Cos } \theta + 3\rho \text{ Sin } \theta = 3$$

Factor out ρ^2 and $-\rho$.

$$\rho^2 (\text{Cos}^2 \theta + \text{Sin}^2 \theta) - \rho (\text{Cos } \theta - 3\text{Sin } \theta) = 3$$

Apply the identity Cos$^2 \theta$ + Sin$^2 \theta$ = 1. Then,

$$\rho^2 - \rho(\text{Cos } \theta - 3 \text{ Sin } \theta) = 3$$

● **PROBLEM** 972

Transform the equation $\rho = 2 \cos \theta$ to rectangular coordinates.

Solution:

$$\rho = \sqrt{x^2 + y^2}, \qquad \cos \theta = \frac{x}{\rho} = \frac{x}{\sqrt{x^2 + y^2}}$$

$$\rho = 2 \cos \theta$$

$$\sqrt{x^2 + y^2} = \frac{2x}{\sqrt{x^2 + y^2}}$$

$$x^2 + y^2 = 2x$$

707

Transform the equation $xy = 4$ to polar coordinates.

<u>Solution:</u>

$$x = \rho \cos \theta$$
$$y = \rho \sin \theta$$
$$xy = 4$$
$$\rho \cos \theta \cdot \rho \sin \theta = 4$$
$$\rho^2 \cos \theta \sin \theta = 4$$

Transform the equation $r = 4/(2 - 3 \sin \theta)$ to an equation in cartesian coordinates.

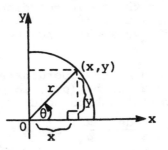

<u>Solution:</u> The given equation is in polar coordinates. This is another system of coordinates where a point (x, y) lies on a circle of radius r whose center is the origin. (see Figure.)

We want to replace r and $\sin \theta$ by rectangular co-ordinates. Observe the following needed substitutions which can be derived from the diagram.

$$\sin \theta = \frac{\text{opposite side}}{\text{hypotenuse}} = \frac{y}{r}$$

Pythagorean Identity $x^2 + y^2 = r^2$

Solving for r: $\sqrt{x^2 + y^2} = r$

Then we proceed as follows:

$r = \dfrac{4}{2 - 3y/r}$ \qquad replacing $\sin \theta$ by y/r

$= \dfrac{4r}{2r - 3y}$ \qquad simplifying the complex fraction

$2r - 3y = 4$ \qquad multiplying each member by $(2r - 3y)/r$

$2\sqrt{x^2 + y^2} = 4 + 3y$ \qquad replacing r by $\sqrt{x^2 + y^2}$ and adding $3y$ to each member

$4x^2 + 4y^2 = 9y^2 + 24y + 16$ equating the squares of each member

$4x^2 - 5y^2 - 24y = 16$ adding $-9y^2 - 24y$ to each member

● **PROBLEM** 975

Convert the equation $r = \tan \theta + \cot \theta$ to an equation in cartesian coordinates.

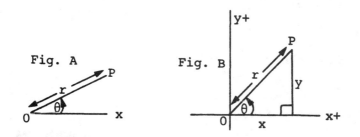

Fig. A   Fig. B

<u>Solution:</u> The given equation is expressed in polar co-ordinates (r, θ) where r is the radius vector, OP, and θ is the angle that r makes with the polar axis, OX. O is the fixed point called the pole. See figure A.

Since $\tan \theta \neq -\cot \theta$, then $r \neq 0$, and the graph of

$$r = \tan \theta + \cot \theta$$

does not pass through the pole. If r were equal to zero, then the curve would pass through $(0,0)$. Therefore in the transformation of this equation to cartesian co-ordinates, we must remember that $(x, y) \neq (0,0)$. Now we must convert all expressions of r and θ into rectangular coordinates (x, y). If P is designated by the coordinates (x, y) in rectangular coordinates and by (r, θ) in polar coordinates, then the following relationships hold true: (see Figure B).

$$\tan \theta = \frac{\text{opposite side}}{\text{adjacent side}} = \frac{y}{x}$$

$$\cot \theta = \frac{\text{adjacent side}}{\text{opposite side}} = \frac{x}{y}$$

By the Pythagorean Identity $x^2 + y^2 = r^2$

Solve for r: $r = \sqrt{x^2 + y^2}$

Substitute these values for r, $\tan \theta$, and $\cot \theta$.

$$\sqrt{x^2 + y^2} = \frac{y}{x} + \frac{x}{y}$$

$$xy\sqrt{x^2 + y^2} = x^2 + y^2$$

Divide by $\sqrt{x^2 + y^2}$

$$xy = \frac{x^2 + y^2}{\sqrt{x^2 + y^2}}$$

Rationalize the denominator by multiplying the numerator and denominator by $\sqrt{x^2 + y^2}$

$$xy = \frac{x^2 + y^2}{\sqrt{x^2 + y^2}} \; \frac{\sqrt{x^2 + y^2}}{\sqrt{x^2 + y^2}}$$

$$xy = \frac{x^2 + y^2}{x^2 + y^2} \sqrt{x^2 + y^2}$$

$$xy = \sqrt{x^2 + y^2}$$

Squaring both sides, we obtain:

$$x^2 y^2 = x^2 + y^2$$

where $x \neq 0$ and $y \neq 0$.

● **PROBLEM** 976

Discuss the graph of the equation

$$r = 4 \cos \theta - 2 \sin \theta \; .$$

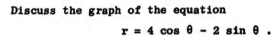

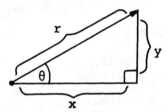

<u>Solution:</u> Instead of plotting points of the form (r, θ), we will write this equation in terms of cartesian coordinates x and y. Cartesian coordinates (x, y) and polar coordinates (r, θ) are related by the following equations: (noting the figure,)

$$\sin \theta = \frac{\text{opposite side}}{\text{hypotenuse}} = \frac{y}{r} \qquad (1)$$

$$\cos \theta = \frac{\text{adjacent side}}{\text{hypotenuse}} = \frac{x}{r} \qquad (2)$$

and by the Pythagorean Theorem,

$$r^2 = x^2 + y^2 \; . \qquad (3)$$

First, we multiply both sides of the given equation by r to obtain the equation

$$r^2 = 4r \cos \theta - 2r \sin \theta \qquad (4)$$

From equations (1), (2), and (3), the new equation (equation (4)) becomes:

$$x^2 + y^2 = 4r\left(\frac{x}{r}\right) - 2r\left(\frac{y}{r}\right)$$

$$x^2 + y^2 = 4x - 2y \qquad (5)$$

Subtract $(4x - 2y)$ from both sides of equation (5):

710

$$x^2 + y^2 - (4x - 2y) = 4x - 2y - (4x - 2y)$$
$$x^2 + y^2 - 4x + 2y = 0$$
$$x^2 - 4x + y^2 + 2y = 0 \qquad (6)$$

Now complete the square in both x and y. This is done by taking half the coefficient of the x term (or y term) and then squaring this value. The result is then added to both sides of equation (6). Completing the square in x:

$$[\tfrac{1}{2}(-4)]^2 = (-2)^2 = 4 \ .$$

Hence, equation (6) becomes:

$$(x^2 - 4x + 4) + y^2 + 2y = 0 + 4$$
or
$$(x^2 - 4x + 4) + y^2 + 2y = 4$$
or
$$(x - 2)^2 + y^2 + 2y = 4 \qquad (7)$$

Completing the square in y:

$$[\tfrac{1}{2}(2)]^2 = (1)^2 = 1.$$

Hence, equation (7) becomes:

$$(x - 2)^2 + (y^2 + 2y + 1) = 4 + 1$$
or
$$(x - 2)^2 + (y^2 + 2y + 1) = 5$$
or
$$(x - 2)^2 + (y + 1)^2 = 5 \qquad (8)$$

Also, note that the equation of a circle is:

$$(x - h)^2 + (y - k)^2 = r^2$$

where (h,k) are the coordinates of the center of the circle and r is the radius of the circle. Equation (8) is in the form for the equation of circle where:

$$x - 2 \quad \text{corresponds to} \quad x - h; \text{ i.e., } h = 2,$$

$$y + 1 \quad \text{corresponds to} \quad y - k; \text{ that is, } y + 1 = y - k$$

$$y + 1 - y = y - k - y$$

$$1 = -k$$

$$-1(1) = (-1)(-k)$$

$$-1 = k;$$

and r^2 corresponds to 5; that is; $r = \sqrt{5}$.

Therefore, the original equation given in polar coordinates (r,θ) represents a circle of center $(h,k) = (2,-1)$ and radius $= r = \sqrt{5}$.

CHAPTER 31

VECTORS AND COMPLEX NUMBERS

> **Basic Attacks and Strategies for Solving Problems in this Chapter. See pages 712 to 741 for step-by-step solutions to problems.**

To find the sum of two vectors A and B, written as $A + B$, we place the initial point of vector B at the terminal point of vector A. The vector with the same initial point as A and the same terminal point as B is the vector sum $A + B$.

Another way to find the sum of two vectors A and B is to use the parallelogram rule. To apply the rule place vectors A and B so that their initial points coincide. Then, complete a parallelogram which has A and B as sides. The diagonal of the parallelogram with the same initial point as A and B is the vector sum $A + B$.

The length of a vector $V = (x, y)$, expressed in rectangular components, is called the magnitude, denoted by $|V|$ and defined by

$$|V| = \sqrt{x^2 + y^2}$$

The magnitude of the resultant sum of two vectors A and B can be determined by using the formula for the Law of Cosines

$$|V|^2 = x^2 + y^2 - 2xy \cos\theta$$
$$|V| = \sqrt{x^2 + y^2 - 2xy \cos\theta},$$

where θ is each of the angles of the parallelogram adjacent to the angle at the initial point of the diagonal vector V.

The standard or rectangular form of a complex number is

$$z = a + bi.$$

The trigonometric polar form of a complex number is

$$z = r(\cos\theta + i \sin\theta),$$

when

$$r = \sqrt{a^2 + b^2}, \quad \cos\theta = \frac{a}{r}, \quad \sin\theta = \frac{b}{r},$$

where r is called the modulus, and θ is the argument of the complex number z. The procedure for finding the modulus and argument of z involves, first substituting the values of a and b in the formula

$$r = \sqrt{a^2 + b^2}$$

and then use

$$\tan \theta = r \cdot \sin \theta / r \cdot \cos \theta = b/a$$

to solve for θ. For example, write

$$z = -2 + 3i$$

in trigonometric form using degrees.

Note that $a = -2$, $b = 3$. Then

$$r = \sqrt{(-2)^2 + 3^2} = \sqrt{13}.$$

To find θ, use the

$$\tan \theta = b/a = 3/(-2) = -1.5.$$

Since a is negative and b is positive, then θ is in the second quadrant, so $\theta = 123.7°$. Then,

$$z = \sqrt{13}\,(\cos 123.7° + i \sin 123.7°) \quad \text{or} \quad z = \sqrt{13}\,\text{cis}\,123.7°.$$

To change a complex number in trigonometric form to its standard form, evaluate $\cos \theta$ and $\sin \theta$. For example,

$$z = 2*(\cos 120° + i \sin 120°) = 2\left(-\frac{1}{2} + \frac{\sqrt{3}}{2}*i\right) = -1 + i\sqrt{3}.$$

The four basic fundamental operations are involved in handling complex numbers. To add or subtract two complex numbers involves simply combining the real number part and the complex numbers parts. For instance, if

$$z_1 = a + bi \quad \text{and} \quad z_2 = c + di,$$

then

$$z_1 + z_2 = (a + c) + (b + d)i \quad \text{or} \quad z_1 - z_2 = (a - c) + (b - d)i.$$

The product of z_1 and z_2 is given by

$$z_1(z_2) = (ac - bd) + (ad + bc)i,$$

while the quotient of z_1 and z_2 is given by

$$\frac{z_1}{z_2} = \frac{a + bi}{c + di} = \frac{a + bi}{c + di} \cdot \frac{c - di}{c - di} = \frac{(ac + bd) + (bc - ad)i}{c(c) + d(d)}.$$

Step-by-Step Solutions to Problems in this Chapter, "Vectors and Complex Numbers"

VECTORS

Which of the following vectors are equal to $\overrightarrow{MN}$ if
M = (2, 1) and N = (3, - 4)?
(a) $\overrightarrow{AB}$, where A = (1, - 1) and B = (2, 3)
(b) $\overrightarrow{CD}$, where C = (- 4, 5) and D = (- 3, 10)
(c) $\overrightarrow{EF}$, where E = (3, - 2) and F = (4, - 7).

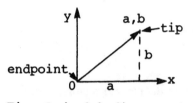

Fig. A:(a-0,b-0) represents the vector.

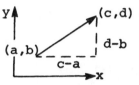

Fig. B:(c-a,d-b) represents the vector.

<u>Solution:</u> With each ordered pair in the plane there can be associated a vector from the origin to that point.

The vector is determined by subtracting the co-ordinates of the endpoint from the corresponding co-ordinates of the tip. As for $\overrightarrow{MN}$, the tip is the point corresponding to the second letter of the alphabetical notation, N, while the endpoint is the point corresponding to the first, M. In this problem the vectors are of a general nature wherein their endpoints do not lie at the origin.

We first find the ordered pair which represents $\overrightarrow{MN}$.

$\overrightarrow{MN}$ = (3 - 2, - 4 - 1) = (1, - 5)

Now, we find the ordered pair representing each vector.

(a) $\overrightarrow{AB}$ = (2 - 1, 3 - (- 1)) = (1, 4)

(b) $\overrightarrow{CD}$ =((- 3) - (- 4), 10 - 5)= (1, 5)

(c) $\overrightarrow{EF}$ = (4 - 3, - 7 - (- 2)) = (1, - 5)

Only $\overrightarrow{EF}$ and $\overrightarrow{MN}$ are equal.

A force of 315 lbs. is acting at an angle of 67° with the horizontal.   What are its horizontal and vertical components?

Solution: Construct the figure shown.

OR = vector force = c.

b = OA = horizontal component.

a = OB = vertical component.

In △OAR: c = 315; α = 67°.		
$\frac{a}{c}$ = sin α, or a = c sin α.	log c = 2.49831 log sin α = 9.96403 − 10 log a = 2.46234	a = 289.96 lbs.
$\frac{b}{c}$ = cos α, or b = c cos α.	log c = 2.49831 log cos α = 9.59188 − 10 log b = 2.09019	b = 123.08 lbs.

Two forces of 50 lbs. and 30 lbs. have an included angle of 60°. Find the magnitude and direction of their resultant.

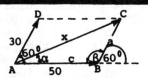

Solution: Construct the parallelogram and label it as in the figure. Since $\overline{AD}$ is parallel to $\overline{BC}$ we have ∠ABC = β = 180° − 60° = 120°. By the law of cosines:

$$x^2 = c^2 + a^2 - 2\,ac\,\cos\beta$$
$$= 2500 + 900 - 2(50)(30)(-\tfrac{1}{2})$$
$$= 2500 + 900 + 1500 = 4900.$$
$$x = 70 \text{ lbs.}$$

$$\cos\alpha = \frac{x^2 + c^2 - a^2}{2xc} = \frac{4900 + 2500 - 900}{2(70)(50)} = \frac{13}{14} = .9286.$$

$$\alpha = 21°47'.$$

Two forces act simultaneously on a body free to move. One force of 112 lbs. is acting due east, while the other of 88 lbs. is acting due north. Find the magnitude and direction of their resultant.

OA = b = 112 lbs.
OB = 88 lbs. = RA = a.

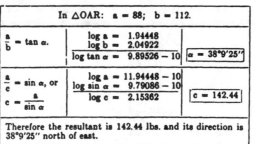

In △OAR: a = 88; b = 112.		
$\frac{a}{b}$ = tan α.	log a = 1.94448 log b = 2.04922 log tan α = 9.89526 − 10	α = 38°9'25"
$\frac{a}{c}$ = sin α, or c = $\frac{a}{\sin α}$	log a = 11.94448 − 10 log sin α = 9.79086 − 10 log c = 2.15362	c = 142.44

Therefore the resultant is 142.44 lbs. and its direction is 38°9'25" north of east.

● **PROBLEM** 981

Find the force required to prevent a 500-pound object from sliding down a 40° incline, disregarding friction.

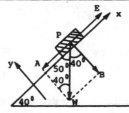

Solution: The weight of the object acts as a 500-pound force vertically downward. The component of the force parallel to the plane tends to force the object down the incline. The component perpendicular to the plane tends to force the object against the plane. The required force is parallel to the plane, equal in magnitude and opposite in direction to the parallel component. Observing the figure, the triangle WPB is similar to the triangle made by the inclined plane and the ground. Angle WPB therefore equals 40°. It follows then that angle APW equals 50°, as it is complementary to angle WPB. By super posing the coordinate axes with the x axis parallel to the inclined plane, observe that the component of the force to be determined lies parallel to the inclined plane and can be found by multiplying the magnitude of weight by the cos of 50°. Here cos 50° can be calculated using the rule cos θ = adjacent/hypotenuse for right triangles. The adjacent side is $\overrightarrow{AP}$ and the hypotenuse is $\overrightarrow{PW}$. Therefore cos 50° = AP/PW or (PW) cos 50° = AP, where Ap and PW represent the magnitudes of the force vectors.

$\overrightarrow{PW}$ = 500, ⦨APW = 90° − 40° = 50°

cos ⦨ APW = $\frac{AP}{PW}$,   where AP = $|\overrightarrow{AP}|$

AP = PW cos ⦨APW

AP = 500 cos 50°

AP = 500(0.6428) = 321

The required force $\overrightarrow{PE}$ is 321 pounds.

Find the magnitude and direction of the force necessary to counteract the effect of a force of 60 pounds and a force of 40 pounds that act on a point at an angle of 60° with each other.

E   Fig. 1   Fig. 2

Solution: The problem requires finding the equilibrant of the two forces. The equilibrant is equal in magnitude, but opposite in direction, to the resultant force. We find the resultant force of two vectors by using the parallelogram law of addition of vectors. Two vectors are drawn and the parallelogram is completed. The resultant is drawn connecting the two opposite vertices (see figure 1).We indicate the magnitude of a vector as the corresponding segment. In the parallelogram, resultant $\overrightarrow{OR}$ may be found by solving triangle OAR.

$$OA = BR = 40$$
$$AR = OB = 60$$
$$\angle A = 180° - 60° = 120° \text{(see figure 2)}$$

By the Law of Cosines:
$$(OR)^2 = (OA)^2 + (AR)^2 - [2(OA)(AR) \cos 120°]$$
$$(OR)^2 = (40)^2 + (60)^2 - [(2)(40)(60)(-½)]$$
$$(OR)^2 = 7,600$$
$$OR = 87$$

From the Law of Sines,
$$\sin\angle AOR = \frac{60 \sin 120°}{87} = \frac{60(0.8660)}{87} = 0.5972$$

$$\angle AOR = 37° \quad \text{(to the nearest degree)}$$

$$\angle AOE = 180° - 37° = 143°, \text{ and the required}$$

force is a force of 87 pounds, 143° from the 40-pound force in the opposite direction from the 87-pound force.

RECTANGULAR AND POLAR/TRIGONOMETRIC FORMS OF COMPLEX NUMBERS

Find the amplitude and the modulus of 5 - 3i.

Solution: The complex number 5 - 3i is expressed in the form a + ib. Here the modulus is the length (distance) r from the origin 0 to the point (a,b) and the amplitude is the angle θ, measured clockwise, that

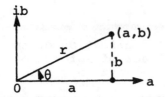

the distance makes with respect to the x axis.

The amplitude θ is determined by the form

$$\tan \theta = \frac{b}{a} \quad \text{or arctan } \frac{b}{a} = \theta.$$

Since a > 0 and b < 0, θ lies in the fourth quadrant, that is 270° < θ < 360°. From the table we find

$$\text{arctan } \frac{3}{5} = \text{arctan } 0.6000 = 31°0´$$

Hence, θ = 360° - 31° = 329°, the amplitude. The

modulus is $r = \sqrt{5^2 + (-3)^2} = \sqrt{34}$.

● PROBLEM 984

Find the polar form of 3 - 4i.

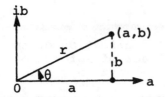

Solution: The given complex number is the cartesian co-ordinate representation. We wish to transform this representation to the polar form, i.e. $r(\cos \theta + i \sin \theta)$. Consult the accompanying figure and notice that r can be determined using the Pythagorean theorem, $r^2 = x^2 + y^2$, and θ can be computed by the formula $\tan \theta = y/x$ or arctan $y/x = \theta$, since we know that x = 3 and y = - 4. Thus,

$$r = \sqrt{x^2 + y^2} = \sqrt{3^2 + (-4)^2} = \sqrt{9 + 16} = 5.$$

θ = arctan - 4/3 = arctan (- 1.3333) = 306°52.25´, since (3, - 4) is a fourth quadrant point. Thus 3 - 4i = 5(cos 306°52.25´+ i sin 306°52.25´).

● PROBLEM 985

Express -6 + 8i in trigonometric form.

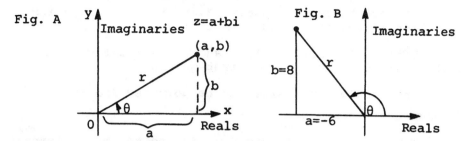

Fig. A

Fig. B

Solution: We are given a complex number z in the form of a + bi, where a and b are real numbers and (a,b) is the corresponding point in the cartesian plane. The value of a is found on the real axis and b is located on the imaginary axis. (See Figure A). Now let r denote the distance between the origin and the point which represents z and let θ be an angle in standard position whose terminal side contains the point z. We want to express a and b in terms of θ.

$$\cos \theta = \frac{a}{r} \quad \Rightarrow \quad a = r \cos \theta$$

$$\sin \theta = \frac{b}{r} \quad \Rightarrow \quad b = r \sin \theta$$

Thus, $a + bi = r \cos \theta + ir \sin \theta = r(\cos \theta + i \sin \theta)$.

In this example, a = -6 and b = 8 (see Figure B). To find r, apply the Pythagorean theorem.

$$r^2 = a^2 + b^2 = (-6)^2 + (8)^2 = 100 \; ;$$

thus, r = 10.

For θ: $\tan \theta = \frac{b}{a} = \frac{8}{-6} = -\frac{4}{3} \approx -1.333 \ldots$. First look up in a table of trigonometric functions the reference angle whose tangent is 4/3. It is 53.1 . But θ is in Quadrant II as we note from Figure B. Hence θ = 180 - 53.1 = 126.9 . Therefore, since $a + bi = r(\cos \theta + i \sin \theta)$,

$$-6 + 8i = 10(\cos 126.9° + i \sin 126.9°).$$

● **PROBLEM** 986

Express each of the following in trigonometric form.
(a) $-\sqrt{2} + \sqrt{2}i$ (b) $3 - 4i$ (c) $2 + i$

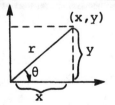

Solution: In the plane, a complex number is represented as x + iy.

Therefore the angle θ can be defined as $\arctan \frac{y}{x} = \theta$ or $\tan \theta = \frac{y}{x}$.

In part a the x coordinate is negative and the y coordinate is positive, therefore θ must lie in the second quadrant.

In part b the x is positive and the y is negative so θ lies in the fourth quadrant.

Finally in part c both x and y are positive implying that θ lies in the first quadrant.

The modulus or radius can be computed from the Pythagorean theorem $r^2 = x^2 + y^2$.

(a)   Tan θ = $\frac{\sqrt{2}}{-\sqrt{2}}$ = - 1, and θ is in the second quadrant.

Since arctan 1 = 45°, θ = 180° - 45° = 135°.

$$r^2 = (-\sqrt{2})^2 + (\sqrt{2})^2 = 4 \quad \text{or} \quad r = \sqrt{4} = 2$$

Therefore, $- \sqrt{2} + \sqrt{2}i$ = 2 (cos 135° + i sin 135°).

(b)   Tan θ = $\frac{-4}{3}$, and θ is in the fourth quadrant.

Since arctan 1.333 = 53° 10′ (to the nearest 10′),

θ = 360° - 53°10′ = 306°50′

$$r^2 = 3^2 + (-4)^2 = 25 \quad \text{or} \quad r = 5.$$

Therefore, 3 - 4i = 5(cos 306°50′ + i sin 306°50′).

(c)   Tan θ = $\frac{1}{2}$, and θ is in the first quadrant.

Since arctan $\frac{1}{2}$ = 26°30′ (to the nearest 10′),

θ = 26°30′

$$r^2 = 2^2 + 1^2 = 5 \quad \text{or} \quad r = \sqrt{5}$$

Therefore, 2 + 1 = $\sqrt{5}$(cos 26°30′ + i sin 26°30′).

● **PROBLEM 987**

Find the value of (4 - 4i) · ($\sqrt{3}$ - i) in polar form.

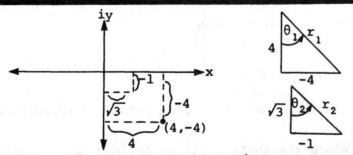

Solution: First change each cartesian representation to its polar representation. That is, we want to transform x + iy to the form r(cos θ + i sin θ). In the figure, notice that r can be determined using the Pythagorean theorem. And θ can be computed using the trigonometric functions.

$$r_1^2 = 4^2 + (-4)^2 = 32$$

$$r_1 = \sqrt{32} = 4\sqrt{2}$$

$$r_2^2 = (\sqrt{3})^2 + (-1)^2 = 3 + 1 = 4$$

$$r_2 = 2$$

Since all three sides of both triangles are known, any of the trigonometric functions can be used to determine θ. i.e. $\sin \theta = x/r$; $\cos \theta = y/r$; $\tan \theta = x/y$

Once θ_1 and θ_2 have been determined the multiplication can be performed according to the formula

$$r_1(\cos \theta_1 + i \sin \theta_1) \cdot r_2(\cos \theta_2 + i \sin \theta_2)$$

$$= r_1r_2\Big(\cos [\theta_1 + \theta_2] + i \sin [\theta_1 + \theta_2]\Big).$$

$\tan \theta_1 = \dfrac{4}{-4} \rightarrow \theta_1 = \tan^{-1}(-1) = 315°$, fourth quadrant.

$\tan \theta_2 = \dfrac{\sqrt{3}}{-1} \rightarrow \theta_2 = \tan^{-1}(-\sqrt{3}) = 330°$, fourth quadrant.

Changing $4 - 4i$ and $\sqrt{3} - i$ to polar form, we obtain $4 - 4i = 4\sqrt{2}$ ($\cos 315° + i \sin 315°$) and $\sqrt{3} - i = 2(\cos 330° + i \sin 330°)$. Thus, $(4 - 4i)(\sqrt{3} - i) = 4\sqrt{2}$ ($\cos 315° + i \sin 315°$) $\cdot$ $2(\cos 330° + i \sin 330°) = 8\sqrt{2}$ ($\cos 645° + i \sin 645°$) $= 8\sqrt{2}(\cos 285° + i \sin 285°)$.

● **PROBLEM** 988

Express each of the following in rectangular form, $a + bi$.   (a) $3(\cos 30° + i \sin 30°)$

(b) $10(\cos 180° + i \sin 180°)$

<u>Solution:</u>   The complex numbers as given are in the trigonometric form

$$r(\cos \theta + i \sin \theta)$$

in part a, $\theta = 30°$,

(a) $3(\cos 30° + i \sin 30°) = \dfrac{3}{2} \sqrt{3} + \dfrac{3}{2} i$

in part b, $\theta = 180°$,

(b) $10(\cos 180° + i \sin 180°) = 10(-1 + i \cdot 0) = -10$

Check: $r^2 = x^2 + y^2$

part a: $(3)^2 = \left(\dfrac{3\sqrt{3}}{2}\right)^2 + \left(\dfrac{3}{2}\right)^2 = \dfrac{27}{4} + \dfrac{9}{4} = 9$

part b: $(10)^2 = (-10)^2$

● **PROBLEM** 989

Find $[2(\cos 30° + i \sin 30°)][8(\cos 60° + i \sin 60°)]$.

Check by converting to rectangular form and multi-plying.

Solution: The two complex numbers are written in the form $r(\cos \theta + i \sin \theta)$ and the rule for multiplying complex numbers in this form is:

$r_1(\cos \theta_1 + i \sin \theta_1)r_2(\cos \theta_2 + i \sin \theta_2)$

$= r_1 r_2 (\cos \theta_1 \cos \theta_2 - \sin \theta_1 \sin \theta_2 + i \sin \theta_1 \cos \theta_2$

$+ i \cos \theta_1 \sin \theta_2)$

$= r_1 r_2 [\cos(\theta_1 + \theta_2) + i \sin(\theta_1 + \theta_2)]$

$[2(\cos 30° + i \sin 30°)][8(\cos 60° + i \sin 60°)]$

$= 16[\cos (30° + 60°) + i \sin (30° + 60°)]$

$= 16 (\cos 90° + i \sin 90°) = 16(0 + i) = 0 + 16i$

Check $2(\cos 30° + i \sin 30°) = 2 (\tfrac{1}{2} \sqrt{3} + \tfrac{1}{2}i) = \sqrt{3} + i$

$8(\cos 60° + i \sin 60°) = 8(\tfrac{1}{2} + \tfrac{1}{2} \sqrt{3}i) = 4 + 4\sqrt{3}i$

$(\sqrt{3} + i)(4 + 4\sqrt{3}i) = 4\sqrt{3} - 4\sqrt{3} + i(4 + 12) = 0 + 16i.$

● **PROBLEM** 990

Compute $\left(\cos \dfrac{3\pi}{2} + i \sin \dfrac{3\pi}{2}\right)^6$.

Solution: To raise the trigonometric representation of a complex number to a power, apply the rule:

$w = r(\cos \theta + i \sin \theta)$

$w^n = \left[r(\cos \theta + i \sin \theta)\right]^n = r^n\left(\cos [n\theta]+ i \sin [n\theta]\right);$

here $n = 6$, $r = 1$. Thus,

$\left(\cos \dfrac{3\pi}{2} + i \sin \dfrac{3\pi}{2}\right)^6 = \cos \dfrac{18\pi}{2} + i \sin \dfrac{18\pi}{2}$

$= \cos 9\pi + i \sin 9\pi.$ Recall the

formula for determining coterminal numbers, $u = u \pm 2n\pi$; $u = 9\pi$, $n = 4$. Then, $9\pi = 9\pi - 2(4)\pi = 9\pi - 8\pi = \pi.$ This means that on the unit circle both π and 9π begin at $(1,0)$ and terminate at π. Thus, we have: $\cos \pi + i \sin \pi$, and since $\cos \pi = -1$ and $\sin \pi = 0$, $= -1 + i(0) = -1$. Note that $\cos 3\pi/2 + i \sin 3\pi/2$ is a number which can be raised to an even power to produce a negative product.

● **PROBLEM** 991

Find $\left[8 \left(\cos \dfrac{\pi}{2} + i \sin \dfrac{\pi}{2}\right)\right] \div \left[2 \left(\cos \dfrac{\pi}{6} + i \sin \dfrac{\pi}{6}\right)\right]$.

Solution: The complex numbers are written in the form $r(\cos\theta + i\sin\theta)$.

Therefore, the division of these two numbers is performed by dividing the first modulus by the second, and subtracting the second angle from the first according to the formula:

$$\frac{r_1(\cos\theta_1 + i\sin\theta_1)}{r_2(\cos\theta_2 + i\sin\theta_2)}$$

$$= \frac{r_1}{r_2}\left[\cos(\theta_1 - \theta_2) + i\sin(\theta_1 - \theta_2)\right]$$

$$r_1 = 8$$

$$r_2 = 2$$

$$\frac{r_1}{r_2} = \frac{8}{2} = 4$$

$$\theta_1 = \pi/2$$

$$\theta_2 = \pi/6$$

$$\theta_1 - \theta_2 = \pi/2 - \pi/6 = \pi/3 = \text{ampltitude}$$

$$8\left[\cos\frac{\pi}{2} + i\sin\frac{\pi}{2}\right] \div 2\cos\left(\frac{\pi}{6} + i\sin\frac{\pi}{6}\right)$$

$$= 4\left(\cos\frac{\pi}{3} + i\sin\frac{\pi}{3}\right)$$

Check $\quad 8\left(\cos\frac{\pi}{2} + i\sin\frac{\pi}{2}\right) = 8(0 + i) = 8i.$

$$2\left(\cos\frac{\pi}{6} + i\sin\frac{\pi}{6}\right) = 2\left(\frac{\sqrt{3}}{2} + \frac{1}{2}i\right) = \sqrt{3} + i$$

$$\frac{8i}{\sqrt{3} + i} = \frac{8i}{\sqrt{3} + i} \cdot \frac{\sqrt{3} - i}{\sqrt{3} - i} = \frac{8 + 8\sqrt{3}i}{3 - i^2}$$

$$= \frac{8 + 8\sqrt{3}i}{4} = 2 + 2\sqrt{3}i$$

$$4\left(\cos\frac{\pi}{3} + i\sin\frac{\pi}{3}\right) = 4\left(\frac{1}{2} + \frac{\sqrt{3}}{2}i\right) = 2 + 2\sqrt{3}i$$

● **PROBLEM** 992

Show $\quad \dfrac{r_1(\cos\theta + i\sin\theta)}{r_2(\cos\phi + i\sin\phi)}$

$$= \frac{r_1}{r_2}\left[\cos(\theta - \phi) + i\sin(\theta - \phi)\right]$$

Solution: The trick is to multiply the original fraction by a fraction that is equivalent to one. The original value of the fraction will remain unchanged, but the complex number in the denominator now becomes a pure real. This is achieved by multiplying by a fraction whose numerator and denominator are the complex conjugate of the complex number in the denominator of the original fraction.

number   conjugate

$$r_2(\cos \phi + i \sin \phi) \qquad\qquad r_2(\cos \phi - i \sin \phi)$$

$$\frac{r_1(\cos \theta + i \sin \theta)}{r_2(\cos \phi + i \sin \phi)}$$

$$= \frac{r_1(\cos \theta + i \sin \theta)}{r_2(\cos \phi + i \sin \phi)} \cdot \frac{r_2(\cos \phi - i \sin \phi)}{r_2(\cos \phi - i \sin \phi)}$$

$$= \frac{r_1 r_2[(\cos \theta \cos \phi + \sin \theta \sin \phi) + i(\sin \theta \cos \phi - \cos \theta \sin \phi)]}{r_2^2(\cos^2 \phi - i^2 \sin^2 \phi)}$$

(Here we recognize the formulae for the cosine and sine of the difference of two angles. Also, we use the fact that $\cos^2 \phi - i^2 \sin \phi = \cos^2 \phi - (-1)\sin^2 \phi = \cos^2 \phi + \sin^2 \phi = 1$.)

$$= \frac{r_1 r_2[\cos(\theta - \phi) + i \sin(\theta - \phi)]}{r_2^2 \cdot 1}$$

$$= \frac{r_1}{r_2} [\cos(\theta - \phi) + i \sin(\theta - \phi)]$$

● **PROBLEM** 993

Find the four fourth roots of 16, including any imaginary roots.

Solution: The fourth roots of 16 are given by the radical $\sqrt[4]{16}$. Let $N = \sqrt[4]{16}$. The equation $N = \sqrt[4]{16}$ is equivalent to $N^4 = 16$, or $(N)(N)(N)(N) = 16$. If $N = \pm 2$, then $N^4 = (\pm 2)^4 = 16$. Hence, two of the four fourth roots of 16 are +2 and -2. Also, note that $i^4 = (i^2)^2 = (-1)^2 = 1$. Hence, if $N = \pm 2i$, then $N^4 = (\pm 2i)^4 = (\pm 2)^4 \cdot i^4 = 16(1) = 16$. Therefore, +2i and -2i are also two fourth roots of 16. Then, the four fourth roots of 16 are +2, -2, +2i, and -2i.

● **PROBLEM** 994

Find the three cube roots of - 1.

Solution: The number - 1 is a pure real number. This can only occur if the amplitude of the angle in the $r(\cos \theta + i \sin \theta)$ representation of - 1 is a value for which $\sin \theta = 0$. This value occurs for either $\theta = \pi$ or π plus integral multiples of π. The real part of this

722

number (equal to minus one) also occurs when $\theta = \pi$ or integral multiples of π.

$$- 1 = \cos \pi + i \sin \pi$$

$$= \cos (\pi + 2k\pi) + i \sin (\pi + 2k\pi)$$

$$(- 1)^{\frac{1}{3}} = \cos \left[\frac{\pi + 2k\pi}{3} \right] + i \sin \left[\frac{\pi + 2k\pi}{3} \right]$$

For $k = 0$, $r_0 = \cos \frac{\pi}{3} + i \sin \frac{\pi}{3}$

For $k = 1$, $r_1 = \cos \pi + i \sin \pi$

For $k = 2$, $r_2 = \cos \frac{5\pi}{3} + i \sin \frac{5\pi}{3}$

Thus, the roots are

$$r_0 = \cos \frac{\pi}{3} + i \sin \frac{\pi}{3}$$

$$r_1 = \cos \pi + i \sin \pi$$

$$r_2 = \cos \frac{5\pi}{3} + i \sin \frac{5\pi}{3}$$

It is possible to check these roots by converting each to rectangular form and raising to the third power.

● **PROBLEM** 995

Find all the fifth roots of 2.

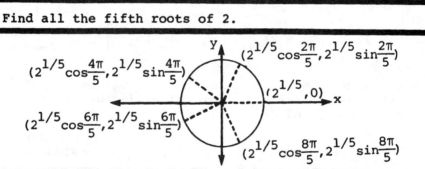

Solution: $2 = 2(\cos 0 + i \sin 0)$

Since 2 is a pure real number, its representation of the form $r(\cos \theta + i \sin \theta)$ has no complex part. Therefore, its amplitude must be equal to zero plus an integral multiple of π radians ($\theta = 0 + 2k\pi$). That is to say, the imaginary part must equal zero ($\sin \theta = 0$) and the real part must equal unity ($\cos \theta = 1$).

Therefore, $2 = 2 [\cos (0 + 2k\pi) + i \sin (0 + 2k\pi)]$

To raise a complex number to an integral power, n, the procedure is as follows:

$$(r[\cos(\theta + 2k\pi) + i \sin (\theta + 2k\pi)])^n$$

$$= r^n [\cos(n(\theta + 2k\pi)) + i \sin (n(\theta + 2k\pi))]$$

For finding n roots just raise the complex number to a rational power which corresponds, i.e. $1/n$:

$$(r[\cos(\theta + 2k\pi) + i \sin(\theta + 2k\pi)])^{1/n}$$

$$= r^{1/n}\left[\cos\left(\frac{\theta + 2k\pi}{n}\right) + i \sin\left(\frac{\theta + 2k\pi}{n}\right)\right]$$

Therefore, for the fifth roots of 2:

$$2^{\frac{1}{5}} = (2[\cos(0 + 2k\pi) + i \sin(0 + 2k\pi)])^{\frac{1}{5}}$$

$$= 2^{\frac{1}{5}}\left[\cos\frac{2k\pi}{5} + i \sin\frac{2k\pi}{5}\right] \quad k = 0, 1, 2, \ldots$$

Now substitute values of k from 0 to 4, inclusive.

For $k = 0$, $\quad r_0 = 2^{\frac{1}{5}} (\cos 0 + i \sin 0)$

For $k = 1$, $\quad r_1 = 2^{\frac{1}{5}}\left[\cos\frac{2\pi}{5} + i \sin\frac{2\pi}{5}\right]$

For $k = 2$, $\quad r_2 = 2^{\frac{1}{5}}\left[\cos\frac{4\pi}{5} + i \sin\frac{4\pi}{5}\right]$

For $k = 3$, $\quad r_3 = 2^{\frac{1}{5}}\left[\cos\frac{6\pi}{5} + i \sin\frac{6\pi}{5}\right]$

For $k = 4$, $\quad r_4 = 2^{\frac{1}{5}}\left[\cos\frac{8\pi}{5} + i \sin\frac{8\pi}{5}\right]$

For $k > 4$, we obtain the same cycle of values. Hence, the fifth roots of 2 are those values designated as $r_0, r_1, r_2, r_3,$ and r_4.

● **PROBLEM** 996

Find the four fourth roots of 9. The polar form of 9 is $9(\cos 0° + i \sin 0°)$.

Solution: 9 is a real number. This implies that its imaginary component equals zero, (i.e., $9 + 0i$). In its polar representation the angle θ must be such that the $\sin \theta = 0$ and the $\cos \theta = 1$, $0°$ satisfies this relationship. Therefore

$$9 = 9(\cos 0° + i \sin 0°).$$

To find the 4th roots of 9, we use the formula

$$w^b = r^b(\cos \theta + i \sin \theta)^b = r^b\left(\cos(b\theta) + i\sin(b\theta)\right)$$

where b is any rational number. We must also keep in mind that since cos and sin are periodic, whenever θ satisfies a given relationship, $\theta + 2k\pi$, where k is an integer, satisfies that relationship also, when we seek the roots of a number, i.e., $w^{1/n}$, we allow k to assume values from $0, \ldots, n-1$. In our problem $\theta = 0°$, $n = 4$ and we evaluate

$$w_k = 9^{1/n}\left[\cos\left(\frac{0 + 2k\pi}{n}\right) + i \sin\left(\frac{0 + 2k\pi}{n}\right)\right]$$

to get:

$$w_0 = 9^{1/4}(\cos 0° + i \sin 0°) = 3^{1/2}(\cos 0° + i \sin 0°)$$

$$w_1 = 9^{1/4}\left(\cos \frac{2\pi}{4} + i \sin \frac{2\pi}{4}\right) = 3^{1/2}\left(\cos \frac{\pi}{2} + i \sin \frac{\pi}{2}\right)$$

$$w_2 = 9^{1/4}\left(\cos \frac{4\pi}{4} + i \sin \frac{4\pi}{4}\right) = 3^{1/2}(\cos \pi + i \sin \pi)$$

$$w_3 = 9^{1/4}\left(\cos \frac{6\pi}{4} + i \sin \frac{6\pi}{4}\right) = 3^{1/2}\left(\cos \frac{3\pi}{2} + i \sin \frac{3\pi}{2}\right)$$

● **PROBLEM** 997

Find the 5th roots of $-1 + i$.

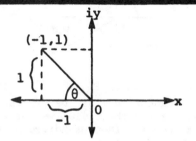

Solution: In the figure notice that we can determine r and θ so that we can transform the complex number from cartesian coordinates to a polar representation. Use the Pythagorean theorem to determine r.

$$r^2 = x^2 + y^2 = (-1)^2 + (1)^2 = 2$$
$$r = \sqrt{2}$$

θ can be computed by using the fact that tan θ = y/x or

$$\arctan \frac{y}{x} = \theta$$
$$\theta = \arctan \left[\frac{1}{-1}\right] = \arctan (-1)$$

i.e. $\frac{\sin \theta}{\cos \theta} = \frac{1}{-1}$; the angle whose sin is 1 and whose cos is -1 is 135° measured counterclockwise from 0°.

The 5th roots are calculated according to the rule

$$w^{1/n} = r^{1/n}\left[\cos \left[\frac{\theta + 2k\pi}{n}\right] + i \sin \left[\frac{\theta + 2k\pi}{n}\right]\right]$$

where $0 \le k \le n - 1$ inclusive

$$-1 + i = \sqrt{2}(\cos 135° + i \sin 135°)$$

$$\sqrt[5]{-1 + i} = \left(2^{\frac{1}{2}}\right)^{1/5}\left[\cos \frac{135° + 2k\pi}{5} + i \sin \frac{135° + 2k\pi}{5}\right]$$

$$k = 0, 1, \ldots, 4$$

725

Thus, $w_0 = 2^{1/10}(\cos 27° + i \sin 27°)$

           $w_1 = 2^{1/10}(\cos 99° + i \sin 99°)$

           $w_2 = 2^{1/10}(\cos 171° + i \sin 171°)$

           $w_3 = 2^{1/10}(\cos 243° + i \sin 243°)$

           $w_4 = 2^{1/10}(\cos 315° + i \sin 315°)$

● **PROBLEM** 998

Use De Moivre's theorem to find the value of $(-\sqrt{3} + i)^7$.

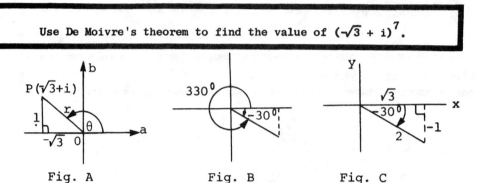

Fig. A   Fig. B   Fig. C

<u>Solution:</u> Let a complex number $a + bi$ be expressed in polar form, $r(\cos \theta + i \sin \theta)$ where r is the radius vector and θ is the angle made between the x-axis and r. Then De Moivre's theorem states that if n is a positive integer,

$$[r(\cos \theta + i \sin \theta)]^n = r^n(\cos n\theta + i \sin n\theta) .$$

We must convert the given complex number to polar form. Plot the point P representing the complex number $(-\sqrt{3} + i)$ which is in the form $a + bi$, (see Figure A).

We need to find the length r of the radius vector of the complex number; $r^2 = 1^2 + (-\sqrt{3})^2 = 1 + 3 = 4$. Since r is always positive, because we cannot have a negative length, $r = 2$. If $r = 2$, then the sine of the reference angle is $1/2$. Thus it is $30°$ and $\theta = 180° - 30° = 150°$. Substitute these two values into $r(\cos \theta + i \sin \theta)$. Hence, $-\sqrt{3} + i = 2(\cos 150° + i \sin 150°)$.

Applying De Moivre's theorem, we obtain
$$(-\sqrt{3} + i)^7 = [2(\cos 150° + i \sin 150°)]^7$$
$$= 2^7(\cos 7 \cdot 150° + i \sin 7 \cdot 150°)$$
$$= 128(\cos 1,050° + i \sin 1,050°)$$
$$= 128[(\cos 2 \cdot 360° + 330°) + i(\sin 2 \cdot 360° + 330°)]$$
$$= 128(\cos 330° + i \sin 330°)$$

The reference angle for $330°$ is $-30°$. (See Figure B). Then,
$$(-\sqrt{3} + i)^7 = 128[\cos(-30°) + i \sin(-30°)]$$
Note that $\cos(-30°) = +\dfrac{\sqrt{3}}{2}$ and $\sin(-30°) = -\dfrac{1}{2}$ since the cosine and sine functions are respectively positive and negative in quadrant IV, (see Figure C). Substitute these values, then:
$$(-\sqrt{3} + i)^7 = 128\left(\dfrac{\sqrt{3}}{2} - \dfrac{1}{2} i\right)$$

$$= 64\sqrt{3} - 64i .$$

Find the equations for $\sin 2\theta$ and $\cos 2\theta$ from the de Moivre equation with $n = 2$.

Solution: Let a complex number be expressed in polar form, $r(\cos \theta + i \sin \theta)$ where r is the radius vector and θ is the angle made between the x-axis and r. Then de Moivre's Theorem states that if n is any rational number, then

$$[r(\cos \theta + i \sin \theta)]^n = r^n(\cos n\theta + i \sin n\theta).$$

Furthermore we can see that:
$$(\cos \theta + i \sin \theta)^n = (\cos n\theta + i \sin n\theta) .$$

Also the complex exponential function is:
$$e^{i\theta} = \cos \theta + i \sin \theta .$$

If we substitute $n = 2$,
$$(\cos \theta + i \sin \theta)^2 = (\cos 2\theta + i \sin 2\theta) = \left(e^{i\theta}\right)^2$$
$$= e^{i\theta} \cdot e^{i\theta} = e^{i\theta+i\theta} = e^{2i\theta}$$

Expand the expression $(\cos \theta + i \sin \theta)^2$.

$$(\cos \theta + i \sin \theta)^2 = (\cos \theta + i \sin \theta)(\cos \theta + i \sin \theta)$$
$$= \cos^2\theta + 2i \sin \theta \cos \theta + i^2\sin^2\theta$$

Noting that $i^2 = -1$, we obtain:
$$(\cos \theta + i \sin \theta)^2 = \cos^2\theta - \sin^2\theta + 2i \sin \theta \cos \theta \qquad (1)$$

Furthermore by de Moivre's Theorem for the case $n = 2$, we have

$$(\cos \theta + i \sin \theta)^2 = \cos 2\theta + i \sin 2\theta \qquad (2)$$

Equate the right side of equations (1) and (2) to obtain:
$$\cos^2\theta - \sin^2\theta + 2i \sin \theta \cos \theta = \cos 2\theta + i \sin 2\theta$$
$$\cos^2\theta - \sin^2\theta + i2 \sin \theta \cos \theta = \cos 2\theta + i \sin 2\theta$$

Equate the real and imaginary parts.
$$\cos^2\theta - \sin^2\theta = \cos 2\theta \qquad (3)$$
$$i2 \sin \theta \cos \theta = i \sin 2\theta \qquad (4)$$

Note that, after dividing both sides by i, equation (4) becomes:
$$2 \sin \theta \cos \theta = \sin 2\theta$$

Therefore, the expressions for $\sin 2\theta$ and $\cos 2\theta$ are:
$$\cos 2\theta = \cos^2\theta - \sin^2\theta$$
$$\sin 2\theta = 2 \sin \theta \cos \theta .$$

OPERATIONS WITH COMPLEX NUMBERS

Express each of the following as the product of i and a real number.

(a) $2i^5$ (b) $\dfrac{-5}{i7}$ (c) $\sqrt{-81}$

Solution: Recalling that $\sqrt{-1} = i$ $\left(\text{or } -1 = i^2\right)$:

(a) $2i^5 = 2 \cdot i^4 \cdot i = 2 \cdot 1 \cdot i = 2i$

(b) $\dfrac{-5}{i^7} = \dfrac{-5}{i^4 \cdot i^2} \cdot \dfrac{1}{i} = \dfrac{-5}{1 \cdot -1} \cdot \dfrac{1}{i} = \dfrac{-5}{1 \cdot -1} \cdot \dfrac{-i^2}{i} = -5i$

Note that $1 = -(-1) = -\left(i^2\right) = -i^2$. Hence $\dfrac{1}{i} = \dfrac{-i^2}{i}$

(c) $\sqrt{-81} = \sqrt{(-1)(81)} = \sqrt{-1} \cdot \sqrt{81} = 9i$

● **PROBLEM 1001**

What is the conjugate of $3 - 2i$ and the conjugate of $5 + 7i$?

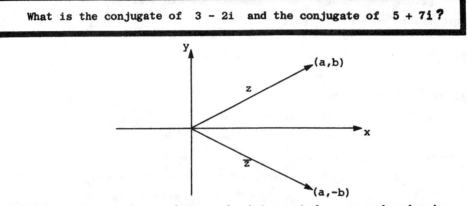

Solution: Any complex number may be interpreted as an ordered pair in the plane with the real component designated by the x value and the imaginary part designated by the y value. The conjugate of a complex number is that number which when multiplied by the original complex number yields a product which is purely real. Geometrically, the complex conjugate is a reflection of the complex number through the x-axis. The complex conjugate of $3 - 2i$ is $3 + 2i$

i.e., $(3 - 2i)(3 + 2i) = 13$.

The conjugate of $5 + 7i$ is $5 - 7i$

$(5 + 7i)(5 - 7i) = 74$.

The conjugate of a pure real number a, which can be written $a + 0i$, is merely itself or $a - 0i$. Geometrically we see that the reflection of a real number is actually itself. The conjugate of a pure imaginary number bi is $-bi$. The conjugate of a complex number $a + bi$ is $a - bi$.

● **PROBLEM 1002**

Add $(3 + 4i)$ and $(2 - 5i)$.

Solution: Numbers of the form $a + bi$, where a and b are real numbers, are called complex numbers. In the complex number $a + bi$, a is called the real part and bi is called the imaginary part. To add two complex numbers, add the real parts and add the pure imaginary parts. Therefore: we have

$(3 + 4i) + (2 - 5i) = (3 + 2) + (4 - 5)i$

$= 5 + (-i)$

$= 5 - i$

Or we may treat the problem as the sum of two binomials:

$(3 + 4i) + (2 - 5i) = 3 + 4i + 2 - 5i$
$= 5 - i$

Perform the indicated operations: $(-2 + 3i) + [5 + (-6)i]$.

<u>Solution:</u> Addition of complex numbers is defined in the following way:

$$(a + bi) + (c + di) = (a + c) + (b + d)i ,$$

where a,b,c,d are real numbers. Thus,

$$(-2 + 3i) + [5 + (-6)i] = (-2 + 5) + [3 + (-6)]i$$
$$= 3 + (-3)i$$
$$= 3 - 3i.$$

Write each of the following in the form a + bi.

    a) $(2 + 4i) + (3 + i)$

    b) $(2 + i) - (4 - 2i)$

    c) $(4 - i) - (6 - 2i)$

    d) $3 - (4 + 2i)$

<u>Solution:</u>

    a) $(2 + 4i) + (3 + i) = 2 + 4i + 3 + i$
$$= (2 + 3) + (4i + i)$$
$$= 5 + 5i$$

    b) $(2 + i) - (4 - 2i) = 2 + i - 4 + 2i$
$$= (2 - 4) + (i + 2i)$$
$$= -2 + 3i$$

    c) $(4 - i) - (6 - 2i) = 4 - i - 6 + 2i$
$$= (4 - 6) + (-i + 2i)$$
$$= -2 + i$$

    d) $3 - (4 + 2i) = 3 - 4 - 2i$
$$= (3 - 4) - 2i$$
$$= -1 - 2i$$

Find the product $(2 + 3i)(-2 - 5i)$.

Solution: Using the following method: product of first elements, + product of outer elements + product of inner elements + product of last elements:

$$(2 + 3i)(- 2 - 5i) = 2(- 2) + 2(- 5i) + 3i(- 2) + 3i(- 5i)$$

$$= - 4 - 10i - 6i - 15i^2$$

$$= - 4 - 16i - 15i^2$$

Recall $i^2 = - 1$, hence, $= - 4 - 16i - 15(- 1)$

$$= - 4 - 16i + 15$$

$$= 11 - 16i$$

The same result is obtained by using the distributive law.

$$(2 + 3i)(- 2 - 5i) = (2 + 3i)(- 2) - (2 + 3i)5i$$

$$= - 4 - 6i - 10i - 15i^2 = 11 - 16i.$$

In other words, if one multiples $2 + 3i$ and $- 2 - 5i$ as if they were polynomials and replaces i^2 by $- 1$, then the correct product is obtained.

● **PROBLEM** 1006

Multiply $(3 + 4i)$ by $(5 + 2i)$.

Solution: To multiply two complex numbers, form the product treating i as an ordinary number and then replace i^2 by -1. Hence,

$$(3 + 4i)(5 + 2i) = 15 + 20i + 6i + 8i^2 = 15 + 26i + 8i^2$$

$$= 15 + 26i - 8 = 7 + 26i$$

Or we may treat the problem as the product of two binomials:

$$(3 + 4i)(5 + 2i) = (3 + 4i)5 + (3 + 4i)2i$$

$$= 15 + 20i + 6i + 8i^2$$

$$= 15 + 26i + 8(-1)$$

$$= 15 + 26i - 8$$

$$= 7 + 26i$$

● **PROBLEM** 1007

Compute the sum and product of the complex numbers $3 + 2i$ and $1 - 3i$.

Solution:

$$(3 + 2i) + (1 - 3i) = 3 + 2i + 1 - 3i = 3+1 +2i - 3i= 4-i$$

$$(3 + 2i)(1 - 3i) = 3(1) +2i(1) + 3(-3i) + 2i(-3i)$$

$$= 3 + 2i - 9i - 6i^2$$

$$(3 + 2i)(1 - 3i) = 3 - 7i - 6i^2 \qquad (1)$$

Since $i^2 = -1$, equation (1) becomes:

$$(3 + 2i)(1 - 3i) = 3 - 7i - 6(-1)$$
$$= 3 - 7i + 6$$
$$= 9 - 7i.$$

● **PROBLEM** 1008

Find the values of the following expressions:

  a. $(2 + 3i) + (6 - 2i)$

  b. $(2 - i)(1 + 3i)$

  c. $i - (2 + 3i)$

<u>Solution:</u> a) $(2 + 3i) + (6 - 2i) = 2 + 3i + 6 - 2i$
$$= 2 + 6 + 3i - 2i$$
$$= 8 + i$$

  b) $(2 - i)(1 + 3i) = 2(1) - i(1) + 2(3i) - i(3i)$
$$= 2 - i + 6i - 3i^2 \qquad (1)$$

Since $i^2 = -1$, equation (1) becomes:

$$(2 - 1)(1 + 3i) = 2 - i + 6i - 3(-1)$$
$$= 2 - i + 6i + 3$$
$$= 2 + 3 - i + 6i$$
$$= 5 + 5i$$

  c) $i - (2 + 3i) = i - 2 - 3i$
$$= -2 + i - 3i$$
$$= -2 - 2i$$

● **PROBLEM** 1009

Express each of the following complex numbers in the form $a + bi$, where a and b are real:

(a) 7   (b) 2i   (c) $\sqrt{-3}$

(d) $\dfrac{1 + \sqrt{-3}}{2}$ (e) $(3 + 2i)(3 - 2i)$ (f) $\dfrac{1 + i}{1 - i}$

<u>Solution:</u>

(a) $7 = 7 + 0i$ $(a = 7, \quad b = 0)$

(b) $2i = 0 + 2i$ $(a = 0, \quad b = 2)$

(c) $\sqrt{-3} = \sqrt{(-1)(3)} = \sqrt{-1} \cdot \sqrt{3} = i\sqrt{3} = 0 + i\sqrt{3}$ $(a = 0, b = \sqrt{3})$

(d) $\dfrac{1 + \sqrt{-3}}{2} = \dfrac{1 + i\sqrt{3}}{2} = \dfrac{1}{2} + \dfrac{\sqrt{3}}{2} i$ $\left(a = \dfrac{1}{2}, \quad b = \dfrac{\sqrt{3}}{2} \right)$

(e) $(3 + 2i)(3 - 2i) = 9 - 6i + 6i - 4i^2 = 9 - 4i^2 = 9 - 4(-1)$

$$= 9 + 4 = 13 = 13 + 0i \qquad (a = 13, \ b = 0)$$

(f)   In order to evaluate $\frac{1 + i}{1 - i}$, we want to eliminate the complex expression in the denominator.   Therefore, we multiply numerator and denominator by the complex conjugate of the denominator, $1 + i$:

$$\frac{1 + i}{1 - i} = \frac{1 + i}{1 - i} \cdot \frac{1 + i}{1 + i} = \frac{1 + 2i + i^2}{1 - i^2} = \frac{1 + 2i + (-1)}{1 - (-1)}$$

$$= \frac{2i}{2} = i = 0 + 1i \quad (a = 0, \ b = 1).$$

● **PROBLEM** 1010

Divide (6 + 3i) by (2 + 4i).

Solution: The division of one complex number by another is accomplished by transforming the fraction into a fraction with a real denominator. In order to do this, multiply the numerator and the denominator of the given fraction by the conjugate of the complex denominator. Note that the numbers (a+bi) and (a-bi) are complex conjugates. Hence, the conjugate of the denominator, 2+4i, is 2-4i. Then, multiplying both the numerator and the denominator:

$$\frac{6 + 3i}{2 + 4i} = \frac{6 + 3i}{2 + 4i} \cdot \frac{2 - 4i}{2 - 4i}$$

$$= \frac{(6 + 3i)(2 - 4i)}{(2 + 4i)(2 - 4i)}$$

$$= \frac{12 + 6i - 24i - 12i^2}{4 + 8i - 8i - 16i^2}$$

$$= \frac{12 - 18i - 12(-1)}{4 - 16(-1)}$$

$$= \frac{12 - 18i + 12}{4 + 16}$$

$$= \frac{24 - 18i}{20}$$

$$= \frac{2(12 - 9i)}{2(10)}$$

$$= \frac{12 - 9i}{10}$$

$$= \frac{12}{10} - \frac{9}{10}i = \frac{6}{5} - \frac{9}{10}i$$

● **PROBLEM** 1011

Find (6 - 5i) ÷ (3 + 4i).

Solution: Write (6 - 5i) ÷ (3 + 4i) as $\frac{6 - 5i}{3 + 4i}$. We then rationalize the denominator. Rationalize the denominator by multiplying the original fraction by a fraction equivalent to unity, consisting of the complex conjugate of the denominator of the original fraction in the numerator and denominator of the new fraction. When a complex number a $\underset{=}{+}$ ib is multiplied by its complex conjugate a $\overline{+}$ ib the result is a real number. i.e.

$$(a + ib)(a - ib) = a^2 - (ib)^2 = a^2 + b^2 \left(\text{note: } i^2 = -1 \right)$$

from the rule for the difference of two squares. Note:

either number is the complex conjugate of the other.

$$\frac{6 - 5i}{3 + 4i} \cdot \frac{3 - 4i}{3 - 4i} = \frac{18 - 15i - 24i + 20i^2}{9 + 16} = \frac{-2 - 39i}{9 + 16}$$

$$= -\frac{2 + 39i}{25}.$$

Check: We check by multiplying the quotient by the divisor. The product must be the dividend.

$$\left(-\frac{2 + 39i}{25}\right)(3 + 4i) = -\frac{1}{25}(6 + 125i - 156)$$

$$= -\frac{1}{25}(125i - 150)$$

$$= 6 - 5i.$$

● **PROBLEM** 1012

Find the real and imaginary parts of

$$(2 + 3i) \div (3 + 4i)$$

Solution: In order to divide one complex number by another, the denominator must be converted to a real number. This can be done by multiplying the numerator and the denominator by the conjugate of the denominator. The complex numbers a + bi and a - bi are conjugates of each other. Therefore, the conjugate of the denominator 3 + 4i is 3 - 4i. Then,

$$\frac{2 + 3i}{3 + 4i} = \frac{(2 + 3i)(3 - 4i)}{(3 + 4i)(3 - 4i)}$$

$$= \frac{6 + 9i - 8i - 12i^2}{9 + \cancel{12i} - \cancel{12i} - 16i^2}$$

$$\frac{6 + i - 12(-1)}{9 - 16(-1)}, \qquad \text{since } i^2 = -1$$

$$= \frac{6 + i + 12}{9 + 16}$$

$$= \frac{18 + i}{25}$$

$$= \frac{18}{25} + \frac{1}{25}i$$

Hence, the real part of the quotient is $\frac{18}{25}$ and the imaginary part is $\frac{1}{25}i$.

● **PROBLEM** 1013

Divide $2 - \sqrt{2}i$ by $2 - i$:

Solution: This problem involves dividing one complex number by another. To perform this division, the numerator and the denominator are multiplied by the conjugate of the denominator. The conjugate of the denominator 2 - i is 2 + i. Then,

$$\frac{2 - \sqrt{2}\,i}{2 - i} = \frac{(2 - \sqrt{2}\,i)(2 + i)}{(2 - i)(2 + i)}$$

$$= \frac{4 - 2\sqrt{2}\,i + 2i - \sqrt{2}\,i^2}{4 - 2i + 2i - i^2}$$

$$= \frac{4 + (2 - 2\sqrt{2})i - \sqrt{2}\,(-1)}{4 - (-1)} \text{,since } i^2 = -1$$

$$= \frac{4 + (2 - 2\sqrt{2})i + \sqrt{2}}{4 + 1}$$

$$= \frac{4 + \sqrt{2} + (2 - 2\sqrt{2})i}{5}$$

$$= \frac{4 + \sqrt{2}}{5} + \frac{(2 - 2\sqrt{2})}{5}\,i$$

● **PROBLEM** 1014

Simplify $\dfrac{3 - 5i}{2 + 3i}$.

Solution: To simplify $\dfrac{3 - 5i}{2 + 3i}$ means to write the fraction without an imaginary number in the denominator. To achieve this, we multiply the fraction by another fraction which is equivalent to unity, (so that the value of the original fraction is unchanged) which will transform the expression in the denominator to a real number. A fraction with this property must have the complex conjugate of the expression in the denominator of the original fraction as its numerator and denominator. The complex conjugate must be chosen because of its special property that when multiplied by the original complex number the result is real.

Note: a + bi; its complex conjugate is a - bi or they can be said to be conjugates of each other. To multiply notice that (a + bi)(a - bi) is the factored form of the difference of two squares. Thus we obtain

$$(a)^2 - (bi)^2; \quad i^2 = -1; \quad (a)^2 - (-1)(b)^2 \text{ or } a^2 + b^2.$$

$$\frac{3 - 5i}{2 + 3i} \cdot \frac{2 - 3i}{2 - 3i} = \frac{6 - 9i - 10i + 15i^2}{4 - 9i^2}$$

$$= \frac{6 - 19i - 15}{4 + 9}$$

$$= \frac{-9 - 19i}{13} \text{ or } \frac{-9}{13} - \frac{19}{13}i$$

Since the resulting fraction has a rational number in the denominator, we have rationalized the denominator.

Simplify: (a) $4i - 7i^3$ (b) $\dfrac{2- 3i}{5i}$

Solution: (a) Factor i in the expression to obtain,

$$i\left(4 - 7i^2\right)$$

$$i^2 = - 1$$

and we obtain,

$$i(4 - 7[-1]) = i(4 + 7) = 11i$$

(b) Rationalize the denominator by multiplying the original fraction by a fraction equivalent to unity which will cause the imaginary expression in the denominator of the original fraction to be eliminated.

$\dfrac{i}{i}$ is suitable because $\dfrac{i}{i} = 1$ and $i^2 = -1$.

$$\frac{2 - 3i}{5i} = \frac{2 - 3i}{5i} \cdot \frac{i}{i} = \frac{2i - 3i^2}{5i^2} = \frac{2i + 3}{-5}$$

Expand $(2 + 3i)^3$.

Solution: The identity $(a + b)^3 = a^3 + 3a^2b + 3ab^2 + b^3$ is still valid in the case of complex numbers. Thus, replacing a by 2 and b by 3i we obtain

$$(2 + 3i)^3 = 2^3 + 3 \cdot 2^2(3i) + 3 \cdot 2(3i)^2 + (3i)^3$$

$$= 8 + 3 \cdot 4 \cdot 3 \cdot i + 3 \cdot 2(3^2i^2) + 3^3i^3$$

$$= 8 + 36i + 6(9)i^2 + 27i^3$$

$$= 8 + 36i + 54i^2 + 27i^3.$$

Recalling that $i^2 = - 1$, since $i = \sqrt{- 1}$ and $i^2 = \sqrt{- 1}\sqrt{- 1} = - 1$; and $i^3 = i^2(i) = (- 1)i = - i$, we obtain:

$$= 8 + 36i + 54(- 1) + 27(- i)$$

$$= 8 + 36i - 54 - 27i$$

$$= 9i - 46.$$

Evaluate $x^2 - 2x + 6$ for $x = 3 + 2i$.

Substituting the given value, we get

$$x^2 - 2x + 6 = (3 + 2i)^2 - 2(3 + 2i) + 6$$
$$= (3 + 2i)(3 + 2i) - 6 - 4i + 6$$

Since $\quad\quad (a + b)(c + d) = ac + ad + bc + bd$

$$x^2 - 2x + 6 = (3)(3) + 6i + 6i + (2i)(2i) - 6 - 4i + 6$$
$$= 9 + 12i + (2i)^2 - 6 - 4i + 6$$

Since $\quad\quad (ab)^2 = a^2 b^2$

$$(2i)^2 = 2^2 (i)^2$$

By definition $i^2 = -1$, hence $(2i)^2 = 4(-1) = -4$ and $x^2 - 2x + 6$

$$= 9 + 12i - 4 - 6 - 4i + 6$$

Combine terms, $\quad = 5 + 8i$.

● **PROBLEM** 1018

Show that $\left(-\dfrac{1}{2} + \dfrac{\sqrt{3}i}{2}\right)^3 = 1$.

Solution: Factor out $\dfrac{1}{2}$:

$$\left(-\frac{1}{2} + \frac{\sqrt{3}i}{2}\right)^3 = \left[\frac{1}{2}(-1 + \sqrt{3}i)\right]^3$$
$$= \left(\frac{1}{2}\right)^3 (-1 + \sqrt{3}i)^3 = \frac{1}{8}(-1 + \sqrt{3}i)^3.$$

Then we apply the identity $(a + b)^3 = (a + b)(a + b)(a + b)$
$= \left(a^2 + 2ab + b^2\right)(a + b) = a^3 + 2a^2 b + ab^2 + a^2 b + 2ab^2 + b^3$
$= a^3 + 3a^2 b + 3ab^2 + b^2$. Let $a = -1$, $b = \sqrt{3}i$, then

$$(-1 + \sqrt{3}i)^3 = (-1)^3 + 3(-1)^2 \sqrt{3}i + 3(-1)(\sqrt{3}i)^2 + (\sqrt{3}i)^3$$
$$= -1 + 3\sqrt{3}i - 9i^2 + 3\sqrt{3}i^3$$
$$= -1 + 3\sqrt{3}i + 9 - 3\sqrt{3}i = 8.$$

since $1^2 = -1$ and $i^3 = i^2 \ i = (-1)i = -i$. Hence

$$\left(-\frac{1}{2} + \frac{\sqrt{3}i}{2}\right)^3 = \frac{1}{8} \cdot 8 = 1.$$

● **PROBLEM** 1019

Show that $\left(\dfrac{1}{\sqrt{2}} + \dfrac{1}{\sqrt{2}}i\right)^4 = -1$.

Solution: Factor out $\dfrac{1}{\sqrt{2}}$:

$$\left(\frac{1}{\sqrt{2}} + \frac{1}{\sqrt{2}}i\right)^4 = \left[\frac{1}{\sqrt{2}}(1 + i)\right]^4 = \left(\frac{1}{\sqrt{2}}\right)^4 (1 + i)^4 = \frac{1}{4}(1 + i)^4$$

Note: $\left(\dfrac{1}{\sqrt{2}}\right)^4 = \left(\dfrac{1}{\sqrt{2}}\right)\left(\dfrac{1}{\sqrt{2}}\right)\left(\dfrac{1}{\sqrt{2}}\right)\left(\dfrac{1}{\sqrt{2}}\right) = \dfrac{1}{2} \cdot \dfrac{1}{2} = \dfrac{1}{4}$

Now we apply the identity $(a + b)^4 =$

$(a + b)^2(a + b)^2 = (a + b)(a + b)(a + b)(a + b)$

$$= \left(a^2 + 2ab + b^2\right)\left(a^2 + 2ab + b^2\right).$$

Then,

$$a^2 + 2ab + b^2$$
$$\underline{a^2 + 2ab + b^2}$$
$$a^4 + 2a^3b + a^2b^2$$
$$+ 2a^3b + 4a^2b^2 + 2ab^3$$
$$\underline{\qquad + a^2b^2 + 2ab^3 + b^4}$$

$(a+b)^4 = a^4 + 4a^3b + 6a^2b^2 + 4ab^3 + b^4.$

Thus, $a = 1$ $b = i.$

We obtain

$$(1 + i)^4 = 1^4 + 4i + 6i^2 + 4i^3 + i^4$$

substitute: $i^2 = -1$ and $i^3 = i^2 \cdot i = (-1)i = -i$ and $i^4 = i^2 \cdot i^2 = (-1)^2 = 1.$ Then,

$$(1 + i)^4 = 1 + 4i - 6 - 4i + 1 = -4.$$

Hence

$$\left(\dfrac{1}{\sqrt{2}} + \dfrac{1}{\sqrt{2}}i\right)^4 = \dfrac{1}{4}(-4) = -1.$$

● **PROBLEM** 1020

Find $\sqrt{3 + 4i}$

<u>Solution:</u> Let $\sqrt{3 + 4i} = x + yi$ where x and y are real numbers. Square both sides of this equation:

$$(\sqrt{3 + 4i})^2 = (x + yi)^2$$
$$3 + 4i = (x + yi)(x + yi) \qquad\qquad (1)$$
$$3 + 4i = x^2 + 2xyi + y^2i^2 \qquad\qquad (2)$$

Since $i^2 = -1$, equation (2) becomes:

$$3 + 4i = x^2 + 2xyi + y^2(-1)$$
$$3 + 4i = x^2 + 2xyi - y^2 \text{ or, by the commutative}$$

property of addition,

$$3 + 4i = x^2 - y^2 + 2xyi$$

Equate the real and imaginary parts of both members:

$$3 = x^2 - y^2 \qquad\qquad (3)$$

$$4i = 2xyi \qquad\qquad (4)$$

Dividing both sides of equation (4) by i:

$$\frac{4i}{i} = \frac{2xyi}{i}$$

$$4 = 2xy \qquad\qquad (5)$$

Therefore, our equations are:

$$3 = x^2 - y^2 \qquad\qquad (3)$$

$$4 = 2xy \qquad\qquad (5)$$

Solving equation (5) for x:
dividing both sides by 2y,

$$\frac{4}{2y} = \frac{2xy}{2y}$$

$$\frac{2}{y} = x$$

Note that the above operation assumes that $y \neq 0$ since division by 0 is undefined. (Also, in our original expression $\sqrt{3 + 4i} = x + yi$, there is assumed to be an an imaginary part; namely, yi. If y were equal to 0, then there would be no imaginary part since $yi = 0(i) = 0$. Hence, y cannot equal 0.)

Substituting the expression for x in equation (3):

$$3 = \left(\frac{2}{y}\right)^2 - y^2$$

$$3 = \frac{4}{y^2} - y^2$$

Obtaining a common denominator of y^2 for the two terms on the right side of this equation:

$$3 = \frac{4}{y^2} - \frac{y^2(y^2)}{y^2}$$

$$3 = \frac{4}{y^2} - \frac{y^4}{y^2}$$

$$3 = \frac{4 - y^4}{y^2}$$

Multiplying both sides by y^2:

$$y^2(3) = y^2\left(\frac{4 - y^4}{y^2}\right)$$

$$3y^2 = 4 - y^4$$

Subtracting $\left(4 - y^4\right)$ from both sides:

$$3y^2 - \left(4 - y^4\right) = 4 - y^4 - \left(4 - y^4\right)$$

$$3y^2 - 4 + y^4 = 0$$

or $y^4 + 3y^2 - 4 = 0$

Factoring the left side of this equation as a product of two polynomials:

$$\left(y^2 + 4\right)\left(y^2 - 1\right) = 0$$

Whenever a product $ab = 0$ where a and b are any two numbers, either $a = 0$ or $b = 0$. Therefore,

either $y^2 + 4 = 0$ or $y^2 - 1 = 0$

$\qquad\qquad y^2 = -4$ or $\qquad y^2 = 1$

$\qquad\qquad\qquad\qquad\qquad\qquad y = \pm\sqrt{1}$

$\qquad\qquad\qquad\qquad\qquad\qquad y = \pm 1$

Note that there is no real solution to $y^2 = -4$ since there is no real number y whose square is -4.

Substituting $y = -1$ in equation (3):

$$3 = x^2 - (-1)^2$$
$$3 = x^2 - (1)$$
$$3 = x^2 - 1$$

Add 1 to both sides:

$$3 + 1 = x^2 - 1 + 1$$
$$4 = x^2$$

Take the square root of both sides:

$$\pm\sqrt{4} = x$$
$$\pm 2 = x.$$

Hence, the two solutions appear to be $(-2,-1)$ and $(2,-1)$. Substituting $y = 1$ in equation (3):

$$3 = x^2 - (1)^2$$
$$3 = x^2 - 1$$

Add 1 to both sides:

$$3 + 1 = x^2 - 1 + 1$$
$$4 = x^2$$

Take the square root of both sides:

$$\pm\sqrt{4} = x \quad \text{or} \quad \pm 2 = x$$

Hence, the two additional solutions appear to be $(-2,1)$ and $(2,1)$. For the four solutions obtained:

when $(x,y) = (-2,-1)$, $\sqrt{3 + 4i} = -2 + (-1)i = -2-i$,

when $(x,y) = (2,-1)$, $\sqrt{3 + 4i} = 2 + (-1)i = 2 - i$,

when $(x,y) = (-2,1)$, $\sqrt{3 + 4i} = -2 + 1i = -2 + i$,

when $(x,y) = (2,1)$, $\sqrt{3 + 4i} = 2 + 1i = 2 + i$.

Checking the four solutions of $\sqrt{3 + 4i}$; namely, $-2-i$, $2-i$, $-2+i$, $2+i$, using equation 1:

for $-2-i$, $3 + 4i = (-2-i)(-2-i)$

$\qquad\qquad 3 + 4i = 4 + 2i + 2i + i^2$

$\qquad\qquad 3 + 4i = 4 + 4i - 1.$ Note that $i^2 = -1$.

$3 + 4i = 3 + 4i$ ✓

Therefore, $-2-i$ is a solution to $\sqrt{3 + 4i}$.

For $2-i$, $3 + 4i = (2 - i)(2 - i)$

$\qquad 3 + 4i = 4 - 2i - 2i + i^2$

$\qquad 3 + 4i = 4 - 4i - 1$ since $i^2 = -1$

$\qquad 3 + 4i \neq 3 - 4i$

Therefore, $2 - i$ is not a solution to $\sqrt{3 + 4i}$.

For $-2+i$, $3 + 4i = (-2 + i)(-2 + i)$

$\qquad 3 + 4i = 4 - 2i - 2i + i^2$

$\qquad 3 + 4i = 4 - 4i - 1$ since $i^2 = -1$

$\qquad 3 + 4i \neq 3 - 4i$

Therefore, $-2 + i$ is not a solution to $\sqrt{3 + 4i}$.

For $2 + i$, $3 + 4i = (2 + i)(2 + i)$

$\qquad 3 + 4i = 4 + 2i + 2i + i^2$

$\qquad 3 + 4i = 4 + 4i - 1$ since $i^2 = -1$

$\qquad 3 + 4i = 3 + 4i$ ✓

Therefore, $2 + i$ is a solution to $\sqrt{3 + 4i}$.

Hence, the only two solutions to $\sqrt{3 + 4i}$ are: $-2-i$ and $2+i$.

● **PROBLEM** 1021

Show that $(a + bi) + (c + di) = (c + di) + (a + bi)$.

Solution: Use the assoiative, distributive, and commutative laws. Associate the corresponding components of the complex numbers, i.e., associate the real and imaginary parts respectively.

$$(a + bi) + (c + di) = (a + c) + (b + d)i$$

$$= a + c + bi + di$$

$$= a + bi + c + di$$

$$= (c + di) + (a + bi)$$

We would suspect that zero is still the additive identity, but zero is a real number. Recall that the real number 5 and the complex number $5 + 0i$ represent the same number. Then the additive identity should be $0 + 0i$. Let us see whether it is. Recall that adding the additive identity to a number does not change the number. Applying the definition of addition.

$$(a + bi) + (0 + 0i) = (a + 0) + (b + 0)i$$

$$= a + bi$$

This verifies that $0 + 0i$ is the additive identity.

Given $f(x) = x^3 + x + 1$, evaluate $f(1 + i)$.

Solution: $f(1 + i)$ indicates that $1 + i$ should be subsituted for x.

$$f(1 + i) = (1 + i)^3 + (1 + i) + 1$$

$$= (1 + i)(1 + i)(1 + i) + (1 + i) + 1$$

$$= \left(1 + 2i + i^2\right)(1 + i) + (1 + i) + 1$$

$$= 1 + 2i + i^2 + i + 2i^2 + i^3 + (1 + i) + 1$$

$$= 1 + 3i + 3i^2 + i^3 + (1 + i) + 1$$

$$= 1 + 3i + 3i^2 + 1^3 + 1 + i + 1$$

Note that $i^2 = -1, i^3 = i^2(i) = (-1)(i) = -i$. Then

$$f(1 + i) = 1 + 3i + 3(-1) + (-1)i + 1 + i + 1$$

$$= 3i.$$

Find the real numbers a and b such that

$$(a + bi) + (2 - 3i) = 2(-2 + i).$$

Solution: Subtract $(2 - 3i)$ from both sides of the given equation.

$$(a+bi) + (2-3i) - (2-3i) = 2(-2+i) - (2-3i)$$

$$a+bi = -4 + 2i - 2 + 3i$$

$$= -4 - 2 + 2i + 3i$$

$$a+bi = -6 + 5i$$

Therefore, $a = -6$ and $b = 5$.

CHAPTER 32

ANALYTIC GEOMETRY

> **Basic Attacks and Strategies for Solving Problems in this Chapter. See pages 742 to 765 for step-by-step solutions to problems.**

In an xy-coordinate system there is a need to be able to determine a point that is halfway between two given points $P_1(x_1, y_1)$ and $P_2(x_2, y_2)$ on a line segment. The procedure for finding this midpoint between P_1 and P_2 is given by the ordered pair $(x_3\, y_3)$, where

$x_3 = (x_1 + x_2)/2$ and $y_3 = (y_1 + y_2)/2$.

The distance between two points in a geometrical configuration in the xy–plane can be determined by using the distance formula which is developed through the application of the well-known Pythagorean Theorem. The distance formula is given by

$$d = \sqrt{(x_2 - x_1)^2 + (y_2 - y_1)^2}\,,$$

where $P(x_1, y_1)$ and $P_2(x_2, y_2)$ are the points. For example, the distance between the points

$(x_1, y_1) = (3, -1)$ and $(x_2, y_2) = (-7, 5)$ is given by:

$d = \sqrt{(-7-3)^2 + [5-(-1)]^2}$

$d = \sqrt{100 + 36}$

$d = \sqrt{136} = 2\sqrt{34}$.

A circle is the set of all points in a plane that are some fixed distance (called the radius) from a given point (called the center). The standard form of the equation of a circle in the xy-plane with center at $C(h, k)$ and a radius r is given by:

$(x - h)^2 + (y - k)^2 = r^2$.

Unfortunately, the equation of a circle will frequently be encountered in the same arrangement other than standard form. When this occurs, the technique for putting the equation in standard form involves applying completion of the square for each variable. For example, in order to put the equation

$$x^2 + y^2 - 4x + 6y + 9 = 0$$

in the standard form of a circle, first rewrite the equation as

$$x^2 - 4x + y^2 + 6y = -9 \tag{1}$$

Then, complete the square for each of the variables by adding the appropriate expressions to both sides of equation (1). Thus,

$$(x^2 - 4x + 4) + (y^2 + 6y + 9) = -9 + 4 + 9.$$

Factoring the left-hand side and simplifying the right-hand side, we get

$$(x - 2)^2 + (y + 3)^2 = 4$$

which is the equation of the circle with center $(2, -3)$ and radius 2.

The measure of the central angle θ (in radians) in a circle of radius r gives rise to a procedure for finding the arc length of a circle. In particular, the length of the arc, denoted by s, of a circle with a central angle θ (in radians) is given by the formula

$$s = r * \theta.$$

For example, the arc length of a circle of radius 10 inches with a central angle that measures 30 degrees or $\pi/6$ radians) is given by

$$s = (\pi/6) \cdot (10)$$

or approximately 5.233 inches.

The area of the sector of the circle with radius r and a central angle θ is given by the following formula:

$$A = (1/2)\, r^2\, \theta.$$

Using the above example, the area of the sector of a circle whose radius is 10 inches and central angle is 30 degrees is as follows:

$$A = (1/2)\, (10)^2\, (0.523) = 26.15 \text{ square inches.}$$

POINTS ON LINE SEGMENTS

● **PROBLEM** 1024

Find the midpoint of the segment from R(-3,5) to S(2,-8).

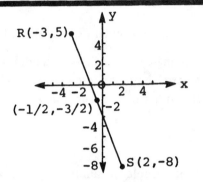

Solution: The midpoint of a line segment from (x_1, y_1) to (x_2, y_2) is given by

$$\left(\frac{x_1 + x_2}{2} , \frac{y_1 + y_2}{2} \right),$$

the abscissa being one half the sum of the abscissas of the endpoints and the ordinate one half the sum of the ordinates of the endpoints. Let the coordinates of the midpoint be $P(x_0, y_0)$. Then,

$$x_0 = \tfrac{1}{2}(-3 + 2) = -\tfrac{1}{2} \qquad y_0 = \tfrac{1}{2}[5 + (-8)] = \tfrac{1}{2}(-3) = -\tfrac{3}{2}.$$

Thus, the midpoint is $P\left(-\tfrac{1}{2}, -\tfrac{3}{2}\right)$.

● **PROBLEM** 1025

What are the coordinates of the midpoint of a line segment joining P(-2,1) and Q(6,4)?

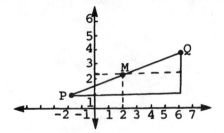

Solution: Let $M(\overline{x},\overline{y})$ be the midpoint of a line segment joining $P(-2,1)$ and $Q(6,4)$. The x-coordinate of M is the average of the x-coordinates of P and Q. The y-coordinate of M is the average of the y-coordinates of P and Q:

$$M(\overline{x},\overline{y}) = M\left(\frac{-2+6}{2}, \frac{1+4}{2}\right) = M\left(2, 2\tfrac{1}{2}\right).$$

Therefore, the coordinates of the midpoint M are $\left(2, 2\tfrac{1}{2}\right)$. Plot the points P and Q, as illustrated in the figure.

● **PROBLEM** 1026

Determine the coordinates of the midpoint of the line segment joining the points (3, -8) and (-7, 5).

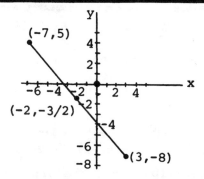

Solution: The coordinates of the desired midpoint are given by one half the sum of the abscissa and one half the sum of the ordinates. Thus,

$$x = 1/2 \ [3 + (-7)] = 1/2 \ (-4) = -2$$

$$y = 1/2 \ [(-8) + 5)] = 1/2 \ (-3) = -3/2$$

Hence the coordinates of the midpoint of the line segment joining these points is (-2, -3/2), as seen in our diagram.

● **PROBLEM** 1027

A line segment AB is $7\tfrac{1}{2}$ in. long. Locate the point C between A and B so that AC is 3/2 in. shorter than twice CB.

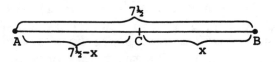

Solution: See the accompanying figure.

Let x = the length of CB in inches. Then $7\tfrac{1}{2}$ - x is the length of AC. We are told AC is 3/2 in. shorter than twice CB. Thus, AC = 2x - 3/2. Therefore

$$7\tfrac{1}{2} - x = 2x - \frac{3}{2}$$

$$\frac{15}{2} - x = 2x - \frac{3}{2}$$

Multiplying both members by 2,

$$15 - 2x = 4x - 3$$

$$-6x = -18$$

$$x = 3$$

Therefore $CB = 3$ and $AC = 7\tfrac{1}{2} - 3 = 4\tfrac{1}{2}$. Hence, C is located $4\tfrac{1}{2}$ in. from A and 3 in. from B.

● **PROBLEM** 1028

Find the point Q that is 3/4 of the way from the point P(- 4, - 1) to the point R(12, 11) along the segment PR.

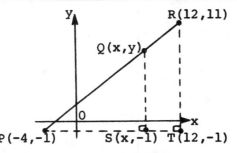

<u>Solution:</u> The figure illustrates the situation; we are to find the numbers x and y, the coordinates of Q. To find these two numbers, we might write the two equations $\overline{PQ} = 3/4\ \overline{PR}$ and $\overline{QR} = \tfrac{1}{4}\ \overline{PR}$ in terms of x and y and solve. Although this method will work, it is easier to use a little geometry. If we introduce the auxiliary points S(x, - 1) having the same x value as Q and same y value as

P, and T(12, - 1) having the same x value as R and the same y value as P, (shown in the figure), we obtain the similar triangles PSQ and PTR. We know these triangles are similar because $\angle$ PSQ and $\angle$ PTR are right angles, and $\triangle$PSQ and $\triangle$PTR have $\angle$ RPS in common. If 2 angles of 2 triangles are equal, their 3rd angles are equal. If two triangles have the same angles they are similar and their corresponding sides are proportional. Therefore,

$$\frac{\overline{PS}}{\overline{PT}} = \frac{\overline{PQ}}{\overline{PR}} \quad \text{and} \quad \frac{\overline{QS}}{\overline{RT}} = \frac{\overline{PQ}}{\overline{PR}} \ . \tag{1}$$

From our figure, we see that $\overline{PS} = x - (- 4) = x + 4$, $\overline{PT} = 12 - (- 4) = 12 + 4 = 16$, $\overline{QS} = y - (- 1) = y + 1$, and $\overline{RT} = 11 - (- 1) = 11 + 1 = 12$, and it is a condition of the problem that $\overline{PQ}/\overline{PR} = 3/4$, since $\overline{PQ} = 3/4\ \overline{PR}$. Hence replacing $\overline{PS}/\overline{PT}$ by $\dfrac{x + 4}{16}$ and $\overline{QS}/\overline{RT}$ by $\dfrac{y + 1}{12}$ in (1), we obtain

$$\frac{x + 4}{16} = \frac{3}{4} \quad \text{and} \quad \frac{y + 1}{12} = \frac{3}{4} \ .$$

744

Cross multiplying,

$$4(x + 4) = 3(16) \quad \text{and} \quad 4(y + 1) = 12(3)$$

$$4x + 16 = 48 \qquad\qquad 4y + 4 = 36$$

$$4x = 32 \qquad\qquad 4y = 32$$

$$x = 8 \qquad\qquad y = 8$$

Thus, the point Q that is 3/4 of the way from P(- 4, - 1) to R(12, 11) is (8,8).

DISTANCES BETWEEN POINTS AND IN GEOMETRICAL CONFIGURATIONS

● **PROBLEM 1029**

What is the distance between the points P(-4,5) and Q(1,-7)? (Observe the accompanying figure).

Solution: Observe the accompanying figure. P being 4 units to the left of the Y-axis and Q being 1 unit to the right, the horizontal distance between P and Q is 5 units. Similarly the vertical distance between P and Q is 12 units. The Pythagorean Theorem states that the sum of the squares of the legs of a right triangle equals the square of the hypotenuse. Thus, in right triangle PQR,

$$(\overline{PR})^2 + (\overline{RQ})^2 = (\overline{PQ})^2 .$$

$$(\overline{PQ})^2 = (12)^2 + (5)^2 = 144 + 25 = 169$$

Taking the square root of both sides, $\overline{PQ}$ = 13. Thus, the distance between (-4,5) and (1,-7), $\overline{PQ}$, is 13.

● **PROBLEM 1030**

What is the distance between the points (2,3) and (7,11)?

Solution: Observe Figure A. The horizontal distance between (2,3) and (7,3) is 7 - 2 = 5. Thus $\overline{BC}$ = 5. Similarly, the vertical distance between (7,11) and (7,3) is 11 - 3 = 8, and $\overline{AC}$ = 8. The Pythagorean Theorem states that the sum of the squares of the legs of a right triangle equals the square of the hypotenuse. Thus, in right triangle

$$ABC, \quad (\overline{BC})^2 + (\overline{AC})^2 = (\overline{AB})^2 .$$

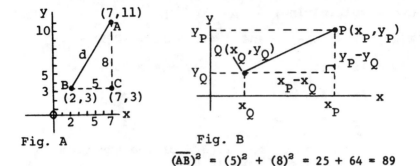

Fig. A   Fig. B

$$(\overline{AB})^2 = (5)^2 + (8)^2 = 25 + 64 = 89$$

Taking the square root of both sides, $\overline{AB} = \sqrt{89}$. Thus, the distance between (2,3) and (7,11), $\overline{AB}$, is $\sqrt{89}$.

Generalizing, suppose we replace these points by P and Q with coordinates (x_P, y_P) and (x_Q, y_Q), respectively (see Figure B). Then what we have done in this problem would amount to using the following general formula for the distance between P and Q:

$$d(P,Q) = \sqrt{(x_P - x_Q)^2 + (y_P - y_Q)^2}$$

This formula continues to hold true in all possible positions of P and Q.

● **PROBLEM** 1031

The point P_1 has coordinates $(-2,-1)$ and the point P_2 has coordinates $(2,2)$. Find the distance $\overline{P_1P_2}$ between these points.

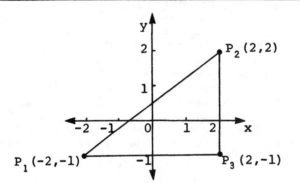

Solution: The points P_1 and P_2 are plotted in the accompanying figure. Let P_3 be the point $(2,-1)$. It is apparent that the points P_1, P_2 and P_3 are the vertices of a right triangle, P_3 being the vertex of the right angle. Since the points P_2 and P_3 lie in the same vertical line, you can easily see that the distance between them is 3 units. Similarly, the distance $\overline{P_1P_3} = 4$. Now according to the Pythagorean Theorem which states that the sum of the squares of the legs of a right triangle is equal to the square of the hypotenuse,

$$\overline{P_1P_2}^2 = \overline{P_1P_3}^2 + \overline{P_3P_2}^2 = (4)^2 + (3)^2 = 16 + 9 = 25.$$

Taking the square root of both sides, $\overline{P_1P_2} = 5$.

The concept of the distance between two points is so important that a formula has been developed for it. The distance between 2 points (x_1, y_1) and (x_2, y_2) is given by

$$d = \sqrt{(x_2 - x_1)^2 + (y_2 - y_1)^2}\,.$$

If we apply this formula to our case, replacing (x_1, y_1) by (-2,-1) and (x_2, y_2) by (2,2) we obtain

$$d = \sqrt{[2-(-2)]^2 + [2-(-1)]^2} = \sqrt{(4)^2 + (3)^2} = \sqrt{16 + 9} = \sqrt{25} = 5$$

as before.

● **PROBLEM** 1032

Find the distance between P(5,3) and Q(8,7).

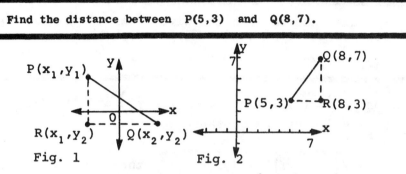

Fig. 1   Fig. 2

<u>Solution:</u> Any two points $P(x_1, y_1)$ and $Q(x_2, y_2)$ not on a line parallel to either axis can be used to locate a third point. We do this by selecting for x-ordinate the x-ordinate of one of the points and y-ordinate the y-ordinate of the other point. The third point shown in Figure 1 is $R(x_1, y_2)$. The choice of R is such that triangle PQR is a right triangle with right angle at R. The distance

$$PR = |y_1 - y_2|; \ RQ = |x_1 - x_2|.$$

Since $\overline{PQ}$ is the hypotenuse of the right triangle, the distance is given as

$$PQ = \sqrt{(PR)^2 + (RQ)^2} = \sqrt{(y_1 - y_2)^2 + (x_1 - x_2)^2}$$

The point R(8,3) is opposite the hypotenuse determined by the points P(5,3) and Q(8,7) (see Figure 2). PR = $|5 - 8|$, RQ = $|3 - 7|$. Hence

$$PQ = \sqrt{|5 - 8|^2 + |3 - 7|^2} = \sqrt{|-3|^2 + |-4|^2}$$

$$= \sqrt{(3)^2 + (4)^2} = \sqrt{9 + 16} = \sqrt{25} = 5$$

We have proved the following theorem:

THEOREM: The distance d between any two points (x_1, y_1) and (x_2, y_2) is given by the formula

$$d = \sqrt{(y_1 - y_2)^2 + (x_1 - x_2)^2}$$

● **PROBLEM** 1033

Find the distance from the origin to the point (x,y).

747

<u>Solution:</u> If P_1 is $(0,0)$ and P_2 is (x,y), then to find the distance from the origin, which is point $(0,0)$, and the point (x,y), apply the distance formula:

$$d = \sqrt{(x_2 - x_1)^2 + (y_2 - y_1)^2}$$

$$d = \sqrt{(x - 0)^2 + (y - 0)^2}$$

$$d = \sqrt{x^2 + y^2}.$$

● **PROBLEM** 1034

Find the distance between the given pair of points, and find the slope of the line segment joining them.

$(3, -5), (2, 4)$

<u>Solution:</u> Let $(3, -5)$ be P_1: $\left(x_1, y_1\right)$ and let $(2, 4)$, be P_2: $\left(x_2, y_2\right)$. By the distance formula,

$$d = \sqrt{\left(x_2 - x_1\right)^2 + \left(y_2 - y_1\right)^2}, \text{ the}$$

the distance between the points $(3, -5)$ and $(2, 4)$ is:

$$d = \sqrt{(2 - 3)^2 + \left(4 - (-5)\right)^2}$$

$$= \sqrt{(-1)^2 + (4 + 5)^2}$$

$$= \sqrt{1 + (9)^2}$$

$$= \sqrt{1 + 81}$$

$$= \sqrt{82}$$

The slope of the line joining the points $(3, -5)$ and $(2, 4)$ is given by the formula:

$$\text{slope} = m = \frac{y_2 - y_1}{x_2 - x_1}$$

Again, let $(3, -5)$ be P_1: $\left(x_1, y_1\right)$ and let $(2, 4)$ be P_2: $\left(x_2, y_2\right)$. Then the slope is:

$$m = \frac{4 - (-5)}{2 - 3}$$

$$= \frac{4 + 5}{-1}$$

$$= \frac{9}{-1}$$

748

= -9

Given the three points P(4,3), Q(4,7), and R(7,3). Find the lengths of $\overline{PQ}$ and $\overline{PR}$.

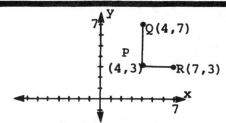

Solution: Points P and Q have the same x-coordinate and lie along a line parallel to the y-axis. Therefore the length of $\overline{PQ} = |y_P - y_Q|$. P and R have the same y-coordinate and lie along a line parallel to the x-axis. Hence the length of $\overline{PR} = |x_P - x_R|$.

$$\overline{PQ} = |3 - 7| = 4 \quad \text{and} \quad \overline{PR} = |4 - 7| = 3.$$

We could have used the distance formula

$$d = \sqrt{(x_1 - x_2)^2 + (y_1 - y_2)^2}$$

Then

$$\overline{PQ} = \sqrt{(4 - 4)^2 + (3 - 7)^2} = 4$$

$$\overline{PR} = \sqrt{(4 - 7)^2 + (3 - 3)^2} = 3$$

Suppose $f = \{(x, 2x - 3)\}$. Choose any three points of the graph of f and show that they lie in a line.

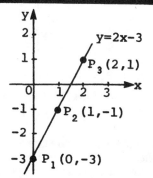

Solution: We are asked to choose any three points, so let us arbitrarily take x = 0, x = 1, and x = 2, and find their corresponding f(x) values:

x	f(x) = 2x-3	f(x)		x	f(x) = 2x-3	f(x)
0	f(0) = 2(0)-3 = -3	-3		2	f(2) = 2(2)-3 = 4 - 3 = 1	1
1	f(1) = 2(1)-3 = 2 - 3 = -1	-1				

Thus, we have three points $P_1(0,-3)$, $P_2(1,-1)$, and $P_3(2,1)$ shown on the accompanying graph of f (see figure). The points are collinear; that is, lie in a line, if the distance $\overline{P_1P_3}$ is equal to the sum of the distances $\overline{P_1P_2}$ and $\overline{P_2P_3}$. (The shortest path between 2 points is a line). Applying the formula for the distance between 2 points (x_1,y_1) and (x_2,y_2), $d = \sqrt{(x_2-x_1)^2 + (y_2-y_1)^2}$, we find

$$\overline{P_1P_3} = \sqrt{(2-0)^2 + [1-(-3)]^2} = \sqrt{2^2+4^2} = \sqrt{4+16}$$

$$= \sqrt{20} = \sqrt{4\cdot5} = \sqrt{4} \cdot \sqrt{5} = 2\sqrt{5}$$

$$\overline{P_1P_2} = \sqrt{(1-0)^2 + [-1-(-3)]^2} = \sqrt{1^2 + (2)^2} = \sqrt{1+4} = \sqrt{5}$$

$$\overline{P_2P_3} = \sqrt{(2-1)^2 + [1-(-1)]^2} = \sqrt{(1)^2+(2)^2} = \sqrt{1+4} = \sqrt{5}$$

and therefore $\overline{P_1P_3} = \overline{P_1P_2} + \overline{P_2P_3} = 2\sqrt{5}$. Therefore we have shown that 3 randomly selected points on the graph $f = \{(x, 2x-3)\}$ lie in a line.

● **PROBLEM** 1037

Use the distance formula to determine whether the points A(0,-3), B(8,3), and C(11,7) are collinear.

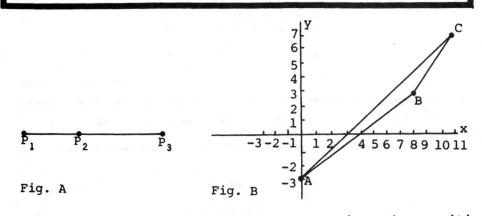

Fig. A   Fig. B

Solution: If three points P_1, P_2, P_3 are in such a position that $\overline{P_1P_2} + \overline{P_2P_3} = \overline{P_1P_3}$ then the three points lie on a straight line and we say that the points are collinear (Figure A).

Thus we find the distances between points A and B, A and C, and B and C to determine whether the sum of any two of these is equivalent to the third, making A, B, and C collinear. Using the formula for the distance between two points (x_1,y_1) and (x_2,y_2),

$$d = \sqrt{(x_2-x_1)^2 + (y_2-y_1)^2}$$

The distance between (0,-3) and (8,3) is

$$d_1 = \sqrt{(8-0)^2 + [3-(-3)]^2} = \sqrt{8^2 + 6^2} = \sqrt{64+36} = \sqrt{100} = 10$$

The distance between (0,-3) and (11,7) is

$$d_2 = \sqrt{(11-0)^2 + [7-(-3)]^2} = \sqrt{(11)^2 + (10)^2} = \sqrt{121 + 100}$$
$$= \sqrt{221} \simeq 14.74$$

The distance between (8,3) and (11,7) is

$$d_3 = \sqrt{(11-8)^2 + (7-3)^2} = \sqrt{(3)^2 + (4)^2} = \sqrt{9 + 16} = \sqrt{25} = 5$$

Since 5 + 10 = 15 and 15 > 14.74, the three points form a triangle as opposed to a straight line. Thus the points are not collinear.

Plotting the points on a graph, and attaching them we also observe that the points form a triangle, not a line. (Figure B).

● **PROBLEM** 1038

Show that the triangle with (-3, 2), (1, 1), and (-4, -2) as vertices is an isosceles triangle.

Solution: If we can show that two sides of the triangle are equal in length, then the triangle is isosceles. This can be done by applying the formula for the distance between two points, (x_1, y_1) and (x_2, y_2):

$$d = \sqrt{(x_1 - x_2)^2 + (y_1 - y_2)^2}$$

Let the given points be designated as A, B, and C respectively. Then

$$|AB| = \sqrt{(1 + 3)^2 + (1 - 2)^2} = \sqrt{17},$$
$$|AC| = \sqrt{(-4 + 3)^2 + (-2 - 2)^2} = \sqrt{17}.$$

Hence $|AB| = |AC|$, and the triangle is isosceles. Furthermore,

$$|BC| = \sqrt{(-4 - 1)^2 + (-2 - 1)^2} = \sqrt{34}.$$

Since $|BC|^2 = |AB|^2 + |AC|^2$ ($\sqrt{34}^2 = \sqrt{17}^2 + \sqrt{17}^2$ or 34 = = 17 + 17), the Theorem of Pythagoras holds, and ABC is a right triangle, with the right angle at A. (See figure.)

Show that the points A (-2, 4), B(-3, -8), and C(2,2) are vertices of a right triangle.

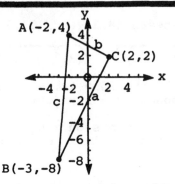

Solution: If triangle ABC is a right triangle, then $a^2 + b^2 = c^2$; that is, the sum of the square of the legs equals the square of the hypotenuse by the Pythagorean Theorem.

Thus we compute the distance from B to C which is side a,

the distance from C to A which is side b,

and the distance from A to B which is side c.

The formula for the distance between two points (x_1, y_1) and (x_2, y_2) is

$$\sqrt{(x_2 - x_1)^2 + (y_2 - y_1)^2}$$

Thus the distance from B to C, from (-3, -8) to (2,2), is

$$\sqrt{[2 - (-3)]^2 + [2 - (-8)]^2} = \sqrt{(2+3)^2 + (2+8)^2}$$
$$= \sqrt{5^2 + 10^2}$$
$$= \sqrt{25 + 100}$$
$$= \sqrt{125}$$

Hence side $a = \sqrt{125}$

The distance from C to A, from (2,2) to (-2,4), is

$$\sqrt{(-2 - 2)^2 + (4 - 2)^2} = \sqrt{(-4)^2 + 2^2}$$
$$= \sqrt{16 + 4}$$
$$= \sqrt{20}$$

Hence side $b = \sqrt{20}$

The distance from A to B, from (-2,4) to (-3, -8), is

$$\sqrt{[-3 - (-2)]^2 + (-8 - 4)^2} = \sqrt{(-3 + 2)^2 + (-12)^2}$$

$$= \sqrt{(-1)^2 + (-12)^2}$$

$$= \sqrt{1 + 144}$$

$$= \sqrt{145}$$

Hence side $c = \sqrt{145}$

Now, if triangle ABC is a right triangle, $a^2 + b^2 = c^2$. Replacing,

a by $\sqrt{125}$, b by $\sqrt{20}$, and c by $\sqrt{145}$

we obtain,

$$\left(\sqrt{125}\right)^2 + \left(\sqrt{20}\right)^2 = \left(\sqrt{145}\right)^2$$

Since, $\left(\sqrt{a}\right)^2 = \sqrt{a}\,\sqrt{a} = \sqrt{a}\cdot\sqrt{a} = \sqrt{a^2} = a$

$$\left(\sqrt{125}\right)^2 = 125$$

$$\left(\sqrt{20}\right)^2 = 20$$

and $\left(\sqrt{145}\right)^2 = 145$

Thus $a^2 + b^2 = c^2$ is equivalent to,
$$125 + 20 = 145$$
$$145 = 145$$

Therefore, triangle ABC is indeed a right triangle.

● **PROBLEM** 1040

Find the equation for the set of points the sum of whose distances from (4,0) and from (-4,0) is 10.

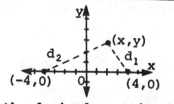

<u>Solution:</u> We find the desired equation by choosing an arbitrary point (x,y) and computing the sum of its distances from (4,0) and (-4,0) (see accompanying figure). Applying the distance formula for the distance between two points (a_1, b_1) and (a_2, b_2), $d = \sqrt{(a_1 - a_2)^2 + (b_1 - b_2)^2}$, we find that the distance from (x,y) to (4,0) is

$$d_1 = \sqrt{(x - 4)^2 + y^2}$$

and the distance from (x,y) to (-4,0) is

$$d_2 = \sqrt{(x + 4)^2 + y^2}.$$

We are given that the sum of the distances $d_1 + d_2 = 10$. Hence, the required equation for the set of points is

$$\sqrt{(x - 4)^2 + y^2} + \sqrt{(x + 4)^2 + y^2} = 10$$

$$\sqrt{(x - 4)^2 + y^2} = 10 - \sqrt{(x + 4)^2 + y^2}.$$

Squaring both sides,

$$\left(\sqrt{(x - 4)^2 + y^2}\right)^2 = \left(10 - \sqrt{(x + 4)^2 + y^2}\right)^2.$$

Since $(\sqrt{a})^2 = \sqrt{a}\,\sqrt{a} = \sqrt{a \cdot a} = \sqrt{a^2} = a$,

$$\left(\sqrt{(x - 4)^2 + y^2}\right)^2 = (x - 4)^2 + y^2.$$

Thus $(x-4)^2 + y^2 = 100 - 20\sqrt{(x+4)^2 + y^2} + (x+4)^2 + y^2$

$x^2 - 8x + 16 + y^2 = 100 - 20\sqrt{(x+4)^2 + y^2} + x^2 + 8x + 16 + y^2$

Adding $- \left(100 + x^2 + 8x + 16 + y^2\right)$ to both members,

$$- 16x - 100 = - 20\sqrt{(x + 4)^2 + y^2}$$

Dividing both sides by -4, $4x + 25 = 5\sqrt{(x + 4)^2 + y^2}$

Squaring again,

$$(4x + 25)(4x + 25) = \left(5\sqrt{(x + 4)^2 + y^2}\right)^2$$

$$16x^2 + 200x + 625 = 25\left[\sqrt{(x + 4)^2 + y^2}\right]^2$$

$$16x^2 + 200x + 625 = 25\left[(x + 4)^2 + y^2\right]$$

$$16x^2 + 200x + 625 = 25\left(x^2 + 8x + 16 + y^2\right)$$

$$16x^2 + 200x + 625 = 25x^2 + 200x + 400 + 25y^2$$

Adding $- (16x^2 + 200x + 400)$ to both members,

$$225 = 9x^2 + 25y^2.$$

Dividing both members by 225, we can write the last equation in the form

$$\frac{9x^2}{225} + \frac{25y^2}{225} = \frac{225}{225}$$

or

$$\frac{x^2}{25} + \frac{y^2}{9} = 1 ,$$

which is the standard form of the equation of an ellipse. This is the desired equation.

● **PROBLEM** 1041

Find the equation for the set of points the difference of whose distances from (5,0) and (-5,0) is 6 units.

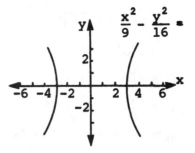

$$\frac{x^2}{9} - \frac{y^2}{16} =$$

Solution: We find the desired equation by choosing an arbitrary point (x,y) and computing the difference of its distance from $(5,0)$ and $(-5,0)$. Applying the distance formula for the distance between two points (a_1,b_1) and (a_2,b_2), $d = \sqrt{(a_1-a_2)^2 + (b_1-b_2)^2}$. From (x,y) to $(5,0)$ is d_1, from (x,y) to $(-5,0)$ is d_2.

$$d_1 = \sqrt{(x-5)^2 + y^2}; \quad d_2 = \sqrt{(x+5)^2 + y^2} \; ;$$

We are told that the difference of these distances, d_2-d_1, is 6. Hence, $\sqrt{(x+5)^2 + y^2} - \sqrt{(x-5)^2 + y^2} = 6$.

$$\sqrt{(x+5)^2 + y^2} = 6 + \sqrt{(x-5)^2 + y^2}$$

Squaring both sides:

$$\left(\sqrt{(x+5)^2 + y^2}\right)^2 = \left(6 + \sqrt{(x-5)^2 + y^2}\right)^2$$

$$\left(\sqrt{(x+5)^2 + y^2}\right)^2 = 36 + 12\sqrt{(x-5)^2 + y^2} + \left(\sqrt{(x-5)^2 + y^2}\right)^2$$

Since $\left(\sqrt{a}\right)^2 = \sqrt{a}\sqrt{a} = \sqrt{a \cdot a} = \sqrt{a^2} = a$,

$$\left(\sqrt{(x+5)^2 + y^2}\right)^2 = (x+5)^2 + y^2 \quad \text{and} \quad \left(\sqrt{(x-5)^2 + y^2}\right)^2 = (x-5)^2 + y^2 \; .$$

Thus we obtain,

$$(x+5)^2 + y^2 = 36 + 12\sqrt{(x-5)^2 + y^2} + (x-5)^2 + y^2$$

$$x^2 + 10x + 25 + y^2 = 36 + 12\sqrt{(x-5)^2 + y^2} + x^2 - 10x + 25 + y^2$$

Adding $-\left(x^2 + 25 + y^2\right)$ to both sides,

$$x^2 + 10x + 25 + y^2 - \left(x^2 + 25 + y^2\right) = 36 + 12\sqrt{(x-5)^2 + y^2} + x^2$$
$$- 10x + 25 + y^2 - \left(x^2 + 25 + y^2\right)$$

$$10x = 36 + 12\sqrt{(x-5)^2 + y^2} - 10x$$

Adding $-36 + 10x$ to both sides,

$$10x - 36 + 10x = 36 + 12\sqrt{(x-5)^2 + y^2} - 10x - 36 + 10x$$

$$20x - 36 = 12\sqrt{(x-5)^2 + y^2}$$

Dividing both sides by 4,

$$5x - 9 = 3\sqrt{(x-5)^2 + y^2}$$

Squaring both sides,

$$(5x-9)^2 = \left(3\sqrt{(x-5)^2 + y^2}\right)^2$$

$$(5x-9)(5x-9) = 3^2\left(\sqrt{(x-5)^2 + y^2}\right)^2$$

$$25x^2 - 90x + 81 = 9[(x-5)^2 + y^2]$$

$$25x^2 - 90x + 81 = 9\left(x^2 - 10x + 25 + y^2\right)$$

$$25x^2 - 90x + 81 = 9x^2 - 90x + 225 + 9y^2$$

Adding $-\left(9x^2 - 90x + 9y^2\right)$ to both sides,

$$25x^2 - 90x + 81 - \left(9x^2 - 90x + 9y^2\right) = 9x^2 - 90x + 225 + 9y^2$$
$$- \left(9x^2 - 90x + 9y^2\right)$$

$$25x^2 - 9x^2 - 9y^2 + 81 = 225$$

$$16x^2 - 9y^2 + 81 = 225$$

Adding -81 to both sides,

$$16x^2 - 9y^2 = 144.$$

Dividing both sides by 144,

$$\frac{16x^2}{144} - \frac{9y^2}{144} = \frac{144}{144}$$

or

$$\frac{x^2}{9} - \frac{y^2}{16} = 1 ,$$

which is the standard form for the equation of a hyperbola.

From the form of the equation we can determine its graph. When the center is at the origin and its vertices are at (a,0) and (-a,0), the equation of the hyperbola is

$$\frac{x^2}{a^2} - \frac{y^2}{b^2} = 1 .$$

In this case, the vertices are (3,0) and (-3,0) since $a^2 = 9$ and $a = \pm 3$ (see figure).

However, when the vertices lie on the y-axis, the equation of the hyperbola is

$$\frac{y^2}{a^2} - \frac{x^2}{b^2} = 1.$$

The vertices would then be (0,a) and (0,-a).　　● **PROBLEM 1042**

Plot the points (1, - 2) and (5, 1) in the xy- plane. What ordered pair corresponds to point C in the Figure? If points A, B, and C of the Figure are three vertices of a parallelogram, what are the coordinates of the fourth vertex in the third quadrant?

Solution: In the Figure, the point (1, -2) is 1 unit to the right of the origin and 2 units below the x-axis. Therefore it is located at point A. The point (5, 1) is located at point B, 5 units to the right of the origin and 1 unit above the x-axis.

The abscissa of point C is 0 and its ordinate is 2. Therefore, the ordered pair is (0, 2).

There are three possible locations for a fourth vertex. In quadrant III the vertex is the intersection of lines parallel to AB and BC, respectively. Since point C is 5 units to the left of point B and 1 unit above it, the required vertex D will be 5 units to the left of point A and 1 unit above it. Its coordinates are (- 4, - 1).

CIRCLES, ARCS, AND SECTORS
● **PROBLEM** 1043

Write equations of the following circles:

    (a) With center at (-1, 3) and radius 9.

    (b) With center at (2, -3) and radius 5.

Solution: The equation of the circle with center at (a,b) and radius r is

$$(x - a)^2 + (y - b)^2 = r^2.$$

(a) Thus, the equation of the circle with center at (-1, 3) and radius 9 is

$$[x - (-1)]^2 + (y - 3)^2 = 9^2$$

$$(x + 1)^2 + (y - 3)^2 = 81$$

(b) Similarly the equation of the circle with center at (2, -3) and radius 5 is

$$(x - 2)^2 + [y - (-3)]^2 = 5^2$$

$$(x - 2)^2 + (y + 3)^2 = 25$$

● **PROBLEM** 1044

Find the center and radius of the circle
$$x^2 - 4x + y^2 + 8y - 5 = 0 \qquad (1)$$

Solution: We can find the radius and the coordinates of the center by completing the square in both x and y. To complete the square in either variable, take half the coefficient of the variable term (i.e., the x term or the y term) and then square this value. The resulting number is then added to both sides of the equation. Completing the square in x:

$$[\tfrac{1}{2}(-4)]^2 = [-2]^2 = 4$$

Then equation (1) becomes:

$$(x^2 - 4x + 4) + y^2 + 8y - 5 = 0 + 4,$$

or

$$(x - 2)^2 + y^2 + 8y - 5 = 4 \qquad (2)$$

Before completing the square in y, add 5 to both sides of equation (2):

$$(x - 2)^2 + y^2 + 8y - 5 + 5 = 4 + 5$$
$$(x - 2)^2 + y^2 + 8y = 9 \qquad (3)$$

Now, completing the square in y:

$$[\tfrac{1}{2}(8)]^2 = [4]^2 = 16$$

Then equation (3) becomes:

$$(x - 2)^2 + \left(y^2 + 8y + 16\right) = 9 + 16,$$

or

$$(x - 2)^2 + (y + 4)^2 = 25 \qquad (4)$$

Note that the equation of a circle is:

$$(x - h)^2 + (y - k)^2 = r^2 ,$$

where (h,k) is the center of the circle and r is the radius of the circle. Equation (4) is in the form of the equation of a circle. Hence, equation (4) represents a circle with center (2,-4) and radius = 5.

● **PROBLEM** 1045

Find the equation for the set of points 5 units from the point (3,4).

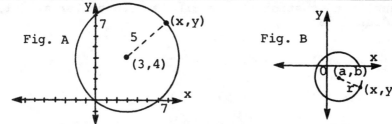

Fig. A   Fig. B

Solution: Apply the distance formula to obtain the distance from the point (3,4) to any point (x,y) 5 units away.

$$d = \sqrt{\left(x_2 - x_1\right)^2 + \left(y_2 - y_1\right)^2}.$$

Thus, replace $\left(x_1, y_1\right)$ by (3,4), $\left(x_2, y_2\right)$ by (x,y) and d by 5 to obtain

$$5 = \sqrt{(x - 3)^2 + (y - 4)^2} \qquad \text{Square both sides,}$$

$$25 = (x - 3)^2 + (y - 4)^2.$$

The equation $(x - 3)^2 + (y - 4)^2 = 25$ is the standard form

758

generally used for the equation of a circle with center at (3,4) and radius 5. See Figure A.

We can generalize the result of this example and find the equation for any circle with center (a,b) and radius r. The required circle is the set of points (x,y) at a distance r from (a,b),

$$\sqrt{(x-a)^2 + (y-b)^2} = r$$

$$(x-a)^2 + (y-b)^2 = r^2$$

as shown in Figure B.

● **PROBLEM** 1046

If θ is an angle of 30°, what is its width in radians?

Solution: If an angle θ is A degrees wide and also t radians wide, then the numbers A and t are related by the equation:

$$\frac{A}{180^\circ} = \frac{t}{\pi} \tag{1}$$

If the given angle of 30° is t radians wide, then by using equation (1):

$$\frac{30^\circ}{180^\circ} = \frac{t}{\pi}$$

$$\frac{1}{6} = \frac{t}{\pi}$$

Multiplying both sides by π,

$$\pi\left(\frac{1}{6}\right) = \cancel{\pi}\left(\frac{t}{\cancel{\pi}}\right)$$

$$\frac{1}{6}\pi = t$$

Hence, the angle's width in radians is $\frac{1}{6}\pi$.

● **PROBLEM** 1047

Find the area of a sector in which the measure of the central angle is 60° and the radius of the circle is 2.

Solution: Note the diagram.

The central angle, 60° is $\frac{1}{6} \times (360^\circ)$; that is, $60^\circ = \frac{1}{6} \times$ (circumference of circle). Therefore, one-sixth of the area of the circle is covered. Using the fact that the area of the total circumference of the circle is πr^2, or $A = \pi r^2$: the area covered by the sector given in this problem is

$$A = \frac{1}{6}\pi r^2 = \frac{1}{6}\pi(2)^2 = \frac{1}{6}\pi(4) = \frac{4\pi}{6}$$
$$= \frac{2\pi}{3}.$$

● **PROBLEM** 1048

Find the length of the minor arc of a circle of radius 1 whose central angle has a measure of 90°.

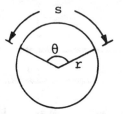

Solution: For problems dealing with arc length of a circle, the following information is helpful. On a circle of radius r, a central angle of θ radians intercepts an arc of length

$$s = r\theta;$$

that is, arc length = radius x central angle in radians. (See Figure). Therefore, in this problem:

$$\text{length of minor arc} = s = (1)(90°)$$
$$= (1)\left(\frac{\pi}{2}\text{ radians}\right)$$
$$= \frac{\pi}{2}\text{ radians}$$

We could have obtained this answer by noting that a central angle of 90° yields an arc equal to $\frac{1}{4}$ the circumference, since

$$90° = \frac{1}{4}(360°) = \frac{1}{4} \times \text{(circumference of circle)}$$
$$= \frac{1}{4}(2\pi\text{ radians})$$
$$= \frac{2\pi}{4}\text{ radians}$$
$$= \frac{\pi}{2}\text{ radians}$$

● **PROBLEM** 1049

What length of arc is subtended by a central angle of 75° on a circle 13.7 inches in radius?

Solution: Let θ denote a central angle in a circle of radius r, and let s be the length of the intercepted arc, measured in the same units as the radius. Then if θ is an angle measured in radians, the length of the arc, s, is s = rθ.

We must express the given angle in radians. Since 2π radians = 360°, then 1° = 2π/360 radians = π/180 radians. Thus, 75° = 75 · π/180 radians = 5/12 π radians.

Substituting this value into the relation s = rθ, we have

$$s = 13.7 \left(\frac{5}{12}\pi\right) \text{ inches} = (5.71\ \pi) \text{ inches}$$

$$= (5.71)(3.14) \text{ inches}$$

$$\approx 17.9 \text{ inches.}$$

● **PROBLEM** 1050

If the central angle of a circle of radius 5 in. is 30°, what is the length of the intercepted arc, and, what is the area of the sector?

__Solution:__ To find the length of the intercepted arc, use the formula

(1) $s = r\theta$, where s is the length of the arc of a circle whose radius is r and whose central angle θ is expressed in radians. We are given the radius and the angle which is measured in degrees. We first convert the given angle, 30°, to radian measure; we have

$$\frac{\theta}{2\pi} = \frac{30°}{360°} \ ,$$

$$\theta = \frac{\pi}{6} \ .$$

Hence equation (1) gives

$$s = r\theta = 5 \times \frac{\pi}{6} = 2.618 \text{ in., approx.,}$$

for the length of arc. To find the area of the Sector, A, apply the formula

(2) $$A = \tfrac{1}{2} r^2 \theta \ .$$

Thus equation (2) yields for the area

$$A = \tfrac{1}{2} \times 5^2 \times \frac{\pi}{6} = 6.545 \text{ in.}^2 \text{ approx.}$$

● **PROBLEM** 1051

Give a geometric description of the following sets of pairs of numbers.

(a) $\{(x,y) \mid x > 0\}$ (d) $\{(x,y) \mid 0 \le x \le 1, 0 \le y \le 1\}$

(b) $\{(x,y) \mid xy = 0\}$

(c) $\{(x,y) \mid x^2 + y^2 < 4\}$

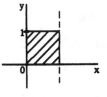

__Solution:__ a) For the set $\{(x,y) \mid x > 0\}$, all points (x,y) in the first and fourth quadrants will be included in the set since $x > 0$ in these quadrants, (not including pts. along the y-axis).

b) For the set $\{(x,y) \mid xy = 0\}$, all points (x,y) will be included for which $xy = 0$. This equation is satisfied whenever $x = 0$ or $y = 0$. Since any point on the y-axis satisfies the equation $x = 0$ and any point on the x-axis satisfies the equation $y = 0$, all points on the x-axis and y-axis satisfy the given set $\{(x,y)\,|\,xy = 0\}$.

c) For the set $\{(x,y)\,|\,x^2 + y^2 < 4\}$, note that the equation $x^2 + y^2 = 4$ is a circle of radius = 2 and center $(0,0)$. Hence, the set $\{(x,y)\,|\,x^2 + y^2 < 4\}$ consists of all points inside the circle of radius = 2 (including center $(0,0)$, since $0^2 + 0^2 < 4$). The points on the circumference of the circle are not included.

d) For the set $\{(x,y)\,|\,0 \le x \le 1,\ 0 \le y \le 1\}$, all points bounded by the vertical lines $x = 0$ and $x = 1$ and by the horizontal lines $y = 0$ and $y = 1$ will be included in the set. Note the figure. All points of the square are in the set $\{(x,y)\,|\,0 \le x \le 1,\ 0 \le y \le 1\}$, including the boundary points.

SPACE RELATED PROBLEMS

● **PROBLEM** 1052

Find the distance of the point (x, y, z) from the origin 0.

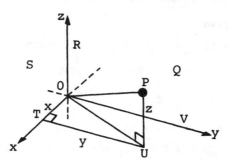

Solution: From the given diagram we see that point (x, y, z) is labeled P. Then OP is the distance of (x, y, z) from the origin 0. Thus, we wish to find OP. Referring to the figure, consider triangle OUP, in which the angle OUP is a right angle. From Pythagoras' theorem,

$$OP^2 = OU^2 + UP^2$$

Consider triangle OTU in which the angle OTU is a right angle. Using Pythagoras' theorem again,

$$OU^2 = OT^2 + TU^2$$

Substituting this value of OU^2 in the first equation

$$OP^2 = \left(OT^2 + TU^2\right) + UP^2$$

But OT = x, TU = y, UP = z, and so

$$OP^2 = x^2 + y^2 + z^2$$

The distance of the points (x, y, z) from the origin 0 is therefore

$$OP = \sqrt{x^2 + y^2 + z^2}.$$

Find the distance from (6, 3, -6) to (10, 0, 6).

Solution: The distance between any two points (x_1, y_1, z_1) and (x_2, y_2, z_2) in xyz-space is given by the formula:

$$d = \sqrt{\left(x_2 - x_1\right)^2 + \left(y_2 - y_1\right)^2 + \left(z_2 - z_1\right)^2}$$

Let (6, 3, -6) be P_1: (x_1, y_1, z_1) and let (10, 0, 6) be P_2: (x_2, y_2, z_2). Therefore, the distance between

(6, 3, -6) and (10, 0, 6) is:

$$d = \sqrt{(10 - 6)^2 + (0 - 3)^2 + \left(6 - (-6)\right)^2}$$

$$= \sqrt{(4)^2 + (-3)^2 + \left(6 + 6\right)^2}$$

$$= \sqrt{(4)^2 + (-3)^2 + (12)^2}$$

$$= \sqrt{16 + 9 + 144}$$

$$= \sqrt{25 + 144}$$

$$= \sqrt{169}$$

$$= 13$$

Find the solution(s) with the given components of the equation in xyz-space.

$$x^2 + 2y^2 - z^2 = 4; \quad (0, 3, ?)$$

Solution: Substituting 0 for x and 3 for y yields

$$(0)^2 + 2(3)^2 - z^2 = 4$$

$$0 + 2(9) - z^2 = 4$$

$$0 + 18 - z^2 = 4$$

$$18 - z^2 = 4$$

Add z^2 to both sides of this equation.

$$18 - \not{z}^2 + \not{z}^2 = 4 + z^2$$

$$18 = 4 + z^2$$

Subtract 4 from both sides of this equation.

$$18 - 4 = \not{4} + z^2 - \not{4}$$

$$14 = z^2$$

Take the square root of both sides of this equation.

$$\pm\sqrt{14} = \sqrt{z^2}$$

$$\pm\sqrt{14} = z$$

Therefore,

$$z = \sqrt{14} \quad \text{or} - \sqrt{14}.$$

Hence, the solutions are $(0, 3, \sqrt{14})$ and $(0, 3, -\sqrt{14})$.

● **PROBLEM** 1055

Find the solutions of the equation in xyz-space where the

   a. first and second components are zero,

   b. first and third components are zero, and

   c. second and third components are zero.

$$x^2 + 3y - z = 4$$

<u>Solution:</u> a) In this case, the first and second components are zero. In xyz-space, the point described is (0, 0, z). Once the third component, z, is found, a solution will be found. Letting x = 0 and y = 0 in the given equation, since the first and second components are zero:

$$(0)^2 + 3(0) - z = 4$$

$$0 + 0 - z = 4$$

$$- z = 4$$

Multiplying both sides of this equation by -1:

$$-1(-z) = -1(4)$$

$$z = -4$$

Therefore, the solution in xyz-space is: (0, 0, -4)

b) In this case, the first and third components are zero. In xyz-space, the point described is $(0, y, 0)$. Once the second component, y, is found, a solution will be found. Letting $x = 0$ and $z = 0$ in the given equation, since the first and third components are zero:

$$(0)^2 + 3y - 0 = 4$$

$$0 + 3y - 0 = 4$$

$$3y = 4$$

Dividing both sides of this equation by 3:

$$\frac{3y}{3} = \frac{4}{3}$$

$$y = \frac{4}{3}$$

Therefore, the solution in xyz-space is: $\left(0, \frac{4}{3}, 0\right)$.

c) In this case, the second and third components are zero. In xyz-space, the point described is $(x, 0, 0)$. Once the first component, x, is found, a solution will be found. Letting $y = 0$ and $z = 0$ in the given equation, since the second and third components are zero:

$$x^2 + 3(0) - 0 = 4$$

$$x^2 + 0 - 0 = 4$$

$$x^2 = 4$$

Taking the square root of both sides:

$$\sqrt{x^2} = \sqrt{4}$$

$$x = \pm 2$$

Therefore, there are two solutions in xyz-space and they are $(2, 0, 0)$ and $(-2, 0, 0)$.

CHAPTER 33

PERMUTATIONS

Basic Attacks and Strategies for Solving Problems in this Chapter. See pages 766 to 775 for step-by-step solutions to problems.

The basic counting procedure for objects is the multiplication principle. It is given by the following statement:

> If one event can occur in m ways and a second event can occur in n ways, then both events can occur in mn ways, provided the outcome of the first event does not influence the outcome of the second.

For instance, to apply this counting principle, consider this example.

How many different meals can be selected if a restaurant offers 3 salads, 5 main dishes, and 2 desserts?

Since the first event can occur in 3 ways, the second event can occur in 5 ways, and the third event can occur in 2 ways, then there are

$$3 \cdot 5 \cdot 2 = 30$$

different meals to select from.

Often we are concerned with the arrangement of a set of distinct objects in some specific order. Such an arrangement of a group of objects is called a permutation of the objects. The number of permutations of n distinct things taken all at a time, denoted by

$$P(n, n),$$

is equal to $n!$. The number of permutations of n distinct things taken r at a time, where

$$0 \le r \le n,$$

is given by

$$P(n, r) = n!/(n - r)!.$$

For instance, the number of ways 12 distinct objects can be arranged taken 3 at a time is given by

$$P(12, 3) = 12!/(12 - 3)!$$

$$= 12!/9!$$

$$= (12 * 11 * 10 * 9!)/9!$$

$$= 12 * 11 * 10$$

$$= 1320 \text{ ways.}$$

Step-by-Step Solutions to Problems in this Chapter, "Permutations"

● **PROBLEM** 1056

Find $_9P_4$.

Solution: Using the general formula for permutations of b different things taken a at a time, $_bP_a = \dfrac{b!}{(b-a)!}$, we substitute 9 for b and 4 for a. Hence $_9P_4 = \dfrac{9!}{(9-4)!} = \dfrac{9!}{5!}$. Evaluating our factorials, we obtain:

$$_9P_4 = \frac{9\cdot 8\cdot 7\cdot 6\cdot(5\cdot 4\cdot 3\cdot 2\cdot 1)}{(5\cdot 4\cdot 3\cdot 2\cdot 1)}$$

cancelling 5! in the numerator and denominator:

$$_9P_4 = 9\cdot 8\cdot 7\cdot 6$$
$$= 3,024 \quad .$$

● **PROBLEM** 1057

Evaluate each of the following symbols:

(a) 5!   (b) $\dfrac{7!}{4!}$ (c) P(6,2)   (d) P(9,2)

Solution:

(a) Recalling $n! = n\cdot(n-1)\cdot(n-2)\cdot(n-3)\ \cdots\ 1$,
$$5! = 5\cdot 4\cdot 3\cdot 2\cdot 1 = 120$$

(b) Recalling $n! = n\cdot(n-1)! = n\cdot(n-1)\cdot(n-2)! = \cdots$
$$\frac{7!}{4!} = \frac{7\cdot 6\cdot 5\cdot \cancel{4!}}{\cancel{4!}} = 210$$

(c) Recalling $P(n,r) = \dfrac{n!}{(n-r)!}$
$$P(6,2) = \frac{6!}{(6-2)!} = \frac{6!}{4!} = \frac{6\cdot 5\cdot \cancel{4!}}{\cancel{4!}} = 30$$

(d) Similarly $P(9,2) = \dfrac{9!}{(9-2)!} = \dfrac{9!}{7!} = \dfrac{9\cdot 8\cdot \cancel{7!}}{\cancel{7!}} = 72$

● **PROBLEM** 1058

Calculate the number of permutations of the letters a,b,c,d taken two at a time.

Solution: The first of the two letters may be taken in 4 ways (a,b,c, or d). The second letter may therefore be selected from the remaining three letters in 3 ways. By the fundamental principle the total number of ways of selecting two letters is equal to the product of the number of ways of selecting each letter, hence

$$p(4,2) = 4 \cdot 3 = 12.$$

The list of these permutations is:

$$ab \quad ba \quad ca \quad da$$

$$ac \quad bc \quad cb \quad db$$

$$ad \quad bd \quad cd \quad dc.$$

● **PROBLEM** 1059

Calculate the number of permutations of the letters a,b,c,d taken four at a time.

<u>Solution:</u> The number of permutations of the four letters taken four at a time equals the number of ways the four letters can be arranged or ordered. Consider four places to be filled by the four letters. The first place can be filled in four ways choosing from the four letters. The second place may be filled in three ways selecting one of the three remaining letters. The third place may be filled in two ways with one of the two still remaining. The fourth place is filled one way with the last letter. By the fundamental principle, the total number of ways of ordering the letters equals the product of the number of ways of filling each ordered place, or $4 \cdot 3 \cdot 2 \cdot 1 = 24 = P(4,4) = 4!$ (read 'four factorial').

In general, for n objects taken r at a time,

$$P(n,r) = n(n-1)(n-2)...(n-r+1) = \frac{n!}{(n-r)!} \quad (r < n).$$

For the special case where $r = n$,

$$P(n,n) = n(n-1)(n-2)...(3)(2)(1) = n!,$$

since $(n-r)! = 0!$ which $= 1$ by definition.

● **PROBLEM** 1060

How many permutations of two letters each can be formed from the letters a,b,c,d,e? Actually write these permutations.

<u>Solution:</u> We recall the general formula for the number of permutations of n different things taken r at a time $_nP_r = n!/(n-r)!$. The number of permutations of 2 letters that can be formed from the 5 given letters is $_5P_2$.

$$_5P_2 = \frac{5!}{(5-2)!} = \frac{5!}{3!} = \frac{5 \cdot 4 \cdot 3!}{3!} = 20$$

Thus, the 20 permutations are:

$$ab \quad ac \quad ad \quad ae$$

$$ba \quad bc \quad bd \quad be$$

$$ca \quad cb \quad cd \quad ce$$

da	db	dc	de
ea	eb	ec	ed

● **PROBLEM** 1061

Determine the number of permutations of three elements
taken from a set of four elements {a, b, c, d}.

Solution:

Method A

In this example we can use the formula for permutations

$$P_a^b = \frac{b!}{(b - a)!} :$$

hence, $P_3^4 = \frac{4!}{(4 - 3)!} = \frac{4!}{1!} = \frac{4 \cdot 3 \cdot 2 \cdot 1}{1} = 24$

Method B

If you do not recall the formula for permutations
you may determine the number of possible permutations of
3 elements taken from a set of four elements by recalling
the fundamental principle: If an act can be performed in m
ways and if, after this first act has been performed, a
second act can be performed in n ways then the number of
ways in which both acts can be performed, in the order
given is m × n ways.

Thus, there are 4 ways of filling our first box
 of 3 elements × 3 ways of filling our second box
              ₓ 2 ways of filling our third box

4 (a or b or c or d)	×	3 (the 3 remaining letters)	×	2 (the 2 re- maining letters)	= 24.

Method C

We can also determine the number of permutations
using a tree diagram:

Hence our permutations are:

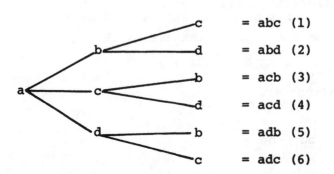

= abc (1)

= abd (2)

= acb (3)

= acd (4)

= adb (5)

= adc (6)

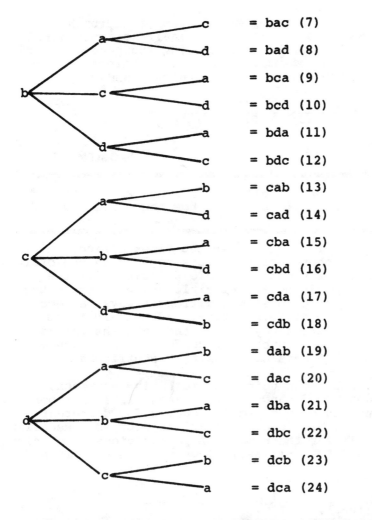

c	= bac	(7)
d	= bad	(8)
a	= bca	(9)
d	= bcd	(10)
a	= bda	(11)
c	= bdc	(12)
b	= cab	(13)
d	= cad	(14)
a	= cba	(15)
d	= cbd	(16)
a	= cda	(17)
b	= cdb	(18)
b	= dab	(19)
c	= dac	(20)
a	= dba	(21)
c	= dbc	(22)
b	= dcb	(23)
a	= dca	(24)

● **PROBLEM** 1062

In how many ways may 3 books be placed next to each other on a shelf?

Solution: We construct a pattern of 3 boxes to represent the places where the 3 books are to be placed next to each other on the shelf:

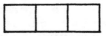

Since there are 3 books, the first place may be filled in 3 ways. There are then 2 books left, so that the second place may be filled in 2 ways. There is only 1 book left to fill the last place. Hence our boxes take the following form:

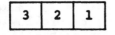

The Fundamental Principle of Counting states that if one thing can be done in a different ways and, when it is done in any one of these ways, a second thing can be done in b different ways, and a third thing can be done in c ways, ... then all the things in succession can be done in a×b×c ... different ways. Thus the books can be arranged in 3·2·1 = 6 ways.

This can also be seen using the following approach. Since the arrangement of books on the shelf is important, this is a permutations problem. Recalling the general formula for the number of permutations of n things taken r at a time, $_nP_r = n!/(n-r)!$, we replace n by 3 and r by 3 to obtain

$$_3P_3 = \frac{3!}{(3-3)!} = \frac{3!}{0!} = \frac{3 \cdot 2 \cdot 1}{1} = 6$$

● **PROBLEM** 1063

Candidates for 3 different political offices are to be chosen from a list of 10 people. In how many ways may this be done?

Solution: There are 10 choices for the first office, and to go with each choice there are 9 choices for the second office, and to go with each of these there are 8 choices for the third office. The Fundamental Principle of Counting states that if one thing can be done in a different ways, a second thing can be done in b different ways, and a third thing can be done in c different ways ..., then all the things in succession can be done in a×b×c ... different ways. Hence, there are 10·9·8 = 720 ways of choosing the officers.

This can also be seen using the following approach. Since the arrangement of candidates on each slate is important (each arrangement represents people running for different political offices), this is a permutations problem. Recalling the general formula for the number of permutations of n things taken r at a time, $_nP_r = n!/(n-r)!$, we replace n by 10 and r by 3 to obtain

$$_{10}P_3 = \frac{10!}{(10-3)!} = \frac{10!}{7!} = \frac{10 \cdot 9 \cdot 8 \cdot \not{7}}{\not{7!}} = 720$$

● **PROBLEM** 1064

A club wishes to select a president, vice-president and treasurer from five members. How many possible slates of officers are there if no person can hold more than one office?

Solution: There are five choices for president, and to go with each choice there are four choices for vice-president, and to go with each president-vice-president choice there are three choices for treasurer. The Fundamental Principle of Counting states that if one thing can be done in a different ways, a second thing can be done in b different ways, and a third thing can then be done in c ways ..., then all the things in succession can be done in a×b×c ... different ways. Hence, there are 5·4·3 = 60 choices. This can also be seen using the following approach. Since the arrangement of people on each slate is important, this is a permutations problem. Recalling the general formula for the number of permutations of n things taken r at a time, $_nP_r = n!/(n-r)!$, we replace n by 5 and r by 3 to obtain

$$_5P_3 = \frac{5!}{(5-3)!} = \frac{5 \cdot 4 \cdot 3 \cdot 2 \cdot 1}{2 \cdot 1} = 60$$

> If a group of 26 members is to elect a president and a
> secretary, in how many ways could the 2 officers be
> elected?

Solution: The group consists of 26 members, anyone of the
26 can serve as president. After the president has been
elected, there are still 25 other members that could be
elected as the secretary. The Fundamental Principle of
Counting states that, if one thing can be done in a
different ways, and a second thing can be done in b
different ways, then the two things in succession can be
done in a · b different ways. Therefore the number of
ways the two officers can be chosen is (26)(25) or 650
ways.

The fundamental principle can be extended to more than
two events. The total number of ways the successive events
could be performed is the product of the numbers of ways
each of the events could be performed.

This can also be seen by using the following approach.
Since the arrangement of officers is important (x serving as president and y serving as secretary is different
than y serving as president and x serving as secretary),
this is a permutations problem. Recalling the general
formula for the number of permutations of n things taken
r at a time, $_nP_r = n!/(n - r)!$, we replace n by 26 and

r by 2 to obtain

$$_{26}P_2 = \frac{26!}{(26 - 2)!} = \frac{26!}{24!} = \frac{26 \cdot 25 \cdot \cancel{24!}}{\cancel{24!}} = 650$$

Thus, once again we find there are 650 ways to elect
a president and secretary from the 26 members.

> How many telephone numbers of four different digits each can be
> made from the digits 0,1,2,3,4,5,6,7,8,9?

Solution: A different arrangement of the same four digits produces a
different telephone number. Since we are concerned with the order in
which the digits appear, we are dealing with permutations.

There are ten digits to choose from and four different ones are to
be chosen at a time. The general formula for the number of permutations
of n things taken r at a time is

$$P(n,r) = \frac{n!}{(n - r)!} \cdot$$

Here n = 10, r = 4, and the desired number is

$$P(10,4) = \frac{10!}{(10 - 4)!} = \frac{10!}{6!} = \frac{10 \cdot 9 \cdot 8 \cdot 7 \cdot 6!}{6!}$$

$$= 5040$$

Thus 5040 telephone numbers of four digits each can be made from the
10 digits.

In how many ways can six books be arranged on a shelf?

Solution: Since we are concerned with the order that the books appear in, we are dealing with permutations. We are to determine the number of permutations of six objects taken six at a time. The solution is obtained by substituting n = 6, r = 6 in the general formula for permutations,

$$p(n,r) = \frac{n!}{(n-r)!}$$

$$p(6,6) = \frac{6!}{(6-6)!}$$

$$= \frac{6!}{0!}$$

0! = 1 by definition, thus

$$= \frac{6!}{1}$$

$$= 6 \cdot 5 \cdot 4 \cdot 3 \cdot 2 \cdot 1$$

$$= 720 \text{ ways.}$$

In how many ways can the letters in the word "Monday" be arranged?

Solution: The word Monday contains 6 different letters. Since different letter arrangements yield different "words", we seek the number of permutations of 6 different objects taken 6 at a time.

Recall the general formula for the number of permutations of n things taken r at a time:

$$_nP_r = \frac{n!}{(n - r)!}$$. Thus, $_6P_6 = \frac{6!}{(6 - 6)!} = \frac{6!}{0!}$.

Since 0! = 1 by definition,

$$_6P_6 = \frac{6!}{1} = 6 \cdot 5 \cdot 4 \cdot 3 \cdot 2 \cdot 1 = 720.$$

Thus, the letters in the word "Monday" may be arranged in 720 ways. We can arrive at the same conclusion using the fundamental theorem of counting, which states that for a given sequence of n events E_1, E_2, ... E_n, if for each i, E_i can occur m_i ways,

then the total number of distinct ways the event may take place is $m_1 \cdot m_2 \cdot m_3 \cdot \ldots \cdot m_n$.

Thus, the first of the 6 letters may be chosen 6 ways
      the second of the 6 letters may be chosen 5 ways
      the third of the 6 letters may be chosen 4 ways
      the fourth of the 6 letters may be chosen 3 ways
      the fifth of the 6 letters may be chosen 2 ways
      the sixth of the 6 letters may be chosen 1 way

Hence the total number of ways the letters may be arranged is $6 \times 5 \times 4 \times 3 \times 2 \times 1 = 720$.

● **PROBLEM** 1069

Determine the number of permutations of the letters in the word BANANA.

Solution: In solving this problem we use the fact that the number of permutations P of n things taken all at a time [P(n,n)], of which n_1 are alike, n_2 others are alike, n_3 others are alike, etc. is

$$P = \frac{n!}{n_1!\,n_2!\,n_3!\ldots} \quad , \text{ with } n_1 + n_2 + n_3 + \ldots = n.$$

In the given problem there are six letters (n = 6), of which two are alike, (there are two N's so that $n_1 = 2$), three others are alike (there are three A's, so that $n_2 = 3$), and one is left (there is one B, so $n_3 = 1$). Notice that $n_1 + n_2 + n_3 = 2 + 3 + 1 = 6 = n$; thus,

$$P = \frac{6!}{2!\,3!\,1!} = \frac{6 \cdot 5 \cdot \overset{2}{\cancel{4}} \cdot 3!}{\cancel{2} \cdot 1 \cdot \cancel{3!} \cdot 1} = 60$$

Thus, there are 60 permutations of the letters in the word BANANA.

● **PROBLEM** 1070

Find the number of permutations of the seven letters of the word "algebra."

Solution: A permutation is an ordered arrangement of a set of objects. For example, if you are given 4 letters a,b,c,d and you choose two at a time, some permutations you can obtain are: ab, ac, ad, ba, bc, bd, ca, cb.

For n things, we can arrange the first object in n different ways, the second in n-1 different ways, the third can be done in n-2 different ways, etc. Thus the n objects can be arranged in order in

$$n! = n \cdot n-1 \cdot n-2 \cdots 1 \text{ ways}$$

Temporarily place subscripts, 1 and 2, on the a's to distinguish them, so that we now have $7! = 5040$ possible permutations of the seven distinct objects. Of these 5040 arrangements, half will contain the a's in the order a_1, a_2 and the other half will contain them in the order a_2, a_1. If we assume the two a's are indistinct, then we apply the following theorem. The number P of distinct permutations of n objects taken at a time, of which n_1 are alike, n_2 are alike of another kind,. . . .,n_k are alike of still another kind, with $n_1 + n_2 + \ldots + n_k = n$ is $P = \frac{n!}{n_1!\,n_2! \ldots n_k!}$ Then, here in this example, the 2 a's are alike so

$$P = \frac{7!}{2!} = 2520 \text{ permutations of the letters of}$$

the word algebra, when the a's are indistinguishable.

● **PROBLEM** 1071

In how many ways can the letters of the word "Tennessee" be arranged?

Solution: The number of permutations P of n things taken all at a time, of which n_1 are alike, n_2 others are alike, n_3 others are alike, etc., is

$$P = \frac{n!}{n_1!n_2!n_3!\ldots} \text{ where } n_1 + n_2 + n_3 + \ldots = n.$$

In this problem we have 4 e's, 2n's, 2s's, and 1 t, and there are 9 letters in all. Consequently, the number of ways the letters can be arranged is

$$\frac{9!}{4!2!2!1!} = \frac{9 \cdot 8 \cdot 7 \cdot 6 \cdot 5 \cdot 4!}{4!2 \cdot 2 \cdot 1} = 3,780$$

● **PROBLEM** 1072

In how many ways may a party of four women and four men be seated at a round table if the women and men are to occupy alternate seats?

Solution: If we consider the seats indistinguishable, then this is a problem in circular permutations, as opposed to linear permutations. In the standard linear permutation approach each chair is distinguishable from the others. Thus, if a woman is seated first, she may be chosen 4 ways, then a man seated next to her may be chosen 4 ways, the next woman can be chosen 3 ways and the man next to her can be chosen in 3 ways ... Our diagram to the linear approach shows the number of ways each seat can be occupied.

4	4	3	3	2	2	1	1

By the Fundamental Principle of Counting there are thus $4 \cdot 4 \cdot 3 \cdot 3 \cdot 2 \cdot 2 \cdot 1 \cdot 1 = 576$ ways to seat the people.

However, if the seats are indistinguishable then so long as each person has the same two people on each side, the seating arrangement is considered the same. Thus we may suppose one person, say a woman is seated in a particular place, and then arrange the remaining three women and four men relative to her. Because of the alternate seating scheme, there are three possible places for the remaining three women, so that there are $3! = 6$ ways of seating them. There are four possible places for the four men, whence there are $4! = 24$ ways in which the men may be seated. Hence the total number of arrangements is $6 \cdot 24 = 144$. In general, the formula for circular permutations of n things and n other things which are alternating is $(n - 1)!n!$. In our case we have

$$(4 - 1)!4! = 3!4! = 3 \cdot 2 \cdot 4 \cdot 3 \cdot 2 = 144.$$

● **PROBLEM** 1073

Prove this identity: $P(n,n-1) = P(n,n)$.

Solution: The general formula for the number of permutations of x objects taken r at a time is

$$P(x,r) = \frac{x!}{(x - r)!}$$

Thus, evaluating the left side of the given identity, we obtain

$$P(n,n-1) = \frac{n!}{[n-(n-1)]!} = \frac{n!}{(n-n+1)!} = \frac{n!}{1!} = \frac{n!}{1} = n!$$

Evaluating the right side of the identity we obtain

$$P(n,n) = \frac{n!}{(n-n)!} = \frac{n!}{0!}$$

$0! = 1$, by definition, hence $\quad \frac{n!}{0!} = \frac{n!}{1} = n!$

Thus $P(n,n-1) = n! = P(n,n)$. Therefore, by the transitive property (If $a = b$ and $b = c$, $a = c$),

$$P(n,n-1) = P(n,n).$$

CHAPTER 34

COMBINATIONS

Basic Attacks and Strategies for Solving Problems in this Chapter. See pages 776 to 789 for step-by-step solutions to problems.

In many instances we are interested in a collection of items, but the order in which they are arranged is not important. In such cases we are dealing with combinations rather than permutations. The number of combinations of n distinct things taken r at a time is given by

$$\binom{n}{r} \quad \text{or} \quad c(n,r) = \frac{n!}{(n-r)!r!}$$

For example, the number of four-member committees that can be formed from a group of nine people is found by using the above formula as follows:

$$c(9, 4) = 9! / (9 - 4)! \, 4!$$
$$= 9! / 5! \, 4!$$
$$= 9 * 8 * 7 * 6 * 5! / 5!(4 * 3 * 2 * 1)$$
$$= 126 \text{ four-member committees}$$

In some applied situations, both knowledge of the fundamental counting principle and the use of combinations are necessary to solve the problem.

Step-by-Step Solutions to Problems in this Chapter, "Combinations"

● PROBLEM   1074

Find the value of C(n,0).

Solution: Starting with the formula for combinations:

$$C(n,r) = \frac{n!}{(n-r)! \, r!}$$

and substituting r = 0, we have

$$C(n,0) = \frac{n!}{(n-0)! \, 0!}$$

Hence,
$$C(n,0) = \frac{n!}{n! \, 0!}$$

Recall that 0! = 1 by definition. Then

$$C(n,0) = \frac{n!}{n!}$$

$$= 1.$$

● PROBLEM   1075

Find $_9C_4$.

Solution: By definition, combinations of b different things taken a at a time, $_bC_a = \frac{b!}{a!(b-a)!}$; hence by substitution, $_9C_4 = \frac{9!}{4!(9-4)!}$.

$$= \frac{9!}{4! \, 5!}$$

$$= \frac{9 \cdot 8 \cdot 7 \cdot 6 \cdot (5 \cdot 4 \cdot 3 \cdot 2 \cdot 1)}{4 \cdot 3 \cdot 2 \cdot 1 \cdot (5 \cdot 4 \cdot 3 \cdot 2 \cdot 1)}$$

Cancelling 5! = (5.4.3.2.1) out of numerator and denominator, we multiply to obtain:

$$_9C_4 = \frac{3024}{24} = 126 \ .$$

● PROBLEM   1076

Evaluate each of the following symbols:
(a)   C(6,3)   (b)   C(18,16)

Solution: Recalling the general formula for the number of combinations of n different things taken r at a time

$$C(n,r) = \frac{n!}{r! \, (n-r)!}:$$

(a) $C(6,3) = \frac{6!}{3!(6-3)!} = \frac{6!}{3! \, 3!} = \frac{\overset{2}{6} \cdot 5 \cdot 4 \cdot 3!}{3 \cdot 2 \cdot 1 \cdot 3!} = 20.$

(b) $\quad C(18,16) = \dfrac{18!}{16!(18 - 16)!} = \dfrac{18!}{16!2!} = \dfrac{\overset{9}{\cancel{18}} \cdot 17 \cdot \cancel{16!}}{\cancel{2} \cdot 1 \cdot \cancel{16!}} = 153$

● **PROBLEM** 1077

How many different five-card hands can be obtained from a fify-two card deck?

<u>Solution:</u> Since the order of the cards is unimportant, we are dealing with a combination problem, as opposed to a permutation problem. Recall our general formula for combinations $\quad {}_bC_a = \dfrac{b!}{a!(b-a)!}$.

We want to know how many $\underline{5}$ card hands can be obtained from a $\underline{52}$ card deck, hence $a = 5$, $b = 52$. Substituting 5 for a and 52 for b in our general formula we obtain:

$$ {}_{52}C_5 = \dfrac{52!}{5!(52-5)!} = \dfrac{52!}{5!(47)!} \; . $$

Recall that $n! = n \cdot (n-1). = n \cdot (n-1) \cdot (n-2). = n \cdot (n-1)(n-2)(n-3)\ldots.$ hence $52! = \dfrac{52 \cdot 51 \cdot 50 \cdot 49 \cdot 48 \cdot (47).}{5.(47).}$

Cancelling (47)! from numerator and denominator, and evaluating 5!:

$$ \dfrac{52 \cdot 51 \cdot 50 \cdot 49 \cdot 48}{5 \cdot 4 \cdot 3 \cdot 2 \cdot 1} \; . $$

Performing the necessary multiplications and divisions:

$$ = 2,598,960 \; . $$

Thus 2,598,960 different five-card hands can be obtained from a fifty-two card deck.

● **PROBLEM** 1078

In how many different ways may a pair of dice fall?

<u>Solution:</u> The Fundamental Principle of Counting states that if one thing can be done in a different ways, and when it is done in any one of these ways, a second thing can be done in b different ways, then both things in succession can be done in a×b different ways. A die has 6 sides, thus it may land in any of six ways. Since each die may land in 6 ways, by the Fundamental Principle both die may fall in 6×6 = 36 ways. We can verify this result by enumerating all the possible ordered pairs of dice throws:

1,1	1,2	1,3	1,4	1,5	1,6
2,1	2,2	2,3	2,4	2,5	2,6
3,1	3,2	3,3	3,4	3,5	3,6
4,1	4,2	4,3	4,4	4,5	4,6

```
5,1    5,2    5,3    5,4    5,5    5,6

6,1    6,2    6,3    6,4    6,5    6,6
```

● **PROBLEM** 1079

How many different bridge hands are possible?

Solution: Since the order in which the cards are dealt
is immaterial, we are dealing with combinations, thus we
are interested in determining the number of combinations
of the 52 cards taken 13 at a time. Recall the general
formula for the number of combinations of n items taken
r at a time,

$$C(n,r) = \frac{n!}{r!(n-r)!} \qquad \text{and}$$

$$C(52,13) = \frac{52!}{13!(52-13)!}$$

$$= \frac{52!}{13!39!}$$

$$= \frac{(52)(51)(50)(49)(48)(47)(46)(45)(44)(43)(42)(41)(40)\cancel{39!}}{(13)(12)(11)(10)(9)(8)(7)(6)(5)(4)(3)(2)(1)\;\cancel{39!}}$$

$$= 635,013,559,600.$$

● **PROBLEM** 1080

In how many ways can we select a committee of 3 from a group
of 10 people?

Solution: The arrangement or order of people chosen is
unimportant. Thus this is a combinations problem. Recalling the general formula for the number of combinations of n
different things taken r at a time

$$C(n,r) = \frac{n!}{r!(n-r)!} \quad ,$$

the number of committees of 3 from a group of 10 people is

$$C(10,3) = \frac{10!}{3!(10-3)!} = \frac{10!}{3!7!} = \frac{10 \cdot \overset{3}{\cancel{9}} \cdot \overset{4}{\cancel{8}} \cdot \cancel{7!}}{\cancel{3} \cdot \cancel{2} \cdot 1 \cdot \cancel{7!}} = 120.$$

● **PROBLEM** 1081

How many committees of four members each can be formed from
a group of seven persons?

Solution: This is a problem in combinations, rather than
permutations, since the order is of no consequence. Thus,
a committee consisting of Smith, Jones, Young, and Robin-
son is the same as a committee consisting of Smith, Robin-
son, Jones, and Young. The number of combinations of n
different objects taken r at a time is equal to:

$$\frac{n(n-1)\ldots(n-r+1)}{1 \cdot 2 \cdots r}.$$

In this example, n = 7, r = 4, therefore

$$c(7,4) = \frac{7 \cdot 6 \cdot 5 \cdot 4}{1 \cdot 2 \cdot 3 \cdot 4} = 35.$$

Alternately: The first member may be selected from the seven persons in 7 ways. The second member may be selected from the remaining six people in 6 ways. The third member may be selected in 5 ways from the remaining five people. The fourth member may be selected from the remaining four people in 4 ways. By the fundamental principle the total number of ways of picking the four members is equal to the product of the number of ways of picking each member, or 7 · 6 · 5 · 4 ways. This is a permutation of 7 people selected 4 at a time. To account for the number of ways in which the same four-person committee is selected, but in a different order, divide 7 · 6 · 5 · 4 by the number of ways in which the same committee of four can be selected. This equals a permutation of 4 people selected 4 at a time. This, by the fundamental principle, applied as above, equals 4 · 3 · 2 · 1. Then

$$\frac{7 \cdot 6 \cdot 5 \cdot 4}{4 \cdot 3 \cdot 2 \cdot 1} = 35.$$

● **PROBLEM** 1082

How many baseball teams of nine members can be chosen from among twelve boys, without regard to the position played by each member?

Solution: Since there is no regard to position, this is a combinations problem (if order or arrangement had been important it would have been a permutations problem). The general formula for the number of combinations of n things taken r at a time is

$$C(n,r) = \frac{n!}{r!(n-r)!}.$$

We have to find the number of combinations of 12 things taken 9 at a time. Hence we have

$$C(12,9) = \frac{12!}{9!(12-9)!} = \frac{12!}{9!3!} = \frac{12 \cdot 11 \cdot 10 \cdot 9!}{3 \cdot 2 \cdot 1 \cdot 9!} = 220$$

Therefore, there are 220 possible teams.

● **PROBLEM** 1083

A manufacturer produces 7 different items. He packages assortments of equal parts of 3 different items. How many different assortments can be packaged?

Solution: Since we are not concerned with the order of the items, we are dealing with combinations. Thus the number of assortments is the number of combinations of 7 items taken 3 at a time. Recall the general formula for the number of combinations of n items taken r at a time,

$$C(n,r) = \frac{n!}{r!(n-r)!}$$

$$C(7,3) = \frac{7!}{3!(7-3)!}$$

$$= \frac{7!}{3!\,4!}$$

$$= \frac{7 \cdot \cancel{6} \cdot 5 \cdot \cancel{4!}}{\cancel{3} \cdot \cancel{2} \cdot \cancel{4!}}$$

$$= 35$$

Thus, 35 different assortments can be packaged.

● **PROBLEM 1084**

A man and his wife decide to entertain 24 friends by giving 4 dinners with 6 guests each. In how many ways can the first group be chosen?

<u>Solution:</u> In the first group we are considering one dinner and there are 6 people out of 24 friends to be invited. We must find the number of ways to choose 6 out of 24. We are dealing with combinations. To select r things out of n objects, we use the definition of combinations:

$$\binom{n}{r} = \frac{n!}{r!(n-r)!} = c(n,r)$$

$$c(24,6) = \binom{24}{6} = \frac{24!}{6!\,18!} = \frac{24 \cdot 23 \cdot 22 \cdot 21 \cdot 20 \cdot 19 \cdot 18!}{6!\ 18!}$$

$$= \frac{24 \cdot 23 \cdot 22 \cdot 21 \cdot 20 \cdot 19}{6 \cdot 5 \cdot 4 \cdot 3 \cdot 2 \cdot 1}$$

$$= 134,596$$

● **PROBLEM 1085**

A lady has 12 friends. She wishes to invite 3 of them to a bridge party. How many times can she entertain without having the same 3 people again?

<u>Solution:</u> Since no reference has been made to order or arrangement (for example, the order in which the guests arrive or their seating arrangement at the bridge table), the problem is one of combinations rather than permutations. Recall the general formula for the number of combinations of n different things taken r at a time,

$$C(n,r) = \frac{n!}{r!(n-r)!} \quad .$$

Thus the number of ways of selecting 3 friends from 12 is

$$C(12,3) = \frac{12!}{3!(12-3)!} = \frac{12!}{3!\,9!} = \frac{\overset{4}{\cancel{12}} \cdot 11 \cdot \overset{5}{\cancel{10}} \cdot \cancel{9!}}{\cancel{3} \cdot \cancel{2} \cdot 1 \cdot \cancel{9!}} = 220$$

In evaluating C(12,3), observe that the numerator and denominator of the fraction are first divided by the larger factorial in the denominator. (9! cancels in our fraction.) Thus, the lady can entertain 220 times without having the same 3 people.

● **PROBLEM 1086**

A Sunday school class of 12 members is to be seated on seven chairs and a bench that accommodates five persons. In how many ways can the bench be occupied?

<u>Solution:</u> If we are concerned with the order of people on the bench (so that we consider the same five people sitting in different arrangements as distinct ways), then this is a permutations problem. Recalling the general formula for the number of permutations of n elements taken r at a time

$$p(n,r) = \frac{n!}{(n-r)!}$$

we find the number of permutations of 12 elements taken 5 at a time, or p(12,5). Thus

$$p(12,5) = \frac{12!}{(12-5)!} = \frac{12!}{7!} = 12 \cdot 11 \cdot 10 \cdot 9 \cdot 8 = 95,040$$

If we are not concerned with the order of the people on the bench this becomes a combinations problem. Recalling the general formula for the number of combinations of n elements taken r at a time

$$c(n,r) = \frac{n!}{r!(n-r)!}$$

we find the number of combinations of 12 elements taken 5 at a time, or c(12,5).

$$c(12,5) = \frac{12!}{5!(12-5)!} = \frac{12!}{5!7!} = \frac{12 \cdot 11 \cdot 10 \cdot 9 \cdot 8 \cdot 7!}{5 \cdot 4 \cdot 3 \cdot 2 \cdot 7!} = 792.$$

● **PROBLEM 1087**

How many different sums of money can be obtained by choosing two coins from a box containing a penny, a nickel, a dime, a quarter, and a half dollar?

<u>Solution:</u> The order makes no difference here, since a selection of a penny and a dime is the same as a selection of a dime and a penny, insofar as a sum of money is concerned. This is a case of combinations, then, rather than permutations. Then the number of combinations of n different objects taken r at a time is equal to:

$$\frac{n(n-1)\ldots(n-r+1)}{1 \cdot 2 \cdots r} \ .$$

In this example, n = 5, r = 2, therefore

$$C(5,2) = \frac{5 \cdot 4}{1 \cdot 2} = 10.$$

As in the problem of selecting four committee members from a group of seven people, a distinct two coins can be selected from five coins in

$$\frac{5 \cdot 4}{1 \cdot 2} = 10 \text{ ways (applying the fundamental principle).}$$

● **PROBLEM 1088**

In how many ways can 5 prizes be given away to 4 boys, when each boy is eligible for all the prizes?

<u>Solution:</u> Any one of the prizes can be given in 4 ways; and then any one of the remaining prizes can also be given in 4 ways, since it may be obtained by the boy who has already received a prize. Thus two prizes can be given away in 4^2 ways, three prizes in 4^3 ways, and so on. Hence the 5 prizes can be given away in 4^5, or 1024 ways.

How many "words" each consisting of two vowels and three consonants, can be formed from the letters of the word 'integral'?

Solution: To find the number of ways to choose vowels or consonants from letters, we use combinations. The number of combinations of n different objects taken r at a time is defined to be

$$C(n,r) = \frac{n!}{r!(n-r)!} \ .$$

Then, we first select the two vowels to be used, from among the three vowels in integral; this can be done in $C(3,2) = 3$ ways. Next, we select the three consonants from the five in integral; this yields $C(5,3) = 10$ possible choices. To find the number of ordered arrangements of 5 letters selected five at a time, we need to find the number of permutations of choosing r from n objects. Symbolically, it is $P(n,r)$ which is defined to be

$$P(n,r) = \frac{n!}{(n-r)!}$$

We permute the five chosen letters in all possible ways, of which there are $P(5,5) = 5! = 120$ arrangements. Finally, to find the total number of words which can be formed, we apply the Fundamental Counting Principle which states that if one event can be performed in m ways, another one in n ways, and another in k ways, then the total number of ways in which all events can occur is m × n × k ways. Hence the total number of possible words is, by the fundamental principle

$$C(3,2)C(5,3)P(5,5) = 3 \cdot 10 \cdot 120 = 3600.$$

From 12 books in how many ways can a selection of 5 be made, (1) when one specified book is always included, (2) when one specified book is always excluded?

Solution: Here the formula for combinations is appropriate: the number of combinations of n things taken r at a time:

$$C(n,r) = {}_nC_r = \frac{n!}{r!(n-r)!}$$

where $n = 11$, and $r = 4$.

(1) Since the specified book is to be included in every selection, we have only to choose 4 out of the remaining 11.
Hence the number of ways $= {}^{11}C_4$

$$
{}^{11}C_4 = \frac{11!}{4!(11-4)!}
$$

$$
= \frac{11!}{4!7!}
$$

$$
= \frac{11 \cdot 10 \cdot 9 \cdot 8 \cdot 7!}{4 \cdot 3 \cdot 2 \cdot 1 \cdot 7!}
$$

$$
= \frac{11 \times 10 \times 9 \times 8}{1 \times 2 \times 3 \times 4}
$$

$$
= 330.
$$

(2) Since the specified book is always to be excluded, we have to select the 5 books out of the remaining 11.

Hence the number of ways = $^{11}C_5$

$$^{11}C_5 = \frac{11!}{5!(11-5)!}$$

$$= \frac{11!}{5!6!}$$

$$= \frac{11 \cdot 10 \cdot 9 \cdot 8 \cdot 7 \cdot \cancel{6!}}{5 \cdot 4 \cdot 3 \cdot 2 \cdot 1 \cdot \cancel{6!}}$$

$$= \frac{11 \times 10 \times 9 \times 8 \times 7}{1 \times 2 \times 3 \times 4 \times 5}$$

$$= 462$$

● **PROBLEM** 1091

How many groups can be formed from ten objects taking at least three at a time?

<u>Solution:</u> The number of combinations of n objects taken r at a time is $C(n,r) = n!/r!(n - r)!$. Thus, the number of groups that can be formed from 10 objects taking three at a time is $C(10, 3)$,

from 10 objects taking 4 at a time is $C(10,4)$,

from 10 objects taking 5 at a time is $C(10,5)$,

from 10 objects taking 6 at a time is $C(10,6)$,

.
.
.

from 10 objects taking 10 at a time is $C(10, 10)$.

Therefore, the number of groups that can be formed from 10 objects taking at least three at a time is $C(10, 3) + C(10, 4) + C(10, 5) + C(10, 6) + C(10, 7) + C(10, 8) + C(10, 9) + C(10, 10)$,

$$\frac{10!}{3!7!} + \frac{10!}{4!6!} + \frac{10!}{5!5!} + \frac{10!}{6!4!} + \frac{10!}{7!3!} + \frac{10!}{8!2!} + \frac{10!}{9!1!} + \frac{10!}{10!0!} =$$

$$\frac{10 \cdot 9 \cdot 8 \cdot 7!}{3 \cdot 2 \cdot 7!} + \frac{10 \cdot 9 \cdot 8 \cdot 7 \cdot 6!}{4 \cdot 3 \cdot 2 \cdot 6!} + \frac{10 \cdot 9 \cdot 8 \cdot 7 \cdot 6 \cdot 5!}{5 \cdot 4 \cdot 3 \cdot 2 \cdot 5!} + \frac{10 \cdot 9 \cdot 8 \cdot 7 \cdot 6!}{4 \cdot 3 \cdot 2 \cdot 6!}$$

$$+ \frac{10 \cdot 9 \cdot 8 \cdot 7!}{3 \cdot 2 \cdot 7!} + \frac{10 \cdot 9 \cdot 8!}{2 \cdot 8!} + \frac{10 \cdot 9!}{1 \cdot 9!} + \frac{10!}{10! \cdot 1}$$

$$= 120 + 210 + 252 + 210 + 120 + 45 + 10 + 1$$

$$= 968.$$

● **PROBLEM** 1092

A man has 6 friends; in how many ways may he invite one or more of them to dinner?

<u>Solution:</u> He has to select some or all of his 6 friends. Here the appropriate procedure is to utilize the formula for combinations in which no particular order is important. Since the question specifies

some or all, every combination must be investigated and the sum will give the total number of ways which are possible. The formula for the number of combinations of n things taken r at a time is

$$_nC_r = \frac{n!}{r!(n-r)!} \quad (\text{where } 0! = 1)$$

The six people can be selected one at a time, two at a time, three at a time etc., so r will vary between 1 and 6 inclusive; therefore the number of selections

$$= {^6C_1} + {^6C_2} + {^6C_3} + {^6C_4} + {^6C_5} + {^6C_6}$$

$$= \frac{6!}{1!(6-1)!} + \frac{6!}{2!(6-2)!} + \frac{6!}{3!(6-3)!} + \frac{6!}{4!(6-4)!}$$

$$+ \frac{6!}{5!(6-5)!} + \frac{6!}{6!(6-6)!}$$

$$= \frac{6!}{1!5!} + \frac{6!}{2!4!} + \frac{6!}{3!3!} + \frac{6!}{4!2!} + \frac{6!}{5!1!} + \frac{6!}{6!0!}$$

$$= \frac{6 \cdot 5!}{1 \cdot 5!} + \frac{6 \cdot 5 \cdot 4!}{2 \cdot 1 \cdot 4!} + \frac{6 \cdot 5 \cdot 4 \cdot 3!}{3 \cdot 2 \cdot 1 \cdot 3!} + \frac{6 \cdot 5 \cdot 4!}{2 \cdot 4!} + \frac{6 \cdot 5!}{5! \cdot 1} + \frac{6! \cdot 1}{6! \cdot 1}$$

$$= 6 + 15 + 20 + 15 + 6 + 1$$

$$= 63$$

● **PROBLEM** 1093

A boy has in his pocket a penny, a nickel, a dime, and a quarter. How many different sums of money can he take out if he removes one or more coins?

Solution: In this problem, we are not considering order; that is, we are not concerned whether we choose a penny first and a nickel second or vice versa. (It is still the same arrangement.) Thus, we are considering combinations, not permutations. We consider the following cases to solve this problem:

a) the boy removes one coin
b) the boy removes two coins
c) the boy removes three coins
d) the boy removes four coins

Now, a combination of n things taken r at a time is:

$$C(n, r) = \frac{n!}{r!(n-r)!} \cdot$$

Thus if for a) the boy removes one coin, then we want to find the number of combinations of 4 coins taken one at a time. Similarly for b), c), and d). The total number of combinations of 4 things taken 1, 2, 3, or 4 at a time is

$$C(4,1) + C(4,2) + C(4,3) + C(4,4) = \frac{4!}{1!3!} + \frac{4!}{2!2!} + \frac{4!}{3!1!} + \frac{4!}{4!0!}$$

$$= \frac{4 \cdot 3!}{1 \cdot 3!} + \frac{4 \cdot 3 \cdot 2!}{2 \cdot 1 \cdot 2!} + \frac{4 \cdot 3!}{3! \cdot 1} + \frac{4!}{4! \cdot 1}$$

$$= 4 + \frac{12}{2} + 4 + 1$$

$$= 4 + 6 + 4 + 1$$

$$= 15.$$

● **PROBLEM** 1094

From 10 men and 6 women, how many committees of 5 people

can be chosen:

(a)   If each committee is to have exactly 3 men?
(b)   If each committee is to have at least 3 men?

Solution:

(a)   The order in which the people on the committee are chosen is unimportant, thus this is a problem involving combinations. The general formula for the number of combinations of n different things taken r at a time is $C(n,r) = n!/r!(n-r)!$. Thus, the number of ways to choose 3 men from 10 men is $C(10,3)$

$$= \frac{10!}{3!(10-3)!} = \frac{10!}{3!7!} = \frac{10 \cdot \overset{3}{9} \cdot \overset{4}{8} \cdot 7!}{3 \cdot 2 \cdot 1 \cdot 7!} = 120.$$

The number of ways to choose 2 women from 6 women is $C(6,2)$

$$= \frac{6!}{2!(6-2)!} = \frac{6!}{2!4!} = \frac{\overset{3}{6} \cdot 5 \cdot 4!}{2 \cdot 1 \cdot 4!} = 15.$$

The Fundamental Principle of Counting states that if the first of two independent acts can be performed in a ways, and if the second act can be performed in b ways, then the number of ways of performing the two acts in the order stated is ab. Thus by the fundamental principle, the number of ways to choose the committee is $C(10,3) \cdot C(6,2) = 120 \cdot 5 = 1,800$.

(b)   If the committee is to contain at least 3 men, the possibilities are 3 men and 2 women, 4 men and 1 woman, 5 men and no women.

We have just shown that the number of committees consisting of 3 men and 2 women is 1,800. The number of committees containing 4 men and 1 woman is

$$C(10,4) \cdot \underset{1}{C(6,1)} = \frac{10!}{4!(10-4)!} \cdot \frac{6!}{1!(6-1)!} = \frac{10!}{4!6!} \cdot \frac{6!}{1!5!}$$

$$= \frac{10 \cdot \overset{3}{9} \cdot \overset{}{8} \cdot 7 \cdot 6!}{4 \cdot 3 \cdot 2 \cdot 1 \cdot 6!} \cdot \frac{6 \cdot 5!}{1 \cdot 5!} = 210 \cdot 6 = 1,260 .$$

The number of committees consisting of 5 men is

$$C(10,5) = \frac{10!}{5!(10-5)!} = \frac{10!}{5!5!} = \frac{\overset{2}{10} \cdot \overset{3}{9} \cdot \overset{2}{8} \cdot 7 \cdot \overset{3}{6} \cdot 5!}{5 \cdot 4 \cdot 3 \cdot 2 \cdot 1 \cdot 5!} = 252$$

The probability that any of several mutually exclusive events will occur is the sum of the probabilities of the separate events.

Hence the number of committees containing at least 3 men is

$$1,800 + 1,260 + 252 = 3,312.$$

● **PROBLEM**   1095

From 7 Englishmen and 4 Americans a committee of 6 is to be formed; in how many ways can this be done, (1) when the committee contains exactly 2 Americans, (2) at least 2 Americans?

Solution: (1) : Case (1) is when we choose exactly 2 Americans and thus we need 4 Englishmen in order to have

6 committee members. We use combinations to find the number of ways in which a number of objects may be selected without regard to order. We are not interested in their arrangement in this particular problem.

The number of combinations of n different objects taken r at a time is defined to be:

$$\binom{n}{r} = \frac{n!}{r'(n-r)!}$$

The number of ways in which two out of four Americans can be chosen

$$\binom{4}{2} = \frac{4!}{2!2!} = \frac{4 \cdot 3 \cdot 2!}{2! \; 2 \cdot 1} = 6$$

The number of ways in which 4 out of 7 Englishmen can be chosen is:

$$\binom{7}{4} = \frac{7!}{4!3!} = \frac{7 \cdot 6 \cdot 5 \cdot 4!}{4! \cdot 3 \cdot 2 \cdot 1} = 35$$

To find the number of ways of choosing the 2 Americans and 4 Englishmen, we apply the Fundamental Principle. This states that if an act can be performed in m ways and another act can be performed in n ways, then the number of ways in which both acts can be performed is mXn ways. Hence,

$$\binom{4}{2}\binom{7}{4} = 6 \cdot 35 = 210 = \text{the number of ways of choosing the}$$

Americans and the Englishmen.

(2) A committee of at least 2 Americans chosen from among 4 Americans may contain 2,3, or 4 Americans.
We shall exhaust all the suitable combinations by forming all the groups containing 2 Americans and 4 Englishmen; then 3-Americans and 3 Englishmen; and lastly 4 Americans and 2 Englishmen.

The sum of the three results will give the answer. Hence the required number of ways

$$= \binom{4}{2}\binom{7}{4} + \binom{4}{3}\binom{7}{3} + \binom{4}{4}\binom{7}{2}$$

$$= \frac{4!}{2!2!} \; \frac{7!}{4!3!} + \frac{4!}{3!1!} \; \frac{7!}{3!4!} + \frac{4!}{4!0!} \; \frac{7!}{2!5!}$$

$$= \frac{7 \cdot 6 \cdot 5 \cdot 4!3!}{2 \cdot 1 \cdot 2 \cdot 1 \; 3!} + \frac{7 \cdot 6 \cdot 5 \cdot 4 \cdot 3!}{3! \; 3 \cdot 2 \cdot 1} + \frac{7 \cdot 6 \cdot 3!}{2 \cdot 1 \cdot 3!}$$

$$= 210 + 140 + 21 = 371$$

In this example we have only to make use of the suitable formulae for combinations, for we are not concerned with the possible arrangements of the members of the committee among themselves.

● **PROBLEM** 1096

Two ordinary dice are rolled. In how many different ways can they fall? How many of these ways will give a sum of nine?

Solution: The first die can fall in any one of six positions, and for each of these positions the second die can also fall in six positions; so there are 6 x 6 = 36 ways that the dice can fall. This is shown in the accompanying construction of possible pairs for the faces of the two dice.

```
1,1   1,2   1,3   1,4   1,5   1,6

2,1   2,2   2,3   2,4   2,5   2,6

3,1   3,2   3,3   3,4   3,5   3,6

4,1   4,2   4,3   4,4   4,5   4,6

5,1   5,2   5,3   5,4   5,5   5,6

6,1   6,2   6,3   6,4   6,5   6,6
```

Of these 36 ways, there are four ways of obtaining a sum
of nine: (6,3), (5,4),(4,5), (3,6) (circled in the figure).

● **PROBLEM** 1097

How many four-letter "words" (we use "word" to mean any
sequence of letters) which begin and end with a vowel may
be formed from the letters a, e, i, p, q.
    (a)   if no repetitions are allowed?
    (b)   if repetitions are allowed?

<u>Solution</u>:   (a) We construct a pattern of four boxes to
represent the four letters of the word to be formed:

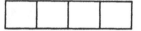

      Since the word is to begin with a vowel, the first
letter may be chosen in 3 ways (since there are three
vowels in the given letters):

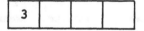

      Since the word must end with a vowel, and repetition of
letters is not allowed, after the first place has been
chosen only 2 letters remain from which to choose the
letter in last place:

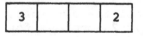

      Having chosen first and last letters, thus using up two
letters of the original five, 3 letters are left from which
to choose the letter in second position:

      Finally, 3 letters having been used, there remain 2
from which to choose the letter in third position:

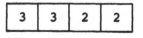

      The Fundamental Principle of Counting states that if one
thing can be done in a different ways and, when it is done in
any one of these ways, a second thing can be done in b different ways, and a third thing then can be done in c ways, ...
then all the things in succession can be done in a×b×c···

different ways. Thus if no repetitions are allowed, then
3×3×2×2 = 36 words can be formed.

(b) Now we consider the case where repetitions are
allowed. Once again we construct a pattern of four boxes to
represent the four letters of the word to be formed:

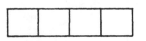

Since the word is to begin with a vowel, the first letter
may be chosen in 3 ways:

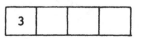

Since the word must end with a vowel, and repetition is
allowed, the last letter may also be chosen in 3 ways:

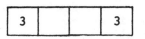

Since there are no specifications regarding the second and
third letters, and repetition is allowed, the second and
third letters may both be chosen in 5 ways. Thus we arrive
at the following array:

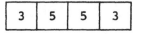

Using the Fundamental Principle of Counting, there are
thus 3×5×5×3 = 225 words which may be formed if repetitions
are allowed.

● **PROBLEM** 1098

If in a series of license plates the letters I and O are
not used and if four successive zeroes can not be used as
the four digits, how many different license plates can be
made?

Solution: A license plate consists of two letters and
four digits. Of the 26 letters of the alphabet, since I
and O are not used, there are now 24 choices for each of
the two letters. Since four successive zeroes can not be
used as the four digits, the digits would form the numbers
from 0001 to 9999; so there are 9999 choices for the four
digits. The fundamental principle states that if one
thing can be done in m different ways and, when it is done
in any one of these ways, a second thing can be done in n
different ways, and if a third thing can then be done in
p ways, ... then the successive things can be done in
mnp... different ways. Therefore,

Number of license plates = (24)(24)(9999)

= 5,759,424

● **PROBLEM** 1099

A baseball manager after determining his starting players
now must determine his batting order. If the pitcher is

to bat last, how many different ways can the manager turn in his batting order?

Solution: Since the pitcher is to bat last only the order of the other eight players must be determined. The fundamental principle states that if one thing can be done in m different ways and, when it is done in any one of these ways, a second thing can be done in n different ways, and if after it has been done in any one of these ways, a third thing can be done in p different ways, . . .,the several things can be done in m·n·p·...different ways. Using the fundamental principle, he has 8 choices for the lead-off man. Once this choice is made, he has 7 choices for the second batter, and so on.

Number of batting orders = 8(7)(6)(5)(4)(3)(2)(1)

= 8!

= 40,320

● **PROBLEM** 1100

Two cards are to be drawn in order from a pack of 4 cards (say, an ace, king, queen, and jack), the drawn card not being replaced before the second card is drawn. How many different drawings are possible?

Solution: The first card can be drawn in four ways, and then the second card in three ways. By the fundamental principle of counting, which states that if the first of two independent acts can be performed in x ways, and if the second act can be performed in y ways, then the number of ways of performing the two acts, in the order stated, is xy, there are 4 · 3 = 12 different drawings. (We regard here ace first, king second as a different drawing from king first, ace second.)

If the first card were to be replaced before the second is drawn in this example, then the answer would be 4 · 4 = 16 different drawings.

CHAPTER 35

PROBABILITY

> **Basic Attacks and Strategies for Solving Problems in this Chapter. See pages 790 to 810 for step-by-step solutions to problems.**

The set S of all possible outcomes of a given experiment is called the sample space for the experiment. Any subset of the sample space S is an event. The probability of an event E, denoted by $P(E)$, is the ratio of the number of outcomes in the sample space S which satisfies event E compared to the total number of outcomes in sample space S. Thus,

$$P(E) = \frac{n(e)}{n(s)} = \frac{\text{number of outcomes satisfying event } E}{\text{number of outcomes in sample space } S}$$

This definition is the procedure for calculating the probability of an event E. For instance, consider the experiment of rolling a single die and finding the probability of the event E = "the number of spots showing on the upper face is greater than three." This is done by first noting that

$n(E) = 3$ since $E = \{4, 5, 6\}$

and $n(S) = 6$ since $S = \{1, 2, 3, 4, 5, 6\}$.

Thus,

$P(E) = n(E)/n(S) = 3/6 = 1/2.$

Calculating probabilities by listing and counting the elements of a sample space, as well as the event, is not always practical. Instead, one should use the counting principles developed in Chapters 33 and 34 to determine the number of elements in the sample space and event, respectively.

Some guidelines for understanding a probability that involves two or more independent events A and B are given below. In a probability statement:

    (1)   The word "or" usually means to add the probabilities of each event, say A and B, if the events are mutually exclusive, that is,

          $P(A \text{ or } B) = P(A) + P(B).$

On the other hand, if the events are not mutually exclusive, then

$P(A \text{ or } B) = P(A) + P(B) - P(A \text{ and } B).$

(2) The word "and" usually means to multiply the probabilities of each event, that is,

$P(A \text{ and } B) = P(A) * P(B).$

(3) The phrase "at least n" means n or more elements of the event.

(4) The phrase "at most n" means n or less elements of the event.

(5) The phrase "exactly n" means just that.

(6) The phrase "not event A" or "A'" means all other elements not in the event A. Note that

$P(A') = 1 - P(A).$

In the case that one event A is conditional to the occurrence of the second event B, denoted by $P(A / B)$, then

$P(A \mid B) = P(A \text{ and } B) / P(B).$

Consider the experiment of drawing a single card from a deck of 52 cards with event A = "spade" and event B = "face card." Then, "A and B" consists of the three face cards that are also spades, and so

$P(A \text{ and } B) = 3/52.$

Since it is clear that $P(B) = 12/52$, then the probability of event A, given event B, is

$P(A \mid B) = P(A \text{ and } B) / P(B)$

$= (3/52) / (12/52) = 3/12 = 1/4.$

Step-by-Step Solutions to Problems in this Chapter, "Probability"

● **PROBLEM** 1101

What is the probability of throwing a "six" with a single die?

<u>Solution:</u> The die may land in any of 6 ways:

    1, 2, 3, 4, 5, 6

The probability of throwing a six,

$$P(6) = \frac{\text{number of ways to get a six}}{\text{number of ways the die may land}}$$

Thus $P(6) = \frac{1}{6}$.

● **PROBLEM** 1102

A deck of playing cards is thoroughly shuffled and a card is drawn from the deck. What is the probability that the card drawn is the ace of diamonds?

<u>Solution:</u> The probaility of an event occurring is

$$\frac{\text{the number of ways the event can occur}}{\text{the number of possible outcomes}}$$

In our case there is one way the event can occur, for there is only one ace of diamonds and there are fifty two possible outcomes (for there are 52 cards in the deck).Hence the probability that the card drawn is the ace of diamonds is 1/52.

● **PROBLEM** 1103

A box contains 7 red, 5 white, and 4 black balls. What is the probability of your drawing at random one red ball? One black ball?

<u>Solution:</u> There are 7 + 5 + 4 = 16 balls in the box. The probability of drawing one red ball,

$$P(R) = \frac{\text{number of possible ways of drawing a red ball}}{\text{number of ways of drawing any ball}}$$

$$P(R) = \frac{7}{16}.$$

Similarly, the probability of drawing one black ball

$$P(B) = \frac{\text{number of possible ways of drawing a black ball}}{\text{number of ways of drawing any ball}}$$

Thus,
$$P(B) = \frac{4}{16} = \frac{1}{4}$$

Calculate the probability of the Dodgers' winning the world series if the Yankees have won the first three games.

Solution:

We assume the probability of each team winning any specific game is 1/2.

The only way the Dodgers can win is if they can take the next four games. Since the games can be considered independent events,

P(DDDD) = 1/2 x 1/2 x 1/2 x 1/2 by the multiplication rule.

P(DDDD)= $(1/2)^4$ = 1/16

In a single throw of a single die, find the probability of obtaining either a 2 or a 5.

Solution: In a single throw, the die may land in any of 6 ways:

1 2 3 4 5 6.

The probability of obtaining a 2,

$$P(2) = \frac{\text{number of ways of obtaining a 2}}{\text{numbers of ways the die may land}}$$

$$P(2) = \frac{1}{6}.$$

Similarly, the probability of obtaining a 5,

$$P(5) = \frac{\text{number of ways of obtaining a 5}}{\text{number of ways the die may land}}$$

$$P(5) = \frac{1}{6}.$$

The probability that either one of two mutually exclusive events will occur is the sum of the probabilities of the separate events. Thus the probability of obtaining either a 2 or a 5, P(2) or P(5), is

$$P(2) + P(5) = \frac{1}{6} + \frac{1}{6} = \frac{2}{6} = \frac{1}{3}$$

If a card is drawn from a deck of playing cards, what is the probability that it will be a jack or a ten?

<u>Solution:</u> The probability that an event A or B occurs, but not both at the same time, is $P(A \cup B) = P(A) + P(B)$. Here the symbol "$\cup$" stands for "or."

In this particular example, we only select one card at a time. Thus, we either choose a jack "or" a ten. $P(\text{a jack or a ten}) = P(\text{a jack}) + P(\text{a ten})$.

$$P(\text{a jack}) = \frac{\text{number of ways to select a jack}}{\text{number of ways to choose a card}} = \frac{4}{52} = \frac{1}{13}.$$

$$P(\text{a ten}) = \frac{\text{number of ways to choose a ten}}{\text{number of ways to choose a card}} = \frac{4}{52} = \frac{1}{13}.$$

$$P(\text{a jack or a ten}) = P(\text{a jack}) + P(\text{a ten}) = \frac{1}{13} + \frac{1}{13} = \frac{2}{13}.$$

Suppose that we have a bag containing two red balls, three white balls, and six blue balls. What is the probability of obtaining a red or a white ball on one withdrawal?

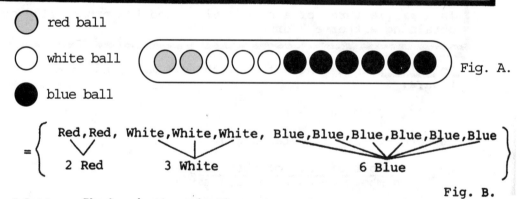

Fig. A.

Fig. B.

<u>Solution:</u> The bag is shown in figure A.

There are two favorable possibilities: drawing a red ball and drawing a white ball. The universal set of equally likely possibilities contains eleven elements; the number of red balls is two, and the number of white balls is three. Refer to figure B.

Hence the probability of drawing a red ball is $\frac{2}{11}$ and the probability of drawing a white ball is $\frac{3}{11}$.

In the set notation, the probability of A, P(A), or the probability of B, P(B), is P(A) + P(B) therefore the probability of drawing a red, <u>or</u> a white ball is the probability of drawing a red ball plus the probability of drawing a white ball

$$P(\text{red}) = \frac{2}{11} \qquad\qquad P(\text{white}) = \frac{3}{11}$$

$$P(\text{red or white}) = \frac{2}{11} + \frac{3}{11} = \frac{5}{11} .$$

Note: In the above example the probability of drawing a blue ball would be $\frac{6}{11}$. Therefore the sum of the probability of a red ball, the probability of a white ball, and the probability of a blue ball is $\frac{2}{11} + \frac{3}{11} + \frac{6}{11} = 1$.

If there are no possible results that are considered favorable, then the probability P(F) is obviously 0. If every result is considered favorable, then P(F) = 1. Hence the probability P(F) of a favorable result F always satisfies the inequality

$$0 \leq P(F) \leq 1.$$

● **PROBLEM** 1108

A bag contains 4 white balls, 6 black balls, 3 red balls, and 8 green balls. If one ball is drawn from the bag, find the probability that it will be either white or green.

Solution: The probability that it will be either white or green is:
P(a white ball or a green ball) = P(a white ball) + P(a green ball).
This is true because if we are given two mutually exclusive events A or B, then P(A or B) = P(A) + P(B). Note that two events, A and B, are mutually exclusive events if their intersection is the null or empty set. In this case the intersection of choosing a white ball and of choosing a green ball is the empty set. There are no elements in common.

P (a white ball) = $\dfrac{\text{number of ways to choose a white ball}}{\text{number of ways to select a ball}}$

$$= \frac{4}{21}$$

P(a green ball) = $\dfrac{\text{number of ways to choose a green ball}}{\text{number of ways to select a ball}}$

$$= \frac{8}{21}$$

Thus,
P(a white ball or a green ball) = $\frac{4}{21} + \frac{8}{21} = \frac{12}{21} = \frac{4}{7}$.

● **PROBLEM** 1109

An urn contains 6 white, 4 black, and 2 red balls. In a single draw, find the probability of drawing: (a) a red ball; (b) a black ball; (c) either a white or a black ball. Assume all outcomes equally likely.

Solution: The urn contains 6 white balls, 4 black balls, and 2 red balls, or a total of 12 balls.

(a) The probability of drawing a red ball,

P (R) = $\dfrac{\text{number of ways of drawing a red ball}}{\text{number of ways of selecting a ball}}$

P (R) = $\frac{2}{12} = \frac{1}{6}$.

(b) The probability of drawing a black ball,

$$P(B) = \frac{\text{number of ways of drawing a black ball}}{\text{number of ways of selecting a ball}}$$

$$P(B) = \frac{4}{12} = \frac{1}{3}.$$

(c) The probability that either one of two mutually exclusive events will occur is the sum of the probabilities of the separate events. Thus the probability of drawing either a white [P(W)] or a black ball [P(B)] is P(W) + P(B).

$$P(W) = \frac{\text{number of ways of drawing a white ball}}{\text{number of ways of selecting a ball}}$$

$$= \frac{6}{12} = \frac{1}{2}.$$

$$P(B) = \frac{1}{3} \text{ [shown in part (b)]}.$$

Thus, $P(W \text{ or } B) = P(W) + P(B) = \dfrac{6}{12} + \dfrac{4}{12}$

$$= \frac{10}{12}$$

$$= \frac{5}{6}.$$

● **PROBLEM** 1110

Determine the probability of getting 6 or 7 in a toss of two dice.

Solution: Let A = the event that a 6 is obtained in a toss of two dice

B = the event that a 7 is obtained in a toss of two dice.

Then, the probability of getting 6 or 7 in a toss of two dice is

$$P(A \text{ or } B) = P(A \cup B).$$

The union symbol "$\cup$" means that A and/or B can occur. Now $P(A \cup B) = P(A) + P(B)$ if A and B are mutually exclusive. Two or more events are said to be mutually exclusive if the occurrence of any one of them excludes the occurrence of the others. In this case, we cannot obtain a six and a seven in a single toss of two dice. Thus, A and B are mutually exclusive.

To calculate P(A) and P(B), use the following table.

Note: There are 36 different tosses of two dice.

A = a 6 is obtained in a toss of two dice

$$= \left\{ (1,5), (2,4), (3,3), (4,2), (5,1) \right\}$$

B = a 7 is obtained in a toss of two dice

= $\left\{ (1,6), (2,5), (3,4), (4,3), (5,2), (6,1) \right\}$.

$P(A) = \dfrac{\text{number of ways to obtain a 6 in a toss of two dice}}{\text{number of ways to toss two dice}}$

$= \dfrac{5}{36}$

$P(B) = \dfrac{\text{number of ways to obtain a 7 in a toss of two dice}}{\text{number of ways to toss two dice}}$

$= \dfrac{6}{36} = \dfrac{1}{6}$.

Therefore, $P(A \cup B) = P(A) + P(B) = \dfrac{5}{36} + \dfrac{6}{36} = \dfrac{11}{36}$.

● **PROBLEM** 1111

A penny is to be tossed 3 times. What is the probability there will be 2 heads and 1 tail?

Solution: We start this problem by constructing a set of all possible outcomes:

We can have heads on all 3 tosses:	(HHH)	
head on first 2 tosses, tail on the third:	(HHT)	(1)
head on first toss, tail on next two:	(HTT)	
●	(HTH)	(2)
●	(THH)	(3)
	(THT)	
●	(TTH)	
	(TTT)	

Hence there are eight possible outcomes (2 possibilities on first toss x 2 on second x 2 on third = 2 x 2 x 2 = 8).

We assume that these outcomes are all equally likely and assign the probability 1/8 to each. Now we look for the set of outcomes that produce 2 heads and 1 tail. We see there are 3 such outcomes out of the 8 possibilities (numbered (1), (2), (3) in our listing). Hence the probability of 2 heads and 1 tail is 3/8.

● **PROBLEM** 1112

Find the probability of throwing two sixes in one toss of a pair of dice.

Solution: To find the probability of throwing two sixes in one toss of a pair of dice, first we express it symbolically.

P(throwing two sixes in one toss of a pair of dice) =

P(throwing a six in one toss of a die) X P(throwing a six in one toss of a die).

This is true because the event of tossing a die is independent of tossing another die. That is, the occurrence of one event has no effect upon the occurrence or non-occurrence of the other. Now,

P(throwing a six in one toss) =

$$\frac{\text{number of ways to obtain a six}}{\text{number of ways to obtain any face value of a die}} = \frac{1}{6}.$$

Hence, the probability of obtaining two sixes is $\left(\frac{1}{6}\right)\left(\frac{1}{6}\right) = \frac{1}{36}$.

● **PROBLEM** 1113

What is the probability of obtaining two aces on two successive throws of a die?

Solution: The throw of a die results in 6 different but equally likely face values. An ace can be obtained only when a certain 1 of the 6 faces shows. Therefore the probability of obtaining an ace in one throw is $\frac{1}{6}$. In a toss of two dice, the fall of either does not affect the fall of the other. Thus the two events, consisting of two sucessive throws, are independent. The probability that all of a set of independent events will happen in a single trial is the product of their separate probabilities. Therefore, the probability is $\frac{1}{6} \cdot \frac{1}{6} = \frac{1}{36}$.

● **PROBLEM** 1114

If a pair of dice is tossed twice, find the probability of obtaining 5 on both tosses.

Solution: We obtain 5 in one toss of the two dice if they fall with either 3 and 2 or 4 and 1 uppermost, and each of these combinations can appear in two ways. The ways to obtain 5 in one toss of the two dice are:

$$(1,4),(4,1),(3,2), \text{ and } (2,3).$$

Hence we can throw 5 in one toss in four ways. Each die has six faces and there are six ways for a die to fall. Then the pair of dice can fall in 6·6 = 36 ways. The probability of throwing 5 in one toss is:

$$\frac{\text{the number of ways to throw a 5 in one toss}}{\text{the number of ways that a pair of dice can fall}} = \frac{4}{36} = \frac{1}{9}.$$

Now the probability of throwing a 5 on both tosses is:

P(throwing five on first toss and throwing five on second toss).

"And" implies multiplication if events are independent, thus p(throwing 5 on first toss and throwing 5 on second toss)

= p(throwing 5 on first toss) X p(throwing 5 on second toss)

Since the results of the two tosses are independent. Consequently, the probability of obtaining 5 on both tosses is

$$\left(\frac{1}{9}\right)\left(\frac{1}{9}\right) = \frac{1}{81}.$$

● **PROBLEM** 1115

A bag contains 4 black and 5 blue marbles. A marble is drawn and then replaced, after which a second marble is drawn. What is the probability that the first is black and second blue?

Solution: Let C = event that the first marble drawn is black.

D = event that the second marble drawn is blue.

The probability that the first is black and the second is blue can be expressed symbolically:

$$P(C \text{ and } D) = P(CD).$$

We can apply the following theorem. If two events A and B, are independent, then the probability that A and B will occur is,

$$P(A \text{ and } B) = P(AB) = P(A) \cdot P(B).$$

Note that two or more events are said to be independent if the occurrence of one event has no effect upon the occurrence or non-occurrence of the other. In this case the occurrence of choosing a black marble has no effect on the selection of a blue marble and vice versa; since, when a marble is drawn it is then replaced before the next marble is drawn. Therefore, C and D are two independent events.

$$P(CD) = P(C) \cdot P(D)$$

$$P(C) = \frac{\text{number of ways to choose a black marble}}{\text{number of ways to choose a marble}}$$

$$= \frac{4}{9}.$$

$$P(D) = \frac{\text{number of ways to choose a blue marble}}{\text{number of ways to choose a marble}}$$

$$= \frac{5}{9}.$$

$$P(CD) = P(C) \cdot P(D) = \frac{4}{9} \cdot \frac{5}{9} = \frac{20}{81}.$$

● **PROBLEM 1116**

A box contains 4 black marbles, 3 red marbles, and 2 white marbles. What is the probability that a black marble, then a red marble, then a white marble is drawn without replacement?

<u>Solution:</u> Here we have three dependent events. There is a total of 9 marbles from which to draw. We assume on the first draw we will get a black marble. Since the probability of drawing a black marble is the

$$\frac{\text{number of ways of drawing a black marble}}{\text{number of ways of drawing 1 out of (4+3+2) marbles}},$$

$$P(A) = \frac{4}{4 + 3 + 2} = \frac{4}{9}$$

There are now 8 marbles left in the box.

On the second draw we get a red marble. Since the probability of drawing a red marble is

$$\frac{\text{number of ways of drawing a red marble}}{\text{number of ways of drawing 1 out of the 8 remaining marbles}}$$

$$P(B) = \frac{3}{8}$$

There are now 7 marbles remaining in the box.

On the last draw we get a white marble. Since the probability of drawing a white marble is

$$\frac{\text{number of ways of drawing a white marble}}{\text{number of ways of drawing 1 out of the 7 remaining marbles}},$$

$$P(C) = \frac{2}{7}$$

When dealing with two or more dependent events, if P_1 is the probability of a first event, P_2 the probability that, after the first has happened, the second will occur, P_3 the probability that, after the first and second have happened, the third will occur, etc., then the probability that all events will happen in the given order is the product $P_1 \cdot P_2 \cdot P_3 \ldots$

Thus, $P(A \cap B \cap C) = P(A) \cdot P(B) \cdot P(C)$

$$= \frac{4}{9} \cdot \frac{3}{8} \cdot \frac{2}{7}$$

$$= \frac{1}{21}.$$

● **PROBLEM 1117**

There is a box containing 5 white balls, 4 black balls, and 7 red balls. If two balls are drawn one at a time from the box and neither is replaced, find the probability that
(a)   both balls will be white.
(b)   the first ball will be white and the second red.
(c)   if a third ball is drawn, find the probability that the three balls will be drawn in the order white, black, red.

Solution: This problem involves dependent events. Two or more events are said to be dependent if the occurrence of one event has an effect upon the occurrence or non-occurrence of the other. If you are drawing objects without replacement, the second draw is dependent on the occurrence of the first draw. We apply the following theorem for this type of problem. If the probability of occurrence of one event is p and the probability of the occurrence of a second event is q, then the probability that both events will happen in the order stated is pq.
(a) To find the probability that both balls will be white, we express it symbolically.
p (both balls will be white) =
p (first ball will be white and the second ball will be white) =
p (first ball will be white) p(second ball will be white) =

$$= \left(\frac{\text{number of ways to choose a white ball}}{\text{number of ways to choose a ball}} \right) \left(\frac{\text{number of ways to choose a second white ball after removal of the first white ball}}{\text{number of ways to choose a ball after removal of the first ball}} \right)$$

$$= \frac{\overset{1}{\cancel{5}}}{\underset{4}{\cancel{16}}} \cdot \frac{\overset{1}{\cancel{4}}}{\underset{3}{\cancel{15}}} = \frac{1}{12}$$

(b) p (first ball will be white and the second red)

= p (first ball will be white) p(the second ball will be red)

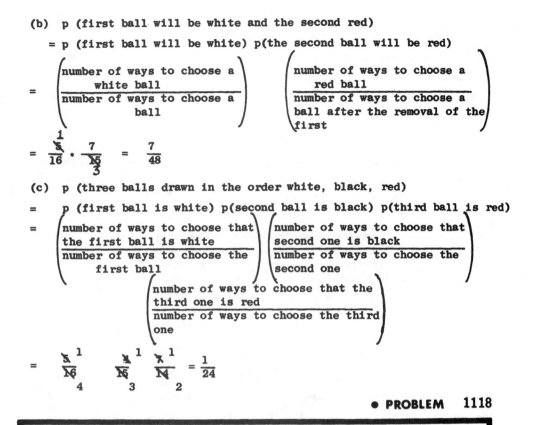

(c) p (three balls drawn in the order white, black, red)

= p (first ball is white) p(second ball is black) p(third ball is red)

● **PROBLEM** 1118

What is the chance of throwing a number greater than 4 with an ordinary die whose faces are numbered from 1 to 6?

Solution: If an event can happen in s ways and fail to happen in f ways, and if all these ways (s + f) are assumed to be equally likely, then the probability (p) that the event will happen is

$$p = \frac{s}{s + f} = \frac{\text{(successful ways)}}{\text{(total ways)}}$$

In our case there are 6 possible ways in which the die can fall (1,2,3,4,5, or 6). Of these, two are favorable to the event required, 5 or 6, therefore the required chance = $\frac{2}{6} = \frac{1}{3}$.

● **PROBLEM** 1119

A bag contains 10 red, 15 green, and 5 yellow beads. If a single bead is drawn from the bag what is the probability (a) that the bead is red, and (b) that the bead is not red?

Solution: (a) If an event can happen in s ways and fail to happen in f ways, and if all these ways (s + f) are assumed to be equally likely, then the probability (p) that the event will happen is

$$p = \frac{s}{s + f} = \frac{\text{(successful ways)}}{\text{(total ways)}}$$

There are 10 + 15 + 5 = 30 beads in the bag, any one of which could be drawn from the bag. Ten beads are red, therefore s = 10. The total of nonred beads is 20, f = 20.

$$p = \frac{s}{s + f}$$

$$= \frac{10}{10 + 20}$$

$$= \frac{1}{3}$$

(b) If the probability that an event will happen is $\frac{a}{b}$ then the probability that this event will not happen is $1 - \frac{a}{b}$. Since the probability that the bead is red is $\frac{1}{3}$, the probability that the bead is not red is $1 - \frac{1}{3} = \frac{2}{3}$.

● **PROBLEM** 1120

A traffic count at a highway junction revealed that out of 5,000 cars that passed through the junction in 1 week, 3,000 turned to the right. Find the probability that a car will turn (a) to the right, and (b) to the left.

<u>Solution:</u> (a) If an event can happen in s ways and fail to happen in f ways, and if all these ways (s + f) are assumed to be equally likely, then the probability (p) that the event will happen is

$$p = \frac{s}{s + f} = \frac{(successful\ ways)}{(total\ ways)}$$

In this case s = 3,000 and s + f = 5,000. Hence, $p = \frac{3,000}{5,000} = \frac{3}{5}$.

(b) If the probability that an event will happen is $\frac{a}{b}$ then the probability that this event will not happen is $1 - \frac{a}{b}$. Thus the probability that a car will not turn right, but left, is $1 - \frac{3}{5} = \frac{2}{5}$. This same conclusion can also be arrived at using the following reasoning: Since 3,000 cars turned to the right, 5,000 - 3,000 = 2,000 cars turned to the left. Hence, the probability that a car will turn to the left is

$$\frac{2,000}{5,000} = \frac{2}{5} .$$

● **PROBLEM** 1121

From 20 tickets marked with the first 20 numerals, one is drawn at random: find the chance that it is a multiple of 3 or of 7.

<u>Solution:</u> If an event can happen in s ways and fail to happen in f ways, and if all these ways (s + f) are assumed to be equally likely, then the probability (p) that the event will happen is

$$p = \frac{s}{s + f} = \frac{(successful\ ways)}{(total\ ways)}$$

There are 6 multiples of 3 in the first 20 numerals (3,6,9,12,15,18), and there are 2 multiples of 7 (7,14). Thus

$$p(multiple\ of\ 3) = \frac{6}{20}$$

and

$$p(multiple\ of\ 7) = \frac{2}{20} .$$

Since the probability that either of two mutually exclusive events will occur is the sum of the probabilities of the separate events, the chance that it is a multiple of 3 or of 7 is

$$\frac{6}{20} + \frac{2}{20} = \frac{8}{20} = \frac{2}{5} .$$

But if the question had been: find the chance that the number is a multiple of 3 or of 5, it would have been incorrect to reason as follows: The chance that the number is a multiple of 3 is $\frac{6}{20}$, and the chance that the number is a multiple of 5 is $\frac{4}{20}$, therefore the chance that it is a multiple of 3 or 5 is $\frac{6}{20} + \frac{4}{20}$, or $\frac{1}{2}$. This is erroneous for the number on the ticket might be a multiple both of 3 and of 5, so that the two events considered are not mutually exclusive.

● **PROBLEM** 1122

What is the probability that the sum 11 will appear in a single throw of 2 dice?

<u>Solution:</u> There are 6 ways the first die may be tossed and 6 ways the second die may be tossed. The Fundamental Principle of Counting states that if the first of two independent acts can be performed in x ways, and if the second act can be performed in y ways, then the number of ways of performing the two acts, in the order stated, is xy. Thus there are 6 X 6 = 36 ways that two dice can be thrown (see accompanying figure).

1,1	1,2	1,3	1,4	1,5	1,6
2,1	2,2	2,3	2,4	2,5	2,6
3,1	3,2	3,3	3,4	3,5	3,6
4,1	4,2	4,3	4,4	4,5	4,6
5,1	5,2	5,3	5,4	5,5	5,6
6,1	6,2	6,3	6,4	6,5	6,6

The number of possible ways that an 11 will appear are circled in the figure. Let us call this set A. Thus,

$$A = \{(5,6),(6,5)\}.$$

The probability that an 11 will appear,

$$p(11) = \frac{\text{number of possible ways of obtaining an 11}}{\text{number of ways that 2 dice can be thrown}}$$

Therefore

$$p(11) = \frac{2}{36} = \frac{1}{18} .$$

● **PROBLEM** 1123

What is the probability of making a 7 in one throw of a pair of dice?

<u>Solution:</u> There are 6 X 6 = 36 ways that two dice can be thrown, as shown in the accompanying figure.

1,1	1,2	1,3	1,4	1,5	1,6
2,1	2,2	2,3	2,4	2,5	2,6
3,1	3,2	3,3	3,4	3,5	3,6
4,1	4,2	4,3	4,4	4,5	4,6
5,1	5,2	5,3	5,4	5,5	5,6
6,1	6,2	6,3	6,4	6,5	6,6

The number of possible ways that a 7 will appear are circled in the figure. Let us call this set B. Thus,

$$B = \{(1,6),(2,5),(3,4),(4,3),(5,2),(6,1)\}.$$

The probability that a 7 will appear,

$$p(7) = \frac{\text{number of possible ways of obtaining a 7}}{\text{number of ways that 2 dice can be thrown}}$$

$$p(7) = \frac{6}{36} = \frac{1}{6} .$$

● **PROBLEM 1124**

If two dice are cast, what is the probability the sum will be less than 5?

Solution: If A, B, and C are mutually exclusive events, that is, their intersection is the null set, then $P(A \cup B \cup C) = P(A) + P(B) + P(C)$. Since the obtaining of sums of 2, 3, and 4 are mutually exclusive events, the probability of obtaining a sum less than 5 is the sum of the probabilities of obtaining a sum of 2,3, and 4. To obtain the sum of 2 with 2 die, we have the following possibilities: (1,1).
Similarly for the sum of 3, we have: (1,2) and (2,1).
For the sum of 4, we obtain: (1,3), (3,1), and (2,2).
Thus P_1 = probability of obtaining a sum of 2

$$= \frac{\text{number of ways to obtain a sum of 2}}{\text{number of ways to throw 2 dice}}$$

$$= \frac{1}{36}$$

P_2 = probability of obtaining a sum of 3

$$= \frac{\text{number of ways to obtain a sum of 3}}{\text{number of ways to throw 2 dice}}$$

$$= \frac{2}{36} = \frac{1}{18}.$$

P_3 = probability of obtaining a sum of 4

$$= \frac{\text{number of ways to obtain a sum of 4}}{\text{number of ways to throw 2 dice}}$$

$$= \frac{3}{36} = \frac{1}{12}.$$

The probability of obtaining a sum less than 5 is

$$P_1 + P_2 + P_3 = \frac{1}{36} + \frac{1}{18} + \frac{1}{12}.$$

$$= \frac{1}{36} + \frac{2}{36} + \frac{3}{36} = \frac{6}{36} = \frac{1}{6}$$

● **PROBLEM 1125**

Find the probability that when a pair of dice are thrown, the sum of the two up faces is greater than 7 or the same number appears on each face.

Solution: The sample space consists of 36 equally likely outcomes as shown in the accompanying figure. Those out-comes that give a sum greater than 7 are

G = {(6,2),(6,3),(6,4),(6,5),(6,6),(5,3),(5,4),(5,5),
(5,6),(4,4),(4,5),(4,6),(3,5),(3,6),(2,6)}

$$S = \begin{array}{cccccc}
1,1 & 1,2 & 1,3 & 1,4 & 1,5 & 1,6 \\
2,1 & 2,2 & 2,3 & 2,4 & 2,5 & 2,6 \\
3,1 & 3,2 & 3,3 & 3,4 & 3,5 & 3,6 \\
4,1 & 4,2 & 4,3 & 4,4 & 4,5 & 4,6 \\
5,1 & 5,2 & 5,3 & 5,4 & 5,5 & 5,6 \\
6,1 & 6,2 & 6,3 & 6,4 & 6,5 & 6,6
\end{array}$$

Then $P(G) =$

$= \dfrac{\text{number of ways the two faces will be greater than 7}}{\text{total number of ways that the dice may fall}}$

$= \dfrac{15}{36}$.

Those outcomes where each die is the same are

$D = \{(1,1),(2,2),(3,3),(4,4),(5,5),(6,6)\}$

Then $P(D) =$

$= \dfrac{\text{number of ways that the same number can appear on each face}}{\text{total number of ways that the dice may fall}}$

$= \dfrac{6}{36}$.

The probability of G or D is $P(G \cup D)$. Recall that

$P(G \cup D) = P(G) + P(D) - P(G \cap D)$.

But $G \cap D = \{(4,4),(5,5),(6,6)\}$.

Hence, $\quad P(G \cap D) = \dfrac{3}{36}$

and $P(G \cup D) = \dfrac{15}{36} + \dfrac{6}{36} - \dfrac{3}{36} = \dfrac{18}{36} = \dfrac{1}{2}$.

● **PROBLEM 1126**

In a single throw of a pair of dice, find the probability of obtaining a total of 4 or less.

Solution: Each die may land in 6 ways. By the Fundamental Principle of Counting the pair of dice may thus land in 6 X 6 = 36 ways:

$$\begin{array}{cccccc}
1,1 & 1,2 & 1,3 & 1,4 & 1,5 & 1,6 \\
2,1 & 2,2 & 2,3 & 2,4 & 2,5 & 2,6 \\
3,1 & 3,2 & 3,3 & 3,4 & 3,5 & 3,6 \\
4,1 & 4,2 & 4,3 & 4,4 & 4,5 & 4,6 \\
5,1 & 5,2 & 5,3 & 5,4 & 5,5 & 5,6 \\
6,1 & 6,2 & 6,3 & 6,4 & 6,5 & 6,6
\end{array}$$

Let us call the possible outcomes which are circled above set A. Then the elements of set A, A = { (1,1), (1,2), (1,3), (2,1), (2,2), (3,1)} are all the possible ways of obtaining four or less.

The probability of obtaining 4 or less,

$$P[(x,y) \leq 4] = \frac{\text{number of ways of obtaining 4 or less}}{\text{number of ways the dice may land}} \quad \left(\begin{array}{l} \text{number of} \\ \text{elements} \\ \text{in set A} \end{array} \right)$$

$$= \frac{6}{36} = \frac{1}{6}.$$

● **PROBLEM** 1127

The probability that A wins a certain game is $\frac{2}{3}$. If A plays 5 games, what is the probability that A will win (a) exactly 3 games? (b) at least 3 games?

Solution: We shall apply the following theorem. If P is the probability that an event will happen in a single trial and q is the probability that this event will fail in this trial, then $_nC_r \, p^r q^{n-r}$ is the probability that this event will happen exactly r times in n trials. $_nC_r$, the number of combinations of n different objects taken r at a time, is

$$_nC_r = \frac{n!}{r!(n-r)!}.$$

Note that p + q = 1.

(a) We are given the probability of a success, p, which is winning a game: $p = \frac{2}{3}$. Therefore from p + q = 1, $q = 1 - p = 1 - \frac{2}{3} = \frac{1}{3}$. The number of ways of winning 3 games out of 5 is

$$_5C_3 = \binom{5}{3} = \frac{5!}{3!2!} = \frac{5 \cdot \cancel{4}^{2} \cdot \cancel{3!}}{\cancel{3!} \, \cancel{2} \cdot 1} = 10.$$

Thus, the probability of A winning 3 games is

$$_nC_r \, p^r q^{n-r} = 5C_3 \left(\frac{2}{3}\right)^3 \left(\frac{1}{3}\right)^2 = 10 \, \frac{2}{3} \cdot \frac{2}{3} \cdot \frac{2}{3} \cdot \frac{1}{3} \cdot \frac{1}{3} = \frac{80}{243}$$

(b) To win at least 3 games A must win either exactly 3 or exactly 4 or all 5 games. In order that A will win at least 3 games, we must calculate the probability that A will win three games, four games, and five games.

$$P = \, _5C_3 \left(\frac{2}{3}\right)^3 \left(\frac{1}{3}\right)^2 + \, _5C_4 \left(\frac{2}{3}\right)^4 \left(\frac{1}{3}\right)^1 + \, _5C_5 \left(\frac{2}{3}\right)^5 \left(\frac{1}{3}\right)^0.$$

$$= \frac{5!}{3!2!} \left(\frac{2}{3}\right)^3 \left(\frac{1}{3}\right)^2 + \frac{5!}{4!1!} \left(\frac{2}{3}\right)^4 \left(\frac{1}{3}\right)^1 + \frac{5!}{5!0!} \left(\frac{2}{3}\right)^5$$

$$= \frac{5 \cdot \cancel{4}^{2} \cdot \cancel{3}}{\cancel{3!} \, \cancel{2} \cdot 1} \frac{8}{243} + \frac{5 \cdot \cancel{4!}}{\cancel{4!} \, 1!} \frac{16}{243} + \frac{32}{243}$$

$$= 10 \cdot \frac{8}{243} + 5 \cdot \frac{16}{243} + \frac{32}{243} = \frac{192}{243} = \frac{64}{81}.$$

Find the chance of throwing at least one ace in a single
throw with two dice.

$$
S =
\begin{matrix}
1,1 & 1,2 & 1,3 & 1,4 & 1,5 & 1,6 \\
2,1 & 2,2 & 2,3 & 2,4 & 2,5 & 2,6 \\
3,1 & 3,2 & 3,3 & 3,4 & 3,5 & 3,6 \\
4,1 & 4,2 & 4,3 & 4,4 & 4,5 & 4,6 \\
5,1 & 5,2 & 5,3 & 5,4 & 5,5 & 5,6 \\
6,1 & 6,2 & 6,3 & 6,4 & 6,5 & 6,6
\end{matrix}
$$

<u>Solution:</u> The chance of throwing at least one ace is the
chance of throwing an ace on either the first die, or the
second die , or both.

The sample space, S, consists of $6 \times 6 = 36$ equally
likely outcomes (see figure). Those outcomes that give an
ace on the first die are shown in the first row of the
figure. Let us call this row A. Thus,

A = {(1,1),(1,2),(1,3),(1,4),(1,5),(1,6)}.

Then $P(A) = \dfrac{\text{number of ways of getting ace on first die}}{\text{total number of ways that the dice may fall}}$

$= \dfrac{6}{36}$

Those outcomes that give an ace on the second die
are shown in the first column of the figure. Let us call
this column B. Thus,

B = {(1,1),(2,1),(3,1),(4,1),(5,1),(6,1)}.

Then $P(B) = \dfrac{\text{number of ways of getting ace on second die}}{\text{total number of ways that the dice may fall}}$

$= \dfrac{6}{36}$

The probability of an ace on the first die or on the
second die is P(A) or P(B), P(A∪B). Recall that

P(A∪B) = P(A) + P(B) - P(A∩B).

Now, A∩B = (1,1);
then P(A∩B)

$= \dfrac{\text{number of ways of getting ace on both dice}}{\text{total number of ways that the dice may fall}} = \dfrac{1}{36}$

Therefore, $P(A∪B) = \dfrac{6}{36} + \dfrac{6}{36} - \dfrac{1}{36} = \dfrac{11}{36}$.

Thus, the chance of throwing at least one ace in a single
throw of 2 dice is 11/36.

A die is tossed five times. What is the probability that an ace will appear: (a) at least twice; (b) at least once?

Solution: This is a problem involving repeated trials of an experiment. The experiment is "tossing a die five times". Apply the following theorem: If p is the probability that an event will happen in a single trial and q is the probability that this event will fail in this trial, then

$$_nC_r p^r q^{n-r}$$

is the probability that this event will happen exactly r times in n trials.

(a) To find the probability that an ace will occur at least twice, find the probability that it will occur twice, or three times, or four times, or five times. The sum (the word "or" implies addition in set notation) of these probabilities will be the probability that an ace will happen at least twice. p = probability that an ace will occur in a given trial

$$= \frac{\text{number of ways to obtain an ace}}{\text{number of ways to obtain any face of a die}}$$

$$= \frac{1}{6}$$

An experiment can only succeed or fail, hence the probability of success, p, plus the probability of failure, q, is one; p+q = 1. Then q = 1-p = 1 - 1/6 = 5/6. Therefore, using $_nC_r p^r q^{n-r}$, p (at least two aces) =

$$_5C_2 (1/6)^2 (5/6)^3 + {}_5C_3 (1/6)^3 (5/6)^2$$

$$+ {}_5C_4 (1/6)^4 (5/6)^1 + {}_5C_5 (1/6)^5 (5/6)^0$$

$_nC_r$ is a symbol for a combination of n things, r at a time, where r objects are chosen from n objects.

$$_nC_r = \frac{n!}{r! \, (n-r)!}$$

Apply this formula. Then,

$$_5C_2 (1/6)^2 (5/6)^3 + {}_5C_3 (1/6)^3 (5/6)^2$$

$$+ {}_5C_4 (1/6)^4 (5/6)^1 + {}_5C_5 (1/6)^5 (5/6)^0$$

$$= \frac{5!}{2!3!} \left(\frac{125}{6^5} \right) + \frac{5!}{2!3!} \left(\frac{25}{6^5} \right) + \frac{5!}{4!1!} \left(\frac{5}{6^5} \right) + \frac{5!}{5!0!} \left(\frac{1}{6^5} \right)$$

$$= \frac{5 \cdot 4 \cdot 3!}{2 \cdot 1 \cdot 3!} \left(\frac{125}{6^5} \right) + \frac{5 \cdot 4 \cdot 3!}{2 \cdot 1 \cdot 3!} \left(\frac{25}{6^5} \right) + \frac{5 \cdot 4!}{4! 1!} \left(\frac{5}{6^5} \right) + \frac{1}{6^5}$$

$$= 10 \left(\frac{125}{6^5} \right) + 10 \left(\frac{25}{6^5} \right) + 5 \left(\frac{5}{6^5} \right) + \frac{1}{6^5}$$

$$= \frac{1250 + 250 + 25 + 1}{6^5} = \frac{1526}{7776} = \frac{763}{3888}$$

Therefore, the probability that an ace will appear at least twice is

$$\frac{763}{3888}.$$

(b) An ace can be obtained at least once by tossing one ace, 2 aces, 3 aces,..., or 5 aces. Hence, the probability of obtaining at least one ace is the sum of the individual probabilities of obtaining one, two, three,..., up to five aces. Apply the same method as in part (a).

$$p(\text{at least one ace}) = {}_5C_1(1/6)^1(5/6)^4 + {}_5C_2(1/6)^2(5/6)^3$$

$$+ {}_5C_3(1/6)^3(5/6)^2 + {}_5C_4(1/6)^4(5/6)^1$$

$$+ {}_5C_5(1/6)^5(5/6)^0$$

$$= \frac{5!}{1!4!}\left(\frac{625}{6^5}\right) + \frac{5!}{2!3!}\left(\frac{125}{6^5}\right) + \frac{5!}{3!2!}\left(\frac{25}{6^5}\right) + \frac{5!}{4!1!}\left(\frac{5}{6^5}\right)$$

$$+ \frac{5!}{5!0!}\left(\frac{1}{6^5}\right) = \frac{5 \cdot 4!}{1 \cdot 4!}\left(\frac{625}{6^5}\right) + \frac{5 \cdot 4 \cdot 3!}{2 \cdot 1 \cdot 3!}\left(\frac{125}{6^5}\right) + \frac{5 \cdot 4 \cdot 3!}{3 \cdot 2!}\left(\frac{25}{6^5}\right)$$

$$+ \frac{5 \cdot 4!}{4! \cdot 1}\left(\frac{5}{6^5}\right) + \left(\frac{1}{6^5}\right) = 5\left(\frac{625}{6^5}\right) + 10\left(\frac{125}{6^5}\right) + 10\left(\frac{25}{6^5}\right)$$

$$+ 5\left(\frac{5}{6^5}\right) + \frac{1}{6^5}$$

$$= \frac{3125 + 1250 + 250 + 25 + 1}{6^5} = \frac{4651}{7776}$$

An alternate, shorter method, is to calculate the probability of failure, (obtaining no aces) and subtract this from one. This is true because $q+p = 1$, hence $q = 1-p$.

$$p \text{ (at least one ace)} = 1 - p \text{ (no aces)}$$
$$p \text{ (no aces)} = {}_5C_0(1/6)^0(5/6)^5 = \frac{5!}{0!5!}\left(\frac{5^5}{6^5}\right)$$

$$= \frac{3125}{7776}$$

Thus,

$$p \text{ (at least one ace)} = 1 - p \text{ (no aces)}$$
$$= 1 - \frac{3125}{7776} = \frac{4651}{7776}$$

Therefore, the probability that an ace appears at least once is

$$\frac{4651}{7776}.$$

● **PROBLEM** 1130

A coin is tossed 3 times, and 2 heads and 1 tail fall. What is the probability that the first toss was heads?

<u>Solution:</u> This problem is one of conditional probability. Given two events, p_1 and p_2, the probability that event p_2 will occur on the condition that we have event p_1 is

$$P\left(p_2/p_1\right) = \frac{P\left(p_1 \text{ and } p_2\right)}{P\left(p_1\right)} = \frac{P\left(p_1p_2\right)}{P\left(p_1\right)}$$

Define

p_1: 2 heads and 1 tail fall,

p_2: the first toss is heads.

$$P\left(p_1\right) = \frac{\text{number of ways to obtain 2 heads and 1 tail}}{\text{number of possibilities resulting from 3 tosses}}$$

$$= \left(\{H,H,T\},\{H,T,H\},\{T,H,H\}\right) / \left(\{H,H,H\},\{H,H,T\},\{H,T,T\},\{H,T,H\},\right.$$
$$\left.\{T,T,H\},\{T,H,T\},\{T,H,H\},\{T,T,T\}\right)$$

$= 3/8$

$P\left(p_1 p_2\right) = P(2 \text{ heads and 1 tail and the first toss is heads})$

$= \dfrac{\text{number of ways to obtain } p_1 \text{ and } p_2}{\text{number of possibilities resulting from 3 tosses}}$

$= \dfrac{\left(\{H,H,T\}, \{H,T,H\}\right)}{8} = 2/8 = 1/4$

$P\left(p_2/p_1\right) = \dfrac{P\left(p_1 p_2\right)}{P\left(p_1\right)} = \dfrac{1/4}{3/8} = 2/3$

A coin is tossed 3 times. Find the probability that all 3 are heads,
   (a) if it is known that the first is heads,
   (b) if it is known that the first 2 are heads,
   (c) if it is known that 2 of them are heads.

Solution: This problem is one of conditional probability. If we have two events, A and B, the probability of event A given that event B has occurred is

$$P(A/B) = \frac{P(AB)}{P(B)}.$$

(a) We are asked to find the probability that all three tosses are heads given that the first toss is heads. The first event is A and the second is B.

$P(AB)$ = probability that all three tosses are heads given that the first toss is heads

$= \dfrac{\text{the number of ways that all three tosses are heads given that the first toss is a head}}{\text{the number of possibilities resulting from 3 tosses}}$

$= \dfrac{\{H, \ HH\}}{\{\{H,H,H\}, \ \{H,H,T\}, \ \{H,T,H\}, \ \{H,T,T\}, \ \{T,T,T\}, \ \{T,T,H\}, \ \{T,H,T\}, \ \{T,H,H\}}$

$= \dfrac{1}{8}.$

$P(B) = P(\text{first toss is a head})$

$= \dfrac{\text{the number of ways to obtain a head on the first toss}}{\text{the number of ways to obtain a head or a tail on the first of 3 tosses}}$

$= \dfrac{\{H,H,H\}, \ \{H,H,T\}, \ \{H,T,H\}, \ \{H,T,T\}}{8}$

$= \dfrac{4}{8}$

$= \dfrac{1}{2}.$

$$P(A/B) = \frac{P(AB)}{P(B)} = \frac{\frac{1}{8}}{\frac{1}{2}} = \frac{1}{8} \cdot \frac{2}{1} = \frac{1}{4}.$$

To see what happens, in detail, we note that if the first toss is heads, the logical possibilities are HHH, HHT, HTH, HTT. There is only one of these for which the second and third are heads. Hence,

$$P(A/B) = \frac{1}{4}.$$

(b) The problem here is to find the probability that all 3 tosses are heads given that the first two tosses are heads.

P(AB) = the probability that all three tosees are heads given that the first two are heads

= the number of ways to obtain 3 heads given that the first two tosses are heads / the number of possibilities resulting from 3 tosses

$$= \frac{1}{8}.$$

P(B) = the probability that the first two are heads

= number of ways to obtain heads on the first two tosses / number of possibilities resulting from three tosses

$$= \frac{\{H,H,H\}, \; \{H,H,T\}}{8} = \frac{2}{8} = \frac{1}{4}.$$

$$P(A/B) = \frac{P(AB)}{P(B)} = \frac{\frac{1}{8}}{\frac{1}{4}} = \frac{4}{8} = \frac{1}{2}.$$

(c) In this last part, we are asked to find the probability that all 3 are heads on the condition that any 2 of them are heads.
Define:

A = the event that all three are heads

B = the event that two of them are heads

P(AB) = the probability that all three tosses are heads knowing that two of them are heads

$$= \frac{1}{8}.$$

P(B) = the probability that two tosses are heads

= number of ways to obtain at least two heads out of three tosses / number of possibilities resulting from 3 tosses

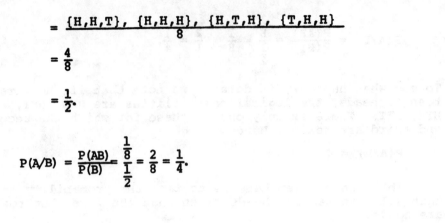

$$= \frac{\{H,H,T\}, \{H,H,H\}, \{H,T,H\}, \{T,H,H\}}{8}$$

$$= \frac{4}{8}$$

$$= \frac{1}{2}.$$

$$P(A/B) = \frac{P(AB)}{P(B)} = \frac{\frac{1}{8}}{\frac{1}{2}} = \frac{2}{8} = \frac{1}{4}.$$

CHAPTER 36

SERIES

> **Basic Attacks and Strategies for Solving Problems in this Chapter. See pages 811 to 817 for step-by-step solutions to problems.**

A series is associated with any sequence. It is the indicated sum of the terms of a finite or infinite sequence. If the sequence is finite and is an arithmetic progression, then the finite series is given by

$$S_n = (n/2) [2a + (n - 1)d],$$

where a is the first term, n is the last term, and d is the common difference. If the sequence is finite and is a geometric progression, then the finite series is given by

$$S_n = (a - ar^n) / (1 - r),$$

where $r \neq 1$, a is the first term, n is the last term, and r is the common ratio.

A series for which the general term is known can be represented by the sigma or summation symbol. For example,

$$S_n = 4 + 7 + 10 + ... + (3n + 1)$$

can be written as

$$S_n = \sum_{j=1}^{n} (3j + 1).$$

For an infinite series, the question of whether it converges or diverges is pertinent. For an infinite series,

$$S_\infty = \sum_{j=1}^{\infty} s_j$$

converges if and only if

$$S_1, S_2, ..., S_n, ...,$$

the corresponding sequence of partial sums, converges. If the sequences of partial sums converge to the number L

$$(\lim_{n \to \infty} S_n = L)$$

then L is said to be the sum of the infinite series and thus

$$S_\infty = \sum_{j=1}^{\infty} s_j = L.$$

For example,

$$\sum_{j=1}^{\infty} [1/j - 1/(j+1)]$$

converges to

$$L = 1 - 1/(n + 1)$$

because the nth partial sum of this series is given by

$$S_n = (1/1 - 1/2) + (1/2 - 1/3) + \ldots + [1/(n - 1) - 1/n] + [1/n - 1/(n + 1)]$$
$$= 1 - 1/(n + 1).$$

An infinite series that does not converge is said to diverge (or, if the sequence of partial sums diverges, then the series is said to diverge). For example, the series

$$\sum_{j=1}^{\infty} a^j$$

diverges because the nth partial sum of this series,

$$S_n = 1 + 2^1 + 2^2 + \ldots + 2^n = (1 - 2^{n+1}) / (1 - 2) = 2^{n+1} - 1$$

does not yield a real number value L as n approaches ∞.

811 – B

Step-by-Step Solutions to Problems in this Chapter, "Series"

● **PROBLEM** 1132

Find the numerical value of the following:

a) $\displaystyle\sum_{j=1}^{7} (2j + 1)$

b) $\displaystyle\sum_{j=1}^{21} (3j - 2)$.

Solution: If $A(r)$ is some mathematical expression and n is a positive integer, then the symbol $\displaystyle\sum_{r=0}^{n} A(r)$ means "Successively replace the letter r in the expression $A(r)$ with the numbers $0,1,2,\ldots,n$ and add up the terms. The symbol Σ is the Greek letter sigma and is a shorthand way to denote "the sum". It avoids having to write the sum $A(0) + A(1) + A(2) + \ldots + A(n)$.

a) For a) successively replace j by $1,\ldots,7$ and add up the terms.

$$\sum_{j=1}^{7} (2j+1) = \Big(2(1)+1\Big) + \Big(2(2)+1\Big) + \Big(2(3)+1\Big) + \Big(2(4)+1\Big) + \Big(2(5)+1\Big)$$
$$+ \Big(2(6)+1\Big) + \Big(2(7)+1\Big)$$
$$= (2+1) + (4+1) + (6+1) + (8+1) + (10+1) + (12+1) + (14+1)$$
$$= 3 + 5 + 7 + 9 + 11 + 13 + 15$$
$$= 63 .$$

b) For b) successively replace j by $1,2,3,\ldots,21$ and add up the terms.

$$\sum_{j=1}^{21} (3j-2) = \Big(3(1)-2\Big) + \Big(3(2)-2\Big) + \Big(3(3)-2\Big) + \Big(3(4)-2\Big) + \Big(3(5)-2\Big)$$
$$+ \Big(3(6)-2\Big) + \Big(3(7)-2\Big) + \Big(3(8)-2\Big) + \Big(3(9)-2\Big)$$
$$+ \Big(3(10)-2\Big) + \Big(3(11)-2\Big) + \Big(3(12)-2\Big) + \Big(3(13)-2\Big)$$
$$+ \Big(3(14)-2\Big) + \Big(3(15)-2\Big) + \Big(3(16)-2\Big) + \Big(3(17)-2\Big)$$
$$+ \Big(3(18)-2\Big) + \Big(3(19)-2\Big) + \Big(3(20)-2\Big) + \Big(3(21)-2\Big)$$
$$= (3-2) + (6-2) + (9-2) + (12-2) + (15-2) + (18-2) + (21-2)$$
$$+ (24-2) + (27-2) + (30-2) + (33-2) + (36-2)$$
$$+ (39-2) + (42-2) + (45-2) + (48-2) + (51-2)$$
$$+ (54-2) + (57-2) + (60-2) + (63-2)$$
$$= 1 + 4 + 7 + 10 + 13 + 16 + 19 + 22 + 25 + 28 + 31 + 34$$
$$+ 37 + 40 + 43 + 46 + 49 + 52 + 55 + 58 + 61$$
$$= 651 .$$

● **PROBLEM** 1133

Determine the general term of the sequence:

$$\frac{1}{2}, \ \frac{1}{12}, \ \frac{1}{30}, \ \frac{1}{56}, \ \frac{1}{90}, \ \ldots$$

Solution: To determine the general term, it is necessary to find how the adjacent terms differ. In this example, it is sufficient to consider the denominator because the numerator is the same for all the terms. The difference between the first two terms is 10. For the second and third terms, the difference is 18. By continuing this process, the results are tabulated as:

$$10, 18, 26, 34, \ldots$$

Note that each difference is larger by 8 **than for the preceding term.**

Now we try to write an expression that generates the series. By inspection, each term is the product of 2 successive integers, for example:

$$\frac{1}{2} = \frac{1}{1} \cdot \frac{1}{2}, \quad \frac{1}{12} = \frac{1}{3} \cdot \frac{1}{4},$$

$$\frac{1}{30} = \frac{1}{5} \cdot \frac{1}{6}, \quad \frac{1}{56} = \frac{1}{7} \cdot \frac{1}{8}$$

This fact can be expressed as

$$\frac{1}{(2n - 1)(2n)} \quad ,$$

and this is the desired answer.

● **PROBLEM   1134**

Determine the general term of the sequence:

$$\frac{1}{5^3}, \frac{3}{5^5}, \frac{5}{5^7}, \frac{7}{5^9}, \frac{9}{5^{11}}, \ldots$$

Solution: The numerators of the terms in the series are consecutive odd numbers beginning with 1. An odd number can be represented by $2n - 1$.

In the denominators, the base is always 5, and the power is a consecutive odd integer beginning with 3.

The general term can therefore be expressed by

$$\frac{2n - 1}{5^{2n+1}} \quad ,$$

and the series is generated by replacing n with n = 1, 2, 3, 4, ...

Establish the convergence or divergence of the series:

$$\frac{1}{1 + \sqrt{1}} + \frac{1}{1 + \sqrt{2}} + \frac{1}{1 + \sqrt{3}} + \frac{1}{1 + \sqrt{4}} + \ldots .$$

Solution: To establish the convergence or divergence of the given series we first determine the nth term of the series. By studying the law of formation of the terms of the series we find the nth term to be

$\frac{1}{1 + \sqrt{n}}$. To determine whether this series is convergent or divergent we use the comparison test. We choose $\frac{1}{n}$, which is a known divergent series since it is a p-series, $\frac{1}{n^p}$, with p = 1. If we can show $\frac{1}{1 + \sqrt{n}} > \frac{1}{n}$, then $\frac{1}{1 + \sqrt{n}}$ is divergent. But we can see this is true, since $1 + \sqrt{n} < n$ for n > 1. Therefore the given series is divergent.

Establish the convergence or divergence of the series:

$$\sin \frac{\pi}{2} + \frac{1}{4} \sin \frac{\pi}{4} + \frac{1}{9} \sin \frac{\pi}{6} + \frac{1}{16} \sin \frac{\pi}{8} + \ldots .$$

Solution: To establish the convergence or divergence of the given series, we first determine the nth term of the series. By studying the law of formation of the terms of the series, we find the nth term to be: $\frac{1}{n^2} \sin \frac{\pi}{2n}$. To determine whether this series is convergent or divergent, we use the comparison test. We choose $\frac{1}{n^2}$, which is a known convergent series, since it is a p-series, $\frac{1}{n^p}$, with p = 2. If we can show $\frac{1}{n^2} \sin \frac{\pi}{2n} < \frac{1}{n^2}$, then

$\frac{1}{n^2} \sin \frac{\pi}{2n}$ is convergent. But we can see this is true since $\sin \frac{\pi}{2n}$ is less than 1 for n > 1. Therefore, the given series is convergent.

Test the series:

$$1 + \frac{2!}{2^2} + \frac{3!}{3^3} + \frac{4!}{4^4} + \ldots$$

by means of the ratio test. If this test fails, use another test.

Solution: To make use of the ratio test, we find the nth term of the given series, and the $(n+1)$th term. If we let the first term, $1 = u_1$, then $\frac{2!}{2^2} = u_2$,

$\frac{3!}{3^3} = u_3$, etc., up to $u_n + u_{n+1}$. We examine the terms of the series to find the law of formation, from which we conclude:

$$u_n = \frac{n!}{n^n} \quad \text{and,} \quad u_{n+1} = \frac{(n+1)!}{(n+1)^{n+1}}.$$

Forming the ratio $\dfrac{u_{n+1}}{u_n}$ we obtain:

$$\frac{(n+1)!}{(n+1)^{n+1}} \times \frac{n^n}{n!}$$

$$\frac{(n+1)(n!)}{(n+1)^n(n+1)} \times \frac{n^n}{n!} = \frac{n^n}{(n+1)^n}.$$

Now, we find $\lim\limits_{n\to\infty}\left|\dfrac{n^n}{(n+1)^n}\right|$. This can be rewritten as:

$$\lim_{n\to\infty} \frac{n^n}{\left[n\left(1+\frac{1}{n}\right)\right]^n} = \lim_{n\to\infty} \frac{n^n}{n^n \cdot \left(1+\frac{1}{n}\right)^n} = \lim_{n\to\infty} \frac{1}{\left(1+\frac{1}{n}\right)^n}.$$

We now use the definition:

$$e = \lim_{x\to 0}(1 + x)^{\frac{1}{x}}.$$

If we let $x = \frac{1}{n}$ in this definition, we have:

$$\lim_{\frac{1}{n}\to 0}\left(1 + \frac{1}{n}\right)^{\frac{1}{\frac{1}{n}}} = \lim_{n\to\infty}\left(1 + \frac{1}{n}\right)^{n}, \text{ which is what we have}$$

above. Therefore, $\lim\limits_{n\to\infty} \dfrac{1}{\left(1 + \frac{1}{n}\right)^n} = \dfrac{1}{e}$. Since $e \approx 2.7$,

$\frac{1}{e} \approx \frac{1}{2.7}$ which is less than 1.

Hence, by the ratio test, the given series in convergent.

● **PROBLEM 1138**

Test the series:

$$1 - \frac{3^2}{2^2} + \frac{3^4}{2^2 \cdot 4^2} - \frac{3^6}{2^2 \cdot 4^2 \cdot 6^2} + \cdots$$

by means of the ratio test. If this test fails, use another test.

Solution: To make use of the ratio test, we find the nth term of the given series, and the $(n+1)$th term. If we let the first term, $1 = u_1$, then

$$\frac{3^2}{2^2} = u_2, \quad \frac{3^4}{2^2 \cdot 4^2} = u_3, \text{ etc. up to } u_n \pm u_{n+1}.$$ We

examine the terms of the series to find the law of formation, from which we conclude:

$$u_n = \frac{3^{2n-2}}{2^2 \cdot 4^2 \cdot \ldots (2n-2)^2},$$

and

$$u_{n+1} = \frac{3^{2(n+1)-2}}{2^2 \cdot 4^2 \cdot \ldots (2n-2)^2 [2(n+1)-2]^2}$$

$$= \frac{3^{2n}}{2^2 \cdot 4^2 \cdot \ldots (2n-2)^2 (2n)^2}.$$

Forming the ratio $\dfrac{u_{n+1}}{u_n}$, we obtain:

$$\frac{3^{2n}}{2^2 \cdot 4^2 \cdot \ldots (2n-2)^2 (2n)^2} \times \frac{2^2 \cdot 4^2 \cdot \ldots (2n-2)^2}{3^{2n-2}}$$

$$= \frac{3^{2n}}{(2n)^2 \times 3^{2n-2}} = \frac{3^{2n-(2n-2)}}{(2n)^2} = \frac{3^2}{4n^2}.$$

Now, we find:

$$\lim_{n \to \infty} \left| \frac{3^2}{4n^2} \right| = 0.$$

Since $\lim\limits_{n \to \infty} \left| \dfrac{u_{n+1}}{u_n} \right| = 0$ and $0 < 1$, the given series converges.

APPLIED PROBLEMS

CHAPTER 37

DECIMAL/FRACTIONAL CONVERSIONS SCIENTIFIC NOTATION

Basic Attacks and Strategies for Solving Problems in this Chapter. See pages 818 to 821 for step-by-step solutions to problems.

The properties of exponents provide a compact method of writing and computing very large or very small numbers. The method is called scientific notation. The procedure for writing a number greater than 10 in scientific notation is to move the decimal point to the position to the right of the first nonzero digit and multiply the results by a power of 10 equal to the number of places the decimal point was moved.

For example,

$437000 = 4.37 \times 10^5.$

For numbers less than 1, move the decimal point to the right of the first nonzero digit and multiply the results by a power of 10 equal to the negative of the number of places the decimal point was moved. For example,

$.000546 = 5.46 \times 10^{-4}.$

If the expression is a decimal fraction conversion to scientific notation, then the numerator and denominator are first converted separately to scientific notation. Next, we divide out the fraction part of the result and write the remaining result in scientific notation. Then, divide out the powers of 10 part of the result using rules of exponents. Finally, multiply the power of 10 factors using rules of exponents to get the quotient in scientific notation.

In the event we wish to

(1) convert a fraction to a decimal that results in a repeating decimal or

(2) convert a repeating decimal to a fraction,

the following procedures should be used.

In the first case, divide the denominator into the numerator and continue dividing until a second repetition occurs. Then, the result is given by placing a bar over the first repetition and deleting all other places in the decimal. For example,

$$\frac{2}{7} = 2.857142857142... = .\overline{285714}.$$

In the second case, suppose we wish to write 0.4242... as a common fraction. The procedure is as follows:

First, let x denote the repeating decimal, that is,

$$x = 0.4242..., \tag{1}$$

and then multiply throughout the equation by 100 to obtain

$$100x = 42.42... \tag{2}$$

and subtract equation (1) from equation (2) to obtain

$$100x = 42.42... \tag{2}$$

$$x = .42... \tag{1}$$

$$99x = 42.0 \tag{3}$$

Then, divide both sides of the equation (3) by 99 to obtain

$$\frac{99x}{99} = \frac{42}{99}$$

$$x = \frac{42}{99} = \frac{14}{33},$$

the common fraction.

Note that it may be necessary to multiply both sides of the first equation by a power of 10 that is equivalent to the number of digits before a repetition begins in order to find the common fraction.

● **PROBLEM** 1139

Use scientific notation to express each number.
(a) 4,375   (b) 186,000   (c) 0.00012   (d) 4,005

Solution: A number expressed in scientific notation is written as a product of a number between 1 and 10 and a power of 10. The number between 1 and 10 is obtained by moving the decimal point of the number (actual or implied) the required number of digits. The power of 10, for a number greater than 1, is positive and is one less than the number of digits before the decimal point in the original number. The power of 10, for a number less than 1, is negative and is one more than the number of zeros immediately following the decimal point in the original number. Hence,

(a) $4,375 = 4.375 \times 10^3$ (b) $186,000 = 1.86 \times 10^5$

(c) $0.00012 = 1.2 \times 10^{-4}$ (d) $4,005 = 4.005 \times 10^3$

● **PROBLEM** 1140

Express $\dfrac{6,400,000}{400}$ in scientific notation.

Solution: In order to solve this problem, we express the numerator and denominator as the product of a number between 1 and 10 and a power of 10. This is known as scientific notation. Thus

$$6,400,000 = 6.4 \times 1,000,000 = 6.4 \times 10^6$$
$$400 = 4 \times 100 = 4 \times 10^2$$

Thus,

$$\frac{6,400,000}{400} = \frac{6.4 \times 10^6}{4.0 \times 10^2}$$

Since $\dfrac{ab}{cd} = \dfrac{a}{c} \cdot \dfrac{b}{c}$: $= \dfrac{6.4}{4.0} \times \dfrac{10^6}{10^2}$

Since $\dfrac{a^x}{a^y} = a^{x-y}$: $= 1.6 \times 10^4$

● **PROBLEM** 1141

Write $\dfrac{2}{7}$ as a repeating decimal.

Solution: To write a fraction as a repeating decimal divide the numerator by the denominator, until a pattern of repeated digits appears.

$$2 \div 7 = .285714285714...$$

Identify the entire portion of the decimal which is repeated. The repeating decimal can then be written in the shortened form:

$$\frac{2}{7} = .\overline{285714}$$

● PROBLEM   1142

Find the common fraction form of the repeating decimal 0.4242....

Solution: Let x represent the repeating decimal.

$$x = 0.4242...$$

$$100x = 42.42...   \text{ by multiplying by 100}$$

$$\underline{x = 0.42...}$$

$$99x = 42   \text{ (1) by subtracting x from 100x}$$

Divide both sides of equation (1) by 99.

$$\frac{99x}{99} = \frac{42}{99}$$

$$x = \frac{42}{99} = \frac{14}{33}$$

The repeating decimal of this example had 2 digits that repeated. The first step in the solution was to multiply both sides of the original equation by the 2nd power of 10 or 10^2 or 100. If there were 3 digits that repeated, the first step in the solution would be to multiply both sides of the original equation by the 3rd power of 10 or 10^3 or 1000.

● PROBLEM   1143

Find $0.25\overline{25}$ as a quotient of integers.

Solution: Let $x = 0.25\overline{25}$.(1) Multiply both sides of this equation by 100:

$$100x = 100(0.25\overline{25})$$

Multiplying by 100 is equivalent to moving the decimal two places to the right, and since the digits 25 are repeated we have:

$$100x = 25.25\overline{25}   \text{ (2)}$$

Now subtract equation (1) from equation (2):

$$100x = 25.25\overline{25}$$
$$\underline{- \; x = \; 0.25\overline{25}}$$
$$99x = 25.0000$$

or   99x = 25   (3)

(Note that this operation eliminates the decimal)
Dividing both sides of equation (3) by 99:

$$\frac{99x}{99} = \frac{25}{99}$$

$$x = \frac{25}{99}$$

Therefore,

$$0.25\overline{25} = x = \frac{25}{99}$$

Also, note that the given repeating decimal, $0.25\overline{25}$, was multiplied by 100 or 10^2, where the power of 10 (which is 2) is the same as the number of repeating digits (namely, 2) for this problem. In general, for problems of this type, if the repeating decimal has n repeating digits, then the repeating decimal should be multiplied by 10^n.

● **PROBLEM** 1144

Write the repeating decimal $14.\overline{23}$ as a quotient of two integers, $\frac{p}{q}$.

<u>Solution:</u> Let $x = 14.\overline{23}$. (1) Multiply both sides of this equation by 100:

$$100x = 100(14.\overline{23})$$

$$100x = 1423.\overline{23} \tag{2}$$

Subtract equation (1) from equation (2):

$$100x = 1423.\overline{23}$$

$$-\ \underline{x =\ \ \ \ 14.\overline{23}}$$
$$99x = 1409.00$$

or   99x = 1409   (3)

(Note that this operation eliminates the decimal.) Dividing both sides of equation (3) by 99:

$$\frac{99x}{99} = \frac{1409}{99}$$

$$x = \frac{1409}{99}.$$

Therefore,

$$14.\overline{23} = x = \frac{1409}{99}.$$

Also, note that the given repeating decimal, $14.\overline{23}$, was multiplied by 100 or 10^2, where the power of 10 (which is 2) is the same as the number of repeating digits (namely, 2) for this problem. In general, for problems of this type, if the repeating decimal has n repeating digits, then the repeating decimal should be multiplied by 10^n.

CHAPTER 38

AREAS AND PERIMETERS

> **Basic Attacks and Strategies for Solving Problems in this Chapter. See pages 822 to 833 for step-by-step solutions to problems.**

The area of a figure simply means finding the number of square units it contains. For a particular figure the use of a formula that expresses the procedure for the area is usually helpful in finding the area. Formulas for areas of some well-known figures are as follows:

Figure	Formula
Square	$A = s^2$
Rectangle	$A = lw$
Parallelogram	$A = bh$
Rhombus	$A = (1/2)\, dd'$
Trapezoid	$A = (1/2)\, h(b + b')$
Triangle	$A = (1/2)\, bh$
Circle	$A = \pi r^2$

To determine the perimeter of a figure means to find the linear distance around the figure. If the figure has straight line sides, then the perimeter is the sum of the lengths of its sides. Formulas for perimeters of some well-known figures with straight line sides are as follows:

Figure	Formula
Rectangle	$P = 2l + 2w$
Triangle	$P = a + b + c$
Square	$P = 4s$
Parallelogram	$P = 2l + 2w$

If the figure is a circle, then the circumference, a concept related to perimeter, is used together with diameter to find the distance around the circle. In particular,

$$C = \pi d$$

is the formula for the circumference of a circle.

Step-by-Step Solutions to Problems in this Chapter, "Areas and Perimeters"

● **PROBLEM** 1145

Find the area of a rectangle with base of length 12 and diagonal of length 13.

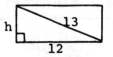

Solution: First find h, the length of the altitude of the rectangle.

In a right triangle, the square of the hypotenuse equals the sum of the squares of the legs.

Hence, $(\text{leg})^2 + (\text{leg})^2 = (\text{hypotenuse})^2$

$$h^2 + 12^2 = 13^2$$
$$h^2 + 144 = 169$$
$$h^2 = 25, \ h = 5$$

Since the area of a rectangle is the product of the base and altitude, the area of the rectangle = 12(5) = 60.

● **PROBLEM** 1146

One side of a rectangle is twice the length of the other side, and the perimeter is 36. Find the area of the rectangle.

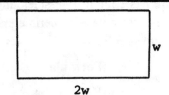

Solution: Since one side of the rectangle is twice the other side, w and 2w can be used to represent the width and the length of the rectangle.

Since the perimeter = 36,

$$2w + 2w + w + w = 36$$
$$6w = 36, \ w = 6$$

Hence, the width = 6 and the length = 12.

Since the area of a rectangle is the product of the length and width, the area = 12(6) = 72.

The area of a square is 9. The square and an equilateral triangle have equal perimeters. Find the length of a side of the triangle.

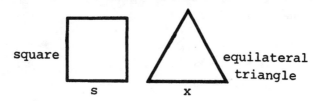

square   equilateral
                triangle
      s   x

<u>Solution:</u>   The area of a square is the square of a side; hence, if s is the side of the square,

$$s^2 = 9$$

Take the square root of each side.

$$s = 3$$

The perimeter of the square is 4 times a side. Hence, the perimeter of the square = 4(3) = 12.

The perimeter of the equilateral triangle is 3 times a side, that is, 3x.

Since the perimeters are equal, 3x = 12 and x = 4. Hence the length of a side of the triangle is 4.

The diagonals of a rhombus are represented by n and n + 3. Express the area of the rhombus in terms of n.

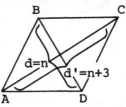

<u>Solution:</u> The area of a rhombus equals one-half the product of the diagonals.

If K = area of rhombus,

K = ½dd'

Substitute n for d and (n+3) for d'.

$$K = \tfrac{1}{2}n(n + 3) \quad \text{or} \quad \tfrac{1}{2}(n^2 + 3n) \quad \text{or} \quad \frac{n^2 + 3n}{2}.$$

The area of a rhombus is equal to the area of a square whose side is 6. If the length of one diagonal of the rhombus is 8, how long is the other diagonal?

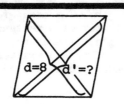

s=6

Since the side, s, of the square is 6, the area of the square $= s^2 = 36$. Therefore,

$$\text{area of rhombus} = 36.$$

The area of a rhombus equals one-half the product of its diagonals. Hence, $\frac{1}{2}dd' = 36$.

Substitute 8 for d.

$$\frac{1}{2}(8)d' = 36$$
$$4d' = 36$$
$$d' = 9$$

● **PROBLEM   1150**

A square parcel of land has twice the area of a rectangular parcel whose length is 9 feet less and whose width is 40 feet less. What are the dimensions of the square parcel of land?

<u>Solution:</u> Let x = the side of the square parcel of land. Then x - 9 is the length of the rectangular parcel, and x - 40 is the width of the rectangular parcel. Since the area of the square parcel is x^2, and the area of the rectangular parcel is $(x - 9)(x - 40)$, the equation relating the two areas is then

$$x^2 = 2(x - 9)(x - 40) \qquad\qquad (1)$$

$$x^2 = 2x^2 - 98x + 720 \qquad\qquad (2)$$

or $\quad x^2 - 98x + 720 = 0 \qquad\qquad (3)$

In factored form, we have

$$(x - 90)(x - 8) = 0 \qquad\qquad (4)$$

So that x = 90 and x = 8.

From the statement of the problem x = 8 must be rejected as the side of the square parcel, because the dimensions of the rectangular parcel are 9 feet and 40 feet less than the side of the square parcel. That is, if x = 8 were accepted as the side of the square, then the rectangular parcel would have negative dimensions, which is impossible.

● **PROBLEM   1151**

Find the dimensions of a rectangular piece of metal whose area is 35 square inches and whose perimeter is 24 inches.

<u>Solution:</u> Let x = width and y = length. The area of a rectangle is equal to width times length. The perimeter of a rectangle is equal to 2 times the width plus 2 times the length. Therefore, the area may be expressed as xy = 35, while the perimeter can be written as 2x + 2y = 24. Since area and perimeter are simultaneous properties of a rectangle we have the system:

$$xy = 35$$
$$2x + 2y = 24$$

The computation can be arranged in the following manner: From $xy = 35$ we obtain
$$y = \frac{35}{x} .$$

Using this expression for y, and substituting, $2x + 2y = 24$ becomes,
$$2x + 2\left(\frac{35}{x}\right) = 24$$

Multiplying both sides of the equation by x: (this is allowed since $x \neq 0$ because width can't = 0)
$$2x^2 + 70 = 24x$$

Put in standard quadratic form and solve for x by factoring:
$$2x^2 - 24x + 70 = 0$$
$$x^2 - 12x + 35 = 0$$
$$(x - 5)(x - 7) = 0$$
$$x - 5 = 0 \quad \text{or} \quad x - 7 = 0$$
$$x = 5 \quad \text{or} \quad x = 7$$

Solve for y using the area equation, $xy = 35$:

when $x = 5$, $5(y) = 35$; thus, $y = 7$

when $x = 7$, $7y = 35$; thus, $y = 5$. Then check using the perimeter equation:
$$x = 5, y = 7 : \quad 2(5) + 2(7) = 24$$
$$x = 7, y = 5 : \quad 2(7) + 2(5) = 24$$

Hence the solution set is $\{(5,7),(7,5)\}$. Thus the dimensions of the piece of metal are 5 by 7 inches.

● **PROBLEM** 1152

The bases of a trapezoid are 10 inches and 20 inches. If the area of the trapezoid is 60 square inches, find the number of inches in the length of the altitude of the trapezoid.

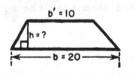

Solution: Find altitude h as follows:

The area of a trapezoid equals one-half the product of the altitude and the sum of the bases.

If K = area of the trapezoid,
$$K = \frac{h}{2}(b + b')$$

Substitute 60 for K, 20 for b, and 10 for b'.
$$60 = \frac{h}{2}(20 + 10) = \frac{h}{2}(30)$$
$$60 = 15h, \quad h = 4; \text{ thus the altitude is 4 inches.}$$

● **PROBLEM** 1153

In parallelogram ABCD, AB = 8 inches, AD = 6 inches, and angle $A = 30°$. Find the number of square inches in the area of the parallelogram.

Solution: In right $\triangle ADE$, h is opposite $30°$. Since a leg opposite $30°$

825

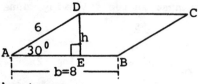

equals one-half the hypotenuse,

$$h = \frac{6}{2} = 3$$

The area of a parallelogram equals the product of the base and the altitude.

Area of $\square$ABCD = bh

Substitute 8 for b and 3 for h.

Area of $\square$ABCD = 8(3) = 24

● **PROBLEM** 1154

The area of a circle is 49π. Find the circumference of the circle in terms of π.

Area = 49π

<u>Solution:</u> First find the radius, r, as follows:

The area of a circle equals the product of π and the square of the radius.

If K = the area of the circle, then

$$K = \pi r^2$$

Substitute 49π for K.

$$49\pi = \pi r^2$$

Divide each side by π and then take the square root of each side.

$$7 = r$$

Find C, the circumference, as follows:

The circumference of a circle equals the product of 2π and the radius.

Hence, C = 2πr

$$= 2\pi(7) = 14\pi .$$

● **PROBLEM** 1155

The circumference of a circle is 8π. What is the area of the circle in terms of π ?

C = 8π

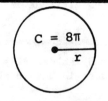

Solution: First find the radius, r, as follows:

The circumference of a circle equals the product of 2π and the radius.

If C = the circumference of the circle, then
$$2\pi r = C$$

Substitute 8π for C.
$$2\pi r = 8\pi$$

Divide each side by 2π.
$$r = 4$$

Now, since area = πr^2, area = $\pi 4^2 = 16\pi$.

● **PROBLEM 1156**

Corresponding sides of two similar polygons are 6 and 8. If the perimeter of the smaller is 27, find the perimeter of the larger.

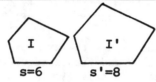

s=6   s'=8

Solution: Find p', the perimeter of the larger polygon, by applying the following principle:

In similar polygons, the ratio of the perimeters equals the ratio of any two corresponding sides.

Hence, $\dfrac{\text{perimeter of I'}}{\text{perimeter of I}} = \dfrac{s'}{s}$.

Substitute 27 for the perimeter of I, 8 for s', and 6 for s.

$$\frac{p'}{27} = \frac{8}{6}$$

Multiply each side by 27.

$$p' = 27\left(\frac{8}{6}\right)$$

$$= 36 .$$

● **PROBLEM 1157**

In the accompanying figure, the large rectangle has been divided into a square and three smaller rectangles. If the areas of the square and two of the rectangles are k^2, 4k, and 8k, respectively, what is the numerical value of the area of the shaded rectangle?

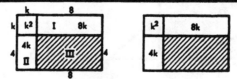

Solution: Since the area of the square is k^2, the length of each side is k.

The area of rectangle I is 8k, and its altitude is k, a side of the square. Hence, its base is 8.

The area of rectangle II is 4k and its base is k, a side of the square. Hence, its altitude is 4.

Since the base and altitude of rectangle III are 8 and 4, its area is the product of 8 and 4, or 32.

● **PROBLEM** 1158

Given quadrilateral ABCD with vertices at A(-3,0), B(9,0), C(9,9), and D(0,12). Find the area of quadrilateral ABCD.

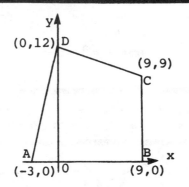

Solution: As shown in the figure the area of quadrilateral ABCD is the sum of the areas of △AOD and trapezoid OBCD.

Recall that the area of a triangle is ½bh, where b = the base and h = the altitude. In the figure note that the base of △AOD is 3 units and h = 12. Thus,

$$\text{area of triangle} = \tfrac{1}{2}bh$$

$$= \tfrac{1}{2}(3)(12) = 18$$

Now recall that the area of a trapezoid is $\frac{h}{2}(b + b')$, where h = altitude and b, b' = bases. In the figure note that the height of trapezoid OBCD = 9 units, and b and b' = 12 and 9. Thus,

$$\text{area of trapezoid} = \frac{h}{2}(b + b')$$

$$= \frac{9}{2}(12 + 9) = \frac{9}{2}(21) = 94\tfrac{1}{2}$$

Area of quadrilateral ABCD = area of triangle + area of trapezoid

= 18 + 94½ = 112½.

● **PROBLEM** 1159

Side AB of △ABC is 5 inches and side AC is 6 inches. If the number of degrees in angle A varies, what is the largest possible area, in square inches, of △ABC?

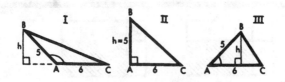

Solution: If the size of ∠A varies, it may be an obtuse angle, a right angle, or an acute angle. Note the three triangles in the diagram. In △I, ∠A is an obtuse angle; in △II, ∠A is a right angle;

828

and in ΔIII, ∠A is an acute angle. In each case, ∠A is included between sides of 5 and 6 inches.

Determine the maximum area by applying the following principle:

The area of a triangle equals one-half the product of a side and the altitude drawn to that side.

The largest altitude that can be drawn to side AC occurs when ∠A is a right angle, as in ΔII. In ΔI and ΔIII, note that the altitude, h, is less than 5.

Hence, as ∠A varies, the largest possible area is obtained when ∠A is a right angle. The area of ΔII = ½(6)(5) = 15.

● **PROBLEM** 1160

A rectangle 4 in. by 8 in. is completely bordered by a strip x in. wide. If the perimeter of the larger rectangle is twice that of the smaller rectangle, what is the value of x?

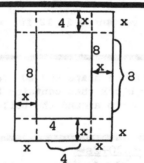

Solution: Observe the accompanying figure. The perimeter of a rectangle is the sum of the lengths of its sides. Thus, the perimeter of the larger rectangle is

(4+x+x) + (8+x+x) + (4+x+x) + (8+x+x) =24 + 8x,

and the perimeter of the smaller or inner rectangle is 8 + 4 + 8 + 4 = 24. We are told that the perimeter of the larger rectangle is twice that of the smaller, thus

$$24 + 8x = 2 \cdot 24$$

$$24 + 8x = 48$$

$$8x = 24$$

$$x = 3 \text{ in.}$$

● **PROBLEM** 1161

A rectangle is twice as long as it is wide. If it is bordered by a strip 2 ft. wide, its area is increased by 160 sq. ft. What are its dimensions?

Solution: Observe the accompanying figure.

Let x = the width of the inner rectangle
Let 2x = the length of the inner rectangle.

The area of a rectangle is its length multiplied by its width. Thus, the area of the inner rectangle is $2x \cdot x = 2x^2$.

The length of the outer rectangle is 2 + 2x + 2 = 2x + 4 and the width of the outer rectangle is 2 + x + 2 = x + 4.

Thus, the area of the outer rectangle is (2x+4)(x+4). We are told

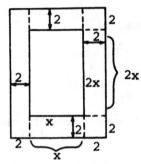

that the outer rectangle has an area of 160 sq. ft. greater than the inner, thus

$$(x+4)(2x+4) = 2x^2 + 160$$
$$2x^2 + 12x + 16 = 2x^2 + 160$$
$$12x + 16 = 160$$
$$12x = 144$$
$$x = 12, \quad 2x = 24.$$

Thus, the width of the inner rectangle is 12 ft. and the length is 24 ft.

● **PROBLEM** 1162

A supermarket, rectangular in shape and 200 feet by 300 feet, is to be built on a city block that contains 81,600 square feet. There will be a uniform strip around the building for parking. How wide is the strip?

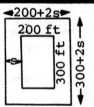

Solution: If the strip is s feet wide, the dimensions of the supermarket will be 200 + 2s by 300 + 2s (see figure). Its area, the product of the width and length, is (200 + 2s)(300 + 2s) square feet. But the area is given as 81,600 square feet. Thus, we have

$$(200 + 2s)(300 + 2s) = 81,600$$
$$60,000 + 1000s + 4s^2 = 81,600$$
$$4s^2 + 1000s - 21,600 = 0 \quad \text{Standard Quadratic Form}$$
$$s^2 + 250s - 5400 = 0 \quad \text{Dividing by 4}$$

Using the quadratic formula

$$s = \frac{-b \pm \sqrt{b^2 - 4ac}}{2a} \quad \text{with } a = 1, \ b = 250, \quad \text{and}$$

c = -5400, we have:

$$s = \frac{-250 \pm \sqrt{250^2 + 21,600}}{2}$$
$$s = \frac{-250 \pm 290}{2}$$

830

s = 20   or   s = -270

The strip is 20 feet wide, since it is impossible for a strip to be a negative width.

Check: If the strip is 20 feet wide, then the block is 340 by 240 feet, and its area must be (340)(240) = 81,600 square feet.

● **PROBLEM** 1163

A rectangle has its length 2 feet greater than its width. If the length is increased by 3 feet and the width by one foot, the area of the new rectangle will be twice the area of the old. What is the length and width of the original rectangle?

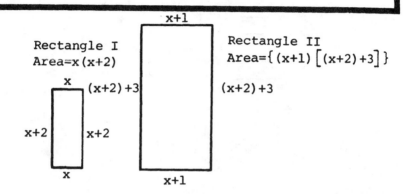

Rectangle I
Area=x(x+2)

Rectangle II
Area=$\{(x+1)[(x+2)+3]\}$

$(x+2)+3$

__Solution:__ We designate the original rectangle by Rectangle I, and the new rectangle by Rectangle II. Let x = the number of feet in the width of Rectangle I.

If x represents the number of feet in the width of the original rectangle, the number of feet in the length is x + 2. (This is obtained from the first and last sentence.) From the second sentence we gain the basic equation

(area of new rectangle) = 2(area of original rectangle)   (1)

Since the area of a rectangle is the product of the width and the length the area of the new rectangle is (x + 1)[(x + 2) + 3] = (x + 1)(x + 5), and the area of the old rectangle is x(x + 2). (See Figures). Substituting these values into equation (1), we have

$$(x + 1)(x + 5) = 2[x(x + 2)].$$

$$x^2 + 6x + 5 = 2(x^2 + 2x)$$

$$x^2 + 6x + 5 = 2x^2 + 4x$$

$$6x + 5 = x^2 + 4x$$

$$5 = x^2 - 2x$$

$$x^2 - 2x - 5 = 0.$$

To find the roots of an equation in the form $ax^2 + bx + c = 0$ we use the quadratic formula

$$x = \frac{-b \pm \sqrt{b^2 - 4ac}}{2a}.$$ In our case a = 1, b = -2, c = -5. Re-

placing these values in the quadratic formula we have

$$x = \frac{-(-2) \pm \sqrt{(-2)^2 - 4(1)(-5)}}{2}$$

$$= \frac{2 \pm \sqrt{4 + 20}}{2}$$

$$= \frac{2 \pm \sqrt{24}}{2}$$

$$= \frac{2 \pm \sqrt{4 \cdot 6}}{2}$$

$$= \frac{2 \pm \sqrt{4}\,\sqrt{6}}{2}$$

$$= \frac{2 \pm 2\sqrt{6}}{2}$$

$$= \frac{2(1 \pm \sqrt{6})}{2}$$

$$= 1 \pm \sqrt{6}$$

Since $\sqrt{6} \approx 2.45$

x = 1 + 2.45 + 3.45 and x = 1 - 2.45 = -1.45.

We reject the negative value, as there are no rectangles with negative sides. Thus, the width of the original rectangle, x, is 3.45 ft., and the length, x + 2, is 5.45 ft.

Check: The area of rectangle I is x(x + 2). Replacing x by 3.45 we have

x(x + 2) = 3.45(3.45 + 2) = (3.45)(5.45) = 18.80.

The area of rectangle II is (x + 1)[(x + 2) + 3]. Re-placing x by 3.45 we have

(x + 1)[(x + 2) + 3] = (3.45 + 1)(3.45 + 5) = (4.45)(8.45)

$$= 37.60$$

$$37.60 = 2(18.80).$$

Thus area rectangle II = 2(area rectangle I), and 3.45 is the correct value of x.

● **PROBLEM** 1164

The corners of a 2-inch square are cut off so as to form a regular octagon. What is the length of the piece to be cut off?

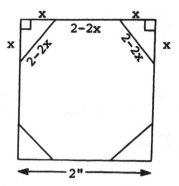

Solution: In the figure we let x equal the length
to be cut off. From the Figure it is seen that the side
of the octagon is equal to the side of the square minus
the lengths of two of the legs of the triangles that are
cut off. Therefore, the side of the octagon is equal to
2 - x - x or 2 - 2x. Since the small triangles formed at
each corner are right triangles, we have, from the
Pythagorean theorem,

$$x^2 + x^2 = (2 - 2x)^2 \qquad\qquad (1)$$

Equation 1 reduces to

$$x^2 - 4x + 2 = 0 \qquad\qquad (2)$$

Using the quadratic formula $x = \dfrac{-b \pm \sqrt{b^2 - 4ac}}{2a}$

with a = 1, b = - 4, c = 2, the roots of Equation 2 are

$$x = 2 - \sqrt{2}, \quad \text{and} \quad x = 2 + \sqrt{2}.$$

We must reject the second root, $2 + \sqrt{2}$, since the
part cut off cannot be greater than the entire side.
Therefore, the size of the piece cut off is $2 - \sqrt{2}$, or
approximately 0.586 inches.

CHAPTER 39

ANGLES OF ELEVATION, DEPRESSION, AND AZIMUTH

> **Basic Attacks and Strategies for Solving Problems in this Chapter. See pages 834 to 837 for step-by-step solutions to problems.**

The solutions of right triangles are frequently involved in problems arising in building construction, navigation, engineering, and science. For some of these application problems, there is a need to know some terminology used in line-of-sight type problems. In particular, angles of elevation and depression are measured with reference to a horizontal and a line-of-sight. If the object being sighted is above the observer, then the angle formed by the line-of-sight and the horizontal line is called an angle of elevation. If the object being sighted is below the observer, then the angle formed by the line-of-sight and the horizontal line is called an angle of depression.

To solve for the angle of elevation, or depression, construct a figure, set up an equation involving the appropriate trigonometric function of the angle being solved for or the unknown side of the right triangle.

When the angle of depression is given and an unknown angle and side of the right triangle are being solved for, it is usually necessary to first apply the alternate interior angles property to find the unknown angle. The next step is to

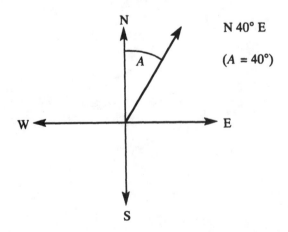

determine the equation involving the appropriate trigonometric function in order to find the unknown side of the right triangle.

In certain navigation problems, the direction or bearing (azimuth) of a ship may be given by stating the size of an acute angle that is measured with respect to a north-south line. For example, a bearing (azimuth) of N40°E denotes an angle whose side points north and whose other side points 40 degrees east of north, as indicated by the figure above.

In order to find the bearing (azimuth) in a problem, first construct an appropriate figure, and then set up and solve an appropriate trigonometric equation.

● PROBLEM   1165

From a point 338 feet from the base of a monument, and in a horizontal plane, the angle of elevation to the top is 27°30'. Find the height of the monument.

Solution: In the figure, side a represents the height of the monument. Side b represents the horizontal distance (given as 338 feet) from the base of the monument. Angle BAC represents the angle of elevation, 27°30'. Since we are given an adjacent side of the angle, side b, and are asked to find an opposite side, side a, we use the Tangent Function. Thus,

$$\tan A = \frac{a}{b}; \quad \tan 27°30' = \frac{a}{338}$$

Multiply both sides by 338. Then,

$$a = 338 \tan 27°30'$$

Looking at our trig table we find

$$\text{Tan } 27°30' = 0.5206$$

therefore, $a = 338(0.5206) = 176$.

The monument is 176 feet tall.   ● PROBLEM   1166

From a point 5 ft. above the horizontal ground, and 30 ft. from the trunk of a tree, the line of sight to the top of the tree is measured as 52° with the horizontal. Find the height of the tree.

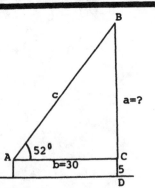

Solution: See the accompanying figure. A is taken at the observer's eye, B is the top of the tree, and the required height is BD = a + 5 ft. One relation connecting the known angle A = 52°, the known distance b = 30 ft., and the unknown length a is

$$\tan A = \frac{\text{opposite side}}{\text{adjacent side}} = \frac{a}{b}$$

We therefore have, with the aid of trigonometric tables,.

$$a = b \tan A = 30 \tan 52°.$$

We find in a table of trigonometric functions that tan 52° is 1.280. Then

$$a = 30 \times 1.280 = 38.4 \text{ ft.};$$

and the height of the tree is

$$BD = a + 5 = 43.4 \text{ ft.}$$

This result may be checked by first finding c from the relation cos A =

$$\cos A = \frac{\text{adjacent side}}{\text{hypotenuse}} = \frac{b}{c}$$

and then getting a from

$$\sin A = \frac{\text{opposite side}}{\text{hypotenuse}} = \frac{a}{c}$$

$$\cos A = \frac{b}{c}$$

$$\cos 52° = \frac{30}{c}$$

$$c = \frac{30}{\cos 52°} = \frac{30}{.6157} \approx 48.7$$

$$\sin A = \frac{a}{c}$$

$$\sin 52° = \frac{a}{48.7}$$

$$a = 48.7 (\sin 52°) = 48.7(.7880)$$

$$a \approx 38.4$$

● **PROBLEM** 1167

From the cockpit of an airplane flying over level ground, the pilot sights the runway of a small airport, and the angle of depression of the runway is 35°41'. If the plane is 14,000 ft above the surface, what is the distance to a point 14,000 ft directly above the runway? (Neglect the curvature of the surface of the earth.

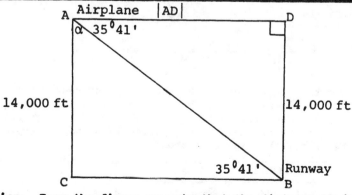

Solution: From the figure we note that the distance AD is required. Triangle ADB is a right triangle; and since AD

is parallel to CB, DB = 14,000, and we have a right triangle with one side and one acute angle given. The solution of the entire triangle is not needed- only one side.
Thus, the side |AD| can be found by computing the tan of
the angle opposite the given side.

$$\tan = \frac{\text{side opposite the angle}}{\text{side adjacent the angle}} \ .$$

In this case the unknown is the adjacent side.

$$\tan 35°41' = \frac{|DB|}{|AD|}$$

$$|AD| = \frac{|DB|}{\tan 35°41'} = \frac{14,000'}{0.71813} = 19,495'$$

● **PROBLEM** 1168

If the angle of elevation of the sun is 49°27' find the height
of a flagpole whose horizontal shadow is 83.59 feet.

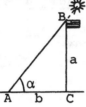

Solution: Construct the figure shown. BC represents the flagpole;
AC, the shadow. The angle of elevation of the sun is at A; we represent it by α. We let AC = b and solve for BC = a.

In △ABC: b = 83.59; α = 49°27'.		
$\frac{a}{b}$ = tan α, or	log b = 1.92215	
	log tan α = 0.06773	
a = b tan α	log a = 1.98988	a = 97.70 ft.

● **PROBLEM** 1169

From the top of a lighthouse 212 feet above a lake, the keeper
spots a boat sailing directly towards him. He observes the angle
of depression of the boat to be 6°13' and then later to be 13°7'.
Find the distance the boat has sailed between the observations.

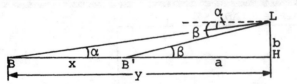

Solution: Construct the figure shown. B and B' represent the two
positions of the boat. LH represents the lighthouse; angle α, the
first angle of depression and β, the second. The line BH is parallel
to the horizontal and therefore the angle at B equals α and the angle
at B' equals β. We letter the figure as shown.

In △B'LH: b = 212; β = 13°7'.		
$\frac{a}{b}$ = cot β, or	log b = 2.32634	
	log cot β = 0.63262	
a = b cot β.	log a = 2.95896	a = 909.8

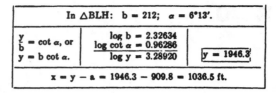
● **PROBLEM   1170**

A ship is 67 mi west and 40 mi north of a port. What is the distance and bearing of the ship from the port? (See Figure).

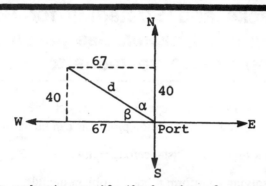

Solution: In order to specify the bearing of some traveling vehicle, the angle of the direction in which the vehicle is traveling must be discovered. To do this we superimpose our direction upon the coordinate axes by extending a line from the origin to the point (x,y) where x is the distance traveled east or west, and y is the distance traveled north or south. The distance in the direction of travel is determined using the Pythagorean theorem: $d^2 = x^2 + y^2$. The angle that the direction makes with the east/west line is determined using the form

$$\tan \beta = \frac{y}{x} = \frac{\text{side opposite } \beta}{\text{side adjacent } \beta} \ .$$

The angle that the direction makes with a north/south line is determined using the form

$$\cot \beta = \frac{\text{side adjacent } \beta}{\text{side opposite } \beta} = \frac{x}{y}$$

or

$$\tan \alpha = \frac{\text{side opposite } \alpha}{\text{side adjacent } \alpha} = \frac{x}{y}$$

$$\tan \alpha = \frac{67}{40} = 1.6750 = \tan 59°9.7'$$

Hence

$$\alpha = 59°9.7'$$

and the bearing (azimuth) of the ship is N59°9.7'W.

$$d = \sqrt{(67)^2 + (40)^2} = \sqrt{4489 + 1600} = \sqrt{6089}$$

$$d = 78.032$$

CHAPTER 40

MOTION

> ## Basic Attacks and Strategies for Solving Problems in this Chapter. See pages 838 to 847 for step-by-step solutions to problems.

Uniform motion means that the speed of an object does not change. The basic equation used to solve uniform motion problems can be written as

Distance = rate × time   or   Distance/rate = time.

The strategy for solving a uniform motion problem includes:

(1) For each object in the problem, write a numerical or variable expression for the distance, rate, and time. The results can be recorded in a table.

(2) Finally, determine how the distances traveled by each object are related.

For example, the total distance traveled by both objects may be known or it may be known that the objects traveled the same distance.

The solution of motion problems which involve an object moving with or against a wind or current normally require two variables. The strategy is to choose one variable to represent the rate of the object in calm conditions and a second variable to represent the rate of the wind or current. Using these variables, express the rate of the object with and against the wind or current. Use the equation

$d = rt$

to write expressions for the distance traveled by the object. Finally, determine how the expressions for distance are related in equations and solve for the unknowns. For example, we solve the following problem which states:

Flying with the wind, a small plane can fly 600 miles in 3 hours. Against the wind, the plane can fly the same distance in 4 hours. Find the rate of the plane in calm air and the rate of the wind.

Let

p = rate of the plane in calm air

w = rate of the wind.

In table format, the following relations can be made using the basic formula

$d = rt$:

	rate	×	time	=	distance
With the wind	$p + w$		3		$3(p + w)$
Against the wind	$p - w$		4		$4(p - w)$

Since the distance traveled is the same, the following system of equations can be written:

$3(p + w) = 600$

$4(p - w) = 600$

The solution of this system is $p = 175$ and $w = 25$. So, the rate (speed) of the plane in calm air is 175 miles per hour and the rate of the wind is 25 miles per hour.

● PROBLEM 1171

● **PROBLEM** 1171

Mike can throw a football exactly 36 yards and Danny can throw a football exactly 30 yards. If they start to throw their football from the same spot, what is the minimum number of throws for each of them so that the footballs will have been thrown the same distance? What is this distance?

Throws:	1st.	2nd	3rd	4th	5th	6th
Mike	36	72	108	144	180	
Danny	30	60	90	120	150	180

Solution: Each time Mike throws the football, it travels a multiple of 36 yards and Danny's football travels a multiple of 30 yards. The figure illustrates what happens.

Multiples of 36 yards are the following:

$$(1)(36) = 36$$
$$(2)(36) = 72$$
$$(3)(36) = 108$$
$$(4)(36) = 144$$
$$(5)(36) = 180$$

Multiples of 30 yards are the following:

$$(1)(30) = 30$$
$$(2)(30) = 60$$
$$(3)(30) = 90$$
$$(4)(30) = 120$$
$$(5)(30) = 150$$
$$(6)(30) = 180$$

From the information in the figure, Mike must throw the football a minimum number of 5 times and Danny must throw it a minimum number of 6 times so that the footballs will be thrown the same distance, this same distance being 180 yards.

● **PROBLEM** 1172

Two cars traveled the same distance. One car traveled at 50 mph and the other car traveled at 60 mph. It took the slower car 50 minutes longer to make the trip. How long did it take the faster car to make the trip?

Solution: Step. 1. Read problem again. Step 2. If we let t represent the number of hours the faster car travels, we can construct the following table from the given statements.

Note: The 50 minutes must be converted to $\frac{5}{6}$ hr.

This is done by the following:

$$\frac{50 \text{ minutes}}{60 \text{ minutes}} \times 1 \text{ hour} = \frac{50}{60} \text{ hours} = \frac{5}{6} \text{ hour}.$$

Step 3

	Distance	Rate	Time
Faster car	D	60 mph	t
Slower car	D	50 mph	t + 5/6

Note: Since the two cars travelled the same distance, the distance for both cars is D, as indicated in the above table.

Step 4

$$\text{Formula } D = rt,$$

where D is distance, r is rate, and t is time.

$$D = 60t$$
$$D = 50 \left(t + \frac{5}{6} \right)$$

Since the distances are the same, we can set the two expressions for D equal as in Step 5.

Step 5
$$60t = 50 \left(t + \frac{5}{6} \right) \qquad (1)$$

Multiply each term within the parentheses by 50 to eliminate the parentheses.

$$60t = 50t + \frac{250}{6} \qquad (2)$$

Subtract 50t from both sides of equation (2).

$$60t - 50t = 50t + \frac{250}{6} - 50t$$

Therefore: $60t - 50t = \frac{250}{6}.$

Therefore: $10t = \frac{250}{6} \qquad (3),$

Multiply both sides of equation (3) by $\frac{1}{10}$.

$$\frac{1}{10} (10t) = \frac{1}{10} \left(\frac{\overset{25}{\cancel{250}}}{6} \right)$$

$$t = \frac{25}{6} = 4 \frac{1}{6} \text{ hours} = 4 \text{ hours } 10 \text{ minutes}.$$

Thus, it took the faster car 4 hours 10 minutes to make the trip.

● **PROBLEM** 1173

Two cars are traveling 40 and 50 miles per hour, respectively.

First Car

A •————————————————————• C

$\qquad\qquad$ 40x miles

Second Car

B • 5 • A ————————————————————• C
$\quad$ miles $\qquad$ 40x miles

$\qquad\qquad$ 50x miles

	Rate (in mph)	Time (hours)	Distance (miles)
First Car	40	x	40x
Second Car	50	x	50x

Solution: Let x = number of hours it takes the second car to overtake the first car. See table.

$$\text{Distance} = \text{rate} \times \text{time}$$

Then $\qquad$ 50x = distance second car travels in x hours,

and $\qquad$ 40x = distance first car travels in x hours.

Since the second car must travel an additional 5 miles (from B to A in diagram).

$$40x + 5 = 50x$$

Simplify,

$$-10x = -5$$

Divide by –10, $\qquad x = \dfrac{1}{2}$, number of hours it takes the second car to overtake the first car.

Check: $\qquad 40\left(\dfrac{1}{2}\right) + 5 = 50\left(\dfrac{1}{2}\right),$

$$25 = 25.$$

● **PROBLEM** 1174

Two trackmen are running on a circular race track 300 feet in circumference. Running in opposite directions, they meet every 10 seconds. Running in the same direction, the faster passes the slower every 50 seconds. Find their rates in feet per second.

10x

starting point

10y

one runner covers this distance, 10x, after 10 seconds

another runner covers this distance, 10y, after 10 seconds

Solution: Let x = rate of the faster runner and y = rate of the slower runner. Using the formula

$$\text{distance} = \text{rate} \times \text{time}$$

the first equation is:

$$10x + 10y = 300 \qquad (1)$$

This is true because both runners together cover the entire distance. (See figure.)

When they run in the same direction, the faster runner must gain 300 feet each time he overtakes the slower runner. Therefore, the distance that the faster runner covers in 50 sec, 50x, is equal to the circumference plus the distance of the slower runner, 50y. That is

$$50x = 300 + 50y$$

$$50x - 50y = 300 \qquad (2)$$

Equations 1 and 2 may be simplified, so that we have

$$x + y = 30$$

$$x - y = 6$$

Add these two equations and solve for x. The solution of this system is

$$x = 18$$

$$y = 12$$

The rates are therefore 18 feet per second for the faster runner and 12 feet per second for the slower runner.

To check the solution, we note that in 10 seconds the runners travel a total of 180 feet + 120 feet, or 300 feet, the length of the track. When they run in the same direction, in 50 seconds the faster travels 900 feet and the slower travels 600 feet, so that the faster runner runs 300 feet more than the slower, and therefore catches up with him every 50 seconds.

● **PROBLEM** 1175

One plane flies at a ground speed 75 miles per hour faster than another. On a particular flight, the faster plane requires 3 hours and the slower one 3 hours and 36 minutes. What is the distance of the flight?

Solution: If the velocity of the fast plane is v_1 and that of the slow plane is v_2, we have

$$v_1 = 75 + v_2 \qquad (1)$$

since the fast plane is 75 miles per hour faster. The distance is the same for each plane. The distance is $d = v \cdot t$, where v is the velocity and t is the time. Note that the time for the slower plane expressed in hours is

$$3 \frac{36}{60} = 3 \frac{3}{5} \text{ hours.}$$

Hence,

$$3v_1 = 3 \frac{3}{5} v_2 \qquad (2)$$

Equations (1) and (2) constitute a system of two equations in two variables. Solving equation (2) for v_1, we have

$$v_1 = \frac{6}{5} v_2$$

Substituting this value in equation (1), we have

$$\frac{6}{5} v_2 = 75 + v_2$$

$$\frac{1}{5} v_2 = 75$$

$$v_2 = 375$$

$$v_1 = 75 + 375 = 450$$

Hence, the length of the trip is $3 \cdot 450 = 1,350$ miles.

Check: The fast plane, velocity 450 miles per hour, is 75 miles per hour faster than the slow one, velocity 375 miles per hour. In 3 hours the fast plane travels $3 \cdot 450 = 1,350$ miles. In 3 hours 36 minutes the slow plane travels the same distance; that is, $3 \frac{3}{5} \cdot 375 = 1,350$ miles.

● **PROBLEM** 1176

An airplane with an air speed of 300 miles per hour flies against a head wind, and flies 900 miles in 4 hours. What is the speed of the head wind?

Solution: Let x = number of miles per hour in the rate of the head wind.

Then since the head wind is detracting from the speed of the airplane, let

$300 - x$ = number of miles per hour in the ground speed of the airplane.

Applying the formula, Rate X Time = Distance, with rate = $(300 - x)$,

Time = 4, Distance = 900, we obtain:

$$(300 - x)(4) = 900$$

Distributing 4, we have:

$$1200 - 4x = 900.$$

Subtract 1200 from both sides:

$$-4x = -300.$$

Divide by -4, $\qquad x = 75$, number of miles per hour in the rate of the head wind.

Check: $\qquad 4(300 - 75) = 900,$

$$900 = 900.$$

● **PROBLEM** 1177

Two planes with speeds of 600 mph (in still air) each make a trip of 990 miles. They take off at the same time and fly in opposite directions. One has a head wind and the other a tail wind. The plane flying with a tail wind lands 20 minutes before the other plane. What is the wind velocity?

Solution: If we let r represent the wind velocity in miles per hour, then the speeds of the planes are:

600 + r   speed of plane with tail wind (faster plane)

600 - r   speed of plane with head wind (slower plane).

Solving the distance formula,

distance = (rate)(time) for time, we obtain

time = $\dfrac{\text{distance}}{\text{rate}}$. We know distance is 990, rate

for slower plane is (600 - r), and rate for faster is (600 + r). Thus, the time for the slower plane is

$$\frac{\text{distance}}{\text{rate}} = \frac{990}{600 - r}$$

and for the faster plane the time is

$$\frac{\text{distance}}{\text{rate}} = \frac{990}{600 + r}$$

The difference in time is 20 minutes. In hours it is $\dfrac{1}{3}$ hour. The time equation is:

$$\frac{990}{600 - r} - \frac{990}{600 + r} = \frac{1}{3}$$

Eliminate the fractions by multiplying both sides of the equation by 3(600 + r)(600 - r).

$$3(990)(600 + r) - 3(990)(600 - r)$$

$$= \frac{1}{3} (3)(600 + r)(600 - r)$$

$$3(990)[(600 + r) - (600 - r)] = (600 + r)(600 - r)$$

$$3(990)(2r) = 360,000 - r^2$$

$$r^2 + 5940r - 360,000 = 0$$

$$(r - 60)(r + 6000) = 0$$

r - 60 = 0	r + 6000 = 0
r = 60 mph	r = - 6000 mph

The value - 6000 mph cannot be considered a solution since it was implied in the problem that the wind velocity was positive. Also a wind speed of 6000 mph is extremely unlikely.

Check: If the wind speed is 60 mph., the planes
will have velocities of 600 - 60 = 540 mph and
600 + 60 = 660 mph. The difference of the times of
flight is

$$\frac{990}{600-r} - \frac{990}{600+r} = \frac{1}{3}, \text{ and substituting:}$$

$$\frac{990}{540} - \frac{990}{660} = \frac{11}{6} - \frac{3}{2}$$

$$= \frac{11}{6} - \frac{9}{6}$$

$$= \frac{1}{3} \text{ hour.}$$

● **PROBLEM** 1178

Two airfields A and B are 400 miles apart, and B is due east of A.
A plane flew from A to B in 2 hours and then returned to A in 2½ hours.
If the wind blew with a constant velocity from the west during the
entire trip, find the speed of the plane in still air and the speed of
the wind.

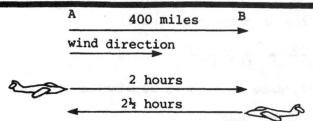

Solution: The essential point in solving this problem is that the wind
helps the plane in flying from A to B and hinders it in flying from B
to A. We therefore have the basis for two equations that involve the
speed of the plane, the speed of the wind, and the time for each trip.
We let

$$x = \text{speed of plane in still air, in miles per hour}$$
$$y = \text{speed of wind, in miles per hour}$$

Then, since the wind blew constantly from the west,

$$x + y = \text{speed of plane from A to B (wind helping)}$$
$$x - y = \text{speed of plane from B to A (wind hindering)}$$

The distance traveled each way was 400 miles, and so we have the following equations based on the formula distance/rate = time:

$$\frac{400}{x+y} = 2 = \text{time required for eastward trip} \qquad (8)$$

$$\frac{400}{x-y} = 2\frac{1}{2} = \text{time required for westward trip} \qquad (9)$$

We solve these equations simultaneously for x and y

$400 = 2x + 2y$	multiplying (8) by $x + y$	(10)
$800 = 5x - 5y$	multiplying (9) by $2(x - y)$	(11)
$2,000 = 10x + 10y$	multiplying (10) by 5	(12)
$1,600 = 10x - 10y$	multiplying (11) by 2	(13)
$3,600 = 20x$	adding equations (12) and (13)	

$$x = 180 \qquad \text{solving for } x$$

$$400 = 360 + 2y \quad \text{replacing } x \text{ by } 180 \text{ in (10)}$$

$$2y = 40$$

$$y = 20$$

Therefore the solution set of equations (8) and (9) is $\{(180,20)\}$, and it follows that the speed of the plane in still air is 180 miles per hour and the speed of the wind is 20 miles per hour.

Check:

$$\frac{400}{180 + 20} = \frac{400}{200} = 2 \qquad \text{from (8)}$$

$$\frac{400}{180 - 20} = \frac{400}{160} = \frac{5}{2} = 2\tfrac{1}{2} \qquad \text{from (9)}$$

● **PROBLEM** 1179

A plane is heading in a direction 45° east of north and flying at an air speed (speed in still air) of 350 miles per hour. The wind is blowing in a direction 10° east of north at the rate of 70 miles per hour. Find the velocity (speed and direction) of the plane.

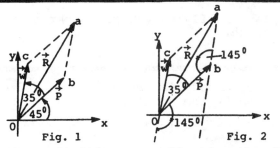

Fig. 1   Fig. 2

Solution: Two velocities are affecting the flight of the plane, that due to the plane itself and that due to the wind.

Velocity is a vector quantity and therefore the resultant or sum of two vectors is obtained using the parallelogram law for addition of vectors.

Let $\overline{P}$ represent the velocity of the plane

Let $\overline{W}$ represent the velocity of the wind.

Draw $\vec{P}$ at a 45° angle to the x axis and $\vec{W}$ 10° to the right of north or at an 80° angle to the x axis.

Now complete the parallelogram by drawing parallel sides to $\vec{P}$ and $\vec{W}$. Then connect the opposite vertices with a diagonal. Call this resultant $\vec{R}$. (See fig. 1)

R can now be determined using the law of cosines

$$c^2 = a^2 + b^2 - 2ab \cos \text{ (included angle)}.$$

The included angle (oba) is determined as follows: $\vec{W}$ makes an angle of 80° and $\vec{P}$ makes an angle of 45° (with the x axis) so subtracting the angle of $\vec{P}$ from the angle of $\vec{W}$ gives the angle of (cob). Together, cob and oba must add up to 180° because they are supplemenarty angles. Angle cob equals 35° and 35° + oba = 180° or oba = 145° (see fig.2). We now have

$R^2 = (P)^2 + (W)^2 - 2PW \cos 145°$ where the symbols without arrows above them represent the magnitudes of the corresponding vectors.

$$(R)^2 = (P)^2 + (W)^2 - [(2)(P)(W) \cos 145°]$$

$$(R)^2 = (350)^2 + (70)^2 - [(2)(350)(70)(-0.8192)]$$

Thus   R = 295

By the law of sines, $\dfrac{\sin \angle aob}{.70} = \dfrac{\sin 145°}{295}$

$$\sin \angle aob = \frac{70(0.5736)}{295} = 0.1361$$

$$\angle aob = 8° \quad \text{(nearest degree)}$$

Hence, the plane is flying 295 miles per hour in direction 37° east of north.

● **PROBLEM** 1180

The distance, s, a body falls in t seconds from a position of rest is given by the equation

$$s = 16t^2$$

where t is time in seconds, and s is measured in feet. A stone is dropped into a well and the sound of the splash is heard 3 seconds later. Taking the velocity of sound to be 1100 feet per second, write and solve the equation that determines the depth of the well.

Solution: From the formula, $s = 16t^2$, the time t_1 for the stone to reach the bottom of the well is

$$(t_1)^2 = \frac{s}{16}, \quad t_1 = \frac{\sqrt{s}}{4}$$

From the formula, distance = rate X time, the time t_2 for the sound of the splash to reach the top of the well is

$$t_2 = \frac{\text{distance}}{\text{rate}} = \frac{s}{1100}$$

Since the total time is 3 seconds, we have the equation

$$t_1 + t_2 = \frac{\sqrt{s}}{4} + \frac{s}{1100} = 3$$

If we multiply each term of equation 4 by 1100, we obtain

$$275 \sqrt{s} + s = 3300$$

This equation may be solved as follows. Let $u = \sqrt{s}$. Then $u^2 = s$ and substituting these values of s and $\sqrt{s}$ in the equation, we get

$$u^2 + 275u - 3300 = 0$$

By means of the quadratic formula

$$u = \frac{-b \pm \sqrt{b^2 - 4ac}}{2a} \;, \quad \text{we find } u = 11.5$$

rejecting the negative root because it has no physical meaning. Since from the above equation, $s = u^2$,

$$s = 132.25$$

Therefore, the depth of the well is approximately 132.25 feet.

CHAPTER 41

MIXTURES/FLUID FLOW

> **Basic Attacks and Strategies for Solving Problems in this Chapter. See pages 848 to 858 for step-by-step solutions to problems.**

A value mixture problem involves combining two ingredients which have different prices into a single blend. The basic strategies for solving such problems include:

(1) For each ingredient in the mixture, write a numerical or variable expression for the amount of the ingredient used, the unit cost of the ingredient, and the value of the amount used.

(2) For the blend, write a numerical or variable expression for the amount, the unit cost of the blend, and the value of the amount.

(3) Display the results in a table.

(4) Determine how the values of each ingredient are related.

Use the fact that the sum of the values of each ingredient is equal to the value of the blend to set up an appropriate equation.

The solution of a percent mixture problem is based upon the equation

$Q = Ar,$

where Q is the quantity of a substance in the solution, r is the percent of concentration, and A is the amount of solution. We use this information together with the strategy outlined above to find the solution. For example, we solve the following problem which states:

How many gallons of a 20% salt solution must be mixed with 6 gallons of a 30% salt solution to make a 22% salt solution?

First, let x = the unknown quantity of 20% solution. Then, using the formula

$Ar = Q,$

we can summarize the relations given in the problem in the following table:

	A ×	r =	Q
	Amt of soln, A	Percent of conc, r	Quantity of substance, Q
20% solution	x	0.20	$0.20x$
30% solution	6	0.30	$0.30(6)$
22% solution	$x + 6$	0.22	$0.22(x + 6)$

Note that the sum of the quantities of the substances in the 20% solution and the 30% solution is equal to the quantity of the substance in the 22% solution. Thus, we can write the equation,

$$0.20x + 0.30(6) = 0.22(x + 6)$$

whose solution is $x = 24$. Thus, 24 gallons of the 20% solution are required.

Step-by-Step Solutions to
Problems in this Chapter,
"Mixtures/Fluid Flow"

How many gallons of a liquid that is 74 percent alcohol must be combined with 5 gallons of one that is 90 percent alcohol in order to obtain a mixture that is 84 percent alcohol?

Solution: If we let x represent the number of gallons needed of the first liquid and remember that 74 percent of x is 0.74x, then the table (see table) shows all the data given in this problem.

	Number of gallons	Percentage of alcohol	Number of gallons of alcohol
First liquid	x	74	0.74x
Second liquid	5	90	0.90(5) = 4.5
Mixture	x + 5	84	0.84(x + 5)

We are told that we are combining the number of gallons of alcohol in the 74 percent alcohol (0.74x) with the number of gallons of alcohol in the 90 percent alcohol (4.5) to obtain the number of gallons of alcohol in the 84 percent alcohol [0.84(x + 5)]. Thus

$$.74x + 4.5 = .84(x + 5)$$

Multiplying both sides by 100,

$$74x + 450 = 84 (x + 5)$$
$$74x + 450 = 84x + 420$$
$$30 = 10x$$
$$x = 3$$

Therefore, 3 gallons of liquid that is 74 percent alcohol must be combined with 5 gallons of one that is 90 percent alcohol to obtain a mixture of 84 percent alcohol.

How much water must be added to 500 gallons of alcohol that is 70 per cent pure to make a mixture that is 60 per cent pure?

Solution: Let x = number of gallons of water to be added.

Let x + 500 = number of gallons of liquid in the 60% pure mixture.

We can thus formulate the following chart:

AMOUNT OF LIQUID	× % ALCOHOL	= NUMBER OF GALLONS OF ALCOHOL
500 gallons	.70	500(.70)
500 + x gallons	.60	(500+x)(.60)

We form the equation:

$$(0.60)(500 + x) = (0.70)(500)$$

and we wish to solve for x.

Distribute, $300 + 0.60x = 350.$

Subtract 300 from both sides,

$$0.60x = 350 - 300.$$

Collect terms, $0.60x = 50.$

Divide by 0.60, $x = 83\frac{1}{3}$, number of gallons of water
to be added.

● **PROBLEM** 1183

A grocer mixes two grades of coffee which sell for 70 cents and 80 cents per pound, respectively. How much of each must he take to make a mixture of 50 pounds which he can sell for 76 cents per pound?

Solution: Let x = the number of pounds of 70-cent coffee. Since the mixture is to contain 50 pounds and there are x pounds of 70 cent coffee, then 50 - x = number of pounds of 80 cent coffee. [Thus the total number of pounds in the mixture is x pounds (of 70-cent coffee) + (50 - x) pounds (of 80-cent coffee) = x - x + 50 = 50 lbs, our desired amount]. Using our formula:

Number of Pounds	×Price per Pound (in cents)	= Total Price
70¢/lb. coffee	x	70	70x
80¢/lb. coffee	50-x	80	80(50-x)
76¢/lb.	50	76	76(50)

The total price of the 70¢ coffee, 70x and total price of the 80¢ coffee, 80(50 - x) equals total price of the 76¢ coffee, (76)(50).

Therefore $70x + 80(50 - x) = (50)(76).$

Using the distributive law, we obtain:

$$70x + 4000 - 80x = 3800.$$

Subtract 4000 from both sides,

$$70x - 80x = 3800 - 4000.$$

Collect terms, $- 10x = -200.$

Divide by -10, $x = 20$, number of pounds of
                                        70-cent coffee.

Then $50 - x = 30$, number of pounds of
                                        80-cent coffee.

Check: $(70)(20 + (80)(30) = (50)(76),$

              $1400 + 2400 = 3800,$

              $3800 = 3800.$

● **PROBLEM 1184**

How much water must be added to 5 gallons of 90% ammonia
solution to reduce the solution to a 60% solution?

Solution: The volume of the original solution is 5
gallons. If we let x represent the number of gallons
of water added to the original solution, then the final
solution will have a volume of (5 + x) gallons. No
ammonia is added so the only ammonia in the final solution
is the ammonia in the original solution. We can now
record our given information in tabular form:

	Total Volume (Gallons)	Percent Concentration	Gallons of 100% Ammonia
Original solution	5	90	4.5
Final solution	5 + x	60	4.5

Note: The gallons of 100% ammonia for the original
solution was obtained in the following way: multiply the
total volume of the original solution by the percent
concentration of the original solution. Therefore,

$$5 \times 90\% = \cancel{5} \times \frac{90}{\cancelto{20}{100}} = \frac{90}{20} = \frac{9}{2} = 4.5 \text{ gallons of } 100\%$$

ammonia. Also, the gallons of 100% ammonia for the final
solution is the same as the gallons of 100% ammonia for
the original solution; that is, 4.5 gallons, since no
ammonia was added.

The formula for the amount of substance in
solution for the final solution is (total volume) X
(percent concentration) = gallons of 100% ammonia.
Therefore: (5 + x) X 60% = 4.5
or
$$(5 + x)\frac{60}{100} = 4.5$$

or $(5 + x) \ .60 = 4.5$

or $.60(5 + x) = 4.5$

            $3 + .60x = 4.5$

850

$$.6x = 4.5 - 3$$

$$.6x = 1.5$$

$$x = \frac{1.5}{.6} = 2.5 \text{ gallons}$$

Therefore, 2.5 gallons of water must be added to reduce the concentration to 60%.

Check: If 2.5 gallons of water are added to the original solution, we will have 7.5 gallons of solution. The amount of 100% ammonia remains the same 4.5 gallons. The concentration equals

$$\frac{4.5}{7.5} \times 100 = \frac{450}{7.5} = 60\%.$$

● **PROBLEM** 1185

How many gallons of a mixture containing 80% alcohol should be added to 6 gallons of a 25% solution to give a 30% solution?

Solution: Here we have

(1) How many gallons of the 80% mixture are needed?

(2) (alcohol in 80% solution) + (alcohol in 25% solution) = (alcohol in 30% solution).

These sentences become, in turn,

(1') Let x = number of gallons of the 80% solution;

(2') 0.80x + (0.25)(6) = 0.30(x + 6);

since the amount of alcohol in any solution is obtained by taking the total amount of the solution times the percentage of alcohol contained in it.

Solving for x from equation (2):

Distributing, 0.80x + (0.25)(6) = 0.30x + (0.30)(6)

0.80x - 0.30 x = (0.30)(6) - (0.25)(6)

0.50x = (0.30 - 0.25)6

0.50x = (0.05)6

$$x = \frac{(0.05)6}{0.50} = \frac{(5)6}{50} = \frac{6}{10} = 0.6$$

Thus, 0.6 gallons of the 80% mixture are needed.

● **PROBLEM** 1186

How many ounces of silver alloy which is 28% silver must be mixed with 24 ounces of silver which is 8% silver to produce a new alloy which is 20% silver?

Solution: Let x = number of ounces of 28% silver to be used. The relationship used to set up the equation is

851

Volume of 28% silver + Volume of 8% silver = Volume of silver in mixture.

$$.28x + .08(24) = .20(x + 24)$$
$$28x + 8(24) = 20(x + 24)$$
$$8x = 288$$
$$x = 36 \text{ ounces of silver}$$

Check: Volume of 28% silver = (.28)(36) = 10.08
Volume of 8% silver = (.08)(24) = 1.92
Total amount of silver = 12 ounces
The total mixture contains 24 + 36 = 60 ounces, and 12 ounces is 20% of 60 ounces.

● **PROBLEM** 1187

How much water must be evaporated from 120 pounds of solution which is 3% salt to make a solution of 5% salt?

Solution: Let x = number of pounds of salt to be evaporated.
The relationship used to set up the equation is

Amount of salt in new mixture = Amount of salt in old mixture.

$$.05(120 - x) = .03(120)$$
$$5(120 - x) = 3(120)$$
$$600 - 5x = 360$$
$$x = 48 \text{ pounds of water}$$

Check: Water in new mixture is 120 - 48 = 72 pounds. Of this

3% of 120 or 3.6 pounds is salt.

$$\frac{3.6}{72} = 5\%.$$

● **PROBLEM** 1188

Of 24 quarts of a mixture, 8% is iodine. Of another mixture, 4% is iodine. How many quarts of the second mixture should be added to the first mixture to obtain a mixture that is 5% iodine?

Solution: Let x = the number of quarts to be taken from second mixture.
The relationship used to set up the equation is:

Volume of iodine in the first mixture + Volume of iodine in the second mixture = Volume of iodine in the resulting mixture.

$$.08(24) + .04x = .05(24 + x)$$
$$1.92 + .04x = 1.20 + .05x$$
$$.01x = .72$$
$$x = 72 \text{ quarts of iodine}$$

Check: Volume of iodine in the first mixture is (.08)(24) or 1.92 quarts. Volume of iodine in the second mixture is (.04)(72) or 2.88 quarts. Total volume of iodine is 4.80 quarts is equal to 5% of 96 quarts.

● **PROBLEM** 1189

A chemist has 24 ounces of a 25% solution of argyrol. How much water must he add to reduce the strength of the argyrol to 20%?

Solution: Let x = number of ounces of water to be added.
The relationship used to set up the equation is (since only water is added)

Amount of argyrol in new mixture = Amount of argyrol in old mixture.

$$.20(24 + x) = .25(24)$$
$$4.8 + .2x = 6$$
$$2x = 12$$
$$x = 6 \text{ ounces of water}$$

Check: The new mixture has 30 ounces. Of this, 6 ounces or 20% of the mixture is argyrol.

● **PROBLEM** 1190

How many quarts of pure alcohol must be added to 40 quarts of a mixture that is 35% alcohol to make a mixture that will be 48% alcohol?

Solution: Let x = number of quarts of pure alcohol to be added. The relationship used to set up the equation is

Amount of alcohol in new mixture = Amount of alcohol in old mixture + Amount of alcohol added

$$.48(40 + x) = (.35)(40) + x$$
$$19.2 + .48x = 14 + x$$
$$x = 10 \text{ quarts of alcohol}$$

Check: Amount of alcohol in new mixture = 14 quarts + 10 quarts = 24 quarts. New mixture contains a total of 40 + 10 = 50 quarts.

$$\frac{24}{50} = 48\% .$$

● **PROBLEM** 1191

A truck radiator contains 32 quarts of a 20% solution of anti-freeze. How much of the original solution must be drawn off and replaced by pure anti-freeze to obtain a solution of 45% anti-freeze?

Solution: Let x = number of quarts of solution drawn off. The relationship used to set up the equation is

Volume of anti-freeze left in 20% solution after drawing off x quarts + Volume of anti-freeze added = Volume of anti-freeze in 45% solution.

$$.20(32 - x) + x = .45(32)$$
$$6.4 - .20x + x = 14.40$$
$$.80x = 8$$
$$x = 10 \text{ quarts of anti-freeze added}$$

Check: The original mixture contained 32 quarts. From this 10 quarts were drawn off. Of this 20% or 2 quarts were anti-freeze. Since the original mixture contained 20% of 32 quarts or 6.4 quarts of anti-freeze, this left 4.4 quarts of anti-freeze. To this was added 10 quarts of anti-freeze making a total of 14.4 quarts of anti-freeze. 14.4 quarts is 45% of 32.

● **PROBLEM** 1192

A storekeeper has two kinds of cookies, one worth $.75 a pound and the other worth $.50 a pound. How many pounds of each should he use to make a mixture of 60 pounds worth $.55 a pound?

Solution: Let x = number of pounds of cookies at 75 cents. Then (60 - x) = the number of cookies at 50 cents a pound.

The relationship used to set up the equation is:

Value of 75 cent cookies + Value of 50 cent cookies = Value of mixture.

$$75x + 50(60 - x) = 60(55)$$
$$75x + 3000 - 50x = 3300$$
$$25x = 300$$
$$x = 12 \text{ pounds of 75 cent cookies}$$
$$60 - x = 48 \text{ pounds of 50 cent cookies}$$

Check: The value of 12 pounds of cookies at 75 cents a pound is $9.00. The value of 48 pounds of cookies at 50 cents a pound is $24.00. The resulting mixture contains 60 pounds worth $33.00. If 60 pounds of cookies are sold for $33.00 then each pound is sold for $.55.

● PROBLEM 1193

If a container contains a mixture of 5 gallons of white paint and 11 gallons of brown paint, how much white paint must be added to the container so that the new mixture will be two-thirds white paint?

Solution: Let x = number of gallons of white paint to be added then x + 5 = number of gallons of white paint in the final mixture. x + 16 = total number of gallons of paint in the final mixture.

The new mixture will be two-thirds white paint if the ratio of the final number of gallons of white paint to the final total number of gallons of paint is equal to $\frac{2}{3}$. The proportion is

$$\frac{x + 5}{x + 16} = \frac{2}{3}$$

$3x + 15 = 2x + 32$ by cross-multiplying

$3x - 2x = 32 - 15$ by isolating x terms

$x = 17$ gallons by combining like terms.

Check: The final mixture would contain x + 5 = 22 gallons of white paint, and the total number of gallons of paint would be x + 16 = 33 gallons.

$$\frac{\text{final number of gallons of white paint}}{\text{final total number of gallons of paint}} = \frac{22 \text{ gal}}{33 \text{ gal}} = \frac{2}{3}.$$

● PROBLEM 1194

A chemist has an 18% solution and a 45% solution of a disinfectant. How many ounces of each should be used to make 12 ounces of a 36% solution?

Solution: Let x = Number of ounces from the 18% solution
         And y = Number of ounces from the 45% solution

(1) $x + y = 12$

(2) $.18x + .45y = .36(12) = 4.32$

Note that .18 of the first solution is pure disinfectant and that .45 of the second solution is pure disinfectant. When the proper

854

quantities are drawn from each mixture the result is 12 gallons of mixture which is .36 pure disinfectant, i.e., the resulting mixture contains 4.32 ounces of pure disinfectant.

When the equations are solved, it is found that

$$x = 4 \quad \text{and} \quad y = 8.$$

● **PROBLEM** 1195

A man wants to obtain 15 gallons of a 24% alcohol solution by combining a quantity of 20% alcohol solution, a quantity of 30% alcohol solution and 1 gallon of pure water. How many gallons of each of the alcohol solutions must be used?

Solution: Let x = The number of gallons of the 20% solution used
And y = The number of gallons of the 30% solution used

(1)   x + y + 1 = 15

(2)   .20x + .30y = .24(15) = 3.6

When these equations are solved it is found that 6 gallons of the 20% solution and 8 gallons of the 30% solution are used.

● **PROBLEM** 1196

In a chemical laboratory one carboy contains 12 gallons of acid and 18 gallons of water. Another carboy contains 9 gallons of acid and 3 gallons of water. How many gallons must be drawn from each carboy and combined to form a solution that is 7 gallons acid and 7 gallons water?

Solution: Let x = Number of gallons taken from first carboy
And y = Number of gallons taken from second carboy

(1)   x + y = 14

(2) $\frac{12x}{30} + \frac{9y}{12} = 7$

In forming the second equation, it should be observed that $\frac{12}{30}$ of the liquid drawn from the first carboy is acid and $\frac{9}{12}$ of the liquid drawn from the second carboy is acid. The two quantities of liquid drawn from the two carboys yield 7 gallons of acid in the mixture.

When the equations are solved, it is found that

$$x = 10 \quad \text{and} \quad y = 4$$

● **PROBLEM** 1197

Sand and gravel have been mixed in two separate piles. In the first pile the ratio of sand to gravel is 1:1, and in the second pile the ratio of sand to gravel is 1:4. A third pile, in which the ratio of sand to gravel is 1:3, is to be formed from the first two piles. If the third pile is to contain 15 cubic yards, how many cubic yards must be taken from each of the first two piles?

Solution: Let x = Number of cubic yards to be taken from first pile
And y = Number of cubic yards to be taken from second pile

(1)   x + y = 15

(2) $\frac{1}{2} x + \frac{1}{5} y = \frac{1}{4}(15)$

In forming the second equation it should be observed that when the ratio of sand to gravel in the first pile is 1:1 then $\frac{1}{2}$ of the material taken from the first pile is sand. Likewise, the ratio of sand to gravel in the second pile is 1:4 and therefore 1/5 of the material taken from the second pile is sand. Since the ratio of sand to gravel in the third pile is 1:3 then $\frac{1}{4}$ of the 15 cubic yards in the third pile is sand.

When the equations are solved it is found that $x = 2\frac{1}{2}$ and $y = 12\frac{1}{2}$.

● **PROBLEM** 1198

What quantities of silver 60 per cent and 82 per cent pure must be mixed together to give 12 ounces of silver 70 per cent pure?

Solution: Let x = number of ounces of 60 per cent silver, and

y = number of ounces of 82 per cent silver.

We use the following table to describe the given information:

	Number of ounces	% Pure Silver	Number of Ounces of Pure Silver
Silver (60%)	x	60	.60x
Silver (82%)	y	82	.82y
Silver (70%)	12	70	.70(12)

From the information obtained in the table we have the following

equations: $\qquad .60x + .82y = .70(12) \qquad (1)$

$$x + y = 12 \qquad (2)$$

Multiplying each term of equation (1) by 100, we obtain:

$$60x + 82y = 70(12) \qquad (3)$$

Equation (2) multiplied by 60 gives:

$$60x + 60y = (12)(60). \qquad (4)$$

Then equation (3)-(4) gives:

$$60x + 82y = 840$$
$$\underline{-60x - 60y = -720} \quad ;$$
$$22y = 120$$

dividing both sides by 22, $\qquad y = \dfrac{120}{22} = \dfrac{60}{11} = 5\dfrac{5}{11}$.

Substituting $5\dfrac{5}{11}$ for y in (2) gives

$$x + 5\frac{5}{11} = 12, \text{ or } \quad x + \frac{60}{11} = \frac{132}{11}$$

Therefore, $\qquad x = \dfrac{72}{11} = 6\dfrac{6}{11}$.

856

Thus, we must mix $6 \frac{6}{11}$ ounces of 60 per cent pure silver and $5 \frac{5}{11}$ ounces of 82 per cent pure silver to obtain 12 ounces of silver 70 per cent pure.

Check: Substituting $6 \frac{6}{11}$ for x and $5 \frac{5}{11}$ for y in (3) gives

$$\left(6 \frac{6}{11}\right)(60) + \left(5 \frac{5}{11}\right)(82) = (70)(12)$$

Convert $6 \frac{6}{11}$ and $5 \frac{5}{11}$ to fractions, $\frac{72}{11}(60) + \frac{60}{11}(82) = 840$.

Multiply, $\frac{4320}{11} + \frac{4920}{11} = 840$

Add fractions, $\frac{9240}{11} = 840$

$$840 = 840 \quad .$$

Substituting in (2) gives

$$6 \frac{6}{11} + 5 \frac{5}{11} = 12$$

$$12 = 12.$$

● **PROBLEM** 1199

A tobacco dealer mixed 12 pounds of one grade of tobacco with 10 pounds of another grade to obtain a blend worth $54. He then made a second blend worth $61 by mixing 8 pounds of the first grade with 15 pounds of the second grade. Find the price per pound of each grade.

Solution: In this problem we have two basic relations that we can use to form two equations. We let

x = price per pound of the first grade, in dollars

y = price per pound of the second grade, in dollars

The relationship used to set up the equations is:

Value of the first grade + value of the second grade = Value of the mixture; that is,

(Number of pounds of the first grade)(Price per pound of the first grade) + (Number of pounds of the second grade)(Price per pound of the second grade) = The total cost of the mixture. Then

$12x + 10y = 54$ (1) using the numbers of pounds as coefficients and the values of the blends as constant terms

$8x + 15y = 61$ (2)

We eliminate y by subtraction:

$36x + 30y = 162$ multiplying (1) by 3

$\underline{16x + 30y = 122}$ multiplying (2) by 2

$20x \qquad = 40$ equating the differences of the members

$$x = 2 \qquad \text{solving for } x$$

$$16 + 15y = 61 \qquad \text{replacing } x \text{ by 2 in (2)}$$

$$15y = 45$$

$$y = 3$$

Therefore the solution set of equations (1) and (2) is {(2, 3)}, and it follows that the prices of the two grades are \$2 and \$3 per pound.

Check: To check the solution of a verbal problem, we re-read the problem, substituting the values found and verify if they make the statements true.

$$12(2) + 10(3) = 24 + 30 = 54$$

$$8(2) + 15(3) = 16 + 45 = 61$$

CHAPTER 42

NUMBERS, DIGITS, COINS, AND CONSECUTIVE INTEGERS

Basic Attacks and Strategies for Solving Problems in this Chapter. See pages 859 to 871 for step-by-step solutions to problems.

A strategy for solving a consecutive integer problem includes:

(1) Let a variable represent one of the integers.

(2) Express each of the other integers in terms of that variable.

(3) Remember that consecutive even or consecutive odd integers differ by 2.

(4) Determine the relationship among the integers and then solve for the unknown.

In solving problems dealing with coins, the following procedure is useful:

(1) For each denomination of coin, write a numerical or variable expression for the number of coins, the value of the coin, and the total value of the coins in cents.

(2) Write the results in a table.

(3) Determine the relationship between the total values of the coins.

(4) Use the fact that the sum of the total values of each denomination of coin is equal to the total value of all the coins.

(5) Solve the equation.

The general plan for solving number, digit, coin, consecutive integer, or any other type of word problem is as follows:

(1) **Read** the problem carefully. Take note of what is being asked and what information is given.

(2) **Plan** a course of action. Represent the unknown number by a letter. If there are two or more unknowns, represent one of them by a letter and

express the others in terms of the letter.

(3) **Create** an equation from the given information.

(4) **Solve** this equation.

(5) **Check** your solution against the information given in the original problem.

(6) **Answer** the original question. Read again the problem to make sure you have answered the question.

● **PROBLEM** 1200

The sum of two numbers is 23. One of the numbers is 7 more than the other number. What are the numbers?

<u>Solution:</u> Let⁻ x = one of the numbers, and x + 7 = the other number. Since we are given that the sum of the two numbers is 23,

$$x + (x + 7) = 23.$$

By the associative law of addition:

$$x + (x + 7) = 23$$

is the same as $(x + x) + 7 = 23,$

or $2x + 7 = 23.$

Subtract 7 from both sides:

$$2x = 23-7.$$

Collect terms, $2x = 16.$

Divide by 2, $x = 8$, one of the numbers.

Then solving for our other number x + 7, we substitute 8 for x.

Hence, $x + 7 = 8 + 7 = 15$, the other number.

Therefore, the two numbers are 8 and 15. We can verify this result by observing that the sum of the two numbers is indeed 23, and 15 is 7 more than 8, $8 + 7 = 15$.

● **PROBLEM** 1201

The sum of a number and its reciprocal is $\frac{65}{28}$. What is the number?

<u>Solution:</u> Given a number x. Its reciprocal is written as $\frac{1}{x}$. The sum of x and its reciprocal $\frac{1}{x}$, that is $x + \frac{1}{x}$, equals $\frac{65}{28}$. We have: $x + \frac{1}{x} = \frac{65}{28}$.

Multiplying both sides of this equation by 28x, the least common denominator, we obtain:

$$28x^2 + 28 = 65x$$
$$28x^2 - 65x + 28 = 0$$
$$(7x - 4)(4x - 7) = 0$$

859

$$7x - 4 = 0 \qquad\qquad 4x - 7 = 0$$
$$x = \frac{4}{7} \qquad\qquad\qquad x = \frac{7}{4}$$

Note: The two possible values of the number are reciprocals so a single check will suffice to show the number could be either $\frac{4}{7}$ or $\frac{7}{4}$.

Check: $\quad \frac{4}{7} + \frac{7}{4} = \frac{16}{28} + \frac{49}{28} = \frac{65}{28}$.

● **PROBLEM** 1202

The sum of two numbers is 24; one number is 3 more than twice the other. Find the numbers.

Solution: Let x = one of the numbers
Let y = the other number
Since the sum of the two numbers is 24,
$$x + y = 2x \qquad\qquad (1)$$
and since one of the numbers is 3 more than twice the other,
$$x = 2y + 3 \qquad\qquad (2)$$
Thus we have 2 equations in 2 unknowns and we solve for x and y:
Since $x = 2y + 3$, we may replace x by $(2y + 3)$ in equation (1),
$$(2y + 3) + y = 24$$
$$3y + 3 = 24$$
$$3y = 21$$
$$y = 7$$
To solve for x we replace y by 7 in equation (2)
$$x = 2(7) + 3$$
$$x = 14 + 3$$
$$x = 17$$
Thus the two numbers are 17 and 7.

Check: The sum of the two numbers is 24:
$$x + y = 17 + 7 = 24$$
One of the numbers is 3 more than twice the other:
$$17 = 2(7) + 3$$
$$17 = 14 + 3$$
$$17 = 17$$

● **PROBLEM** 1203

The sum of two numbers is 25 and the difference of their squares is 225. Find the numbers.

Solution: Let x = one of the two numbers.
Let y = the other number.
Since the sum of the two numbers is 25,
$$x + y = 25 \qquad\qquad (1)$$

and since the difference of their squares is 225,

$$x^2 - y^2 = 225 \qquad (2)$$

Thus we now have 2 equations in 2 unknowns:

$$x + y = 25 \qquad (1)$$
$$x^2 - y^2 = 225 \qquad (2)$$

Solving equation (1) for y we obtain

$$y = 25 - x \qquad (3)$$

To solve for x, we substitute this value of y into equation (2),

$$x^2 - (25 - x)^2 = 225$$
$$x^2 - (625 - 50x + x^2) = 225$$
$$x^2 - 625 + 50x - x^2 = 225$$
$$50x - 625 = 225$$
$$50x = 850$$
$$x = 17$$

To solve for y, we substitute this value of x into equation (1)

$$17 + y = 25$$
$$y = 8$$

Thus the two numbers are 17 and 8.

Check: Replace x and y by 17 and 8 in equations (1) and (2):

$$x + y = 25 \qquad (1)$$
$$17 + 8 = 25$$
$$25 = 25$$
$$x^2 - y^2 = 225 \qquad (2)$$
$$(17)^2 - (8)^2 = 225$$
$$289 - 64 = 225$$
$$225 = 225$$

● **PROBLEM 1204**

Find the number which, increased by its reciprocal, is equal to 37/6.

Solution: Let x = The number

Then $\frac{1}{x}$ = The reciprocal

$$x + \frac{1}{x} = \frac{37}{6}$$
$$6x^2 + 6 = 37x$$
$$6x^2 - 37x + 6 = 0$$
$$(6x - 1)(x - 6) = 0$$
$$6x - 1 = 0$$
$$x = \frac{1}{6}$$
$$x - 6 = 0$$
$$x = 6$$

The number may be taken as 1/6 and the reciprocal as 6, or the number may be taken as 6 and the reciprocal as 1/6.

Check: If the number is 6, the reciprocal is 1/6. Then

$$6 + \frac{1}{6} = \frac{37}{6} .$$

● PROBLEM 1205

Find two numbers such that twice the first added to the second equals 19, and three times the first is 21 more than the second.

Solution: Let x = the first number and y = the second number. The equations are

$2x + y = 19$ (twice the first added to the second equals 19)

$3x = y + 21$ (three times the first is 21 more than the second)

To solve this system

$$2x + y = 19$$

$$3x = y + 21$$

obtain all the variables on one side of the equations.

$2x + y = 19$	(1)
$3x - y = 21$	(2)

Add (2) to (1)

$2x + y = 19$	(1)
$\underline{3x - y = 21}$	(2)
$5x \quad\ = 40$	(3)

Divide by 5 to obtain x

$$x = 8$$

Substitute $x = 8$ into (1) or (2).

(1) $2x + y = 19$

$2(8) + y = 19$

$16 + y = 19$

$y = 3$

The solution of this system is

$x = 8$, the first number

$y = 3$, the second number

To check the solution, show that the two numbers satisfy the conditions of the problem.

Twice the first number is $2(8) = 16$. Add this result

to the second is 16 + 3 = 19. Thus 19 = 19. Then three
times the first number is 3(8) =24 which is 21 more than
3. That is 24 = 21 + 3; 24 = 24.

● PROBLEM 1206

Find two numbers such that the sum of twice the larger and the
smaller is 64. But, if 5 times the smaller be subtracted from four
times the larger the result is 16.

Solution: Let x = the larger number
           And y = the smaller number
           Then 2x + y = 64
                 4x - 5y = 16

When these equations are solved simultaneously the larger number
is found to be 24 and the smaller number 16.

● PROBLEM 1207

Separate 120 into two parts such that the larger exceeds three
times the smaller by 12.

Solution: Let x = the larger number
           And y = the smaller number
           Then x + y = 120
                 x = 3y + 12

When these equations are solved simultaneously the larger number is
found to be 93 and the smaller 27.

● PROBLEM 1208

Find two real numbers whose sum is 10 such that the sum of the
larger and the square of the smaller is 40.

Solution: Let x = the smaller number
           Let y = the larger number

The sum of the numbers is 10, therefore

$$x + y = 10 \qquad (1)$$

The sum of the larger and the square of the smaller is 40, therefore

$$y + x^2 = 40 \qquad (2)$$

Solving for y in equation (1) by adding (-x) to both sides we
obtain

$$y = 10 - x \qquad (3)$$

Replacing this value of y in equation (2) we obtain

$$(10 - x) + x^2 = 40$$

Adding -40 to both sides,

$$10 - x + x^2 - 40 = 0$$

$$x^2 - x - 30 = 0$$

Factoring,

$$(x - 6)(x + 5) = 0$$

Whenever the product of two numbers ab = 0, either a = 0 or
b = 0. Thus, either

$$x - 6 = 0 \quad \text{or} \quad x + 5 = 0$$

and

$$x = 6 \qquad \text{or} \quad x = -5.$$

To find the corresponding y values we replace x by each of these

values in equation (3):
    Replacing x by 6:
$$y = 10 - 6$$
$$y = 4$$

    Replacing x by -5:
$$y = 10 - (-5)$$
$$y = 10 + 5$$
$$y = 15$$

Thus the two possible solutions are (6,4) and (-5,15).
Since we assumed x to be the smaller number, and 6 is greater
than 4, not smaller than it, we reject (6,4). To check if (-5,15)
fits the conditions of this problem, we replace (x,y) by (-5,15) in
equations (1) and (2):

$$x + y = 10 \qquad\qquad (1)$$
$$-5 + 15 = 10$$
$$10 = 10$$

$$y + x^2 = 40 \qquad\qquad (2)$$
$$15 + (-5)^2 = 40$$
$$15 + 25 = 40$$
$$40 = 40$$

Thus the pair of numbers whose sum is ten such that the sum of the
larger and the square of the smaller is 40 is (-5,15).

● **PROBLEM** 1209

If 3 is subtracted from the numerator of a certain fraction,
the value of the fraction becomes 3/5. If 1 is subtracted from the
denominator of the same fraction then the value of the fraction becomes 2/3. Find the original fraction.

Solution: Let x = the numerator of the fraction
           And y = the denominator of the fraction

Then (1) $\dfrac{x - 3}{y} = \dfrac{3}{5}$, $5x - 15 = 3y$

(2) $\dfrac{x}{y - 1} = \dfrac{2}{3}$, $3x = 2y - 2$

(3) $5x - 3y = 15$

(4) $3x - 2y = -2$

When these equations are solved simultaneously x is found to be 36
and y is found to be 55. Therefore, the original fraction is 36/55.

● **PROBLEM** 1210

The units digit of a two digit number is two larger than
the tens digit. When the digits are reversed, the new
two digit number is equal to seven times the sum of the
digits. What is the original number?

Solution: When dealing with number problems that are
concerned with the digits of a number, we must utilize
position values and write a two digit number as
10a + b, a three digit number as 100x + 10y + z, and
so on.

    Let x represent the digit in the tens position,

then the digit in the units position must be x + 2.
The original two digit number is of the form 10a + b
where a = digit in tens position and b = digit in the
units position. Therefore, the original number is

10x + (x + 2)   or 10x + x + 2   or 11x + 2.

When the digits are reversed, the new number can
be expressed as

10(x + 2) + x   or 11x + 20.

The sum of the digits is x + (x + 2) = 2x + 2.
Setting the new number equal to seven times the sum of
the digits, we have:

11x + 20 = 7(2x + 2)

11x + 20 = 14x + 14

20 - 14 = 14x - 11x

6 = 3x

x = 2, x + 2 = 4

The original number 10x + (x + 2) = 10(2) + 4 = 24.

If the digits are reversed, we get 42, and this is
the product of seven and the sum of the digits (six).

● **PROBLEM** 1211

The sum of the digits of a two-digit number is 9. The number is
equal to 9 times the units' digit. Find the number.

Solution: (1) t + u = 9

(2) 10t + u = 9u

If these two equations are solved simultaneously, t = 4 and
u = 5 and the number is 45.

● **PROBLEM** 1212

The units' digit of a two-digit number is 4 less than 3 times
the tens' digit. If the digits are reversed, a new number is formed
which is 12 less than twice the original number. Find the number.

Solution: (1) u = 3t - 4

(2) 10u + t = 2(10t + u) - 12

If these equations are solved simultaneously t = 4 and u = 8
and the number is 48.

● **PROBLEM** 1213

A purse contains 19 coins worth $3.40. If the purse contains only
dimes and quarters, how many of each coin are in the purse?

Solution: Let x = the number of dimes in the purse
Then 19 - x = the number of quarters in the purse
10x = the value of the dimes

$$25(19 - x) = \text{the value of quarters}$$

The relationship used in setting up the equation is:
The value of the dimes + the value of the quarters = \$3.40
$$10x + 25(19 - x) = 340$$
$$10x + 475 - 25x = 340$$
$$x = 9$$

There are 9 dimes and 10 quarters in the purse.

Check: The dimes are worth \$.90 and the quarters are worth \$2.50, making a total of \$3.40.

● **PROBLEM** 1214

A toy savings bank contains \$17.30 consisting of nickels, dimes, and quarters. The number of dimes exceeds twice the number of nickels by 3 and the number of quarters is 4 less than 5 times the number of nickels. How many of each coin are in the bank?

Solution: Let x = the number of nickels
Then 2x + 3 = the number of dimes
And 5x - 4 = the number of quarters
The relationship used in setting up the equation is:

Value of nickels + Value of dimes + Value of quarters = 1730.

$$5x + 10(2x + 3) + 25(5x - 4) = 1730; \ x = 12$$

There are 12 nickels, 27 dimes and 56 quarters in the bank.

Check: The nickels are worth \$.60, the dimes are worth \$2.70, and the quarters are worth \$14.00, making a total of \$17.30.

● **PROBLEM** 1215

Max has \$1.45 in coins. He has fourteen coins in nickels, dimes, and quarters. There are two more nickels than dimes and quarters combined. How many of each kind of coin does he have?

Solution: Let the number of nickels be n, dimes d, and quarters q. The total value of the coins yields one equation.

$$0.05n + 0.10d + 0.25q = 1.45 \qquad (1)$$

The total number of coins yields another.

$$n + d + q = 14 \qquad (2)$$

The number of nickels yields a third.

$$n = d + q + 2 \qquad (3)$$

Substituting the value of n from (3) in (2),

$$d + q + 2 + d + q = 14$$
$$2d + 2q + 2 = 14$$
$$d = 6 - q$$

Substituting this value of d in (2),

$$n + 6 - q + q = 14$$
$$n = 8$$

Substituting n = 8 and d = 6 - q in (1),
$$0.05(8) + 0.10(6 - q) + 0.25q = 1.45$$

$$0.40 + 0.60 - 0.10q + 0.25q = 1.45$$

$$0.15q = 0.45$$
$$q = 3$$

Since $d = 6 - q$, $d = 3$. Therefore, Max has $q = 3$ quarters, $d = 3$ dimes, and $n = 8$ nickels.

● **PROBLEM** 1216

The three angles of a triangle are together equal to $180°$. The smallest angle is half as large as the largest one, and the sum of the largest and smallest angles is twice the third angle. Find the three angles.

Solution: Let x = the smallest angle,

y = the largest angle,

z = the third angle.

From the given information we formulate the following equations:

$$x + y + z = 180 \qquad (1)$$
$$x = \frac{y}{2} \qquad (2)$$
$$x + y = 2z \qquad (3)$$

We wish to solve for x, y and z. Multiply both sides of equation (2) by 2. Thus, $2x = 2\left(\frac{y}{2}\right)$

$$2x = y$$

Now, subtract y from both sides. Thus,

$$2x - y = y - y,$$

or $2x - y = 0$. (4)

Subtract $2z$ from both sides of equation (3):

$$x + y - 2z = 2z - 2z ,$$

or $x + y - 2z = 0$. (5)

Equation (1) multiplied by 2 gives

$$2x + 2y + 2z = 360 .   (6)$$

Equation (5) + (6) gives

$$x + y - 2z = 0$$
$$\underline{+ (2x + 2y + 2z = 360)}$$
$$3x + 3y = 360 \qquad (7)$$

Equation (4) multiplied by 3 gives

$$6x - 3y = 0 .   (8)$$

Equation (7) + (8) gives

$$3x + 3y = 360$$
$$\underline{+ (6x - 3y = 0)}$$
$$9x = 360 \qquad ;$$

therefore, $x = 40°$

Substituting $40°$ for x in (2) gives $40 = \frac{y}{2}$. Multiply both sides by 2: $80° = y$. Substituting $40°$ for x and $80°$ for y in (1) gives

$$40° + 80° + z = 180° .$$

Subtract 120 from both sides: $z = 60°$.

Thus, the angles are $40°$, $80°$, and $60°$.

Check: Substituting $40°$ for x, $80°$ for y, $60°$ for z gives:
$$40° + 80° + 60° = 180°$$
$$40° = \frac{80°}{2}$$
$$40° + 80° = 2(60°) .$$

● **PROBLEM** 1217

Show that the sum of any positive number and its reciprocal cannot be less than 2.

Solution: Express the given example as a mathematical statement, recalling the reciprocal of $x = 1/x$. Write "the sum of any positive number and its reciprocal cannot be less than 2" as
$$a + \frac{1}{a} \not< 2,$$

where a represents the positive number. Thus, the sum of any positive number and its reciprocal can be greater than or equal to 2. We are to show that
$$a + \frac{1}{a} \geq 2, \quad a > 0.$$

If this relation is true, the following inequalities will also hold:
$$a^2 + 1 \geq 2a, \quad \text{Multiply both members of the inequality by a ,}$$
$$a^2 - 2a + 1 \geq 0, \quad \text{Transpose 2a}$$
$$(a - 1)^2 \geq 0, \quad \text{Factor}$$

Now this simple relation is easily shown to be true, for, whether a - 1 is positive, negative, or zero, its square must be non-negative. This is therefore a suitable starting point, and our synthesis, constituting the actual proof, is as follows. Since
$$(a - 1)^2 \geq 0,$$

for the reason just stated, expansion of the left member gives us the equivalent relations
$$a^2 - 2a + 1 \geq 0 , \quad \text{Expanding}$$
$$a^2 + 1 \geq 2a , \quad \text{Transposing 2a}$$
$$a + \frac{1}{a} \geq 2 , \quad \text{Divide both sides of the inequality by a, with } a > 0 .$$

We see, incidentally, that the equality holds only if a = 1; if 0 < a < 1, or if a > 1, the inequality holds.

CONSECUTIVE INTEGERS

● **PROBLEM** 1218

The sum of three consecutive odd numbers is 45. Find the numbers.

Solution: If we look at a series of odd numbers 1,3,5,7,9,11... we note that each consecutive odd number is 2 more than the one before it.

i.e.,
$$3 = 1 + 2$$
$$5 = 3 + 2$$
$$7 = 5 + 2 \quad \text{etc.}$$

Therefore, if x is the first odd number,

$x + 2$ is the second odd number,

and $(x + 2) + 2 = x + (2 + 2) = x + 4$

is the third odd number. Therefore

Let x = the first odd number,

$x + 2$ = the second odd number,

$x + 4$ = the third odd number.

Then, since the sum of these numbers is 45, we have:

$$x + (x + 2) + (x + 4) = 45.$$

Using the Associative and Commutative Laws of Addition:

$$3x + 6 = 45.$$

Subtract 6 from both sides:

$$3x = 45 - 6.$$

Collect terms, $\qquad 3x = 39.$

Divide by 3, $\qquad x = 13$, the first odd number.

Replace x by 13, $\quad x + 2 = 15$, the second odd number.

Replace x by 13, $\quad x + 4 = 17$, the third odd number.

Check: $\qquad 13 + 15 + 17 = 45,$

$$45 = 45.$$

● **PROBLEM** 1219

The sum of four consecutive even numbers is 140. What are the numbers?

Solution: An even number can be represented by $2n$, where n is an integer. Consecutive even integers (or odd integers) differ by 2. Therefore, four consecutive even integers can be represented by the following:

the first even integer is $2n$,

the next even integer is $2n + 2$

the next even integer is $(2n + 2) + 2 = 2n + 4$, and

the next or fourth even integer is $(2n + 4) + 2 = 2n + 6$.

Hence, the four consecutive even integers are $2n$, $2n + 2$, $2n + 4$, and $2n + 6$. Then

$$2n + (2n + 2) + (2n + 4) + (2n + 6) = 140$$

$$8n + 12 = 140$$

$$8n = 128$$

$$n = 16$$

Therefore:

the first even integer = 2n = 2(16) = 32
the next even integer = 2n + 2 = 2(16) + 2 = 32 + 2 = 34
the next even integer= 2n + 4 = 2(16) + 4 = 32 + 4 = 36
and the next or fourth even integer
$$= 2n + 6 = 2(16) + 6 = 32 + 6 = 38$$

Note: The sum of these four consecutive even integers=

$$32 + 34 + 36 + 38 = 66 + 36 + 38 = 102 + 38 = 140.$$

● **PROBLEM** 1220

The product of two consecutive odd integers is 35.
Find the integers.

Solution: Let x represent the first odd integer. Then $x + 2$ represents the next odd integer. The product $x(x + 2)$ equals 35.
Thus, we solve the following equation.

$$x(x + 2) = 35.$$

$x^2 + 2x = 35$	Distributing
$x^2 + 2x - 35 = 0$	Subtracting 35 from both members
$(x + 7)(x - 5) = 0$	Factoring

In order for this product to = 0, either $(x + 7) = 0$ or $(x - 5) = 0$.
Therefore,

$$x = -7 \quad \text{or} \quad x = 5$$

Since x represents the first odd integer, we see that there are two solutions.
   When $x = -7$, $x + 2 = -5$, and $(-7) \cdot (-5) = 35$.
   When $x = 5$, $x + 2 = 7$, and $5 \cdot 7 = 35$.
The two integers are 5 and 7 or -7 and -5.

● **PROBLEM** 1221

Find 3 consecutive positive integers such that when 5 times the largest be subtracted from the square of the middle one the result exceeds three times the smallest by 7.

Solution: Let x = The smallest number

   Then $x + 1$ = The next larger number

      And $x + 2$ = The largest number

$$(x + 1)^2 - 5(x + 2) = 3x + 7$$
$$x^2 + 2x + 1 - 5x - 10 = 3x + 7$$
$$x^2 - 6x - 16 = 0$$
$$(x - 8)(x + 2) = 0$$
$$x = -2 \text{ reject}$$

$x = 8$
$x + 1 = 9$
$x + 2 = 10$

Check: $(9)^2 - 5(10) = 3(8) + 7$
      $81 - 50 = 24 + 7$, $31 = 31$

If the largest of four consecutive odd integers is represented by n, what is the smallest of these integers represented by?

Solution: Smaller consecutive odd integers are obtained by subtracting 2 from each successive odd integer. For example, the sequence 17, 15, 13 may be extended to include 11 and 9.

Hence, if n is the largest of four consecutive odd integers, then the three smaller ones found by subtracting 2 are (n - 2), (n - 4), and (n - 6). The smallest of these is (n - 6).

Can the sum of three consecutive odd integers be (a) 25? (b) 45?

Solution: Notice that all consecutive odd integers differ by 2:

1, 1 + 2 = 3, 3 + 2 = 5, 5 + 2 = 7, 7 + 2 = 9, ...

Thus, if we let x = the first consecutive odd integer

$\qquad$ x + 2 = the 2nd consecutive odd integer

and

(x+2) + 2 = x + 4 = the 3rd consecutive odd integer,

(a) We take the sum of these three numbers and determine if it can be 25:

$$x + (x+2) + (x+4) = 25$$
$$3x + 6 = 25$$
$$3x = 19$$
$$x = \frac{19}{3} .$$

Since 19/3 is not an integer, there are no such odd integers.

(b) If 25 is replaced by 45, the equation takes the form

$$x + (x+2) + (x+4) = 45$$
$$3x + 6 = 45$$
$$3x = 39$$
$$x = \frac{39}{3} = 13$$
$$x + 2 = 13 + 2 = 15$$
$$x + 4 = 13 + 4 = 17$$

Thus, the three consecutive odd integers are 13, 15, 17.

In general, if the sum of three consecutive odd integers is to be the number N, then N must be an integral multiple of 3.

CHAPTER 43

AGE AND WORK

> **Basic Attacks and Strategies for Solving Problems in this Chapter. See pages 872 to 876 for step-by-step solutions to problems.**

A strategy for solving a work problem is as follows:

(1) For each person or machine described in the problem, write a numerical or variable expression for the rate of work, the time worked, and the part of the task completed. The results can be recorded in a table.

(2) Determine how the parts of the task completed are related. Use the fact that the sum of the parts of the task completed must equal 1 in order to write an equation.

(3) Solve the equation and check the results.

For example, if a painter can paint a wall in 20 minutes and his apprentice can paint the same wall in 30 minutes, how long will it take to paint the wall when they work together? Let t denote the time worked. Then, the part of the task completed by the painter is t times his rate of work, 1/20, or $t/20$. Similarly, the part of the task completed by the apprentice is $t/30$. Thus, the equation may be stated as follows:

$$t/20 + t/30 = 1,$$

the completed job. After solving the equation, we find the answer to be

$$t = 12 \text{ minutes.}$$

This is the time needed to complete the task working together.

A strategy for solving an age problem includes the following:

(1) Represent the ages in terms of numerical or variable expressions. To represent a past age, subtract from the present age. The results can be recorded in a table.

(2) Determine the relationship among the ages in order to write an equation.

(3) Solve the equation and check the results.

For example: a car is 20 years old and its owner is 10 years old. How many years ago was the car three times as old as its owner was then?

Let x represent the age a number of years ago. Then, the past age of the car is

$20 - x$ and $10 - x$

for the owner. These past ages are related as follows:

$20 - x = 3(10 - x)$.

The solution is $x = 5$. So, five years ago the car was three times as old as its owner.

Step-by-Step Solutions to Problems in this Chapter, "Age and Work"

● **PROBLEM** 1224

John is 4 times as old as Harry. In six years John will be twice as old as Harry. What are their ages now?

<u>Solution:</u> Let x represent Harry's age now. John's age now is then represented by 4x. In six years their respective ages will be (x + 6) and (4x + 6). In tabular form our data is:

	Now	In Six Years
John's Age	4x	4x + 6
Harry's Age	x	x + 6

From the statement six years from now John will be twice as old as Harry, we can write the equation necessary to solve the problem. To form an equation from this information we must multiply Harry's age in six years by 2, to account for John's age being 2 times Harry's. Thus, we have:

$$4x + 6 = 2(x + 6), \text{ or}$$
$$4x + 6 = 2x + 12.$$

We now want to solve for x to obtain Harry's age. To do this we proceed as follows: Subtract 2x from both sides of the equation, 4x + 6 = 2x + 12. We obtain:

$$4x - 2x + 6 = 2x - 2x + 12$$
$$2x + 6 = 12. \tag{1}$$

Now, subtracting 6 from both sides of Equation (1), we have:

$$2x + 6 - 6 = 12 - 6$$
$$2x = 6. \tag{2}$$

Finally, dividing both sides of Equation (2) by 2, we obtain:

$$\frac{2x}{2} = \frac{6}{2} \text{ or, } x = 3.$$

Therefore, Harry's age now is 3 years, and John's is 4(3) = 12 years.
We can verify these values by observing that in six years Harry will be (3 + 6) or 9 years old and John will be [4(3) + 6] or 18 years old. Therefore, John will then be twice as old as Harry, which was given above.

● **PROBLEM** 1225

A mother is now 24 years older than her daughter. In 4 years, the mother will be 3 times as old as the daughter. What is the present age of each?

Solution: Let x = the age of the daughter
Then x + 24 = the age of the mother
x + 4 = the age of the daughter in 4 years
x + 24 + 4 = x + 29 = the age of the mother in 4 years

The relationship used in setting up the equation is:
In 4 years, the mother will be 3 times as old as the daughter.

$$x + 28 = 3(x + 4); \quad x = 8$$

Check: The daughter is now 8 years old and the mother is 32 years old.
In 4 years, the daughter will be 12 years old and the mother will be
36 years old. At this time, the mother will be 3 times as old as the
daughter.

● **PROBLEM** 1226

A man is now 6 times as old as his son. In two years, the man
will be 5 times as old as his son. Find the present ages of the man
and his son.

Solution: Let x = the present age of the son.
Then let 6x = the present age of the father.
x + 2 = the son's age in 2 years.
6x + 2 = the father's age in 2 years.

The relationship used in setting up the equation is:
In two years the father will be five times as old as the son.
$$6x + 2 = 5(x + 2)$$
$$6x + 2 = 5x + 10$$
$$x = 8$$

Check: The son is now 8 and the father is now 48.
In two years, the son will be 10 and the father will be 50.
At this time, the father will be 5 times as old as the son.

● **PROBLEM** 1227

John is now 18 years old and his brother, Charles, is 14 years
old. How many years ago was John twice as old as Charles?

Solution: Let x = the number of years ago John was twice as old as
Charles.
Then 18 - x = John's age x years ago.
And 14 - x = Charles' age x years ago.
The relationship used in setting up the equation is:

x years ago, John was twice as old as Charles

$$18 - x = 2(14 - x); \quad x = 10$$

Check: 10 years ago, John was 8 and Charles was 4. At this time,
John was twice as old as Charles.

● **PROBLEM** 1228

How long will it take Jones and Smith working together to plow a
field which Jones can plow alone in 5 hours and Smith alone in 8 hours?

Solution: Here we have

(1) How long will it take Jones and Smith working together to plow
a field?
(2) Jones and Smith working together plow a field.
(3) Jones can plow it alone in 5 hours.

873

(4)   Smith can plow it alone in 8 hours.

These sentences become, in turn:

(1')   Let x = number of hours that it takes Jones and Smith to plow the field.

(2')   (Jones's fractional part of the work) + (Smith's fractional part of the work) = 1, since they do 1 job together

(3')   Since Jones does the job alone in 5 hours, we can write 1 job = 5 hours work.   Divide both sides of this equation by 5:

$$\frac{1}{5} \text{ job} = \frac{5 \text{ hours work}}{5}$$

$$\frac{1}{5} \text{ job} = 1 \text{ hour work.} \qquad (1)$$

Hence, Jones does 1/5 of the job in 1 hour.   Note that the numerator of the fraction on the left side of equation (1) is equal to the number of hours; i.e., 1 = 1.   Therefore, in x hours, Jones does x/5 of the job.

(4')   Similarly, in one hour Smith does 1/8 of the job and, in x hours, x/8 of the job. Hence, we have

$$\frac{x}{5} + \frac{x}{8} = 1 \qquad (2)$$

Obtaining a common denominator of 40 for the two fractions on the right side of equation (2):

$$\frac{8(x)}{8(5)} + \frac{5(x)}{5(8)} = 1$$

$$\frac{8x}{40} + \frac{5x}{40} = 1$$

$$\frac{8x + 5x}{40} = 1$$

$$\frac{13x}{40} = 1 \qquad (3)$$

Multiply both sides of equation (3) by 40/13,

$$\frac{40}{13}\left(\frac{13x}{40}\right) = \frac{40}{13} \qquad (1)$$

$$x = \frac{40}{13} \text{ hours} = 3\frac{1}{13} \text{ hours} \approx 3 \text{ hours 5 min-}$$

utes. Therefore, it takes Jones & Smith 3 hours 5 minutes to plow the field.

● **PROBLEM** 1229

If A can do a job in 8 days and B can do the same job in 12 days, how long would it take the two men working together?

<u>Solution:</u> Let x = the number of days it would take the two men working together.

Then $\frac{x}{8}$ = the part of the job done by A

and $\frac{x}{12}$ = the part of the job done by B

The relationship used in setting up the equation is:

Part of job done by A + Part of job done by B = 1 job

$$\frac{x}{8} + \frac{x}{12} = 1$$

$$3x + 2x = 24$$
$$x = 4\frac{4}{5} \text{ days}$$

Check: $\dfrac{4\frac{4}{5}}{8} + \dfrac{4\frac{4}{5}}{12} = 1$

$$\dfrac{\frac{24}{5}}{8} + \dfrac{\frac{24}{5}}{12} = 1$$

$$\frac{3}{5} + \frac{2}{5} = 1$$

● **PROBLEM** 1230

A mechanic and his helper can repair a car in 8 hours. The mechanic works 3 times as fast as his helper. How long would it take the helper to make the repair, working alone?

Solution: Let x = Number of hours it would take the mechanic working alone.

Then 3x = Number of hours it would take the helper working alone.

The relationship used in setting up the equation is:

Part of job done by mechanic + Part of job done by helper = 1 job.

$$\frac{8}{x} + \frac{8}{3x} = 1$$
$$x = 10\frac{2}{3} \text{ hours by mechanic}$$
$$3x = 32 \text{ hours by helper .}$$

Check: $\dfrac{8}{10\frac{2}{3}} + \dfrac{8}{32} = 1$

$$\frac{24}{32} + \frac{1}{4} = 1$$

$$\frac{3}{4} + \frac{1}{4} = 1$$

● **PROBLEM** 1231

A man can do a job in 9 days and his son can do the same job in 16 days. They start working together. After 4 days the son leaves and the father finishes the job alone. How many days did the man take to finish the job alone?

Solution: Let x = the number of days it takes the man to finish the job.

Note that the man actually works (x + 4) days, and the son actually works 4 days.

The relationship used to set up the equation is:
Part of job done by man + Part of job done by boy = 1 job

$$\frac{x + 4}{9} + \frac{4}{16} = 1$$

$$16(x + 4) + 4(9) = 144$$

$$x = 2\frac{3}{4} \text{ days}$$

Check: $\dfrac{2\frac{3}{4} + 4}{9} + \dfrac{4}{16} = 1$

$\dfrac{3}{4} + \dfrac{1}{4} = 1$

A tank can be filled in 9 hours by one pipe, in 12 hours by a second pipe, and can be drained when full, by a third pipe, in 15 hours. How long would it take to fill the tank if it is empty, and if all pipes are in operation?

<u>Solution:</u> Let x = the number of hours the pipes are in operation

Then $\dfrac{x}{9}$ = part of tank filled by first pipe

and $\dfrac{x}{12}$ = part of tank filled by second pipe

and $\dfrac{x}{15}$ = part of tank emptied by third pipe

The relationship used in setting up the equation is:

Part of tank filled by first pipe + Part of tank filled by second pipe - Part of tank emptied by third pipe = 1 Full tank.

$$\dfrac{x}{9} + \dfrac{x}{12} - \dfrac{x}{15} = 1$$

$$20x + 15x - 12x = 180$$

$$x = 7\dfrac{19}{23} \text{ hours}$$

Check: $\dfrac{7\frac{19}{23}}{9} + \dfrac{7\frac{19}{23}}{12} - \dfrac{7\frac{19}{23}}{15} = 1$

$$\left(\dfrac{180}{23} \cdot \dfrac{1}{9}\right) + \left(\dfrac{180}{23} \cdot \dfrac{1}{12}\right) - \left(\dfrac{180}{23} \cdot \dfrac{1}{15}\right) = 1$$

$$\dfrac{20}{23} + \dfrac{15}{23} - \dfrac{12}{23} = 1$$

CHAPTER 44

RATIOS, PROPORTIONS, AND VARIATIONS

> **Basic Attacks and Strategies for Solving Problems in this Chapter. See pages 877 to 891 for step-by-step solutions to problems.**

A proportion is an equation framed by equating two ratios, say, a/b and c/d, as follows:

$a/b = c/d.$

The procedure for solving a proportion referred to as the cross multiplication or means-extremes property. This property states that the proportion

$a/b = c/d$ yields $ad = bc,$

where b and d are not 0.

Once this property is applied, then we solve for the unknown variable in the equation.

If $y = kx$, where k is a constant, then y is said to vary directly as x. If x is not zero, then the procedure for finding k, the constant of variation, is to divide both sides of $y = kx$ by x to obtain $y/x = k$. For instance, if a car travels at a speed of 50 miles per hour for t hours, then the distance covered is $d = 50t$. Thus, d varies directly as t and 50 is the constant of variation.

If $y = k/x$, where k is a constant and x is not zero, then y varies inversely as x. The procedure for finding the constant of variation is to multiply the values of x and y to obtain $xy = k$.

For example: If a rectangle has the relation

$xy = 12$ square units,

then dividing by x (not zero), we get $y = 12/x$. So y varies inversely as x since the product of x and y is the constant 12.

Step-by-Step Solutions to Problems in this Chapter, "Ratios, Proportions, and Variations"

RATIOS AND PROPORTIONS

● **PROBLEM** 1233

Solve the proportion $\frac{x + 1}{4} = \frac{15}{12}$.

Solution: Cross multiply to determine x; that is, multiply the numerator of the first fraction by the denominator of the second, and equate this to the product of the numerator of the second and the denominator of the first.

$$(x + 1)12 = 4 \cdot 15$$

$$12x + 12 = 60$$

$$x = 4.$$

● **PROBLEM** 1234

If a/b = c/d, a + b = 60, c = 3, and d = 2, find b.

Solution: We are given $\frac{a}{b} = \frac{c}{d}$. Cross multiplying we obtain ad = bc. Adding bd to both sides, we have ad + bd = bc + bd, which is equivalent to d(a + b) = b(c + d) or

$$\frac{a + b}{b} = \frac{c + d}{d}.$$

Replacing (a + b) by 60, c by 3 and d by 2 we obtain

$$\frac{60}{b} = \frac{3 + 2}{2}$$

$$\frac{60}{b} = \frac{5}{2}.$$

Cross multiplying, 5b = 120

$$b = 24.$$

● **PROBLEM** 1235

On a map, $\frac{3}{16}$ inch represents 10 miles. What would be the length of a line on the map which represents 96 miles?

Solution: The lengths of line segments on the map are pro-portional to the actual distances on the earth. If L re-presents the length of the line segment on the map corresponding to a distance of 96 miles, then

$$\frac{\frac{3}{16} \text{ inches}}{\text{L inches}} = \frac{10 \text{ miles}}{96 \text{ miles}}$$

$$\frac{3}{16}(96) = 10L \qquad \text{by cross multiplying}$$

$$L = \frac{(3)(96)}{(16)(10)} = \frac{3(6)}{10}$$

$$L = \frac{18}{10}$$

$$L = 1\frac{4}{5} \text{ inches.}$$

● **PROBLEM** 1236

Find the height of a tree which casts a shadow 20 feet
long at the same time a vertical yard stick casts a
shadow 30 inches long.

Solution: Set up this proportion by realizing that each
height is proportional to the length of its shadow.
Since this is the case, the ratio of the height of the
tree and its shadow can be equated with the ratio of
the height of the yardstick and its shadow.

Let: H represent the height of the tree
h represent the height of the yardstick = 36 in
S represent the length of the shadow
of the tree = 20 ft
s represent the length of the shadow
of the yardstick = 30 in.

We now have: $\dfrac{H}{S} = \dfrac{h}{s}$

This equation is solved by multiplying both sides of the
Least Common Denominator, LCD, which is found by multiplying the two denominators together.

$$LCD = Ss$$

Our equation becomes

$$Ss\left(\frac{H}{S}\right) = Ss\left(\frac{h}{s}\right), \text{ or } Hs = hS.$$

Dividing by s: $H = \dfrac{hS}{s}$

$$H = \frac{36 \text{ in } X \text{ 20 ft}}{30 \text{ in}} \text{ ; inches cancel; and}$$

$$H = \frac{6}{5} X \text{ 20 ft} = 24 \text{ ft.}$$

Hence, the height of the tree is 24 feet.

Find the ratios of x : y : z from the equations

$$7x = 4y + 8z, \quad 3z = 12x + 11y.$$

Solution: By transposition we have

$$7x - 4y - 8z = 0$$

$$12x + 11y - 3z = 0.$$

To obtain the ratio of x : y we convert the given system into an equation in terms of just x and y. z may be eliminated as follows: Multiply each term of the first equation by 3, and each term of the second equation by 8, and then subtract the second equation from the first. We thus obtain:

$$21x - 12y - 24z = 0$$

$$- \underline{(96x + 88y - 24z = 0)}$$

$$- 75x + 100y = 0$$

Dividing each term of the last equation by 25 we obtain,

$$- 3x - 4y = 0 \quad \text{or,}$$

$$- 3x = 4y.$$

Dividing both sides of this equation by 4, and by - 3, we have the proportion:

$$\frac{x}{4} = \frac{y}{-3} \, .$$

We are now interested in obtaining the ratio of y : z. To do this we convert the given system of equations into an equation in terms of just y and z, by eliminating x as follows: Multiply each term of the first equation by 12, and each term of the second equation by 7, and then subtract the second equation from the first. We thus obtain:

$$84x - 48y - 96z = 0$$

$$- \underline{(84x + 77y - 21z = 0)}$$

$$- 125y - 75z = 0.$$

Dividing each term of the last equation by 25 we obtain

$$- 5y - 3z = 0, \quad \text{or,}$$

$$- 3z = 5y$$

Dividing both sides of this equation by 5, and by - 3, we have the proportion:

$$\frac{z}{5} = \frac{y}{-3} \, .$$

From this result and our previous result we obtain:

$\frac{x}{4} = \frac{y}{-3} = \frac{z}{5}$ as the desired ratios.

If $(2ma + 6mb + 3nc + 9nd)(2ma - 6mb - 3nc + 9nd)$

$= (2ma - 6mb + 3nc - 9nd)(2ma + 6mb - 3nc - 9nd)$,

prove that a, b, c, d are proportionals.

Solution: Dividing both sides of the given equation by $(2ma - 6mb - 3nc + 9nd)$, and then by $(2ma - 6mb + 3nc - 9nd)$ we have

$$\frac{2ma + 6mb + 3nc + 9nd}{2ma - 6mb + 3nc - 9nd} = \frac{2ma + 6mb - 3nc - 9nd}{2ma - 6mb - 3nc + 9nd} .$$

Since the above two ratios are of the form $\frac{a}{b} = \frac{c}{d}$, we can use the Law of Proportions which states that

$$\frac{a + b}{a - b} = \frac{c + d}{c - d} .$$

Doing this we obtain,

$$\frac{2ma + 6mb + 3nc + 9nd + (2ma - 6mb + 3nc - 9nd)}{2ma + 6mb + 3nc + 9nd - (2ma - 6mb + 3nc - 9nd)} =$$

$$\frac{2ma + 6mb - 3nc - 9nd + (2ma - 6mb - 3nc + 9nd)}{2ma + 6mb - 3nc - 9nd - (2ma - 6mb - 3nc + 9nd)}$$

or, $\frac{4ma + 6nc}{12mb + 18nd} = \frac{4ma - 6nc}{12mb - 18nd}$; and factoring gives us,

$$\frac{2(2ma + 3nc)}{2(6mb + 9nd)} = \frac{2(2ma - 3nc)}{2(6mb - 9nd)} \quad \text{or,}$$

$$\frac{2ma + 3nc}{6mb + 9nd} = \frac{2ma - 3nc}{6mb - 9nd} .$$

Now, since $\frac{a}{b} = \frac{c}{d}$ can be alternately written as $\frac{a}{c} = \frac{b}{d}$, we write:

$$\frac{2ma + 3nc}{2ma - 3nc} = \frac{6mb + 9nd}{6mb - 9nd} .$$

Now, rewriting this last proportion as,

$$\frac{2ma + 3nc + (2ma - 3nc)}{2ma + 3nc - (2ma - 3nc)} = \frac{6mb + 9nd + (6mb - 9nd)}{6mb + 9nd - (6mb - 9nd)}$$

we obtain: $\frac{4ma}{6nc} = \frac{12mb}{18nd} .$

Again, using the fact that $\frac{a}{b} = \frac{c}{d}$ can be rewritten as $\frac{a}{c} = \frac{b}{d}$, we write:

$$\frac{4ma}{12mb} = \frac{6nc}{18nd} \qquad \text{or,}$$

$$\frac{a}{3b} = \frac{c}{3d} \; .$$

Thus, $\frac{a}{b} = \frac{c}{d}$ or, $a : b = c : d$.

DIRECT VARIATION

If y varies directly with respect to x and y = 3 when x = -2, find y when x = 8.

Solution: If y varies directly as x then y is equal to some constant k times x; that is, $y = kx$ where k is a constant. We can now say $y_1 = kx_1$ and $y_2 = kx_2$ or

$\frac{y_1}{x_1} = k$, $\frac{y_2}{x_2} = k$ which implies $\frac{y_1}{x_1} = \frac{y_2}{x_2}$ which is a proportion.

We use the proportion $\frac{y_1}{x_1} = \frac{y_2}{x_2}$. Thus $\frac{3}{-2} = \frac{y_2}{8}$. Now solve for y_2:

$$8\left(\frac{3}{-2}\right) = 8\left(\frac{y_2}{8}\right)$$

$$-12 = y_2 .$$

When x = 8, y = -12.

According to Hooke's Law, the length of a spring, S, varies directly as the force, F, applied on the spring. In a spring to which Hooke's Law applies, a force of 18.6 lb stretches the spring by 1.27 in. Find k, the proportionality constant.

Solution: The direct variation of the length of the spring, S, and the force applied on it, F, is expressed symbolically as

F = kS, where k is the constant of proportionality.

We are given that F = 18.6 lb and S = 1.27 in. Thus, it is necessary merely to substitute the given values in the equation F = kS, obtaining

$$18.6 = k(1.27),$$

from which k = 18.6/1.27 = 14.65 lb/in.

The area of a sphere is directly proportional to the square of the radius. If a sphere of radius 3 inches has an area of 36π square inches, deduce the formula for the area of a sphere.

Solution: If a quantity x is directly proportional to another quantity y, x is the product of a constant k and y; that is, $x = ky$. Let a = the area of a sphere and let r = the radius of the sphere. Since the area a of a sphere is directly proportional to the square of the radius r, the following equation can be written:

$$a = kr^2 \qquad\qquad (1)$$

When the area a is 36π square inches, the radius is 3 inches. Then, using equation (1):

$$36\pi = k(3)^2$$

$$36\pi = k(9)$$

$$36\pi = 9k$$

Divide both sides of this equation by 9:

$$\frac{36\pi}{9} = \frac{9k}{9}$$

$$4\pi = k$$

Substituting this value for the constant k in equation (1): $a = (4\pi)r^2$, is the equation which represents the formula for the area of a sphere.

A falling body strikes the ground with a velocity v which varies directly as the square root of the distance s it falls. If a body that falls 100 feet strikes the ground with a velocity of 80 feet per second, with what velocity will a ball dropped from the Washington monument (approximately 550 feet high) strike the ground?

Solution: A variable y is said to vary directly as another variable x if y is equal to some constant k times x, that is, if $y = kx$. Here, the variable v, velocity, varies directly as the square root of the distance s. The general equation for this variation is

$$v = k\sqrt{s} \qquad\qquad (1)$$

Since $v = 80$ when $s = 100$, we have

$$80 = k\sqrt{100} \qquad\qquad (2)$$

$$80 = k(10)$$

$$k = \frac{80}{10}$$

$$k = 8. \tag{3}$$

Replacing $k = 8$ in Equation 1

$$v = 8\sqrt{s}. \tag{4}$$

When $s = 550$

$$v = 8\sqrt{550} \tag{5}$$

$$\simeq 8(23.45) \simeq 188.$$

Therefore the velocity is approximately 188 feet per second.

● **PROBLEM 1243**

The resistance R of a given size of wire at constant temperature varies directly as the length ℓ. It is found that the resistance of 100 feet of number 14 copper is 0.253 ohm. Construct a table of values for the given lengths of number 14 copper wire assuming the temperature is constant.

ℓ	25	75	125	175	225
R					

Solution: Direct variation implies a variable y is equal to a constant c times x; that is, $y = cx$. In this particular example, the resistance R varies directly as the length ℓ. Since several values of R are to be found, we use the general equation for this variation. Thus,

$$R = k\ell \tag{1}$$

Since $R = 0.253$ when $\ell = 100$

$$0.253 = 100k \tag{2}$$

Solving for k

$$k = 0.00253 \tag{3}$$

The specific equation is therefore

$$R = 0.00253\ell \tag{4}$$

The values of R corresponding to the given values of ℓ may then be found directly from Equation 4, as follows.

ℓ	25	75	125	175	225
R	0.063	0.190	0.316	0.443	0.569

Notice that if values of ℓ were to be determined from given values R, it would be convenient to solve Equation 4 for ℓ in terms of R. Thus

$$\ell = \frac{R}{0.00253} \tag{5}$$

or

$$\ell = 395.26R$$

● **PROBLEM 1244**

The square of the time of a planet's revolution varies

as the cube of its distance from the Sun; find the time
of Venus' revolution, assuming the distances of the Earth
and Venus from the Sun to be 91 ¼ and 66 millions of
miles, respectively.

Solution: Since x varies directly as y, x = ky, where k
is a constant, called the constant of proportionality,
or the constant of variation. We are given that
(time of revolution)^2 = k (distance from Sun)^3.

Let P be the periodic time measured in days, D the distance in millions of miles; we have

$$P^2 = kD^3,$$

where k is some constant.

We are given D for the Earth, and since we know
that P for Earth is 365 days, we can use this information to find k. We have:
$(365)^2 = k(91¼)^3$. This can be rewritten as:

$365 \times 365 = k \times 91¼ \times 91¼ \times 91¼$

Now, since $91¼ = \dfrac{365}{4}$, we write:

$365 \times 365 = k \left[\dfrac{365}{4} \times \dfrac{365}{4} \times \dfrac{365}{4}\right]$ or,

$365 \times 365 = k \left[\dfrac{365 \times 365 \times 365}{4 \times 4 \times 4}\right]$. Solving for
k we have:

$\left(\dfrac{4 \times 4 \times 4}{365 \times 365 \times 365}\right)$ $365 \times 365 = k.$ Thus,

$k = \dfrac{4 \times 4 \times 4}{365}$.

We can now use this value of k to solve for the time
of Venus' revolution. We are given that D = 66. Thus, we
have:

$P^2 = kD^3$ and, substituting we obtain:

$P^2 = \left(\dfrac{4 \times 4 \times 4}{365}\right) (66)^3$; therefore,

P = the time of Venus' revolution =

$$\sqrt{\dfrac{4 \times 4 \times 4}{365} \cdot (66)^3}$$

INVERSE VARIATION

● PROBLEM 1245

Express y in terms of x if y is inversely proportional

to x and y = 2 when x = 3.

Solution: y is inversely proportional to x means there exists a number k such that y = k/x. Our problem will be solved when we determine the number k. If we substitute 3 for x and 2 for y in the equation y = k/x, we find 2 = k/3. Multiplying both sides by 3 we obtain k = 6. Thus, the equation relating x and y is y = 6/x.

● PROBLEM 1246

If y varies inversely as the cube of x, and y = 7 when x = 2, express y as a function of x.

Solution: The relationship "y varies inversely with respect to x" is expressed as,

$$y = \frac{k}{x}.$$

The inverse variation is now with respect to the cube of x, x^3 and we have,

$$y = \frac{k}{x^3}$$

Since y = 7 and x = 2 must satisfy this relation, we replace x and y by these values,

$$7 = \frac{k}{2^3} = \frac{k}{8},$$

and we find k = 7 · 8 = 56. Substitution of this value of k in the general relation gives,

$$y = \frac{56}{x^3},$$

which expresses y as a function of x. We may now, in addition find the value of y corresponding to any value of x. If we had the added requirement to find the value of y when x = 1.2, x = 1.2 would be substituted in the function so that for x = 1.2, we have,

$$y = \frac{56}{(1.2)^3} = \frac{56}{1.728} = 32.41$$

Other expressions in use are "is proportional to" for "varies directly," and "is inversely proportional to" for "varies inversely."

● PROBLEM 1247

The cube root of x varies inversely as the square of y; if x = 8 when y = 3, find x when y = 1½.

Solution: If x varies inversely as y, then $x = \frac{m}{y}$, and in this case $\sqrt[3]{x}$ varies inversely as y^2, thus,

$\sqrt[3]{x} = \frac{m}{y^2}$, where m is constant.

Putting x = 8, y = 3, we have $\sqrt[3]{8} = \frac{m}{3^2}$ or, $2 = \frac{m}{9}$;

therefore,   m = 18.

Now, we want to find x when $y = 1\frac{1}{2} = \frac{3}{2}$. We know

from above that m = 18, and since $\sqrt[3]{x} = \frac{m}{y^2}$, by sub-

stitution we obtain:

$$\sqrt[3]{x} = \frac{18}{\left(\frac{3}{2}\right)^2}$$

$$\sqrt[3]{x} = \frac{18}{\frac{9}{4}}$$

$\sqrt[3]{x} = 18 \left(\frac{4}{9}\right) = 8$. Now cubing both sides we obtain:

x = 8^3 or x = 512.

● **PROBLEM** 1248

The weight W of an object above the earth varies inversely
as the square of the distance d from the center of the
earth. If a man weighs 180 pounds on the surface of the
earth, what would his weight be at an altitude of 1000
miles? Assume the radius of the earth to be 4000 miles.

Solution: W varies inversely with d^2; therefore $W = \frac{k}{d^2}$

where k is the proportionality constant. Similarly,

$W_1 = \frac{k}{d_1^2}$, $W_2 = \frac{k}{d_2^2}$ and, solving these two equations for k,

$W_1 d_1^2 = k$ and $W_2 d_2^2 = k$.   Hence,

$$k = W_1 d_1^2 = W_2 d_2^2$$

or $$\frac{W_1 d_1^2}{W_2} = \frac{W_2 d_2^2}{W_2}$$

$$\frac{W_1 d_1^2}{W_2} = d_2^2$$

$$\frac{W_1 d_1^2}{W_2 d_1^2} = \frac{d_2^2}{d_1^2}$$

$$\frac{W_1}{W_2} = \frac{d_2^2}{d_1^2} \qquad (1)$$

Letting d_1 = radius of the earth, 4000, then d_2 = 4000 +

1000 = 5000.

Substituting the given values in Equation (1):

$$\frac{180}{W_2} = \frac{5000^2}{4000^2} = \frac{(5 \times 1000)^2}{(4 \times 1000)^2} = \frac{5^2 \times \cancel{1000}^2}{4^2 \times \cancel{1000}^2}$$

$$= \frac{5^2}{4^2}$$

$$= \frac{25}{16}$$

$$\frac{180}{W_2} = \frac{25}{16}$$

$$W_2\left(\frac{180}{W_2}\right) = W_2\left(\frac{25}{16}\right)$$

$$180 = \frac{25}{16}W_2$$

$$\frac{16}{\cancel{25}}\overset{36}{(\cancel{180})} = \frac{\cancel{16}}{\cancel{25}}\left(\frac{\cancel{25}}{\cancel{16}}\, W_2\right)$$

$$\frac{576}{5} = W_2$$

$$115\tfrac{1}{5} \text{ pounds} = W_2$$

$$\text{or } 115.2 \text{ pounds} = W_2$$

JOINT AND COMBINED DIRECT - INVERSE VARIATIONS

● **PROBLEM** 1249

If y varies jointly as x and z, and 3x:1 = y:z, find the constant of variation.

__Solution:__ A variable s is said to vary jointly as t and v if s varies directly as the product tv; that is, if s = ctv where c is called the constant of variation.

Here the variable y varies jointly as x and z with k as the constant of variation.

$$y = kxz$$
$$3x:1 = y:z$$

Expressing these ratios as fractions.

$$\frac{3x}{1} = \frac{y}{z}$$

Solving for y by cross-multiplying,

$$y = 3xz$$

Equating both relations for y we have:

$$kxz = 3xz$$

Solving for the constant of variation, k, we divide both sides by xz, $k = 3$

The pressure of wind on a sail varies jointly as the area of the sail and the square of the wind's velocity. When the wind is 15 miles per hour, the pressure on a square foot is one pound. What is the velocity of the wind when the pressure on a square yard is 25 pounds?

Solution: Let p = pressure of the wind, in pounds
v = the velocity of the wind, in miles per hour
a = the area of the sail, in square feet.

Pressure, p, varies jointly as the area of the sail, a, and the square of the wind's velocity, v^2. Therefore p varies directly as the product av^2 times a proportionality constant, k. k must be determined before we can proceed to find v as desired. Use the given information a = 1 and p = 1 when v = 15 to determine the proportionality constant, k.

$$p = kav^2.$$

$$1 = k(1)(225).$$

$$k = \frac{1}{225}, \text{ value of the proportionality constant.}$$

Now we can find v using $k = \frac{1}{225}$ when p = 25 and a = 9 (1 yard = 3 feet, 1 square yard = 9 square feet).

$$p = \frac{1}{225}av^2.$$

$$25 = \frac{1}{225}(9)v^2.$$

$$v^2 = \frac{(25)(225)}{9}$$

$$v = \sqrt{\frac{(25)(225)}{9}} = \frac{(5)(15)}{3}$$

$$v = 25, \text{ number of miles per hour.}$$

The elongation, E, of a steel wire when a mass, m, is hung from its free end varies jointly as m and the length, x, of the wire and inversely as the cross sectional area A, of the wire. Given that E = 0.001 inches when m = 20 pounds, x = 10 inches, and A = 0.01 square inches, find E when m = 40 pounds, x = 15.5 inches, and A = 0.015 square inches.

Solution: If E is directly proportional to m and x and inversely proportional to A, our equation is

$$E = \frac{kmx}{A}$$

where k is a constant which we can determine from the given information.

Since, when m = 20, x = 10, and A = .01, E = .001, we can use this information to solve for k. Substituting we obtain:

$$0.001 = \frac{k(20)(10)}{.01} \quad , \text{ and multiplying both}$$

sides by $\frac{.01}{(20)(10)}$ we obtain:

$$k = \frac{(.001)(.01)}{200} = \frac{.00001}{200} .$$

Multiplying numerator and denominator by 100,000 (move the decimal 5 places to the right) we obtain:

$$\frac{1}{20,000,000}$$

Hence, $\quad E = \frac{1}{20,000,000} \frac{mx}{A}$

and when m = 40, x = 15.5, and A = 0.015, we have:

$$E = \frac{1}{20,000,000} \frac{40(15.5)}{0.015}$$

= 0.00207 inches (approximately).

● **PROBLEM** 1252

The current, i, in amperes in an electric circuit varies directly as the electromotive force, E, in volts and inversely as the resistance, R, in ohms. If, in a certain circuit, i = 30 amperes when R = 15 ohms and E = 450 volts, find i when R = 20 ohms and E = 200 volts.

Solution: The relation "i varies directly as E and inversely as R" may be expressed as $i = \frac{kE}{R}$ where k is the proportionality constant. k must be determined before i can be found as desired.

i = kE/R and so

$$30 = \frac{k \cdot 450}{15} \quad \text{or, } k = 30\left(\frac{15}{450}\right)$$

$$= \frac{15}{15}$$

$$= 1.$$

Hence i = E/R and when R = 20 and E = 200, we have i =

= 200/20 = 10 amperes.

A certain beam L ft. long has a rectangular cross section b in. in horizontal width and d in. in vertical depth. It is found that, when the beam is supported at the ends, the deflection D at the center varies directly as the fourth power of L, inversely as b, and inversely as the cube of d. If the length is decreased by 10 per cent but the width kept the same, by how much should the depth be changed in order that the same deflection D be obtained?

Solution: A quantity m varies directly as another q quantity n if m equals the product of a constant and n; that is, m = cn where c is the constant. Also, a quantity p varies inversely as another quantity q if p equals the product of a constant and the reciprocal of q; that is, p = c(1/q) = c/q where c is the constant. From the statement of the problem, we see that the combined variation is given by

$$D = \frac{kL^4}{bd^3} \quad , \text{ where } k \text{ is a constant.}$$

Since corresponding values of the four variables are not known, we cannot determine the value of the constant of proportionality k. But if one set of variables is designated with the subscript 1, and the new set with the subscript 2, we have

$$D_1 = \frac{kL_1^4}{b_1 d_1^3} \quad , \quad D_2 = \frac{kL_2^4}{b_2 d_2^3} \quad .$$

Since it is desired to obtain a relationship between d_1 and d_2, carry out the following procedure in order to isolate d_1 and d_2 on different sides of the equation.

Divide D_2 by D_1:

$$\frac{D_2}{D_1} = \frac{kL_2^4/b_2 d_2^3}{kL_1^4/b_1 d_1^3} \tag{1}$$

Since division by a fraction is equivalent to multiplication by that fraction's reciprocal, equation (1) becomes:

$$\frac{D_2}{D_1} = \frac{kL_2^4}{b_2 d_2^3} \frac{b_1 d_1^3}{kL_1^4} = \frac{L_2^4 b_1 d_1^3}{b_2 d_2^3 L_1^4} = \frac{L_2^4 b_1 d_1^3}{L_1^4 b_2 d_2^3}$$

or

$$\frac{D_2}{D_1} = \frac{L_2^4 b_1 d_1^3}{L_1^4 b_2 d_2^3}$$

In this problem we have $D_2 = D_1$, $L_2 = 0.9L_1$, and $b_2 = b_1$. Therefore we get

$$\frac{D_1}{D_1} = \frac{\left(0.9L_1\right)^4 b_1 d_1^3}{L_1^4 b_1 d_2^3}$$

$$1 = \frac{(0.9)^4 L_1^4 b_1 d_1^3}{L_1^4 b_1 d_2^3}$$

$$1 = (0.9)^4 \frac{d_1^3}{d_2^3}$$

Multiplying both sides by d_2^3:

$$d_2^3(1) = d_2^3(0.9)^4 \frac{d_1^3}{d_2^3}$$

$$d_2^3 = (0.9)^4 d_1^3$$

Take the cube root of each side:

$$\sqrt[3]{d_2^3} = \sqrt[3]{(0.9)^4 d_1^3}$$

$$d_2 = \sqrt[3]{(0.9)^4} \sqrt[3]{d_1^3}$$

$$d_2 = \sqrt[3]{(0.9)^4} \, d_1$$

$$d_2 = (0.9)^{4/3} d_1$$

$$d_2 = \left[(0.9)^{1/3}\right]^4 d_1$$

$$d_2 = \left[\sqrt[3]{0.9}\right]^4 d_1$$

$$d_2 \approx (.966)^4 d_1$$

$$d_2 \approx .871 \, d_1$$

Hence, the new depth, d_2 is only $.871 \, d_1$. Subtracting $.871 \, d_1$ from $1 \, d_1$:

$$
\begin{array}{r}
1.000 \ d_1 \\
- \ 0.871 \ d_1 \\
\hline
0.129 \ d_1
\end{array}
$$

Therefore, the depth has been decreased by an amount of .129 or

$$\frac{129}{1000} = \frac{129}{10} \times \frac{1}{100}$$

$$= \frac{129}{10} \text{ \% since "hundredths means per cent"}$$

$$= 12.9\%.$$

Hence, the depth has been decreased approximately 13%.

CHAPTER 45

COSTS

Basic Attacks and Strategies for Solving Problems in this Chapter. See pages 892 to 897 for step-by-step solutions to problems.

Cost, the price which a business pays for a product, is related to selling price and markup (profit). Selling price is the price a business sells a product to a customer, which includes a markup (profit) that is added to the cost. Thus,

Cost = Selling Price – Markup (Profit)   or $C = S - M$.

Usually, the markup (profit) is expressed in terms of a percent of the business cost. Thus, markup (profit),

$M = r \cdot C$,

where r is the markup rate and C is the cost.

The procedure for solving problems involving cost focuses upon first representing the various unknown quantities and then determining their relation and expressing it using the basic equation

$S = C + M$ or $S = C + rC$.

Then, solve the equation.

For example: A store employee uses a markup rate of 40% on all items. The selling price of a lawnmower is $105. Then the cost of the lawnmower is given by

$S = C + rC$.

Thus, we have

$105 = C + 0.4 \cdot C$ or $C = \$75$.

Many cost problems involve the use of discount, which is the amount by which a retailer reduces the regular price of a product for a promotional sale. This is usually given as a percent, the discount rate. The basic discount equation for solving cost problems involving discount is as follows:

Sale Price = Regular Price − Discount

$$S = R - D,$$

where

Discount = Discount Rate × Regular Price

$$D = r \times R.$$

For example: The sale price of a sprayer is $23.40. This price is 35% off the regular price. Then, by the above equations, the regular price is given by

$$S = R - r \cdot R \quad \text{or}$$

$$23.40 = R - 0.35 \cdot R \quad \text{or}$$

$$\$36 = R, \text{ the regular price.}$$

● **PROBLEM** 1254

Reserved seat tickets to a football game are $6 more than general admission tickets. Mr. Jones finds that he can buy general admission tickets for his whole family of five for only $3 more than the cost of reserved seat tickets for himself and Mrs. Jones. How much do the general admission tickets cost?

<u>Solution:</u> Let x = the cost of general admission tickets.

Let x + 6 = the cost of reserved seat tickets.

Thus, 5x is the cost of five general admission tickets and 2(x+6) is the cost of two reserved seat tickets. Since the five tickets cost three dollars more than the two reserved tickets,

$$5x = 2(x + 6) + 3$$
$$5x = 2x + 12 + 3$$
$$5x = 2x + 15$$
$$3x = 15$$
$$x = 5,$$

and $x + 6 = 5 + 6 = 11$

Thus, general admission tickets are $5.00, and reserve tickets are $11.00.

● **PROBLEM** 1255

A television set cost a dealer $102. At what price should he mark the set so that he can give a discount of 15% from the marked price and still make a profit of 20% on the selling price?

<u>Solution:</u> Let x = The marked price

Then .85x = The selling price

The relationship used to set up the equation is:

Selling price = Cost + Profit

$$.85x = 102 + (.20)(.85x)$$
$$.85x = 102 + .17x$$
$$.68x = 102$$
$$x = \$150$$

Check: The marked price = $150. The selling price is 15% less or $127.50. Since the cost is $102, the profit is $25.50. The profit ($25.50) is 20% of the selling price ($127.50).

● **PROBLEM** 1256

At a movie showing there were 356 paid admissions. The total receipts were $287.40. If orchestra seats sold for $.90 and balcony seats for $.65, how many of each kind were sold?

<u>Solution:</u>   Let x = The number of $.90 seats sold

Then (356 - x) = The number of $.65 seats sold

The relationship used to set up the equation is:

Orchestra receipts + Balcony receipts = Total receipts or $287.40

$$.90x + .65(356 - x) = 287.40$$

$$90x + 65(356 - x) = 287.40$$

$$90x + 23140 - 65x = 287.40$$

$$25x = 5600$$

$$x = 224 \text{ orchestra seats sold}$$

$$356 - x = 132 \text{ balcony seats sold}$$

Check: 224 at $.90 = $201.60
132 at $.65 = $85.80
356 tickets for $287.40

● **PROBLEM   1257**

A merchant paid $1,800 for a group of men's suits. He sold all but 5 of the suits at $20 more per suit than he paid, thereby making a profit of $200 on the transaction. How many suits did the merchant buy?

<u>Solution:</u>   Let x = The number of suits the merchant bought

Then $\dfrac{1,800}{x}$ = The cost of each suit

The relationship used to set up the equation is:

The number of suits sold X The selling price of each suit

$$= \$1,800 + \$200$$

$$(x - 5) \left(\frac{1,800}{x} + 20 \right) = 2,000$$

$$1,800 + 20x - \frac{9,000}{x} - 100 = 2,000$$

$$20x - \frac{9,000}{x} = 300$$

$$20x^2 - 300x - 9,000 = 0$$

$$x^2 - 15x - 450 = 0$$

$$(x - 30)(x + 15) = 0$$

$$x = 30, \quad x = -15 \text{ reject}$$

The merchant bought 30 suits.

Check: The merchant bought 30 suits at $60 each. He sold 25 suits at $80 each, thus taking in $2,000.

● **PROBLEM   1258**

A dealer can buy a certain number of ties for $30. If 5 more could be bought for the same money, the price would be $3.60 less per dozen. What is the price per dozen?

<u>Solution:</u>   Let x = The number of ties the dealer can buy for $30.

Then $\frac{30}{x}$ = The cost per tie

And $\frac{30}{x + 5}$ = The supposed cost per tie

The relationship used to set up the equation is:
The supposed cost per dozen = The old cost per dozen minus $3.60

$$12\left(\frac{30}{x + 5}\right) = 12\left(\frac{30}{x}\right) - \frac{18}{5}$$

$$12(30)(5x) = 12(30)(5)(x + 5) - 18(x)(x + 5)$$

$$1,800x = 1,800x + 9,000 - 18x^2 - 90x$$

$$18x^2 + 90x - 9,000 = 0$$

$$x^2 + 5x - 500 = 0$$

$$(x + 25)(x - 20) = 0$$

$$x = -25 \text{ reject}, \quad x = 20$$

The man bought 20 ties for $30, which is at the rate of $18 per dozen.

Check: The man bought 20 ties for $30, paying $1.50 per tie or $18 per dozen. Had he bought 25 ties for $30 he would have paid $1.20 per tie or $14.40 per dozen. This would have been a saving of $3.60 per dozen.

● PROBLEM 1259

A haberdasher sold 3 shirts and 4 ties to one customer for $18.70. Another customer bought 4 shirts and 7 ties of the same quality for $27.75. What was the price per shirt and price per tie?

Solution: Let x = Cost of one shirt
And y = Cost of one tie

(1) $3x + 4y = 1870$
(2) $4x + 7y = 2775$

When the equations are solved simultaneously it is found that $x = 398$, i.e., the price of a shirt is $3.98 and $y = 169$, i.e., the price of a tie is $1.69.

● PROBLEM 1260

A real estate dealer received $1,200 in rents on two dwellings last year, and one of the dwellings brought $10 per month more than the other. Find the monthly rental on each if the more expensive house was vacant for 2 months.

Solution: On inspecting the problem we see that there are two basic relations involved, the relation between the separate rentals, and the relation between the monthly rentals and the income per year. Since the monthly rentals differ by $10, we let

x = monthly rental of the more expensive house in dollars
y = monthly rental of the less expensive house in dollars

and we write $\qquad x = y + 10$

or $\qquad x - y = 10 \qquad$ (1)

Furthermore, since the first of the two houses was rented

for 10 months and the other was rented for 12 months, we know that 10x + 12y is the total annual income. Hence

$$10x + 12y = 1,200 \qquad (2)$$

We now solve Eq. (1) and (2) simultaneously by eliminating y:

12x - 12y = 120	multiplying (1) by 12
10x + 12y =1200	(2) recopied
22x = 1,320	equating the sums of corresponding members
x = 60	solving for x
60 - y = 10	replacing x by 60 in (1)
y = 50	solving for y

Therefore, the solution set {(x, y)} = {(60, 50)}, and it follows that the monthly rentals are $60 and $50, respectively.

We check the obtained values by substituting in equations (1) and (2). Thus

$$x - y = 10 \qquad (1)$$
$$60 - 50 = 10$$
$$10 = 10$$

$$10x + 12y = 1200 \qquad (2)$$
$$10(60) + 12(50) = 1200$$
$$600 + 600 = 1200$$
$$1200 = 1200$$

● **PROBLEM** 1261

The income of a business is at the rate of $1000 a week and the expenses are at the rate of $900 a week. If the business was worth $8000 at the start, obtain the function showing the value of the business at the end of t weeks, assuming that the profits are retained in the business.

Solution: Since income exceeds expenses by $100 a week, the rate of increase of the function is m = 100; the value of the function at t = 0 is 8000, that is, b = f(0) = 8000, since the original value was 8000. If t represents the number of weeks, the desired function of time will be linear with slope 100 and y-intercept 8000; and f(t) = 100t + 8000.

The time at which the business is worth a specified amount is of particular interest. For example, the time at which the business is worth $10,000 is seen to be t = 20 weeks, for, with t = 20, we have

$$100 \cdot 20 + 8000 = 10,000$$

This value of t may also be thought of as the value of t for which the function

$$(100t + 8000) - 10,000, \quad \text{or } 100t - 2000$$

vanishes.

More generally, if we set the function mx + b equal to c, we get

$$mx + b = c, \quad \text{or} \quad mx + b - c = 0.$$

The value of x which satisfies the equation in either of the two forms is then $x = (c - b)/m$. From the second form of the equation it is seen that for this value of x the value of the function mx + b - c is zero.

● **PROBLEM** 1262

A wholesale outlet has room in its radio and television section for not more than 150 radio and television sets. A radio set weighs 50 pounds and a television set weighs 100 pounds, and the floor is limited by the city in-spector to a total weight of 10,000 pounds. The profit on a radio set is $50, and the profit on a television set is $75. In order to realize a maximum profit, how many of each shall be stocked? We shall assume, of course, that radio sets and television sets sell equally well.

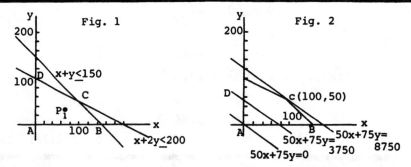

Solution: Let x = the number of radio sets and y = the number of television sets. Since each radio weighs 50 pounds and each television weighs 100 pounds: Let 50x= total weight contributed by the radios and 100y = total weight contributed by the television sets. Because of the limitation on weight, we have the first constraint

$$50x + 100y \leq 10,000 \tag{1}$$

Since at most 150 sets can be stocked, the second constraint is

$$x + y \leq 150 \tag{2}$$

Since the number of sets of either kind cannot be less than zero, we have two more constraints

$$x \geq 0 \tag{3}$$
$$y \geq 0 \tag{4}$$

The graph of this system (noting that inequality 1 can be reduced to x + 2y < 200) is shown in Figure 1. It is clear that only points such as P_1 which lie within or on the boundary of the polygon will satisfy the given constraints, and are therefore called feasible points.

We now consider the profit equation also

$$50x + 75y = P \qquad (5)$$

where $50x$ = total profit on all radio sets and $75x$ = total profit on all television sets.

For all values of P, Equation 5 represents a system of parallel lines, as shown in Figure 2. We note further that as P increases, the lines move to the right. Of particular significance is the fact that the maximum profit occurs at the vertex C, since no feasible points exist beyond this point. This can be shown by substituting other points of the polygon into the profit equation. For example, given the point (40,8), then P = $8000. The coordinates of C are found by solving the system of equations, derived from the first two constraints, since this is their point of intersection. The equations are:

$$x + y = 150$$
$$x + 2y = 200 \qquad (6)$$

Subtracting the first equation from the second we obtain:

$x = 100$, the number of radio sets
$y = 50$, the number of television sets

Thus, the maximum profit is

$$P = 100(50) = 50(75) = \$8750.$$

CHAPTER 46

INTEREST AND INVESTMENTS

> **Basic Attacks and Strategies for Solving Problems in this Chapter. See pages 898 to 901 for step-by-step solutions to problems.**

The annual simple interest which an investment earns is given by the equation

$I = Prt,$

where I is the simple interest, P is the principle or the amount invested, and r is the simple interest rate. To find I, the interest, simply substitute in the formula and simplify.

The strategy for solving a problem involving an investment in two simple interest accounts includes the following:

(1) For each amount invested, write a numerical or variable expression for the principal, the interest rate, and the interest earned. The results can be displayed in a table.

(2) Determine how the amounts of interest earned on each investment are related. For example, the total interest earned by both investments may be known or it may be known that the interest earned on one investment is equal to the interest earned by the other investment.

(3) Use the relations determined in Step 2 to write an equation. Solve the equation and check the results.

For example: An investor has a total of $10,000 to deposit into two simple interest accounts. On one account, the annual simple interest rate is 7%. On the second account, the annual simple interest rate is 11%. How much should be invested in each account so that the total interest is earned is $1,000?

Let x denote the amount invested at 7%. Then use the table below to summarize the representation of all other information.

	P	$\times$	r	$\times$	t	$=$	I
	P		r		t		Interest earned, I
Amount at 7%	x		0.07		1		$0.07x$
Amount at 11%	$10{,}000 - x$		0.11		1		$0.11(10{,}000 - x)$

Since the total interest earned is $1,000, then the equation is formed by setting the sum of the interest earned column in the table equal to $1,000. Thus,

$$0.07x + 0.11(10{,}000 - x) = 1{,}000$$

The solution is $x =$ $2,500. So, the amount invested at 7% is $2,500 and the amount invested at 11% is $10,000 – $2,500 = $7,500.

● **PROBLEM** 1263

$10,000 is to be invested for one year. A part of the $10,000 is invested at 6% and the remainder at 7%. The income from the $10,000 is $640 per year. Find the amount invested at each rate.

Solution: Let x represent the amount invested at 6%, then the remaining amount $10,000 - x would be invested at 7%. The money is invested for one year, therefore, time= t = 1.

This problem involves four different items: 1) the interest or income, I, that accumulates from an investment over a period of time, 2) the amount of money invested, P, 3) the rate, r, at which an amount of money is invested, (the rate is given in percents), and 4) the time, t, for which an amount of money is invested.

The following formula relates these four items together: I = Prt.

	P	r	t	I = Prt
Investment No. 1	x	.06	1	.06x
Investment No. 2	10,000 - x	.07	1	.07(10,000 - x)

The income $640 from the $10,000 invested is the sum of the amount of interest from the two investments. If I_1 is the interest from the first investment and I_2 is the income from the second investment, then

$$I_1 + I_2 = \$640 \qquad\qquad (3)$$

From the table above, the interest from the first investment, I_1, is .06x and the interest from the second investment, I_2, is .07(10,000 - x). Therefore,

$$.06x + .07(\$10,000 - x) = \$640$$

$$.06x + \$700 - .07x = \$640$$

Note: $.07(\$10,000) = \dfrac{7}{\cancel{100}} \times \$\cancel{10},000 \overset{100}{\underset{1}{}} = \$700$

A man wishes to invest a part of $4200 in stocks earning 4% dividends and the remainder in bonds paying 2½%. How much must he invest in stocks to receive an average return of 3% on the whole amount of money?

Solution: Let x = amount invested at 4%

Then 4200 - x = amount invested at 2½%

The relationship used to set up the equation is:

Income from 4% investment + Income from 2½% investment = 3% of $4200, or $126.

$$.04x + .025(4200 - x) = 126$$
$$40x + 25(4200 - x) = 126,000$$
$$40x + 105,000 - 25x = 126,000$$
$$15x = 21,000$$
$$x = 1,400$$
$$4200 - x = 2,800$$

Check: 4% of $1,400 = $56
        2½% of $2,800 = $70
        Total income = $126 which is 3% of $4,200

A man made two investments amounting to a total of $5000. On the first he gained 8% and on the second he lost 3%. His net gain on the two investments was $70. What was the amount of each investment?

Solution: Let x = number of dollars in the first investment, and

5000 - x = number of dollars in the second investment.

We comprise the following table:

Investment	×	%Gained	=	Net Gain
x		8		.08x
5000-x		-3		-.03(5000-x)

Then 0.08x = number of dollars gained, and -0.03(5000 - x) = number of dollars lost. Since the man had a net gain of $70. :

$$0.08x - 0.03(5000 - x) = 70.$$

Distribute, $$0.08x - 150 + 0.03x = 70.$$

Add 150 to both sides, $$0.08x + 0.03x = 70 + 150.$$

Collect terms, $$0.11x = 220.$$

Divide by 0.11; then   x = 2000, number of dollars in the first investment.

Then   5000 - x = 3000, number of dollars in

the second investment.

Check: 0.08(2000) - 0.03(3000) = 150 - 90 = 70.

● PROBLEM 1266

A man derives an income of $309 from some money invested at 3%
and some at 4½%. If the amounts of the respective investments were
interchanged, he would receive $336. How much did he originally in-
vest at each rate?

Solution: Let x = Amount originally invested at 3%
           And y = Amount originally invested at 4½%
    (1)   .03x + .045y = 309
    (2)   .045x + .03y = 336

When the equations are solved simultaneously it is found that
x = $5,200 and y = $3,400.

● PROBLEM 1267

A man invested $7,800, part at 6% and part at 4%. If the income
of the 4% investment exceeded the investment at 6% by $92, how much
was invested at each rate?

Solution: Let x = Amount invested at 4%

           And y = Amount invested at 6%

    (1)   x + y = 7800

    (2)   .04x = .06y + 92

When the equations are solved it is found that x = $5,600, i.e., the
amount invested at 4%, and y = $2,200, i.e., the amount invested at
6%.

● PROBLEM 1268

A man invests $9,000, part at 3% and rest at 4%. If his total
income from the two investments is $295, how much did he invest at
each rate?

Solution: Let x = amount invested at 3%.
           Then ($9,000 - x) = amount invested at 4%.
    The relationship used to set up this equation is

    Income from 3% investment + Income from 4% investment = $295

$$
\begin{aligned}
.03x + .04(9,000 - x) &= 295 \\
3x + 4(9,000 - x) &= 29,500 \\
3x + 36,000 - 4x &= 29,500 \\
x &= 6,500 \\
9,000 - x &= 2,500
\end{aligned}
$$

Check: $6,500 at 3% for 1 year = $195
        $2,500 at 4% for 1 year = $100
              Total income = $295

● PROBLEM 1269

If $100 is invested at 3% interest compounded annually, how much

will it amount to at the end of 18 yrs.?

900

<u>Solution:</u> We apply the formula to find the amount of a sum of money invested at compound interest. Note that P is the principal which is the sum of money upon which the interest is paid.

Use the general formula: $P at interest rate r compounded j times a year will amount in t years to A_{jt} where $A_{jt} = P(1 + r/j)^{jt}$

$$P = 100.00$$
$$r = 3\%$$
$$j = 1$$
$$t = 18$$

$$A_{jt} = 100.00 \left(1 + \frac{.03}{1}\right)^{18} = 100(1.03)^{18}$$

Now solve for A_{jt} by using logs. That is;

$$\log A_{jt} = \log\left[100(1.03)^{18}\right] = \log(100) + 18 \log(1.03)$$
$$= 2 + 18(.0128) = 2 + .2304$$
$$\log\left(A_{jt}\right) = 2.2304$$

A_{jt} can now be determined by taking the antilog of both sides (the antilog of the log of x yields x itself)

$$\text{antilog}\left[\log(A_{jt})\right] = \text{antilog}[2.2304]$$
$$A_{jt} = \text{antilog}[2.2304]$$

Here the characteristic is 2 and the mantissa is .2304. Using a table of common logs, look up the value 2304 and see that the number whose log is .2304 is 1.70 and since the characteristic is 2 the number is then multiplied by $10^2 = 100$ and the final answer is:

$$A_{jt} = 10^2(1.70) = 170$$

Thus at the end of 18 years the $100. will amount to $170. with 3% interest compounded annually.

CHAPTER 47

PROBLEMS IN SPACE

Basic Attacks and Strategies for Solving Problems in this Chapter. See pages 902 to 907 for step-by-step solutions to problems.

Algebra and trigonometry concepts are useful in solving problems in space. Space science is based on a mathematical description of the universe. This mathematical description is in turn based on defining physical quantities clearly and precisely so that all observers can agree on any measurement of these quantities. Many relations between quantities in space are expressed in terms of formulas that can be solved using algebra, analytic geometry, and trigonometry. For instance, the altitude of an object in space can be found by first understanding conic sections and orbits and then using the circular orbital velocity equation,

$$V = \sqrt{GM/r},$$

and the equation

$$V = 2\pi r/t,$$

where G is the constant of proportionality (called the universal gravitational constant), M is the mass of the object, r is the radius, and t is the time. The complete solution is given in Problem #1270.

In addition to the use of algebra and trigonometry in the exploration and solution of space problems, there is a natural use of the tools and techniques of physics, astronomy, and higher mathematics.

• PROBLEM   1270

A synchronous Earth satellite is one which is placed in a west-to-east orbit over the Equator at such an altitude that its period of revolution about Earth is 24 hours, the time for one rotation of Earth on its axis. Thus the orbital motion of the satellite is synchronized with Earth's rotation, and the satellite appears from Earth to remain stationary over a point on Earth's surface below. Such communication satellites as Syncom, Early Bird, Intelsat, and ATS are in synchronous orbits. Find the altitude for a synchronous Earth satellite.

Solution   The velocity can be found from the equation for circular orbital velocity. It can also be found by dividing the distance around the orbit by the time required; that is, $v = \dfrac{2\pi r}{t}$. Because the two velocities are equal,

$$\frac{2\pi r}{t} = \sqrt{\frac{GM}{r}}$$

$$\left(\frac{2\pi r}{t}\right)^2 r = GM$$

$$r^3 = \frac{GMt^2}{4\pi^2}$$

$$r = \sqrt[3]{\frac{GMt^2}{4\pi^2}}.$$

It is apparent that $t = 24$ hours. Substituting the other values yields

$$r = \sqrt[3]{\frac{1.24 \times 10^{12} \times 24^2}{4 \times 3.14^2}} = 10^4 \sqrt[3]{\frac{178.56}{9.86}}$$

$$= 10^4 = \sqrt[3]{18.1} = 26{,}260 \text{ mi.}$$

$$\text{Altitude} = 26{,}260 - 3{,}960 = 22{,}300 \text{ mi.}$$

$$v = \frac{2 \times 3.14 \times 26{,}260}{24} = 6{,}870 \text{ mi/hr.}$$

To understand orbits, we must know something of the nature and properties of the conic sections. They get their name, of course, from the fact that they may be formed by cutting or sectioning a complete right circular cone (of two nappes) with a plane. Any plane perpendicular to the axis of the cone cuts a section that is a circle. Incline the plane a bit, and the section formed is an ellipse. Tilt the plane still more until it is parallel to a ruling of the cone and the section is a parabola. Let the plane cut both nappes, and the section is a hyperbola, a curve with two branches. It is apparent that closed orbits are circles or ellipses. Open or escape orbits are parabolas or hyperbolas.

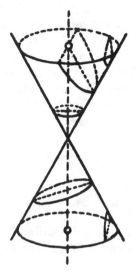

Another way of classifying the conic sections is by means of their eccentricity. If we represent the eccentricity by e, then a conic section is

A circle if $e = 0$,
An ellipse if $0 < e < 1$,
A parabola if $e = 1$,
A hyperbola if $e > 1$.

In actual practice, orbits that are exactly circular or parabolic do not exist because the eccentricity is never exactly equal to 0 or 1.

● **PROBLEM** 1271

A typical altitude for manned spacecraft about Earth is 100 miles because this is the lowest altitude at which air resistance becomes small enough to make a stable orbit possible. Because the speed in a circular orbit at this altitude is about 17,500 miles per hour, this speed is sometimes quoted as a typical one for space travel. How many years would it take for a spaceship to travel 1 light year if its rate is 17,500 miles per hour?

Solution   Solving the distance-time-rate equation, $d = vt$, for t, we obtain

$$t = \frac{d}{v}$$

$$= \frac{5.88 \times 10^{12} \text{ mi}}{1.75 \times 10^4 \text{ mi/hr}} = 3.36 = 10^8 \text{ hr}.$$

Converting 3.36×10^8 hours to years yields

$$3.36 \times 10^8 \text{ hr} = 3.36 \times 10^8 \text{ hr} \times \frac{1 \text{ day}}{24 \text{ hr}} \times \frac{1 \text{ yr}}{365 \text{ days}}$$

$$= 3.36 \times 10^8 \text{ hr} \times \frac{1 \text{ yr}}{8,760 \text{ hr}}$$

$$= 3.84 \times 10^4 \text{ yr}$$

$$= 38,400 \text{ yr}.$$

This problem uses an interesting application of a binomial expansion to investigate the relationship between Newton's and Einstein's formulas for kinetic energy. The exploration of space outside our solar system will be feasible only if we can produce spacecraft that will travel nearly as fast as light. (Even if we could travel at the speed of light, it would require a little more than 4 years to reach Alpha Centauri, the closest star outside the solar system.) At speeds close to that of light, the theory of relativity changes the formula for kinetic energy E_K. Whereas in Newtonian mechanics we have

$$E_K = \frac{1}{2}m_0 v^2,$$

the relativistic formula is

$$E_K = (m_r - m_0)c^2.$$

At first sight these formulas look quite different. We shall see, however, that the Newtonian formula can be regarded as an approximation to the relativistic one.

Verify that the binomial expansion of

$$\frac{1}{\sqrt{1-x^2}} = (1-x^2)^{-1/2}$$

is

$$1 + \frac{1}{2}x^2 + \frac{1}{2}\cdot\frac{3}{4}x^4 + \frac{1}{2}\cdot\frac{3}{4}\cdot\frac{5}{6}x^6 + \cdots.$$

Solution   Using the binomial expansion for $(a + b)^n$, namely

$$(a+b)^n = a^n + na^{n-1}b + \frac{n(n-1)}{2\cdot 1}a^{n-2}b^2 + \cdots$$

with $a = 1$, $b = -x^2$, and $n = -\frac{1}{2}$, we find

$$(1-x^2)^{-1/2} = (1)^{-1/2} + \frac{\left(-\frac{1}{2}\right)}{1}(1)^{-3/2}(-x^2)$$

$$+ \frac{\left(-\frac{1}{2}\right)\left(-\frac{3}{2}\right)}{1\cdot 2}(1)^{-5/2}(-x^2)^2$$

$$+ \frac{\left(-\frac{1}{2}\right)\left(-\frac{3}{2}\right)\left(-\frac{5}{2}\right)}{1\cdot 2\cdot 3}(1)^{-7/2}(-x^2)^3 + \cdots$$

$$= 1 + \frac{1}{2}x^2 + \frac{1}{2}\cdot\frac{3}{4}x^4 + \frac{1}{2}\cdot\frac{3}{4}\cdot\frac{5}{6}x^6 + \cdots.$$

Use the first two terms of the expansion as an approximation to $(1-x^2)^{-1/2}$ and set $x = v/c$. Show that the relativistic kinetic energy formula reduces to the Newtonian one.

Solution We are given that

$$m_v = \frac{m_0}{\sqrt{1-(v/c)^2}}$$

Hence

$$\frac{m_v}{m_0} = \frac{1}{\sqrt{1-(v/c)^2}}$$

Using $1+1/2(v/c)$ as an approximation to $\dfrac{1}{\sqrt{1-(v/c)^2}}$, we have

$$\frac{m_v}{m_0} \doteq 1+1/2(v/c)^2$$

$$m_v \doteq m_0+1/2m_0(v/c)^2$$

$$m_v-m_0 \doteq 1/2m_0(v/c)^2$$

$$(m_v-m_0)c^2 \doteq 1/2m_0v^2.$$

(We are using $x\doteq y$ to mean "x is approximately equal to y.")
Thus we have shown that the relativistic kinetic energy formula
reduces approximately to the Newtonian one when v is small
compared with c.

● **PROBLEM** 1273

The average angle subtended by the Sun for an observer on
the surface of Earth is 0.533⁰. Assuming that the diameter of
the Sun is 866,000 miles, find the distance from the surface
of Earth to the center of the Sun. Assume OCT is a right
triangle.

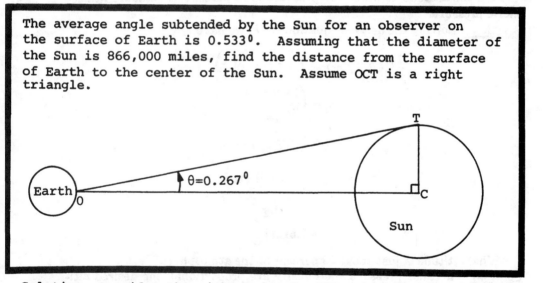

Solution Consider the right triangle OCT. Because the total
angle subtended by the Sun as viewed by an Earth observer is
0.533⁰, angle TOC is one-half the angle of 0.267⁰.

The distance OC between Earth's surface and the center of the
Sun may be calculated by using the tangent function:

$$\tan\theta = \frac{TC}{OC} \quad \text{or} \quad OC = \frac{TC}{\tan\theta}$$

$$= \frac{433,000\,\text{mi}}{0.00466} = 92,900,000\,\text{mi}$$

An approximate rule for atmospheric pressure at altitudes less than 50 miles is the following: Standard atmospheric pressure, 14.7 pounds per square inch, is halved for each 3.25 miles of vertical ascent.

Write a simple exponential equation to express this rule.

Solution Letting P denote the atmospheric pressure at altitudes less than 50 miles and h the altitude. we have

$$P = 14.7 \text{ lb/in}^2 \left(\frac{1}{2}\right)^{h/3.25 \text{ mi}}$$

Compute the atmospheric pressure at an altitude of 19.5 miles.

Solution Using the equation derived in part a,

$$P = (14.7 \text{ lb/in}^2)\left(\frac{1}{2}\right)^{19.5 \text{ mi}/3.25 \text{ mi}}$$

$$= (14.7 \text{ lb/in}^2)\left(\frac{1}{2}\right)^6$$

$$= (14.7 \text{ lb/in}^2)\frac{1}{64}$$

$$= 0.23 \text{ lb/in}^2.$$

Find the altitude at which the pressure is 20 percent of standard atmospheric pressure.

Solution Solving the derived equation for h, we have

$$\frac{1}{5}(14.7 \text{ lb/in}^2) = (14.7 \text{ lb/in}^2)\left(\frac{1}{2}\right)^{h/3.25 \text{ mi}}$$

$$\frac{1}{5} = \left(\frac{1}{2}\right)^{h/3.25 \text{ mi}}$$

$$\log\frac{1}{5} = \left(\frac{h}{3.25 \text{ mi}}\right)\log\frac{1}{2}$$

$$h = \frac{(3.25 \text{ mi}) \log\frac{1}{5}}{\log\frac{1}{2}}$$

$$= 7.54 \text{ mi}.$$

What altitude is just above 99 percent of the atmosphere?

Solution Because pressure and density are proportional, the desired altitude is the point at which the pressure is 1 percent of standard atmospheric pressure. Hence

$$(0.01)(14.7 \text{ lb/in}^2) = (14.7 \text{ lb/in}^2)\left(\frac{1}{2}\right)^{h/3.25 \text{ mi}}$$

$$0.01 = \left(\frac{1}{2}\right)^{h/3.25 \text{ mi}}$$

$$\log 0.01 = \left(\frac{h}{3.25 \text{ mi}}\right)\log\frac{1}{2}$$

$$h = \frac{(3.25 \text{ mi}) \log 0.01}{\log\frac{1}{2}}$$

$$= \frac{(3.25 \text{ mi})(-2)}{-0.301}$$

$$= 21.6 \text{ mi}.$$

A sweeping light beam is used with a light-source detector to determine the height of clouds directly above the detector, as in the following diagram.

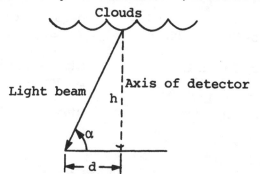

With the axis of the detector vertical, the light beam is allowed to sweep from the horizontal ($\alpha=0$) to the vertical ($\alpha=90^0$). When the beam illuminates the base of the clouds directly above the detector, as in the figure, the angle α is read, and with d known, the height h can be computed.

Express h in terms of an appropriate trigonometric function of α and d.

Solution   Applying the definition of the tangent function gives

$$h=d(\tan \alpha).$$

If the light source is 1,000 feet from the detector and the angle is 45^0, compute the height of the cloud.

Solution   Using the equation from part a gives

$$h=(1,000 \text{ ft})(\tan 45^0)$$
$$=1,000 \text{ ft}$$

If the height of the cloud is 2,050 feet and the distance d is 1,000 feet, compute the angle α.

Solution   Using the same equation, we find

$$2,050 \text{ ft} =(1,000 \text{ ft})(\tan \alpha)$$
$$2.050 =\tan \alpha$$
$$\alpha =64^0.$$

Find the angle α when clouds are 1,000 feet high and the light source is located 100 feet from the detector.

Solution   Applying the same equation again gives

$$1,000 \text{ ft} =(100 \text{ ft})(\tan \alpha)$$
$$10 =\tan \alpha$$
$$\alpha =84.29^0 \text{ or about } 84^0.$$

INDEX

Numbers on this page refer to **PROBLEM NUMBERS,** not page numbers

REA'S
PROBLEM
SOLVERS®

The PROBLEM SOLVERS® are comprehensive supplemental textbooks designed to save time in finding solutions to problems. Each PROBLEM SOLVER® is the first of its kind ever produced in its field. It is the product of a massive effort to illustrate almost any imaginable problem in exceptional depth, detail, and clarity. Each problem is worked out in detail with a step-by-step solution, and the problems are arranged in order of complexity from elementary to advanced. Each book is fully indexed for locating problems rapidly.

Accounting
Advanced Calculus
Algebra & Trigonometry
Automatic Control Systems/Robotics
Biology
Business, Accounting & Finance
Calculus
Chemistry
Differential Equations
Economics
Electrical Machines
Electric Circuits
Electromagnetics
Electronics
Finite & Discrete Math
Fluid Mechanics/Dynamics

Genetics
Geometry
Linear Algebra
Mechanics
Numerical Analysis
Operations Research
Organic Chemistry
Physics
Pre-Calculus
Probability
Psychology
Statistics
Technical Design Graphics
Thermodynamics
Topology
Transport Phenomena

If you would like more information about any of these books,
complete the coupon below and return it to us or visit your local bookstore.